PRINCIPLES and PRACTICES
in
PLANT ECOLOGY

Allelochemical Interactions

Edited by

Inderjit
Department of Agricultural Sciences (Weed Science),
The Royal Veterinary and Agricultural University,
Denmark

K.M.M. Dakshini
Department of Botany, University of Delhi, India

Chester L. Foy
Department of Plant Pathology, Physiology, and Weed Science,
Virginia Polytechnic Institute and State University,
Blacksburg, Va, USA

CRC Press
Boca Raton   London   New York   Washington, D.C.

Library of Congress Cataloging-in-Publication Data

Principles and practices in plant ecology : allelochemical
    interactions / Inderjit, K.M.M. Dakshini, Chester L. Foy, editors.
        p. cm.
    Includes index.
    ISBN 0-8493-2116-6 (alk. paper)
      1. Allelochemicals. 2. Allelopathy. 3. Allelopathic agents.
    4. Plant chemical ecology. 5. Plant-microbe relationships.
    I. Inderjit. II. Dakshini, K. M. M., 1934–   III. Foy, Chester L.
    QK898.A43P75 1999
    581.7—dc21   98-44681
                                                             CIP

© 1999 by CRC Press LLC

No claim to original U.S. Government works
International Standard Book Number 0-8493-2116-6
Library of Congress Card Number 98-44681
Printed in the United States of America 1 2 3 4 5 6 7 8 9 0
Printed on acid-free paper

072999–11436B

Preface

Allelochemical interactions, during the last three decades, have evolved as an important branch of plant ecology. In this book, in general, the effects of chemical compounds released from plants (including microorganisms), on other plants in their vicinity are considered under the term "allelopathy." The term "allelochemical" is used in a wider context in the field of ecology where it includes, but is not limited to, plant and microorganism interactions. Allelochemicals released from plants (including microorganisms) have multifaceted influences on ecosystems; these also influence soil microbial ecology, soil nutrients, and physical, chemical and biological soil factors. We believe that it is extraordinarily difficult to separate the influence of allelochemicals on each of these components of an ecosystem. Effects on any one of these components, due to allelochemicals, may influence growth, distribution, and survival of plant species.

The aim of this book is to provide insight and recent progress on allelochemical research from this multifaceted standpoint. Research articles—reporting results of substantially completed work, and review articles—presenting novel and critical appraisals of specific topics of interest, are included. Yet it may not be a comprehensive treatise on the subject. The sequence of chapters in the book starts with an overview followed by 34 chapters contributed by scientists around the world, thus presenting a global perspective on allelochemical research. Section I—Methodologies (Chapters 2–8), discusses important aspects of methodology in the study of allelopathy, shortcomings of bioassays for allelopathy, bioassays for different plant groups, extraction of allelochemicals from soil, sampling procedures, and an outline of analytical methods for different classes of allelochemicals. Section II—Interactions Among Plant and Microbial Systems (Chapter 9–15), presents allelochemical research in aquatic and terrestrial ecosystems, and includes other important subjects like pollen allelopathy. Section III—Ecological Aspects (Chapters 16–22), illustrates the significance of ecological studies in allelochemical research, and discusses the important role that the soil environment plays in the functioning of allelochemicals. Section IV—Biochemical, Chemical and Physiological Aspects (Chapters 23–30), discusses biochemical, molecular, and physiological aspects of allelopathy, including information on modes of action of allelochemicals in allelopathy. Allelochemicals have been successfully used in biocontrol of plant pathogens and weeds. This important applied aspect of allelochemistry is discussed under Section V—Biological Control of Plant Disease and Weeds: Applied Aspects (Chapters 31–34). Thus, in totality, the book illustrates the processes, procedures, and applications related to allelochemicals.

The Editors

Inderjit is an Assistant Professor, in the Department of Agricultural Sciences (Weed Science) at The Royal Veterinary and Agricultural University in Denmark. His formal education includes a BSc in Botany from the University of Delhi (1984); an MSc from Panjab University (1986); and a PhD, specializing on weed ecology, from the University of Delhi (1993). His research interests include different aspects of weed biology and ecology, allelopathy, soil chemistry, phytoremediation, and phytochemistry. Dr. Inderjit has investigated the allelopathic effects of different annual and perennial cropland weeds from India and the United States. He has investigated the interference potential of certain boreal forest understory shrubs in certain areas in Canada.

Dr. Inderjit has demonstrated exceptional leadership in the field of chemical ecology, more specifically in the allelochemical interactions of soils and plant products. His work, which explores the agricultural potentials of such interactions, will unquestionably provide the foundation for improved crop yields in the future. His unique interests in the delicate balance between allelochemicals and natural soils, and the likelihood of serious disruption of this critical association by industrial contamination, are at the cutting edge of modern agricultural management techniques. The significant findings of his research on the allelopathic interactions between weeds and their associations have a wide range of applications in environmental manipulations and biological control and other vegetation management strategies.

Dr. Inderjit has published several papers in reputed international journals. He was the senior editor of a book, "Allelopathy: Organisms, Processes, and Applications," which was published by the American Chemical Society in 1995. He visited research centers and gave lectures in different states in the United States, as well as in Canada, Japan, Spain, Italy, France, and the Philippines. He has organized a Symposium on Allelopathy under aegis of the American Institute of Biological Sciences in the Botanical Society of America meeting held at Iowa State University, Ames, from August 1 through August 5, 1993. He was awarded the STA (Science & Technology Agency) fellowship by Japan International Scientific and Technology Exchange Center (JISTEC), Japan. He did some collaboration work with Dr. Chester L. Foy at the Department of Plant Pathology, Physiology & Weeds Science, Virginia Tech, Blacksburg, in 1997. Dr. Inderjit has investigated forest interference problems in Canada (1995–1997), and during that time was located at the Department of Biology, Lakehead University, Thunder Bay, Canada. During 1996, Dr. Inderjit visited the Department Vegetale Biologie, Second University of Napoli as a Visiting Professor.

Other recognitions: Dr. Inderjit has been invited to a joint meeting (April 20–29, 1997) of Indian and Polish experts at the Department of Biotechnology, Government of India, New Delhi. Dr. Inderjit was *Guest Invited Speaker* at the First World Allelopathy Congress, Cadiz, Spain (September 16, 1996 through September 20, 1996). The Citizen Ambassador Program, Washington, D.C., invited him to join the *Soil Science Delegation* traveling to South Africa (July 30, 1996 through August 11, 1996) to participate in bilateral technical exchanges with emerging leaders in South Africa Soil Science. The objective of the delegation was to review the status of soil science in South Africa, with particular reference to those areas important to the region's economy. He was nominated *Active Member* by New York Academy of Sciences in 1995, and appointed *Associate* by the Indian Academy of Sciences in 1995. His biography was published in "Marquis Who's Who in Science and Engineering" in 1992.

K. M. M. Dakshini was born in Bareilly, India on July 25, 1934. Presently he is a Professor of Botany at the University of Delhi, India. He has been teaching ecology at the postgraduate level for the past 40 years. His research interests include allelopathy, swamp forests, plant introduction, ecological implication of biodiversity, soil degradation, restoration of degraded land, ecological adaptation, nutrient dynamics, and phytochemistry. He has published more than 150 research papers, 20 review articles, and has co-edited a book, "Allelopathy: Organisms, Processes, and Applications" which was published by the American Chemical Society in 1995. He has presented over 30 invited lectures at international meetings and over 40 invited lectures at national meetings and seminars. Dr. Dakshini has directed (as Major Advisor) more than 30 MPhil and PhD students. Dr. Dakshini has traveled extensively in North America, Europe, Asia, Australia, and the USSR.

Chester Larrimore Foy (Larry) was born in Tennessee on July 8, 1928. His formal education includes: involvement in the Army Specialized Training Reserve Program; Pre-engineering at Clemson Agricultural College, Virginia Military Institute Institute, and Texas A & M University (1945–1946); a Certificate in Agriculture from the University of Tennessee Junior College (1952); a BS in Agronomy (and Soils) from the University of Tennessee (1952); an MS in Field Crops (Chemical Weed Control) from the University of Missouri (1953); and a PhD in Plant Physiology (Weed Science-Botany) from the University of California—Davis (1958).

Dr. Foy is now Professor of Plant Physiology/Weed Science at Virginia Polytechnic Institute and State University, Blacksburg. His primary responsibilities are research (basic and applied) and teaching in the areas of weed science and other related subjects.

Dr. Foy's recognitions include election to membership in the following academic honorary societies: Alpha Zeta, Phi Kappa Phi, Gamma Sigma Delta, Gamma Alpha, Sigma *Xi*, Epsilon Sigma *Xi*, Phi Beta Delta, as well as several "Who's Who" listings. He also received several other significant awards for professional achievements as follows: National Academy of Sciences (NRC) Resident Research Associateship (1964–1965); WSSA Outstanding Paper Award (1968); Gamma Sigma Delta Faculty Research Award (1977); WSSA Fellow (1980); SWSS's first *Weed Scientist of the Year* Award; WSSA's prestigious *Outstanding Researcher* Award (also, a concomitant industry-sponsored *Ag Recognition Award,* a 10-day visit to CIBA-GEIGY's world corporate headquarters in Basle, Switzerland; IWSS's *Outstanding Achievement Award—Developed Countries* (1996), etc. Dr. Foy's affiliations with other professional organizations past and present include: American Society of Agronomy, American Association for the Advancement of Science, American Institute of Biological Sciences, American Society of Plant Physiologists, Council of Agricultural Science and Technology (CAST), International Weed Science Society (IWSS), International Congress of Plant Protection, International Congress of Pesticide Chemistry, Plant Growth Regulator Society of America (PGRSA), Southern Weed Science Society (SWSS); as well as other regional and state conferences, Torch International, Virginia Academy of Sciences, and Weed Science of America (WSSA). Currently, he is a Founding Trustee, member of the Science Advisory Council, and Co-Editor of the newly incorporated international Association of Formulation Chemists.

Dr. Foy is a charter member of WSSA and IWSS and is a regular contributor to the affairs of these societies. He served as president of IWSS for a two-year term (1991–1993). In WSSA activities he served as vice-president, president–elect (program chairman), and president, with four years on the executive committee; he chaired committees on Industry Financed Awards, EPA Relations, Liaison, Nonherbicidal Aspects of Plant Growth Regulation (which contributed to the formation of the PGRSA), and Outstanding Paper Award. He was chairman (editor) for the standing committee on *Reviews of Weed Science* for four years and is currently serving as editor of *Weed Technology* for a third three-year term.

Dr. Foy conducts and/or directs field, greenhouse, and laboratory studies in the following areas: crop production and protection; vegetation management in agronomic and fruit crops, and control of specific perennial weeds; routes and mechanisms of adsorption, translocation, accumulation, and exudation of herbicides and growth regulators, and surfactants and other adjuvants; metabolism and fate of these substances; physiological, biochemical, and morphological changes induced by exogenous chemicals; modes of action and selectivity of herbicides and growth regulators; minimizing pesticide residues in the biosphere; allelopathy; and parasitic weeds. He is a prolific author in the plant sciences (weed science, physiology, crop protection) and related fields. Publications (and manuscripts prepared) since 1952 include: 2 books; 24 book chapters; 5 book reviews; 111 peer-reviewed scientific journal papers; 5 technical research bulletins; 372 contributions to conference proceedings, research reports, abstracts, or scientific papers and special technical research articles; and over 215 semi-technical or extension publications.

Dr. Foy has taught graduate and/or undergraduate courses in field crops, botany, weed science, herbicidal action, pesticide usage, and the use of radioisotopes in biological research. He has directed (as major advisor) 40 MS and PhD students from 14 countries and served on numerous other committees in weed science–plant physiology. Dr. Foy has traveled extensively and presented invited lectures in several countries, including the United States, Canada, Puerto Rico, England, Germany, Denmark, Australia, Korea, Kenya, Egypt, Israel, India, the Netherlands, and France.

Contributors

T. Mitchell Aide
Department of Biology
University of Puerto Rico
San Juan, PR

Giovanni Aliotta
Dipartmento di Scienze della Vita
Facultad di Science MM.FF.NN.
Seconda Universita di Napoli
Caserta, Italy

Oz Barazani
Department of Botany
George S. Wise Faculty of Sciences
Tel Aviv University
Tel Aviv, Israel

Mark A. Berhow
Bioactive Constituents Research
USDA, ARS, Midwest Area
National Area for Agricultural
  Utilization Research
Peoria, IL

Udo Blum
Department of Botany
North Carolina State University
Raleigh, NC

Carolee T. Bull
USDA, ARS, Pacific West Area
U. S. Agricultural Research Station
Salinas, CA

Gennaro Cafiero
Universita Degli Studi Napoli Federico II
Centro Interdipartmentale
  di Ricerca Sulle
  Ultrastructure Biologiche
Napoli, Italy

Natividad Chaves
Dpto. de Ecología
Facultad de Ciencias
Universidad de Extremadura
Badajoz, Spain

H. H. Cheng
Department of Soil, Water
and Climate
College of Agricultural, Food
  and Environmental Sciences
University of Minnesota
St. Paul, MN

Olaf Christen
Institute of Crop Science
University of Kiel
Kiel, Germany

Horace G. Cutler
Southern School of Pharmacy
Mercer University
Atlanta, GA

Randy A. Dahlgren
University of California
Davis, CA

K. M. M. Dakshini
Department of Botany
University of Delhi
Delhi, India

Barry R. Dalton
Department of Biological Sciences
Natural Science Center
Northern Kentucky University
Highland Heights, KY

Frank A. Einhellig
Southwest Missouri State University
Springfield, MO

Stella D. Elakovich
Department of Chemistry and Biochemistry
University of Southern Mississippi
Hattiesburg, MS

J. C. Escudero
Dpto. de Ecología
Facultad de Ciencias
Universidad de Extremadura
Badajoz, Spain

M. C. Feng
School of Chemical Sciences
Universiti Sains Malaysia
Penang, Malaysia

W. Feucht
Institute of Fruit Growing
Faculty of Agriculture and Horticulture
Technical University of Munich
Freising-Weihenstephan, Germany

Chester L. Foy
Department of Plant Pathology, Physiology
  and Weed Science
Virginia Polytechnic and State University
Blacksburg, VA

Annette Friebe
Institut fur Landwirtschaftliche Botanik
Der Rheinischen Friedrich-Wilhelms-
  Universitat Bonn
Bonn, Germany

Jacob Friedman
Department of Botany, George S. Wise
  Faculty of Sciences
Tel Aviv University
Tel Aviv, Israel

Y. Fujii
Allelopathy Laboratory
National Institute of Agro-Environmental
  Sciences
Tsukuba, Ibaraki, Japan

Daniel W. Gilmore
Aspen/Larch Genetics Cooperative
University of Minnesota
North Central Experimental Station
Grand Rapids, MN

Elizaeth M. Gross
Limnologisches Institut
Universität Konstanz
Konstanz, Germany

Robert E. Hoagland
Southern Weed Science Lab
USDA ARS
Stoneville, MS

P. M. Huang
Department of Soil Science
University of Saskatchewan
Saskatoon, SK, Canada

Inderjit
Department of Agricultural Sciences
  (Weed Science)
The Royal Agricultural and Veterinary
  University (KVL)
Frederiksberg C, Copenhagen
Denmark

Philippe Jeandet
UFR Sciences Exactes et Naturelles
Université de Reims
Reims, France

Takao Katase
Laboratory of Chemistry for Bioresources
  and Environment
Graduate School of Bioresource Sciences
Nihon University
Kanagawa, Japan

Kathleen I. Keating
Department of Environmental Sciences
Rutgers—The State University
Cook Campus
New Brunswick, NJ

Bong-Seop Kil
Department of Biology Education
Wonkwang University
Iri, Republic of Korea

R. E. Ley
Department of Environmental, Population
  and Organisimic Biology
University of Colorado at Boulder
Boulder, CO

John V. Lovett
Grain Research and Development Corporation
ACT 2600, Australia

B. Lutz-Bruning
Mainz, Germany

Francisco A. Macías
Department of Organic Chemistry,
 Faculty of Science
University of Cadiz
Puerto Real (Cadiz), Spain

Eugenia V. Melnikova
Russian Academy of Sciences
Institute of Cell Biophyscics
Pushchino, Moscow Region, Russia

José M. G. Molinillo
Department of Organic Chemistry
Faculty of Science
University of Cadiz
Puerto Real (Cadiz), Spain

Stephen D. Murphy
Department of Environment
 and Resource Studies
University of Waterloo
Waterloo, ON, Canada

Chandrashekhar I. Nimbal
Department of Horticulture
 and Landscape Architecture
University of Kentucky
Lexington, KY

Hiroyuki Nishimura
Department of Bioscience and Technology,
 School of Engineering
Hokkaido Tokai University
Sapporo, Japan

Robert R. Northup
University of California
Davis, CA

Wieslaw Oleszek
Department of Biochemistry
Institute of Soil Science and
 Plant Cultivation
Pulawy, Poland

Victoria V. Roshchina
Russian Academy of Sciences
Institute of Cell Biophysics
Pushchino, Moscow Region, Russia

T. Schmeller
Pharmaceuticals
Karlsruhe, Germany

Stephen K. Schmidt
Department of Environmental, Population
 and Organisimic Biology
University of Colorado at Boulder
Boulder, CO

Margot Schulz
Institut fur Landwirtschaftliche Botanik
Der Rheinischen Friedrich-Wilhelms-
 Universitat Bonn
Bonn, Germany

Christiane Theuer
Institute of Crop Science
University of Kiel
Kiel, Germany

Ascensión Torres
Department of Organic Chemistry
Faculty of Science
University of Cadiz
Puerto Real (Cadiz), Spain

D. Treutter
Institute of Fruit Growing
Faculty of Agriculture and Horticulture
Technical University of Munich
Freising-Weihenstephan, Germany

Rosa M. Varela
Department of Organic Chemistry
Faculty of Science
University of Cadiz
Puerto Real (Cadiz), Spain

Steven F. Vaughn
Bioactive Constituents Research
USDA, ARS, Midwest Area
National Area for Agricultural
 Utilization Research
Peoria, IL

George R. Waller
Department of Biochemistry
 and Molecular Biology
Oklahoma State University
Stillwater, OK

M. C. Wang
Department of Soil Science
National Chung Hsing University
Taichung, Taiwan, Republic of China

M. K. Wang
Department of Agricultural Chemistry
National Taiwan University
Taipei, Taiwan, Republic of China

Leslie Weston
Department of Floriculture
 and Ornamental Horticulture,
Cornell University
Ithaca, NY

Rick Willis
School of Botany
University of Melbourne
Parkville, Victoria, Australia

Michael Wink
Institut fur Pharmazeutische Biologie
Ruprecht-Karls-Universitat Heidelberg
Heidelberg, Germany

Robert M. Zablotowicz
Southern Weed Science Lab
Stoneville, MS

Jess K. Zimmerman
Department of Biology
San Juan, PR

Alicja M. Zobel
Department of Chemistry
Trent University
Peterborough, ON, Canada

Acknowledgment

We are indebted to authors for submitting their valuable work for this book in a given time frame. We are grateful to Professors Udo Blum and Frank Einhellig, and to Kathleen Keating for their advice and help throughout the preparation of this book. We also extend our appreciation to John Sulzycki, Acquisitions Editor, Marsha Baker, former CRC Acquisitions Editor, Carolyn Spence, Editorial Assistant, Susan Zeitz, and Pat Roberson of CRC Press for their cooperation and help throughout the project. We express our gratitude to Professors N. N. Bhandari, K. G. Mukerji and M. R. Parthasarthy, and to Satish Handa, Jasleen Kaur, Surinder Kaur, Reetu Sharma, and Harold Witt for their help in various ways. We trust that the information presented in this volume will truly provide an ecological perspective on allelochemical research, and will help the scientific community toward a better understanding of the subject. Equally we hope that it will stimulate young scientists to take up studies on allelochemical interactions.

Inderjit, K. M. M. Dakshini, and Chester L. Foy
April, 1998

From the Editors

Each chapter included in this book has been peer reviewed by two or more referees. The editors and authors are highly grateful to reviewers for their thorough, critical, and constructive reviews. We extend our deepest appreciation for the time and efforts that the following referees have spent on reviewing different chapters.

A. L. Anaya
M. A. Berhow
Dan Binkley
Udo Blum
Paul Brown
Jack Butler
Larry G. Butler*
John Cardina
H. H. Cheng
C. H. Chou
Olaf Christen
Gallian A. Cooper-Driver
Horace G. Cutler
Randy A. Dahlgren
Barry R. Dalton
Roger Del Moral
Kelsey L. Downum
Stephen O. Duke
Frank A. Einhellig
Stella D. Elakovich
W. Feucht
Nikolaus H. Fischer
Richard F. Fisher
G. E. Fogg
Jonathan Gershenzon
Daniel Gilmore
Stephen R. Gliessman
Jeffrey B. Harborne
Rod M. Heisey
Robert E. Hoagland

Stephen B. Horsley
P. M. Huang
Gorden Kayahara
Kathleen I. Keating
Edna Levy
John V. Lovett
Dean F. Martin
Darrell A. Miller
Dieter Muller-Dombois
Stephen D. Murphy
David Netzly
W. Oleszek
David T. Patterson
James Rasmussen
Stephen K. Schmidt
Donn G. Shilling
Helen A. Stafford
C. S. Tang
John R. Teasdale
David O. TeBeest
George R. Waller
David Wardle
Jeffrey D. Weidenhamer
Leslie A. Weston
Rick Willis
M. Wink
C. Peter Wolk
A. D. Worsham
Olle Zackrisson
Robert L. Zimdahl

* DECEASED.

Dedication

We feel honored to dedicate this book to Dr. Elroy L. Rice, David Ross Boyd Professor Emeritus of Botany at the University of Oklahoma. Most of the present research on allelopathy can be attributed to Dr. Rice.

Dr. Rice earned his PhD degree in December 1947 from the University of Chicago. In January 1948, he joined the Department of Botany & Microbiology at the University of Oklahoma, Norman, OK. He was elected president of the Southwestern Association of Naturalists and the Oklahoma Academy of Science in 1970 and 1981, respectively. Dr. Rice was named Outstanding Scientist in Oklahoma in 1984 by the membership of the Oklahoma Academy of Sciences.

He was given the Distinguished Service Award for a lifetime of teaching and pioneering work in allelopathy research by the Medical and Biological Science Alumni Association of the University of Chicago in 1995. He was granted a Professional Achievement Citation as one of the world's top authorities in allelopathy by the University of Chicago Alumni Association, also in 1995. After retirement in 1984, Dr. Rice was appointed David Ross Boyd Professor Emeritus of Botany.

Dr. Rice has supervised 33 doctoral students, and several MS and MNS students. He has published eight books and over 100 research articles in refereed journals of international repute. His classical textbook "Allelopathy," published in 1984, is cited in almost every paper published on allelopathy. Inderjit would like to extend his special gratitude to Dr. Rice as he has helped him throughout his career in various ways. It is with pride that we dedicate this book to the *Father of Allelopathy*, Dr. Elroy L. Rice.

Contents

Overview

1 Allelopathy: One Component in a Multifaceted Approach to Ecology

K. M. M. Dakshini, Chester L. Foy, and Inderjit

CONTENTS

1.1 ABSTRACT

The growth of allelopathy as a discipline is an indication of its ubiquitous occurrence. Allelopathy is expressed through the release of allelochemicals by the donor plant in the vicinity of a receptor plant species. Aside from their many roles in allelopathy—influencing soil microbial ecology, nutrient dynamics, and other abiotic and biotic factors—allelochemicals play key roles in structuring of other trophic levels, especially affecting predators and pests and mediating competitive circumstances. Data presented in the various chapters included in this book amply illustrate the significance of applying multifaceted approaches to problems in allelopathy that encompass interactions from different aspects of the environment. In this chapter, the need for a multifaceted approach in allelochemical studies is argued.

1.2 INTRODUCTION

A successful plant/organism has the capacity to interfere with other organisms in its surroundings. Furthermore, the successful plant should be able to create conditions in which its interference outweighs the resilience potential of the other plants, the survival of which is affected. Interference can be accomplished through competition, allelopathy, or facilitation (Goldberg, 1990). The intermediaries involved in competition and facilitation are resources of the habitat and natural enemies. In allelopathy, however, the medium of interference is allelochemicals.

In nature it is not possible to separate many of the interference mechanisms (e.g., allelopathy, resource competition, nutrient immobilization, plant/microbe interactions) since they operate sequentially and/or simultaneously, and identifying one mechanism as the source for the observed pattern would be unnatural. However, it is true that through the use of each of these mechanisms, the individual tries to maximize its inherent potential to survive, and during this process the entire ecological realm is unfolded in response (Harborne, 1977).

Large numbers of secondary metabolites (allelochemicals) have been reported from different groups of plant species (Whittaker and Feeny, 1971). Aside from their role in plant–plant interactions (allelopathy), allelochemicals may also influence abiotic and biotic components of the substratum. Ecological roles of allelochemicals are discussed in this chapter, and a need for a multifaceted approach of allelochemical interaction is argued.

1.3 ALLELOCHEMICAL INTERACTIONS

A classical definition of allelopathy includes plant–plant chemical interactions in which the term 'plant' also includes microorganisms, and interactions include both inhibitory and stimulatory effects (Rice, 1984). As evident from the other reports included in this volume, and our experience on the subject, we are reasonably convinced that allelochemicals released from plants also affect other components of the environment. We use the term 'allelochemical interaction' in a wider perspective encompassing (1) allelopathy, (2) effect of allelochemicals released from plants on abiotic (inorganic and organic components) and biotic constituents of soil, and (3) the regulation of the production and ultimate release of allelochemicals by abiotic and biotic components of the ecosystem.

1.3.1 ALLELOPATHY: PAST ACHIEVEMENTS AND THE FUTURE APPROACH

Williamson (1990) stated, "Allelopathy as a field of research had grown albeit sluggishly, to the stage where it required integration into general ecological theory." He discussed the progress of allelopathy research in terms of published papers on allelopathy. He compared the number of journal articles listed in *Biological Abstracts* relative to the total annual listing, and found that progress was relatively low until the 1970s. However, after this time point there was an exponential increase in the number of papers published from 1975 to 1985, especially from 1980 to 1985. Growing interest in allelopathy is also evident from the foundation of the *International Allelopathy Society* in 1994 at New Delhi. Presently this society has over 300 members representing over 38 countries.

The growth pattern of the progress of allelopathic research followed the usual three phase phenomenon: (1) initial (initiation), (2) lag, and (3) developmental phase. The taxonomist Augustin de Candolle (1832) was the first to popularize the idea of the chemical interaction of plants. He did no experimental work of his own, but was substantially influenced by the simple experiments of Brugmans and Coulon in 1785 on root exudation in *Lolium* (*c.f.* Willis, 1985), and those by his colleague Macaire on various species, published in the 1830s (Willis, 1985). The concept of allelopathy languished in the second half of the 19th century, but was revitalized in the early 20th century largely due to the influence of Pickering in England, and Schreiner and Shorey in the United States (Willis, 1997).

From 1925 to 1972, the progress in allelopathic research was slow despite the advancement made during the World War II period, 1939 to 1945, when allelopathic activities were discovered accidentally in plants growing in California deserts. In the late 1960s and early 1970s, the pioneer work of Keating (1977, 1978), Muller (1969), and Rice (1984) placed this branch of science on firm footing. This was the developmental phase of the subject. The involvement of ecophysiologists and microbiologists widened the scope of allelopathic research. Although during this phase there was some classical ecological work on allelopathy by Keating (1977, 1978), Rice (1984), and Muller (1966, 1969), the bulk of the studies used preliminary bioassays (e.g., leachate or extract bioassay in absence of soil: see Rice, 1984, 1995). A large number of these studies did not focus on the importance of abiotic and biotic ecological components in bringing about allelopathic effects (Inderjit and Dakshini, 1995). As a result, allelopathy did not receive the attention of many ecologists, and those who studied it even faced criticism (Harper, 1977; Connell, 1990). During this phase, numerous papers were published on identification, isolation, and characterization of allelopathic compounds, and their role in interference was demonstrated. Thus, even though there was appreciable advancement of phytochemical aspects of allelopathy, ecological aspects remained largely neglected.

During the last decade, the importance of natural soils in allelopathy has been noted. This led to shifting the approach from laboratory studies to field studies. The importance of abiotic and biotic stress factors (e.g., nutrients, shade, herbicides, plant diseases, and herbivores), plant density, habitat, and climate has been realized (Einhellig, 1989, 1995, 1996; Weidenhamer, 1996; Weidenhamer et al., 1989; Blum et al., 1987, 1991, 1992, 1993, Dalton et al., 1989a, b; Fischer et al., 1994; Inderjit and Dakshini, 1996; Inderjit et al., 1996; Einhellig, this volume). These studies have been, in some way, follow-ups of some classical studies conducted between 1960 and 1970 (Rice, 1964, 1968, 1971; Rice and Pancholy, 1973, 1974; Muller, 1965, 1966, 1969; Muller and Muller, 1964; Del Moral and Cates, 1971; Del Moral and Muller, 1969; Del Moral et al., 1978; Jackson and Willemsen, 1976).

A multifaceted approach to allelochemical interaction is the principal challenge since a plant cannot be isolated from the environment. Entire ecophysiology adaptations and modifications of a plant's life-cycle processes are controlled by abiotic and biotic components of the ecosystem. This dynamic relationship influences not only the allelopathic potential of a plant, but also the pattern of synthesis, storage and release of allelochemicals, their persistence, and their fate in the environment (see other chapters in this volume by Huang et al.; Katase; Northup et al.; Schmidt and Ley).

1.3.2   ALLELOPATHY: A MULTIFACETED APPROACH

In nature many mechanisms of interference are operating, such as resource competition, allelopathy, microbial nutrient immobilization, and mycorrhizal activity. Separating these mechanisms, and identifying one as a probable cause of the observed patterning of vegetation, would be unnatural (Inderjit and Del Moral, 1997). Some investigators have argued that resource competition and allelopathy may operate simultaneously or sequentially in nature, and it is extraordinarily difficult, if not impossible, to separate these mechanisms. There are situations in which allelopathy evolved because of competition and vice-versa (Williamson, 1990).

1.3.2.1   Soil Nutrient Dynamics

Plants can make allelochemicals available through residue/debris/litter incorporation, natural release (mostly allelopathic compounds), litter decomposition, lignin degradation, etc. (Rice, 1984; Facelli and Pickett, 1991). Northup et al. (1995) reported that polyphenols released from leaf litter of *Pinus muricata* D. Don influenced the release of dissolved organic nitrogen and mineral nitrogen in soils. Northup et al. (this volume) have convincingly demonstrated how polyphenols influence soil nutrient dynamics. So far, however, not much emphasis has been given to the study of allelopathic compounds on soil nutrient dynamics. Supporting the lobby of ecologists who firmly believe that soil is a living dynamic biological system and realizing its significance in terms of gathering ecologically relevant data, we strongly encourage the study of the effects of allelopathic compounds on soil nutrient dynamics. We are confident that this approach will lead the way to receiving some plaudits from ecologists who are still critical about allelopathic research. For example, Harper (1977) believed that observed growth inhibition may be due to a temporary depletion of nitrogen as a result of higher microbial activity, which in turn is a result of the leaching of organics by plants. However, even if this is the case, the initial stage of this chain of events (organic molecules → higher microbial activity → temporary depletion of N → poor growth) begins only after the release of organic compounds into the soil. Therefore, if a multifaceted approach to allelopathy (i.e., the effect of allelopathic compounds on plants, soil microbes, nutrients, etc.) is adopted, the allelopathic interactions can be better understood.

Inderjit and Mallik (1997) reported the significance of the effects of phenolic compounds, protocatechuic, *p*-coumaric, *p*-hydroxybenzoic, and ferulic acids and catechol on soil inorganic ions in relation to allelopathy. The concentration of certain phenolics is correlated with soil podzolization (Vance et al., 1986). Plant phenolics have the potential to influence soil nutrients and rates of nutrient cycling, which ultimately influence the growth pattern (Lyu and Blum, 1990; Schlesinger, 1991). Phenolic compounds may influence the accumulation and availability of phosphate as they compete for anion absorption sites on clay and humus. This results in their binding to Al, Fe, and Mn, which may otherwise bind to phosphate, affecting its availability (Appel, 1993; Tan and Binger, 1986). To gain a better insight into the ecological roles of allelochemicals, their influence on soil nutrients must be addressed.

1.3.2.2   Soil Ecology and Allelochemical Expression

The abiotic and biotic components of soil may toxify or detoxify allelochemicals. The importance of soil microorganisms in allelopathy is well documented (Blum and Shafer, 1988; Schmidt and Ley, this volume). Juglone, for example, has been reported as a potent allelochemical released from black walnut (*Juglans nigra* L.), and is often cited as a classical example of allelopathy (Rice, 1984). *Pseudomonas putida,* a bacterium isolated from walnut-infested soils, has the potential to catabolize juglone (Rettenmaier et al., 1983; Schmidt, 1988). Schmidt (1990) suggests that the allelopathic influence of juglone depends on its interaction with abiotic and biotic conditions, as juglone is not always present at a phytotoxic threshold. Willis (this volume) reported that mature trees of certain species of *Eucalyptus* (e.g., *E. pilularis, E. delegetensis, E. regens*) accumulate allelochemicals in soil that results in a soil environment that favors pathogens and eventually causes the death of mature trees.

The sorption and retention of phenolic allelochemicals is well documented (Huang et al., 1977; Shindo and Kuwatsuka, 1975; Huang et al., this volume). While studying the differential sorption of exogenously applied ferulic, *p*-hydroxybenzoic, and vanillic acids in Cecil Portsmouth and White Store soils, Dalton et al. (1989b) reported that recovery of these phenolics varies with soil type, soil horizon, and type of functional group present on the aromatic ring of phenolic compound. Wang and Huang (1989) reported the catalytic transformation of pyrogallol, an important degradation product of plant debris/litter, by oxides of Mn, Fe, Al, and Si. While discussing the fate of allelochemicals in soils, Cheng (1989) opined that abiotic oxidation of allelochemicals is not directly related to Mn and Fe content of the soil, but rather to the degree of exposure of mineral surface to the organic molecules. He suggested that soil organic matter may prevent oxidation of allelochemicals by coating mineral surfaces. This aspect should be considered when assessing bioassays for allelopathy and in mulch studies. Agricultural soils often have a higher input of herbicides; additive effects of allelopathic compounds and herbicides have been suggested (Einhellig, 1996, and this volume).

1.3.2.3 Allelopathy in Relation to Ecological Factors

1.3.2.3.1 Environmental and Climatic Factors

Environmental and climatic factors influence production and release of allelochemicals (Einhellig, 1996; Weidenhamer, 1996). Jonasson et al. (1986) suggested that fluctuations in the levels of allelochemicals might be due to climatic variations. They state, ". . . The carbon nutrient balance could be altered by climatic variation that changes carbohydrate production. When carbohydrates are produced in excess of growth demands, excess photosynthates might be used for the production of phenolics or terpenes. Conversely, when photosynthesis is more limiting than nutrient absorption, carbohydrate concentrations in plants decline, as do phenolic or terpene levels, making plants more susceptible to herbivores or pathogens."

Inderjit and Dakshini (1994a, 1994b, 1996) demonstrated the allelopathic interference of *Pluchea lanceolata* (DC.) C. B. Clarke. The variation in different seasons and sites, in the allelopathic interference by *P. lanceolata* in terms of phenolic content has also been shown (Inderjit, 1998; Inderjit et al., 1996, 1998). The phenolic content of topsoil infested with *P. lanceolata* was maximal during October in Site 3 (Fig 1.1a); those of subsoil were maximal in July at Site 1 (Fig 1.1b).

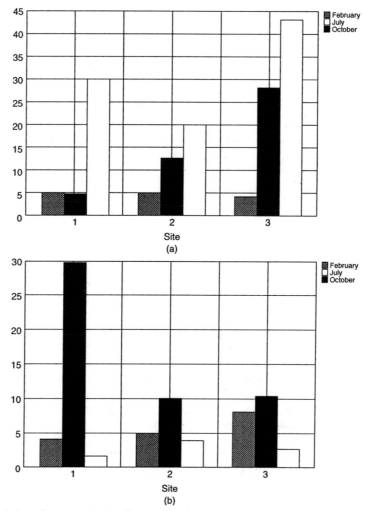

FIGURE 1.1 Total phenolic content during February, July, and October of (a) top soil, and (b) subsoil from three different sites infested with *Pluchea lanceolata*.

Similarly, influences of climate on allelochemical production have been reported elsewhere (Lodhi 1975, 1976; Weidenhamer and Romeo, 1989; Jalal and Read, 1983; Rice, 1984). In other words, climate, habitat, and environmental factors are important determinants of allelopathic interference in nature.

The significance of soil texture has been suggested in allelopathy (Del Moral and Muller, 1970; Oleszek and Jurzysta, 1987; Inderjit and Dakshini, 1994a). As discussed by Rice (1984), a complete inhibition of *Sorghum vulgare* L. crop, following the same crop, occurred on sandy soils in Senegal (Africa), but no such inhibition was observed on clay soils with high montmorillonite content due to rapid decomposition of allelochemicals. The influence of radiation, mineral deficiencies, water stress, and temperature on the production of allelopathic compounds has often been discussed (Del Moral, 1972; Koeppe et al., 1970; Rice, 1984; Inderjit and Del Moral, 1997). Additionally, the influence of habitat on the persistence and fate of allelochemicals has been suggested (J. R. Teasdale, Beltsville, MD, 1998 personal communication). Del Moral and Muller (1970) reported that degradation and decomposition of phenolic allelochemicals is delayed in arid/semiarid environments compared with humid conditions.

1.3.2.3.2   Fire
The role of allelopathy has been noted in relation to natural fire. Fischer et al. (1989, 1994) studied the allelopathic mechanism in two Florida scrub communities: (1) a sand pine scrub community dominated by *Pinus clausa* (Chapm. ex Engelm.) Vasey ex Sarg., and (2) a sandhill community dominated by *P. palustris* Mill. Surface fire occurs every 3 to 8 years in the sandhill community, and every 20 to 25 years in the sand pine community. Richardson and Williamson (1988) hypothesized that allelochemicals released from the scrub community inhibit fire-prone sandhill grasses. They reported that scrub community members allelopathically interfere with the sandhill grasses such as *Schizachyrium scoparium* (Michx.) Nash, *Andropogon gyrans* Ashe, and *Leptochloa dubia* (HBK.) Nees. There is a need to understand such ecological phenomena in relation to allelochemical interactions.

1.3.2.3.3   Ecological Succession
Rice and his students (see Rice, 1984) made in-depth studies to understand the probable involvement of allelopathy in old-field succession. The succession in old fields in Central Oklahoma included four stages: (1) pioneer weeds, (2) annual grasses, (3) perennial bunch grasses, and (4) true prairie. These investigators demonstrated the role of allelopathy in replacing one stage by another in old-field succession (Rice, 1964, 1968, 1971; Rice and Pancholy, 1973, 1974; Rice and Parenti, 1978). They suggested that the rate of nitrification is slow as succession proceeds toward climax in many ecosystems. However, this theory may not be universally applicable (Vitousek and Reiners, 1975; Robertson and Vitousek, 1981). In their review on changes in the rates of nitrification during primary, secondary, and old-field succession, Vitousek et al. (1989) reported that primary succession has low nitrogen availability and nitrification during early succession, but the rate of nitrification increases with succession. According to these authors, succession followed by chronic disturbances may control any increase or decrease in nitrification. The role of allelopathy in succession has also been suggested by Jackson and Willemsen (1976), Quinn (1974), and Gant and Clebsch (1975).

1.4   THE FUTURE OF ALLELOPATHY

An evaluation of earlier literature, published books (Rice, 1974, 1984, 1995; Thompson, 1985; Putnam and Tang, 1986; Waller, 1987; Chou and Waller, 1989; Rizvi and Rizvi, 1992; Inderjit et al., 1995), and international symposia on allelopathy clearly demonstrates that the study of allelopathy is a major discipline of ecology. The literature also shows that studies on allelopathy entered a logarithmic phase of growth since 1970. Since allelochemicals released have been shown to influence, in diverse ways, cultivated and natural ecosystems, efforts to understand and utilize allelopathy will continue.

Allelochemical interactions, in fact, are challenges to plant physiologists, ecologists, microbiologists, natural product chemists, soil scientists, and botanists (Putnam, 1987). The fact that allelopathic research involves a multifaceted approach is demonstrated by the variety of reports included in this volume. We are confident that the information included in the chapters will be a step forward to bring together increasing interaction of professionals from different disciplines to understand the complexity of the ecosystem. Ecologists, in particular, have a responsibility for demonstrating the importance of allelochemicals in plant–plant interactions, soil nutrient dynamics, soil microbial ecology, and rhizosphere ecology.

In the Conclusion and Recommendations section of the book edited by Chou and Waller (1989), Gliessman stated, "We need to develop a greater understanding of the role phytochemicals play at the ecosystem level. Team projects, several years long and interdisciplinary in nature, are becoming more important. We need to continue to move out of the laboratory and into the field and to develop tools for managing natural and agricultural ecosystems."

The information included in this book supports the need for future studies that provide evidence on the following assumptions:

1. Allelopathic potential varies with climatic, environmental, and habitat factors.
2. Abiotic and biotic soil components decide the fate (and ultimately biological activity) of allelochemicals.
3. In addition to influencing other plants present in the vicinity of donor plants, allelochemicals may influence soil nutrient dynamics, microbial ecology, and other abiotic and biotic components of the ecosystem.

1.4.1 AREAS NEEDING IN-DEPTH INVESTIGATIONS

Appreciating the significance of a multifaceted ecological approach to understanding allelochemical interactions, we suggest that the following areas should be given priority to better understand the scope of allelopathy and allelochemical interactions:

1. Significance of designing bioassays under more natural conditions (chapters in this volume by Blum; Foy; Elakovich; Dalton).
2. Production and release of allelochemicals in response to abiotic and biotic stress (see Einhellig; Chaves and Escudero, this volume). This aspect remains to be studied in most plants.
3. Influence of allelochemicals on soil nutrient dynamics (see Northup et al.; Inderjit and Dakshini, this volume).
4. Role of soil ecology in deciding the fate of allelochemicals (see Huang et al.; Katase; Schmidt and Ley, this volume).
5. Additive effects of allelochemicals as modified by other organics (e.g., herbicides) present in the substratum (Einhellig, this volume).
6. Use of allelochemicals in biocontrol of specific weedy taxa, especially in cultivated areas (Aliotta; Bull; Cutler; Macias, this volume).
7. Role of allelochemicals to improve degraded soils. If successful at the field level, it will be of significant value to phytoremediation programs. Mixing and loading of chemicals for application to crop plants has resulted in higher concentrations of herbicides and other

pesticides in agricultural soils (Kruger et al., 1997). Allelochemicals may help in degradation of certain synthetic chemicals responsible for soil degradation by increasing microbial activity, by influencing their availability and solubility in the medium, that is, creating conditions suitable for their degradation. The remedial potential of allelochemicals should be explored in terms of investigations and protection practices.

8. Whether allelopathic effects are species (receptor) specific, and how they vary in soils with different clay, silt, and loam content (Foy, this volume).
9. Role of allelochemicals in the evolution of donor plant and receptor species.
10. Impact of usual agricultural and silvicultural practices on allelochemical interactions.
11. Understanding of allelochemical mechanisms using the latest techniques available in molecular biology and biotechnology (see Section IV, this book).
12. Site/soil specific mode of release, transportation of allelochemicals from donor to receptor plant.

1.4.2 Basic Problems Related to Allelopathy Research

1. As discussed above, investigation related to allelochemical interactions is multifaceted and, therefore, to obtain meaningful data, active collaboration among natural products chemists, ecologists, microbiologists, soil scientists, and plant physiologists is required. Such collaborations are not always realistic and many times are difficult to achieve due to differing viewpoints or practical events as simple as differences in schedules. Putnam (1987) stated that plant ecologists and agricultural scientists in discussing plant interactions seldom agree on anything, except on some degree of competition. Furthermore, phytochemists may be interested only in new structures, which may not always be considered very important by ecologists or microbiologists. In any case, the need for such collaboration cannot be avoided, and concerted effort is needed to achieve it.
2. Presently worldwide research on allelopathy is not favored by funding agencies. The reasons are many, but mostly are due to lack of communication between scientists and bureaucrats, as well as the relative values given at present to basic and applied aspects of such investigations. Only recently has the costly damage caused by toxic algal blooms (Keating, this volume) drawn the attention of funding agencies to that single facet of allelochemical study.
3. There is insufficient dissemination of knowledge regarding allelopathy and allelochemical interactions. Since few textbooks mention or discuss this topic, a large section of the student community and general public are unaware of these phenomena. For the success of any discipline, a large participation is necessary and, therefore, scientists involved in such investigations have to work diligently to disseminate the importance of these fascinating phenomena.

1.5 CONCLUSION

This science of allelopathy has the potential to contribute greatly to agricultural production and stability (see Einhellig, this volume). C. S. Tang (1998 personal communication) states, "The greatest challenge in allelopathy as a science, perhaps, is the utilization of our existing knowledge in promoting sustainable agriculture and protecting natural plant resources. Only if allelopathy makes credible contribution to these objectives, can it catch the imagination of most researchers and gain prominence." Topics within this book address not only this aspect, but also findings regarding the role of ecological factors of the habitats on allelochemical interactions and other such issues. Collectively, they provide valuable information on allelopathy and the importance of allelochemical interactions.

ACKNOWLEDGMENTS

We sincerely thank Drs. Udo Blum, Frank A. Einhellig, Kathleen I. Keating, Roger del Moral, and Rick Willis for their comments on the manuscript.

REFERENCES

Aliotta, G. and Cafiero, G., Biological properties of rue *(Ruta graveolens* L.*)*: potential use in sustainable agricultural systems, 1999 (this volume), Chapter 34.

Appel, H. M., Phenolics in ecological interactions: the importance of oxidation. *J. Chem. Ecol.* 19, 1521, 1993.

Blum, U., Designing laboratory plant debris soil bioassays: some reflections, 1999 (this volume), Chapter 2.

Blum, U. and Shafer, S. R., Microbial populations and phenolic acids in soil. *Soil Biol. Biochem.* 20, 793, 1988.

Blum, U., Weed, S. B., Dalton, B. R., Influence of various soil factors on the effects of ferulic acid on leaf expansion of cucumber seedlings. *Plant Soil* 98, 111, 1987.

Blum, U., Wentworth, T. R., Klein, K., Worsham, A. D., King, L. M., Gerig, T. M., and Lyu, S. W., Phenolic acid content of soils from wheat no-till, wheat-conventional till, and fallow-conventional till soybean cropping systems. *J. Chem. Ecol.* 17, 1045, 1991.

Blum, U., Gerig, T. M., Worsham, A. D., Holappa, L. D., King, L. D., Allelopathic activity in wheat-conventional and wheat no-till soils: development of soil extract bioassays. *J. Chem. Ecol.* 18, 2191, 1992.

Blum, U., Gerig, T. M., Worsham, A. D., and King, L. D., Modification of allelopathic effects of *p*-coumaric acid on morning-glory seedling biomass by glucose, methionine, and nitrate. *J. Chem. Ecol.* 19, 2791, 1993.

Bull, C. T., Allelopathic interactions in the biological control of postharvest diseases of fruit, 1999 (this volume), Chapter 32.

Chaves, N. and Escudero, J. C., Variation of flavonoid synthesis induced by ecological factors, 1999 (this volume), Chapter 17.

Cheng, H. H., Assessment of the fate and transport of allelochemicals in the soil, *in Phytochemical Ecology: Allelochemicals, Mycotoxins and Insect Pheromones and Allomones,* Chou, C. H. and Waller, G. R., Eds., Institute of Botany, Academia Sinica Monogr. Ser. No. 9, Taipei, ROC, 1989, 209.

Chou, C. H. and Waller, G. R., Eds., *Phytochemical Ecology: Allelochemicals, Mycotoxins, and Insect Pheromones and Allomones,* Academia Sinica, Monogr. Ser. No. 9, Acad. Sinica, Taipei, ROC, 1989.

Connell, J. H., Apparent versus "real" competition in plants, *in Perspectives in Plant Competition,* Grace, J. B. and Tilman, D., Eds., Academic Press, San Diego, CA, 1990, 9.

Cutler, H. G., Potentially useful natural product herbicides from microorganisms, 1999 (this volume), Chapter 31.

Dalton, B. R., The occurrence and behavior of plant phenolic acids in soil environment and their potential involvement in allelochemical interference interactions: methodological limitations in establishing conclusive proof of allelopathy, 1999 (this volume), Chapter 6.

Dalton, B. R., Blum, U., and Weed, S. B., Differential sorption of exogenously applied ferulic, *p*-coumaric, *p*-hydroxybenzoic, and vanillic acids in soil. *Soil Sci. Soc. Am. J.* 53, 757, 1989a.

Dalton, B. R., Blum, U., and Weed, S. B., Plant phenolic acids in soils: sorption of ferulic acid by soil and soil component sterilized by different techniques. *Soil Biol. Biochem.* 21, 1011, 1989b.

de Candolle, M. A. P., *Physiologie Vegetale.* III. Bechet Jeune, Lib. Fac. Med. 1474, Paris, 1832.

Del Moral, R., On the availability of chlorogenic acid concentration. *Oecologia* 15, 65, 1972.

Del Moral, R. and Muller, C. H., Fog drip: a mechanism of toxin transport from *Eucalyptus globulus. Bull. Torrey Bot. Club* 96, 467, 1969.

Del Moral, R. and Muller, C. H., The allelopathic effects of *Eucalyptus camaldulensis. Am. Midl. Nat.* 83, 254, 1970.

Del Moral, R. and Cates, R. G., Allelopathic potential of the dominant vegetation of western Washington. *Ecology* 52, 1030, 1971.

Del Moral, R., Willis, R. J., and Ashton, D. H., Suppression of coastal heath vegetation by *Eucalyptus baxteri. Aust. J. Bot.* 26, 203, 1978.

Einhellig, F. A., Interactive effects of allelochemicals and environmental stress, *in Phytochemical Ecology: Allelochemicals, Mycotoxins, and Insect Pheromones and Allomones,* Chou, C. H. and Waller, G. R., Eds., Institute of Botany, Academia Sinica Monogr. Ser. No. 9, Taipei, ROC, 1989, 101.

Einhellig, F. A., Allelopathy: current status and future goals, *in Allelopathy: Organisms, Processes, and Applications,* Inderjit, Dakshini, K. M. M., and Einhellig, F. A., Eds., American Chemical Society, Washington, D.C., 1995, 1.

Einhellig, F. A., Interactions involving allelopathy in cropping systems. *Agron. J.* 88, 886, 1996.

Einhellig, F. A., An integrated view of allelochemicals and multiple stresses, 1999 (this volume), Chapter 30.

Elakovich, S. D., Bioassays applied to allelopathic herbaceous vascular hydrophytes, 1999 (this volume), Chapter 5.

Facelli, J. M. and Pickett, S. T. A., Plant litter: its dynamics and effects on plant community structure. *Bot. Rev.* 57, 1, 1991.

Fischer, N. H., Williamson, G. B., Weidenhamer, J. D., Tanrisever, N., de la Pena, A., Jordan, E. D., and Richardson, D. R., Allelopathic mechanisms in the Florida scrub community, *in Phytochemical Ecology: Allelochemicals, Mycotoxins, and Insect Pheromones and Allomones,* Chou, C. H., and Waller, G. R., Eds., Institute of Botany, Academia Sinica Monogr. Ser. No. 9, Taipei, ROC, 1989, 183.

Fischer, N. H., Williamson, G. B., Weidenhamer, J. D., and Richardson, D. R., In search of allelopathy in the Florida scrub: the role of terpenoids. *J. Chem. Ecol.* 20, 1355, 1994.

Foy, C. L., Some suggestions on designing bioassays for allelopathy under controlled conditions, 1999 (this volume) Chapter 3.

Gant, R. E. and Clebsch, E. E. C., The allelopathic influences of *Sassafras albidum* in old-field succession in Tenessee. *Ecology* 56, 604, 1975.

Goldberg, D. E., Components of resource competition in plant communities, *in Perspectives on Plant Competition,* Grace, J. B., and Tilman, D., Eds., Academic Press, New York, 1990, 27.

Harborne, J. B., *Introduction to Ecological Chemistry,* Academic Press, London, 1977.

Harper, J. L., *Population Biology of Plants,* Academic Press, London, 1977.

Huang, P. M., Wang, T. S. C., Wang, M. K., Wu, M. H., and Hsu, N. W., Retention of phenolic acids by noncrystalline hydroxy-aluminum and -iron compounds and clay mineral of soil. *Soil Sci.* 123, 213, 1977.

Huang, P. M., Wang, M. C., and Wang, M. K., Catalytic transformation of phenolic compounds in soils, 1998. (this volume)

Inderjit, Influence of *Pluchea lanceolata* on selected soil properties. *Am. J. Bot.* 85, 303, 1998.

Inderjit and Dakshini, K. M. M., Allelopathic effect of *Pluchea lanceolata* (Asteraceae) on characteristics of four soils and tomato and mustard growth. *Am. J. Bot.* 81, 799, 1994a.

Inderjit and Dakshini, K. M. M., Allelopathic potential of phenolics from the roots of *Pluchea lanceolata.* *Physiol. Plant.* 92, 571, 1994b.

Inderjit and Dakshini, K. M. M., On laboratory bioassays in allelopathy. *Bot. Rev.* 61, 28, 1995.

Inderjit and Dakshini, K. M. M., Allelopathic potential of *Pluchea lanceolata:* comparative study of cultivated fields. *Weed Sci.* 44, 393, 1996.

Inderjit and Dakshini, K. M. M., Bioassays for allelopathy: interactions of soil organic and inorganic constituents, 1999 (this volume), Chapter 4.

Inderjit and Mallik, A. U., Effect of phenolic compounds on selected soil properties. *For. Ecol. Manage.* 92, 11, 1997.

Inderjit and Del Moral, R., Is separating resource competition from allelopathy realistic? *Bot. Rev.* 63, 221, 1997.

Inderjit, Dakshini, K. M. M., and Einhellig, F. A., Eds., *Allelopathy: Organisms, Processes, and Applications.* American Chemical Society, Washington, D.C., 1995, 389.

Inderjit, Surinder Kaur, and Dakshini, K. M. M., Determination of allelopathic potential of a weed *Pluchea lanceolata* through a multi-faceted approach. *Can. J. Bot.* 74, 1445, 1996.

Inderjit, Foy, C. L., and Dakshini, K. M. M., *Pluchea lanceolata:* A noxious weed. *Weed Technol.* 12, 190, 1998.

Jackson, J. R. and Willemsen, R. W., Allelopathy in the first stages of secondary succession on the piedmont of New Jersey. *Am. J. Bot.* 63, 1015, 1976.

Jalal, M. A. and Read, D. J., The organic acid decomposition of *Calluna* heathland soil with special reference to phyto- and fungitoxicity. II. Monthly quantitative determination of the organic acid content of *Calluna* and spruce-dominated soil. *Plant Soil* 70, 273, 1983.

Jonasson, S., Bryant, J. P., Chapin, F. S., III, and Anderson, M., Plant phenols and nutrients in relation to variation in climate and rodent grazing. *Am. Nat.* 128, 394, 1986.

Katase, T., Lignin-related phenolic acids in peat soils and implications in tropical rice sterility problems, 1999 (this volume) Chapter 21.

Keating, K. I., Allelopathic influence on blue-green blooms sequence in a eutrophic lake. *Science* 196, 885, 1977.

Keating, K. I., Blue green algal inhibition of diatom growth: Transition from mesotrophic to eutrophic community structure. *Science* 199, 971, 1978.

Keating, K. I., Allelochemistry in plankton communities, 1999 (this volume), Chapter 11.

Koeppe, D. E., Rohrbaugh, L. M., Rice, E. L., and Wender, S. H., The effect of age and chilling temperature on the concentration of scopolin and caffeoylquinic acids in tobacco. *Physiol. Plant.* 23, 258, 1970.

Kruger, E. L., Anderson, T. A., and Coats, J. R., Eds., *Phytoremediation of Soil and Water Contaminants.* ACS Symposium Series 664, American Chemical Society, Washington, D.C., 1997.

Lodhi, M. A. K., Soil-plant toxicity and its significance in patterning of herbaceous vegetation in bottomland forest. *Am. J. Bot.* 62, 618, 1975.

Lodhi, M. A. K., Role of allelopathy as expressed by dominating trees in a lowland forest in controlling productivity and pattern in herbaceous growth. *Am. J. Bot.* 63, 1, 1976.

Lyu, S. W. and Blum, U., Effect of ferulic acid, an allelopathic compound, on net P, K, and water uptake in cucumber seedling in a split-root system. *J. Chem. Ecol.* 16, 2429, 1990.

Macias, F. A., Potentiality of cultivar sunflower (*Helianthus annuus* L.) as a source of natural herbicide models, 1999 (this volume) Chapter 33.

Muller, C. H., Inhibitory terpenes volatilized from *Salvia* shrubs. *Bull. Torrey Bot. Club* 92, 38, 1965.

Muller, C. H., The role of chemical inhibition (allelopathy) in vegetational composition. *Bull. Torrey Bot. Club* 93, 332, 1966.

Muller, C. H., Allelopathy as a factor in ecological process. *Vegetatio* 18, 348, 1969.

Muller, W. H. and Muller, C. H., Volatile growth inhibitors produced by *Salvia* species. *Bull. Torrey Bot. Club* 91, 327, 1964.

Northup, R. R., Yu, Z., Dahlgren, R. A., and Vogt, K. A., Polyphenol control of nitrogen release from pine litter. *Nature* 377, 227, 1995.

Northup, R. R., Dahlgren, R. A., Aide, T. M., and Zimmerman, J. K., Effect of plant polyphenols on the nitrogen cycle and implications for community structure, 1999 (this volume), Chapter 22.

Oleszek, W. and Jurzysta, M., The allelopathic potential of alfalfa root medicagenic acid glycosides and their fate in soil environment. *Plant Soil* 98, 67, 1987.

Putnam, A. R., Introduction, *in Allelochemicals: Role in Agriculture and Forestry,* Waller, G. R., Ed., ACS Sym. Ser. 330, American Chemical Society, Washington, D.C., 1987, xiii.

Putnam, A. R. and Tang, C. S., Allelopathy: state of the science, *in The Science of Allelopathy,* Putman, A. R. and Tang, C. S., Eds., John Wiley & Sons, New York, 1986, 1.

Quinn, J. A., *Convolvulus sepia* in old field succession on the New Jersey Piedmont. *Bull. Torrey Bot. Club* 101, 89, 1974.

Rettenmaier, H., Kupas, U., and Lingens, F., Degradation of juglone by *Pseudomonas putida* J 1. *FEMS Microbiol. Lett.* 19, 193, 1983.

Rice, E. L., Inhibition of nitrogen-fixing and nitrifying bacteria by seed plants. I,. *Ecology* 45, 824, 1964.

Rice, E. L., Inhibition of nodulation of inoculated legumes by pioneer plant species from abandoned fields. *Bull. Torrey Bot. Club* 95, 346, 1968.

Rice, E. L., Inhibition of nodulation of inoculated legumes by leaf leachates from pioneer plant species from abandoned fields. *Am. J. Bot.* 58, 368, 1971.

Rice, E. L., *Allelopathy.* Academic Press, New York, 1974.

Rice, E. L., *Allelopathy.* Academic Press, Orlando, FL, 1984.

Rice, E. L., *Biological Control of Weeds and Plant Diseases: Advances in Applied Allelopathy,* University of Oklahoma Press, Norman, OK, 1995.

Rice, E. L. and Pancholy, S. K., Inhibition of nitrification by climax ecosystems. II. Additional evidence and possible role of tannins. *Am. J. Bot.* 60, 691, 1973.

Rice, E. L. and Pancholy, S. K., Inhibition of nitrification by climax ecosystems. III. Inhibitors other than tannins. *Am. J. Bot.* 61, 1095, 1974.

Rice, E. L. and Parenti, R. L., Causes of decreases in productivity in undisturbed tall grass prairie. *Am. J. Bot.* 65, 1091, 1978.

Richardson, D. R. and Williamson, G. B., Allelopathic effects of shrubs of sand pine scrub on pines and grasses of the sandhills. *For. Sci.* 34, 592, 1988.

Rizvi, S. J. H. and Rizvi, V., Eds., *Allelopathy: Basic and Applied Aspects,* Chapman and Hall, London, 1992.

Robertson, G. P. and Vitousek, P. M., Nitrification potentials in primary and secondary succession. *Ecology* 63, 1561, 1981.

Schlesinger, W. H., *Biogeoochemistry: An Analysis of Global Change,* Academic Press, San Diego, CA, 1991.

Schmidt, S. K., Degradation of juglone by soil bacteria. *J. Chem. Ecol.* 14, 1561, 1988.

Schmidt, S. K., Ecological implications of the destruction of juglone (5-hydroxy-1,4-naphthoquinone) by soil bacteria. *J. Chem. Ecol.* 16, 3547, 1990.

Schmidt, S. K. and Ley, R., Microbial competition and soil structure limit the expression of allelochemicals in nature, 1999 (this volume) Chapter 20.

Shindo, H. and Kuwatsuka, S., Behaviour of phenolic substances in the decaying process of plants. III. Degradation pathway of phenolic acids. *Soil Sci. Plant Nutr.* 21, 227, 1975.

Tan, K. H. and Binger, A., Effect of humic acid on aluminum toxicity in corn plants. *Soil Sci.* 141, 20, 1986

Thompson, A. C. (Ed.), *The Chemistry of Allelopathy,* ACS Symposium Series 268, American Chemical Society, Washington, D.C., 1985.

Vance, G. F., Mokma, D. L., and Boyd, S. A., Phenolic compounds in soils of hydrosequences and developmental sequences of podzols. *Soil Sci. Soc. Am. J.* 50, 992, 1986.

Vitousek, P. M. and Reiners, W. A., Ecosystem succession and nutrient retention: a hypothesis. *Bioscience* 25, 376, 1975.

Vitousek, P. M., Matson, P. A., and Van Cleve, K., Nitrogen availability and nitrification during succession: primary, secondary, and old-field seres. *Plant Soil* 115, 229, 1989.

Waller, G. R. (Ed.), *Allelochemicals: Role in Agriculture and Forestry,* ACS Symposium Series 330, American Chemical Society, Washington, D.C., 1987.

Wang, M. C. and Huang, P. M., Pyrogallol transformations as catalyzed by oxides of Mn, Fe, Al and Si, *in Phytochemical Ecology: Allelochemicals, Mycotoxins and Insect Pheromones and Allomones,* Chou, C. H. and Waller, G. R., Eds., Institute of Botany, Academia Sinica Monogr. Ser. No. 9, Taipei, ROC, 1989, 195.

Weidenhamer, J. D., Distinguishing resource competition and chemical interference: Overcoming the methodological impasse. *Agron. J.* 88, 866, 1996.

Weidenhamer, J. D. and Romeo, J. T., Allelopathic properties of *Polygonella myriophylla:* field evidence and bioassays. *J. Chem. Ecol.* 15, 1957, 1989.

Weidenhamer, J. D., Hartnett, D. C., and Romeo, J. T., Density-dependent phytotoxicity: distinguishing resource competition and allelopathic interference in plants. *J. Appl. Ecol.* 26, 613, 1989.

Whittaker, R. H. and Feeny, P. P., Allelochemicals: chemical interactions between plant species. *Science* 171, 757, 1971.

Williamson, G. B., Allelopathy, Koch's postulates and the neck riddle, *in Perspectives in Plant Competition,* Grace, J. B. and Tilman, D., Eds., Academic Press, New York, 1990, 143.

Willis, R. J., The historical basis of the concept of allelopathy. *J. Hist. Biol. 18,* 71, 1985.

Willis, R. J., The history of allelopathy, 2. The second phase (1900–1920): the era of S.U. Pickering and the U.S.D.A. bureau of soils. *Allelo J.* 4, 7, 1997.

Willis, R. J., Australian studies of allelopathy in *Eucalyptus:* a review, 1999 (this volume) Chapter 13.

Section I

Methodologies

2 Designing Laboratory Plant Debris–Soil Bioassays: Some Reflections

Udo Blum

CONTENTS

2.1 ABSTRACT

Concentration-dependent inhibition or stimulation of germination or seedling growth in soils with surface or incorporated plant debris has been interpreted frequently as evidence for allelopathic interactions. However, allelopathic interactions may or may not be the actual causal agents because one can never be certain whether the observed effects are actually due to allelopathic interactions or due to physicochemical (e.g., nutritional, soil temperature, and moisture) and biotic (e.g., microbial) modifications associated with increasing plant debris on or in the soil. Therefore, experimental designs necessary to establish allelopathic interactions as causative agents must include appropriate controls for such physicochemical and biotic factors. Furthermore, putative allelopathic agents must be isolated, and their mode of transmission and action must be characterized before an allelopathic interaction can be identified with confidence.

0-8493-2116-6/99/$0.00+$.50

2.2   INTRODUCTION

Plant debris–soil bioassays in the laboratory have been and continue to be an important tool for investigating allelopathic interactions between plants. In most instances, bioassays consist of monitoring seed germination and/or seedling growth in soils with surface debris or containing incorporated plant debris. Soils without debris or with biologically inert debris are used as controls. Concentration-dependent inhibition or stimulation of germination or seedling growth is considered by most investigators working in the area of allelopathy as evidence for allelopathic activity.

Superficially such bioassays appear to be simple and straightforward, but in fact they are not. Laboratory bioassays are simplified representations of managed or natural systems. Bioassays consist of black boxes in which the levels of debris (inputs) are related to rates of germination or seedling growth (outputs). Establishing such relationships are extremely important to the study of allelopathy, but the relevance of the resulting relationships to the system under study will depend upon how well a laboratory bioassay reflects the true environment of the system under consideration. Furthermore, the observed relationships may or may not be a result of allelopathic interactions. For example, one cannot be certain that the observed behavior of seedlings in such bioassays is a result of chemical, physical, or biotic modifications resulting from the increasing levels of surface or incorporated debris (Qasem and Hill, 1989). Thus, laboratory debris bioassays are, as a general rule, only a first step in identifying mechanisms of plant behavior that may appear to involve allelopathy. At the minimum, allelopathic agents must be isolated, their mode of transmission and action must be identified, and other soil factors (e.g., soil moisture and temperature) must be eliminated as causative agents before allelopathy can be implicated as the causative agent (Willis, 1985). In some situations, however, it may not be possible to identify individual allelopathic agents, because responses of bioassay species may be due to complex mixtures of toxic and nontoxic organic compounds, each at very low concentrations (Blum, 1996). Consequently, care must be taken in designing bioassay experiments and in interpreting their results. Bioassays should not be initiated unless and until clear objectives have been formulated, because the experimental design of any debris bioassay must be consistent with its objectives. There are, however, a series of common experimental considerations that must be addressed when designing any plant debris bioassay.

2.3   LIST OF EXPERIMENTAL CONSIDERATIONS

2.3.1   Type of Plant Material (or Debris) to Be Tested

What should be collected:

1. Living or partially decomposed or weathered material?
2. Leaf, root, stem, reproductive structure, or whole-plant material?
3. Material from a single species or from a multiple-species mixture?

2.3.2   Handling and Storage of Plant Material

Should the collected material be:

1. Used immediately, frozen, freeze-dried, air-dried, and/or oven-dried?
2. Ground or stored whole?
3. Stored in an open or closed container, in a desiccator, at room temperature and/or in a freezer?

How long can frozen or dried material be stored before its chemical nature is modified?

2.3.3 ADDITION OF PLANT MATERIAL TO SUBSTRATE

Should the plant material or debris be:

1. Ground, cut, or added whole?
2. Added to the surface or mixed into the substrate?
3. Mixed only in the upper part of a container or throughout the entire substrate in the container?

How much material should be added to each container? What type of substrate should be used?

2.3.4 BIOASSAY SPECIES

What type of bioassay species should be used? How many seeds or seedlings should be used in each experimental unit (e.g., petri dish or pot)? When should the bioassay species be introduced into the system? Should seeds or seedlings be introduced immediately after addition of plant material or after some aging/decomposition of the debris? When should the experiment be terminated and the bioassay species measured or harvested? What parameters should be measured?

2.3.5 GROWTH ENVIRONMENT

What type of container should be used (e.g., size, open, or closed on bottom)? Should the bioassay be conducted under a light bank, in a growth chamber, in the greenhouse, or outdoors? How and when should the bioassay system be fertilized and watered? Should substrate be limed and/or autoclaved before use?

2.3.6 ADEQUATE CONTROLS

What type of controls should be included to identify and characterize the effects of the various soil physicochemical and biotic factors resulting from the addition of plant materials or debris? When and how should physicochemical soil factors be regulated or monitored?

2.4 EXPERIMENTAL APPROACH

The approach below describes how the phytotoxicity of plant materials and debris can be maximized in laboratory bioassays. This approach may or may not be very realistic or appropriate for a given system under study, but it does provide an opportunity to list experimental approaches that should be utilized with extreme care when bioassays are conducted in the laboratory.

2.4.1 HOW TO MAXIMIZE THE INPUT OF PHYTOTOXINS INTO THE SOIL

Collect living tissues (leaf tissue frequently is the best) of healthy but slightly nutrient or water stressed plants from the field prior to extended rain events and freeze-dry them. This will ensure the highest amount of soluble organic matter, including phytotoxins, in the tissue. If tissue must be stored before use, store it intact (not ground) because oxidation occurs rapidly after grinding. Cut or grind the tissue neither too fine nor too coarse (this may require some preliminary experimentation) immediately before adding tissue to soil (Ells and McSay, 1991). This will provide an extended maximum dosage (assuming agents do not have to be produced by microbial activity) of soluble organic matter, including toxins, instead of a single pulse. A broad range of tissue concentrations should also be used to maximize the range of input of phytotoxins.

2.4.2   How to Maximize the Detection of Phytotoxic Effects

Test seedling emergence or growth (the part of the life cycle most likely to be affected) and use species that are very sensitive (e.g., species such as lettuce [*Lactuca sativa* L.], cucumber [*Cucumis sativus* L.], etc.) instead of more tolerant species. Use only individual seeds or seedlings to maximize the concentration of the phytotoxins interacting with the bioassay species. Increasing density has been shown to reduce the effects of toxins (Weidenhamer et al., 1987, 1989; Thijs et al., 1994). Use a growth medium and container which in combination will minimize leaching losses of and maximize seed or root contact with soluble organic matter. Soil is generally better than sand unless the substrate is contained in a closed-bottom vessel. Mix the cut or ground tissue throughout the soil so that all roots of the bioassay seedling will be in continuous contact with the soluble organic toxins released from the tissue. Provide moisture and nutrition for optimizing the release and/or microbial production of phytotoxins from plant tissue, with the understanding that too much moisture and nutrition can lead to rapid detoxification of phytotoxins by microorganisms.

2.4.3   How Appropriate Is This Approach?

Any of these techniques to maximize allelopathic interactions may be used, but the implications of utilizing these techniques should be clearly understood by the investigator. For example, testing soil incorporated freeze-dried living plant tissue is inappropriate if the observed effects in the field occur under herbicide desiccated surface debris, but may be quite appropriate if the observed effects in the field occur when green manure (living plant material) is incorporated into the soil.

Obviously the type and the timing of collection of tissue or debris must be consistent with the suspected occurrences of allelopathic interactions in the field. Storage and manipulation of collected debris should be minimized, and the debris should be placed on the surface of or incorporated into field soil in a way that is consistent with the field environment. The use of surrogate species (i.e., sensitive species) is appropriate as long as the actual field species are also included for comparison. Allelopathic effects are concentration-dependent, so most researchers will add a range of debris concentrations. These concentrations should span the range found in the field. Unfortunately, changes in soil pH, water holding capacity, microbial populations, aeration, nutrition, soil temperature, etc., are common with increasing concentrations of debris on or in the soil. The co-linearity of many of these factors makes conclusive identification of causative agents difficult. However, attempts have been made to deal with this vexing problem.

2.5   CASE HISTORIES OF EXPERIMENTAL APPROACHES

The first case history illustrates how biologically inactive materials, in this instance *Populus* wood shavings, can be used to help separate the physical and chemical effects of rye (*Secale cereale* L.) surface debris on weed biomass (Putnam and DeFrank, 1983). *Populus* wood shavings, which did not have apparent toxic or nutritional effects on weeds being tested, were applied to the surface of soil in flats at rates equal to applications of rye debris. Wood shavings reduced the light intensity by approximately 97 percent and soil temperature by about 1°C. This was similar to effects of the rye surface debris. A bare soil treatment was included as control. Weed fresh weight in the flats was suppressed 44 percent by the shavings (physical effects) and 80 percent by the rye (physical and allelopathic effects). When debris was mixed throughout the soil, these effects did not occur (DeFrank and Putnam, 1978; Putnam et al., 1983). Apparently, debris on the soil surface created an inhibition zone near the surface of the soil, the site of seed germination, which was not formed when debris was mixed into the soil. A number of phytotoxins have been identified for rye (Chou and Patrick, 1976; Barnes and Putnam, 1987; Barnes et al., 1986; Gagliardo and Chilton, 1992; Yenish et al., 1995), although it has not been established which of these phytotoxins (alone or in combination) is really important in the field soil environment.

The second case history illustrates how co-linearity of the soil physicochemical factors associated with increasing debris levels can be minimized by mixing the same amount of debris, but with different concentrations of phytotoxins, into the soil. Hall et al. (1982) grew sunflower (*Helianthus annuus* L.) plants under various nutrient levels to create sunflower tissues with different total phenolic acid (TPA) contents. The resulting sunflower tissue allowed these researchers to mix identical amounts of these tissues into soil. Hall et al. (1982) observed that germination of redroot pigweed (*Amaranthus retroflexus* L.) declined as the TPA content of the sunflower tissue mixed into soil increased. Addition of nutrient solution to the debris–soil mixture brought about a decline in the inhibition of redroot pigweed germination. The inhibitory effects of TPA on redroot pigweed seedling growth, however, were significant only when total sunflower tissue nitrogen, phosphorus, and potassium were included in regression models (Hall et al., 1983). TPA and total nitrogen were related negatively, while total phosphorus and potassium were related positively to redroot pigweed seedling biomass. The addition of nutrient solution eliminated the inhibitory effects of sunflower tissue on redroot pigweed seedling growth. Thus nutrient content of the soil as well as that in the sunflower tissue added to the soil (i.e., representing debris) substantially influenced the response of pigweed seed germination and seedling growth to phytotoxins. The most abundant phenolic acids in sunflower are the chlorogenic acids, so chlorogenic acid was substituted for sunflower tissue in another bioassay. Germination of redroot pigweed was not affected by chlorogenic acid, but seedling biomass was inversely related to chlorogenic acid levels applied to the soil. As with sunflower tissue, the effects of chlorogenic acid on seedling biomass were eliminated when nutrient solution was added. A number of phytotoxins, other than chlorogenic acids, have also been identified for sunflower tissue (Wilson and Rice, 1968; Einhellig et al., 1970; Lehman and Rice, 1972; Macias et al., 1993).

The third case history illustrates how a physical factor, such as soil moisture, can mislead the unwary researcher. Lehman and Blum (1996) incorporated into soil different quantities of ground clover crimson (*Trifolium incarnatum* L. CV Tibbee) and subterranean clover (*Trifolium subterraneum* L. CV Mt. Barker) and small grain [rye (*Secale cereale* L. CV Wrens Abruzzi) and wheat (*Triticum aestivum* L. CV Coker 983] surface debris collected at various times after glyphosate [*N*-(phosphomethyl)glycine] desiccation of cover crops and monitored subsequent pigweed seedling emergence. Soil nitrogen, temperature, and moisture were controlled. Inhibition of pigweed emergence by surface debris collected after glyphosate desiccation declined rapidly for all cover crops except wheat. Nitrogen up to 14 μgN g soil^{-1} did not modify pigweed seedling emergence. Redroot pigweed emergence was related directly to temperature and inversely related to increasing concentrations of toxic debris, but the effects of temperature and debris on seedling emergence were independent of each other. Soil moisture, however, modified the effects of debris. Inhibition of pigweed emergence by phytotoxic debris was not evident when soil moisture levels were greater than or equal to 0.147 g H_2O g soil^{-1} (i.e., the inhibitory effects of soil moisture were dominant over the phytotoxins). Emergence was stimulated in the presence of nonphytotoxic debris when the soil moisture level was 0.173 g H_2O g soil^{-1}, the highest level tested. Nontoxic debris appeared to be binding soil water. Although levels of soil moisture in this study were above field capacity (0.095 g H_2O g soil^{-1}), they were below the highest average seasonal moisture observed under surface rye debris for 1992 and 1993 (0.20 g H_2O g soil^{-1}). A variety of phytotoxins have also been identified for these small grain and clover cover crops (Chou and Patrick, 1976; Liebl and Worsham, 1983; Shilling et al, 1985, 1986a, b; Barnes et al., 1986).

Finally, research by Weidenhamer et al. (1989) and Thijs et al. (1994) indicated that phytotoxic effects are density, as well as, concentration dependent. They observed that phytotoxicity of known toxins decreased as density of plants increased. They attributed this decline of toxicity to a reduced availability (i.e., dilution effect) of toxins to individual plants as density increased. This deviation from the expected plant-density response, they proposed, could be used as a tool to identify resource competition from allelopathy. Although not experimentally determined it may be possible, for example, to use this deviation to rule out competition for resources by plants and microbes in plant debris–soil bioassays.

These experiments suggest that the observed effects of debris bioassays using increasing amounts of surface or soil incorporated debris result from physical (e.g., temperature, moisture), chemical (e.g., debris toxin, nutrient content), and/or associated biological (e.g., microbial) factors that must be characterized before the role of any one factor as a causative agent can be identified. This may be done by the inclusion of an appropriate control, such as *Populus* shavings (Putman and DeFrank, 1983), utilization of constant amounts of debris with different chemical compositions (Hall et al., 1982; Lehman and Blum, 1996), with orthogonal factorial experiments that keep physical factors independent of debris levels (Lehman and Blum, 1996), and/or with density studies (Weidenhamer et al., 1989). In all cases, however, every attempt should be made to include experimental conditions which are within the range of the field environment under study (Inderjit and Dakshini, 1995). Finally, it should be recognized that laboratory bioassays can never prove that allelopathy is operational in the field, they can only demonstrate the possibility.

ACKNOWLEDGMENTS

The author wishes to thank R. C. Fites, M. E. Lehman, S. R. Shafer, T. R. Wentworth, and J. D. Weidenhamer for their review and thoughtful comments.

REFERENCES

Barnes, J. P. and Putnam, A. R., Role of benzoxazinones in allelopathy by rye, *J. Chem. Ecol.*, 4, 889, 1987.

Barnes, J. P., Putnam, A. R., Burke, B. A., and Aasen, A. J., Isolation and characterization of allelochemicals in rye herbage, *Phytochemistry*, 26, 1385, 1986.

Blum, U., The use of plant-microbe-soil model systems for characterizing allelopathic interactions involving mixtures of phenolic acids and/or other compounds, *J. Nematol.*, 28, 259, 1996.

Chou, C. H. and Patrick, Z. A., Identification and phytotoxic activity of compounds produced during decomposition of corn and rye residues in soil, *J. Chem. Ecol.*, 2, 369, 1976.

DeFrank, J. and Putnam, A. R., Weed and crop response to allelopathic crop residues, *North Cent. Weed Control Conf. Proc.*, 33, 44, 1978.

Einhellig, F. A., Rice, E. L., Risser, P. G., and Wender, S. H., Effects of scopoletin on growth, CO_2 exchange rates, and concentration of scopoletin, scopolin, and chlorogenic acid in tobacco, sunflower and pigweed. *Bull. Torrey Bot. Club*, 97, 22, 1970.

Ells, J. E. and McSay, A. E., Allelopathic effects of alfalfa plant residues on emergence and growth of cucumber seedlings, *Hortscience*, 26, 368, 1991.

Gagliardo, R. W. and Chilton, W. S., Soil transformation of 2(3H)-benzoxazolone of rye into phytotoxic 2-amino-3H-phenoxazin-3-one, *J. Chem. Ecol.*, 18, 1683, 1992.

Hall, A. B., Blum, U., and Fites, R. C., Stress modification of allelopathy of *Helianthus annuus* L. debris on seed germination, *Amer. J. Bot.*, 69, 776, 1982.

Hall, A. B., Blum, U., and Fites, R. C., Stress modification of allelopathy of *Helinathus annuus* L. debris on seedling biomass production of *Amaranthus retroflexus* L., *J. Chem. Ecol.,* 9, 1213, 1983.

Inderjit and Dakshini, K. M. M., On laboratory bioassays in allelopathy, *Bot. Rev.,* 61, 28, 1995.

Lehman, M. E. and Blum, U., Debris effects on weed emergence as modified by environmental factors, *Allelopathy J.,* 4, 69, 1996.

Lehman, R. H. and Rice, E. L., Effect of deficiencies of nitrogen, potassium and sulfur on chlorogenic acids and scopolin in sunflower, *Am. Midl. Nat.,* 87, 71, 1972.

Liebl, R. A. and Worsham, A. D., Inhibition of morning-glory (*Ipomoea lacunosa* L.) and certain other weed species by phytotoxic components of wheat (*Triticum aestivum* L.) straw, *J. Chem. Ecol.,* 9, 1027, 1983.

Macias, F. A., Varela, R. M., Torres, A., and Molinillo, J. M. G., Potential allelopathic quaianolides from cultivar sunflower leaves var SH-222, *Phytochemistry,* 34, 669, 1993.

Putnam, A. R. and DeFrank, J., Use of phytotoxic plant residues for selective weed control, *Crop Prot.,* 2, 173, 1983.

Putnam, A. R., DeFrank, J., and Barnes, J. P., Exploitation of allelopathy for weed control in annual and perennial cropping systems, *J. Chem. Ecol.,* 9, 1001, 1983.

Qasem, J. R. and Hill, T. A., On difficulties with allelopathy methodology, *Weed Res.,* 29, 345, 1989.

Shilling, D. G., Liebl, R. A., and Worsham, A. D., Rye (*Secale cereale* L.) and wheat (*Triticum aestivum* L.) mulch: the suppression of certain broad-leaved weeds and the isolation and identification of phytotoxins, in *The Chemistry of Allelopathy: Biochemical Interactions Among Plants,* Thompson, A. C., Ed., ACS Symp. Ser. 268, American Chemical Society, Washington, D.C., 1985, 243.

Shilling, D. G., Jones, L. A., Worsham, A. D., Parker, C. E., and Wilson, R. F., Isolation and identification of some phytotoxic compounds from aqueous extracts of rye (*Secale cereale* L.), *Agric. Food Chem.,* 34, 633, 1986a.

Shilling, D. G., Worsham, A. D., and Danehower, D. A., Influence of mulch, tillage, and diphenamid on weed control, yield, and quality in no-till flue-cured tobacco (*Nicotiana tabacum* L.), *Weed Sci.,* 34, 738, 1986b.

Thijs, H., Shann, J. R., and Weidenhamer, J. D., The effects of phytotoxins on competitive outcome in a model system, *Ecology,* 75, 1959, 1994.

Weidenhamer, J. D., Morton, T. C., and Romeo, J. T., Solution volume and seed number: often overlooked factors in allelopathic bioassays, *J. Chem. Ecol.,* 13, 1481, 1987.

Weidenhamer, J. D., Hartnett, D. C., and Romeo, J. T., Density-dependent phytotoxicity distinguishing resource competition and allelopathic interference in plants, *J. Appl. Ecol.,* 26, 613, 1989.

Willis, R. J., The historical bases of the concept of allelopathy, *J. Hist. Biol.,* 18, 71, 1985.

Wilson, R. E. and Rice, E. L., Allelopathy as expressed by *Helianthus annuus* and its role in old-field succession, *Bull. Torrey Bot. Club.,* 95, 432, 1968.

Yenish, J. P., Worsham, A. D., and Chilton, W. S., Disappearance of DIBOA-glucoside, DIBOA, and BOA from rye (*Secale cereale* L.) cover crop residue, *Weed Sci.,* 43, 18, 1995.

3 How to Make Bioassays for Allelopathy More Relevant to Field Conditions with Particular Reference to Cropland Weeds

Chester L. Foy

CONTENTS

3.1 ABSTRACT

Increased attention has been given to the role and potential use of allelopathy as a management strategy for crop protection against weeds and other pests. Allelopathy has been implicated and reported for many crop and weed species, but conclusive proof is often lacking. Laboratory bioassay is the first step to investigate probable involvement of allelopathy, and the use of soil in this bioassay is essential. Bioassays with plant leachates and/or extracts in the absence of soil provide only limited information on the allelopathic potential of plants and may have little or no correspondence to interactions in the field. Some aspects that need to be considered when designing bioassays for allelopathy (for cropland weed species only) are discussed.

3.2   INTRODUCTION

Molisch (1937) coined the term "allelopathy" which includes chemical interactions among plants, including microorganisms. Rice (1984) defined allelopathy as the effect(s) of one plant (including microorganisms) on other plants through the release of chemical compounds in the environment. Although the term was coined in 1937, the phenomenon was observed 2000 years ago (Willis, 1997a, 1997b). Scientists from Virginia Polytechnic Institute reported the allelopathy phenomenon in the 1920s. Massey (1925) reported that black walnut (*Juglans nigra* L.) and butternut walnut (*J. cinerea* L.) caused wilting and dying of alfalfa (*Medicago sativa* L.), tomato (*Lycopersicon esculentum* Mill.), and potato (*Solanum tuberosum* L.). Davis (1928) associated the toxicity of black walnut with synthetic juglone (5-hydroxy-alpha-naptha-quinone) and reported its toxic effects on alfalfa and tomato.

Allelopathy has been evidenced, indicated, or implied for numerous crop and weed species, but conclusive proof is often lacking. Laboratory bioassay in allelopathic research is of great significance (Inderjit and Dakshini, 1995). In nature, many factors are interacting (simultaneously and sequentially) and constantly changing, and there are many co-linearities (e.g., debris increase, soil temperature decline, increased water content of soil, and increased phenolic content). Laboratory bioassay is the first step to investigate probable involvement of allelopathy. Some aspects that need to be considered, while designing bioassays for allelopathy, are discussed in this chapter. The discussion, however, is limited to cropland weed species only.

A study on allelopathy can be divided into:

1. Laboratory and greenhouse studies, that is, under controlled environment.
2. Field studies.
3. Phytochemical analyses of alleged allelopathic candidates.

A plant species with allelopathic potential is referred to as the donor plant. The term receptor plant is used hereafter for the plant species being affected by the allelopathic compounds released by the donor plant. Any unnatural experimental condition to ease the release of allelopathic compound should be avoided. It is important to simulate field conditions in laboratory bioassay, and carefully design an allelopathy bioassay having correspondence to field conditions. Although it is essentially impossible to duplicate exact field conditions, some of the points being discussed here might decrease the gap between laboratory bioassays and field interactions.

3.3   LABORATORY AND GREENHOUSE STUDY

Bioassays with plant leachates and/or extracts in the absence of soil do not provide much information on the allelopathic potential of the donor plant, and may have little or no correspondence to field interactions (Inderjit and Dakshini, 1995). Aqueous leachate of any plant is certain to contain some organic molecules. However, not all plants have the potential to interfere allelopathically with other plants in the vicinity. Therefore, leachate or extract bioassays for allelopathy are not discussed here. As discussed by several workers (Inderjit and Dakshini, 1995; Schmidt, 1990; Blum, this volume; Dalton, this volume), involvement of soil in laboratory bioassay for allelopathy is essential to demonstrate weed allelopathy in a terrestrial agroecosystem.

Soil is often amended with plant debris, residues, and leachates to evaluate the effects of organic molecules leached from donor plant species (Selleck, 1972; Alsaadawi et al., 1990; Anaya et al., 1990; Inderjit and Dakshini, 1991, 1992, 1994a, 1994b; Sahid and Sugau, 1993; Blum, this volume). The following points should be considered in laboratory plant debris/leachate–soil amending experiments.

3.3.1   COLLECTION OF SOIL SAMPLES

Care must be taken while collecting soil samples (Cheng, 1995). Soil samples should be air-dried, and stored in dark under nonhumid conditions to avoid microbial degradation of allelopathic compounds. Soil should be sampled from the rhizosphere and bulk zones. It is important to differentiate between rhizosphere and bulk soil, particularly in the case of annual weed species. Cheng (1995) viewed that sampling of rhizosphere soil is a difficult job because the soil may contain small fragments of roots and root hairs. He discussed the significance of spatial and temporal scale of the rhizosphere in relation to residence time, retention, transport, and transformation of allelopathic compounds in the rhizosphere. He emphasized the short distance movement of allelopathic compounds in the rhizosphere, and this as an important factor in soil sampling. The rhizosphere zone has higher microbial activity and biomass at root-shoot interfaces (Curl and Truelove, 1986). Most allelochemicals are prone to microbial degradation (Blum, 1998), thus a difference in allelopathic activity in the rhizosphere and bulk soil is likely. Many perennial weed species, for example, *Pluchea lanceolata* (DC) C. B. Clarke, have a deep and dense rhizomatous subterranean system (Inderjit et al., 1998) and a large rhizosphere zone when compared to some shallow rooted annual weeds such as rabbitfoot polypogon [*Polypogon monspeliensis* (L.) Desf.]. Since the zone of the rhizosphere is mostly in the order of millimeters, care must be taken to collect soils from the rhizosphere zone. Perennial weeds, in general, have many pockets of the rhizosphere owing to their dense and deep subterranean parts. When collecting soil from the rhizosphere, fine root hair should be carefully removed. The problems related to soil sampling in allelopathic studies are discussed in detail by Dalton (this volume).

3.3.2   ANALYSIS OF NATURAL SOIL INFESTED WITH THE DONOR PLANT

Before starting an amending experiment, the natural soil infested with the donor plant should be collected from several weed-infested fields from different sites, and analyzed for its soil texture, pH, organic matter, and certain inorganic ions (e.g., Na^+, K^+, Cu^{2+}, Zn^{2+}, Mg^{2+}, Ca^{2+}, PO_4, NO_3^-, NH_4^+). In order to get some information on the ecological amplitude of the donor plant species, sampling must be large and repeated in different seasons and at different locations. The investigator may use this information to calculate soil/plant ratios for plant debris–soil amendments. Inderjit and Dakshini (1994a) suggested that weed-free natural soil should be amended with weed plant parts in different amounts, and the amendment level having least/no significantly different soil characteristics when compared to weed-infested natural soil should be selected for bioassay. This means that amended soil should have chemical characteristics within the range of those of natural soil infested with the weed. Data on soil chemistry are important in terms of abiotic soil factors influencing residence time, persistence, and fate of allelopathic compounds (Inderjit and Dakshini, this volume; Huang et al., this volume).

3.3.3   DONOR PLANT MATERIAL

The next important question concerns the amount of donor plant material to be added for amendments. The range of biomass to be incorporated can be calculated on the basis of a clipped quadrat. Blum (this volume) recommends the use of different amounts of donor plant material for amending soil, as long as adequate controls are included. The plant material should not be chopped or ground. This practice may result in leaching of many organic molecules which are otherwise never released in nature (Inderjit and Dakshini, 1995). Furthermore, use of any kind of organic solvents should be avoided for growth experiments and for making extracts (Schmidt, 1990).

3.3.4   SELECTION OF AN APPROPRIATE CONTROL

The next question which must be addressed is the selection of an appropriate control. Williamson and Richardson (1988) argued for the significance of measuring treatment responses with independent controls (see also Blum, this volume). Inert organic matter or peat moss has been added to control soil to maintain a similar organic matter status of the amended soil (Pardales et al., 1992; Wardle et al., 1993; Ahmed and Wardle, 1994; Rice, 1995). Inderjit and Mallik (1997) reported the potential of peat moss to alter chemical characteristics of soil. In some studies, no peat moss was added to control soils (Inderjit and Dakshini, 1992, 1994a, 1994b). K. M. M. Dakshini (1998 personal communication) argued that under field conditions, allelopathic plant parts are plowed into the soil, and the effect on growth of a crop species should be analyzed after comparison with fields not amended with phytotoxic substances. In natural systems, no peat moss is added to the fields free from donor plants. Whether the organic matter of control and amended soils should be kept similar is still debatable. To identify allelopathy as a mechanism, controls are required to separate physical factors from chemical factors (i.e., allelochemicals). An investigator should keep these points in mind while designing a control for amending experiments.

Natural soil not infested with the donor plant, but in the immediate vicinity, should be used in plant debris–soil amending experiments. Since physical, chemical, and biological characteristics of soil influence the expression of allelopathy, it is important that abiotic (inorganics and other organics) and biotic (microbial population) characteristics of natural weed-free soil are considered. Variation in abiotic and biotic soil characteristics may influence the extraction of allelopathic compounds, and hence, modify allelopathic activities. Any difference in microbial ecology may cause significant differences in the fate of allelopathic compounds (Blum, 1998b; Schmidt and Ley, this volume). Therefore, the use of artificial soils, vermiculite, and/or sand should be avoided in plant debris–soil bioassays for allelopathy.

Amended soil should be analyzed for chemical characteristics. Amending soil with plant debris will result in changes in the organic and inorganic ion status of soil. Leachate from donor plants also has organic and inorganic ions (Tukey, 1970). It is important to find out which component is the main discriminating factor between amended and nonamended soils. Inderjit and Dakshini (1994a, 1994b) suggested that plant debris should be added to soil in different amounts followed by chemical analysis, and statistical comparison of amended and natural soil. Amended soils having the least differences in chemical characteristics in comparison to natural soils should be selected for bioassays. The observed effects on growth of receptor plants may be owing to their altered inorganic ion status, and may not be owing to organic chemical substances. The study confirmed the importance of an appropriate control in demonstrating allelopathy of ecological relevance. Any amended soil with unrealistically high amounts of plant debris will not give meaningful results. This aspect is discussed in detail by Inderjit and Dakshini (this volume). It would be worthwhile to study the interactions of organic and inorganic molecules, and their influence on growth patterns.

3.3.5   MOISTURE LEVELS OF AMENDED SOILS

In the case of cropland weeds, water often acts as the medium of leaching allelopathic compounds from weed species. To simulate such field conditions, amended soil is moistened with water to release plant leachates (including both inorganic and organic molecules) into soil. It is, however, important to keep moisture levels essentially similar to those of field conditions. Unnecessarily flooding amended soil may bring out quantitative and qualitative changes in organic molecules (Dao, 1987; Inderjit et al., 1996). Flooding, with poor drainage, may result in an increase in facultative microbial populations, and thus change the type of microorganisms involved (Dao, 1987).

3.3.6   Concentrations of Allelopathic Substances in Soil

The next step should be the estimation of the concentration of allelopathic substances in soil. Cultivated soils generally do not have homogeneous mixing of plant debris or residues, and estimating concentrations of organic molecules is relatively difficult. There are pockets of toxins in the natural fields (Liebl and Worsham, 1983). Inderjit and Dakshini (1996) reported that while preparing fields for cultivation, farmers plowed plant debris of *Pluchea lanceolata* into the soil. There may be significant variation in concentrations of allelopathic substances in nature, and this aspect must be taken into consideration. Soil samples should be collected from different fields at different locations having weed infestation. Even within the same field, soil samples should be randomly collected from various places. This would help in collecting data on a range of concentrations of allelopathic compounds.

Allelochemicals may exist in three forms:

1. Free.
2. Reversibly bound.
3. Irreversibly bound.

Most of the studies consider the first two forms to be allelopathically active forms (Whitehead et al., 1981; Inderjit and Dakshini, 1992, 1994a, 1994b; Inderjit, 1996). In allelopathic interactions, free and reversibly bound forms are important, but the role of irreversibly bound organics cannot be ruled out (Rice, 1984). Novak et al. (1995) reported that fungal hyphae can penetrate the organic matrix or inner clay layer which results in the release of bound forms into soil solution. Blum et al. (1994) suggested EDTA extraction for analyzing reversibly-bound phenolic acids. Blum (1997) reported the benefits of citrate over EDTA for extracting phenolic acids from soils. He reported that unlike EDTA, citrate did not interfere with the Folin Ciocalteu's phenol reagent during estimation of total phenolic acid content. Selection of standards is another important aspect of analyzing total phenolics from soil. It is important to identify suspected allelochemicals to understand the mechanism of the action of allelopathy. It is, however, equally important to identify the role of mixtures. Pue et al. (1995) stated, "soil bioassays of individual toxins identified as allelopathic compounds may not provide adequate information on the concentration required for the inhibition of plant germination and growth in soils."

The abiotic and biotic soil factors significantly influence the recovery of allelochemicals from soil (Dalton et al., 1983; Blum, 1998b; Schmidt and Ley, this volume). Dalton et al., (1983) found that recovery of phenolic acids from soil decreases with increasing acidity of soil at the time of extraction. Since the identification of promotor-inhibitor complexes is difficult, most studies are focused on individual phytotoxins. Most allelopathic activities in nature are owing to either additive or partially antagonistic activities of phytotoxins (Einhellig, 1995; Inderjit, 1996), and allelopathic effect can occur when the concentrations of individual toxins in the mixture is far below their individual toxic levels. (Blum, 1996).

3.3.7   What Species Should Be Selected for the Bioassay?

Use of artificially sensitive species, for example lettuce (*Lactuca sativa* L.), as a test species should be avoided (Inderjit and Dakshini, 1995). Plant species naturally occurring or cultivated in association with allelopathic plants should be given preference. Selection of several test species, at least in laboratory bioassays, will give more meaningful results. Microbial populations associated with particular crop species might help in degrading allelopathic compounds or other organic molecules present in the rhizosphere. This is particularly important because allelopathic effects are owing to additive effects of allelopathic compounds and other organic molecules present in the rhizosphere (Blum, 1997). Many pesticides are known to be degraded by microbes associated with roots of crop species (Anderson and Coats, 1995). Microbial communities associated with wheat (*Triticum aestivum* L.) roots are capable of degrading mecoprop [2-(4-chloro-2-methylphe-noxy)propanoic acid], 2,4-D [(2,4-dichlorophenoxy)acetic acid], and MCPA [(4-chloro-2-methylphenoxy)acetic acid] (Lappin et al., 1985). Microbes in the rhizosphere of sugarcane (*Saccharum officinarum* L.) are capable of degrading 2,4-D and those associated with corn (*Zea mays* L.) have potential to degrade atrazine [6-chloro-N-ethyl-N'-(1-methylethyl)-1,3,5-triazine-2,4-diamine] (Seibert et al., 1981). The rhizosphere microbes, specific to a particular crop actually cultivated in the field infested with the donor allelopathic weed, are important in determining allelopathic effects.

Leather and Einhellig (1985) reported that selection of bioassay species depends on the quantity of allelopathic compound available and the growth parameter to be tested. They found that

1. *Lemna* bioassays are particularly useful in determining the biological activity of fractions collected from HPLC of plant or soil extracts.
2. When the quantity of allelopathic compounds is not limited, sorghum [*Sorghum bicolor* (L.) Moench] bioassays are of particular help in determining uptake and metabolism of allelopathic compounds.

3.3.8 GROWTH RESPONSES

Seed germination and seedling growth are the main parameters to monitor growth responses owing to amendment. Most growth experiments are terminated after a period of one or two weeks. There is nothing wrong in this approach, but sometimes additional data might give more relevant results. Referring to the hypothesis that there may be a threshold which is dependent on the concentration of allelochemicals, Putnam and Tang (1986) opined that allelopathy may also have startling effects. They cited an example of a tomato which suddenly dies when growing in close vicinity to a black walnut. Putnam and Weston (1986) reported that soil amended with quackgrass [*Elytriga repens* (L.) Nevski] inhibits the development of root hairs in snap bean (*Phaseolus vulgaris* L.). They found that untreated control snapbean root had numerous root hairs. However, root hairs were absent in snap-bean treated with extracts of quackgrass shoots, and treated roots showed flaking of epidermal tissues. Sometimes studies such as transmission and scanning electron microscopy provide interesting evidence for allelopathy. Membranes, in fact, are likely to be the primary site(s) of allelopathic compounds, and the effect may be short-term and reversible. This fact should be considered as it may better help to identify allelopathy as a primary mechanism.

Weidenhamer (1996) discussed allelopathy as a density-dependent phenomenon. He stated, "Density-dependent phytotoxic effects may be defined as the differences in the degree of inhibition observed when plants are grown at different densities in soil containing a phytotoxic substance." He and his coworkers (Thijs et al., 1994; Weidehamer, 1996) reported that when compared to plants growing at higher density, plants growing at lower density have more toxin available in their surroundings. These authors recommend that density of assay species must be taken into account in designing bioassays.

3.4 SUMMARY

Studies conducted along the previously mentioned lines will tell whether there are any growth responses from receptor species after amendment. Also, factors other than organic molecules can be known. Sometimes, nonsignificant concentrations of organic molecules might bring out significant effects on growth of receptor species owing to their additive effects (Blum, 1996; Einhellig, this volume). Although it is essentially impossible to duplicate field conditions, one should not create artificial situations in controlled experiments in order to ease the release of organic compounds. In any event, it may not be possible to prove that *only* allelopathy is operative in a given situation, and it is the *only* responsible factor for the observed pattern. However, one can prove that allelopathy is the best explanation for the observed pattern (Willis, 1985). While demonstrating allelopathy, an investigator should exercise extreme caution in attempting to separate allelopathy from other mechanisms of interference such as nutrient limitation, microbial nutrient immobilization, plant-microbe interactions, etc. It is extraordinarily difficult, if not impossible, to separate these mechanisms of interference at the field level (Inderjit and del Moral, 1997). However, one can certainly conduct experiments under controlled conditions to understand some particular aspect(s) of allelopathy. To demonstrate allelopathy as a mechanism in bioassays, there is a need to separate allelopathic effects from other mechanisms of interference such as resource competition (Weidenhamer et al., 1989; Nilsson, 1994; Weidehamer, 1996).

In natural systems, particularly in perennial weeds, periodic replenishment of allelopathic compounds occurs owing to the evergreen nature of the weed (e.g., *Pluchea lanceolata* (DC) C. B. Clarke, *Cyperus rotundus* L., *Cyperus esculentus* L.). Even if allelopathic compounds are sorbed onto soil particles and/or degraded/transformed in soil, their fresh release is maintained owing to the presence of the weed all through the year. This characteristic of perennial weeds is difficult to maintain in plant debris/leachate–soil amending experiments. Such amending studies with perennial weeds may be good for short-term growth experiments, but may not be equally convincing for long-term (e.g., three- to four-month) growth bioassays. Investigators should consider experiments designed to maintain a fresh supply of allelopathic compounds in the case of perennial weeds.

Establishing the role of allelopathic interference in the field is essential before using allelopathy in successful weed management (Inderjit and Foy, 1998). Various ecological factors such as habitat, climate, soil properties (physical, chemical, and biological), and agricultural practices greatly influence the fate, persistence, and quantitative and qualitative availability of allelopathic compounds in the rhizosphere (Einhellig, this volume).

ACKNOWLEDGMENTS

Sincere appreciation is expressed to Dr. Inderjit for suggesting the topic and generously providing most of the reference materials for this chapter. Special thanks are also owing to Harold L. Witt for valuable assistance in preparing and proofreading the manuscript.

REFERENCES

Ahmed, M. and Wardle, D. A., Allelopathic potential of vegetative and flowering ragwort (*Senecio jacobaea* L.) plants against associated pasture species, *Plant Soil,* 164, 61, 1994.

Alsaadawi, I. S., Sakeri, I. A. K., and Al-Dulaimy, S. M., Allelopathic inhibition of *Cynodon dactylon* (L.) Pers. and other plant species by *Euphorbia prostrata* L., *J. Chem. Ecol.,* 16, 2747, 1990.

Anaya, A. L., Calera, M. R., and Pereda-Miranda, R., Allelopathic potential of compounds isolated from *Ipomoea tricolor* Cav. (Convolvulaceae), *J. Chem. Ecol.,* 16, 2145, 1990.

Anderson, T. A. and Coats, J. R., An overview of microbial degradation in the rhizosphere and its implications for bioremediation, in *Bioremediation: Science and Applications,* Skipper, H. D. and Tureo, F. R., Eds., Soil. Sci. Soc. Am. Pub. No. 43, Madison, WI, 1995.

Blum, U., Allelopathic interactions involving phenolic acids, *J. Nematol.,* 28, 259, 1996.

Blum, U., Benefits of citrate over EDTA for extracting phenolic acids from soils and plant debris, *J. Chem. Ecol.*, 23, 347, 1997.

Blum, U., Designing laboratory plant debris–soil bioassays: some reflections, 1999 (this volume), Chapter 2.

Blum, U., Effects of microbial utilization of phenolic acids and their phenolic acid breakdown products on allelopathic interactions, *J. Chem. Ecol.*, 24, 685, 1998.

Blum, U., Worsham, A. D., King, D. L., and Gerig, T. M., Use of water and EDTA extraction to estimate available (free and reversibly bound) phenolic acids in Cecil soils, *J. Chem. Ecol.*, 20, 341, 1994.

Cheng, H. H., Characterization of the mechanisms of allelopathy: Modeling and experimental approaches, in *Allelopathy: Organisms, Processes and Applications,* Inderjit, Dakshini, K. M. M., and Einhellig, F. A., Eds., American Chemical Society, Washington, D.C., 1995, 132.

Curl, E. A. and Truelove, B., *The Rhizosphere,* Springer-Verlag, Berlin, 1986.

Dalton, B. R., The occurrence and behavior of plant phenolic acids in soil environment and their potential involvement in allelochemical interference interactions: methodological limitations in establishing conclusive proof of allelopathy, 1999 (this volume), Chapter 6.

Dalton, B. R., Blum, U., and Weed, S. B., Allelopathic substances in ecosystems: effectiveness of sterile soil components in altering recovery of ferulic acid, *J. Chem. Ecol.*, 9, 1185, 1983.

Dao, T. H., Sorption and mineralization of plant phenolic acids in soils, in *Allelochemicals: Role in Agriculture and Forestry,* Waller, G. R., Ed., American Chemical Society, Washington, D.C., 1987, 358.

Davis, E. F., The toxic principle of *Juglans nigra* as identified with synthetic juglone, and its toxic effects on tomato and alfalfa plants, *Am. J. Bot.*, 15, 620.

Einhellig, F. A., Allelopathy: current status and future goals, in *Allelopathy: Organisms, Processes, and Applications,* Inderjit, Dakshini, K. M. M., and Einhellig, F. A., Eds., American Chemical Society, Washington, D.C., 1995, 1.

Einhellig, F. A., An integrated view of allelochemicals amid multiple stresses, 1999 (this volume), Chapter 30.

Huang, P. M., Wang, M. C., and Wang, M. K., Catalytic transformation of phenolic compounds in soil, this volume, 1998.

Inderjit, Plant phenolics in allelopathy, *Bot. Rev.*, 62, 182, 1996.

Inderjit and Mallik, A. U., Effects of phenolic compounds on selected soil properties, *For. Ecol. Manage.*, 92, 11, 1997.

Inderjit and Dakshini, K. M. M., Investigations on some aspects of chemical ecology of cogongrass, *Imperata cylindrica* (L.) Beauv., *J. Chem. Ecol.*, 17, 343, 1991.

Inderjit and Dakshini, K. M. M., Interference potential of *Pluchea lanceolata* (Asteraceae): growth and physiological responses of asparagus bean, *Vigna unguiculata* var. *sesquipedalis.*, *Am. J. Bot.*, 79, 977, 1992.

Inderjit and Dakshini, K. M. M., Allelopathic effect of *Pluchea lanceolata* (Asteraceae) on characteristics of four soils and tomato and mustard growth, *Am. J. Bot.*, 81, 798, 1994a.

Inderjit and Dakshini, K. M. M., Allelopathic potential of phenolics from the roots of *Pluchea lanceolata, Physiol. Plant.*, 92, 571, 1994b.

Inderjit and Dakshini, K. M. M., On laboratory bioassays in allelopathy, *Bot. Rev.*, 61, 28, 1995.

Inderjit and Dakshini, K. M. M., Allelopathic potential of *Pluchea lanceolata*: comparative studies of a cultivated field, *Weed Sci.*, 44, 393, 1996.

Inderjit and Dakshini, K. M. M., Bioassays for allelopathy: interactions of soil organic and inorganic constituents, 1999 (this volume), Chapter 4.

Inderjit and Del Moral, D., Is separating resource competition from allelopathy realistic?, *Bot. Rev.*, 63, 221, 1997.

Inderjit and Foy, C. L., An ecological update on weed allelopathy, *Weed Sci. Soc. Am. Abstr.*, 38, 40, 1998.

Inderjit, Foy, C. L., and Dakshini, K. M. M., *Pluchea lanceolata*: a noxious weed, *Weed Technol.*, 12, 190, 1998.

Inderjit, Kaur, S., and Dakshini, K. M. M., Determination of allelopathic potential of a weed *Pluchea lanceolata* through a multifaceted approach, *Can. J. Bot.*, 74, 1445, 1996.

Lappin, H. M., Greaves, M. P., and Slater, J. H., Degradation of the herbicide mecoprop [2-(2-methyl-4-chlorophenoxy)propionic acid] by a synergistic microbial community, *Appl. Environ. Microbiol.*, 49, 429, 1995.

Leather, G. R. and Einhellig, F. A., Mechanism of allelopathic action in bioassay, in *The Chemistry of Allelopathy*, Thompson, A. C., Ed., ACS Symp. Ser. 268, American Chemical Society, Washington, D.C., 1985, 197.

Liebl, R. H. and Worsham, A. D., Inhibition of pitted morningglory (*Ipomoea lacunosa* L.) and certain other weeds by phytotoxic components of wheat (*Triticum aestivum* L.) straw, *J. Chem. Ecol.,* 9, 1027, 1983.

Massy, A. B., Antagonism of walnuts, *Phytopathology,* 15, 773, 1925.

Molisch, H., *Der Einfluss einer Pflanze auf die andere-Allelopathige,* Fischer, Jena, Germany, 1937.

Nilsson, M.-C., Separation of allelopathy and resource competition by the boreal dwarf shrub *Empetrum hermaphroditum* Hagerup, *Oecologia,* 98, 1, 1994.

Novak, J. M., Jayachandra, K., Moorman, T. B., and Weber, J. B., Sorption and binding of organic compounds in soils and their relation to bioavailability, in *Bioremediation: Science and Applications,* Skipper, H. D. and Turco, R. F., Eds., Soil. Sci. Soc. Am. Pub. No. 43, Madison, WI, 1995, 13.

Pardales, J. R., Jr., Kono, Y., Yamauchi, A., and Iijima, M., Seminal root growth in sorghum (*Sorghum bicolor*) under allelopathic influences from residues of taro (*Colocasia esculenta*), *Ann. Bot.,* 69, 493, 1992.

Pue, K. J., Blum, U., Gerig, T. M., and Shafer, S. R., Mechanism by which noninhibitory concentrations of glucose increase inhibitory activity of *p*-coumaric acid in morning glory seedling biomass accumulation, *J. Chem. Ecol.,* 21, 833, 1995.

Putnam, A. R. and Weston, L. A., Adverse impacts of allelopathy in agricultural systems, in *The Science of Allelopathy,* Putnam, A. R. and Tang, C. S., Eds., John Wiley & Sons, New York, 1986, 43.

Putnam, A. R. and Tang, C. S., Allelopathy: state of science, in *The Science of Allelopathy,* Putnam, A. R. and Tang, C. S., Eds., John Wiley & Sons, New York, 1986, 1.

Rice, E. L., *Allelopathy,* Academic Press, Orlando, FL, 1984.

Rice, E. L., *Biological Control of Weeds and Plant Diseases: Advances in Applied Allelopathy*, University of Oklahoma Press, Norman, OK, 1995.

Sahid, I. B. and Sugau, J. B., Allelopathic effects of lantana (*Lantana camara*) and siam weed (*Chromolaena odorata*) on selected crops, *Weed Sci.,* 41, 303, 1993.

Schmidt, S. K., Ecological implication of destruction of juglone (5-hydroxy-1-,4-napthoquinone) by soil bacteria, *J. Chem. Ecol.,* 16, 3547, 1990.

Schmidt, S. K. and Ley, R., Microbial competition and soil structure limit the expression of allelochemicals in nature, 1999 (this volume), Chapter 20.

Seibert, K., Fuehr, F., and Cheng, H. H., Experiments on the degradation of atrazine in maize rhizosphere, in *Proceedings of the Theory and Practical Use of Soil Applied Herbicide Symposium*, European Weed Resource Society, Paris, France, 1981, 137.

Selleck, G. W., The antibiotic effects of plants in laboratory and field, *Weed Sci.,* 20, 189, 1972.

Thijs, H., Shann, J. D., and Weidenhamer, J. D., The effect of phytotoxins on competitive outcome in a model system, *Ecology,* 75, 1959, 1994.

Tukey, H. B., Jr., The leaching of substances from plants, *Ann. Rev. Plant Physiol.,* 21, 305, 1970.

Wardle, D. A., Nicholson, K. S., and Rahman, A., Influence of plant age on the allelopathic potential of nodding thistle (*Carduus nutans* L.) against pasture grasses and legumes, *Weed Res.,* 33, 69, 1993.

Weidenhamer, J. D., Distinguishing resource competition and chemical interference: overcoming the methodological impasse, *Agron. J.,* 88, 866, 1996.

Weidenhamer, J. D., Hartnett, D. C., and Romeo, J. T., Density-dependent phytotoxicity: distinguishing resource competition and allelopathic interference in plants, *J. Appl. Ecol.,* 26, 613, 1989.

Whitehead, D. C., Dibb, H., and Hartley, R. D., Extractant pH and the release of phenolic compounds from soil, plant roots, and leaf litter, *Soil Biol. Biochem.,* 13, 343, 1981.

Williamson, J. B. and Richardson, D., Bioassay for allelopathy: measuring treatment responses with independent controls, *J. Chem. Ecol.,* 14, 181, 1988.

Willis, R. J., The historical basis of the concept of allelopathy, *J. Hist. Biol.,* 18, 71, 1985.

Willis, R. J., The history of allelopathy, 1. The first phase 1785–1845: the era of A. P. de Candolle, *Allelo. J.,* 4, 164, 1997a.

Willis, R. J., The history of allelopathy, 2. The second phase (1900–1920): the era of S. U. Pickering and the U.S.D.A. bureau of soils, *Allelo. J.,* 4, 7, 1997b.

4 Bioassays for Allelopathy: Interactions of Soil Organic and Inorganic Constituents

Inderjit and K. M. M. Dakshini

CONTENTS

4.1   ABSTRACT

Little attention has been paid to investigating the role(s) of soil attributes, particularly soil chemistry, in bioassays for allelopathy. Physical, chemical, and biological soil components influence biotic and static availability, bioactive concentration, persistence, and fate of allelopathic compounds in the rhizosphere. The influence of allelopathic compounds on soil inorganic ions, however, has received less attention. It is suggested that bioassay protocols be designed to take into account both the effects of allelopathic compounds on soil chemistry and the influence of soil chemistry on availability of allelopathic compounds.

4.2   INTRODUCTION

Many ecologists have expressed their concern over the way allelopathic research is done (Willis, 1985; Connell, 1990; Williamson, 1990; Weidenhamer, 1996; Inderjit and Del Moral, 1997). Harper (1975) stated, "Demonstrating this (allelopathy) has proved extraordinarily difficult—it is logically impossible to prove that it doesn't happen and perhaps nearly impossible to prove that it does." This statement may be true for conclusively demonstrating the existence of competition as well. Two species may compete whenever there is a limited supply of resources. Competition can also occur when resources are not in limited supply, but two species interfere with each other's use of the resource(s) (Kimmins, 1997). The phenomena of *annidation* (complementary use of resources) should be important to investigate in interference studies.

Most laboratory bioassays for allelopathy have been conducted without considering the evolutionary context of the organism; yet, it is difficult to conclude through current laboratory bioassays that allelopathy is, or is not, a main force influencing competition in natural systems (Lewis,

1986). Mere presence of allelopathic compounds in plant parts does not demonstrate allelopathy (Heisey, 1990). Grümmer (1961) reported several phenolic compounds, including 4-hydroxybenzoic and vanillic acids from roots and rhizomes of couch grass (*Agropyron repens*). However, he states, ". . . it is difficult to believe that an effect specific to couch grass should depend on substances so common in the plant kingdom." Henn et al. (1988) studied the interference between mouseear hawkweed (*Hieracium pilosella* L.) and tall oatgrass (*Arrhenatherum elatius* Beauv.) in colliery spoils from the north of France. They identified phytotoxins such as umbelliferon, apigenin glucoside, and skimin from the roots of mouseear hawkweed, and this species was reported to be allelopathic earlier (Rice, 1984). However, they observed no toxicity in soils underneath mouseear hawkweed and no allelopathic compounds were detected in its rhizosphere.

We have previously discussed the significance of soil texture, associated species as test species, bioassay design, growth parameters, critical age of donor plant, and critical stage of allelochemical release (Inderjit and Dakshini, 1995a). In this review, we will analyze some of the concerns about laboratory and greenhouse bioassays, and importance of soil chemistry in relation to plant debris and soil interactions. Following the analysis, some modest proposals have been suggested for improving the protocols of allelopathic research.

4.3   LABORATORY BIOASSAYS

Despite the significant upsurge in allelopathic research and an unwavering belief in its importance, the study of plant–plant chemical interactions (allelopathy) often has not employed methods that consistently demonstrate interference by chemical compounds in natural systems. However, laboratory bioassays are an important methodology. While reviewing laboratory bioassays for allelopathy, we (Inderjit and Dakshini, 1995a) concluded that laboratory bioassays should allow researchers to eliminate all possible alternative interferences through carefully controlled experimental designs and manipulations of as many experimental parameters as is possible. In this way investigators can simulate complex field conditions, one factor at a time, to search for mechanistic interactions. In the last 40 years, many bioassay techniques have been proposed to demonstrate allelopathy (Dekker et al. 1983; Einhellig et al. 1985, Williamson and Richardson, 1988; Dornbos and Spencer, 1990; Grakhov et al. 1993; Inderjit and Dakshini, 1994a). Many bioassays, however, have been criticized because they have little correspondence to field interactions (May and Ash, 1990).

Much of the criticism of research focusing on allelopathy has been based on the lack of recognition of natural system functions in laboratory bioassays. Allelopathic compounds, directly or indirectly, are released into the rhizosphere by leaching, residue incorporation and decomposition, and volatilization, or via root exudates. This suggests that a growth response with leachate and extract bioassays may be of limited ecological relevance. Stowe (1979) concludes that it is too easy (and misleading) to demonstrate an allelopathic potential for almost any species with some allelopathic bioassay procedures. Harper (1977) states, "Almost all species can, by appropriate digestion, extraction and concentration, be persuaded to yield a product that is toxic to one species or another." We have discussed previously the unsuitability of leachate bioassays in allelopathic studies and strongly discourage the use of leachate and extract bioassays to demonstrate allelopathy in laboratory bioassays (Inderjit and Dakshini, 1995a).

4.4   PLANT DEBRIS–SOIL BIOASSAYS: EFFECT OF ORGANIC
MOLECULES ON SOIL CHEMISTRY

Allelopathic chemicals, after their entry into the host environment, are largely responsible for direct (rapid) or indirect (delayed) allelopathic reactions (Figure 4.1). The direct effects of allelochemicals will largely depend upon their bioactive concentration, fate, and residence time in soil (Figure 4.1).

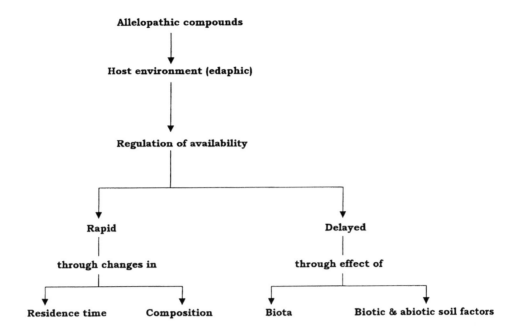

FIGURE 4.1   Schematic representation of the host environment in the availability of allelopathic chemicals.

Indirect effects could be caused by degradation/transformation of released allelochemicals via both abiotic and biotic soil factors (Blum and Shafer, 1988; Chase et al. 1991; Dalton, 1989; Dec and Bollag, 1988; Cheng, 1995; Schmidt, 1988; Shindo and Huang, 1982, 1984; Inderjit et al., this volume). Inderjit and his co-workers (1997) reported the differential recovery of certain terpenoids and phenolics from the soil. Pal and co-workers (1994) report the ability of abiotic (clay minerals) and biotic (enzymes) catalysts to oxidize phenolic compounds such as catechol and 2,6-dimethoxyphenol. They also report that for both abiotic and biotic agents, the polyvalent elements Cu^{2+}, Mn^{2+}, and Fe^{3+} facilitate the transformation of phenolic compounds. Delayed effects of allelochemicals could also be related to their effects on soil biota and on the physicochemical characteristics of soils (Figure 4.1).

Many allelopathic bioassays are concerned with evaluating the effects of plant debris, litter, or organic molcules directly on the growth of test species (Rice, 1984, 1995; Blum, this volume). Organic molecules, however, also influence accumulation of soil inorganic ions. Phenolics have significant effects on nutrient cycling, and by forming complexes with nutrients they can influence rates of nutrient turnover in soils (Appel, 1993). While studying the effects of phenolic compounds on selected soil properties such as pH, organic matter, and inorganic ions, Inderjit and Mallik (1997) found that various phenolic compounds influence the accumulation of soil inorganic ions. Both phenolics and terpenoids have been shown to influence mineralization rates (Horner and co-workers, 1988). However, not all phenolics form complexes with nutrients (Kuiters and Mulder, 1992). Furthermore, most of the aromatic compounds oxidize to catechol before the cleavage of a benzene ring (Gibson, 1968; Cheng et al. 1983).

Protocatechuic acid has been widely studied as an allelopathic compound. It originates in soils through plant debris, residue decomposition, root exudation (Rice, 1995), or biological and chemical degradation of p-hydroxybenzoic and vanillic acids (Haider and Martin, 1975). Shindo and Kuwatsuka (1977) report that as amounts of protocatechuic acid increase more Al^{3+} and Fe^{3+} can be recovered from soils. Vance and co-workers (1986) reported that owing to an orthohydroxy functional group, protocatechuic acid can form complexes with Fe^{3+} and Al^{3+} resulting in increased solubility and mobility of these metal ions.

Rice and his co-workers (see Rice, 1984) reported inhibition of nitrification in the climax vegetation. They concluded that phenolic compounds such as caffeic and ferulic acids, myricetin, tannins, and tannin derivative compounds inhibit oxidation of NH_4^+ to NO_3^- by *Nitrosomonas*. White (1986, 1994) tested the validity of the hypothesis that terpenoids may play an important role in the inhibition of nitrification. In contrast, Bremner and McCarty (1988) disagreed with this conclusion drawn by Rice (1984) and White (1986), and reported that terpenoids enhanced immobilization of ammonium N by soil organisms rather than by inhibition of nitrification. Northup and co-workers (1995) report that leaf litter of bishop pine (*Pinus muricata* D. Don) influences the release of dissolved organic nitrogen and mineral nitrogen in soils through the production of polyphenols. Certainly, organic molecules influence nutrient cycling and availability in soil. Many of the phenolics, terpenoids, and C ring-based plant secondary metabolites reduce the rate of decomposition and nitrogen mineralization, thus affecting soil N availability. Soil adsorption of an allelopathic compound is an important factor in soil nutrient dynamics. Nitrification is impeded in the forest floor (litter layer) of forest soil owing to phenolics. Once phenolics are sorbed to mineral surfaces in the mineral soil horizons (e.g., A horizons), nitrification is observed to occur (Olson and Reiners, 1983). Thus, there is a spatial scale of allelopathic effects within a soil profile owing to sorption of allelopathic compounds. A bioassay must therefore be designed to capture these spatial factors as they occur in the natural soil profile (Randy A. Dahlgren, 1998 personal discussion).

To produce ecologically relevant results, the influence of allelopathic compounds on soil chemistry needs to be considered when designing laboratory bioassays for allelopathy. Plant debris–soil bioassays are widely employed to study allelopathic interference (Rice, 1984, 1995). In allelopathic interactions, the release of chemical compounds into the host environment is the initial step, not the ultimate step. The significance of rhizosphere chemistry cannot be ignored. Amending soils with plant debris could alter physical (soil moisture, temperature, aeration, etc.), biotic (microbial population), and/or chemical (organic and inorganic constituents, pH, nutrient availability) factors of the soil (Figure 4.2). In many plant debris–soil bioassays, amending soils with plant debris is followed by irrigation with water to ease the release of allelochemicals (Rice, 1984). However, under flooded conditions free phenolic acids may undergo anaerobic degradation because with the increase of soil

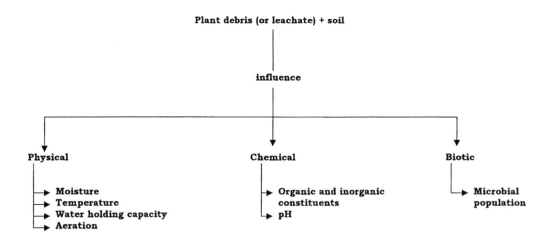

FIGURE 4.2   Schematic representation of how biotic and abiotic soil factors can be changed in plant debris–soil bioassays.

water content, the microbial community becomes dominated by facultative anaerobic organisms that degrade phenolic compounds through fermentation (Dao, 1987).

Any effect of plant debris–soil amendment on growth of test species does not necessarily demonstrate the allelopathic potential of the donor plant. The addition of plant debris and leachate to soil results in higher microbial biomass and activity. This tends to produce a temporary depletion of available pools of N and P and has adverse effects on plant growth (Michelsen and co-workers, 1995). Even cellulose and glucose can have similar effects. In plant debris–soil bioassays, it is important to demonstrate not only the natural release of chemical compounds in bioactive concentrations, but also their persistence in the rhizosphere for sufficient time to cause the observed effects. Plant debris and litter, directly or indirectly, may change the chemical and/or physical environment of the rhizosphere. Decomposing plant litter and debris may release both nutrients and chemical compounds. In a grassland ecosystem, accumulation of litter may influence soil pH, reduce ammonia losses, decrease wet nitrogen deposition, and change the chemical composition of rainfall reaching the soil (Knapp and Seastedt, 1986; Facelli and Pickett, 1991). Soil pH is an important factor for uptake of allelopathic compounds (Blum, 1996). Incorporation of rice straw decreases both available N and concentrations of soil cations such as Zn, Ca, Cu, Mn, and Na (Chou and Chiou, 1979).

In addition to direct effects of allelopathic compounds on plant growth, added plant debris may directly influence nutrient availability in soil (Figure 4.3). This, in turn, could have adverse effects on plant growth. Any adverse effect on growth from altered soil nutrient status would negate invoking allelopathy. The allelopathic compounds released by plant debris may, directly or indirectly, influence nutrient accumulation in soil through degraded by-products. This, in turn, may affect plant growth. The significance of transport, transformation, and retention of allelopathic compounds in the rhizosphere should be appreciated in plant debris and soil bioassays for allelopathy (Cheng, 1995).

We studied the allelopathic potential of the weed *Pluchea lanceolata* (DC.) C. B. Clarke by amending soils with different amounts of weed leaves and leaf leachate (Inderjit and Dakshini, 1994a). We concluded that plant debris should be added to the soil in different amounts followed by chemical analysis, and statistical comparison of amended soils with natural soils should be made.

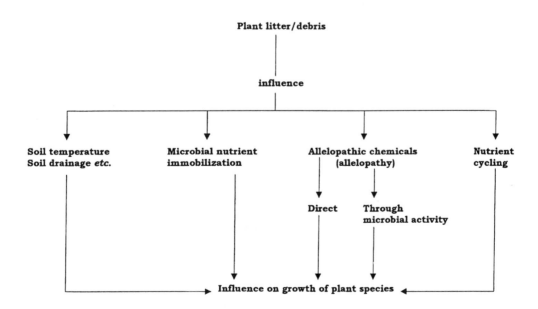

FIGURE 4.3   Schematic representation of effects of plant litter/debris on the soil nutrient and allelopathic chemical pool.

Thereafter, dilution levels having the least differences in chemical characteristics to those of natural soils should be selected for bioassays. Our study confirmed the importance of an appropriate control and soil texture as well as demonstrating allelopathy of ecological relevance.

There is an extensive literature on the contribution of phytotoxins from plant debris/litter (Rice 1995). However, many of these studies were conducted under unnatural conditions (see reviews by Inderjit and Dakshini, 1995a; Inderjit, 1996). Mulching, tilling, and reclamation of degraded habitat may also affect the impact of allelopathy. Natural mulching could result in

1. Reduction of weed infestation.
2. Prevention of soil freezing.
3. Reduction of evaporation.
4. Soil erosion (Facelli and Pickett, 1991).

Also, nutrient immobilization has been neglected in plant debris–soil bioassays. Depletion of oxygen in the soil and toxicity of carbon dioxide produced by microorganisms should also be considered.

4.5   ADDITIONAL CONSIDERATIONS

The plant density of target species greatly influences the response to allelopathic compounds (Weidenhamer and his co-workers, 1989; Thijs and her co-workers, 1994). That bioassays must be designed to detect density-dependent phytotoxic effects has been recently argued by Weidenhamer (1996). While investigating the effects of weed density on herbicide absorption and bioactivity, Winkle and co-workers (1981) report that activities of two herbicides, alachlor and atrazine, decrease with increased weed populations. Hoffman and Lavy (1978) report that high plant density is not as effective as low plant population in detecting low levels of atrazine in soil.

Life-cycle patterns (annual, biennial, or perennial) should also be considered in all plant debris–soil bioassays. The question of whether annual (mainly monocarpic) weeds are allelopathic under field conditions has been raised in earlier investigations (Bhowmik and Doll, 1984; Inderjit and Dakshini, 1995b). While investigating the allelopathic potential of the monocarpic annual weed rabbitfoot polypogon (*Polypogon monspeliensis* (L.) Desf.), we found that this weed has allelopathic potential expressed through incorporation of its straw (Inderjit and Dakshini, 1995b). It does not affect the same-season crop, but rather interferes with the next season's crop after incorporation of its straw. We suggest that if annual weeds appear at the time of crop seed sowing or after crop seed sowing, it is unlikely that the weeds will be allelopathic to the same-season crop.

In most forest ecosystem interference problems, it is difficult to separate the effects of mycorrhizae, nutrient limitations, root competition, microbial activity, litter quality, and allelopathy from each other (Kimmins, 1997). Nilsson (1994) investigated the separation of resource competition and allelopathic interference by the boreal shrub, crowberry *Empetrum hermaphroditum* (Lunge) Hagerup. She reports that both below-ground competition and allelopathic interference by crowberry are important factors retarding growth of Scots pine (*Pinus sylvestris* L.).

The pervasive scepticism regarding allelopathy results because allelopathy as a mechanism responsible for plant interference is not considered unless

1. There is convincing evidence supporting it.
2. Other mechanisms have been eliminated.

This invokes not only isolating the compound(s), but also demonstrating that a toxic effect is the primary function of a compound. Additionally, when other factors (such as resource limitations) are controlled or eliminated, the allelopathic effects must persist. We do not agree that this is the only acceptable approach. The approach may be valid only in cases of direct effects of allelopathic compounds. However, this may not always be true. If allelopathic compounds are influencing the availability of soil nutrients, allelopathic effects are not likely to persist when factors such as resource limitation are eliminated (Figure 4.1).

Plant phenolics and terpenoids are the two major classes of secondary compounds most often implicated in allelopathy. Investigating probable involvement of one (phenolics) does not eliminate the possible involvement of the other (terpenoids). Many question the criteria for selection of a particular class of the secondary compound. Often selection of a particular class of the secondary compound depends upon the expertise of the investigators and the analytical facilities available in their laboratory. Sometimes workers invoke the water-soluble nature of phenolics as a basis for their selection as probable allelopathic candidates. Many highly water-soluble compounds have low toxicity while many slightly water-soluble compounds are highly toxic (J. D. Weidenhamer, 1996 personal communication). Weidenhamer et al. (1993) reported that many allelopathic compounds have biological activity at concentrations well below the aqueous solubilities of these compounds. They reported that two monoterpenes, borneol and camphor, have biological activity at 10 and 25 ppm, respectively. These concentrations are well below the actual aqueous solubility of borneol (274 ppm) and camphor (550 ppm). Also, most phytochemical analyses focus on compounds with relatively high concentrations. The compounds with relatively low concentrations are often ignored. From the standpoint of allelopathy, as long as the solubility exceeds concentrations required for biological activity, the compound needs to be regarded as potentially able to exert allelopathic effects either by themselves or in interactions with other compounds (Weidenhamer et al., 1993).

After isolation and identification of potential allelopathic compounds, growth tests are performed to confirm their allelopathic activity (Inderjit et al., 1995). Investigators usually searched for the most toxic compound and ignored the role of complex mixtures. Under field conditions, additive or partially antagonistic effects may become more influential even at low concentrations, when compared to individual compounds (Inderjit, 1996). Blum (1996) states, ". . .the field mixture of phenolic acids and other organic compounds can cause inhibitory effects even though the concentrations of individual compounds are well below their inhibitory levels." In such cases allelopathic interactions would be the result of mixtures of allelopathic compounds (e.g., phenolic acids) and other organic compounds rather than of a single compound (Inderjit and Dakshini, 1994b; Einhellig, 1996; Inderjit, 1996; Blum, 1997). Einhellig (1995) states that all cases of allelopathic growth inhibition are the result of the combined effect of several compounds. It is therefore desirable to perform growth bioassays with a mixture of all identified and unidentified compounds, in addition to bioassays with single compounds. The significance of the uptake of the allelopathic compound by an affected plant has also been argued (Putnam, 1985; Willis, 1985; Horsley, 1991). Direct uptake of the allelochemical may be one of the mechanisms of allelopathic action, but it may not be the required factor for growth inhibition or promotion.

Allelopathic interference thus is a multifaceted chain of chemical reactions involving diverse facets of edaphic systems. Allelopathic compounds released into the host environment may also cause growth responses through their effect on

1. Soil microbial ecology.
2. Availability of nutrient ions.
3. Abiotic and biotic soil components.

4.6   RECOMMENDATIONS

To demonstrate allelopathy, it is therefore recommended that the following steps be considered in addition to those of earlier protocols (Putnam, 1985; Willis, 1985; Horsley, 1991; Inderjit and Dakshini, 1995a).

1. The effects of allelopathic compounds on soil chemistry.
2. Influence of soil chemistry on the quantitative and qualitative availability of allelopathic compounds.
3. To distinguish between direct or indirect (on soil biota, nutrient cycling) effects of allelopathic chemicals released into the environment.
4. Life-cycle patterns to be considered prior to planning debris–soil bioassays.
5. To confirm whether plant density-dependent phytotoxic effects exist.

It should be stated that following the bioassay protocols recommended previously, do not in any way exclude the possibility of the operation of other ecophysiological phenomenon, but highlight only that allelopathy offers the most reasonable explanation for the observed pattern (Inderjit and Del Moral, 1997; Blum, this volume).

ACKNOWLEDGMENTS

We sincerely thank Professors Randy Dahlgren, Frank Einhellig, Daniel Gilmore, Stephen Horsley, Kathleen Keating, Stephen Murphy, and Jeff Weidenhamer for their comments and suggestions to improve the quality of the paper.

REFERENCES

Appel, H. M., Phenolics in ecological interactions: the importance of oxidation. *J. Chem. Ecol.,* 19, 1521, 1993.
Bhowmik, P. C. and Doll, J. D., Allelopathic effects of annual weed residues on growth and nutrient uptake of corn and soybean, *Agron. J.,* 76, 383, 1984.
Blum, U., Allelopathic interactions involving phenolic acids, *J. Nemato.,* 28, 259, 1996.
Blum, U., Benefits of citrate over EDTA for extracting phenolic acids from soils and plant debris, *J. Chem. Ecol.,* 23, 347, 1997.
Blum, U., Designing laboratory plant debris-soil bioassays: some reflections, 1999 (this volume), Chapter 2.
Blum, U. and Shafer, S. R., Microbial populations and phenolic acids in soil, *Soil Biol. Biochem.,* 20, 793, 1988.
Bremner, J. M. and McCarty, G. W., Effects of terpenoids on nitrification in soil, *Soil Sci. Soc. Am. J.,* 52, 1630, 1988.
Chase, W. R., Nair, M. G., Putnam, A. R., and Mishra, S. K., 2,2-oxo-1,1-azobenzene: microbial transformation of rye (*Secale cereale* L.) allelochemicals in field soils by *Acinetobacter calcoaceticus*: III, *J. Chem. Ecol.,* 17, 1575, 1991.
Cheng, H. H., Characterization of the mechanisms of allelopathy: modeling and experimental approaches, in *Allelopathy: Organisms, Processes and Applications,* Inderjit, Dakshini, K. M. M., and Einhellig, F. A. Eds., American Chemical Society, Washington, D.C., 1995, 132.
Cheng, H. H., Haider, K., and Harper, S. S., Catechol and chlorophenols in soil: degradation and extractability, *Soil Biol. Biochem.,* 15, 311, 1983.
Chou, C. H. and Chiou, S. J., Autointoxication mechanism of *Oryza sativa* II. Effects of cultural treatment on the chemical nature of paddy soil and on rice productivity, *J. Chem. Ecol.,* 5, 539, 1979.
Connell, J. H., Apparent versus "real" competition in plants, in *Perspectives in Plant Competition,* Grace, J. B. and Tilman, D., Eds., Academic Press, San Diego, CA, 1990, 9.
Dalton, B. R., Physiochemical and biological processes affected the recovery of exogenously applied ferulic acid from tropical forest soils, *Plant Soil,* 115, 13, 1989.
Dao, T. H., Sorption and mineralization of plant phenolic acids in soil, in *Allelochemicals: Role in Agriculture and Forestry,* Waller, G. R., Ed., American Chemical Society, Washington, D.C., 1987, 358.
Dec, J. and Bollag, J. M., Microbial release and degradation of catechol and chlorophenols bound to synthetic humic acid, *Soil Sci. Soc. Am. J.,* 52, 1366, 1988.
Dekker, J. H., Meggitt, W. F., and Putnam, A. R., Experimental methodologies to evaluate allelopathic plant interaction: the *Abutilon theophrasti–Glycine max* model, *J. Chem. Ecol.,* 9, 945, 1983.
Dornbos, D. L. and Spencer, G. F., Natural products phytotoxicity: a bioassay suitable for small quantities of slightly water-soluble compounds, *J. Chem. Ecol.,* 16, 339, 1990.

Einhellig, F. A., Allelopathy: current status and future goals, in *Allelopathy: Organisms, Processes, and Applications,* Inderjit, Dakshini, K. M. M., and Einhellig, F. A., Eds., American Chemical Society, Washington, D.C., 1995.

Einhellig, F. A., Interactions involving allelopathy in cropping systems, *Agron. J.,* 88, 886, 1996.

Einhellig, F. A., Leather, G. R., and Hobbs, L. L., Use of *Lemna minor* L. as bioassay in allelopathy, *J. Chem. Ecol.,* 11, 65, 1985.

Facelli, J. M. and Pickett, S. T. A., Plant litter: its dynamics and effects on plant community structure, *Bot. Rev.,* 57, 1, 1991.

Gibson, D. T., Microbial degradation of aromatic compounds: metabolic pathways reveal a general formula for the degradation of several compounds, *Science,* 161, 1093, 1968.

Grakhov, V. P., Kozeko, V. G., and Golovko, E. A., Modeling of allelopathic interactions in laboratory tests, *Ukra. Bot. Zurn.,* 50, 86, 1993.

Grümmer, G., The role of toxic substances in the interrelationships between higher plants, in *Mechanisms in Biological Competition,* Milthorpe, E. L., Ed., Academic Press, New York, 1961, 219.

Haider, K. and Martin, J. P., Decomposition of specifically carbon-14 labeled benzoic and cinnamic acid derivatives in soil, *Soil Sci. Soc. Am. Proc.,* 39, 657, 1975.

Harper, J. L., Allelopathy, *J. Biol. Q.,* 50, 493, 1975.

Harper, J. L., *Population Biology of Plants,* Academic Press, London, 1977.

Heisey, R. M., Evidence for allelopathy by tree-of-heaven (*Ailanthus altissima*), *J. Chem. Ecol.,* 16, 239, 1990.

Henn, H., Petit, D., and Vernet, P., Interference between *Hieracium pilosella* and *Arrhenatherum elatius* in colliery spoils of north of France: allelopathy or competition?, *Oecologia,* 76, 268, 1988.

Hoffman, D. W. and Lavy, T. L., Plant competition for atrazine, *Weed Sci.,* 26, 94, 1978.

Horner, J. D., Gosz, J. R., and Cates, R. G., The role of carbon-based plant secondary metabolites in decomposition in terrestrial ecosystems, *Amer. Nat.,* 132, 869, 1988.

Horsley, S. B., Allelopathy, in *Biophysical Research for Asian Agroforestry,* Avery, M. E., Cannell, M. G. R., and Ong, C. K., Eds., Winrock International, USA, 1991, 167.

Inderjit, Cheng, H. H., and Nishimura, H., Plant phenolics and terpenoids: transformation, degradation and potential for allelopathic interactions, 1999 (this volume), Chapter 16.

Inderjit, Plant phenolics in allelopathy, *Bot. Rev.,* 62, 182, 1996.

Inderjit and Dakshini, K. M. M., Allelopathic effect of *Pluchea lanceolata* (Asteraceae) on characteristics of four soils and tomato and mustard growth, *Am. J. Bot.,* 81, 799, 1994a.

Inderjit and Dakshini, K. M. M., Allelopathic potential of phenolics from the roots of *Pluchea lanceolata, Physiol. Plant.,* 92, 571, 1994b.

Inderjit and Dakshini, K. M. M., On laboratory bioassays in allelopathy, *Bot. Rev.,* 61, 28, 1995a.

Inderjit and Dakshini, K. M. M., Allelopathic potential of an annual weed, *Polypogon monspeliensis,* in crops in India, *Plant Soil,* 173, 251, 1995b.

Inderjit and Del Moral, R., Is separating resource competition from allelopathy realistic?, *Bot. Rev.,* 63, 221, 1997.

Inderjit and Mallik, A. U., Effect of phenolic compounds on selected soil properties, *For. Ecol. Manage.,* 92, 11, 1997.

Inderjit, Dakshini, K. M. M., and Einhellig, F. A., Eds., *Allelopathy: Organisms, Processes, and Applications.* American Chemical Society, Washington, D.C., 1995.

Inderjit, Muramatsu, M., and Nishimura, H., On allelopathic potential of terpenoids and phenolics and their recovery in soil, *Canad. J. Bot.,* 75, 888, 1997.

Kimmins, J. P., *Forest Ecology: A Foundation for Sustainable Management,* 2nd ed., Prentice-Hall, NJ, 1997.

Knapp, A. K and Seasedt, T. R., Detritus accumulation limits productivity of tallgrass prairie, *BioScience,* 36, 622, 1986.

Kuiters, A. T. and Mulder, W., Gel permeation chromatography and C-binding of water-soluble organic substances from litter and humus layers of forest soils, *Geoderma,* 52, 1, 1992.

Lewis, W. M., Jr., Evolutionary interpretations of allelochemicals interactions in phytoplankton algae, *Amer. Nat.,* 127, 184, 1986.

May, F. E. and Ash, J. E., An assessment of allelopathic potential of *Eucalyptus, Aust. J. Bot.,* 38, 245, 1990.

Michelsen, A., Schmidt, I. K., Jonasson, S., Dighton, J., Jones, H. E., and Callaghen, T. V., Inhibition and growth, and effects on nutrient uptake of arctic graminoids by leaf leachates extracts—allelopathy or resource competition between plants and microbes?, *Oecologia,* 103, 407, 1995.

Nilsson, M.-C., Separation of allelopathy and resource competition by the boreal dwarf shrub *Empetrum hermaphroditum* Hagerup, *Oecologia,* 98, 1, 1994.

Northup, R. R., Yu, Z., Dahlgren, R. A., and Vogt, K. A., Polyphenol control of nitrogen release from pine litter, *Nature,* 377, 227, 1995.

Olsen, R. K., and Reiners, W. A., Nitrification in subalpine balsam fir soils: tests for inhibitory factors, *Soil Biol. Biochem.,* 15, 413, 1983.

Pal, S., Bollag, J. M., and Huang, P. M., Role of abiotic and biotic catalysts in the transformations of phenolic compounds through oxidative coupling reactions, *Soil Biol. Biochem.,* 26, 813, 1994.

Putnam, A. R., Weed allelopathy, in *Weed Physiology.* Vol. I. *Reproduction and Ecophysiology,* Duke, S. O., Ed., CRC Press, Boca Raton, FL, 1985, 131.

Rice, E. L., *Allelopathy,* Academic Press, Orlando, FL, 1984.

Rice, E. L., *Biological Control of Weeds and Plant Diseases: Advances in Applied Allelopathy,* University of Oklahoma Press, Norman, OK, 1995.

Schmidt, S. K., Degradation of juglone by soil bacteria, *J. Chem. Ecol.,* 14, 1561, 1988.

Shindo, H. and Huang, P. M., Role of Mn(IV) oxide in abiotic formation of humic substances in the environment, *Nature,* 298, 363, 1982.

Shindo, H. and Huang, P. M., Significance of Mn(IV) oxide in abiotic formation of organic nitrogen complexes in natural environment, *Nature,* 308, 87, 1984.

Shindo, H. and Kuwatsuka, S., Behavior of phenolic substances in decaying process of plants. III. Degradation pathway of phenolic acids, *Soil Sci. Plant Nut.,* 21, 227, 1977.

Stowe, L. G., Allelopathy and its influence on distribution of plants in Illinois old-field, *J. Ecol.,* 67, 1065, 1979.

Thijs, H., Shann, J. D., and Weidenhamer, J. D., The effect of phytotoxins on competitive outcome in a model system, *Ecology,* 75, 1959, 1994.

Vance, G. F., Mokma, D. L., and Boyd, S. A., Phenolic compounds in soils of hydrosequences and developmental sequences of sodzols, *Soil Sci. Soc. Am. J.,* 50, 992, 1986.

Weidenhamer, J. D., Distinguishing resource competition and chemical interference: overcoming the methodological impasse, *Agron. J.,* 88, 866, 1996.

Weidenhamer, J. D., Hartnett, D. C., and Romeo, J. T., Density-dependent phytotoxicity: distinguishing resource competition and allelopathic interference in plants, *J. Appl. Ecol.,* 26, 613, 1989.

Weidenhamer, J. D., Macias, F. A., Fischer, N. H., and Williamson, G. B., Just how insoluble are monoterpenes?, *J. Chem. Ecol.,* 19, 1799, 1993.

White, C. S., Volatile and water-soluble inhibitors of nitrogen mineralization and nitrification in a ponderosa pine ecosystem, *Biol. Fertil. Soils,* 2, 97, 1986.

White, C. S., Monoterpenes: their effect on ecosystem nutrient cycling, *J. Chem. Ecol.,* 20, 1381, 1994.

Williamson, G. B., Allelopathy, Koch's postulates and the neck riddle, in *Perspectives in Plant Competition,* Grace, J. B. and Tilman, D., Eds., Academic Press, New York, 1990, 143.

Williamson, G. B. and Richardson, D., Bioassay for allelopathy: measuring treatment responses with independent control, *J. Chem. Ecol.,* 14, 181, 1988.

Willis, R. J., The historical basis of the concept of allelopathy, *J. Hist. Biol.,* 18, 71, 1985.

Winkle, M. E., Leavitt, J. R. C., and Burnside, O. C., Effects of weed density on herbicide absorption and bioactivity, *Weed Sci.,* 29, 405, 1981.

5 Bioassays Applied to Allelopathic Herbaceous Vascular Hydrophytes

Stella D. Elakovich

CONTENTS

5.1 ABSTRACT

Germination and growth target systems used in laboratory bioassays for exploring aquatic and wetland plant allelopathic interactions include lettuce (*Lactuca sativa*), *Lemna* species, hydrilla (*Hydrilla verticillata*), algae and microalgae, ferns, bacteria, as well as some other types of systems. A few large scale and field experiments have also been applied to allelopathic interactions of hydrophytes. Strengths and limitations of these assay systems and a review of their applications to herbaceous hydrophyte allelopathic interactions are presented.

5.2 INTRODUCTION

Bioassays are crucial to the investigation of allelopathic interactions. In spite of the large number of allelopathic studies, few adequately address the separation of allelopathic interactions from other plant–plant interactions such as those involving competition. This is largely owing to the complexity of the system. Several mechanisms of interaction may be operative in sequence or simultaneously, making the separation of these mechanisms virtually impossible (Putnam and Weston, 1986). Many bioassays have been used to probe allelopathic potential, but few show any correspondence to plant interactions in the field (Stowe, 1979; Leather and Einhellig, 1986). Although he compiled the information 19 years ago, Stowe (1979) noted that of the 96 qualitatively different bioassays reported in the literature, few if any are of value in relating the results to the demonstration of allelopathy under natural or field conditions. In spite of their limitations, laboratory bioassays allow the investigator to manipulate interactions and vary one parameter at a time to probe allelopathic potential. What must be remembered is that the mere presence of an allelochemical does not prove that it is ecologically relevant. Laboratory bioassay evidence for allelopathy of many species of weeds may not hold true upon further testing (Putnam, 1985). Conclusive proof of allelopathy must show that the allelochemicals are released at high enough concentrations to influence associated plant species (Putnam and Tang, 1986).

Many assays involve extraction of ground plant material with hot water or organic solvents. High temperatures or grinding may result in the release of enzymes, salts, amino acids, and nitrogen compounds which may not be released under natural conditions (Chou and Muller, 1972). Stowe (1979) found that bioassays involving whole plant extracts, foliar washings, and decomposed litter inhibited the germination and growth of plant species even though the plant was not allelopathic under natural conditions. Cheng (1992) suggested that after chemicals enter the soil, a number of processes take place which affect the fate of these chemicals and hence their allelopathic potential. Organic solvents can extract chemicals from litter, soil organic matter, humic acid, and microbe membranes along with the plant constituents (Schmidt, 1990). Some workers extract allelochemicals in aqueous phase at room temperature (Rice, 1984), which provides a bias for water soluble compounds. Water-soluble phenolics have historically been preferred as primary allelochemicals because of their relatively easy identification and ready availability (Inderjit and Dakshini, 1990, 1992; Alsaadawi et al., 1985). Inderjit (1996) has recently reviewed the role of plant phenolics in allelopathy and found overwhelming evidence of their significant role. Fischer et al.(1987) earlier proposed the possibility of transport of water insoluble compounds through micelle formation with natural sterols and fatty acids, but they more recently (Weidenhamer et al., 1993) have reconsidered this hypothesis. Their studies on the water solubilities of 31 biologically active monoterpenes show that these compounds are often biologically active at concentrations below their aqueous solubilities. The allelopathic activity of some monoterpenes exceeded that of the well known allelopathic compound juglone, and the aqueous solubility also exceeded that of juglone. Thus, water solubility must be approached in a new light. What is important is the solubility relative to the concentration necessary for biological activity (Weidenhamer et al., 1993). Putnam and Tang (1986) claim that all cases of alleged allelopathy appear to involve a complex of chemicals rather than one specific phytotoxin.

The purpose of selected bioassays may range from simply demonstrating allelopathic potential of isolated compounds, to defining the concentration at which that activity is expressed, to determining the fate of allelochemicals in the environment (Inderjit, 1996), to probing ecologically meaningful interactions in order to further the understanding of allelopathy. In their discussion of bioassays in the study of allelopathy, Leather and Einhellig (1985) conclude: "There is no perfect assay that will meet all the requirements for detecting bioactivity of allelochemicals and it would be prudent to use several for each case of suspected allelopathic interaction." Ten years later, Inderjit and Dakshini (1995) reiterate this sentiment and further suggest that beyond growth parameters, physiological and biochemical parameters of affected target plants should also be examined. This paper discusses the range of bioassay methods that have been applied by investigators who report on allelopathic herbaceous vascular hydrophytes with an emphasis on those assay methods which use aquatic plants as the target species.

In a review we published of the literature from 1970 to 1994 (Elakovich and Wooten, 1995), we found 67 genera and 97 species of herbaceous vascular hydrophytes reported to be allelopathic. A total of 302 allelopathic plant–target plant interactions were found. Of these, as summarized in Table 5.1, 54 involved lettuce as the target plant; 23 involved *Lemna* (duckweed) species; and 15 involved the aquatic plant hydrilla. Hydrophyte-algae interactions were reported in 35 cases; hydrophyte-fern

TABLE 5.1
Summary of Target Plants Used in Bioassays Involving 302 Allelopathic Interactions of Herbaceous Vascular Hydrophytes (Elakovich and Wooten, 1995)

Target Plant	Frequency of Use for Bioassay
Lettuce	54
Duckweed (*Lemna* species)	23
Hydrilla	15
Algae and micro-algae	35
Fern	13
Bacteria	10

interactions were reported 13 times. Of the 238 hydrophyte-spermatophyte interactions, 107 involved hydrophyte-hydrophyte interactions. Ten target species were bacteria, three were fungi; an unusual target was tomato cell cultures (Ashton et al., 1985). The following discusses those assay systems listed in Table 5.1 as well as large scale assays which have been applied to aquatic and wetland plants.

5.3 LETTUCE SEEDLING BIOASSAY

Although the mere counting of the number of times a particular assay is used may result more in a reflection of the number of species tested using a given assay by a single investigator, it is, nevertheless, clear that lettuce is the most commonly selected target species for examining hydrophyte allelopathy. Lettuce is used both in germination assays and plant growth assays. Bonasera et al. (1979) compared lettuce sensitivity to that of cucumber (*Cucumber sativus* L.), radish (*Raphanus sati vus* L.), and tomato (*Lycopersicon esculentum* Mill.), all used in other studies as bioassay species, and found lettuce to be the most sensitive to plant extracts of four hydrophytes tested. They also found lettuce radicle growth to be more sensitive than germination. Lettuce is an excellent laboratory assay species as its seeds readily germinate within 24 h, the plant has linear growth which is insensitive to pH values over a wide range, and insensitive to osmotic potentials of less than 70 milliosmol/kg (Cheng and Riemer, 1988). Growth of lettuce radicles exposed to plant extracts can be compared to growth in control plates after only 72 h. Thus, the assay is simple, reproducible, and relatively quick. It shares with all other bioassays the problem of testing water-insoluble compounds or extracts and the necessity of using sterile techniques because of the opportunity for fungal contamination. We have found that small amounts of ethanol or DMSO may be added to aid in sample solubility. Lettuce seeds may be sterilized by successive 30 s washes in detergent, 70 percent alcohol, and 10 percent bleach followed by thorough rinsing in sterile distilled water. Assays are reproducible to within 10 percent when repeated on different days. Comparison of three cultivated varieties of lettuce seeds at three concentrations suggests that seed cultivated variety is unimportant in lettuce seedling bioassay (Boyd and Elakovich, 1989). Tang and Young (1982) published their seminal paper describing a hydrophobic root exudate trapping system as a solution to the tedious isolation of allelopathic compounds from an undisturbed system and using lettuce germination as their target plant system. Although the lettuce assay is easy to perform and provides a sensitive measure of both seed germination and growth inhibition, it has no ecological relevance when applied to hydrophytes. If the purpose of a selected bioassay is to identify ecologically meaningful interactions, then use of artificially sensitive species should be avoided (Inderjit, 1996). Some hydrophytes which have been reported to be allelopathic according to lettuce seedling bioassay are listed in Table 5.2.

TABLE 5.2
Hydrophytes Reported to Possess Allelopathic Activities According to Lettuce Seedling Bioassays

Common Name	Scientific Name	Reference
Giant ragweed	*Ambrosia trifida*	Bonasera et al., 1979
Burmarigold	*Bidens laevis*	Bonasera et al., 1979
Water shield	*Brasenia schreberi*	Elakovich and Wooten, 1987
Purple cabomba	*Cabomba caroliniana*	El-Ghazal and Riemer, 1986; Elakovich and Wooten, 1989
Coontail	*Ceratophyllum demersum*	Elakovich and Wooten, 1989
Purple nutsedge	*Cyperus esculentus*	Tames et al., 1973
Water nutgrass	*C. serotinus*	Komai et al., 1981
	Elodea nuttallii	El-Ghazal and Riemer, 1986
	Eleocharis acicularis	Wooten and Elakovich, 1991
	E. equisetoides	Wooten and Elakovich, 1991
	E. flavescens	Wooten and Elakovich, 1991
	E. montana, E. obtusa	Wooten and Elakovich, 1991
	E. quadrangulata	Wooten and Elakovich, 1991
	E. tuberculosa	Wooten and Elakovich, 1991
	Hemarthria altissima	Tang and Young, 1982
	Heracleum laciniatum	Junttila, 1975
Hydrilla	*Hydrilla verticillata*	Elakovich and Wooten, 1989
	Juncus repens	Elakovich and Wooten, 1989
	Limnobium spongia	Elakovich and Wooten, 1989
Parrot feather	*Myriophyllum aquaticum*	Elakovich and Wooten, 1989
Eurasian watermilfoil	*M. spicatum*	Elakovich and Wooten, 1989
	Myriophyllum sp.	El-Ghazal and Riemer, 1986
Southern naiad	*Najas guadalupensis*	Elakovich and Wooten, 1989
Yellow water lily	*Nuphar lutea*	Elakovich and Wooten, 1991
Fragrant water lily	*Nymphaea odorata*	Elakovich and Wooten, 1989
	Nymphoides cordata	Elakovich and Wooten, 1989
Green arum	*Peltandra virginica*	Bonasera et al., 1979
	Potamogeton foliosus	Elakovich and Wooten, 1989
Bur-reed	*Sparganium americanum*	Cheng and Riemer, 1988; Elakovich and Wooten, 1989
	Spirodela polyrhiza	El-Ghazal and Riemer, 1986
Common cat-tail	*Typha latifolia*	Bonasera et al., 1979
American eelgrass	*Vallisneria americana*	Cheng and Riemer, 1988; Elakovich and Wooten, 1989; El-Ghazal and Riemer, 1986

5.4 *LEMNA* (DUCKWEED) BIOASSAYS

Leather and Einhellig (1985, 1986), and Einhellig et al. (1985a) developed assays using floating aquatic *Lemna* (duckweed) species. Their assays are rapid, sensitive, and require only small volumes of test solution. They found that duckweed cultured in 1.5 ml of growth medium for 7 days in 24-well tissue culture cluster plates maintained linear growth. This ability to limit test volumes to only 1.5 ml is a distinct advantage of duckweed bioassays. They measured growth rate by counting frond production, growth by dry weight, and, for obscure duckweed (*Lemna obscura* (Austin) Daubs), chlorophyll and anthocyanin production. They found anthocyanin production by obscure duckweed to be the most sensitive attribute of those measured. Martin and Norris (1988) have measured growth rate of obscure duckweed as the area of monolayer plant growth which was directly related to plant fresh weight. The assay has also been used to follow bioactivity during bioassay-directed isolation procedures (Saggese et al., 1985). We employed common duckweed (*Lemna minor* L.) in the testing of extracts from some 25 hydrophytes (Elakovich and Wooten, 1989, 1991; Wooten and Elakovich, 1991). The frond number was the most reproducible attribute; our attempts to measure dry weights of such small amounts of plant material met with little success. Each cluster plate allows four treatments, three test samples and a control, with six replications in a randomized block design. The assay requires a minimum of seven days for growth, and also requires maintenance of aseptic Lemna cultures. We have found that common duckweed cultures can be maintained in agar slants for months on the lab bench with no special light or temperature considerations. Common duckweed is not sensitive to pH changes in the range pH 4 to 6 (Einhellig et al., 1985b) nor to osmotic potential below 143 milliosmols/kg (Elakovich and Wooten, 1991). Up to 0.3 percent v/v of ethanol may be added for better sample solubility.

Sutton and Portier (1989) used a similar assay system employing tropical duckweed (*Lemna paucicostata* Hegelm.) which had been maintained in axenic culture for more than 10 y. They selected those fronds in the L4 and R4 stage of development for use and counted the fronds produced after 10 days growth. In this system they tested four hydrophytes, Carolina mosquitofern (*Azolla caroliniana* Willd.), Illinois pondweed (*Potamogeton illinoensis* Moeong.), gulfcoast spikerush (*Eleocharis cellulosa* Torr.), and spikerush (*E. interstincta* R. Br.), as well as four different isolated allelochemicals. Low concentrations of extracts of Carolina mosquitofern and Illinois pondweed stimulated *Lemna* growth, but higher concentrations reduced its growth. Extracts of both *Eleocharis* species inhibited *Lemna* growth.

Wolek (1974) carried out 14 day mixed culture experiments in 500 ml flasks in an environmental chamber using common duckweed and inflated duckweed (*Lemna gibba*) in combination with rootless duckweed (*Wolffia arrhiza* Wimm.) and giant duckweed (*Spirodela polyrrhiza* (L.) Schleid) to establish the character of interaction among these four floating aquatic species. He examined growth of each in pure sterile control cultures, in uncrowded bi-part cultures designed to eliminate competition for space and light, and in pure culture with added extract from each of the four species. Because of the similar appearance of common duckweed and inflated duckweed, these were not tested in mixed culture. Rootless duckweed growth rate initially rose in the bi-part cultures as compared to the control monoculture, but soon fell. The largest decrease in growth rate was with giant duckweed, next was common duckweed, and least, and not significantly different from the control, was common duckweed. However, there was a significant decrease of rootless duckweed growth rate in cultures to which common duckweed extract had been added. On the other hand, in the same rootless duckweed/common duckweed bi-part culture, common duckweed growth rate was significantly higher than the control. In the common duckweed/giant duckweed bi-part cultures, common duckweed growth rate decreased. Inflated duckweed bi-part culture growth rates mirrored those of common duckweed. There was stimulation by all plant extracts at low concentrations, and inhibition or indifference at higher extract concentrations. Among Wolek's conclusions from this work were that in uncrowded mixed bi-part cultures, plant morphological characteristics play a decisive role in determining the better competing plant; species with longer roots or stronger developed underwater parts that reach deeper into the nutrient solution are the better competitors. Also, a stimulating allelopathic influence may be concealed by the presence of a stronger competitor because the competition for nutrient salts represents a more important factor.

5.5 *HYDRILLA VERTICILLATA* AND *POTAMOGETON* BIOASSAYS

Anderson (1985) developed assays using explants of the nuisance aquatic plant hydrilla and vegetative propagules of the pondweeds American pondweed (*Potamogeton nodosus* Poir.) and sago pondweed (*Potamogeton pectinatus* L.). Hydrilla and *Potamogeton* spp. are among the ten most noxious aquatic weeds. Anderson showed that these systems responded to many plant growth regulators and herbicides much as do whole plants. He suggested that many attributes could be measured: frequency of new shoot production, elongation of new shoots, chlorosis, root production, necrosis of explant, and inhibition of sprouting. Culture and whole-plant testing of these aquatic weeds requires more space, time and cost than *Lemna* bioassays.

Ashton et al. (1985) applied both of these aquatic plant assays to fractions obtained from the leachate of axenic cultures of dwarf spikerush (*Eleocharis coloradoensis* (Britt.) Gilly), a plant which had been suggested as an aquatic weed control agent. The crude spikerush extract at 500 ppm concentration inhibited hydrilla adventitious root formation to 36 percent of the control. Portions of the spikerush extract reduced the number of shoots of sago pondweed (*Potamogeton pectinatus* L.) producing adventitious roots to 38 percent of the control.

Dooris and Martin (1984) used a hydrilla assay in their search for naturally-occurring hydrilla inhibitors. Field collected plants were grown under controlled conditions in closed systems and growth was measured by either plant wet weight or by a change in the dissolved oxygen concentration of the growth medium.

Frank and Dechoretz (1980) subjected winterbuds and tubers of American pondweed and sago pondweed to co-planting with dwarf spikerush sod and found significant reduction of new shoots. Similar reductions were observed when dwarf spikerush sod leachate was applied to the pondweeds.

Jones (1993) applied a modification of the hydrilla bioassay to 11 aquatic species including purple cabomba, coontail, parrotfeather, Eurasian watermilfoil (*Myriophyllum spicatum*), southern naiad (*Najas gudalupensis* Magnus), American lotus (*Nelumbo lutea* Willd.), fragrant waterlily (*Nymphaea odorata* Ait.), lance-leaved sweet-flag (*Pontederia lanceolata* Wall ex Kunth), American pondweed, lance-leaf arrowhead (*Sagittaria lancifolia* L.), and American eelgrass (*Vallisneria americana* Michx.). Axenic cultures of hydrilla were harvested from tubers and turions from greenhouse grown plants. Cultures required 6 l of medium in 8 l cylinders aerated with compressed air. A 2 cm long apical tip explant was placed in an 80 ml test tube with artificial lake water medium and the sample to be tested. Test tubes were aerated with compressed air. For each treatment 5 replicates and 10 controls were used. Plants were grown for 14 days, dried and weighed. Results were that weights differed from 3 to 8 mg, and plant length was an unreliable estimate of plant growth. The ranking of inhibition among the test species was dependent on the extract concentration. At the highest tested concentrations, 167 ppt (parts per thousand), the most inhibitory eight species, purple cabomba, coontail, Eurasian watermilfoil, southern naiad, American lotus, fragrant waterlily, American pondweed, and lanceleaf arrowhead inhibited hydrilla growth by 49 to 63 percent. The degree of inhibition was not significantly different among any of this group. Five of these eight, purple cabomba, coontail, Eurasian watermilfoil, southern naiad, and fragrant waterlily had previously been subjected to lettuce seedling and Common duckweed bioassays, both of which were more discriminating than the hydrilla assay (Elakovich and Wooten, 1989). Of these five species, in the lettuce seedling assay, fragrant waterlily was the most inhibitory, inhibiting 95 percent of lettuce seedling radicle growth at the highest concentration tested, 250 ppt. Coontail was the most inhibitory at 25 ppt, the lowest concentration tested, where it caused a 66 percent radicle growth inhibition. All five caused greater than 50 percent inhibition at 250 ppt, and all except purple cabomba caused greater than 50 percent inhibition at 125 ppt. In the common duckweed assay, fragrant waterlily was also the most inhibitory, and all five species caused greater than 50 percent inhibition at 200 ppt. All except purple cabomba caused greater than 50 percent inhibition at 100 ppt. Thus, these three assays are not in disagreement, although the rank of inhibition differs, and in all three assay systems, the rank of inhibition is dependent on the extract concentration tested.

The hydrilla assay system has the disadvantages of extensive facilities required for growth of cultures, large sample testing volumes, 14 day growth periods, and measurements over a small (5 mg) range. This latter disadvantage contributes greatly to the lack of discrimination of the assay. Except for this latter disadvantage, most higher plant assays share these disadvantages.

Large scale assays involving hydrilla are discussed in Section 5.9.

5.6   ALGAL BIOASSAYS

Both liquid and agar-plate cultures of algae and micro-algae have served as the target species in bioassays of aqueous extracts, organic solvent extracts, and isolated compounds from potentially allelopathic hydrophytes. In investigations of the dynamic interactions in the eelgrass ecosystem, the water-soluble fraction of dead leaves of the eelgrass (*Zostera marina* L.) were tested on the growth of eight species of micro-algae. The extracts of leaves which had been dead less than two weeks were lethal at concentrations of as little as 0.25 mg dry leaf/ml. In contrast, extracts of leaves which had aged and dried several months in the field had no effect until concentrations of 13 mg dry leaf/ml were reached (Harrison and Chan, 1980). Aliotta et al. (1991) tested several unsaturated C^{18} fatty acids, sterols, and the phenylpropanoid α-asarone isolated from the floating-leaved macrophyte *Pistia stratiotes* L. against liquid cultures of 19 algal strains and found 17 of them to be affected. Ethyl ether extracts of *P. stratiotes* were also active, but the chloroform and methanol extracts were inactive. Organic solvent extracts of watershield were inhibitory to agar-plate cultures of eukaryotic and prokaryotic alga (Elakovich and Wooten, 1987). Phenylpropanes isolated from grassy-leaved pickerel weed (*Acorus gramineus* Ait.) were inhibitory to green and blue-green algae, some at the same level as the well known algicide copper sulphate (Della Greca et al., 1989).

5.7   BACTERIA BIOASSAYS

Bacteria have been used as bioassays for the detection of allelopathic interactions, but usually in combination with some other bioassay system. Watershield extracts were found active against nine bacteria in an agar-plate, paper disk assay. Three of the test bacteria were isolated from the lake where the watershield was collected, but from an area where the plants were not growing. The same extracts were also inhibitory in agar-plate, paper disk antialgal assays (see Section 5.6), and inhibitory at 100 ppm in the lettuce seedling assay (Elakovich and Wooten, 1987). Extracts of fresh, beginning to decay, and dried leaves of eelgrass were subjected to an agar-plate, paper disk assay against the bacterium *Staphylococcus aureus* (Harrison and Chan, 1980). The fresh and older leaves were always inhibitory while the dried leaves had little or no effect. These extracts were also tested against eight species of micro-algae with similar effects (see Section 5.6) (Harrison and Chan, 1980). Cariello and Zanetti (1979) showed that leaf sections and leaf extracts of sea-grass (*Posidonia ocean-ica*) stimulated growth of *S. aureus*. Plant growth stimulation, as well as inhibition, was included in Molisch's (1937) original definition of allelopathy.

5.8   FERNS AS BIOASSAY TARGET PLANTS

Use of ferns as target species in bioassay of hydrophytes is essentially limited to examination of fern–fern interactions. Peterson and Fairbrothers (1980) performed initial survey experiments with six fern species. These six species were sown on agar plates in adjacent paired strips in all combinations. Control plates contained spores from only one species. Plates were examined daily for percent germination or growth; a clear area at the interface between species indicated inhibition. Follow-up experiments involved preparation of control plates and experimental plates containing a mixture of the species being examined. To differentiate between nutrient competition and allelopathy, separate liquid cultures were grown for two weeks. The gametophytes were removed from these cultures by filtration and the supernatants were conserved. Subsequent experimental plates were prepared and sprayed with 1 ml of the conserved supernatant. Control plates consisted of the spores of one species sprayed with the supernatant of that species. Experimental plates contained spores of one species sprayed with the supernatant of the other. This reciprocal supernatant experiment showed that the gametophytes of each species were suppressing cell division of the gametophytes of the other species through the release of inhibitory compounds, not by competition. They determined that the hydrophyte, cinnamon fern (*Osmunda cinnamonea* L.) inhibits the growth and development of the fancy fern (*Dryopteris intermedia*).

Assays involving the effects of aqueous leaf extracts, leaf leachates, and litter infusions of the cinnamon fern and interrupted fern (*Osmunda claytoniana*) on spore germination of three other fern species showed the aqueous leaf extracts to be most inhibitory; the leaf leachates were inactive, perhaps owing to the small amount of material leached in the procedure (Munther and Fairbrothers, 1980).

Wagner and Long (1991) examined the allelopathic effect of sporophytes of cinnamon fern on the number and growth rate of gametophytes of tooth wood fern (*Dryopteris carthusiana*), crested wood fern (*D. cristata*), and goldie's wood fern (*D. goldiana*), three ferns which grow in the same habitat as cinnamon fern. Leachate from the cinnamon fern fronds was used to water 48-cell tissue culture test plates; control plates were watered with deionized water. Each of 16 cells of each plate contained 50 spores of each of the three target species. After four weeks, the number of gametophytes in each cell was counted, and gametophyte size was measured. The leachate did not affect the number of spores which germinated or the number of gametophytes which survived, but it did reduce the size of the gametophytes of goldie's wood fern (*D. goldiana*). It is interesting to note that all three target fern species occur in the same habitat as cinnamon fern, but only tooth wood fern and crested fern grow adjacent to cinnamon fern.

5.9   LARGE SCALE ASSAYS AND FIELD TRIALS

Few large scale assays or field trials investigating hydrophyte allelopathy have been conducted. During the 1970s workers at the Institute of Ecology, Polish Academy of Sciences, carried out pot experiments on the interrelations of aquatic plants (Szczepanska, 1971; Szczepanski, 1977; Szczepanska and Szczepanski, 1973, 1976a, 1976b). Experiments were outdoors in 5 l plastic buckets containing soil that was sand, peat, garden soil, or profundal sediment from an eutrophic lake. Seedlings of common reed (*Phragmites communis* Trin.) and common cattail (*Typha latifolia* L.) were planted separately and together in each of the four soil types. Growth was allowed for 4.5 months. A second three month experiment included a common culture of various plant species on lake mud. A third three month experiment included 101 pots planted with year old common reed plants, with the same size clumps of *Carex hudsoni,* and with both species. A fourth 2.5 month experiment involved 80 containers of common reed seedlings to which dried, milled plant material from six different aquatic plants was added. Among the conclusions from these many experiments were that introduction of nonliving plant material had much stronger influence than the mixed cultures with plants. Also, the influence of various plant species is strongly connected with the environmental conditions, including soil type (Szczepanska, 1971).

Extensive field surveys in India showed that neither coontail nor *Ceratophyllum muricatum* grow in the presence of hydrilla. To examine the interaction of these species (Kulshreshtha and Gopal, 1983), each was grown separately, and bi-part cultures of hydrilla with each *Ceratophyllum* species were grown in 45 × 45 × 60 cm cement tanks. In the bi-part cultures, the two species were separated by a wire netting so that the plants were not in direct competition. Initially, all species grew well, but after 30 days the growth of the *Ceratophyllum* grown with the hydrilla declined, and after 70 days, the *Ceratophyllum* was dead whereas the monocultures were healthy.

In an examination of the affect of other aquatic plants on the growth and survival of hydrilla, outdoor experiments were conducted in circular, plastic-lined pools, 0.9 m in height by 3.6 m in diameter (Sutton, 1986). The aquatic plants *Eleocharis geniculata* and slender arrowhead (*Sagittaria graminea* Michx) were established in 19 cm in diameter circular pans elevated 20 cm from the bottom of the pool. Cultures of *E. geniculata* were established in two months, but cultures of slender arrowhead required six months to become established. Hydrilla tubers were planted in each established culture and allowed to grow for eight weeks after which dry weight of hydrilla and the number of tubers produced were both measured. Hydrilla dry weight was reduced by both species, but slender arrowhead was more effective than *E. geniculata*.

Efforts to evaluate the ability of established native plant populations to resist reinvasion by nuisance species focused on the competitive interactions of Eurasian watermilfoil with American eelgrass (*Vallisneria americana* Michx.), American pondweed, or American lotus (Doyle and Smart, 1995). The experiments took place in Chisenhall Embayment, a small cove on Guntersville Reservoir, AL, a part of the Tennessee River system. Native species were established in 1.5 × 1.5 m plots within a larger 20 × 30 m enclosure. This process was hindered by common herbivores such as slider turtles, muskrat, and grass carp, which plagued the entire study, and the plant establishment stage required replantings over two years. Established populations of American eelgrass and American lotus significantly slowed the growth of *M. spicatum*. In spite of the great time, expense, and effort of this long field study, the authors stressed in their conclusions the correspondence with earlier, much simpler laboratory assays.

Experimental times of months or years rather than days, extensive space requirements, interference from herbivores, adverse weather conditions, and the need for personnel to initiate large scale and field experiments clearly illustrate the great disadvantages of such experiments. Their advantage is that the results may have much greater ecological significance than do the simpler laboratory experiments.

5.10 MISCELLANEOUS BIOASSAY SYSTEMS

Few aquatic species other than *Lemna* species, hydrilla, and *Potamogeton* species have been used as the target of hydrophyte bioassays. Hollyleaf najad (*Najas marina* L.) was used to evaluate allelopathic relationships with Eurasian watermilfoil (Agami and Waisel, 1985). Cultivated crop plants including barley (*Hordeum vulgare* L.) (Del Moral and Cates, 1971), mustard (*Brassica campestris* L.), rice (*Oryza sativa* L.) (Paria and Mukherjee, 1981; Komai et al., 1981; Kobayashi et al., 1980), cucumber, radish, tomato (Bonasera et al., 1979; Tames et al., 1973), and wheat (*Triticum aestivum* L.) (El-Ghazal and Riemer, 1986), have frequently been used in the bioassay of hydrophytes. Weed species such as common ragweed (*Ambrosia artemisiifolia* L.) (Kobayashi et al., 1980), barnyardgrass (*Eichinochloa crus-galli* (L.) Beauv.), and ivyleaf morninglory (*Ipomoea hederacea* Jacq.) (El-Ghazal and Riemer, 1986), have also been used as the target species. Additional target species include pearl millet (*Pennisetum typhoideum* Rich.) (Singhvi and Sharma, 1984), beet (*Beta vulgaris* L.), birdsfoot trefoil (*Lotus corniculatus* L.), perennial ryegrass (*Lolium perenne* L.), pea *(Pisum sativum* L.), white clover *(Trifolium repens* L.) (Tames et al., 1973), and watercress (*Nasturtium officinale* R. Br.) (Stevens and Merrill, 1980). In some studies, target species were carefully selected to probe the interactions occurring in selected plant communities. Drifmeyer and Zieman (1979) examined the effect of leachates from salt marsh species on one another. Germination of seeds of the marsh species saltgrass (*Distichlis spicata* (L.) Greene) and *Scirpus robustus* was not significantly inhibited by aqueous washings from the rhizospheres of other marsh species common reed, needle rush (*Juncus roemerianus*), or narrowleaf cattail (*Typha angustifolia* L.), but saltgrass seed germination increased 2.5 fold when treated with narrowleaf cattail rhizosphere leachate. Carral et al. (1988) examined the effects of leaves and their extracts from the meadow species broadleaf dock (*Rumex obtusifolius* L.) on the germination and root growth of the four meadow species perennial ryegrass, white clover, Kentucky bluegrass *(Poa pratensis* L.), and orchardgrass (*Dactylis glomerata* L.). That broadleaf dock exerts allelopathic control over meadow species is supported not only by these assay results, but by field observations that there is small-scale distribution of the meadow species in the neighborhood of broadleaf dock plants, and the area affected and the intensity of the effect both increase with the size of the individual broadleaf dock plant. In a probe of the dominant factors in secondary succession, Kobayashi *et al.* (1980) examined the effect of a polyacetylene isolated from tall goldenrod (*Solidago altissima* L.) on two species which it succeeds, common ragweed and eulaliograss (*Miscanthus sinensis* Andress.). The isolated polyacetylene inhibited growth of both species at only 1.0 ppm (72 and 85 percent of control, respectively). In contrast, rice (*Oryza sativa*) was only weakly inhibited at 50 ppm (88 percent of control).

5.11   GENERAL CONCLUSIONS

Except for the field studies, few of the reports described in this review address the question of ecological relevance. Wagner and Long's work (1991) on the effect of sporophytes of a fern on the
growth rate of gametophytes of three ferns growing in the same habitat is a notable exception. Of the
laboratory assays reported, lettuce seedling growth bioassay has the advantages that it is a widely
applied assay, it is simple to run, requires minimal space, supplies, and equipment, requires only four
days for growth, and is relatively reproducible. It is also sensitive and discriminating. Its major disadvantage is the total lack of ecological relevance when applied to an aquatic or wetland species.
Common duckweed bioassay eliminates this disadvantage while maintaining most of the positive
attributes of the lettuce seedling growth bioassay. Large scale assays which have more ecological
significance, also, invariably, require much more time, space, supplies, and equipment, and are subject to interference from herbivores, weather, and other outside effects. Most of these assays require
months rather than days to complete. The conclusion must be that there is no one perfect assay.
Rather, the assay used must be selected to fit the investigation. If the goal is to identify phytotoxic
compounds, then the many laboratory assays seem adequate and appropriate. If the goal is to identify plants which might be valuable in an integrated management system of undesired plants, then
larger scale assays or field trials must be completed. The undesired species is the ideal target species.
The more than ten year old wisdom of Leather and Einhellig (1985) bears repeating: "There is no
perfect assay that will meet all the requirements for detecting bioactivity of allelochemicals and it
would be prudent to use several for each case of suspected allelopathic interaction."

REFERENCES

Agami, M. and Waisel, Y., Inter-relationships between *Najas marina* L. and three other species of aquatic
    macrophytes, *Hydrobiologia,* 126, 169, 1985.
Aliotta, G., Monaco, P., Pinto, G., Pollio, A., and Previtera. L., Potential allelochemicals from *Pistia stratiotes*
    L., *J. Chem. Ecol.,* 17, 2223, 1991.
Alsaadawi, I. S., Arif, M.B. and Alrubeaa, A. A. Allelopathic effects of *Citrus aurantium* L. Isolation, charac-
    terization, and biological activities of phytotoxins. *J. Chem. Ecol.* 11, 1527, 1985.
Anderson, L. W. J., Use of bioassays for allelochemicals in aquatic plants, in *The Chemistry of Allelopathy,*
    Thompson, A. C., Ed, American Chemical Society, Washington, D.C. 1985, 351.
Ashton, F. M., DiTomaso, J. M., and Anderson, J. W. J., Spikerush (*Eleocharis* spp.): a source of allelopathics
    for the control of undesirable aquatic plants, in *The Chemistry of Allelopathy,* Thompson, A. C., Ed.,
    American Chemical Society, Washington, D.C., 1985, 401.
Bonasera, J., Lynch, J., and Leck, M. A., Comparison of the allelopathic potential of four marsh species, *Bull.
    Torrey Bot. Club,* 106, 217, 1979.
Boyd, J. L. and Elakovich, S. D., An examination of the effect of seed cultivated variety on allelopathy bio-
    assays, *J. Miss. Acad. Sci.,* 34, 1, 1989.
Cariello, L. and Zanetti, L., Effect of *Posidonia oceanica* extracts on the growth of *Staphylococcus aureus, Bot.
    Mar.,* 22, 129, 1979.
Carral, E., Reigosa, M. J., and Carballeira, A., *Rumex obtusifolius* L.: release of allelochemical agents and their
    influence on small-scale spatial distribution of meadow species, *J. Chem. Ecol.,* 14, 1763, 1988.
Cheng, H. H., A conceptual framework for assessing allelochemicals in soil environment, in *Allelopathy: Basic
    and Applied Aspects,* Rizvi, S. J. H. and Rizvi, V., Eds., Chapman and Hall, London, 1992, 21.
Cheng, T.-S. and Riemer, D. N., Allelopathy in threesquare burreed (*Sparganium americanum*) and American
    eelgrass (*Vallisneria americana*), *J. Aquat. Plant Manage.,* 26, 50, 1988.
Chou, C. H. and Muller, C. H., Allelopathic mechanisms of *Arctostaphylos glandulosa* var *zacaensis, Amer.
    Midl. Nat.,* 88, 324, 1972.
Del Moral, R. and Cates, R. G., Allelopathic potential of the dominant vegetation of western Washington,
    Ecology, 52, 1030, 1971.
Della Greca, M., Monaco, P., Previtera, L., Aliotta, G., Pinto, G., and Pollio, A., Allelochemical activity of
    phenylpropanes from *Acorus gramineus, Phytochemistry,* 28, 2319, 1989.

Dooris, P. M. and Martin, D. F., Naturally occurring substances that inhibit the growth of *Hydrilla verticillata,* in *The Chemistry of Allelopathy*, Thompson, A. C., Ed., American Chemical Society, Washington, D.C., 1984, 381.

Doyle, R. D. and Smart, R. M., Potential use of native aquatic plants for long-term control of problem aquatic plants in Guntersville Reservoir, Alabama; Report 2, Competitive interactions between beneficial and nuisance species, *Tech. Rep. A-93-6,* U.S. Army Engineer Waterways Experiment Station, Vicksburg, MS.

Drifmeyer, J. E. and Zieman, J. C., Germination enhancement and inhibition of *Distichlis spicata* and *Scirpus robustus* seeds from Virginia, *Estuaries,* 2, 16, 1979.

Einhellig, F. A., Leather, G. R., and Hobbs, L. L., Use of *Lemna minor* L. as a bioassay in allelopathy, *J. Chem. Ecol.,* 11, 65, 1985a.

Einhellig, F. A., Stille Muth, M., and Schon, M. K., Effects of allelochemicals on plant-water relationships, in *The Chemistry of Allelopathy,* Thompson, A. C., Ed., American Chemical Society, Washington, D.C., 1985b, 170.

Elakovich, S. D. and Wooten, J. W., An examination of the phytotoxicity of the water shield, *Brasenia schreberi, J. Chem. Ecol.,* 13, 1935, 1987.

Elakovich, S. D. and Wooten, J. W., Allelopathic potential of 16 aquatic and wetland plants, *J. Aquat. Plant Manage.,* 27, 78, 1989.

Elakovich, S. D. and Wooten, J. W., Allelopathic potential of *Nuphar lutea* (L.) Sibth. & Sm. (Nymphaeaceae), *J. Chem. Ecol.,* 17, 707, 1991.

Elakovich, S. D. and Wooten, J. W., Allelopathic, herbaceous, vascular hydrophytes, in *Allelopathy: Organisms, Processes, and Applications*, Inderjit, Dakshini, K. M. M., and Einhellig, F. A., Eds., American Chemical Society, Washington, D.C., 1995, 58.

El-Ghazal, R. A. K. and Riemer, D. N., Germination suppression by extracts of aquatic plants, *J. Aquat. Plant Manage.,* 24, 76, 1986.

Fischer, N. H., Tanrisever, N., de la Pena, L., and Williamson, G. B., The chemistry and allelopathic mechanisms in the Florida scrub community, in *Proceedings of the 14th Annual Plant Growth Regulator Society of America Meeting,* August 2 to 6, 1987, Honolulu, HI, 1987, 192.

Frank, P. A. and Dechoretz, N., Allelopathy in dwarf spikerush (*Eleocharis coloradoensis*), *Weed Sci.,* 28, 499, 1980.

Harrison, P. G. and Chan, A. T., Inhibition of the growth of micro-algae and bacteria by extracts of eelgrass (*Zostera marina*) leaves, *Mar. Biol.,* 61, 21, 1980.

Inderjit, Plant phenolics in allelopathy, *Bot. Rev.,* 62, 186, 1996.

Inderjit and Dakshini, K. M. M., The nature of interference potential of *Pluchea lanceolata* (DC.) C. B. Clark (Asteraceae), *Plant Soil,* 122, 298, 1990.

Inderjit and Dakshini, K. M. M., Interference potential of the weed *Pluchea lanceolata* (Asteraceae): growth and physiological responses of asparagus bean, *Vigna unguiculata* var. *sesquipedalis, Amer. J. Bot.,* 79, 977, 1992.

Inderjit and Dakshini, K. M. M., On laboratory bioassays in allelopathy, *Bot. Rev.,* 61, 28, 1995.

Jones, H. L., Allelopathic Influence of Various Aquatic Plant Extracts on the Growth of Hydrilla (*Hydrilla verticillata* L.f. (Royle)). MS thesis, University of Southern Mississippi, MS, 1993.

Junttila, O., Allelopathy in *Heracleum laciniatum*: inhibition of lettuce seed germination and root growth, *Physiol. Plant,* 33, 22, 1975.

Kobayashi, A., Morimoto, S., Shibata, Y., Yamashita, K., and Nunata, M., C-10 polyacetylenes as allelopathic substances in dominants in early stages of secondary succession, *J. Chem. Ecol.,* 6, 119, 1980.

Komai, K., Sugiwaka, Y., and Sato, S., Plant-growth retardant of extracts obtained from water nutgrass (*Cyperus serotinus* Rottb.), *Kinke Daigaku Nogakubu Kigo,* 14, 57, 1981.

Kulshreshtha, M. and Gopal, B., Allelopathic influence of *Hydrilla verticillata* (L. F.) Royle on the distribution of *Ceratophyllum* species, *Aquat. Bot.,* 16, 207, 1983.

Leather, G. R. and Einhellig, F. A., Mechanisms of allelopathic action in bioassay, in *The Chemistry of Allelopathy*, Thompson, A. C., Ed., American Chemical Society, Washington, D.C., 1985, 197.

Leather, G. R. and Einhellig, F. A., Bioassays in the study of allelopathy, in *The Science of Allelopathy,* Putnam, A. R. and Tang, C.-S., Eds., John Wiley & Sons, New York, 1986, 142.

Martin, D. F. and Norris, C. D., Effect of selected dyes on the growth of duckweed, *J. Environ. Sci. Health,* A23, 765, 1988.

Molisch, H., Der Einfluss einer Planze auf die andere—Allelopathie. G. Fischer, Jena, Germany, 1937.

Munther, W. E. and Fairbrothers, D. E., Allelopathy and autotoxicity in three eastern North American ferns, *Amer. Fern J.,* 70, 124, 1980.

Oborn, E. T., Moran, W. T., Greene, K. T., and Bartley, T. R., Joint Laboratory Report SI-2, U.S. Department of Agriculture, Bureau of Reclamation Engineering Laboratory and U.S. Department of Agriculture, ARS Field Crops Branch, 1954, 16.

Paria, N. and Mukherjee, A., Allelopathic potential of a weed *Alternanthera philoxeroides* (Mart.) Griseb, *Bangladesh J. Bot.,* 10, 86, 1981.

Peterson, R. L. and Fairbrothers, D. E., Reciprocal allelopathy between the gametophytes of *Osmunda cinnamomea* and *Dryopteris intermedia, Amer. Fern J.,* 70, 73, 1980.

Putnam, A. R., Weed allelopathy, in *Weed Physiology,* Vol. I, Duke, S. O., Ed., CRC Press, Boca Raton, FL, 1985, 132.

Putnam, A. R. and Tang, C. S., Allelopathy: state of science, in *The Science of Allelopathy,* Putnam, A. R. and Tang, C. S., Eds., John Wiley & Sons, New York, 1986, 1.

Putnam, A. R. and Weston, L. A., Adverse impacts of allelopathy in agricultural systems, in *The Science of Allelopathy,* Putnam, A. R. and Tang, C. S., Eds., John Wiley & Sons, New York, 1986, 43.

Rice, E. L., *Allelopathy,* 2nd ed., Academic Press, Orlando, FL, 1984.

Saggese, E. J., Foglia, T. A., Leather, G., Thompson, M. P., Bills, D. D., and Hoagland, P. D., Fractionation of allelochemicals from oilseed sunflowers and Jerusalem artichokes, in *The Chemistry of Allelopathy,* Thompson, A. C., Ed., American Chemical Society, Washington, D.C., 1985, 99.

Schmidt, S. K., Ecological implication of the destruction of juglone (5-hydroxy-1,4-naphthoquinone) by soil bacteria, *J. Chem. Ecol.,* 16, 3547, 1990.

Singhvi, N. R. and Sharma, K. D., Allelopathic effects of *Ludwigia adscendens* Linn. and *Ipomoea aquatica* Forsk on seedling growth of pearlmillet (*Pennisetum typhoideum* Rich.), *Trans. Isdt Ucds,* 9, 95, 1984.

Stevens, K. L. and Merrill, G. B., Growth inhibitors from spikerush, *J. Agric. Food Chem.,* 28, 644, 1980.

Stowe, L. G., Allelopathy and its influence on the distribution of plants in an Illinois old-field, *J. Ecol.,* 67, 1065, 1979.

Sutton, D. L., Growth of Hydrilla in established stands of spikerush and slender arrowhead, *J. Aquat. Plant Manage.,* 24, 16, 1986.

Sutton, D. L. and Portier, K. M., Influence of allelochemicals and aqueous plant extracts on growth of duckweed, *J. Aquat. Plant Manage.,* 27, 90, 1989.

Szczepanska, W., Allelopathy among the aquatic plants, *Pol. Arch. Hydrobiol.,* 18, 17, 1971.

Szczepanska, W. and Szczepanski, A., Emergent macrophytes and their role in wetland ecosystems, *Pol. Arch. Hydrobiol.,* 20, 41, 1973.

Szczepanska, W. and Szczepanski, A., Growth of *Phragmites communis* Trin., and *Typha latifolia* L. in relation to the fertility of soils, *Pol. Arch. Hydrobiol.,* 23, 233, 1976a.

Szczepanska, W. and Szczepanski, A., Effect of density on productivity of *Phragmites communis* Trin. and *Typha latifolia* L., *Pol. Arch. Hydrobiol.,* 23, 391, 1976b.

Szczepanski, A. J., Allelopathy as a means of biological control of water weeds, *Aquat. Bot.,* 3, 193, 1977.

Tames, R. S., Gesto, M. D. V., and Vieitez, E., Growth substances isolated from tubers of *Cyperus esculentus* var *aureus, Physiol. Plant,* 28, 195, 1973.

Tang, C. S. and Young, C. C., Collection and identification of allelopathic compounds from the undisturbed root system of bigalta limpograss (*Hemarthria altissima*), *Plant Physiol.,* 69, 155, 1982.

Wagner, H. B. and Long, K. E., Allelopathic effects of *Osmunda cinnamomea* on three species of *Dryopteris, Am. Fern J.,* 81, 134, 1991.

Weidenhamer, J. D., Macias, F. A., Fischer, N. H., and Williamson, G. B., Just how insoluble are monoterpenes?, *J. Chem. Ecol.,* 19, 1799, 1993.

Wolek, J., A preliminary investigation on interactions (competition, allelopathy) between some species of *Lemna, Spirodela,* and *Wolffia, Ber. Geobot.* ETH Stiftung Rubel., 42, 140, 1974.

Wooten, J. W. and Elakovich, S. D., Comparisons of potential allelopathy of seven freshwater species of spikerushes (*Eleocharis*), *J. Aquat. Plant Manage.,* 29, 12, 1991.

6 The Occurrence and Behavior of Plant Phenolic Acids in Soil Environments and Their Potential Involvement in Allelochemical Interference Interactions: Methodological Limitations in Establishing Conclusive Proof of Allelopathy

Barry R. Dalton

CONTENTS

6.1   ABSTRACT

In ecological systems, phenolic compounds undergo continuous cycles of synthesis, deposition, decomposition, transformation, plant uptake, and incorporation into soil organics. Their phytotoxic potential has been clearly demonstrated in model systems, and there has been much speculation about their role in the patterning of native vegetation across landscapes and their potential utility in agrosystems to control undesirable weeds. However, no single study has yet to provide indisputable proof that chemical interference interactions occur under ordinary ecological conditions. Distinguishing allelopathy from resource competition and demonstrating that allelochemicals are present at sufficient concentrations in soil for sufficient periods of time to affect plant growth have not yet been satisfactorily documented. New approaches and techniques to help overcome some of the current methodological limitations in establishing conclusive proof of allelopathy are discussed in this text. Field studies are needed that address soil allelochemical dynamics: peak concentration events, shifts in chemical composition, background concentrations, duration of peak concentration events, seasonal cycles, yearly cycles, etc. More care is needed in reducing chemical alterations of samples during soil collection, soil extraction, and sample analysis. Major advancements will come when increased attention is given to designing studies and model systems that better approximate situations and conditions typically encountered in ecological systems.

6.2   PLANT PHENOLIC ACIDS

Phenolic compounds are one of the most widely distributed classes of secondary metabolites found in plants and much attention has been given to their potential functional role in plant ecology (Kuiters, 1990; Siqueira et al., 1991). Phenolic compounds comprise the bulk of the structural matrix of plants and the pigmentation of flowers. They are believed to function as defensive agents against herbivores and invading microbes and as signal molecules in plant interactions with pathogens, mutualistic microbes, and parasitic angiosperms. Upon degradation and release into the soil environment, phenolics exist as components of soil organic matter. Because some phenolic compounds are toxic at millimolar concentrations to a wide range of species in bioassay experiments, their involvement in allelopathic interactions in many natural and cultivated ecosystems is often suggested (Inderjit, 1996; Rice, 1984; Whitehead et al., 1981, 1982, 1983). Taking into consideration the abundance of phenolic compounds in soil systems and their phytotoxic potential, it is reasoned that certain situations may exist in which allelochemical interference is involved in differential seedling growth and survival, and ultimately, vegetation patterning. However, to date, no study has provided conclusive proof that allelopathic interactions occur under ordinary field conditions (Connell, 1990; Weidenhamer, 1996), and unambiguous proof may not even be attainable for ecological phenomena as complex as allelopathy (Weidenhamer, 1996).

With the multitude of factors affecting ecological systems it is incredibly difficult to isolate any one factor that may be responsible for changes in plant growth and survival. Although there is mounting circumstantial evidence to support the existence and significance of chemical interference interactions between plants, there still remains a void of information regarding the behavior, bioavailability, and flux rates of allelochemical substances in soil systems. That allelochemicals are present at sufficient concentrations in soil for sufficient periods of time to affect plant growth has not yet been satisfactorily documented; however, it is essential in building a strong case to support an allelochemical interference hypothesis. Crucial to this is identifying and monitoring potential allelochemical concentrations that are actually free in the soil environment and available to plants and other organisms growing there. The allelopathy literature is replete with studies where harsh soil extractions have been used to identify potential soil allelochemical levels. This has often resulted in reported bioavailable levels that are over exaggerated (Dalton et al., 1987; Kaminsky and Muller, 1977, 1978; Katase, 1981c, Whitehead et al., 1981). However, over the past decade, much progress has been made in identifying and quantifying potential allelochemicals in soil extracts as well as in improving soil extraction methods to recover potentially bioavailable concentrations. This has proven to be no simple task, and there is still much debate over the development and use of appropriate soil extraction methodology.

This review takes a detailed look at the occurrence, behavior, and potential bioavailability of the most commonly identified group of phenolic compounds encountered in soils of most terrestrial, plant-dominated ecosystems, the phenolic acids. New approaches and techniques to help overcome some of the current methodological limitations in establishing conclusive proof of allelopathy are also discussed.

6.3   SYNTHESIS AND OCCURRENCE

As a group, phenolic acids are relatively simple in structure, are fairly easily identified and quantified in soil extracts, and are toxic to a wide range of vascular plant species. In fact, much of our knowledge about the behavior and potential bioavailability of phenolic compounds in soil is limited to phenolic acids and their common derivatives. A more in-depth understanding of the behavior of simple phenolic compounds in soil may aid in our understanding of more complex phenolic compounds and other allelochemical substances.

The term "phenolic acid" encompasses a variety of water-soluble, monomeric phenolic compounds possessing a singular carboxyl, and at least one hydroxyl group. Typically, the term is used to describe a limited number of simple phenolics, primarily derivatives of benzoic acid (phenylcarboxylic acid) and cinnamic acid (phenylacrylic or hydroxycinnamic acid). Benzoic acids, along with their corresponding aldehydes and alcohols, have been identified in all angiosperms so far examined (Goodwin and Mercer, 1983; Harborne, 1980). Cinnamic acids are thought to be universally present in higher plants, occurring free and in a wide range of esterified forms (Harborne, 1964, 1980).

The majority of phenolic compounds are products of the shikimate and acetate pathways in plants and synthesis (or accumulation) tends to be enhanced by environmental stress (Kuiters, 1990; Siqueira et al., 1991). Phenolic acids and other low-molecular weight phenolics are not necessarily end products, but undergo further degradation, conjugation, and polymerization reactions with other phenolic and nonphenolic substances. In living plant cells, phenolic substances are almost exclusively in the form of glycosides or esters (Harborne, 1964, 1980; Newby et al., 1980; Swain, 1978). Phenolic acids most often identified in extracts from plant tissues include *p*-hydroxybenzoic, protocatechuic, vanillic, syringic (benzoic acid derivatives), and caffeic, *p*-coumaric, sinapic, and ferulic acids (corresponding cinnamic acid derivatives).

Phenolics exist as low-molecular weight compounds as well as structural units in high-molecular weight substances of plants. Lignin, the second most abundant compound in nature, is a polymer of phenylpropanoid units consisting of an aromatic ring and a three-carbon side chain. Lignin comprises 22 to 34 percent of the total solid material of wood and is believed to be synthesized from various cinnamyl alcohols derived from certain phenolic acids (Siqueira et al., 1991). Decaying plant materials are rich sources of phenolic acids because lignin releases, upon physical and biological degradation, these and other low-molecular weight aromatic compounds. Aqueous extracts of decomposing straw and other lignified materials have been found to contain phenolic acids (Flaig, 1964; Kuiters and Sarink, 1986; Nord, 1964; Oglesby et al., 1967; Shindo and Kuwatsuka, 1975). In fact, these compounds can be isolated from decomposing plant material throughout the decomposition process (Flaig, 1964). It is therefore not surprising that phenolic acids and other structurally similar phenolics have been found in varying quantities in soils from many different natural and cultivated ecosystems around the world (Blum, 1996; Blum et al., 1991; Carballeira, 1980; Carballeira and Cuervo, 1980; Chandramohan et al., 1973; Chou and Patrick, 1976; Gallet and Lebreton, 1995; Jalal and Read, 1983; Katase, 1981a,b,c,d; Kuiters and Denneman, 1987; Lodhi,

1975a, b, 1976, 1978; Patrick, 1971; Patrick and Koch, 1958; Shindo et al., 1978; Turner and Rice, 1975; Wang et al., 1967; Whitehead, 1964; Whitehead et al., 1981, 1982, 1983).

There is a point during the decomposition process where soil organic matter reaches a relatively stable intermediate state and very little subsequent degradation occurs. This could be because intermediate compounds are relatively resistant to decomposition or they are protected in the soil environment by forming resistant complexes with metal ions or clays (Haider et al., 1977; Martin and Haider, 1971). These relatively stable soil organics are typically referred to as humic substances (Flaig, 1964). Lignins are generally considered to be the primary source of phenolic units for humic acid synthesis (Burges et al., 1964; Martin and Haider, 1971). Although their exact chemical structure and composition are incompletely known (Bohn et al., 1979), humic acids appear to be complex macromolecules of phenolic units with linked amino acids, peptides, and other organic compounds (Martin and Haider, 1980), and may comprise 50 to 80 percent of the soil humus (Bohn et al., 1979; Martin and Haider, 1971). Although humic substances are regarded as being fairly chemically recalcitrant, it is likely that further, gradual degradation would result in the release of small amounts of monomeric phenolics into soil solution over time.

6.4   RELEASE INTO THE SOIL ENVIRONMENT

Leachates from different plant species can differ quantitatively and qualitatively in their spectra of water-soluble phenolic compounds (Gallet and Lebreton, 1995). Problems of natural regeneration and reforestation are often attributed to the presence of particular phenolic substances deposited in the soil by previous tree species or herbaceous vegetation (Fisher, 1980). Potentially, the differences in phenolic compound input can lead to variations in the development of soil under different plant communities (DeKimpe and Martel, 1976; Kogel and Zech, 1985; Kuiters and Denneman, 1987; Malcolm and McCracken, 1968; Shindo et al., 1978; Whitehead et al., 1982). Kuiters and Denneman (1987) found considerable variation in phenolic acid concentrations in soils supporting different tree species. They concluded that the variation was partially explained by the differences in litter properties among species. Highest concentrations of phenolic acids were released from broad-leaved tree species, particularly hornbeam (*Carpinus betulus* L.), hazelnut (*Corylus avellana* L.), and birch (*Betula pendula* Roth.) (Kuiters and Sarink, 1986). Kogel and Zech (1985) found phenolic acid composition in soil humus layers to be highly correlated with phenolic substances released from leaf litter. Shindo et al. (1978) concluded that higher levels of phenolic acids found in forest soils, as compared with paddy fields, resulted from higher levels of phenolic substances in decomposing wood and leaves of trees than in decomposing straw.

The most common phenolic acids found in soil are ferulic, *p*-coumaric, vanillic, protocatechuic (Chou and Lee, 1991; Katase, 1981c; Kuiters and Denneman, 1987; Li et al., 1992; Shindo et al., 1978; Vance et al., 1985, 1986; Whitehead et al., 1982), *p*-hydroxybenzoic (Katase, 1981c; Shindo et al., 1978; Vance et al., 1985, 1986; Whitehead et al., 1982, 1983), syringic (Kuiters and Denneman, 1987; Shindo et al., 1978; Whitehead et al., 1981), caffeic (Lodhi, 1976, 1978), and salicylic (Jalal and Read, 1983; Shindo et al., 1978; Vance et al., 1985). Typically, there is greater variety and higher levels of benzoic acid derivatives extracted from soil than cinnamic acid derivatives (Jalal and Read, 1983; Katase, 1981c; Vance et al., 1985, 1986; Whitehead et al., 1983).

6.5   BEHAVIOR IN SOILS

One of the main difficulties in obtaining strong supportive evidence for allelopathic interference in field situations is demonstrating that phenolic acids are present and available in soil in sufficient quantities for sufficient periods of time to affect plant growth. This is complicated by the fact that there is as yet no clear understanding or consensus of what actually constitutes bioavailable and non-bioavailable fractions of phenolic acids in soil. To understand the issue of bioavailability, one must first attempt to understand the complexity of soil systems and how different soil components and processes affect the behavior and concentration of such compounds in soils.

6.5.1　Physicochemically Mediated Processes

When phenolic acids enter the soil environment, physicochemical and biological processes immediately begin to affect the solution-phase concentration of these compounds (Blum et al., 1994; Dalton et al., 1989a,b). Certain molecules become sorbed to soil particle surfaces and other molecules remain 'free' in the soil solution-phase (or bulk-phase). Phenolic acids adsorbed onto soil surfaces by short-range retention forces, such as van der Waals-London or hydrogen bonding, are held reversibly; they can be released (desorbed) back into the solution-phase as a result of decreasing ionic strength of the soil solution, or by displacement by a competing ion (Dalton et al., 1987). Some investigators suggest that bioavailable molecules are those in soil solution and those weakly adsorbed to soil particle surfaces (Blum et al., 1994; Dalton et al., 1987; Dao, 1987; Whitehead et al., 1982, 1983). Adsorption–desorption processes are very dynamic, and concentrations of phenolic acids in soil solution may vary substantially even with slight changes in soil moisture, nutrition, temperature, pH, etc. Adsorption of phenolic acids by soil components is greatly affected by pH. At soil pH values below the pKa of phenolic acids (approximately 4.5), most molecules are nonionized and may be adsorbed to organic matter and clay through weak physical adsorption forces. Under slightly acid to basic conditions, phenolic acids become anions that can adsorb to positively charged sites on soil surfaces (Bailey and White, 1964; Weber and Sheets, 1977). At these same soil pHs, polyvalent cations may function as bridges in the binding of negatively charged phenolic acid molecules to negatively charged sites on clay and organic matter (Greenland, 1965, 1971). The binding strength of the resulting complex is determined by the specific polyvalent cation involved. Therefore, the nature of the soil/phenolic acid complex determines how easily it can be disrupted through ion exchange or displacement reactions.

Some reactions involving phenolic acids and soil surfaces are not easily reversible and molecules do not re-enter (or are very slow to re-enter) the soil solution under normal soil conditions. Mechanisms of this type of adsorption might include ligand-exchange or oxidation reactions of phenolic acids with soil mineral surfaces (Lehmann et al., 1987; Lehmann and Cheng, 1988; McBride, 1987), or incorporation of phenolic acids into soil humic material (Haider et al., 1977; Martin et al., 1972; Martin and Haider, 1976). Reactions of this kind probably involve strong organometallic complexes or have some degree of covalent bonding associated with soil particle surfaces (mineral and organic). Soil extractions liberating such compounds (e.g., alkaline hydrolysis) provide phenolic concentration data of questionable ecological value (Dalton et al., 1987; Kaminsky and Muller, 1977, 1978; Katase, 1981c).

In general, cinnamic acid derivatives appear to be more reactive in soils (Dalton et al., 1989a; Lehmann et al., 1987; Haider and Martin, 1975) and more toxic to plants (Blum et al., 1984, 1985a, b) than benzoic acid derivatives bearing the same functional groups. The presence of a methoxy group or acrylic side chain on the aromatic ring of phenolic acids increases molecular sorption in soils (Dalton et al., 1989a). Blum et al. (1994) observed a rapid initial sorption of both cinnamic and benzoic acid derivatives in Cecil soil followed by a slow, long-term sorption of cinnamic acid derivatives only. In fact, in most field studies using mild soil extractants, benzoic acids are found in higher concentrations than cinnamic acids (Jalal and Read, 1983; Katase, 1981c; Vance et al., 1985, 1986; Whitehead et al., 1983). It appears that the very functional groups that tend to promote plant toxicity also promote molecular sorption to soil particles.

6.5.2 Microbially Mediated Processes

Once in the soil environment, phenolic acids are often decomposed to simpler compounds by soil microorganisms (Dagley, 1965; Flaig, 1964; Gibson, 1968; Haider and Martin, 1967; Horvath, 1972; Kunc and Macura, 1966; Martin and Haider, 1971). However, microbial degradation should not be equated with detoxification since some breakdown products have been shown to be more toxic in plant bioassays than the parent compound (Liebl and Worsham, 1983). Microbes can utilize free phenolic acids as carbon and energy sources (Blum and Shafer, 1988; Rahouti et al., 1989; Sugai and Schimel, 1993) and cause polymerization of phenolic acids through the activity of polyphenoloxidases and peroxidases (Haider and Martin, 1975; Kassim et al., 1982). Indigenous soil microbes reportedly have no difficulty metabolizing, degrading, or transforming exogenously applied phenolic acids in model soil systems (Blum et al., 1987; Blum and Shafer, 1988; Dalton, 1989). Such observations, however, are typically made under ideal growth conditions in which microbes are provided with more than adequate temperatures, nutrition, and moisture. In the field, many soil systems experience periods when conditions for rapid microbial growth are lacking (e.g., low temperatures, low nutrition, high levels of toxins, anaerobic conditions, etc.), and physicochemical reactions of phenolic acids with soil particle surfaces may become increasingly important in determining the fate of phenolic acid molecules and their derivatives. In fact, because physicochemical reactions occur rapidly, they may play a significant role in reducing phenolic acid bioavailability, even when microbial activity is high.

  Blum and Shafer (1988) found that ferulic, p-coumaric, p-hydroxybenzoic, and vanillic acids were readily utilized as carbon sources by soil microbes in a clay-loam, bottomland soil material. Soil bacteria and fungi were stimulated by phenolic acids at concentrations of 0.5 mol/g $^{-1}$ and were generally reduced at higher concentrations. Rates of phenolic acid biodegradation were greater in the A_1 horizon (organic), while toxicity to microbial populations were more frequent in the B_1 horizon (clay). In a study of decomposition and incorporation of ^{14}C-labeled glucose and phenolics into biomass on a taiga forest floor, Sugai and Schimel (1993) found that after 4 h, less than 10 percent of the added ^{14}C-compounds typically remained unchanged. Much of the radiolabel was found as $^{14}CO_2$, suggesting microbial degradation and use. More than twice as much glucose was incorporated into biomass than either phenolic acid tested in the study. Sugai and Schimel surmised that glucose was incorporated into new cellular material by microbes, while the phenolics were used predominantly as energy sources. Variation was noted between similar phenolic acids; p-hydroxybenzoic acid was metabolized and used for biomass efficiently, while salicyclic acid essentially was not assimilated.

6.6 EXTRACTION FROM SOILS

In allelopathy studies it is essential to distinguish between bioavailable and nonbioavailable amounts of phenolic acids in soil to identify amounts that are potentially available to interact with organisms living there. However, direct determination of chemical bioavailability in soil presents serious technical challenges. There is much debate over valid ways to monitor ecologically relevant or meaningful concentrations of soil phenolic acids. Parallels to this problem exist in determining the availability of nutrients to plants. Often the assumption is that nutrients in soil solution are available to plants, when in fact, studies have shown that the chemical activity of specific forms correlate more closely with plant uptake than concentration (Sposito, 1984). The problem is compounded with ionizing organics like phenolic acids since additional problems exist in determining accurate stability constants needed to calculate their chemical activity (Baham, 1984). Until these technical problems are somehow resolved, best estimates of available and potentially available phenolic acids in soil may come from measurements of amounts found in soil solution and amounts found weakly sorbed to soil particle surfaces, respectively. The following discussion summarizes the limited amount of information available on the extraction and behavior of phenolic acids in soils.

6.6.1 EXTRACTION FROM NATURAL AND CULTIVATED SYSTEMS

Various extractants and extraction procedures have been used to determine phenolic acid concentrations in soils of natural and cultivated systems (Blum et al., 1991, 1992; Carballeira and Cuervo, 1980; Chou and Patrick, 1976; Guenzi and McCalla, 1966; Kaminsky and Muller, 1977; Katase, 1981b; Shindo and Kuwatsuka, 1976; Turner and Rice, 1975; Wang et al., 1967; Whitehead et al., 1981, 1982). Whitehead et al. found the amounts of phenolic acids extracted by water from roots (1982) and litter (1981) greater than the amounts extracted from their associated soils. They speculated that the levels extracted from soil by water alone were probably lower than levels likely to exert allelopathic effects. However, extraction of the same soil with 0.5 percent $Ca(OH)_2$ recovered levels that have been reported to have adverse effects on plant growth (Whitehead et al., 1982). Such effects might occur in field situations as a consequence of liming. In addition, all soils analyzed by Whitehead et al. (1983) contained both free and bound forms of water-soluble phenolic compounds. Substantial amounts of water-extractable phenolic compounds found in roots and leaf litter occurred as free forms, while more than 50 percent were in bound forms. Glass (1976) used the same mixture and concentration of phenolic acids as identified in field soil solution by Whitehead (1964) and observed that the development of roots of barley (*Hordeum vulgare* L.) plants in hydroponic culture was severely inhibited.

Katase (1981a) extracted phenolic acids from peat soil and identified three forms. Form A, isolated by repeated reflux with hot ethyl acetate, was designated as the free form. Form B, first extracted with hot ethyl acetate, then repeatedly refluxed with 2 M NaOH for 24 h, and adjusted to pH 2, was designated as a combined form. Form C was another combined form not extractable with hot ethyl acetate, but isolated with hot 2 M NaOH by repeated extraction. Katase found form C to be tightly bound to the soil matrix. The A, B, and C forms of *p*-hydroxybenzoic acid were 11, 69, and 980 µg/g soil, respectively; vanillic acid was 85, 130, and 860 µg/g soil, respectively; and ferulic acid was 53, 450, and 2700 µg/g soil, respectively. The combined totals of the three forms for all three phenolic acids were 1200, 1100, and 3200 µg/g soil, respectively. Form A, which Katase considered important for allelopathy studies, ranged from 2 percent (of total ferulic acid) to 10 percent (of total *p*-hydroxybenzoic acid) while form C was more than 80 percent. The combined totals of the three forms for all three phenolic acids occupied approximately 1 percent of the total soil organic matter.

Whitehead et al. (1983) found the amounts of phenolic acids extracted by water from peat and forest soils by Katase (1981a,c) to be considerably lower than amounts that could be removed by ethyl acetate (form A). Water-soluble forms of the phenolic compounds accounted for less than 0.7 percent of the amount extracted by 2 M NaOH (Whitehead et al., 1983). Kuiters and Denneman (1987) found that concentrations of water-extractable phenolic acids in vegetated soil varied from 3 to 26 µg/g soil. The highest amounts of phenolic acids occurred under *Pinus* and *Picea* trees and ferulic acid was highest in concentration (11 to 15 µg/g dry soil). In general, ferulic, *p*-coumaric, syringic, vanillic, and protocatechuic acids were found in greatest abundance, except in soils under *Quercus,* where benzoic acid was higher in concentration.

6.6.2 EXTRACTION FROM MODEL SYSTEMS

Model systems (specific soils spiked with compounds) have also been used to investigate phenolic acid extractability (Blum et al., 1994; Dalton, 1989; Dalton et al., 1987, 1989a,b). Under controlled conditions, Dalton et al. (1987) compared many of the soil extraction procedures previously used to recover allelochemicals from natural and cultivated systems (Carballeira and Cuervo, 1980; Chou and Patrick, 1976; Dalton et al.,1983; Guenzi and McCalla, 1966; Kaminsky and Muller, 1977; Katase, 1981b; Turner and Rice, 1975). Dalton (1989) added ferulic acid (1000 mg/kg) to different soils (two horizons of two different soil types) and allowed them to equilibrate for 90 d under aseptic conditions before extraction. The efficiency of various extractants in recovering ferulic acid were directly compared by allowing the extracting solutions to remain in contact with soil materials for 1, 3, 6, 12, 24, and 48 h before extraction and analysis by high performance liquid chromatography (HPLC). Regardless of soil type, the ability of various extractants to recover ferulic acid was generally in the order: water=methanol<sodium acetate<EDTA=DTPA<sodium hydroxide. Increased ferulic acid extraction by sodium acetate is brought about by the disruption of weak retention forces easily disrupted by sodium and acetate ions. Differences between sodium acetate and chelating extractions were considered to represent the recovery of molecules held by polyvalent cations, either exchangeable or nonexchangeable. Dalton et al. (1987) concluded that phenolic acid molecules recovered by harsh chelating and sodium hydroxide extractions would probably not be readily released under typical ecological conditions.

Basic EDTA and water extractions have also been used (Blum et al., 1994) to extract soil. They appear to provide good estimates of free and reversibly bound cinnamic acid derivatives, but not benzoic acid derivatives. Neutral (pH 7) EDTA and water extractions, however, provide good estimates for both cinnamic and benzoic acid derivatives. By conducting studies with and without microbes, Blum et al. (1994) confirmed that neutral EDTA and water soil extractions provide accurate estimates of bioavailable phenolic acids to soil microbes and, therefore, are potentially available to seeds and roots. Blum (1997) also reported that citrate extractions were equivalent to or better than EDTA in recovering ferulic acid from soil and plant debris. Allelochemical profiling in soil alone does not prove that chemical interference interactions are functioning in a particular system, however, it does provide fundamental knowledge about the soil chemical environment and may help in identifying soil types and environmental conditions in which allelopathic interactions are more likely to occur.

6.6.3 EXTRACTION FOR SPECIFIC PURPOSES OR CONDITIONS

In some cases, it may be desirable to directly use soil extracts in bioassays for the detection of allelopathic activity. Many soil extractants, however, may not be appropriate for use directly in bioassays. For example, EDTA soil extractions should not be used directly in germination bioassays because solutions of EDTA (0.5 M at pH 8) completely inhibit the germination of some test species (Blum et al., 1992). Soil extracts may not always remain stable enough to reflect actual bioavailable levels and forms. Longer soil extraction times (in water and other aqueous solutions) allow soil microbes enough time to utilize and modify phenolic acid concentractions in the extracting solution (Blum and Shafer, 1988; Blum et al., 1991). To avoid such complications, Blum et al. (1992) suggested the use of a water-autoclave procedure. This provides sterile soil extracts that can be used directly in germination bioassays. In addition, the water-autoclave extraction procedure can effectively extract free and reversibly bound phenolic acids from wheat debris (Blum et al., 1992).

If soil sterilization is to be used in allelochemical studies, or as part of a soil extraction process, it is recommended that pilot tests be conducted to determine if such treatments affect the sorption and stability of test compounds. Steam sterilization of soil can potentially affect the physical and chemical properties of soil (Lopes and Wollum, 1976; Salonius et al., 1967; Williams-Linera and Ewel, 1984), thereby changing the way phenolic acids interact with soil particles. Comparisons of immediate extractions of ferulic acid reveal that recoveries from nonsterile soils are more similar to recoveries from autoclaved than from methyl bromide or gamma-irradiated soils (Dalton et al., 1989b). However, changes in phenolic acid recovery brought about by soil sterilization techniques are greatly dependent upon soil type. Dalton (1989) found treatment effects in the immediate recovery of ferulic acid from steam-sterilized and nonsterilized tropical forest soils.

In soil extracts of most natural and cultivated ecosystems studied so far, phenolic acids are the most commonly identified compounds with demonstrated allelochemical activity. Their prevalence may, in part, be a function of the ease of isolation and identification of phenolic acids as compared with the more complex phenolics (e.g., coumarins, flavones, flavonones, isoflavones, flavonols, chalcones, etc.), or a result of degradation of the more complex phenolics (including lignified and humified materials) in extracting solutions. Further investigations in which careful attention is given to reducing chemical alterations of samples during soil collection, soil extraction, and sample analysis are undeniably needed. Labware potentially containing impurities and residual charge may differentially sorb or facilitate the degradation of potential allelochemicals that may be present at trace levels. Sample contact with nonacid-washed plastics and glass should be avoided. In all chemical interference studies, controls should be included to evaluate allelochemical stability at various stages of extracting, processing, storing, bioassaying, etc. In fact, certain laboratory lighting conditions can cause rapid *cis-trans* isomerization of cinnamic acids (Hartley and Jones, 1975; Katase, 1981c, d, 1983). Because of the sensitivity of these compounds to degradation and sorption during handling and extraction, it is essential that sterile, cold, dark, and noncharged conditions be maintained whenever possible.

6.7 INVOLVEMENT IN ALLELOPATHIC INTERACTIONS

Allelopathic interactions specifically involving phenolic acids have been suggested for many species of plants growing in natural systems (Carballeira and Cuervo, 1980; Chou and Lee, 1991; Fischer et al., 1994; Gallet, 1994; Glass, 1976; Jalal and Read, 1983; Kuiters, 1989; Lodhi and Nickell, 1973; Lodhi, 1975a,b, 1976, 1978; Moreland et al., 1966) as well as in cultivated systems (Blum, 1996; Blum et al., 1991, 1992; Borner, 1960; Chandramohan et al., 1973; Chou and Leu, 1992; Chou and Patrick, 1976; Einhellig and Rasmussen, 1979; Li et al., 1992; Liebl and Worsham, 1983; Patrick and Koch, 1958; Putnam and Duke, 1978; Putnam, 1983; Wang et al., 1967). Further supportive evidence from hydroponic and soil culture bioassay studies clearly indicate the phytotoxic potential of many phenolic acids, singularly and in mixtures (Blum, 1996; Blum and Dalton, 1985; Blum et al., 1984, 1985a,b, 1989, 1993; Einhellig and Rasmussen, 1979; Glass, 1976; Kuiters, 1989). However, there still remains no definitive proof that phenolic acids actually affect plant growth under ordinary field conditions (Blum, 1995).

Rapid loss of phenolic acids (in soil) by biodegradation, sorption, and leaching are frequently cited as reasons for dismissing their involvement in allelopathic interactions. The continued detection of low concentrations of bioavailable phenolic acids in soils, and the fact that much higher concentrations are necessary in hydroponic and soil culture bioassays to cause plant growth supression, have furthered this skepticism. This is complicated by the fact that phenolic acids have a low toxicity relative to other compounds (e.g., sesquiterpene lactones) that may be more biologically active by an order of magnitude or more. It is perhaps time to reconsider whether it is reasonable to expect that elevated concentrations of soil allelochemicals are necessary to substantiate allelochemical interference in field situations. It may also be time to re-evaluate some of the bioassay and model systems traditionally used to study allelopathic interference interactions and determine how well such artificial systems represent real ecological conditions and situations.

In natural soil systems, it is possible that continual low levels of phenolic acids, resulting from continuous release by plants, soils, and microbes, are cumulatively effective in inhibiting plant growth over time. It is also possible that there are localizations of allelochemical concentrations in soil that have previously gone undetected. Most soil extraction methods provide average allelochemical concentrations for a particular unit area of soil (Dalton et al., 1987). The chemical composition of the soil environment, however, is heterogeneous and concentrations of phenolic compounds likely exist in gradients around fragments of decomposing plant material, actively growing roots (rhizosphere), and possibly, specifically charged soil components. Therefore, even in systems where low (average) amounts of allelochemicals are detected, there may be localized concentrations that are high enough to affect plant growth and survival. To better understand the potential allelochemical profile in soil, detailed studies and innovative techniques are needed to better identify and monitor bioavailable background concentrations, fluctuations, and localizations of these compounds in soil environments.

Unlike most field studies, bioassays are typically conducted in artificial environments with adequate to optimum growing conditions for plants and soil microbes. Therefore, the high concentrations of phenolic acids necessary to suppress the growth of plants exposed under the conditions of most bioassay studies should not be too surprising. Thriving microbial populations can quickly alter and deplete concentrations of phenolic acids in the growing medium before there is sufficient contact with plant roots to affect growth. Plants in these circumstances are receiving all essential nutrients and are not likely to be especially vulnerable to a chemical stress. However, conditions in natural and cultivated systems are frequently growth limiting, increasing the probability that low levels of phenolic acids could impact plant or microbial systems.

The observed phytotoxicity of phenolic acids has been attributed to the inhibition of mineral uptake, nitrogen fixation, photosynthesis, respiration, and interference with plant growth substances (Rice, 1979, 1984). Some evidence suggests that changes in plant growth may also occur indirectly through effects on mycorrhizal growth and host infection (Perry and Choquette, 1987). In fact, ectomycorrhizal fungi appear to be more sensitive to phenolic compounds than litter decomposing saprotrophs. Olsen et al. (1971) found the saprotroph, *Marasmius,* to be less inhibited by benzoic acid and catechol than the ectomycorrhizal fungus, *Boletus.* The sensitivity of other mycorrhizal fungi to phenolic substances has been corroborated (Rose et al., 1983; Cote and Thibault, 1988) and is further supported by laboratory evidence that the *in-vitro* growth of actinorhizal actinomycetes of the genus *Frankia* is inhibited by the presence of certain phenolic acids, particularly the cinnamic acid derivatives (Perradin et al., 1983; Vogel and Dawson, 1986).

Blum and his colleagues, using well-defined nutrient and soil culture bioassays, have shown that mixtures of phenolic acids can inhibit the growth of cucumber (*Cucumis sativus* L.) seedlings in an additive or antagonistic manner (Blum et al., 1989). As an outcome of such studies, statistical methods have been designed to determine the joint inhibitory action of phenolic acids and similar compounds on plant growth (Gerig et al., 1989). Further studies have shown that differential utilization of compounds in organic mixtures by soil microbes can modify the toxicity of a given concentration of phenolic acid on seedling growth (Blum et al., 1993; Pue et al., 1995). With all that has been learned from this model system, it appears that phytotoxicity by phenolic acids is more likely to occur in acidic soils with low organic matter and slightly nutrient- and moisture-limiting conditions (Blum, 1995). This has interesting ecological implications since the nutrient status of forest soil has been shown to affect phenolic production in plant tissues. Muller et al. (1987, 1989) observed a trend of increasing astringent phenolic concentrations in foliage and fine roots, respectively, along gradients of declining soil fertility in a forested system. It is unknown, however, if increases in root and foliage astringent phenolics directly correlate with increases in bioavailable amounts of soil phenolics.

Major advancements in allelopathy will come when increased attention is given to designing studies that better approximate situations and conditions typically encountered in ecological systems. If allelopathy is ever established as an ecological process, it could have a tremendous influence on the management of natural and cultivated systems.

6.8 NEW DIRECTIONS IN SOIL SAMPLING METHODOLOGY

Weidenhamer (1996) acknowledged that while unambiguous proof may not be attainable for an ecological phenomenon as complex as allelopathy, the challenge to investigators is still to provide convincing evidence for chemical interference. He suggested two key areas in which additional progress is needed to overcome the methodological impasse: (1) distinguishing allelopathy from resource competition and other interference mechanisms in plant growth studies; and (2) developing new analytical techniques that measure the flux rates of allelochemicals in soil. Weidenhamer (1996) provided a substantive discussion on current methodological limitations and progress in distinguishing resource competition and chemical interference. The following discussion elaborates on possible ways to improve techniques to better approximate bioavailable flux rates of allelochemicals in soil.

Blum (1995) suggested that a systems approach will be required to obtain definitive insight concerning the role of phenolic acids in allelopathic interactions. He suggests that model systems should first be studied so that correlations between system parameters and characteristics can be made. Aside from the sensitivity of the target plant to phenolic acids, Blum feels that the rate and timing of phenolic acid input into the bioavailable soil fraction and the rates of root uptake and modification are much more important than static, single-point soil phenolic acid determinations.

Determining bioavailable allelochemical flux rates in soil will require more precise measurements of the soil chemical environment than has been conducted in the past. Much progress can be made by developing sampling techniques that measure *in situ* soil chemical fluctuations. For example, Weidenhamer (1996) suggested the use of resin traps containing XAD or other adsorbents for trapping certain available soil allelochemicals (particularly lipophilic compounds) to provide measurements of flux rates. He conducted sorption studies using polyurethane foam plugs in aqueous solutions of juglone and found rapid sorption (approximately 70 percent) of the compound with moderately high recovery rates (approximately 85 percent). He has also recovered alpha-terthienyl from plugs buried in pots of marigold (*Tagetes erecta* L.), growing both in sand and in potting soil (J. D. Weidenhamer, 1996 personal communication). There is a real need for more innovative approaches using microtechniques to pinpoint allelochemical localizations in soil and to better understand allelochemical gradients.

More studies using ^{14}C-labeled allelochemicals are needed to detect chemical movement and bioavailability in model plant–soil–microbe systems. The expense of ^{14}C-labeled material and the expense of facilities in which to conduct such studies have done much to impede progress. Of particular importance are long-term studies of the release, degradation, transformation, sorption, and polymerization of ^{14}C-labeled phenolic substances in model and contained systems as well as determining the effects that environmental stresses have on allelochemical production, composition, concentration, deposition into the environment, and toxicity.

6.8.1 *In Situ* Soil Sampling

The complexity of the plant–soil–microbe system offers unique challenges with regard to determining soil allelochemical concentrations in allelopathy studies. Because of the dynamic nature of soil systems and the continual changes in the soil chemical environment, peak occurrences of free or bioavailable amounts may be missed with single-point sampling. Moreover, there is considerable potential for allelochemical modification during soil collection, transport, storage, and processing. Data generated using such techniques and conditions should be interpreted accordingly. Ideally, *in situ* soil sampling would allow continued monitoring of potentially bioavailable allelochemical levels and reduce the risk of creating sampling artifacts during soil collection and processing.

In some instances, particularly in wet to mesic situations, soil water samplers (or lysimeters) may be an optional method to measure *in situ* flux rates of phenolics, providing several advantages over traditional soil solution extractions. *In situ* soil sampling would eliminate the potential for chemical modifications associated with the removal of soil from the field and the processing time associated with soil extractions. The use of such devices would allow low-volume samples to be collected at the same location each sampling time. Certain precautionary measures, however, are recommended when using soil water samplers in allelopathy studies. First, it is essential to establish that samplers do not significantly interfere with sample chemistry. Lysimeters should be tested to determine the short-term and long-term potential for sorption of test compounds to the sampler matrix. During installation, soil disturbance around the lysimeter should be minimized, and the addition of substances like silica around sampler tips should be excluded or minimized (and tested) to alleviate the potential for additional sorption reactions. In addition, samplers should be as small as possible, not only to reduce initial soil disturbance, but to reduce the potential for chemical and hydrologic changes in the immediate vicinity of the sampler tip as a result of removing high volumes of water. Vacuum soil samplers may not prove useful in all habitat situations, nor are they applicable for all research objectives. If carefully employed, however, they may provide more realistic assessments of bioavailable allelochemical fluctuations than attainable with current soil extraction methods. Soil water sampler usage in allelochemical studies might be limited to wetlands and mesic habitats or situations in xeric habitats during rainy seasons or immediately following rain events.

Porous ceramic cups are made of heat-fused clay, and as a result, they possess ion-exchange properties that potentially interfere with sampling of charged compounds like phenolic acids (Debyle et al., 1988; Grover and Lamborn, 1970; Hansen and Harris, 1975; Hughes and Reynolds, 1988). Flushing cups with dilute solutions of HCl reduces the charge of the ceramic matrix and consequently reduces the interference with certain nutrients (Debyle et al., 1988; Grover and Lamborn, 1970; Hughes and Reynolds, 1988; Neary and Tomassini, 1985).

The author tested the passage of dilute solutions of phenolic acids through porous ceramic cup vacuum samplers to determine their potential utility in *in situ* field studies (unpublished data, October, 1991). Solutions of protocatechuic, *p*-hydroxybenzoic, vanillic, caffeic, syringic, *p*-coumaric, ferulic, and sinapic acids were passed through non–acid-washed, acid-washed, and 2 y field tested porous ceramic cup water samplers. Ten and 100 M solutions of each phenolic acid (at pH 4.5) were pulled through the ceramic matrix and quantified by HPLC. The greatest amount of interference with phenolic acids occurred with the non–acid-washed ceramic cups. Ten molar concentrations of protocatechuic, caffeic, and sinapic acids were significantly reduced by 39.5, 40.5, and 14.75 percent, respectively. Acid-washed samplers had significant, but dramatically less interference with phenolic acids than non–acid-washed samplers. Ferulic acid (10 M) was only reduced by 1.75 percent; caffeic, ferulic, and sinapic acids (100 M) were reduced by 5, 3.25, and 6 percent, respectively. Samplers previously acid-washed and used in a muck soil for approximately 2 y were rinsed with distilled water and lightly brushed to remove soil particles before testing the passage of phenolic acids. These samplers did not significantly interfere with the passage of phenolic acids at either concentration. There was more than 97 percent passage of all compounds tested. This suggests that during the 2 y field exposure there was no substantial increase in the sorption capacity of the ceramic matrix.

Porous ceramic cup vacuum samplers have been shown to be ideal for sampling field soil solution, both in terms of minimum alteration of soil solution and in terms of low failure rates and adequate volumes of solution (Silkworth and Grigal, 1981). Other investigators have indicated their effectiveness in sampling nutrients in soil water (Debyle et al., 1988; Grover and Lamborn, 1970; Hughes and Reynolds, 1988; Neary and Tomassini, 1985). With proper pre-cleaning, porous ceramic cup vacuum samplers might prove useful in sampling simple organic molecules like phenolic acids and other allelochemicals from field soils. In each case, however, samplers should be painstakingly tested to determine the potential for sorption reactions with substances to be monitored as well as periodically tested during long-term field exposure to determine the degree of interference the ceramic matrix might have on sample chemistry.

Soil water sampling devices composed of less reactive materials (e.g., teflon and glass) might interfere less with sample chemistry than ceramic samplers. Advances in technology will further reduce sample volume needed for chemical analysis and smaller soil sampling devices and microtechniques will likewise be developed. This will no doubt have a profound effect on our ability to monitor allelochemical flux rates and better understand their spatial heterogeneity in soil environments.

6.9   CONCLUDING REMARKS/RECOMMENDATIONS

An enormous amount of information has been published about plant phenolics and their potential involvement in chemical interference interactions. Although no single study has provided indisputable proof that such interactions occur under ordinary ecological conditions, there is mounting evidence from model, cultivated, and natural ecosystem studies clearly indicating the potential for such interactions. Results from model systems and bioassay experiments have provided unequivocal evidence that specific compounds from plants and decomposing plant debris have the ability to affect plant growth, development, and even survival. In natural and cultivated ecosystems these compounds are regularly found in soil extracts in the immediate vicinity of plants and plant debris, and they appear to be widespread, occurring in various soils of most plant-dominated, terrestrial ecosystems. However, at this time it is still uncertain whether compounds recovered from soil are in bioavailable forms or whether they occur at sufficient concentrations for sufficient durations in soil to affect the health of plants and other soil organisms. Because of the complexity of plant–soil–microbe systems and our current limited knowledge and understanding of allelochemical bioavailability in such systems, it is likely that this uncertainty will continue to plague allelopathy studies for some time.

The following suggestions are offered to help overcome some of the current methodological limitations encountered in studying phenolics and other potential allelochemicals in soil environments.

1. Until new soil extraction methods and techniques are developed, best estimates of available and potentially available water-soluble phenolics may come from measurements of amounts found in soil solution and amounts found weakly sorbed to soil particle surfaces, respectively. Soil solution concentrations can be approximated by separating and analyzing soil water or low-volume water extracts (lysimeters might be appropriate in certain situations). Weakly sorbed phenolic compounds can be extracted from soil using competing ions or mild chelating agents. Sequential extraction approaches (using extractants of increasing strength) can provide a valuable profile of the allelopathic potential of a particular soil.

2. There is an undeniable need for field studies addressing long-term, soil allelochemical dynamics: peak concentration events, shifts in chemical composition, background concentrations, duration of peak concentration events, seasonal cycles, yearly cycles, etc.

3. Increased attention is needed in reducing chemical alterations of samples during soil collection, soil extraction, and sample analysis. Because of the sensitivity of some compounds to light, microbial degradation, and sorption, it is recommemded that samples be maintained under sterile, cold, dark and noncharged conditions.

4. If sterilized soils are used in chemical interference studies, or if soil sterilization (e.g., autoclave treatment) is part of a soil extraction process, tests should be conducted to determine if there are sterilization treatment effects on the sorption and stability of test compounds.

5. Traditional model systems have often provided plants and soil organisms with optimum growing conditions not commonly encountered in field situations. Model systems for studying chemical interference interactions should be improved to better approximate ecological conditions.

6. *In situ* allelochemical sampling (or trapping) may provide an improvement over traditional soil extractions used to recover bioabailable compounds. If done correctly, *in situ* methods could greatly reduce the risk of creating sampling artifacts inherent in collecting and processing soil material.

Unraveling the intricate details of this potentially pervasive but subtle ecological phenomenon will most likely demand decades of individual and cooperative efforts from investigators in many different areas of specialization. Major advancements will come when increased attention is given to designing studies that approximate situations and conditions typically encountered in ecological systems. The unequivocal establishment of allelopathy as an ecological process will revolutionize the way we manage certain cultivated and natural ecosystems.

ACKNOWLEDGMENTS

The author thanks Drs. J. D. Weidenhamer, J. O. Luken, R. F. Fisher, H. H. Cheng, and Inderjit for providing constructive comments on the manuscript. Special recognition and kind thanks are given to Drs. U. Blum, S. B. Weed, and J. R. Shann for editorial comments and for many years of collaborative research and thought-provoking discussions regarding this subject.

REFERENCES

Baham, J., Prediction of ion activities in soil solutions: computer equilibrium modeling, *Soil Sci. Soc. Am. J.,* 48, 525, 1984.

Bailey, G. W. and White, J. L., Review of adsorption and desorption of organic pesticides by soil colloids, with implications concerning pesticide bioactivity, *Agric. Food Chem.,* 12, 324, 1964.

Blum, U., The value of model plant–microbe–soil systems for understanding processes associated with allelopathic interactions: one example, in *Allelopathy: Organisms, Processes and Applications,* Inderjit, Dakshini, K. M. M., and Einhellig, A., Eds., American Chemical Society, Washington, D.C., 127, 1995.

Blum, U., Allelopathic interactions involving phenolic acids, *J. Nematol.,* 28, 259, 1996.

Blum, U., Benefits of citrate over EDTA for extracting phenolic acids from soils and plant debris, *J. Chem. Ecol.,* 23, 347, 1997.

Blum, U. and Dalton, B. R., Effects of ferulic acid, an allelopathic compound, on leaf expansion of cucumber seedlings grown in nutrient culture, *J. Chem. Ecol.,* 11, 279, 1985.

Blum, U. and Shafer, S. R., Microbial populations and phenolic acids in soils, *Soil Biol. Biochem.,* 20, 793, 1988.

Blum, U., Dalton, B. R., and Rawlings, J. O., Effects of ferulic acid and some of its microbial metabolic products on radicle growth of cucumber, *J. Chem. Ecol.,* 10, 1169, 1984.

Blum, U., Dalton, B. R., and Shann, J. R., Effects of various mixtures of ferulic acid and some of its microbial metabolic products on cucumber leaf expansion and dry matter in nutrient culture, *J. Chem. Ecol.,* 11, 619, 1985a.

Blum, U., Dalton, B. R., and Shann, J. R., The effects of ferulic and *p*-coumaric acids in nutrient culture on cucumber leaf expansion as influenced by pH, *J. Chem. Ecol.,* 11, 1567, 1985b.

Blum, U., Gerig, T. M., and Weed, S. B., Effects of mixtures of phenolic acids on leaf area expansion of cucumber seedlings grown in different pH Portsmouth A1 soil materials, *J. Chem. Ecol.,* 15, 2413, 1989.

Blum, U., Gerig, T. M., Worsham, A. D., Holappa, L. D., and King, L. D., Allelopathic activity in wheat-conventional and wheat-no-till soils: development of soil extract bioassays, *J. Chem. Ecol.,* 18, 2191, 1992.

Blum, U., Gerig, T. M., Worsham, A. D., and King, L. D., Modification of allelopathic effects of *p*-coumaric acid on morning-glory seedling biomass by glucose, methionine, and nitrate, *J. Chem. Ecol.,* 19, 2791, 1993.

Blum, U., Weed, S. B., and Dalton, B. R., Influence of various soil factors on the effects of ferulic acid on leaf expansion of cucumber seedlings, *Plant Soil,* 98, 111, 1987.

Blum, U., Wentworth, T. R., Klein, K., Worsham, A. D., King, L. D., Gerig, T. M., and Lyu, S.-W., Phenolic acid content of soils from wheat-no till, wheat-conventional till, and fallow-conventional till soybean cropping systems, *J. Chem. Ecol.,* 17, 1045, 1991.

Blum, U., Worsham, A. D., King, L. D., and Gerig, T. M., Use of water and EDTA extractions to estimate available (free and reversibly bound) phenolic acids in Cecil soil, *J. Chem. Ecol.,* 20, 341, 1994.

Bohn, H. L., McNeal, B. L., and O'Connor, G. A., *Soil Chemistry,* John Wiley and Sons, New York, 1979.

Borner, H., Liberation of organic substances from higher plants and their role in the soil sickness problem, *Bot. Rev.,* 26, 393, 1960.

Burges, N. A., Hurst, H. M., and Walkden, B., The phenolic constituents of humic acid and their relation to the lignin of the plant cover, *Geochim. Cosmochim. Acta,* 28, 1547, 1964.

Carballeira, A., Phenolic inhibitors in *Erica australis* L. and in associated soil, *J. Chem. Ecol.,* 6, 593, 1980.

Carballeira, A. and Cuervo, A., Seasonal variation in allelopathic potential of soils from *Erica australis* L. heathland, *Ecol. Plant,* 1, 345, 1980.

Chandramohan, D., Purushothaman, D., and Kothandaraman, R., Soil phenolics and plant growth inhibition, *Plant Soil,* 39, 303, 1973.

Chou, C.-H. and Lee, Y.-F., Allelopathic dominance of *Miscanthus transmorrisonensis* in an alpine grassland community in Taiwan, *J. Chem. Ecol.,* 17, 2267, 1991.

Chou, C.-H. and Leu, L.-L., Allelopathic substances and interactions of *Delonix regia* (BOJ) RAF, *J. Chem. Ecol.,* 18, 2285, 1992.

Chou, C.-H. and Patrick, Z. A., Identification and phytotoxic activity of compounds produced during decomposition of corn and rye residues in soil, *J. Chem. Ecol.,* 2, 369, 1976.

Connell, J. H., Apparent versus "real" competition in plants, in Grace, J. B. and Tilman, D., Eds., *Perspectives on Plant Competition,* Academic Press, San Diego, 1990, 9.

Cote, J.-F. and Thibault, J.-R., Allelopathic potential of raspberry foliar leachates on growth of ectomycorrhizal fungi associated with black spruce, *Am. J. Bot.,* 75, 966, 1988.

Dagley, S., Degradation of the benzene nucleus by bacteria, *Sci. Progr.,* 53, 381, 1965.

Dalton, B. R., Physicochemical and biological processes affecting the recovery of exogenously applied ferulic acid from tropical forest soils, *Plant Soil,* 115, 13, 1989.

Dalton, B. R., Blum, U., and Weed, S. B., Allelopathic substances in ecosystems: effectiveness of sterile soil components in altering recovery of ferulic acid, *J. Chem. Ecol.,* 9, 1185, 1983.

Dalton, B. R., Weed, S. B., and Blum, U., Plant phenolic acids in soils: a comparison of extraction procedures, *Soil Sci. Soc. Am. J.,* 51, 1515, 1987.

Dalton, B. R., Blum, U., and Weed, S. B., Differential sorption of exogenously applied ferulic, *p*-coumaric, *p*-hydroxybenzoic, and vanillic acids in soils, *Soil Sci. Soc. Am. J.,* 53, 757, 1989a.

Dalton, B. R., Blum, U., and Weed, S. B., Plant phenolic acids in soils: sorption of ferulic acid by soil and soil components sterilized by different techniques, *Soil Biol. Biochem.,* 21, 1011, 1989b.

Dao, T. H., Sorption and mineralization of plant phenolic acids in soils, in *Allelochemicals: Role in Agriculture and Forestry,* Waller, G. R., Ed., American Chemical Society, Washington, D.C., 1987, 358.

Debyle, N. V., Hennes, R. W., and Hart, G. E., Evaluation of ceramic cups for determining soil solution chemistry, *Soil Sci.,* 146, 30, 1988.

Dekimpe, C. R. and Martel, Y. A., Effects of vegetation on the distribution of carbon, iron, and aluminum in the B horizons of northern Appalachian Spodosols, *Soil Sci. Soc. Am. J.,* 40, 77, 1976.

Einhellig, F. A. and Rasmussen, J. A., Effects of three phenolic acids on chlorophyll content and growth of soybean and grain sorghum seedlings, *J. Chem. Ecol.,* 5, 815, 1979.

Fischer, N. H., Williamson, G. B., Weidenhamer, J. D., and Richardson, D. R., In search of allelopathy in the Florida scrub: the role of terpenoids, *J. Chem. Ecol.,* 20, 1355, 1994.

Fisher, R. F., Allelopathy: a potential cause of regeneration failure, *J. For.,* 6, 346, 1980.

Flaig, W., Effects of micro-organisms in the transformation of lignin to humic substances, *Geochim. Cosmochim. Acta,* 28, 1523, 1964.

Gallet, C., Allelopathic potential in bilberry-spruce forests: influence of phenolic compounds on spruce seedlings, *J. Chem. Ecol.,* 20, 1009, 1994.

Gallet, C. and Lebreton, P., Evolution of phenolic patterns in plants and associated litters and humus of a mountain forest ecosystem, *Soil Biol. Biochem.,* 27, 157, 1995.

Gerig, T. M., Blum, U., and Meier, K., Statistical analysis of the joint inhibitory action of similar compounds, *J. Chem. Ecol.,* 15, 2403, 1989.

Gibson, D. T., Microbial degradation of aromatic compounds, *Science,* 161, 1093, 1968.

Glass, A. D. M., The allelopathic potential of phenolic acids associated with the rhizosphere of *Pteridium aquilinum, Can. J. Bot.,* 54, 2440, 1976.

Goodwin, T. W. and Mercer, E. I., *Introduction to Plant Biochemistry,* 2nd ed., Pergamon Press, New York, 1983.

Greenland, D. J., Interaction between clays and organic compounds in soils. Part 1. Mechanisms of interaction between clays and defined organic compounds, *Soil Fertil.*, 28, 415, 1965.

Greenland, D. J., Interactions between humic and fulvic acids and clays, *Soil Sci.*, 111, 34, 1971.

Grover, B. L. and Lamborn, R. E., Preparation of porous ceramic cups to be used for extraction of soil water having low solute concentrations, *Soil Sci. Soc. Am. Proc.*, 34, 706, 1970.

Guenzi, W. D. and McCalla, T. M., Phytotoxic substances extracted from soil, *Soil Sci. Soc. Am. Proc.*, 30, 214, 1966.

Haider, K. and Martin, J. P., Synthesis and transformation of phenolic compounds by *Epicoccum nigrum* in relation to humic acid formation, *Soil Sci. Soc. Am. Proc.*, 31, 766, 1967.

Haider, K. and Martin, J. P., Decomposition of specifically carbon-14 labeled benzoic and cinnamic acid derivatives in soil, *Soil Sci. Soc. Am. Proc.*, 39, 657, 1975.

Haider, K., Martin, J. P., and Rietz, E., Decomposition in soil of ^{14}C-labeled coumaryl alcohols; free and linked into dehydropolymer and plant lignins and model humic acids, *Soil Sci. Soc. Am. J.*, 41, 556, 1977.

Hansen, E. A. and Harris, A. R., Validity of soil-water samples collected with porous ceramic cups, *Soil Sci. Soc. Am. Proc.*, 39, 528, 1975.

Harborne, J. B., Phenolic glycosides and their natural distribution, in *Biochemistry of Phenolic Compounds*, Harborne, J. B., Ed., Academic Press, London, 1964, 129.

Harborne, J. B., Plant phenolics, in *Secondary Plant Products*, Bell, E. A. and Charlwood, B. V., Eds., Springer-Verlag, Berlin, 1980, 329.

Hartley, R. D. and Jones, E. C., Effect of ultraviolet light on substituted cinnamic acids and the estimation of their cis and trans isomers by gas chromatography, *J. Chromatogr.*, 107, 213, 1975.

Horvath, R. S., Microbial co-metabolism and the degradation of organic compounds in nature, *Bact. Rev.*, 36, 146, 1972.

Hughes, S. and Reynolds, B., Cation exchange properties of porous ceramic cups: implications for field use, *Plant Soil*, 109, 141, 1988.

Inderjit, Plant phenolics in allelopathy, *Bot. Rev.*, 62, 186, 1996.

Jalal, M. A. F. and Read, D. J., The organic acid composition of *Calluna* heathland soil with special reference to phyto- and fungitoxicity: II. Monthly quantitative determination of the organic acid content of *Calluna* and spruce dominated soils, *Plant Soil*, 70, 273, 1983.

Kaminsky, R. and Muller, W. H., The extraction of soil phytotoxins using a neutral EDTA solution, *Soil Sci.*, 124, 205, 1977.

Kaminsky, R. and Muller, W. H., A recommendation against the use of alkaline soil extractions in the study of allelopathy, *Plant Soil*, 49, 641, 1978.

Kassim, G., Stott, D. E., Martin, J. P., and Haider, K., Stabilization and incorporation into biomass of phenolic and benzenoid carbons during biodegration in soil, *Soil Sci. Soc. Am J.*, 46, 305, 1982.

Katase, T., The different forms in which *p*-hydroxybenzoic, vanillic, and ferulic acids exist in a peat soil, *Soil Sci.*, 132, 436, 1981a.

Katase, T., The different forms in which *p*-coumaric acid exists in a peat soil, *Soil Sci.*, 131, 271, 1981b.

Katase, T., Distribution of different forms of *p*-hydroxybenzoic, vanillic, *p*-coumaric and ferulic acids in forest soils, *Soil Sci. Plant Nutr.*, 27, 365, 1981c.

Katase, T., Stereoisomerization of *p*-coumaric and ferulic acids during their incubation in peat soil extract solution by exposure to fluorescent light, *Soil Sci. Plant Nutr.*, 27, 421, 1981d.

Katase, T., The presence of cis-4-hydroxycinnamic acid in peat soils, *Soil Sci.*, 135, 296, 1983.

Kogel, I. and Zech, W., The phenolic acid content of cashew leaves (*Anacardium occidentale* L.) and of the associated humus layer, Senegal, *Geoderma*, 35, 119, 1985.

Kuiters, A. T., Effects of phenolic acids on germination and early growth of herbaceous woodland plants, *J. Chem. Ecol.*, 15, 467, 1989.

Kuiters, A. T., Role of phenolic substances from decomposing forest litter in plant-soil interactions, *Acta Bot. Neerl.*, 39, 329, 1990.

Kuiters, A. T. and Denneman, C. A. J., Water-soluble phenolic substances in soils under several coniferous and deciduous tree species, *Soil Biol. Biochem.*, 19, 765, 1987.

Kuiters, A. T. and Sarink, H. M., Leaching of phenolic compounds from leaf and needle litter of several deciduous and coniferous trees, *Soil Biol. Biochem.*, 18, 475, 1986.

Kunc, F. and Macura, J., Oxidation of aromatic compounds in soil, *Fol. Microbiol.*, 11, 248, 1966.

Lehmann, R. G. and Cheng, H. H., Reactivity of phenolic acids in soil and formation of oxidation products, *Soil Sci. Soc. Am. J.,* 52, 1304, 1988.

Lehmann, R. G., Cheng, H. H., and Harsh, J. B., Oxidation of phenolic acids by soil iron and manganese oxides, *Soil Sci. Soc. Am. J.,* 51, 352, 1987.

Li, H.-H., Nishimura, H., Hasegawa, K., and Mizutani, J., Allelopathy of *Sasa cernua. J. Chem. Ecol.,* 18, 1785, 1992.

Liebl, R. A. and Worsham, D., Inhibition of pitted morning glory (*Ipomoea lacunosa* L.) and certain other weed species by phytotoxic components of wheat (*Triticum aestivum* L.) straw, *J. Chem. Ecol.,* 9, 1027, 1983.

Lodhi, M. A. K., Soil-plant phytotoxicity and its possible significance in patterning of herbaceous vegetation in a bottomland forest, *Am. J. Bot.,* 62, 618, 1975a.

Lodhi, M. A. K., Allelopathic effects of hackberry in a bottomland forest community, *J. Chem. Ecol.,* 1, 171, 1975b.

Lodhi, M. A. K., Role of allelopathy as expressed by dominating trees in a lowland forest in controlling the productivity and pattern of herbaceous growth, *Am. J. Bot.,* 63, 1, 1976.

Lodhi, M. A. K., Allelopathic effects of decaying litter of dominant trees and their associated soil in a lowland forest community, *Am. J. Bot.,* 65, 340, 1978.

Lodhi, M. A. K. and Nickell, G. L., Effects of leaf extracts of *Celtis laevigata* on growth, water content, and carbon dioxide exchange rates of three grass species, *Bull. Torrey Bot. Club,* 100, 159, 1973.

Lopes, A. S. and Wollum, A. G., Comparative effects of methylbromide, propylene oxide, and autoclave sterilization on specific soil chemical characteristics, *Turrialba,* 26, 351, 1976.

Malcolm, R. L. and McCracken, R. J., Canopy drip: a source of mobile soil organic matter for mobilization of iron and aluminum, *Soil Sci. Soc. Am. Proc.,* 32, 834, 1968.

Martin, J. P. and Haider, K., Microbial activity in relation to soil humus formation, *Soil Sci.,* 111, 54, 1971.

Martin, J. P. and Haider, K., Decomposition of specifically carbon-14-labeled ferulic acid: free and linked into model humic acid-type polymers, *Soil Sci. Soc. Am. J.,* 40, 377, 1976.

Martin, J. P. and Haider, K., A comparison of the use of phenolase and peroxidase for the synthesis of model humic acid-type polymers, *Soil Sci. Soc. Am. J.,* 44, 983, 1980.

Martin, J. P., Haider, K., and Wolf, D., Synthesis of phenols and phenolic polymers by *Hendersonula toruloidea* in relation to humic acid formation, *Soil Sci. Soc. Am. Proc.,* 36, 311, 1972.

McBride, M. B., Adsorption and oxidation of phenolic compounds by iron and manganese oxides, *Soil Sci. Soc. Am. J.,* 51, 1466, 1987.

Moreland, D. E., Egley, G. H., Worsham, A. D., and Monaco, T. J., Regulation of plant growth by constituents from higher plants, *Adv. Chem.,* 53, 112, 1966.

Muller, R. N., Kalisz, P. J., and Kimmerer, T. W., Intraspecific variation in production of astringent phenolics over a vegetation-resource availability gradient, *Oecologia,* 72, 211, 1987.

Muller, R. N., Kalisz, P. J., and Luken, J. O., Fine root production of astringent phenolics, *Oecologia,* 79, 563, 1989.

Neary, A. J. and Tomassini, F., Preparation of alundum/ceramic plate tension lysimeters for soil water collection, *Can. J. Soil Sci.,* 65, 169, 1985.

Newby, V. K., Sablon, R.-M., Synge, R. L. M., Casteele, K. V., and Van Sumere, C. F., Free and bound phenolic acids of lucerne (*Medicago sativa* cv Europe), *Phytochemistry,* 19, 651, 1980.

Nord, F. F., The formation of lignin and its biochemical degradation, *Geochim. Cosmochim. Acta,* 28, 1507, 1964.

Oglesby, R. T., Christman, R. F., and Driver, C. H., The biotransformation of lignin to humus—facts and postulates, *Adv. Appl. Microbiol.,* 9, 171, 1967.

Olsen, R. A., Odham, G., and Lindeberg, G., Aromatic substances in leaves of *Populus tremula* as inhibitors of mycorrhizal fungi, *Physiol. Plant.,* 25, 122, 1971.

Patrick, Z. A., Phytotoxic substances associated with the decomposition in soil of plant residues, *Soil Sci.,* 111, 13, 1971.

Patrick, Z. A. and Koch, L. W., Inhibition of respiration, germination, and growth by substances arising during the decomposition of certain plant residues in the soil, *Can. J. Bot.,* 36, 621, 1958.

Perradin, Y., Mottet, M. J., and Lalonde, M., Influence of phenolics on *in vitro* growth of *Frankia* strains, *Can. J. Bot.,* 61, 2807, 1983.

Perry, D. A. and Choquette, C., Allelopathic effects on mycorrhizae: influence on structure and dynamics of forest ecosystems, in *Allelochemicals: Role in Agriculture and Forestry,* Waller, G. R., Ed., American Chemical Society, Washington, D.C., 1987, 358.

Pue, K. J., Blum, U., Gerig, T. M., and Shafer, S. R., Mechanisms by which noninhibitory concentrations of glucose increase inhibitory activity of *p*-coumaric acid on morning-glory seedling biomass accumulation, *J. Chem. Ecol.,* 21, 833, 1995.

Putnam, A. R., Allelopathic chemicals: nature's herbicides in action, *Chem. Eng. News,* April 4, 34, 1983.

Putnam, A. R. and Duke, W. B., Allelopathy in agroecosystems, *Ann. Rev. Phytopathol.,* 16, 431, 1978.

Rahouti, M., Seigle-Murandi, F., Steiman, R., and Eriksson, K. E., Metabolism of ferulic acid by *Paecilomyces variotii* and *Pestalotia palmarum. Appl. Environ. Microbiol.,* 55, 2391, 1989.

Rice, E. L., Allelopathy—An update, *Bot. Rev.,* 45, 15, 1979.

Rice, E. L., *Allelopathy,* 2nd ed., Academic Press, Orlando, FL, 1984

Rose, S. L., Perry, D. A., Pilz, D., and Schoeneberger, M. M., Allelopathic effects of litter on the growth and colonization of mycorrhizal fungi, *J. Chem. Ecol.,* 9, 1153, 1983.

Salonius, P. O., Robinson, J. B. and Chase, F. E., A comparison of autoclaved and gamma-irradiated soils as media for microbial colonization experiments, *Plant Soil,* 27, 239, 1967.

Shindo, H. and Kuwatsuka, S., Behavior of phenolic substances in the decaying process of plants: II. Changes of phenolic substances in the decaying process of rice straw under various conditions, *Soil Sci. Plant Nutr.,* 21, 215, 1975.

Shindo, H. and Kuwatsuka, S., Behavior of phenolic substances in the decaying process of plants: IV. Adsorption and movement of phenolic acids in soils, *Soil Sci. Plant Nutr.,* 22, 23, 1976.

Shindo, H., Ohta, S., and Kuwatsuka, S., Behavior of phenolic substances in the decaying process of plants: IX. Distribution of phenolic acids in soils of paddy fields and forests, *Soil Sci. Plant Nutr.,* 24, 233, 1978.

Silkworth, D. R. and Grigal, D. F., Field comparison of soil solution samplers, *Soil Sci. Soc. Am. J.,* 45, 440, 1981.

Siqueira, J. O., Nair, M. G., Hammerschmidt, R., and Safir, G. R., Significance of phenolic compounds in plant-soil-microbial systems, *Crit. Rev. Plant Sci.,* 10, 63, 1991.

Sposito, G., The future of an illusion: Ion activities in soil solution, *Soil Sci. Soc. Am. J.,* 48, 531, 1984.

Sugai, S. F. and Schimel, J. P., Decomposition and biomass incorporation of ^{14}C-labeled glucose and phenolics in taiga forest floor: effect of substrate quality, successional state, and season, *Soil Biol. Biochem.,* 25, 1379, 1993.

Swain, T., Phenolics in the environment, *Rec. Adv. Phytochem.,* 12, 617, 1978.

Turner, J. A. and Rice, E. L., Microbial decomposition of ferulic acid in soil, *J. Chem. Ecol.,* 1, 41, 1975.

Vance, G. F., Boyd, S. A., and Mokma, D. L., Extraction of phenolic compounds from a spodosol profile: an evaluation of three extractants, *Soil Sci.,* 140, 412, 1985.

Vance, G. F., Mokma, D. L., and Boyd, S. A., Phenolic compounds in soils of hydrosequences and developmental sequences of spodosols, *Soil Sci. Soc. Am. J.,* 50, 992, 1986.

Vogel, C. S. and Dawson, J. O., *In vitro* growth of five *Frankia* isolates in the presence of four phenolic acids and juglone, *Soil Biol. Biochem.,* 18, 227, 1986.

Wang, T. S. C., Yang, T.-K., and Chuang, T.-T., Soil phenolic acids as plant growth inhibitors, *Soil Sci.,* 103, 239, 1967.

Weber, J. B. and Sheets, T. J., Fate of organic contaminants, in *National Conference on Composting of Municipal Residues and Sludges,* August 23–25, Information Transfer, Rockville, MD, 1978, 81.

Weidenhamer, J. D., Distinguishing resource competition and chemical interference: overcoming the methodological impasse, *Agron. J.,* 88, 866, 1996.

Whitehead, D. C., Identification of *p*-hydroxybenzoic, vanillic, *p*-coumaric and ferulic acids, *Nature* 202, 417, 1964.

Whitehead, D. C., Dibb, H., and Hartley, R. D., Extractant pH and the release of phenolic compounds from soils, plant roots and leaf litter, *Soil Biol. Biochem.,* 13, 343, 1981.

Whitehead, D. C., Dibb, H., and Hartley, R. D., Phenolic compounds in soil as influenced by the growth of different plant species, *J. Appl. Ecol.,* 19, 579, 1982.

Whitehead, D. C., Dibb, H., and Hartley, R. D., Bound phenolic compounds in water extracts of soil, plant roots and leaf litter, *Soil Biol. Biochem.,* 15, 133, 1983.

Williams-Linera, G. and Ewel, J. J., Effect of autoclave sterilization of a tropical andept on seed germination and seedling growth, *Plant Soil,* 82, 263, 1984.

7 Biochemical Analysis of Allelopathic Compounds: Plants, Microorganisms, and Soil Secondary Metabolites

George R. Waller, Meow-Chang Feng, and Yoshiharu Fujii

CONTENTS

7.1 ABSTRACT

Allelopathic scientists need to be aware of a number of metabolic changes in plants, and to a certain extent microorganisms, when they are planning to conduct research. Such changes as diurnal variation, growth, onset of flowering, fruit development, seeds for the future, and throughout the different periods of the life span that are controlled by plant metabolism need to be recognized. Thus, it is important to recognize the stage of development when allelochemicals are to be isolated. These changes also influence the surrounding soil with respect to metabolic composition. The metabolites may be modified by the soil through:

1. Chemical modification.
2. Microbial modification.
3. Reacting with the humus fraction of the soil.
4. Extreme physical factors.

Once the plant has been selected, extraction should be done carefully making sure of the composition of the extracting solution. The extracted mixture of biochemical metabolites need to be concentrated, and partitioned into different solvents and purified following the chromatographic techniques described. Often there is a fraction of compounds that are more active in the bioassay than others; it is quite legitimate to pursue this fraction until the pure compound is isolated. Bioassays must be conducted along with the procedures as they are being developed. The most important step is the recognition of the need to obtain pure compound(s). Some of the techniques for extraction, purification, and identification have been outlined. Biochemical interference between devil's claw plants and cotton has been discussed in detail as a specific example to demonstrate different steps necessary to demonstrate allelopathy.

7.2 INTRODUCTION

The recent advances in natural products chemistry have been made through the use of traditional and modern methods of extraction, isolation, purification, and identification of secondary metabolites. It has been suggested that care should be taken using the appropriate assay to check the loss of compound. By incorporating both classical and molecular biological approaches, the role of secondary metabolites found in plants has been evaluated with reference to allelopathic potential. We are concerned in this paper with the biochemical analysis of secondary metabolites found in plants, microorganisms and in soil. The initial observations may be made in the field, greenhouse experiments, or other laboratory experiments under controlled conditions (Blum, this volume). This is followed by a several step purification process leading to the crystalization of the compound of interest. This compound, after establishing its chemical structure, becomes an authentic standard used in bioassay experiments to evaluate its biological activity.

Plants produce and store large amounts of primary and secondary metabolic products (Table 7.1). In general, the bulk of allelochemicals are secondary in nature of synthesis and they are accumulated in the plant, soil, or in culture medium. These vary in their chemical composition, concentration, and localization and are species-specific. In addition, residence time, persistence, and fate of secondary metabolites may vary in soil, and are greatly influenced by abiotic and biotic soil factors (Huang, this volume; Schmidt and Lvy, this volume). The secondary plant metabolites may influence the growth of associated plant species (allelopathy) or even own growth (autotoxicity). Some such compounds (allelochemicals) may be induced in plants by microorganisms, insects, or higher grazers that affect the plant's development.

In this paper we will discuss some of the methods used for extracting, purifying, and chemically identifying the secondary metabolites produced by plants. These compounds are released into the rhizosphere and may undergo microbial degradation in the rhizosphere.

TABLE 7.1

Classes of Secondary Metabolites

Alkaloids

Anthocyanins

Cyanohydrins, cyanohydrin glycosides

Diterpenes and diterpenoids

Flavanoids, isoflavanoids, biflavanoids, chalcones, aurones, and xanthones

Flavones, flavanols and their glycosides

Humic and fulvic acids

Lignins

Mono- and dicarboxylic aliphatic acids (fatty acids)

Monoterpenes and monoterpenoids

Napthoquinones, anthraquinones, stilbenes, phenanthrenes, and quinones

Non-protein amino acids

Phenols and phenolic acids

Phytoalexins

Polyacetylenes

Polyketides

Saponins (triterpenoid, steroid and steroid alkaloidal types)

Sesquiterpenes and sesquiterpenoids

Sterols and steroids

Sulfur compounds

Tannins

Triterpenes and triterpenoids

Various other chemical types, i.e., with antibiotic or phytoalexin activity

7.3 METABOLIC CHANGES ABOUT WHICH NATURAL PRODUCTS SCIENTISTS SHOULD BE AWARE

With a few exceptions such as those in C_4 photosynthesis, plants use the same primary metabolic processes for growth and development and for production of crops and seed for the next generation. But these plants differ widely in their production of secondary metabolites. So it follows that members of the plant kingdom vary in their abilities to produce allelochemicals and other effects of secondary natural products. The biochemical reactions in which we are interested are pathways for the formation of secondary metabolites, possibly through a metabolic grid, starting with one of the compounds produced via a primary metabolic pathway. The reason for the diversities in plant metabolism is not clear. However, it seems likely that the formation of highly individualized and specialized pathways has resulted from the pressure of natural selection (Waller and Nowacki, 1978).

A metabolic pathway consists of a series of compounds, reactions, and enzyme catalysts (Davis, 1955). An advanced terminology that has been widely accepted is:

1. A *precursor* is any compound, either endogenous or exogenous, that is converted by an organism into a product.
2. An *intermediate* is a compound that is both formed and then altered by the organism under identical conditions.
3. An *obligatory intermediate* is a component of the only path by which the organism can synthesize a given substance from given precursors or degrade a substance to given products.

A *metabolic grid,* according to Bu'Lock (1965), represents the metabolism of any type of compound (phenolic acids, terpenes and terpenoids, alkaloids, saponins, and other compounds listed in Table 7.1) in which a series of parallel reactions and interconversions occur in a polydimensional

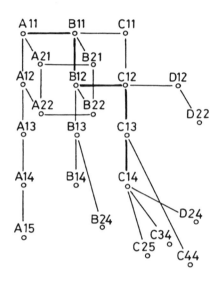

FIGURE 7.1 A metabolic grid representing a pattern of responses occurring in plants (Nowacki and Waller, 1975; Bu'Lock, 1965).

array of the metabolic pathways (Figure 7.1). Thus a compound (**A11**) may be converted to a secondary product (**A22, B22, B12, C12, D12,** or **D22**) by several different pathways, but at different rates. Exclusive or predominant use of one pathway is called *channeling,* as indicated by the heavy line in Figure 7.1 (**A11 → B11 → B12 → C12 → C13 → C14**).

If the sequence **A → B → C** is suspected, and labeled **B** gives better yields of labeled **C** than labeled **A,** this is evidence for the sequence as written. However, as Adelberg (1953) has pointed out, it is difficult to decide whether a certain compound **Y** occupies position **C** or **X** in the metabolic pathway.

$$A \rightarrow B \rightarrow C \rightarrow D \rightarrow \text{end product}$$
$$\downarrow\uparrow$$
$$X$$

The only way to solve this problem is to show that a single enzyme catalyzes the conversion **B → Y,** and another enzyme catalyzes the conversion **Y → D;** this would show that **Y** has position **C** and not position **X.**

One might be tempted to draw conclusions about intermediates from inadequate evidence in the metabolic pathway above. If **B** is incubated with appropriate reagents and, in separate experiments, with each of a series of compounds suspected of being **C;** if one greatly reduces the incorporation of the label into **D,** it is likely to be **C.** In general, however, it is far easier by these methods to rule out pathways than to demonstrate that they are obligatory. Of course, when the enzymes are isolated for each step, then it becomes possible to establish the unambiguous pathway leading to the end product. However, few natural products have received the attention of enzymologists or molecular biologists. The current trend is not to use ^{14}C and ^{3}H labeled substrates, but ^{13}C and ^{2}H labeled substrates can be used to avoid the possible dangers which can occur during the use of radioactivity.

Secondary compounds are metabolically active in plants and microorganisms; their biosynthesis and biodegradation play an important role in the ecology and physiology of the organism in which they occur. Some are accumulated at various stages of growth (Figure 7.2a), while sometimes levels depend on the time of day (Figure 7.2b, Figure 7.3). Coniine and γ-coniceine (Figures 7.2a and 7.2b), secondary compounds produced by western hemlock, vary in concentration in the hemlock fruit in ways that are complementary to each other (diurnal variation). Since the coniine content is much higher than that of the γ-coniceine, all the coniine represented by the maximum in both 1958 and 1959 cannot be produced from the γ-coniceine that disappears concurrently, unless it appears that additional γ-coniceine is synthesized very rapidly and then converted into coniine (Fairbairn and Suwal, 1961). Morphine, which is mostly found in the roots of young poppy (*Papaver somniferum*

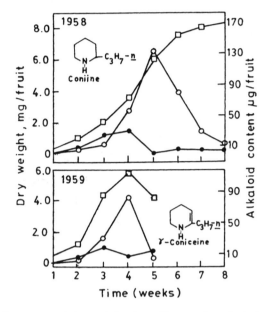

FIGURE 7.2A Alkaloidal content and dry weight of weekly samples of whole *Conium maculatum* plants (Fairbairn and Suwal, 1961); dry weight (○); coniine (•); γ-coniceine (□).

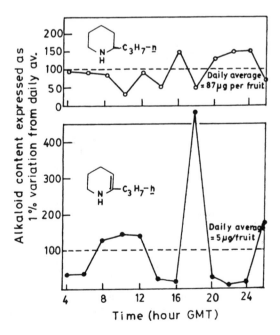

FIGURE 7.2B Diurnal changes in the alkaloid content of *Conium maculatum* in week 4 (Fairbairn and Wassel, 1964); coniine (○); γ-coniceine (•).

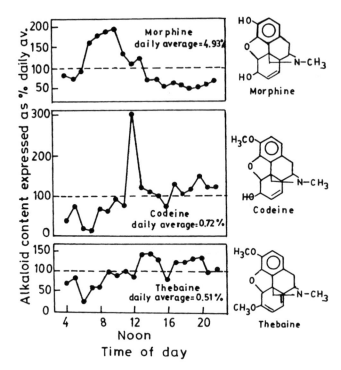

FIGURE 7.3 Diurnal changes in the alkaloid content of whole *Papaver somniferum* plant (Fairbairn and Wassel, 1964).

L.) plants, gradually increases in concentration in leaves. At the stage of fruit formation, morphine disappears from leaves and accumulates in high levels in the fruit capsule (Fairbairn and Wassel, 1964). Diurnal variation of alkaloid concentration in the plant (Figure 7.3) suggests complementary interconversion of thebaine and codeine, but both drop just before the morning rise of morphine. This is consistent with the known biosynthetic pathway: thebaine → codeine → morphine. Morphine is always predominant (like coniine in the hemlock alkaloids), and more morphine is made than can be accounted for by the simple conversion of thebaine and codeine, both of which disappeared. A reasonable hypothesis is the rapid biosynthesis of thebaine and conversion to codeine, and codeine to morphine, followed by the rapid conversion of morphine to nonradioactive substances. Such a process results in the observed daily changes in concentrations of morphine.

We have selected alkaloids for Figures 7.2 and 7.3 because we have some experience with this class of secondary compounds in our laboratories, but parallel processes exist for almost all secondary metabolites. The rates of biosynthesis and biodegradation are reflected in the pool sizes of the particular secondary metabolite in the tissue of the plant; these rates vary with physiological states of development, diurnal variations, and functionally different parts of the plant. There are even different variation patterns for different secondary compounds in the same species.

The soil with its chemical compounds, some derived from microorganisms (both free-living microbes and those associated with the plants), insects, and grazing animals also have an important role in the recovery of secondary compounds. Plant litter drops onto the soil when the leaves and fruit begin to age. In such aging, metabolites that can be useful are translocated to plant parts that remain behind. With annuals all that remain are seeds and some of these are left on the surface or are plowed into the soil. The primary metabolites introduced into the soil are rapidly metabolized by soil microorganisms whereas the secondary metabolites are destroyed slowly because they are of a more complex nature, and indeed may accumulate and be absorbed by other plants. The same applies to metabolites produced by microorganisms. It is this absorption of compounds from the soil by plants that can also give rise to an entirely new set of compounds that occur naturally, and this has not received the attention that it deserves owing to the difficulties associated with the chemical compounds extracted from humic and fulvic acids associated with soil.

Roots and their rhizosphere have been the subject of research for more than 100 years. Soil moisture, temperature, gases, humus, and inorganic (mineral) and organic compounds affect root development and its function. The root system can have symbiotic associations with fungi (mycorrhiza) and bacteria (bacterial nodules) where these microorganisms bring in inorganic nutrients in exchange for some of the organic compounds produced by the plant (i.e., vitamins, carbohydrates, etc.)

The objective of this section is to remind natural products scientists and those working in the biological and pharmaceutical sciences to consider these factors when they sample materials for investigation, and to be cautious about such procedures and variations when they interpret chemical analysis data.

7.4   SOURCES, EXTRACTION, AND ISOLATION OF SECONDARY PLANT METABOLITES

The primary resource books on the subject are *Methods in Plant Biochemistry* (Dey and Harborne, 1989–1994) and *Enzymes of Secondary Metabolites* (Lea, 1993); however, there are many others that may be used. The methods for extraction, isolation, and concentration are simple in one respect. All chemicals have some solubility in water at a certain pH and technically can be extracted with water, although in practice the individual research worker may choose organic solvents instead. After the leaching of the powdered plant material, the worker frequently has to choose among organic solvents for further separation. If the concentration of the chemical in water is high, selection of the organic solvent is easy; but if the concentration is low, the researcher must select a solvent of the highest purity to eliminate possible difficulties (e.g., contaminants in solvent, compound losses, intra- and intermolecular rearrangements). Otherwise, the possibility of introducing an artifact during the isolation and purification process is increased significantly. Some common difficulties encountered in isolation that lead to artifact formation caused by extraction with water are dehydration, hydration, and hydrolysis. Some esterification may occur, with an alcohol as the solvent, and some thermal decomposition. Therefore, it is recommended that mild extraction conditions be adopted. Ende and Spiteller (1982) discussed problems associated with contamination and their elimination. The levels of sophistication vary widely with individual scientists but column, thin-layer, paper, and liquid chromatographic techniques are the routine methods. However, more refined methods such as electrophoresis can then be employed to purify the compound. Final identification may be made by using mass spectrometry and infrared, ultraviolet, visible, and nuclear magnetic resonance spectrometry. If the scientist has isolated a new compound then a further, but often difficult, proof of structure is to synthesize the compound and put it through the same procedure. The compound should be biologically or chemically tested.

Biochemicals that may be found in various natural sources are of different physical and chemical nature. In selecting the choice of a suitable extraction method, all these factors should be considered. The various extraction procedures will be discussed in the following sections.

7.4.1   GENERAL

A general scheme for the extraction is shown in Figure 7.4. This procedure permits the biochemicals to be fractionated into groups of different polarities; each of these fractions should be bioassayed accordingly.

When fresh plant material is used, following two methods can be used:

1. Grind the fresh tissue with a mortar and pestle with sand and methanol, filter, regrind with methylene chloride or methanol, and filter. This removes lipids and the sample is ready for the procedure outlined in Figure 7.4.

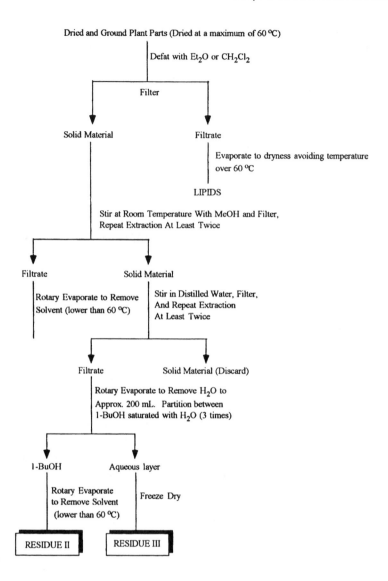

Dried and Ground Plant Parts (Dried at a maximum of 60 °C)

Defat with Et$_2$O or CH$_2$Cl$_2$

Filter

Solid Material   Filtrate

Evaporate to dryness avoiding temperature
over 60 °C

LIPIDS

Stir at Room Temperature With MeOH and Filter,
Repeat Extraction At Least Twice

Filtrate   Solid Material

Rotary Evaporate to Remove   Stir in Distilled Water, Filter,
Solvent (lower than 60 °C)   And Repeat Extraction
                                  At Least Twice

Filtrate   Solid Material (Discard)

Rotary Evaporate to Remove H$_2$O to
Approx. 200 mL. Partition between
1-BuOH saturated with H$_2$O (3 times)

1-BuOH   Aqueous layer

Rotary Evaporate
to Remove Solvent   Freeze Dry
(lower than 60 °C)

RESIDUE II   RESIDUE III

FIGURE 7.4   General procedure for the extraction of nonvolatile secondary plant metabolites.

2. Grind the fresh tissue in liquid nitrogen. When the nitrogen evaporates, the sample can be treated as in step 1 above.

7.4.2   NONVOLATILE COMPOUNDS FROM PLANTS AND THEIR RHIZOSPHERE

Release of secondary compounds (Putnam, 1983) from plants may depend upon the pattern of exudation or leaching of biochemical compounds from living roots, leaves, stems, fruit, rhizomes, seeds, and flowers, and from plant residues. If the plant is dying, soil microbes such as fungi, or bacteria may play an important role in synthesizing or metabolizing a phytotoxic agent that alters the secondary metabolites released from the plant. All chemicals produced by plants, and microorganisms, are ultimately returned to the soil or atmosphere. We are aware of the type of soil being used for growing crops, grasses, etc., and are able to eliminate the live or decaying tissue as a source; however, microbes are always present and their activities have to be considered in the analysis of secondary metabolites.

Chemical extraction can be done on the plant by separating it into its parts, and processing it directly, which is the preferred procedure because this increases the yields. Frequently, it is necessary to dry (up to 60°C maximum), and then grind the material until it can pass through a 20-mesh (0.846 mm^2) screen sample. A sample of the plant should be deposited in the herbarium; the location, time, and date of collection, and age of the plant are essential data to be recorded.

Researchers should be knowledgeable about characteristics of the soil and its history for the last several years, and should be careful to take a statistically significant sample from the plot or field. It is better to work with fresh soil. The preferred route is to extract it quickly without drying. The second best method is to store it at -80°C. The most commonly used method is to air-dry the soil, pass it through a 1.5 cm^2 screen to remove large rock and plant debris, then pulverize and pass it through a 20-mesh (0.846 mm^2) screen. The small roots and stems can be separated from the soil before extraction; however, this method is less desirable because the soil can undergo changes in chemical and physical composition.

The procedure for extracting secondary metabolites from root exudates and plant debris from the soil is outlined in Figure 7.5.

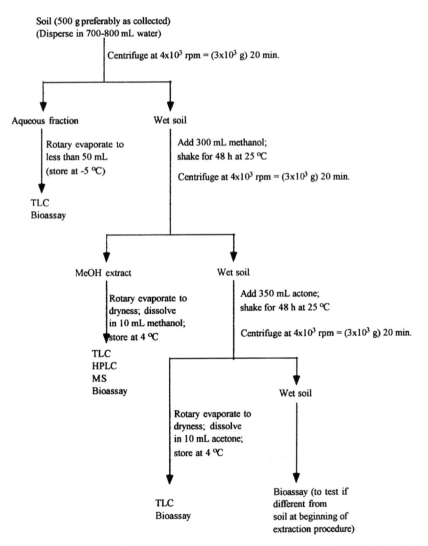

FIGURE 7.5 Extraction of soil containing secondary metabolites from root exudates or plant debris.

7.4.3 PARTS OF THE PLANT THAT CONTAIN VOLATILE COMPOUNDS

7.4.3.1 Volatile Compounds or Essential Oils

If the compounds concerned are highly volatile, it is advisable to trap them in a suitable adsorbing column such as Tenax and analyze them by head space capillary gas chromatography (GC) or gas chromatography/mass spectrography/data analysis (GC/MS/DA). For essential oils, it is possible to subject the samples to steam distillation, saturate the distillate with NaCl, and extract with diethyl ether, dichloromethane, or other low-boiling solvents. To avoid artifact formation the samples should be extracted with the solvent at room temperature. This should be followed by filtering or centrifuging. Alternatively, the sample can be placed in a vacuum system which will allow the volatiles to condense in a cold trap.

7.4.3.2 Extraction Efficiency

Barnes et al. (1986) found that water removed more compounds from dried rye shoot tissue than 50 percent methanol, and little difference was observed between 0.5 and 2 h extractions. However, they showed that sequential partitioning against solvents of increasing polarity resulted in each solvent fraction showing some inhibitory activity toward cress in their bioassay as shown in Table 7.2. The yield, I_{50} (which cannot be generalized for all active natural products), and unit activity were determined and compared with the initial aqueous activity fraction. The ethyl ether fraction was found to have the greatest specific activity with a I_{50} of 150 mg or 100 ppm (w/v) and accounted for 12 percent of the activity and only 1.0 percent of the crude weight.

TABLE 7.2
Activity of Rye Shoot Tissue Extracts on Garden Cress (*Lepidium sativum* L.)

Fraction	Percent of Crude	I_{50} (mg)	Units of Inhibition*
Initial aqueous	100.0	1.90	13,000
Acetone precipitate	17.0	1.64	2,550
Hexane	0.1	0.68	47
Ethyl ether	1.0	0.15	1,580
Dichloromethane	0.2	0.8	144
Ethyl acetate	0.6	0.29	498
Final aqueous	55.0	3.80	3,570

*Determined by weight of dried sample × quantity necessary for 50 percent inhibition.

Source: Reprinted by permission of John Wiley & Sons, Inc., Barnes et al., 1986.

7.4.3.3 Supercritical Fluid Extraction (SFE)

Supercritical Fluid Extraction (SFE) coupled with gas chromatography and mass spectrometry provides an analytical method of separation and quantitation that has only recently become available to laboratory scientists although it has been used in industrial applications for some time. It was introduced in 1962. An analysis of four supercritical fluid extractions of commercially available SFE systems by Lopez-Avila et al. (1992) indicates that the average recoveries varied from 10 to 130 percent for organochlorine pesticides, 46 to 92 percent for diphenyl ethers, and 42 to 89 percent for recovery of synthetic organic compounds. Supercritical fluid chromatography has been used in natural product applications by Foley and Crow (1990); however, they report only a few examples.

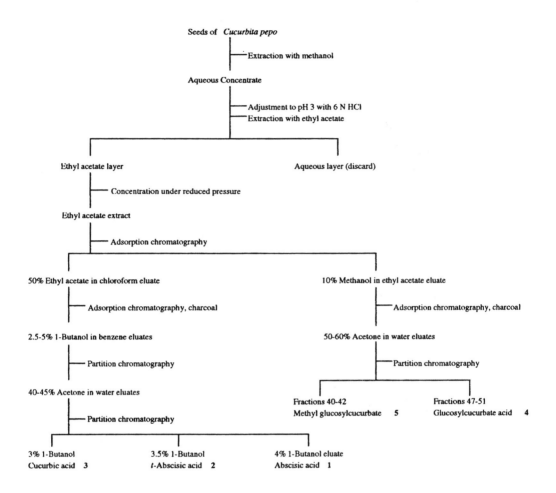

FIGURE 7.6 Isolation procedure for plant growth inhibitors from seeds of *Cucurbita pepo* (Fukui and Koshimizu, 1981).

7.4.3.4 Liquid Chromatography (LC)

In most cases, the isolation of allelochemicals involves considerable effort on chromatographic fractionation. Trace amounts of allelopathically active compounds may be isolated from crude extracts using bioassays. Each step must be considered carefully to avoid degradation of the active compounds. The isolation (Figure 7.6) of plant growth inhibitors from the seeds of *Cucurbita pepo* L. shows that compounds present in a complex mixture cannot be purified by a single method in a single step. It requires a combination of methods arranged in an appropriate order (Fukui and Koshimizu, 1981) to be successful in isolating and proving the structures of *cis*-abscisic acid, *trans*-abscisic acid, cucurbic acid, glycosylcucurbic acid, and methyl glycosylcucurbate.

7.5 PURIFICATION

Because of the complexities of the crude mixture extracted, the purification of secondary metabolites poses a great challenge to scientists. Such purification, like that of many other minor constituents in natural products, involves good technique and skill. It is sometimes considered as an art rather than as a routine technique. There is no straightforward procedure in the work, but sometimes a short cut may be found if one has sufficient knowledge about the chemical nature of the component and the crude extract with which one is dealing.

The following techniques available for separation and purification of secondary metabolites are related to chromatography.

7.5.1 CHROMATOGRAPHIC TECHNIQUES

According to the generalized chromatography nomenclature developed by a special committee of the International Union of Pure and Applied Chemistry (IUPAC, 1974) and Horvath (1980) chromatography is characterized as, "A method, used primarily for separation of the components of a sample, in which the components are distributed between two phases, one of which is stationary, while the other moves. The stationary phase may be a solid, or a liquid supported on a solid, or a gel. The stationary phase may be packed in a column, spread as a layer, or distributed as a film, etc.; in these definitions, a chromatographic bed is used as a general term to denote any of the different forms in which the stationary phase may be used. The mobile phase may be gaseous or liquid." Therefore, any kind of separation method based on differential partitioning of the components to be separated between a stationary and a mobile phase can be considered as a chromatographic method (Englehardt, 1985; Graham, 1991; Grinsburg, 1990; Hancock and Sparrow, 1984; Heftman, 1983a, 1983b; Horvath, 1980; Robards et al., 1994).

7.5.1.1 Column Chromatography

The simplest form of liquid chromatography consists of a column, usually of glass, packed with a suitable stationary phase (adsorbent) which is wetted with a solvent. The mixture to be separated is added at the top and then components eluted with a suitable mobile phase (solvent). Depending on the nature of the interaction between the solute and the stationary phase (packing material), liquid chromatography can be carried out in different modes:

1. Adsorption chromatography.
2. Partition chromatography.
3. Ion exchange chromatography.
4. Exclusion chromatography.

Some of the common types of solid support used in the previously mentioned types of chromatography together with the mode of chromatography are shown in Table 7.3.

TABLE 7.3
Some Common Solid Supports for Chromatography

Mode of Chromatography	Absorbent
Adsorption	Silica gel, florisil, alumina, charcoal
Partition	Cellulose, silica gel
Ion exchange	
cationic	Resins containing -COO, $-SO_3$ and $-PO_3$ groups
anionic	Resins containing $-CH_2N^+(CH_3)_3$ groups
Bonded phase (reversed phase)	Silica gel bonded to C_2, C_8, C_{18}, an alkyl phenyl group

7.5.1.2 Paper Chromatography

Paper chromatography, the oldest of the chromatographic techniques, is the separation of components on a piece of filter paper (e.g., Whatman No. 3) eluted by the ascending or descending method. Although many sophisticated chromatographic techniques are available these days, it is still an important, widely used method for the analysis of mixtures containing phenols, flavonoids, anthocyanins, alkaloids, sugars, etc. For preparative work, the mixture in a solvent is streaked horizontally on a piece of paper (Whatman 3 MM) at a short distance from an edge of the paper and developed as for thin-layer chromatography (TLC). When the solvent has traveled a desired distance, the paper is removed and dried, with the bands corresponding to the components visualized by UV or suitable reagents. The paper bearing the band of interest can be cut out and washed with methanol or water for the components to be separated. Table 7.4 shows a number of the commonly used solvents for paper chromatography.

TABLE 7.4
Solvent Systems for Paper Chromatography

Composition	Preparation	Compound or Group
1-BuOH : AcOH : H_2O	4 : 1 : 5 (upper layer)	Most classes of phenols
1-BuOH : 2N ammonia	1 : 1 (upper layer)	Cinnamic acids
AcOH : conc. HCl : H_2O	30 : 3 : 10	Most flavonoid aglycones
Phenol : C_2H_5COOH : H_2O	2 :2 : 1 (upper layer)	Xanthones, cinnamic acid, coumarins, biflavonyls, and simple phenols
1-BuOH = 1-butanol		C_2H_5COOH:Propionic Acid

7.5.1.3 Thin-Layer Chromatography (TLC)

In TLC, the partition of the sample occurs on a thin layer of adsorbent coated on a glass, metal, or plastic plate. Analytical TLC is usually carried out on a layer of 0.2 to 0.25 mm thickness of adsorbent. For preparative work, a plate of size 20 × 20 cm with a coating of 2.5 mm is normally used. It can be used for separation of about 100 mg of a mixture depending on its complexity. Because of its advantage in speed, cost, and simplicity, TLC is still one of the most commonly used methods for analytical and preparative work in laboratories. For details on the practice of TLC techniques the reader is referred to Grinsburg (1990), Jork et al. (1980), Stahl (1969), and Kalász et al. (1997).

7.5.1.4 High Performance Liquid Chromatography (HPLC)

HPLC has become the most important chromatographic technique for chemical analysis as well as for purification of compounds in the modern laboratories. Its advantages over the conventional open-column liquid chromatography are its speed of elution, suitability for separating thermally labile compounds, high efficiency, and reusability of the column. Details on the instrumentation and theory can be found in many of the monographs on chromatography [Englehardt (1985); Graham (1991); Grinsburg (1990); Hancock and Sparrow (1984); Heftman (1983a); and Snyder and Kirkland (1979)].

A list of the commercially available HPLC columns can be found in many monographs on the subject [Horvath (1980) and Robards et al. (1994)].

7.5.1.5 Capillary Gas-Liquid Chromatography (CGLC)

In gas-liquid chromatography, the components are carried through a column by an inert gas and are separated by differential partitioning between the inert gas and the stationary phase, which is a nonvolatile liquid coated on an inert solid support (Jennings, 1981). The rate of migration is determined by the flow rate of the gas, the nature of the stationary liquid, and the operating temperature.

Identification of the components in the mixture is achieved by comparison of the retention times with those of standard samples. For more accurate determination, the gas chromatograph is connected to a mass spectrometer; this enables the mass spectra of individual components to be recorded, analyzed, and the components identified. Quantitation of a component can be done by measuring the peak size and comparing it with a calibration curve. Sometimes it is also possible to isolate individual components by using a preparative column by collecting the eluted fractions and condensing them in a small trap.

7.5.1.6 Supercritical Fluid Chromatography (SFC)

SFC is an instrumental chromatographic method similar to liquid and gas chromatography, except that it employs a supercritical fluid as the mobile phase. Because supercritical fluids possess properties that are intermediate between those of gases and liquids, their high solvent power (relative to GC) may be exploited, yet retain high solute diffusivities (relative to HPLC). SFC coupled with liquid chromatography may be used for the analysis for thermally labile or nonvolatile compounds, which represent about 60 to 70 percent of the natural products present. SFC/GC or SFC/MS/GC can be used for the analysis of the remaining secondary metabolites (natural products).

7.6 SECONDARY METABOLITE IDENTIFICATION

We shall now illustrate the application of the preceding methods, which has been discussed only very generally, to the isolation, purification, and, in some examples, identification of biochemical compounds from plants and soil. The following selected examples will be listed in the order of relative degree of difficulty of purification.

7.6.1 PAPER CHROMATOGRAPHY

Paper chromatography is a well established technique for the identification of certain groups of polar compounds, in particular the phenolics, isoflavonoids, alkaloids, terpenes, and terpenoids. Chou and Waller (1980) investigated the phytotoxic (allelopathic) substances of coffee seedlings, roots, and fallen leaves by using paper and thin-layer chromatography. Figure 7.7 shows the TLC of the extract from the various materials compared with standard compounds. The identity of these compounds was confirmed by isolating them with preparative paper and TLC and subjecting them to mass spectral analysis.

7.6.2 THIN-LAYER CHROMATOGRAPHY (TLC)

Barnes et al. (1986) used TLC of the ether fraction (Table 7.2) to separate and identify by mass spectrometry the structure of the DIBOA glycoside, DIBOA (2,4-dihydroxy-1,4-(2H)-benzoxazin-3-one), and BOA (2(3H)-benzoxazolinone) isolated from rye residues.

7.6.3 ANALYSIS FOR PHENOLIC ACIDS IN SOIL SAMPLES BY HPLC

Phenolic acids are a class of important allelochemicals. Their presence in the soil has been a great concern of allelopathic scientists. For qualitative identification and quantitative determination, HPLC is the technique of choice. Blum et al. (1991) using HPLC analysis have conducted extensive studies on the phenolic acid contents of soil under various conditions. Graham (1991) examined the profile of allelopathic plant and microbial aromatic secondary metabolites using HPLC (chromatographic conditions: column Merck Lichrosorb RP-18 10 mm column; solvent 0 to 55 percent CH_3CN in pH 3 water for 25 min, followed by a step increase to 100 percent CH_3CN, held for 2 min.).

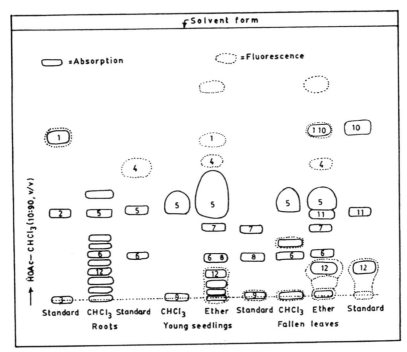

FIGURE 7.7   Identification of allelochemicals present in Coffea arabica by TLC (Chou and Waller, 1980); 1, ferulic acid; 2, *p*-coumaric acid; 3, gallic acid; 4, scopoletin, 5, caffeine; 6, theobromine; 7, theophylline; 8, paraxathine; 9, chlorogenic acid; 10, vanillic acid; 11, *p*-hydroxybenzoic acid; and 12, caffeic acid.

7.6.4   COMBINATION OF TLC, COLUMN CHROMATOGRAPHY, AND HPLC

The metabolites produced by microorganisms in soil from DIBOA (2,6-dihydroxy-1,4-(2*H*)benzox-azin-3-one) and MBOA (6-methoxy-2,3-benzoxazolinone) were analyzed and isolated by initial col-umn chromatography using various solvents followed by repeated TLC, which finally led to the isolation of pure AZOB (2,2′-oxo-1, 1′-azobenzene) as the pure allelopathic compound (Nair et al. 1990).

7.6.5   CAPILLARY GAS CHROMATOGRAPHY/MASS SPECTROMETRY (CGC/MS)

Identification of the components of many naturally occurring compounds can be made using this technique. Essential oils or air surrounding plants can often be analyzed by capillary gas chro-matography/mass spectrometry/data analysis system (CGC/MS/DAS) method. Although individual components may not be isolated, the data obtained can enable the researcher to identify the compo-nents and then carry out bioassays and chemical assays with authentic compounds. Wilt et al. (1993) analyzed the monoterpene concentration in fresh, senescent, and decaying foliage of single leaves of pines by the head space GC/MS method. The concentrations of eight monoterpenes were monitored. This provided evidence that the monoterpene hydrocarbons present in the vapor phase of the single-leaf pinyon pine understory may be toxic to a variety of plant species. This demonstrates the role of the air containing toxic compounds from pines that controls the pattern of vegetation in these forests and is an example of allelopathy.

7.6.6   COMBINATION OF LC, TLC, HPLC, MS AND NMR SPECTROMETRY

Saponins are one of the classes of substances that are difficult to separate by simple column chro-matography or TLC because of the similarities in polarities of many members in this group. The separation usually requires a combination of several chromatographic methods involving typical absorption columns, TLC, and HPLC.

In the isolation of the allelopathic saponin medicagenic acid 3-glucoside and several other saponins from alfalfa roots, Nowacka and Oleszek (1994) removed the sugar and phenols from the crude extract with a C_{18} Sep-Pak cartridge column with methanol as eluent, followed by treatment with aqueous methanol; see Figure 7.8. The fractions corresponding to the saponins which were subjected to repeated HPLC separation with a Lichroprep C_{18} column followed by a Lichroprep Si60 column or by a Eurosphere 80 column as shown in Figure 7.9. Some of the compounds were identified as shown in Figure 7.8. The aglycones, medicagenic acid, and soyasapogenol B are also shown in this figure. Final identification was made by Mass spectrometry (FAB-MS) and Nuclear magnetic resonance spectrometry (NMR) (Oleszek et al., 1990).

In the study of the allelopathy of mungbean (*Vigna radiata* L.) seedlings, Waller et al. (1993) used a Hyperprep 120 ODS column to chromatograph a fraction obtained from the RP-18 column-eluted crude saponins from the extract of 7-day-old mungbean seedlings. A fraction identified by fast atom bombardment and liquid secondary ion mass spectrometry was shown to contain mainly soyasaponin I, with a little soyasaponin II, and a small amount of 3-O-[β-D-galactopyranosyl-(1→2)-β-D-glucoronopyranosyl] sophradiol (Lee et al., 1996a); confirmation was shown by electrospray mass spectrometry (Lee et al., 1996b).

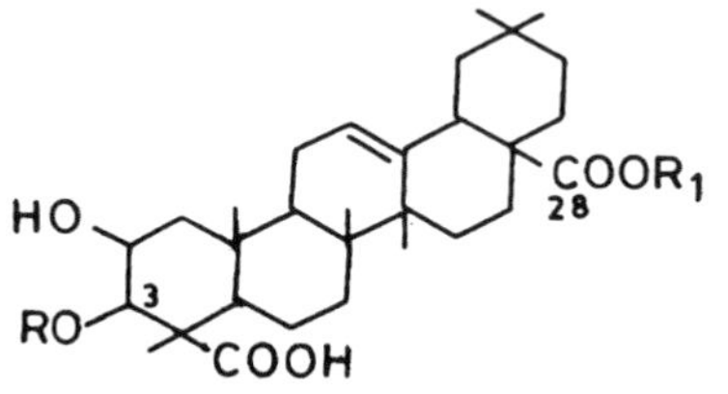

Medicagenic Acid and Saponin Glycosides 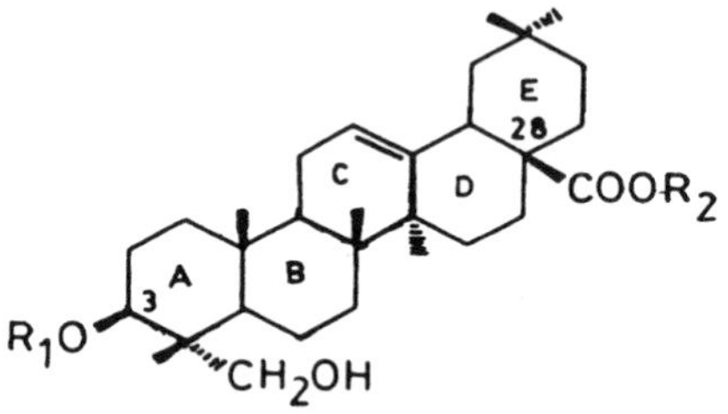 Hederagenin and Saponin Glycosides

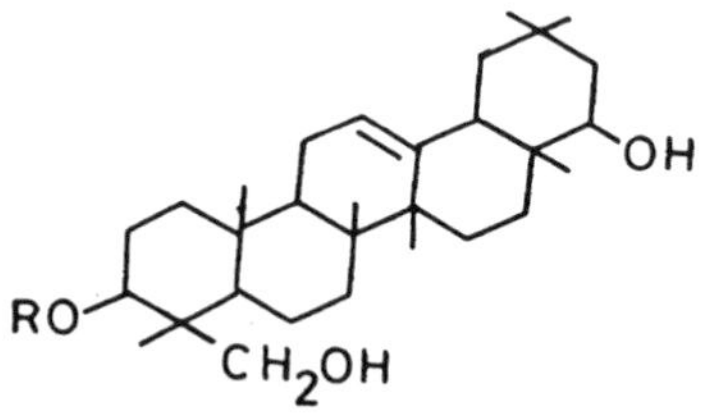

Soyasapogenol B and Saponin Glycosides

Compound	Peak No. from Fig. 9	M/W	-R	-R1
Medicagenic Acid (MA)		502	-H	-H
3-Glc MA	10	664	-Glc	-H
3-Glc,28-Glc MA	2	826	-Glc	-Glc
3-Glc,28-Ara Rha Xyl MA	1	1074	-Glc	-Ara-Rha-Xyl
3-GlcA,28-Ara Rha Xyl MA	7	1236	-GlcA	-Ara-Rha-Xyl
Hederagenin (H)		488	-H	-H
3-Ara Glc Ara (H)	8	898	Ara-Glc-Ara	-H
Soyasapogenol B		458	-H	
3-GlcA Gal Rha SB (Soyasaponin I)	9	942	-GlcA-Gal-Rha	
3-GlcA Ara Rha SB (Soyasaponin II)		926	-GlcA-Ara-Rha	

FIGURE 7.8 Structures of some alfalfa (*Medicago sativa* L.) root saponins and 7-day-old sprouts of mungbeans (*Vigna radiata* L.).

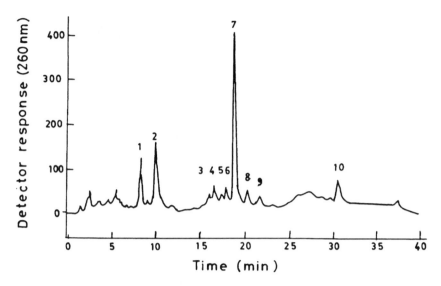

FIGURE 7.9 Analytical determination of alfalfa saponins by HPLC using the Eurosphere 80 C18 column (Nowacka and Oleszek, 1994).

7.6.7 STEAM DISTILLATION APPLIED TO WHEAT SOIL USING CAPILLARY GAS CHROMATOGRAPHY/MASS SPECTROMETRY/DATA ANALYSIS SYSTEM (CGC/MS/DAS)

An example is shown using the combination of CGC/MS/DAS for the analysis of the fatty acids and other compounds isolated from the steam distillation of wheat soil. Soils contain a heterogeneous collection of organic matter of various origins. Waller et al. (1984, 1985, 1987) studied steam distillation, followed by mass spectrometry, for identification in residual soil and some of the compounds identified as allelochemicals The relative proportion of the different fractions depends upon the biological activity of the individual soils. An Oklahoma soil was collected from Altus, OK, in 1985 with the soil type Tillman, clay/loam (fine, mixed, thermic Typic paleustolls). Wheat (*Triticum aestivum* L.) had been grown on the soil for the past decade in no tillage vs. conventional tillage; however, no tillage was for the past three years. Samples of 2 kg each were subjected to successive steam distillations at pH 5.9 (initial conditions) (fraction I), pH was then adjusted to 1.0 in the same batch (fraction A), adjusted to pH 11.0 (fraction B), adjusted to pH 13.5 (fraction BH), adjusted to pH 7.0 (fraction N) and finally adjusted pH to 0.35 (fraction AA). Organic compounds were extracted from each distillate with ethyl ether and subsequently with CH_2Cl_2 or toluene in some cases. Reagent-grade solvents were used. These organics were analyzed by CGC, using a 30m $\times$ 0.32mm I.D. DB-5 J&W Durabond column with a 1m film thickness held at 40°C for 4 min and then programmed to 300°C at 8°C/min and held at 300°C. The compounds were tentatively identified with a GC/MS/DAS (Finnigan Model 4500) as shown in Figure 7.10. Peaks not identified were presumed to be trace natural products, degradation products of pesticides, or they could have been impurities present in the diethyl ether or other extraction solvents.

The mass spectral data were quite distinctive, providing the possibility of establishing identifications of a large number of individual compounds. A data base served to store compound identifications from the different fractions. The SAS* program SORT was then run to group compounds into various classes: *n*-saturated fatty acids from C_4 to C_{18} (Figure 7.11); alcohols (aromatic and aliphatic); phenols; amines (aromatic and aliphatic); polynuclear hydrocarbons; and compounds having the following general composition of CHNO, CHNS and CHS, and CHN_2.

---

*SAS (Statistical Analysis System) is a computer software system, SAS Institute Inc., Box 8000, Cary, NC 27511.

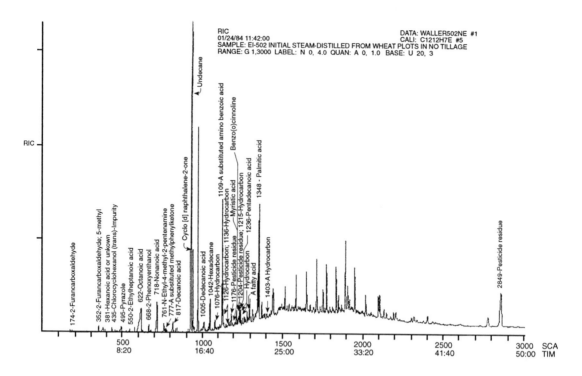

FIGURE 7.10   Reconstructed ion current mass spectral plot of compounds from Oklahoma soil used to grow wheat. Steam distillation at initial pH of 5.9 which was the pH of no-tillage wheat plots, 2 kg sample, air dried before distillation.

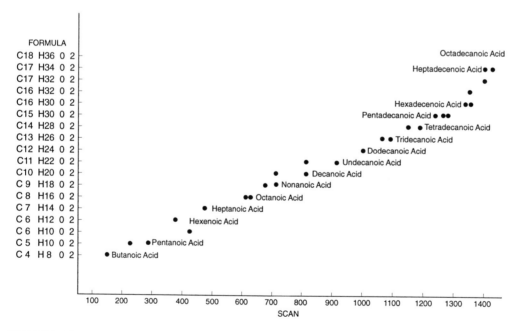

FIGURE 7.11   Plot of scan (GLC retention times) vs. carbon numbers for fatty acids distributed in Oklahoma soil.

TABLE 7.5
Compound Groups Obtained from Soil by Steam Distillation as Identified by CGC/MS/DAS

	Initial (pH 5.9)	Neutral (pH 7.0)	Acidic (pH 1.0, 0.35)	Basic (pH 11, 13.5)
Fatty acids	8	0	20	0
Fatty acid esters	1*	0	3*	0
Alcohols	0	1	2	0
Aldehydes	0	3	6	8
Ketones	0	1	4	2
C-N and other N-containing compounds	1	3	5	17
S-containing compounds	2	0	1	2
Cl-containing compounds	1	1	0	0
Aromatics not otherwise included	1	6	1	23
Aliphatics not otherwise included	4	7	33	16
Totals	18	22	75	75

*All ethyl esters

SAS PLOT procedures were used to plot scans (retention times) against carbon numbers for fatty acids which serve as a support to identification and are shown in Figure 7.11 as well as in Table 7.5. Short- and long-chain fatty acids have been implicated in allelopathy by several investigators (Chou and Waller, 1982, 1989; Waller, 1987).

7.7 BIOCHEMICAL INTERACTION BETWEEN DEVIL'S CLAW PLANT (*PROBOSCIDEA LOUISIANICA*) AND COTTON (*GOSSYPIUM HIRSUTUM* L.): USE OF TLC, CGC/MS-50/DAS

Riffle et al. (1990) reported the allelopathic interference of the devil's claw plant [*Proboscidea louisianica* (Mill.) Thell.] to cotton. The devil's claw plant is an annual broadleaf weed of the family Martyniaceae, found in the cotton-growing areas of Texas and Oklahoma, and has been found to reduce cotton yields up to 83 percent. This viscid, pubescent plant is difficult to control with conventional herbicides, and often hand labor or spot treatment with more concentrated herbicides is used. It can reach heights of 1 m and the canopy spread is approximately 2 m. There is conclusive evidence that the effect of allelochemicals upon the cotton plant can be a contributing factor when the devil's claw plant is grown in the presence of cotton, i.e., essential oils were collected, from the leaves and stems, and from the pods that were inhibitory (27 and 37 percent, respectively) to cotton radicle elongation.

The devil's claw plant is densely covered with glandular hairs, each tipped with a droplet of oil. A strong acrid odor of the plant growing in the field can be detected. Senescence and finally death occur in a portion of the cotton leaves exposed to a devil's claw plant leaf only a few cm away (when the wind blows, actual physical contact is made between the devil's claw plant leaf and the cotton leaf). When the cotton plant is more than 3 to 10 cm away, the volatiles emitted are rapidly dispersed in the surrounding air. Some cotton plants show only a small response to the devil's claw plant so that they are able to rapidly metabolize compounds produced by the weed or are insensitive to them.

To isolate a fraction containing components that might be allelopathically active, essential oil was selected. This was isolated by steam distillation of the leaves, stems, pods, and roots of the devil's claw plant for 5 h. The condensate was treated with NaCl to saturation, extracted with ethyl ether, and dried over anhydrous Na_2SO_4, and the ether evaporated to dryness under N_2. The essential oils were tested for their allelochemical activity against cotton seeds and were analyzed by CGC/MS/DA using an MS-50 (double focusing mass spectrometer) set at a resolution of 2000. The Kratos mass spectrometer was equipped with a Varian model 3700 gas chromatograph containing an OV-1 fused silica column 50 m $\times$ 0.32 mm and the samples were analyzed by using a 1.0 ml injection with the splitter turned off. The oven, first at 50°C, was programmed to rise at 2°C/min to 225°C and held for 5 min at a He flow of 0.5 ml/min. The data were acquired and analyzed with a Kratos DS-55 data system. Identifications were based on comparison of unknown with known spectra and visual interpretation of the fragmentation patterns.

The MS-50 profile of the essential oils required approximately 140 min and 3500 spectra for each sample and indicated that 150 to 220 compounds were present. From this mixture the following compounds were identified from the essential oil: vanillin, perillyl acetate, δ-cadinene, α-bisabolol, traxolide, 2-methyl-1,4-naphthoquinone, (1-hydroxy-2(or 3)-hydroxymethyl-9, 10-anthrace-nedione), and hexadecanoic acid, with small amounts of 6-methyl-5-hepten-2-one and piperitenone. δ-cadinene is the intermediate in cotton phytoalexine by substance which has been identified by Davis and Essenberg (1995) and it is of interest to find it being produced in moderate amounts by devil's claw plants. The remaining compounds which were not identified were mostly terpenes, terpenoids, and other hydrocarbons.

The essential oils of devil's claw extracted from the upper plant parts were inhibitory to cotton radicle elongation (Table 7.6). Piperitenone and α-bisabolol were included for comparison purposes. The essential oil contains up to 220 compounds, so that each compound is acting in only minute quantities.

The bioassays were conducted by dissolving the essential oil in methanol and placing 2 ml of this mixture on two layers of 9 cm Whatman No. 1 filter paper in a petri dish. After methanol evaporation, 10 cotton seeds were placed between the layers of filter paper, 3 ml of distilled water were added, and the covered disk placed in a Ziploc bag to prevent loss of volatiles. Dishes were incubated at 29°C for 72 h, and then measurements of the radicle length were made. Each sample was used to treat at least 40 seeds. Table 7.6 shows the results of bioassay against cotton using devil's claw plant steam distillates and compounds thereof. The concentrations of compounds tested for inhibitory activity against cotton radicle elongation are listed as mg per dish (Table 7.7). The column of data shows that two monoterpenes, p-cymen-9-ol and piperitenone, inhibited radicle elongation by 16 and 13 percent, respectively. The sesquiterpene alcohol, α-bisabolol, and vanillin were 9 percent and 11 percent inhibitory, respectively. These studies proved helpful in evaluating the allelopathic response to cotton; however, they did not show the degree of inhibition that was observed in the field.

TABLE 7.6
Effect of Devil's Claw Essential Oils Piperitenone and α-Bisabolol on Cotton Radicle Elongation

Samples	Concentration (mg/dish)	Inhibition* (%) Cotton
Control	0	0a
Essential oil, 8/14	0.80	7ab
Essential oil, 8/22	0.80	15bc
Essential oil, 8/27	0.80	12bc
Piperitenone (root)	0.60	17c
a-Bisabolol (root)	0.89	12bc

*Means followed by the same letter are not significantly different at the 5 percent level using LSD.

TABLE 7.7
Effects of Authentic Volatile Compounds from Devil's Claws on Cotton Radicle Elongation and the Plant Part Extracted and Relative Amount

Compound	Plant part Extracted	Relative Amount	Concentration (mg/dish)	Inhibition*(%) Cotton
Control		moderate	0	0a
α-Bisabolol	root	moderate	0.44	9bcd
γ-Cadinene	root	low	0.41	2ab
p-Cymen-9-ol	pod	low	0.30	13cd
Phenethyl alcohol	pod	low	0.24	6abc
Piperitenone	root	low	0.30	16d
Vanillin	root, pod	high	0.30	11cd

*Means followed by the same letter are not significantly different at the 5 percent level using a protected LSD.

It is important to recognize the synergistic properties of components of essential oils that are of different types and classes of compounds. These compounds individually may be phytotoxic, non-toxic, or stimulatory. An example is the isolated inactive root essential oil (Table 7.6); it still contains α-bisabolol, γ-cadinene, piperitenone, and vanillin, which are allelochemicals in the pure state. In the allelopathic activity of devil's claw, preliminary evidence has been obtained that vanillin, when added to the essential oil, produced more phytotoxic response. It shows that vanillin, which occurs in the essential oil of the devil's claw, may be one compound out of many that exert some control over its allelochemical response.

7.8 BIOASSAY EXPERIMENTS

The secondary metabolites have been somewhat hampered by lack of suitable techniques for bioassays at varying stages of their purification. Following phytochemical analysis, a need for an appropriate bioassay to check biological activity of isolated compounds, must be performed (see Section I of this book). The significance of natural plant species associated with an allelopathic plant as a target species as well as other soil factors should be considered to get ecologically relevant data on allelopathic potential of isolated compounds (Blum, (this volume); Dalton, this volume; Inderjit, 1996; Inderjit and Dakshini, 1995).

7.9 CONCLUSION

It is our recommendation that all scientists interested in investigations pertaining to allelopathic studies should devote some time to understanding better the types of chemicals involved in producing a biological response. This can be done by extracting the plant material followed by some purification as per the methods previously discussed. These compounds can then be used for assaying their biological activity which may provide some information on the significance of the chemical to isolated compound. The objective should be to produce a pure compound even if several techniques are required further. Of the many thousands of naturally occurring compounds identified every year in plants, microorganisms, and soil, few have been studied for their potential use. The complexities of biochemical and physiological behavior between species and among species still remain little understood. For detailed information on the nature of these compounds, a wide array of reactions may be referred to as a metabolic grid or a biological network that must be considered during the ebb and flow of the life cycle of plants and microorganisms. It is quite possible that through genetic manipulation by molecular biology techniques the pathways of producing secondary metabolites could be modified to produce compounds that have a much higher level of biological activity than those presently found. All of these techniques may not be available to every investigator. However, like many allelopathic scientists they search for collaboration with specially trained individuals for a specialized technique; the specially trained mass spectrometrist or nuclear magnetic resonance spectrometrist can be of enormous help in identifying the mixtures or the pure compounds. Using these techniques will produce results which may lead others to explore the market for the allelochemical of choice.

ACKNOWLEDGMENT

We appreciate the editorial assistance and keen insight of Margaret K. Essenberg, Andrew Mort, and Otis C. Dermer. This is published with the approval of the Oklahoma Agricultural Experiment Station, Oklahoma State University, Stillwater, OK. M.-C. Feng and Y. Fujii express their appreciation to the Department of Biochemistry and Molecular Biology, Oklahoma State University, Stillwater, OK, for their sabbatical leaves.

REFERENCES

Adelberg, E. A., The use of metabolically blocked organisms for the analysis of biosynthetic pathways, *Bacteriol. Rev.*, 17, 253, 1953.

Barnes, J. P., Putnam, A. R., and Burke, B. A., Allelopathic activity of rye (*Secale cereale* L.), in *The Science of Allelopathy*, Putnam, A. R., and Tang, C. S., Eds., Wiley-Interscience, New York, NY, 1986, 271.

Blum, U., Designing laboratory plant debris-soil bioassays: some reflections, 1999 (this volume), Chapter 2.

Blum, U., Wentworth, T. R., Klein, K., Worsham, A. D., King, L. D., Yerig, T. M., and Lyu, S. W., Phenolic acid content of soils from wheat-no till, wheat-conventional till, and fallow-conventional till, soybean cropping system, *J. Chem. Ecol.*, 17, 1045, 1991.

Bu'Lock, J. D., *The Biosynthesis of Natural Products*, McGraw-Hill, New York, 1965.

Chou, C. H. and Waller, G. R., Isolation and identification by mass spectrometry of phytotoxins in *Coffea arabica*, *Bot. Bull. Acad. Sinica*, 21, 25, 1980.

Chou, C. H. and Waller, G. R., Allelochemicals and pheromones, Institute of Botany, Academia Sinica, Taipei, Taiwan, ROC, 1982.

Chou, C. H. and Waller, G. R., Phytochemical ecology, allelochemicals, mycotoxins, and insect pheromones and allomones, Institute of Botany, Academia Sinica, Taipei, Taiwan, ROC, 1989.

Dalton, B. R., The occurrence and behavior of plant phenolic acids in soil environment and their potential involvement in allelochemical interactions: methodological limitations in establishing conclusive proof of allelopathy, 1999 (this volume), Chapter 6.

Davis, B. D., Intermediates in amino acid biosynthesis, *Adv. Enzymol.*, 16, 247, 1955.

Davis, G. P. and Essenberg, M., (+)-s-Cadinene is a product of sesquiterpene cyclase activity in cotton, *Phytochemistry*, 39, 553, 1995.

Dey, P. M. and Harborne, J. B., *Methods in Plant Biochemistry*, Vols. 1–9, Academic Press, London, 1989–1994.

Ende, E. and Spiteller, G., Contaminants in mass spectrometry, *Mass Spectrom. Rev.*, 1, 29, 1982.

Englehardt, H., *Practice of High Performance Liquid Chromatography*, Springer-Verlag, Berlin, Germany, 1985.

Fairbairn, J. W. and Suwal, P. N., The alkaloids of hemlock (*Conium maculatum* L.) II, evidence for a rapid turnover of the major alkaloids, *Phytochemistry*, 1, 38, 1961.

Fairbairn, J. W. and Wassel, G., The alkaloids of *Papaver somniferum* L., evidence for a rapid turnover of the major alkaloids, *Phytochemistry*, 3, 25, 1964.

Foley, J. P. and Crow, J. A., Supercritical fluid chromatography for the analysis of natural products, in *Recent Advances in Phytochemistry, Modern Phytochemical Methods*, Fischer, N. H., Isman, M. B., and Staffoard, H. A., Eds., Plenum Press, New York, NY, 1990, Chapter 4, Vol. 25.

Fukui, H. and Koshimizu, K., Isolation of plant growth inhibitors from fruits, in *Advances in Natural Product Chemistry: Extraction and Isolation of Biologically Active Compounds*, Natori, S., Ikekawa, N., and Suzuki, M., Eds., Kodansha, Ltd., Tokyo, 1981, chap. 15.

Graham, T. L., A rapid high resolution high performance liquid chromatography profile procedure for plant and microbial aromatic secondary metabolites, *Plant Physiol.,* 95, 584, 1991.

Grinsburg, N., Chromatographic Science Series: Modern Thin-Layer Chromatography, 52, 1990.

Hancock, W. S. and Sparrow, J. T., *HPLC Analysis of Biological Compounds: A Laboratory Guide,* Chromatographic Series, Vol. 46, Marcel Dekker, NY, Vol. 46, 1984, 361.

Heftman, E., *Chromatography: Fundamentals and Applications of Cromatographic and Electrophoretic Methods,* Elsevier, New York, Vol. 22a, 1983a, A1.

Heftman, E., *Chromatography: Fundamentals and Applications of Chromatographic and Electrophoretic Methods,* Elsevier, New York, Vol. 22b, 1983b, B1.

Horvath, C., *High-Performance Liquid Chromatography—Advances and Perspective,* Vol. 1, Academic Press, New York, 1980, 88.

Hunang, P. M., Wang, M. G., and Wang, M. K., Catalytic transformation of phenolic compounds in soils, 1999 (this volume), Chapter 18.

Inderjit, Plant phenolics in allelopathy, *Bot. Rev.,* 62, 182, 1996.

Inderjit and Dakshini, K. M. M., On laboratory bioassays in allelopathy, *Bot. Rev.,* 61, 28, 1995.

IUPAC, Analytical Chemistry Division, Commission on Analytical Nomenclature, Recommendations on nomenclature for chromatography, *Pure Appl. Chem.,* 37, 447, 1974.

Jennings, W. G., *Applications of Glass Capillary Gas Chromatography,* Chromatographic Series, Vol. 15, Marcel Dekker, NY, 1981, 629.

Jork, H., Funk, W., Fischer, W., and Wimmer, H., *Thin-Layer Chromatography,* Vol. 1a, 1980, 1.

Kalász, H., Ettre, L. S., and Báthori, M., Past accomplishments, present status and future challenges of thin-layer chromatography, *LC-GC,* 15, 1044, 1997.

Lea, P. J., *Enzymes of Secondary Metabolites,* Academic Press, London, 1993.

Lee, M. K., Ling, Y. C., Jurzysta, M., and Waller, G. R., Saponins from alfalfa, clover, and mungbeans analyzed by electrospray ionization-mass spectrometry as compared with positive and negative FAB-mass spectrometry, in *Saponins Used in Food and Agriculture,* Waller, G. R. and Yamasaki, K., Eds., Plenum Press, New York, 1966a, 353.

Lee, M. R., Lee, J. S.,Wang, J. C., and Waller, G. R., Structural determination of saponins from mungbean sprouts by tandem mass spectrometry, in *Saponins Used in Food and Agriculture,* Waller, G. R. and Yamasaki, K., Eds., Plenum Press, New York, 1966, 331.

Lopez-Avila, V., Dodhiwala, N. S., Benedicto, J., and Beckert, W. F., Evaluation of four supercritical fluid extraction systems for extracting organics from environmental samples, *LC-GC,* 10, 762, 1992.

Nair, M. G., Whitenack, C. J., and Putnam, A. R.,. 2,2′-Oxo-1,1′-azobenzene, a microbially transformed allelochemical from 2,3-benzoxazolinone, *J. Chem. Ecol.,* 16, 353, 1990.

Nowacki, E. K. and Waller, G. R., Use of the metabolic grid to explain the metabolism of quinolizidine alkaloids in *Leguminosae, Phytochemistry,* 14, 165, 1975.

Nowacka, J. and Oleszek, W., Determination of alfalfa saponins by high performance liquid chromatography, *J. Agric. Food Chem.,* 42, 727, 1994.

Oleszek, W., Price, K. R., Colquhoun, I. J., Jurzysta, M., Ploszynski, M., and Fenwick, G. R., Isolation and identification of alfalfa (*Medicago sativa* L.) root saponins: their activity in relation to a fungal bioassay, *J. Agric. Food Chem.,* 38, 1810, 1990.

Putnam, A. R., Allelopathic chemicals: nature's herbicides in action, *Chem. Eng. News,* 61, 34, 1983.

Riffle, M. S., Waller, G. R., Murray, D. S., and Sgaramella, R. P., Devil's-claw (*Proboscidea louisianica*) essential oil and its components, potential allelochemical agents on cotton and wheat, *J. Chem. Ecol.,* 16, 1927, 1990.

Robards, K., Haddad, P. R., and Jackson, P. E., *Principles and Practice of Modern Chromatographic Methods,* Academic Press, San Diego, 1994, 496.

Schmidt, S. K., and Ley, R. E., Microbial competition and soil structure limit the expression of allelochemicals in nature, 1999 (this volume), Chapter 20.

Snyder, L. R. and Kirkland, J. J., *Introduction to High Performance Liquid Chromatography,* 2nd ed., Interscience, New York, 1979, 863.

Stahl, E., *Thin Layer Chromatography: A Laboratory Handbook,* 2nd ed., Springer Verlag, Berlin, 1969, 1041.

Waller, G. R., *Allelochemicals: Role in Agriculture and Forestry,* American Chemical Society, Washington, D.C., 1987, 606.

Waller, G. R. and Nowacki, E. K., *Alkaloid Biology and Metabolism in Plants,* Plenum Press, New York, 1978.

Waller, G. R., West, P. R., Cheng, C. S., Ling, Y. C., and Chou, C. H., The occurrence of soyasaponin I in *Vigna radiata* L. (mungbean) sprouts as determined by fast atom bombardment, liquid secondary ion mass spectrometry, and linked scanning at constant B/E MS/MS, *Bot. Bull. Acad. Sin.,* 34, 323, 1993.

Waller, G. R., Krenzer, E. G., Jr., McPherson, J. F., and McGowan, S. R., Allelopathic compounds in soil from notillage vs. conventional tillage in wheat production, *Plant Soil,* 98, 1987.

Waller, G. R., McPherson, J. K., Ritchey, C. R., Krenzer, E. G., Jr., Smith, G., and Hamming, M., Natural products from wheat soil, book of abstracts, Paper presented at the 190th American Chemical Society Meeting, Chicago, IL, September 9–14, AGFD 110, 1985.

Waller, G. R., Ritchey, C. R., Krenzer, E. G., Smith, G., and Hamming, M., Natural products from soil, 32nd Annual Conference (ASMS) on Mass Spectrometry and Allied Topies, San Antonio, TX, Abstract MPB 14, May 27–June 1, 1984, 144.

Wilt, F. M., Miller, G. C., and Evert, R. L., Measurement of monoterpene hydrocarbon levels in vapor phase surrounding single leaf *(Pinus monophylla* Torr. & Fem: Pinaceae) understory litter, *J. Chem. Ecol.,* 19, 1417, 1993.

8 Microspectrofluorimetry of Intact Secreting Cells, with Applications to the Study of Allelopathy

Victoria V. Roshchina and Eugenia V. Melnikova

CONTENTS

8.1 ABSTRACT

Microspectrofluorimetry has been used for the identification of intact secretory cells filled with allelochemicals, based on their fluorescence characteristics. The technique's possibilities have been shown on certain types of secreting cells: terpenoid-, phenol-, alkaloid-, and amine-containing secretory cells of nongenerative tissues, and secreting cells of generative organs. The fluorescence spectra of secretory cells containing allelochemicals differ in the position of maxima and intensity, which permits a preliminary discrimination of dominating allelopathically active compounds in an analyzed cell type. The dynamics of secretion accumulation, dependent on the secretory structure and the season, have been demonstrated, for instance, by the observation of the development of some glands. This deals with pollen–pistil interactions, pollen–pollen allelopathy in mixtures *in vitro,* and pistil stigma-pollen germination, etc. It has also been shown that surface chemoreceptory processes can be studied by microspectrofluorimetry, based on fast changes in the fluorescence spectra and fluorescence intensity.

8.2 INTRODUCTION

Recent progress in allelopathy leads toward the exploration of the pesticide/herbicide potential of natural plant excretions (Rice, 1984, 1995). It is therefore important to design relevant bioassays to collect data on occurrence, composition, and mechanisms of allelochemical interactions, and to invoke the role(s) of allelochemicals in biological control programs (Inderjit and Dakshini, 1995). The tests may depend on biochemical and biophysical characteristics of the living cell. Biophysical methods are preferred in many cases because they usually do not damage the cell. Among similar methods is a spectral analysis of intact tissues.

The search for new methodological approaches to allelopathic studies may involve the use of optical methods. Allelochemicals such as alkaloids, polyacetylenes, coumarins, and terpenoids fluoresce under ultraviolet (UV) radiation (Wolfbeis, 1985). Their fluorescent spectra have characteristic maxima peculiar to individual allelochemicals. These features may be used for the identification of the compounds in intact secretory cells containing allelochemicals, if there is an appropriate optic system of registration. It should be noted that until recently such secretory structures were studied by electron microscopy (Vasilyev, 1977; Fahn, 1979) but this method does not permit investigation of the allelopathic and physiological activity of a secretory process *in vivo.* There have been attempts to use fluorescence under UV light for the microphotography of secretory cells (Curtis and Lersten, 1990; Zobel and March, 1993). Shapovalov (1973) recommended the use of fluorescent products of plant root excretions for the study of chemical interactions between plants.

The microspectrofluorimetric technique is one of the noninvasive methods used in cellular diagnostics (Karnaukhov, 1972, 1978, 1988), and has also been applied to the study of secreting plant cells and their excretions (Roshchina and Melnikova, 1994, 1995; Roshchina et al., 1995). This method appears to be useful in the study of dynamic physiological responses of the intact plants in allelopathy without damage to either acceptor cells or donor cells.

The aim of this chapter is to discuss the potential use of microfluorimetry in the study of (1) secretory cells containing allelochemicals, their diagnosis, and content; (2) seasonal dynamics of their fluorescence, spreading of the secretion along secretory cells; (3) pollen germination, pollen–pistil interaction (the latter is useful for understanding pollen allelopathy); and (4) histochemical reactions associated with allelochemicals or their related systems.

8.3　MATERIALS AND METHODS

8.3.1　Main Principle of Microspectrofluorimetry Applied to Cellular Diagnostics *In Vivo*

Microspectrofluorimetry is a technique for fluorescence measurement, focusing the optical detector on microobjects that can fluoresce under excitation by UV light. The main principle of similar apparatus for the recording of the fluorescence spectra on living cells was presented by Karnaukhov (1978, 1988). Detailed analysis of the apparatus, approaches, applications, and particular assays have been described in some monographs (Karnaukhov, 1978, 1988). Secretory cells also contain fluorescent substances with allelopathic features, such as phenols, flavins, and quinones (Roshchina and Roshchina, 1989, 1993), that make possible fluorimetric analysis of the intact cell *in vivo*. The microspectrofluorimeter has detectors with optical sounds (zonds) of various diameters that can receive fluorescence from individual cells and even from cell walls, large organelles, and secretions in periplasmic space (extracellular space), as well as from drops of secretions from secretory cells on the cellular surface (Roshchina and Melnikova, 1995). The technique permits fluorescence measurements on biological objects, using both artificial and natural fluorophores. Fluorescence of dyes and substances penetrated into cells is widely used (Karnaukhov, 1978; Kohen et al., 1981; Taylor and Salmon, 1989; Whitaker, 1995), while the autofluorescence of a cell itself is not well understood. The term autofluorescence is used for luminescence of naturally occurring molecules of intact cells induced by UV light (Taylor and Salmon, 1989). Plants have secretory cells with fluorescent content, and this phenomenon can be used in their diagnostics (Roshchina and Melnikova, 1994, 1995). It should be noted that the secretions contain allelochemicals that differ in characteristic maxima of fluorescence. The possibility of plant vegetative tissue to fluoresce under UV irradiation has been associated with chlorophyll. However, most secretory cells are of epidermal origin and located on the surface, whereas chlorophyll-rich mesophyll cells are located under the secreting cells.

The recorded fluorescence of a plant secretory cell is the sum of fluorescence of components of the surface and the interior of the secreting cell. The cell wall is impregnated by polymers from phenolic residues and it weakly fluoresces in the blue region of the spectrum owing to this feature (Frey-Wyssling, 1964). The main light emission appears to be associated with the compartment filled with some kind of secretion. In the event that secretory function prevails in the cells, the contribution of chloroplasts estimated as the fluorescence intensity at 680 nm becomes minimal, unlike that of the nonsecretory cells. It may also be identified by measurements of the fluorescence spectra of the secretion itself.

Pollen grain is a unique complex that is covered by a lipid polymeric product, named sporopollenin. The mature pollen is a multilayered structure that consists of a vegetative tube cell (a pollen tube-forming cell), a male generative cell with two spermiums within, and an envelope from intine on callose base and exine formed by sporopollenin. Figure 8.1 shows the pathway of both fluorescent flow and UV light in the pollen grain. This is accompanied by reflectance and scattering. The main fluorescence is excited in the exine external layer.

Microspectrofluorimetry can be used for recording in two ways: (1) the diagnostics of secretory cells containing allelochemicals, and (2) the dynamics of chemical interaction between secreting cells.

8.3.2　Objects and Assays

8.3.2.1　Objects

Secreting and nonsecreting cells of nongenerative and generative plant organs are used. Lists of the plant species are given in Tables 8.1 and 8.2. Among these were greenhouse-grown plants such as *Hippeastrum hybridum, Epiphyllum* sp., *Alstroemeria aurantiaca* D. Don., *Drosera capensis* L., *Utricularia* sp., and allelopathically active species collected from natural habitats:

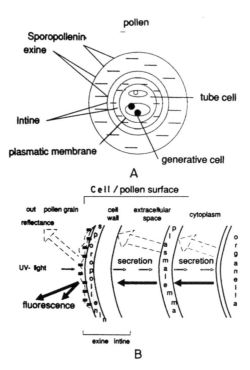

FIGURE 8.1   The diagrammatic representation of pollen grain (A) and possible fluxes of ultraviolet (UV) radiation and fluorescence (B).

1. Weeds—*Achillea millefolium* L., *Artemisia vulgaris* L., *Matricaria chamomilla* L., *Urtica dioica* L., etc.
2. Meadow herbaceous plants—*Alopecurus pratensis* L., *Dactylis glomerata* L., *Trifolium repens.* L., etc.
3. Woody plants—*Larix decidua* L., *Populus balsamifera* L., *Betula verrucosa* Ehrh., etc.

8.3.2.2   Absorbance Measurement

The absorbance spectra of pollen grains and water or acetone extracts from plant material were measured in the first instance on quartz glasses to which pollen grains adhered and in quartz cuvettes in the second one by the use of a Specord M-40 spectrophotometer (Karl Zeiss, Jena, Germany). The sensitivity of the apparatus was 0.001 units of optical density.

8.3.3.3   Fluorescence Registration

Fluorescence from intact cellular surfaces induced by UV light (360 to 380 nm) was observed with a fluorescence microscope Fluoval (Karl Zeiss, Jena, Germany), and cells were photographed on photofilm A-2 used for aerial photography. The fluorescence spectra of the intact cells were recorded with a microspectrofluorimeter of original construction (Karnaukhov, 1978), whereas those of solutions were obtained with a Perkin-Elmer spectrofluorimeter, Coleman 575. The fluorescence intensity was measured in relative units.

8.3.3.4   Histochemical Assay of a Cholinesterase Activity

Cholinesterase in intact *Hippeastrum* pollen on subject glass was determined by incubation in 1 mM acetylthiocholine or butyrylthiocholine in 0.1 M Na/K-phosphate buffer pH 7.4, for 1 h, then dried and stained with a red analog of Ellman reagent DTPDD, ([2, 2''-dithio-bis(p-phenyleneazo-bis

TABLE 8.1
Fluorescence Maxima of Intact Secretory Plant Cells

Plant Species	Wavelength (nm)			
	Secreting Cells	Nonsecreting Cells	Secretion	Individual Fluorescent Component(s)
TERPENOID (OIL)-CONTAINING				
Asteraceae				
Achilea millefolium	465, 500, 550 (a)	460	410 (Oil)	380, 410 (Sesquiterpenes)
Artemisia vulgaris	475 (a)	450, 680	460 (Oil)	380, 410 (Sesquiterpenes)
	450, 680 (b)	450, 680	460 (Oil)	380, 410 (Sesquiterpenes)
Matricaria chamomilla	460, 550, 650 (b)	440	460 (Oil)	380, 410 (Sesquiterpenes)
	460, 550, 675 (b)	450, 680	460 (Oil)	380, 410 (Sesquiterpenes)
Labiatae				
Salvia splendens	465, 550 (a)	460	410 (Oil)	420 (Monoterpenes)
Pinaceae				
Picea excelsa	430 (c)	460, 680	430 (Resin)	430 (Terpenoids)
ALKALOID AND AMINE-CONTAINING				
Asteraceae				
Taraxacum officinale	550 (d)	680	460 (Latex)	460 (Alkaloids)
	465 (e)	—	—	—
Berberidaceae				
Berberis vulgaris	475, 520 (f)	680	—	450, 520 (Berberin, Serotonin)
Euphorbiaceae				
Euphorbia viminalis	540, 680 (d)	680	465 (Latex)	460 (Alkaloids)
Papaveraceae				
Chelidonium majus	480, 560 (d)	680	560 (Latex)	520 (Berberin)
Solanaceae				
Lycopersicon esculentum	540, 680 (g)	680	410 (Oil)	410–420, 460 (Terpenoids, Alkaloids)
Urticaceae				
Urtica dioica	538 (h)	680 (Water leachate)	440 (Biogenic amines)	440–460

PHENOL-CONTAINING

Betulaceae

Betula verrucosa	495–498, 520 (i)		680 (Exudate)	460–470 (Flavonoids) 460–470
	500, 520 (e)	—	—	—

Hippocastanaceae

Aesculus	480 (j)	680	460 (Exudate)	460–480 (Flavonoids)
hippocastanum	500 (e)	—	—	—

Hypericaceae

Hypericum	460(k)	680	440 (Exudates)	460 (Quinones)
perforatum	440, 460 (i), 680	440	—	

Salicaceae

Populus	500 (m)	680	470 (Exudate)	460-480 (Flavonoids)
balsamifera	500 (e)			

SLIME (POLYSACCHARIDE-PROTEIN-PHENOL)-CONTAINING

Droseraceae

Drosera capensis	560 (n)	680	555 (Slime)	460 (Phenols)

Lentibulariaceae

Utricularia sp.	560 (b)	680	545 (Slime)	460 (Phenols)

Scrophulariaceae

Symphytum officinale	535-557 (o)	680	535 (Slime)	460 (Phenols)
	510 (e)	—	—	—

Note: The letters in column two indicate the following: a, floral glands; b, leaf glands; c, resin ducts of needles; d, laticifers; e, secretory hair; f, fruit secretory cells; g, leaf secretory hair; h, stinging hair; i, secretory cell of bud scale; j, bud scale secretory cells; k, glands of flowers; l, glands of leaves; m, bud scale secretory cell; n, leaf glandular hair; o, idioblast.

TABLE 8.2
Absorbance and Fluorescence Maxima in Pollen of Certain Allelopathically Active Species

	Wavelength (nm)	
Plant Species	**Absorbance**	**Fluorescence**
Aceraceae		
Acer campestre	—	440
A. negundo	—	475
Alstroemeridaceae		
Alstroemeria hybrida	—	475, 550
Amaryllidaceae		
Hippeastrum hybridum	380, 434	480
Zephyranthes sp.	—	490, 520

Asteraceae

Artemisia vulgaris	—	500
Calendula officinalis	—	460, 520, 580
Matricaria chamomilla	380, 450, 855	435, 465, 570, 620
Taraxacum officinale	—	520
Tussilago farfara	—	460, 520, 675

Betulaceae

Alnus sp.	380, 434	520
Betula verrucosa	379, 434, 841	440

Cactaceae

Epiphyllum hybridum	No maxima	540
(*Phyllocactus*)		

Caprifoliaceae

Lonicera tatarica	763	—

Caryophyllaceae

Dianthus deltoides	379, 439	—

Clusiaceae

Hypericum perforatum	385, 434, 846	460, 550

Convolvulaceae

Calystegia sepium	—	550

Dipsacaceae

Knautia arvensis	—	475, 515

Fabaceae

Medicago falcata	—	480, 550, 620
Mellilotus albus	—	480, 520
M. officinalis	—	480, 500–510
Trifolium pratense	—	500, 675
T. repens	—	515, 675

Geraniaceae

Geranium pratense	380, 439, 855	430

Hippocastanaceae

Aesculus hippocastanum	379, 430, 870	470, 680

Hydrophyllaceae

Phacelia tanacetifolia	340, 442, 835	430

Iridaceae

Gladiolus sp.	380, 439, 847	470
Crocus vernalis	—	550
Funkia sp. (Hosta)	375, 440, 830	470

Liliaceae

Hemerocallis fulva	379, 437, 850	470, 560, 620, 680
Narcissus pseudonarcissus	379, 430, 830	470, 675
Tulipa sp.	380, 439, 849	470

Onagraceae

Chamaenerium angustifolium	379, 434, 880	460

Papaveraceae

Chelidonium majus	—	540
Papaver orientale	379, 434, 847	470

Pinaceae

Larix decidua		460 (675)
Pinus sylvestris	385, 454, 846	480

Plantaginaceae

Plantago major	380, 843	540

Poaceae

Alopecurus pratensis	—	500, 680
Dactylis glomerata	—	500, (540)

Portulacaceae

Portulaca hybrida	—	480, 520

Rosaceae

Cerasus vulgaris	—	500, 550
Crataegus sp.	—	480
Geum urbanum	—	475-500
Filipendula ulmaria	—	480, 550
Malus domestica	380, 431	540
Rosa davuricata	—	500
Ribus idaeus	749	430

Ranunculaceae

Delphinium consolida	380, 820	480 (560)

Rubiaceae

Gallium borealis	—	475
Gallium verum	—	475

Salicaceae

Populus balsamifera	—	470, 515
Salix virgata	—	470, 515

Solanaceae

Petunia hybrida	438, 842	450, 525, 663

Saxifragaceae

Philadelphus grandiflorus	380, 441, 858	455, 510, 620

Tiliaceae

Tilia cordata	379, 441, 847	470

Umbelliferae

Heracleum sp.	—	480, 520

Urticaceae

Urtica dioica	380, 442, 835	445 (680)

(1-hydroxy-8-chloro-3,6)-disulfonic acid sodium salt]). A 6 mg DTPDD was dissolved in 200 ml of the same buffer. The colored zones of pollen grains were viewed in transmitted light under a Karl Zeiss Microscope and their fluorescence spectra were recorded by microspectrofluorimetry (see Fluorescence Registration).

8.3.3.5 Histochemical Assay of Catecholamines

The histochemical determination of catecholamines in secreting pollen grains was carried out accord-ing to recommendations of Markova et al. (1983) for sexual cells of invertebrates. Dry pollen grains of *Hippeastrum hybridum* were incubated on a subject glass in one to two drops of the fixation medium for up to 3 min. The medium contained 340 mg of sucrose, 160 mg KH_2PO_4, and 50 mg of glyoxalic acid, all mixed and dissolved in 2.5 ml of distilled H_2O; then the solution was titrated with 1 N NaOH to pH 7.4 and diluted with water to a final volume of 5 ml. After the 3-min incubation of the preparation in the fixation medium, it was dried by hot air (80°C) for 5 to 20 min. Then the fluo-rescence of the preparation was analyzed by microspectrofluorimetry in UV light (360 to 380 nm).

8.3.3.6 Pollen Germination

Pollen of *Hippeastrum hybridum* was cultivated on subject glasses in the wet chambers of Petri dishes. The nutrient medium included 10 percent sucrose. Sodium borate (0.001 percent) had no influence on the pollen germination; *Hippeastrum* pollen can germinate in distilled water without any additions (Johri and Vasil, 1961).

8.4 RESULTS AND DISCUSSION

8.4.1 MICROSPECTROFLUORIMETRIC ANALYSIS OF INTACT SECRETORY CELLS

A microspectrofluorimeter can register a fluorescent spectrum from individual cells both of non-generative and generative tissues: leaves, flowers, buds, anthers, pistils, and pollen grains. Allelochemicals present in the secretions from nongenerative and generative tissues possess charac-teristic maxima in the absorbance and fluorescence spectra that are favorable for the identification of the compounds in mixtures.

8.4.1.1 Autofluorescence of Secretory Cells of Leaves, Flowers, and Stems

Many species of genera *Artemisia, Achillea, Matricaria,* and *Urtica* have glandular trichomes, glands, and glandular cells filled with allelochemicals (Rice, 1984). Some of the species belonging to these genera are listed below. Table 8.1 summarizes our data on fluorescence of the intact secre-tory cells from different species in comparison with intact nonsecretory cells, their exudates, and individual substances for which location is established (Roshchina and Roshchina, 1989, 1993). Below, we focus on characteristic types of cellular fluorescence.

8.4.1.1.1 *Terpenoid-containing cells (oil- and resin-containing cells)*

Terpenoid-containing cells of glands and glandular hairs on flowers and leaves differ in the charac-ter of the fluorescence spectra from the nonsecretory cells of epidermis and lower-lying leaf meso-phyll cells (Figure 8.2). In secretory cells, the emission maxima at 680 nm peculiar to chlorophyll is either small or practically absent. Significant differences are seen in the luminescence spectra of secretory cells from various species. For example, under the microscope, red petals of flowers from *Salvia splendens* Dello ex Nees show the glands lightening in the region of 450 to 460 nm with two maxima at 465 and 550 nm, whereas nonglandular cells show no fluorescence here (Figure 8.2). The emission appears to be due to monoterpenes of the essential oils (Table 8.1). Glands and glandular hairs containing sesquiterpenes on the surface of flowers and leaves of *Achillea* and *Artemisia* have

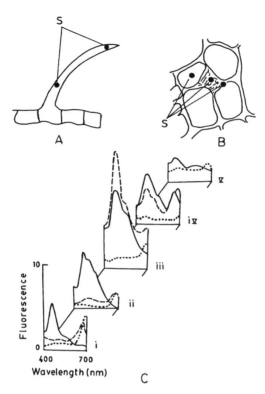

FIGURE 8.2   (A) Glandular hair; (B) Leaf surface, S, position of optical sound (zond); (C) fluorescence spectra of terpenoid-containing cells in glandular hairs, gland on the leaf surface and the secreting cells on the surface of *Salvia splendens* (i), *Achillea millefolium* (ii), *Artemisia vulgaris* (iii), *Matricaria chamomilla* (iv), and *Picea excelsa* (v). Spectrum of secreting cells in plants (i-iv), (_____) petal glands, (---------) leaf surface, and (...............) nonsecretory cells of mesophyll in plant (v). (_____) resin ducts, (---------) nonsecretory cells from needles.

other fluorescence spectra (Figure 8.2) with three significant maxima at 465, 500, and 535 nm for ligulate flower of *Achillea* and two maxima at 460 and 550 nm for *Artemisia*. However, in the leaves of both the plants there is only one maximum, at 500 nm (*Achillea*), and at 460 to 510 nm (*Artemisia*). Unlike these genera, secretory cells of flowers and leaves of *Matricaria chamomilla* L. exhibited more complex fluorescence spectra, with maxima at 460, 550, and 650 to 675 nm. The latter component may be associated with the light emission of chamazulene (prochamazulene) and/or chlorophyll (Table 8.1). Unlike herbaceous species, resin ducts in needles of *Picea excelsa* Wall. ex D. Don have a low intensity of fluorescence and have a maximum at 430 nm, characteristic for monoterpenes (Figure 8.2). The contribution in the fluorescence can be due to phenolic compounds contained in the resin (Roshchina and Roshchina, 1989). On the whole, terpenoid-containing secretory cells fluoresce in blue (450 to 460 nm) and green-yellow (500 to 550 nm) spectral regions (Table 8.1). The first is characteristic for sesquiterpenes and monoterpenes of oils. Azulene-containing species (Figure 8.2) also show maxima in the red region, differing from chlorophyll (max 680 nm), as is known for some azulenes (Roshchina et al., 1995).

8.4.1.1.2   Phenol-, amine- and alkaloid-containing cells

One of the examples among the secretory cells containing phenols, amines, and alkaloids is of stem laticifers of *Chelidonium majus* L. Latex of this species includes phenols, dopamine, and various alkaloids derived from the amine such as berberine and chelerythrine (Roshchina and Roshchina, 1993). As shown in Figure 8.3, the fluorescence of the laticifer of the species has a maximum at 575 to 590 nm while surrounding nonsecretory cells have peaks at 510 nm and 680 nm (chlorophyll). Comparison of exuded latex and laticifer shows that the intensity of latex emission is two to three fold more intensive than that of nondamaged laticifer. The latex fluorescence grows during the first 30 to 40 min after the beginning of exudation (Fig. 8.3A) and achieves a plateau when the orange latex coagulates in the air, changing to a brown viscous mass. The orange fluorescence of latex and laticifer is similar and corresponds to berberine fluorescence at 520 to 550 nm. Comparison of these data with other alkaloid- and amine-containing species (Table 8.1) shows that many types of laticifers have similar maxima in fluorescence spectra in blue and yellow regions, as do the secretory cells of fruit skin (peel) of *Berberis vulgaris* L. Their fresh latex fluorescence is only in the blue region, except for *Chelidonium majus* L. Phenol- and quinone-containing cells fluoresce presumably in the blue or blue-green region, as well as amine-containing cells (Table 8.1). This is demonstrated in some species belonging to Urticaceae and Solanaceae.

8.4.1.1.3 Slime-containing cells

A third type of secretory cell belongs to plants containing slime (Figure 8.4). The cells have a significant fluorescence unlike nonsecretory ones. First of all, secretory cells either have no maxima in the chlorophyll related region at 680 nm or the cells of the mesophyll contribute a small peak. The glands of carnivorous plants releasing trapping mucilage are known to contain enzymes, polysaccharides, phenols, and other substances (Roshchina and Roshchina, 1989, 1993). Perhaps these substances make a contribution to the light emission. Secretory cells of *Drosera capensis* L. (Droseraceae) and *Utricularia vulgaris* L. (Lentibulariaceae) showed a marked fluorescence with

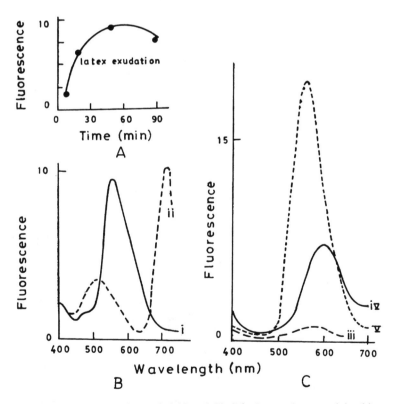

FIGURE 8.3 Fluorescence spectra of stem laticifer of *Chelidonium majus*, containing biogenic amines and alkaloids in latex. (A) Variation in the fluorescence intensity at 550 nm of latex during exudation. (B and C) The fluorescence spectra of stem laticifer (i), nonsecretory cell (ii), surface of flower petal (iii), leaf secretory cell (iv), and latex drop exuding from stem laticifer (v).

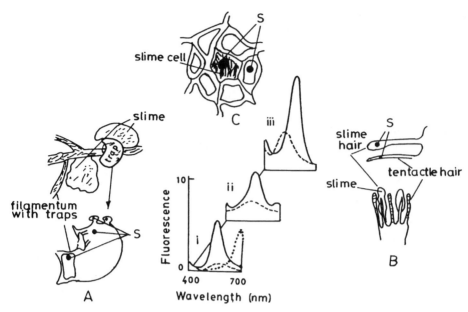

FIGURE 8.4 Fluorescence spectra of intact secretory cells: (_____), slimy cell; (---------), slime; (...............), nonsecretory cell. S, position of optical sound. (i) *Utricularia* sp., (ii) *Drosera capensis,* and (iii) *Symphytum officinalis.* (A) *Utricularia* sp. (Lentibulariaceae) with weakly fluorescent trap on the thread-like leaves that release slime. (B) *Drosera capensis* (Droseraceae) Surface hairs on the trapping leaf tentacle hair, nonflurescent and slime hair with weak yellow fluorescence. (C) *Symphytum officinale* (Boraginaceae) with fluorescent slime cells on the leaf surface.

maximum at 560 nm near the analogous maximum due to released slime of *Drosera* (Figure 8.4). In *Utricularia* mucilage fluorescence is in a shorter wavelength region with maximum at 545 nm. Unlike those of carnivorous plants, the slime cell and released mucilage of *Symphytum officinale* L., lacking digestive glands, luminesce with a maximum at 535 nm, which indicates another component in the secretion.

8.4.1.1.4   Nectar-containing cells

The nectar of some plants is reported to have allelopathic features (Murphy, 1992). Nectar-containing cells are clearly seen on the petals of *Alstroemeria* flowers (Figure 8.5) near red spots formed by anthocyanin-including cells. The fluorescence spectra of different parts of the petals demonstrate bright-lightening secretory cells (max at 500 nm) differing from nectar-containing cells and anthocyanin-containing cells, which show weak fluorescence. None of these cells have the 680 nm maximum peculiar to chlorophyll, unlike nonsecretory cells located on the top of petals.

Microfluorimetric analysis of plant secretory cells on the surface of tissues demonstrates that their autofluorescence differs markedly from that of nonsecretory structures. The character of the fluorescence spectra indicates a chemical composition for study *in vivo.* This technique may be recommended for the study of excretory function of plants *in vivo* and for the identification of secretory cells in microscopic research.

8.4.1.2   Autofluorescence of Pollen and Pistil

The ability of pollen mixed from different species to react with each other, stimulating or inhibiting pollen tube growth, was reported by Bransheidt (1930), Zanoni (1930), and Golubinskii (1946). These authors analyzed both cultivated lawn species of the genera *Lilium, Tulipa,* and *Narcissus* as well as species from natural phytocenosis such as *Corylus avellana, Linaria vulgaris* (L.) Mill., *Papaver somniferum* L., etc. Later, experiments connected with allelopathy were carried out on the

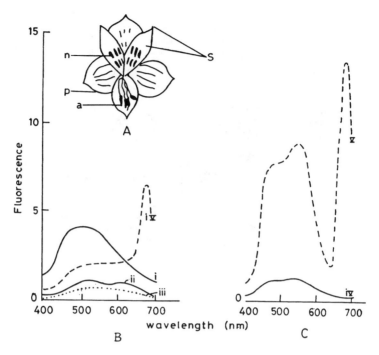

FIGURE 8.5 (A) Diagrammatic representation of flower of *Alstroemeria aurantiaca*. S, position of optical sound; n, nectar-containing cells in red spots; p, petals; a, anthers with pollen. (B and C) Fluorescence spectra. (i) Secretory cells—brightly fluorescent, (ii) red anthocyanin-containing cell, (iii) nectar-containing cells, (iv) nonsecretory cells on the tip of petal, (v) anther, (vi) pollen grain.

basis of pollen germination *in vivo* (Char 1977; Sukhada and Jayachandra, 1980a; Thomson et al., 1982; Murphy, 1992), where fertilization by mixtures of its own and foreign pollen grains was observed. Although pollen autofluorescence of many allelopathically active species was demonstrated (Berger, 1934; Asbeck, 1955; van Gijzel, 1961, 1967, 1971; Willemse, 1971; Driessen et al., 1989) along with stigma fluorescence (Kenrick and Knox, 1981), the phenomenon was never considered in a context of allelopathy until recently. Fast changes in fluorescence may be an indicator of chemical relations.

As shown in Figure 8.6, all generative structures show fluorescence under UV light. Fluorescing pollen grains are clearly observed on the surface of anthers or the stigma, when excited by UV irradiation. Table 8.2 summarizes the absorbance and fluorescence maxima of studied pollen species, and shows the variety of peaks dealing with the different components of their surface and excretion. Among these are mono-, di-, and three-component systems. The absorbance spectra of pollen have from one to three maxima at 380 to 390 nm, 410 to 450 nm, and 760 to 840 nm, while their fluorescence spectra have maxima at 460 to 490 nm, 510 to 550 nm, and 620 to 680 nm (Table 8.2). The first could be connected with phenolic compounds, the second with carotenoids, and the third with chlorophyll (Wolfbeis, 1985; Chappele et al., 1990). One example of the results is analyzed in detail in Figure 8.7. The registration of the fluorescence spectra of pollen grains (i–ix) collected in natural conditions and male sexual cells of mouse (x) demonstrates remarkable differences between functionally related structures of plants and animals (Figure 8.7). The difference is in the higher intensity of light emission in pollen and in the position of maxima in the fluorescence spectra. When pollen grains are mature, they often have no maxima in the red region (1 to 6), whereas in immature pollen grains of catch weed (7) the maximum at 680 nm peculiar to chlorophyll is observed. Unlike the sperm of mouse, pollen shows a clearly seen maximum in the blue-green region at 480 to 500 nm (corb scabious, wild heliotrope, catch weed, hawthorn) and/or a second maximum in the yellow-orange region (celandine poppy, delphinium, and apple tree). Sometimes a third maximum

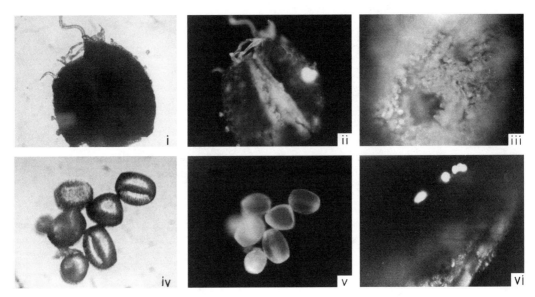

FIGURE 8.6   The fluorescence of anthers, pollen grains, and pistil stigmas under UV microscope × 136. Anther of *Corylus avellana* in (i) transmitted white light and (ii) UV light (360–380 nm). On the surface brightly fluorescent pollen is seen. (iii) The stigma of *Berberis vulgaris,* the papillae are seen. Pollen grains of *Aesculus hippocastanum* in transmitted white light (iv), and in UV light (v). (vi) Fluorescent pollen grains of *Campanula persicifolia* on the stigma of pistil of *Campanula persicifolia.*

in the red region at 600 to 650 nm arises, as for the apple tree. As shown in Table 8.2, there are from one to three maxima at 460 to 490 nm, 510 to 550 nm, and 620 to 680 nm in pollen fluorescence spectra of various species. It depends on the substances included in the sporopollenin and/or exuded on the pollen surface. Van Gijzel (1961, 1967, 1971) and Willemse (1971) recorded pollen fluorescence spectra of forest trees and observed one or two maxima in them, mainly in blue and oranged-red regions.

The surface stigma is composed of specialized receptive glandular cells which are able to recognize and discriminate among pollen grains according to their genotype (Dumas et al., 1988). During interspecific matings it distinguishes "not self," that is, pollen belonging to a species other than that of the pistil, is generally rejected, assuring maintenance of stability of the species. By contrast, in intraspecific mating, "non-self," which corresponds to allopollen, is accepted, while self-pollen is rejected. This later process enforces outbreeding and characterizes "the self-incompatibility phenomenon." Figure 8.8 demonstrates that the use of microfluorimetry permits registering the fluorescence spectra of a pistil stigma and of pollen grains lying on this stigma (see Section 8.4.2.2). The approach to the study of intact surface of plant generative cells could be the measurement of their fluorescence spectra at pollen germination and during pollen–pistil interaction. *Hippeastrum* pistils have marked changes in autofluorescence, mainly in the blue region (Figure 8.8A). Blue fluorescence of pistils was also demonstrated earlier by Kenrick and Knox (1981). The observed changes in fluorescence in our work reflect the excretory and metabolic processes on the surface of pistil and pollen. In the secretion of the stigma, phenolic substances are found (Knox, 1984) that can fluoresce in the same region of the spectra.

8.4.1.3   Contribution of Individual Components in Autofluorescence of Cells

The main problem in the fluorescence analysis is to single out the individual component fluorescing under UV light excitation out of pool of allelochemicals of secreting cells. The observed fluorescence of intact cells is the sum of emissions belonging to several different groups of substances, both excreted and linked on the cellular surface (Tables 8.3 and 8.4).

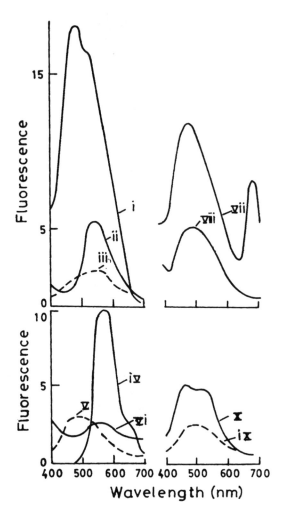

FIGURE 8.7   Fluorescence spectra of pollen grains (i-ix) and sperm of mouse (x): (i) *Knautia arvensis,* (ii) *Chelidonium majus,* (iii) *Delphinium consolida,* (iv) *Malus domestica,* (v) *Phacelia tanacetifolia,* (vi) *Crocus vernalis,* (vii) *Gallium spurium* or *G. aparine,* (viii) *Gallium verum,* (ix) *Craetaegus oxyacanytha,* (x) sperm of mouse.

Among allelochemicals excreted, many substances fluoresce in the blue region of the visible spectral part, such as alkaloids of acridone and tropolonic types, coumarins, cytokinins, flavonoids, furanocoumarins, hydrocinnamic acids and their esters, indole derivatives, some monoterpene alcohols, and azulenes; or in the yellow-orange region, for instance, the isoquinoline type of alkaloids and polyacetylenes (Table 8.3). Azulenes may also have maxima in the red region of the spectra. There are some groups of allelochemicals that fluoresce out of the visible part of the spectra in the UV region (e.g., proteins and amino acids, catecholamines, and some indoles and monoterpenes) and therefore they should not be included in our spectral interpretation.

The main fluorescence of secretory cells of trichomes and glands full of allelochemicals appears owing to monoterpenes and their alcohols in the family Lamiaceae (*Salvia, Mentha*) or to sesquiterpene lactones and their derivatives, such as azulenes, in the family Asteraceae (genera *Achillea, Artemisia, Matricaria*). The components are found in essential oils and are allelopathically active on seed germination; especially menthol, which induces 100 percent inhibition (Garshtya and

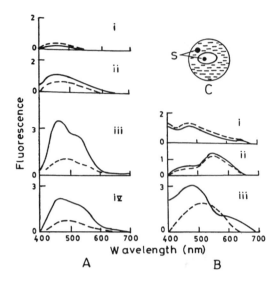

FIGURE 8.8 (A) The fluorescence spectra of stigma of *Hippeastrum* without (- - -) and with pollen grains (____). (i) + *Dactylis glomerata* pollen, (ii) + *Hemerocallis fulva* pollen, (iii) + own pollen (cross - pollination by pollen from other plants of the *Hippeastrum* (Amaryllidaceae), (iv) + own pollen (self - pollination by pollen of the same flower). (B) The fluorescence spectra of pollen (of i, ii, or iii) with (- - -) and without (____) *Hippeastrum* stigma interaction. (i) *Dactylis glomerata,* (ii) *Hemerocallis fulva,* (iii) *Hippeastrum* sp. (C) Outline of pollen showing the position of optical sound, S.

TABLE 8.3
Fluorescence Maxima of Allelochemicals Found in Plant Secretory Cells and Plant Excreta

Class of Allelochemicals	Occurrence	Site of Location	Representative	λ_{ex}/λ_f (fluorophore)[a]
Alkaloids				
Acridone type	Rutaceae	SC	n-methylacridone	381/427 (acridone)
Isoquinoline	Berberidaceae	SC of leaves,	Berberine,	380/510, 540
type	*Berberis vulgaris* L.	roots, stems fruits	chelerythrine	(isoquinoline)
	Chelidonium majus L.	Latex	Berberine, chelerythrine	380/510, 540 (isoquinoline)
Tropolonic type	Iridaceae *Crocus autumnalis* L.	SC	Colchicine	360/435
			Phenylalanine,	260/282
Amino acids	All taxa	PE of roots	tryptophan,	287/348
			tyrosine	275/303
Azulenes	Asteraceae genera	SC (oil glands,	Azulene at 77K,	340/720
(Sesquiterpene	*Artemisia Achillea*	trichomes)	Azulenes of	360–380/420, 620, 725
lactones)	*Matricaria*		pollen bee-collected,	
			chamazulene,	360–380/410, 430, 725
			1,3-	
			dichloroazulene	360/430 (lactone)
Catecholamines	Fruits of many	SC	Adrenaline,	285/325
	plants		dopamine,	285/325
			noradrenaline	285/325 (phenolic ring)
	Many families.			
Coumarins	Hippocastanaceae	SC of woody	Esculetin	360/475 (coumarin)

(Dicoumarins)	*Aesculus hippocastanum*	plant buds. PE		
Cytokinins	Many families	PE	Kinetin	380/410, 430 (adenine)
Flavonoids	Many families	SC of woody plant buds, glands	Galangin, Kaempferol, Quercetin	365/447–461 365/450 365/445
Furanocoumarins	Umbelliferae *Heracleum sibiricum* L., *Psoralea corylifolia*	SC, glands, PE	4-methylpsoralen	360/440–420 (furanocoumarin)
C₆–C₃			Caffeic acid,	365/450
Hydrocinnamic acids and their esters	Many families	PE, SC of buds	ferulic acids	350/440 (phenolic ring)
Indole derivatives	Many families Urticaceae, *Urtica dioica* L.	PE, SC of stinging trichomes	Indole, acetic acid, serotonin	290/350 295/340 360/410–420 (indole)
Monoterpenes and their alcohols	Rutaceae *Mentha piperita*	SC, glands	Menthol, eugenol	360–380/415–420 280/320
Polyacetylenes	Umbelliferae *Cicuta virosa* L.	SC, reservoirs	Cicutotoxin	360–380/580 (triple bonds)

Abbreviations: SC, plant secretory cells; PE, plant excreta.

[a] λ_{ex} and λ_f, wavelengths of excitation light and fluorescence.

Sources: Roshchina and Melnikova, 1995.

TABLE 8.4
Fluorescence Maxima of Surface Components in Secretory Cells

Substance	Occurrence on Surface	λ_{ex}/ λ_f	Reference
Azulenes	Pollen	360–380/410–440; 620; 725	Roshchina et al., 1995
Carotenoids	Pollen	360/500–525	Chapelle et al., 1990
Flavins			
FAD (flavin adenine dinucleotide)	Pollen, leaves everywhere	365/520	Wolfbeis, 1985
FMN (Flavin mononucleotide)	Pollen, leaves everywhere	365/520	Wolfbeis, 1985
Flavoproteins	Plasmalemma of every cell	365/520–540	Wolfbeis, 1985
Riboflavin	Pollen	377/526	Chapelle et al., 1990
Flavonoids	Pollen, leaf exudates, root exudates	365/445–461	Wolfbeis, 1985
Folic acid	Pollen	365/450	Wolfbeis, 1985
Lipofuscin	Seeds and leaves of corn and soybean	365/450–460	Brooks and Csallany, 1978
Lipofuscin	Pea leaves	350–370/410–440	Merzlyak, 1988
NAD(P)H	Everywhere	340/460–470 360/440	Wolfbeis, 1985
Phenolic residues	Cell wall	360/440–460	Wolfbeis, 1985

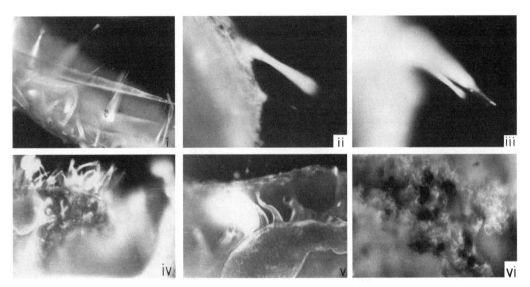

FIGURE 8.9   The fluorescence of the stinging cell of *Urtica dioica* (i, ii, iii) and secretory cell of bud scale from *Betula verrucosa* (iv, v, vi) under the fluorescence microscope. (i-iii), × 68; (iv-v), × 136; (vi), × 272.

Kovalchuk, 1972). In *Picea* (Pinaceae) luminescence has been attributed to the presence of terpenoids and phenols (Roshchina and Roshchina, 1989), and/or of the cuticle due to azulenes (Konovalov, 1995). Fluorescence of intact cells at 600 to 650 nm, differing from that of chlorophyll, is supposed to be connected with the blue pigments (azulenes) occurring in oil cells of leaves and flowers (Roshchina and Roshchina, 1993) and in pollen grains (Roshchina et al., 1995), and/or with lipofuscin, a pigment of aging, found in both retinal pigment epithelia and plant tissue (Merzlyak, 1988). Azulenes are often met within blue-green leaves of many species, but, in particular the blue color of pine *Picea pungens* needles is supposed to be associated with the highest reflectance from surface waxes (Reicosky and Hanover, 1978). Recently red fluorescence at 600 to 650 nm has been found for the intact surfaces of blue-green leaves of azulene-containing species such as *Artemisia,* *Matricaria,* and *Achillea* (see Section 8.4.1.1) and in pollen grains (Roshchina et al., 1995). All studied species exhibiting such red fluorescence contained azulenes. Laticifers also contain phenolic compounds and their derivatives, among which there are also alkaloids (Roshchina and Roshchina, 1989, 1993). Slimes of *Symphytum officinale* L. contain tannins and an alkaloid cynoglossin that can also cause fluorescence at 460 to 480 nm. Blue-green emission of bud secretory cells in the studied tree species appear to be connected with high concentrations of flavonoids, coumarins, and hydrocinnamic acids (Roshchina and Roshchina, 1993). The slime of carnivorous plants with medicinal features includes 1,4-naphthoquinones such as plumbagin and 7-methyljuglone (Blehova et al., 1995), which fluoresce in the blue region. The emission of water-soluble compounds NADPH, flavins FMC and FAD, is in the blue-yellow region (Wolfbeis, 1985). Nectar-containing cells may fluoresce owing to the phenols present (Roshchina and Roshchina, 1993).

As for substances that are not allelochemicals, the fluorescence at 440 to 480 nm may also be owing to phenolic substances. The fluorescence maxima of a number of flavonoid components linked on the cellular surface or in the cellular interior are shown in Table 8.4. The blue fluorescence (460 to 490 nm) of normal green leaves and their water extracts is associated with NAD(P)H and reduced flavins. Chlorophyll luminesces in intact leaf nonsecretory cells at 680 nm. Significant additions to the emission by derivatives of hydrocinnamic acids are found in bud exudates of *Populus* and *Betula* (Wollenweber et al., 1987, 1991). β-Carotene also adds to the blue fluorescence, with a maximum at 470 nm. The fluorescence in the region 500 to 520 nm may be because of riboflavin and some carotenoids (Chappele et al., 1990).

8.4.2   THE STUDY OF AUTOFLUORESCENCE IN DYNAMICS OF SECRETORY PROCESS

Earlier the fluorescence spectra were never recorded in dynamic processes of plant cell secretion. In this section such processes will be considered with examples of secretory cells from nongenerative and generative organs. In the latter cases pollen–pistil and pollen–pollen interactions will be analyzed in connection with fertility and pollen allelopathy.

8.4.2.1   Secretory Cells in Nongenerative Tissue

Microspectrofluorimetry also permits the observation of how secretory cells are filled by secretions, for example, the accumulation of secretion in stinging hairs (trichomes) of *Urtica dioica* and secretory cells of woody plant bud scales, which fluoresce brightly under the microscope (Figure 8.9). In spring the filling of secretory glands (colleters) of bract scales by resin-like secretion leads to the bright luminescence mainly in the blue-yellow region of spectra and, except for *Aesculus,* in the red region at 650 to 690 nm that is peculiar to chlorophyll (Figure 8.10).

   The form of fluorescence spectra of *Betula* and *Populus* changed during this time period, unlike the *Aesculus* bud scales in which only the height of the peak at 500 to 510 nm changed. In *Betula* spectra, there was a shoulder at 480 nm, maximum at 520 to 530 nm, and a chlorophyll maximum

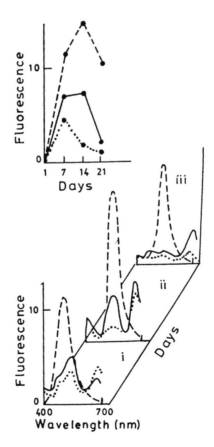

FIGURE 8.10   The fluorescence spectra of secretory cells in bud scales from woody plants showing dynamics of their development through 7 (i), 14 (ii), and 21 (iii) days from the start of the experiment. (_____) *Populus balsamifera,* (- - -) *Aesculus hippocastanum,* and (. . . .) *Betula verrucosa.* At the start of the experiment, fluorescence was not observed.

at 680 nm on April 1; on April 14 the shoulder disappeared, and the height of the 520 to 530 nm maximum sharply decreased. Lastly, on April 21 all maxima decreased. Unlike *Betula*, in *Populus*
spectra the shoulder at 480 nm was absent on April 14, but the maximum at 510 nm did not decrease;
and on April 21 a decrease in the whole blue region fluorescence was observed, whereas the marked
chlorophyll maximum did not disappear.

As the secretion accumulated, the intensity of emitted energy increased up to a peak after two
weeks in April, and decreased to zero at the end of April when the bud scale dropped and the young
leaf opened. At the beginning of May, leaves arose from the buds. The secretory function serves as
a defense of the primordial leaf from damage factors such as pests and late frosts. In the bud secretions, phenolic compounds, mainly flavonoids and esters of aromatic acids, as well as terpenoids,
prevail (Wollenweber et al., 1987, 1991; Roshchina and Roshchina, 1989). In the bud exudates of
Populus deltoides similar components are found (Greenaway et al., 1990).

The spreading of the secretion in secretory cells of *Urtica dioica* L. may be cited as another
example (Figure 8.11). Changes in the intensity of fluorescence along the stinging hair of the common nettle *Urtica* are observed. The intensity of fluorescence is minimal on top of the hair where
there is intensive vacuolation of the secretion. However, it is highest in the base of the hair. Blue
lightening of the secretion is supposed to be determined by the presence of the indole derivative serotonin (Roshchina and Melnikova, 1995).

8.4.2.2   Pollen Germination and Pollen–Pistil Interaction

Experiments have been performed on pollen grains and pistils of intact laboratory model herbaceous
plant *Hippeastrum hybridum* (Amaryllidaceae), which is a suitable object for the experiments
because it has large pollen grains and the longest pistil. Moreover, *Hippeastrum* pollen germinates
directly in water and does not need additional nutrition (Johri and Vasil, 1961).

8.4.2.2.1   Pollen germination
As shown in Figure 8.12, two main types of fluorescent grains differing in intensity were seen:
brightly lightening (A) and weak lightening (B). The former swelled and maintained their emission,

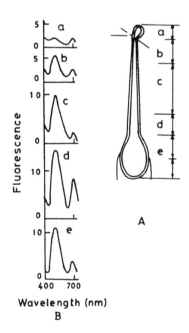

FIGURE 8.11   Outline diagram (A) and fluorescence spectra (B) of the different parts (a, b, c, d, e) of the glandular stinging hair of *Urtica dioica*.

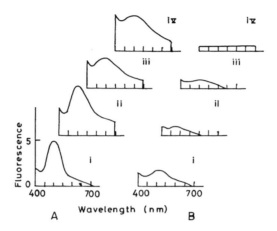

FIGURE 8.12 The fluorescence spectra of *Hippeastrum* pollen grains during the germination *in vitro*. (A) Bright-lightening and (B) weak-lightening pollen. (i) Dry pollen at the beginning; and (ii–v) after 10, 30, 60, and 120 min of moistening, respectively. The pollen began to germinate and pollen tubes had a weak emission or lacked (after 120 min) any fluorescence.

but did not germinate over 24 h. The average intensity of the luminescence was two to three fold higher than for germinating pollen grains. In contrast, the viable pollen grains decreased their auto-fluorescence during germination, and after 60 to 120 min, when pollen tubes arose, the emission was quenched completely. Thus the autofluorescence may be an indicator of pollen viability during the first 10 to 30 min after moistening. The phenomenon has been confirmed by experiments on pollen from other species such as *Aesculus hippocastanum* L., *Sorbus aucuparia* L., and *Betula verrucosa* Ehrh.

8.4.2.2.2 Pollen–pistil interaction

The experiments on pollination included fluorescence measurements 2 to 5 min following germination and 1 month following ovary development. The fluorescence spectra of the *Hippeastrum* stigma was analyzed after pollination by foreign and own pollen (Figure 8.8). When pollen of wind-pollinated species such as *Dactylis glomerata* L. (Figure 8.8), *Populus balsamifera* L., and *Betula verrucosa* were added, the weak own intrinsic fluorescence of the *Hippeastrum* pistil responded to the pollen addition on the stigma. The character of spectra was changed by both cross-pollination and self-pollination, where new peaks arose (Figures 8.8, i–iii), whereas pollination by pollen from insect-pollinated *Hemerocallis fulva* L. only increased the whole fluorescence intensity (Figure 8.10, ii). After 1 month of observation, the ovary formation was at variance with that following cross- or self-pollination. In other cases, fruit maturation was not seen.

Simultaneously with the measurements on the pistil, the fluorescence of the same species pollen added on the *Hippeastrum* stigma was also recorded (Figure 8.8). The pollen grains of foreign pollen weakly or practically did not change the autofluorescence spectra. However, own-pollen fluorescence intensity decreased 2 to 3 min after its addition on the pistil. Its maximum at 480 nm shifted to the long wavelength region (Figure 8.8). This phenomenon appeared to be associated with the chemosensory features of the surfaces of both the stigma and the pollen grains.

8.4.2.3 Pollen–Pollen Interaction in Mixtures

One of the suitable modes of estimating pollen–pollen interactions in mixtures is observation of fast changes in their autofluorescence. The responses are obviously dependent on both the composition of the excretion of donor pollen and the chemosensory peculiarities of the surface component of the plant-acceptor pollen. In this section we will consider the chemical interactions between dry pollen grains, for example, through a communication by volatile excretion, unlike our experiments with moistened pollen and leachates from pollen (Roshchina and Melnikova, 1996).

The present data involve pollen of meadow wind-pollinated species; forest-living wind-polli-nated species; meadow wind-pollinated and insect-pollinated species, grown in flower gardens; and weeds.

The fluorescence spectra of pollen mixtures may reflect initial fast responses in pollen allelopa-thy. One example is shown in Figure 8.13, where the fluorescence spectra of mixtures from wind-pollinated herbaceous meadow plants such as meadow foxtail (*Alopecurus pratensis* L.) and orchardgrass (*Dactylis glomerata* L., Poaceae), and cultural flower garden species poppy (*Papaver orientale* L., Papaveraceae) and day lily (*Hemerocallis fulva* L., Liliaceae) are given. The changes are seen mainly in the fluorescence intensity. The value is 1.5-fold higher for *Alopecurus* in a mix-ture of *Alopecurus* and *Papaver* (var. i, ii), whereas poppy has no visible shifts. In contrast, meadow foxtail decreases its autofluorescence intensity and shifts the main maximum to the long wavelength region in a mixture with day lily pollen (var. iii, iv). The latter shows no changes in its autofluores-cence. When pollen grains of *Hemerocallis* and *Dactylis* (var. v, vi) are mixed, orchardgrass under-goes significant changes and a new maximum at 480 nm arises, major maximum at 525 nm becomes more flat, and red fluorescence with maximum at 650 nm is more obvious, while the intensity of flu-orescence in maxima decreases more than twofold. The autofluorescence of day lily pollen also decreases. If pollen of orchardgrass and poppy are mixed, the fluorescence of both remains unchanged (var. vii, viii).

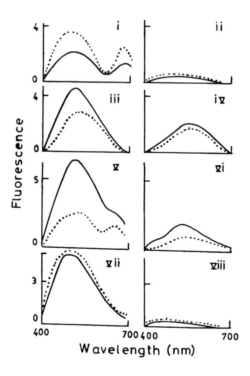

FIGURE 8.13 The fluorescence spectra of pollen grains in mixtures of allelopathically active species. Pollen fluorescence spectra (___) without and (- - -) with the foreign pollen in mixture. (i) Pollen of *Alopecurus pratensis* without and with pollen of *Papaver orientale;* (ii) pollen of *Papaver orientale* without and with pollen of *Alopecurus pratensis;* (iii) pollen of *Alopecurus pratensis* without and with pollen of *Hemerocallis fulva;* (iv) pollen of *Hemerocallis fulva* without and with pollen of *Alopecurus pratensis;* (v) pollen of *Dactylis glom-erata* without and with pollen of *Hemerocallis fulva;* (vi) pollen of *Hemerocallis fulva* without and with pollen of *Dactylis glomerata;* (vii) pollen of *Alopecurus pratensis* without and with pollen of *Dactylis glomerata;* (viii) pollen of *Papaver orientale* without and with pollen of *Dactylis glomerata.*

There were also fast changes (1 to 2 min) in the fluorescence intensity of the pollen grains from weed species *Artemisia vulgaris* L. and *Urtica dioica* and day lily as well as of wind-pollinated, woody species *Larix decidua* and *Betula verrucosa*, which reflected the effects of volatile excretion (Table 8.5). When the microspores of weed species or woody species interacted, in both cases a 60 percent drop in light emission was observed. By contrast, day lily increased its fluorescence by 45 to 80 percent in the presence of weed pollen grains. According to Stanley and Linskens (1974), pollen has a noticeable smell. Terpenoids prevail among the volatile excretion (Egorov and Egofarova, 1971) and may influence the foreign pollen grains as well as attract insects.

Pollen–pollen antagonism, estimated by the ability to germinate, has been known since the 1930s for cultural plants of parterres (Bransheidt, 1930; Zanoni, 1930), whereas a mutual stimulation was first found for natural field plants by Golubinskii in 1946. Unlike the experiments needed for 1 to 24 h for the pollen tube growth, our experiments permitted testing the chemosensitivity of different pollen species in mixtures for 1 to 2 min, and can indicate compatible or incompatible species in phytocenosis. The ability to depress or stimulate fertilization by foreign pollen could be one of the determinants of the mutual existence of the species. Certainly, pollen of *Papaver orientale* and *Hemerocallis fulva*, as parterre species, were either insensitive or weakly sensitive to pollen of field-grown, wind-pollinated *Alopecurus pratensis* L. and *Dactylis glomerata*. Only the fluorescence of *Hemerocallis* pollen was stimulated by pollen of weeds *Artemisia vulgaris* and *Urtica dioica*. In contrast, the pollen grains of wind-pollinated weeds and woody species were extremely sensitive to each other, decreasing their autofluorescence. The sensitivity between pollen of wind-pollinated meadow plants *Alopecurus* and *Dactylis* was marked, but without a strong correlation with their germination.

Fast processes observed as quick shifts in the fluorescence spectra or intensity should be associated with (1) metabolic processes, such as redox changes of NAD(P)H and flavines (Karnauchov 1978); (2) photodynamic changes of some photosensitive pigments of excreta and/or structures of cellular surface (Aucoin et al., 1992); and (3) free radical reactions, both in excreta and on the cellular surface.

Known photodynamic processes are connected with the allelochemicals excreted on the cellular surface such as furanocoumarins (Ceska et al., 1986), or located in surface glands, such as hypericin and juglone and its derivatives (Aucoin et al., 1992). Free radical reactions are most rapid. They are shown in plant excretion containing allelochemicals (Yurin et al., 1972) as spots of fluorescent pigment lipofuscin (Brooks and Csallany, 1978; Merzlyak, 1988), especially when in contact with ozone (Roshchina, 1996), on the pollen surface (Dodd and Ebert, 1971), or on the surface of seeds and leaves tested on final products. Free radicals and the products of free radical processes can contribute to visible light emission. Moreover, some substances such as hypericin, oleic acid, and berberine, may undergo both photodynamic and free radical reactions (Aucoin et al., 1992).

The fast changes in fluorescence at 430 to 470 nm from a plant surface may be due to NAD(P)H fluorescence. For instance, for 25 s the fluorescence at 460 nm in ascitic tumors decreased by 50 percent in the presence of fatty acids (Shwartsburd and Aslanidi, 1991).

TABLE 8.5
Fast Observations of Chemical Interactions of Dry Pollen Grains in Mixtures Estimated by the Fluorescence Intensity

Species of Pollen-Acceptor of Volatile Excreta	Species of Pollen-Donor of Volatile Excreta	Fluorescence Intensity of Pollen-Acceptor at 475 nm (% of control)
Urtica dioica L.	*Artemisia vulgaris* L.	40 ± 1.0
Hemerocallis fulva L.	*Artemisia vulgaris* L.	181 ± 2.3
Hemerocallis fulva L.	*Urtica dioica* L.	145 ± 5.1
Larix decidua L.	*Betula verrucosa* Ehrh.	45 ± 2.4

8.4.3   Fluorescence of Secretory Cells Treated with Histochemicals for Luminescence Measurement

8.4.3.1   Assay of the Cholinesterase Activity

Acetylcholine is one of the nitrogen-containing allelochemicals (Roshchina, 1994). The enzyme cholinesterase hydrolyzing this substance and known as a component of animal cholinergic systems, was also found in plants (Roshchina, 1991). This enzyme is supposed to be essential for cell development and in particular for fertilization. Recently, cholinesterase activity was observed in pollen grains (Roshchina et al., 1994). DTPDD is a red analog of the Ellman reagent that generates a blue product in the reaction with thiocholine (Figure 8.14). By using the histological method based on the red-blue conversion of DTPDD and changes in luminescence, cholinesterase was demonstrated to be located on the exine of pollen grains. After incubation with acetylthiocholine, DTPDD stained the exine blue, while the inner part of the pollen grain turned red, assuming the initial color of the reagent (Figure 8.14). These reactions changed the fluorescence emitted by the pollen; its maximum shifted from 480 nm (untreated pollen) to 510 nm (blue stained pollen zone) and a new maximum appeared at 640 to 650 nm (Figure 8.14). In addition, the fluorescence of the blue-stained zone of pollen grains at 510 nm increased. This effect can be used for qualitative and semi-quantitative analysis of the enzyme; that is important for acetylcholine indication in which cholinesterase serves as a marker for the occurrence of the allelochemical.

8.4.3.2   Assay of Biogenic Amines

8.4.3.2.1   The catecholamines determination

The induction fluorescence in oocytes of sea urchin treated with 1 percent glyoxalic acid is considered to be associated with catecholamines noradrenaline, dopamine, or adrenaline (Markova

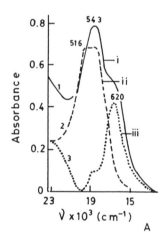
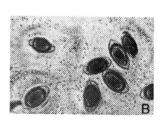
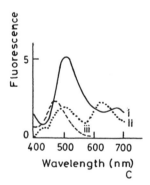

FIGURE 8.14   The histochemical reaction showing cholinesterase activity in pollen. (A) Absorbance spectra of color-generated solutions in the reaction with the cholinesterase activity: (i) DTPDD; (ii) DTPDD + 10^{-3} M thiocholine; (iii) difference spectrum (1 − 2), showing the increase of the blue component. (B) Microphotograph of *Hippeastrum* pollen stained with DTPDD after 1 h of incubation with 10^{-3} M acetylthiocholine (after Roshchina et al., 1994). (C) The fluorescence spectra of *Hippeastrum* pollen (after Roshchina et al., 1994). Blue color due to acetylthiocholine hydrolysis in exine of pollen grains. (i) Pollen, its edge after 1 h of incubation with acetylthiocholine followed by DTPDD; (ii) after blue color developed (in central part of the pollen after the same procedure, red color of the reagent DTPDD is retained); (iii) untreated pollen grains.

et al., 1983). These substances are also found in plants (Roshchina, 1991) and can regulate the seed germination that appears to be connected with allelopathic relations (Roshchina, 1994). The microspectrofluorimetric assay of catecholamines was applied to the pollen grains of *Hippeastrum*. After treatment of intact microspores with 1 percent glyoxalic acid and excitation by UV light (360 to 380 nm), the surface of pollen fluoresced as well as the secretion around it (Figure 8.15). Owing to the light emission, the secretion becomes visible. The fluorescence spectrum of the system has a large maximum at 550 nm, unlike untreated pollen with a maximum at 475 nm (Figure 8.16). The intensity of the fluorescence is enhanced two- to three-fold in comparison with untreated pollen grains. Noradrenaline and adrenaline as markers of catecholamines themselves have no such emission in the visible region of spectra when not treated with 1 percent glyoxalic acid, but after such treatment a maximum of fluorescence at 550 nm is seen. This shows the presence of catecholamines on the surface of pollen grains and in the secretory liquid. A small contribution of other phenol-containing substances in the fluorescence intensity is also not excluded. The occurrence of catecholamines is known in latex of many plants (Roshchina and Roshchina, 1993) where they serve as protective agents upon wounding. A similar defensive function is possible for the secretions of pollen.

8.5 CONCLUSION

Spectral characteristics of individual allelochemicals contained in secretory cells can be measured *in vivo*. These are significant for both their identification and modeling of the allelopathic interactions. Microspectrofluorimetry is especially useful for such studies. The fluorescence spectra of intact secretory cells containing various allelochemicals differ in the positions of maxima and in intensity. This permits a preliminary discrimination of dominating allelochemicals in the secretory cell analyzed. The surface chemoreceptor processes in allelopathy can be studied by microspectrofluorimetry, utilizing the fast changes in the fluorescence. This is owing to pollen–pollen interactions in mixtures of pollen grains *in vitro* and *in vivo*, on the pistil stigma, with pollen germination, and in other circumstances. The observation of fluorescence can be used to identify secretory cells among nonsecretory ones in leaves, seeds, flowers, and fruits and to mark the accumulation of secretions. It may be done on the surface of both herbaceous and woody plants. The method is also useful in indicating the viability of pollen grains and the receptory ability of pistil stigma, discriminating between their own and foreign pollen, that could be applied in the modeling of allelopathic relations in generative organs.

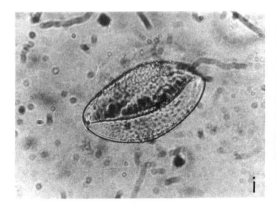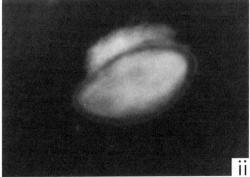

FIGURE 8.15 The view of pollen of *Hippeastrum* in UV light (i) or in transmitted white light (ii), after treatment with 1 percent glyoxalic acid, × 260. The excreta could be seen clearly under UV light.

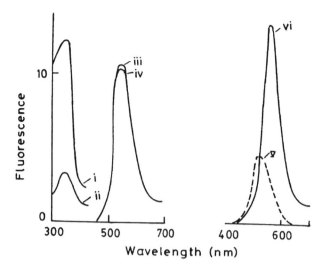

FIGURE 8.16 The fluorescence spectra of *Hippeastrum* pollen after treatment with 1 percent glyoxalic acid and individual substances participating in histochemical reaction on catecholamines: (i) 10^{-5} M adrenaline hydrochloride, (ii) 1 percent glyoxalic acid, (iii) 10^{-5} M adrenaline + 1 percent glyoxalic acid, (iv) 10^{-5} M noradrenaline hydrotartrate + 1 percent glyoxalic acid, (v) untreated pollen of *Hippeastrum*, (vi) pollen of *Hippeastrum* treated with 1 percent glyoxalic acid. λ of excitation: i, ii–280 nm; iii to 6–360 to 380 nm.

ACKNOWLEDGMENTS

This work was supported by a grant from the Russian Foundation of Fundamental Research, N 96-04-48091. The authors would like to thank Dr Valerii N. Karnaukhov for support of the line of the study and Mme Mit'kovskaya for help in computer programming.

REFERENCES

Aucoin, R. R., Schneider, E., and Arnason, J. T., Evaluating the phytotoxicity and photogenotoxicity of plant secondary compounds, in *Plant Toxin Analysis,* Linskens, H. F. and Jackson, J. F., Eds., Springer, Berlin, Heidelberg, 1992, 75.

Asbeck, F., Fluorescezieren der Blütenstaub, *Naturwissenschaften,* 42, 632, 1955.

Berger, F., Das Verhalten der heufiebererregenden Pollen in filtrien ultravioleten Licht, *Beitr. Biologie Pflanzen,* 22, 1, 1934.

Blehova, A., Erdelsky, K., Repcak, M., and Garcar, J., Production and accumulation of 7-methyljuglone in callus and organ culture of *Drosera spathulata* Labill, *Biologia* (Bratislava), 50, 397, 1995.

Branscheidt, P., Zur Physiologie der Pollenkeimung und ihrer experimentellen Beeinflussung, *Planta,* 11, 368, 1930.

Brooks, R. I. and Csallany, A. S., Effects of air, ozone, and nitrogen dioxide exposure on the oxidation of corn and soybean lipids, *J. Agric. Food Chem.,* 28, 1203, 1978.

Ceska, O., Chaudhary, S., Warrington, P., Poulton, G., and Ashwood-Smith, M., Naturally-occurring crystals of photocarcinogenic furanocoumarins on the surface of parsnip roots sold as food, *Experientia,* 42, 1302, 1986.

Chappele, E. W., McMurtrey, J. E., and Kim, M. S., Laser induced blue fluorescence in vegetation, *in Remote Sensing Science for the Nineties,* Vol. 3, Mills, R., Ed., IGARSS'90: 10th Ann. Int. Geosci. and Remote Sensing Symp., Washington, D.C., May 20–24, 1990, University of Maryland, The Institute of Electrical and Electronic Engineers, College Park, MD, 1990, p. 1919.

Char, M. B. S., Pollen allelopathy, *Naturwissenschaften,* 64, 489, 1977.

Curtis, J. D. and Lersten, N. R., Internal secretory structures in *Hypericum* (Clusiaceae), *H. perforatum* L. *and H. balearicum* L., *New Phytol.,* 114, 571, 1977.

Dodd, N. J. F. and Ebert, M., Demonstration of surface free radicals on spore coats by ESR techniques, in *Sporopollenin,* Brooks, I., Grant, P. R., Muir, M., van Gijzel, P. R., and Show, G., Eds., Proc. Symp. Geology Department, Imperial College, London, September 23–25, 1970, Academic Press, London, 1971, 408.

Driessen, M. N. B. M., Willemse, M. T. M., and van Luijn, I. A. G., Grass pollen grain determination by light and UV microscopy, *Grana,* 28, 115, 1989.

Dumas, C., Bowman, R. B., Gaude, T., Gully, C. M., Heizman, Ph., Roeckel, P., and Rougier, M., Stigma and stigmatic secretion reexamined, *Phyton,* 28, 193, 1988.

Egorov, I. A. and Egofarova, R. Kh., The study of essential oils of blossom pollen of grape, *Doklady AN SSSR,* 199, 1439, 1971.

Fahn, A., *Secretory Tissue in Plants,* Academic Press, London, 1979, 302.

Feucht, W. and Treutter, D., Flavan-3-ols in trichomes, pistils, and phelloderm of some tree species, *Ann. Bot.,* 65, 225, 1990.

Frey-Wyssling, A., Ultraviolet and fluorescence optics of lignified cell, in *The Formation of Wood in Forest Trees,* Zimmerman, M. N., Ed., Academic Press, New York, 1964, 153.

Garshtya, L. Ya. and Kovalchuk, Yu. G., On the problem of phytodynamic effect of solutions and vapours of some compounds of essential oils, in *Physiologo-Biochemical Bases of Plant Interactions in Phytocenosis,* Grodzinskii, A. M., Ed., Naukova Dumka, Kiev, 1972, 21.

van Gijzel, P., Autofluorescence and age of some fossil pollen and spores, *Proc. Koninkl. Nederl. Acad. Wet., ser B., Phys. Sci.,* 64, 56, 1961.

van Gijzel, P., Autofluorescence of fossil pollen and spores with special reference to AGE determination and coalification, *Leidse Geol. Mededel.,* 40, 263, 1967.

van Gijzel, P., Review on the UV-fluorescence microphotometry of fresh and fossil exines and exosporia, in *Sporopollenin,* Brooks, J., Grant, P. R., Muir, M., van Gijzel, P., and Shaw, G., Eds., Proc. Symp. Geology Department, Imperial College, London, September 23–25, 1970, Academic Press, London, 1971, 659.

Gillispie, G. D. and Lim, E. C., S_2–S_1 fluorescence of azulene in a Shpol'skii matrix, *J. Chem. Phys.,* 65, 4314, 1976.

Golubinskii, I. N., Mutual influence of pollen grains of different species on their mutual germination in artificial media, *Doklady AN SSSR,* 53, 73, 1946.

Greenaway, W., English, S., and Whatley, F. R., Phenolic composition of bud exudates of *Populus deltoides, Z. Naturforsch.,* 45, 587, 1990.

Griesser, H. I. and Wild, U. P., Energy selection experiments in glassy matrices: The linewidths of the emissions S_2–S_0 and S_2–S_1 of azulene, *J. Chem. Phys.,* 73, 4715, 1980.

Inderjit and Dakshini, K. M. M., On laboratory bioassays in allelopathy, *Bot. Rev.,* 61, 28, 1995.

Johri, B. M. and Vasil, I. K., Physiology of pollen, *Bot. Rev.,* 27, 326, 1961.

Karnaukhov, V. N., Spectral methods for investigations of energy regulation in living cells, in *Biophysics of the Living Cell,* Frank, G. M. and Karnaukhov, V. N., Eds., Institute of Biophysics, Pushchino, 1972, 100.

Karnaukhov, V. N., *Luminescent Spectral Analysis of the Cell,* Moscow, Nauka, 1978, 208.

Karnaukhov, V. N., *Spectral Analysis of the Cell in Ecology and Environmental Protection (Cell Biomonitoring),* Institute of Biophysics, Biological Center of USSR Academy of Sciences, Pushchino, 1988, 124.

Kenrick, J. and Knox, R. B., Structure and histochemistry of the stigma and style of some Australian species of *Acacia, Austral. J. Bot.,* 29, 733, 1981.

Knox, R. B., Pollen-pistil interactions, in *Cellular Interactions. Encyclopedia of Plant Physiology,* New Ser., Vol. 17, Linskens, H. F. and Heslop-Harrison, J., Eds., Springer Verlag, Berlin, 1984, 508.

Kohen, E., Thorell, B., and Hirschberg, I. G., Microspectrofluorimetric procedures and their applications in biological systems, in *Modern Fluorescence Spectroscopy,* Wehry, E. L., Ed., Plenum Press, New York, 1981, 295.

Konovalov, D. A., Natural azulenes, *Rastitelnye Resursi,* 31, 101, 1995.

Markova, L. N., Volina, E. V., and Kovachevich, H., Histochemical determination of biogenic amines in developing sea urchin embryos, *Ontogenez (Russ. J. Develop. Biol.),* 14, 634, 1983.

Merzlyak, M. N., Lipo-soluble fluorescent "aging pigments" in plants, in *Lipofuscin—1987: State of the Art,* Zs.-Nagy, I., Ed., Academiai Kiado/Elsevier, Budapest/Amsterdam, 1988, 451.

Murphy, S. D., The determination of the allelopathic potential of pollen and nectar, in *Plant Toxin Analysis,* Linskens, H. F. and Jackson, I. F., Eds., Springer Verlag, Berlin, 1992, 333.

Reicosky, D. A. and Hanover, J. W., Physiological effects of surface waxes. Light reflectance for glaucous and nonglaucous *Picea pungens, Plant Physiol.,* 62, 101, 1978.

Rice, E. L., *Allelopathy,* 2nd ed., Academic Press, Orlando, FL, 1978, 422.

Rice, E. L., *Pest Control with Nature's Chemicals: Allelochemics and Pheromones in Gardening and Agriculture,* Oklahoma University Oklahoma Press, Norman, OK, 1995, 238.

Roshchina, V. V., *Biomediators in Plants. Acetylcholine and Biogenic Amines. (Biomediatori v rasteniakh. Atsetilkholin I biogennie amini),* Biological Center of USSR Academy of Sciences, Pushchino, 1991, 192.

Roshchina, V. V., Chemosensory mechanisms in allelopathy, in *Allelopathy in Agriculture and Forestry,* Narwal, S. S. and Tauro, P., Eds., Scientific Publishers, Jodhpur, 1994, 273.

Roshchina, V. V., Volatile plant excretions as natural antiozonants and origin of free radicals, in *Allelopathy. Field Observation and Methodology,* Narwal, S. S. and Tauro, P., Eds., Scientific Publishers, Jodhpur, 1996, 233.

Roshchina, V. V. and Melnikova, E. V., Spectral analysis of secretory products and excreta in intact plant cells, in *Abstracts of 2nd National Symposium "Allelopathy in Sustainable Agriculture, Forestry and Environment,"* September 6–8, New Delhi, Indian Society of Allelopathy, New Delhi, 1994, 140.

Roshchina, V. V. and Melnikova, E. V., Spectral analysis of intact secretory cells and excretions of plants, *allelopathy, J.,* 2, 179, 1995.

Roshchina, V. V. and Melnikova, E. V., Microspectrofluorometry: A new technique to study pollen allelopathy, *Allelopathy J.,* 3, 51–58, 1996.

Roshchina, V. V., Melnikova, E. V., Kovaleva, L. V., and Spiridonov, N. A., Cholinesterase of pollen grains, *Dokl. Biol. Sci.,* 337, 424, 1994.

Roshchina, V. V., Melnikova, E. V., Spiridonov, N. A., and Kovaleva, L. V., Azulenes, the blue pigments of pollen, *Dokl. Biol. Sci.,* 340, 93, 1995.

Roshchina, V. D. and Roshchina, V. V., *The Excretory Function of Higher Plants (Videlitelnaja functsia visshikh rastenii),* Moscow, Nauka, 1989, 210.

Roshchina, V. V. and Roshchina, V. D., *The Excretory Function of Higher Plants,* Springer Verlag, Berlin, 1993, 314.

Roshchina, V. V., Solomatkin, V. P., and Roshchina, V. D., Cicutotoxin as an inhibitor of electron transport in photosynthesis, *Fisiol. Rast.* (USSR), 27, 704, 1980.

Shapovalov, A. A., Possibility and perspectives in application of fluorescent method of allelopathic investigations, in *Physiologo-biochemical Base of Plant Interactions in Phytocenosis,* Grodzinskii, A. M., Ed., Naukova Dumka, Kiev, 1973, 128.

Shwartsburd, P. M. and Aslanidi, K. B., Spectrokinetic characteristics of two types of fluorescence of refractive granules in native individual cells from ascitic tumours, *Biomed. Sci.,* 2, 391, 1991.

Stanley, R. G. and Linskens, H. F., *Pollen. Biology, Biochemistry, Managements.* Springer Verlag, Berlin, 1974, 307.

Sukhada, K. D. and Jayachandra, Pollen allelopathy—a new phenomenon, *New Phytol.,* 84, 739, 1980.

Taylor, D. L. and Salmon, E. D., Basic fluorescence microscopy, in *Methods in Cell Biology. Living Cell in Culture,* Wang, J. L. and Taylor, D. L., Eds., Academic Press, San Diego, 1989, 207.

Thomson, I. D., Andrews, B. I., and Plowright, R. C., The effect of a foreign pollen on ovule development in *Dervilla lonicera (Caprifoliaceae), New Phytol.,* 90, 777, 1982.

Vasilyev, A. E., *Functional Morphology of Plant Secretory Cells,* Nauka, Leningrad, 1977, 208.

Whitaker, M., Fluorescence imaging in living cells, in *Cell Biology. A Laboratory Handbook,* Vol. 2, Celis, J. E., Ed., Academic Press, San Diego, 1995, 37.

Willemse, M. T. M., Morphological and fluorescence microscopical investigation on sporopollenin formation at *Pinus sylvestris* and *Gasteria verrucosa,* in *Sporopollenin,* Brooks, J., Grant, P. R., Muir, M., van Gijzel, P., and Shaw, G., Eds., Proc. Symp. Geology Department, Imperial College, London, September 23–25, 1970, Academic Press, London, 1971, 68.

Wolfbeis, O. S., The fluorescence of organic natural products, in *Molecular Luminescence Spectroscopy. Methods and Applications,* Schulman, S. G., Ed., John Wiley & Sons, Chichester, 1985, 167.

Wollenweber, E., Asakawa, Y., Schillo, D., Lehmann, U., and Weigel, H., A novel caffeic acid derivative and other constituents in *Populus* bud excretion and propolis (bee-glue), *Z. Naturforsch.,* 42, 1030, 1987.

Wollenweber, E., Mann, K., and Roitman, J. N., Flavonoid aglycones from the bud exudates of three Betulaceae, *Z Naturforsch.,* 46, 495, 1991.

Yurin, P. V., Shef, R. P., and Chernysheva, V. I., Role of free radicals at the physiologo-biochemical plant interactions in agrophytocenosis, in *Physiologo-Biochemical Base of Plants Interactions in Phytocenosis,* Vol. 3, Grodzinskii, A. M., Ed., Naukova Dumka, Kiev, 1972, 127.

Zanoni, D. G., Antagonismo pollinico, *Revista di Biologia,* 12, 126, 1930.

Zobel, A. M. and March, R. E., Autofluorescence reveals different histological localization of furanocoumarins in fruits of some *Umbelliferae* and *Leguminosae, Ann. Bot.,* 71, 251, 1993.

Section II

Interactions Among Plants and Microbial Systems

9 Pollen Allelopathy

Stephen D. Murphy

CONTENTS

9.1 ABSTRACT

Pollen allelopathy occurs when pollen is transferred interspecifically. Allelochemicals from pollen allelopathic species interfere with pollen germination or tube growth, stigma or style receptivity, ovule development, respiration, seedling germination and/or growth, leaf chlorophyll, seed set, and nonspecialist pollinators or nectarivores. Allelopathic pollen has been detected mainly in Poaceae and Asteraceae. Pollen allelopathy tends to be more common in wind-pollinated species; this probably is because wind-pollinated species are more likely to transfer pollen interspecifically. Testing for pollen allelopathy involves both *in vitro* and field tests. *In vitro* tests are appropriate for mass screening of many species but require care in accounting for potential artifacts related to pH, chemical dilution, pollen nutrients, anoxia, and relative humidity. Field tests require intensive manipulation of habitats where the physiography and microclimate are relatively predictable. The synthesis and mechanisms of pollen allelopathy are not well established. Since all pollen allelochemicals identified to date are phenolics and terpenoids, synthesis is likely similar to general allelochemical production via shikimate or terpenoid pathways. In terms of the mechanism of pollen allelopathy, pollen likely has to rehydrate on heterospecific stigmas in order to exude pollen allelochemicals. The allelochemicals probably are absorbed by target tissues (e.g., pollen, stigmas) and may interfere with mitosis, oxidative phosphorylation, or membrane permeability. From an ecological perspective, pollen allelopathy simply is one of a suite of competitive abilities. As such, individuals exposed repeatedly to pollen allelochemicals likely will develop resistance. The implication of this for agricultural applications of pollen allelopathy is that just because pollen allelochemicals may be bioherbicides does not mean they should be used carelessly. Indeed, that several pollen-allelopathic species are weeds (e.g., *Hieracium* sp., *Parthenium hysterophorus*) means that pollen allelopathy may be a greater risk to agriculture than a potential benefit unless a farming system approach is used.

9.2 THE STATUS OF ALLELOPATHIC RESEARCH

In my experience, research on allelopathy seems to provoke strong reactions among plant scientists. Many are enthusiastic about the concept and its potential implications for physiological and field ecology and agriculture. Many (including myself) also have a skeptical attitude as there is much research needed to elucidate the mechanisms and implications of allelopathy. Others still react with a contempt often reserved only for benighted souls who claim to have been abducted by aliens and taken to see Elvis at the intergalactic doughnut shop. As a doctoral student, I was less immersed in the controversies because I did research into possibly the most obscure form of allelopathy: pollen allelopathy. However, I doubt pollen allelopathy is truly less controversial; as my contribution here will demonstrate, there are still many questions unanswered.

9.3 POLLEN ALLELOPATHY

The first task is to explain the juxtaposition of "pollen" and "allelopathy." Allelopathy is generally described as the detrimental or beneficial effects of chemicals exuded from living or decomposing plant vegetative tissues on other plants (Rice, 1984; Rizvi and Rizvi, 1992). The role of pollen in allelopathic effects on heterospecific individuals is relatively unknown but it may be an important limit to sexual reproduction. Usually, pollen allelopathy is mediated by heterospecific pollen transfer (HPT) (Murphy and Aarssen, 1989, 1995a, b, c, d, 1996; Murphy, 1992, 1993). Pollen allelopathy occurs when pollen chemicals interfere with pollen germination or tube growth (Sukhada and Jayachandra, 1980a; Murphy and Aarssen 1989, 1995a, b), stigma or style receptivity (Sukhada and Jayachandra, 1980a), ovule development (Thomson et al., 1982), respiration (Jimenez et al., 1983; Ortega et al., 1988), seedling germination and/or growth (Anaya et al., 1992a, b), leaf chlorophyll (Sukhada and Jayachandra, 1980a), seed set (Char 1977; Sukhada and Jayachandra, 1980a; Murphy and Aarssen, 1995c, d, 1996), and nonspecialist pollinators or nectarivores (Tepedino et al., 1989; Murphy 1992). The phenomenon appears to have been described first by Char (1977), albeit based wholly on the research of Sukhada and Jayachandra (1980a, b) in India. In North America, Thomson et al. (1982) appear to have been the first to test for allelopathic pollen, but there has been a large body of excellent research on pollen allelopathy based in Mexico (Jimenez et al., 1983; Ortega et al., 1988; Anaya et al., 1987, 1992a, b). Extensive research on pollen allelopathy has been based on studies of eastern Canadian plants completed by myself and Lonnie Aarssen (Murphy and Aarssen, 1989, 1995a, b, c, d, 1996; Murphy, 1992, 1993). These studies began to examine the fundamental questions of identifying and quantifying the pollen allelopathic effect, determining the biochemical and physiological basis of pollen allelopathy, and exploring the implications of pollen allelopathy for evolutionary and applied ecology. These are the questions that this chapter will review in detail.

9.4 ARE SOME SPECIES MORE LIKELY TO BE POLLEN-ALLELOPATHIC?

9.4.1 WHICH SPECIES ARE POLLEN-ALLELOPATHIC?

Allelopathic pollen has been experimentally determined in four Asteraceae (*Hieracium floribundum* Wimmer. & Grab., *H. aurantiacum* L., *H. pratense* Tasuch, *Parthenium hysterophorus* L.) and several Poaceae (*Phleum pratense* L., *Zea mays* L. var. *chalquiñocónico* Hernández) (Char, 1977; Sukhada and Jayachandra, 1980a, b; Thomson et al., 1982; Jimenez et al., 1983; Ortega et al., 1988; Anaya et al., 1987, 1992a, b; Murphy, 1992, 1993; Murphy and Aarssen, 1989, 1995a, b, c, d, 1996; A. L. Anaya, personal communication). Indirect evidence suggests that *Bombax ceiba* L. (Malvaceae), *Brassica oleracea* L. (Brassicaceae), *Cucurbita pepo* L. (Cucurbitaceae), and *Zigadenus* L. (Liliaceae) may be pollen-allelopathic (see Murphy, 1992). I have tested over 150 other angiosperm species from the Asteraceae, Brassicaceae, Cyperaceae, Fabaceae, Juncaceae, Lamiaceae, Liliaceae, Poaceae, Ranunculaceae, and Rosaceae; of these, I have only found other allelopathic species of *Hieracium* (S. D. Murphy, unpublished data).

Of the experimentally verified pollen-allelopathic species, there do not seem to be many taxonomic correlations, save for the cluster of *Hieracium* species. The most striking example of a lack of taxonomic correlation of pollen allelopathy is from *Zea mays*. The pollen-allelopathic effect within *Zea mays* appears unique to Mexican varieties, for example, *Z. mays* var. *chalquiñocónico* (Anaya et al., 1992a; A. L. Anaya, personal communication); however, *Z. mexicana* L. is not pollen-allelopathic (Anaya et al., 1992a) nor is pollen from an unidentified cultivar of *Z. mays* allelopathic to *Citrullus lanatus* (Sedgley and Blessing, 1982; Murphy, 1992). I have recently tested two common Canadian cultivars of corn (*Z. mays* cv. 'Pride 5' and *Z. mays* cv. 'Pioneer 3902'); they are also not pollen-allelopathic (S. D. Murphy, unpublished data). Broadly, the only trends are that three pollen-allelopathic species (*Parthenium hysterophorus*, *Phleum pratense*, *Zea mays* var. *chalquiñocónico*) are wind-pollinated, while the *Hieracium* species are all agamospermous.

9.4.2 POLLEN ALLELOPATHY IN WIND-POLLINATED SPECIES

The stochastic nature of wind pollination often selects for increases in pollen production and pollen to ovule ratio (Murphy, 1992, 1993). As a specific example, the wind-pollinated *Phleum pratense* produces quantities of pollen that may yield ambient concentrations of between 3.9 to 26.4 billion (pollen grains/diurnal pollination cycle) at points along the circumference of an emission source (area = 1052 m^2) (Raynor et al., 1972). It is not surprising that Evans (1916) described the sudden emission of pollen from a field of *Phleum pratense* as a "haze" that could be seen at a distance of approximately 1 km. If anything, he was being conservative with his description.

Because of the increased probability of HPT, Murphy and Aarssen (1989) predicted that pollen allelopathy may be more common and more likely to reduce reproductive success of heterospecific individuals in wind-pollinated species. Since HPT is more probable, the potential for loss of resources (pollen and the energy used to manufacture it) may be greater in a wind-pollinated individual. To compensate for this unavoidable loss, it may be advantageous to sequester chemicals into pollen that can be used to disrupt reproduction in potential competitors.

Much of this depends on the relative ability to disperse pollen; in two of the anemophilous pollen-allelopathic species, *Phleum pratense* and *Parthenium hysterophorus,* this ability is relatively great. In contrast, the anemophilous, pollen-allelopathic *Zea mays* var. *chalquiñocónico* has large, dense pollen that has an extremely leptokurtic pattern of dispersal and so HPT is unlikely to occur (Paterniani and Stort, 1974; Jimenez et al., 1983; Ortega et al., 1988; Anaya et al., 1992a, b). Ambient concentrations of pollen of *Zea mays* may be high in some cases, but this often reflects the practice of planting *Zea mays* in dense monocultures (Paterniani and Stort, 1974). Studies of *Zea mays* var. *chalquiñocónico* have been predicated on the assumption that the high density of pollen (1×10^7 grains/m^2) deposited on the ground near below-canopy crops is what causes pollen allelopathy. This density appears sufficient to inhibit oxidative phosphorylation and/or mitosis in the hypocotyls of seedlings or roots or leaves of mature sympatric plants (Ortega et al., 1988). Though pollen of *Parthenium hysterophorus* does not require HPT to cause a similar effect (i.e., decrease the chlorophyll content of leaves of sub-canopy crops), HPT does mediate pollen-allelopathic effects on pollen and seeds in the sub-canopy crops (Sukhada and Jayachandra, 1980a, b).

9.4.3 POLLEN ALLELOPATHY IN AGAMOSPERMOUS SPECIES

Agamospermous species rarely (or never) reproduce sexually in the sense that pollen delivery of sperm is not needed and fertilization does not occur. Why then would pollen continue to be produced in agamospermous species? Thomson et al. (1982) suggested an answer: agamospermous species (e.g., *Hieracium*) may use allelopathic pollen in competition, perhaps to compensate for the production of otherwise unnecessary pollen. This hypothesis is attractive as it explains why plants would continue allocating energy to produce apparently useless pollen. However, this assumes that pollen production otherwise would be easily eliminated by natural selection and/or there would be strong selection against pollen production to save energy. Additionally, the hypothesis assumes that agamospermous pollen is unnecessary for sexual reproduction. In *Hieracium* species, pollen apparently is needed to supply hormones for agamospermous seed production and fertilization occurs occasionally (see Murphy and Aarssen, 1995a).

For *Hieracium* species, the questions about the selective advantage of pollen allelopathy in agamospermous species appears moot. While *Hieracium* pollen is allelopathic, there is insufficient HPT to sympatric species for pollen allelopathy to occur (Murphy, 1993; Murphy and Aarssen, 1995a; McLernon et al., 1996). I caution that to date I have studied the role of *Hieracium*'s allelopathic pollen in only a half-dozen communities that are dominated by sympatric nonactinomorphic species from the Fabaceae. I suspect that in habitats more dominated by species with actinomorphic flowers (similar to *Hieracium*), pollen allelopathy might be more important.

9.4.4 POLLEN ALLELOPATHY IN ANIMAL-POLLINATED SPECIES

I have not yet detected allelopathic pollen in any species that is truly animal-pollinated; I exclude the agamospermous species of *Hieracium,* though pollen is harvested by floral visitors. *Bombax ceiba* L. (Malvaceae), *Brassica oleracea* L. (Brassicaceae), *Cucurbita pepo* L. (Cucurbitaceae), and

Zigadenus L. (Liliaceae) all may be pollen-allelopathic, but the evidence is indirect (Murphy, 1992). By this, I mean that none of the studies were designed explicitly to test for pollen allelopathy, they do not mention the term "pollen allelopathy," yet results suggest the pollen is allelopathic. I have wondered if searching for pollen allelopathy in animal-pollinated species is not a fool's errand simply because the chances for HPT are much lower. However, HPT does happen frequently in animal-pollinated species because pollinators are more labile in the behavior than is often assumed, though this depends greatly on the taxa of pollinators and plants involved (Murphy and Aarssen, 1989; Murphy, 1992, 1993; McLernon et al., 1996). It is possible that pollen allelopathy then could be important in plant communities where floral structures and heights are similar enough to mediate HPT (Murphy and Aarssen, 1995a).

9.5   TESTING FOR POLLEN ALLELOPATHY

9.5.1   GENERAL REQUIREMENTS FOR *IN VITRO* POLLEN SCREENING

In 1992, I reviewed how to test for pollen allelopathy (Murphy, 1992). I have since refined this method and will address further the main problems with *in vitro* screening that can render results suspect or worthless. Generally, most explicit studies of pollen allelopathy have used an *in vitro* approach as a screening method (Char, 1977; Sukhada and Jayachandra, 1980a, b; Thomson et al., 1982; Jimenez et al., 1983; Ortega et al., 1988; Anaya et al., 1992a, b; Murphy and Aarssen, 1989, 1995a, b, c, d, 1996; Murphy, 1992, 1993). This allows testing of hundreds of pairwise species combinations. The initial screening usually involves whole-pollen extraction in polar solvents (e.g., distilled water) because most allelochemicals tend to be polar, water-soluble chemicals (Murphy, 1992).

Pollen grains exhibit variability in their ability to germinate. *In situ* this may be caused by genotypic and phenotypic differences in the ability to produce a pollen tube, genetic incompatibility, genomic incongruity, pollen selection by stigmas and/or styles, changes in the pH of the germination environment, osmotic effects, effects of ultraviolet (UV) radiation, or effects of pollutants (Murphy, 1992). Pollen is highly sensitive to many chemicals and radiation, hence its extensive use in bioassays for metal toxicity, effects of acid-rain and other atmospheric pollutants, herbicidal, fungicidal, and pesticidal toxicity, and effects of various forms of electromagnetic radiation. The implication is that even broad-spectrum screening methods require care. The challenge is to harvest fresh pollen quickly, extract any allelochemicals, and create an artificial growth medium that will (1) homogenize the micro-environment for pollen germination, (2) not produce toxic effects itself, (3) allow the putative allelochemicals to be added in low (ml) concentrations to media, and (4) be free of contaminants such as bacteria or fungi. Additionally, the pollen from target species (i.e., ones that might be susceptible to pollen allelochemicals) must also be harvested fresh, quickly, and carefully to avoid rupture.

9.5.2   *IN VITRO* SCREENING FOR POLLEN ALLELOPATHY: AVOIDING ARTIFACTS

9.5.2.1   pH

The potential effect of pH on pollen germination provides a good example in contrast to the effects of pollen allelopathy. The response curves of pollen germination, seed set, and ovule development to increased extract concentrations decrease according to functions that describe a rectangular hyperbola (Thomson et al., 1982; Murphy and Aarssen, 1989, 1995a, b, c, d, 1996; Murphy, 1992, 1993). In terms of pollen germination *in vitro* this response curve is unique to pollen allelopathy. The response curve describing the effect of pH on pollen germination may follow a fairly normal distribution in some species but is sigmoidal in others (Murphy, 1992). In many species, there is an optimal pH range bounded by sub-optimal and lethal ranges (Murphy, 1992). Much of this optimality probably reflects adaptation to acidic cytoplasm or may be used as a control mechanism by females to induce pollen competition (Ganeshaiah et al., 1988). This creates a problem for *in vitro* screening methods because the pH will be affected by the germination chemicals used (e.g., $CaCO_3$, H_3BO_4) and this can create sub-optimal conditions for the pollen from the target species. It is important to measure the normal pH of the stigmatic environment of the target species and to buffer germination media appropriately. If the response curve to the putative allelochemicals is anything other than a rectangular hyperbola, the effect is likely caused by an inappropriate pH of the medium or even of the pollen exudate tested for allelochemicals (Murphy, 1992).

9.5.2.2 Dilution of Pollen Germination Factors

Pollen germination may decrease if there is a dilution of a pollen germination factor common to con- or heterospecific individuals (Thomson et al., 1982). In such cases, however, the decrease does not describe the rectangular hyperbola of pollen allelopathy (Murphy, 1992). Nonetheless, Thomson et al. (1982) suggested that dilution of a putative germination promotor could contribute to a possible hyperbolic decrease in ovule development upon exposure to increased relative concentrations of pollen of an allelopathic species. They did not, however, specify the physiological basis for this.

9.5.2.3 Chemicals in Germination Media

In vitro experiments require that a sufficient percentage of pollen germinates in the absence of any allelochemicals, that is, in the control media. Although the use of a standard solution will result in sub-optimal germination conditions for many species, it is relative pollen germination values that are of interest. While it is acknowledged that there are many alternatives for germination media, the fact that a medium may be more conducive to one species over another has few consequences for studies on pollen allelopathy (Murphy, 1992).

Nonetheless, it is important to be aware that some pollen may be extremely susceptible to the chemicals of the germination media (often this includes boron, calcium, and potassium) (Murphy, 1992). Although high concentrations of exogenous boron may be toxic, it appears to be necessary for germination of pollen of most of angiosperms (Leduc et al., 1990). Boron may promote the proper exportation of proteins used in cell wall construction and may regulate the permeability of membranes (Sidhu and Malik, 1986). Notably, high cellular levels of boron appear to inhibit the manufacture of certain phenolics and terpenoids, possibly including allelochemicals. Such inhibition allows carbohydrates (and/or their precursors) to be diverted to the pentose phosphate and tryptophan biosynthesis pathways, which then promote the manufacture of compounds needed for pollen tube biosynthesis (Sidhu and Malik, 1986). It is unclear, however, if this affects the biosynthesis of allelochemicals. Further, it assumes that allelochemicals are synthesized, *de novo,* and are phenolics or terpenoids (Murphy, 1992).

Excessive calcium may prevent *in situ* germination of pollen in Poaceae (at least), possibly by altering the pectic substances of the Zwischenkörper (the outer layer of the intine controlling the operculum covering the pollen aperture, i.e., where the tube emerges) (Heslop-Harrison, 1979a). Nonetheless, calcium may be necessary to initiate, promote, and guide pollen tube growth through the style in some species (Reger et al., 1992). In monocolpate pollen (e.g., Poaceae), calcium also appears to promote extension of the pollen tube wall, that is, deposition of callose and activation of the actin cytoskeleton (Heslop-Harrison, 1979a; Heslop-Harrison and Heslop-Harrison, 1992). In general, structural proteins are released to the media (instead of constructing the pollen tube walls) if boron and calcium are not added to media, indicating the necessity of their inclusion in most germination media (Jackson and Kamboj, 1986).

In contrast to calcium and boron, high external concentrations of potassium are not toxic and may prevent excessive loss of potassium due to minor damage to the pollen membranes (Kim et al., 1985; Hoekstra and Barten, 1986). This would promote the integrity of the membrane and increase the probability of germination of pollen (Kim et al., 1985; Hoekstra and Barten, 1986).

9.5.2.4   Germination Media Solvents

I recommend using 1 percent agar (spread about 1 to 5 mm thick) as the solvent for germination media because it creates a more homogenous environment than other solvents (e.g., distilled water) and it is easier to control for contamination (Murphy, 1992). Agar is also nontoxic (assuming the agar was pure initially), acts as an osmoticum to reduce pollen bursting, and may reduce damage to the phospholipids in the pollen membranes by providing an external source of carbohydrates (Hoekstra et al., 1989, 1991, 1992a). Aniso-osmotic effects are apparent and distinguishable from any effects of allelopathic pollen as pollen simply ruptures in cases of osmotic imbalance (Murphy, 1992).

Nonetheless, agar can create some artifacts. Agar media may create anaerobic pockets into which pollen may sink (Murphy, 1992). Anoxia will affect both germination and pollen tube growth (Van Aelst and van Went, 1989). Similar to the effects of extreme pH levels, anoxic effects are detected easily as they cause pollen tubes to burst (Van Aelst and van Went, 1989). The solution is to spread agar thinly (0.1 mm thickness) to prevent anoxia but allow pollen germination (Murphy and Aarssen, 1989, 1995a, b, c; Murphy, 1992, 1993).

If a uniform thickness is achieved and anoxia is avoided, agar media creates a homogeneous micro-environment for pollen germination *in vitro*. All chemical solutions must be thoroughly mixed and carefully introduced into agar so they do not form discrete layers (Pederson, 1986; Dorbos and Spencer, 1990). From the perspective of controlling experimental error this is an advantage. For example, it is relatively simple to avoid over-concentrating pollen extract to unrealistic values (>100 ml) in some areas of the agar plate. The problem is that such a homogeneous profile may not exist *in situ* and the *in vitro* screens may exaggerate the allelopathic effect that would occur *in situ* (Thomson et al., 1982; Murphy, 1992, 1993; Murphy and Aarssen, 1995b, c, d, 1996). Nonetheless, such homogeneity is necessary in order to quantify the pollen-allelopathic effect *in vitro,* that is, heterogeneity may cause confounding effects of differential concentrations (allelochemicals, calcium, boron) within the media (Murphy, 1992, 1993; Murphy and Aarssen, 1995b, c, d, 1996).

9.5.2.5   Relative Humidity and Rehydration of Pollen Introduced
to Growth Media

For many species, especially the Poaceae, control of the relative humidity is more important than the constituents of the germination medium itself (assuming that none of these constituents is toxic) (Heslop-Harrison and Heslop-Harrison, 1985, 1988; Shivanna et al., 1991). Rehydration of pollen in an environment of high relative humidity (prior to germination on artificial media) is crucial as it allows the gradual restoration of the membranes; these normally are desiccated upon dehydration of pollen in the anther (prior to anthesis) (Hoekstra and Barten, 1986; Heslop-Harrison and Heslop-Harrison, 1988; Hoekstra et al., 1989, 1991, 1992a, b). Failure to rehydrate pollen before introduction to germination media often results in rupture of the pollen grain, possibly from a breach in the membranes (Hoekstra and Barten, 1986; Heslop-Harrison and Heslop-Harrison, 1988). It also is important to germinate the pollen as rapidly as possible as trinucleate pollen metabolizes so rapidly after harvest that delays can result in a lack of significant germination (Mulcahy and Mulcahy, 1988; Murphy, 1992; Murphy and Aarssen, 1995c). I have normally collected mature pollen (i.e., at anther dehiscence), rehydrated it in Petri dishes with wet filter paper, and germinated it within 30 minutes of harvest (Murphy and Aarssen, 1989, 1995a, b, c).

If pollen is not properly rehydrated, then larger compounds can penetrate damaged membranes (Hoekstra and Barten, 1986; Heslop-Harrison and Heslop-Harrison, 1988) and cause artifacts (Murphy,

1992; Murphy and Aarssen, 1995b, c). These larger compounds include proteins (and possibly DNA) that are normally exuded from pollen. These interact with the stigma and usually do not appear to penetrate other pollen grains or tubes (Linskens and Schrauwen, 1969). In contrast, smaller molecular weight compounds, such as phenyl indoles, can penetrate pollen and its tube easily (Heslop-Harrison and Heslop-Harrison, 1988).

9.5.3   FIELD TESTS FOR POLLEN ALLELOPATHY

In collaboration with Lonnie Aarssen, my main experimental contribution has been in successful field testing for pollen allelopathy (Murphy and Aarssen, 1995d, 1996). As mentioned in Section 9.3.3, the studies on *Hieracium* revealed that there was insufficient transfer of *Hieracium* pollen (i.e., mean of two grains or less) to stigmas of sympatric species to cause pollen allelopathy. This conclusion was based on observations of floral visitors and by examining the species composition of pollen found on harvested stigmas (Murphy and Aarssen, 1989). The studies with *Phleum pratense* were more positive but rather labor-intensive (Murphy and Aarssen, 1995d, 1996); glibly, I often have thought that few would be sufficiently masochistic to spend several weeks clipping all of the inflorescences of pollen-allelopathic *Phleum pratense*. These studies examined pollen allelopathy within abandoned pastures in Canada that contain naturalized individuals of a once-popular forage crop. With the exception of the contribal *Agrostis stolonifera* L. (Poaceae: Agrostidae), pollen from naturalized populations of *Phleum pratense* across eastern Canada is allelopathic to all heterospecifics tested to date (Murphy, 1992, 1993; Murphy and Aarssen, 1995b, c). The field studies were the crucial link in demonstrating that pollen allelopathy was more than an interesting phenomenon that existed *in vitro*.

To test whether pollen allelopathy by *Phleum pratense* was important *in situ*, we chose old fields of similar sizes, physiography (including species composition), and, of course, ones that contained *Phleum pratense* and other species of concurrently flowering Poaceae (Murphy and Aarssen, 1995d, 1996). Importantly, we could pair fields because they were separated by a forest tract and, in fact, were surrounded on all sides by forest (Murphy and Aarssen, 1995d, 1996). This is important because in fields surrounded by such forest tracts, nightly surface-based radiation inversions create what is essentially an atmospheric barrier that prevents export of dispersed pollen beyond the field or import of pollen from other fields until sunrise (Chatigny et al., 1979). Most individuals of *Phleum pratense* release pollen before sunrise and most of the pollen would be deposited within the field in which it was originally dispersed prior to the gradual elimination of the radiation inversion (Chatigny et al., 1979; Murphy and Aarssen, 1995d, 1996). This did indeed prevent contamination of control fields where we clipped inflorescences of *Phleum pratense* (see next paragraph) as we found almost no pollen from *Phleum pratense* on stigmas of other species of Poaceae (Murphy and Aarssen, 1995d, 1996).

The field studies mainly involved clipping all of the inflorescences of *Phleum pratense* in a small field to prevent pollination. The only exception was that we bagged *Phleum pratense* inflorescences close to marked individuals of other species of Poaceae (Murphy and Aarssen, 1995d, 1996). This ensured that any responses in these marked individuals were not caused by any change in the ability of neighboring *Phleum pratense* plants to compete for resources (a possible artifact caused by clipping inflorescences) (Murphy and Aarssen, 1995d, 1996). We left the adjacent paired field unmanipulated and compared the number of pollen grains of *Phleum pratense* on stigmas from other Poaceae (*Elytrigia repens* [L.] Nevski and *Danthonia compressa* Aust.) and seed set in those species (Murphy and Aarssen, 1995d, 1996). The result was that in the fields that were not manipulated, there were the usual billions of pollen grains from *Phleum pratense* in the atmosphere. In these fields, *Elytrigia repens* and *Danthonia compressa* typically had 5 to 15 pollen grains from *Phleum pratense* on the stigmas (well within the range of the pollen-allelopathic effect indicated by *in vitro* studies) and the seed set was 5 to 30 times lower than in fields without *Phleum pratense* pollen (Murphy and Aarssen, 1995d, 1996). It appears that the copious production of low-density, small-diameter (30 mm), allelopathic pollen by *Phleum pratense* is an effective means of reducing heterospecific seed set and production of possible future competitors (at least in the short-term; sections 9.7 and 9.8 discuss the longer-term implications) (Murphy and Aarssen, 1995d, 1996).

9.6 HOW MIGHT POLLEN ALLELOCHEMICALS BE SYNTHESIZED IN PLANTS?

9.6.1 POSSIBLE SYNTHESIS MECHANISMS

Pollen allelochemicals appear to be terpenoids or phenolics produced via the terpenoid or shikimic acid pathways (Sukhada and Jayachandra, 1980a, b; Ortega et al., 1988; Anaya et al., 1992a; Murphy, 1992). As such, allelochemicals could be produced for a specific reason and not simply as byproducts of normal metabolic pathways and/or toxins that must be eliminated. Indeed, most secondary compounds do not appear to be mere byproducts and, since autotoxicity does not seem to occur, this suggests that allelochemicals would not be considered toxic waste (Murphy and Aarssen, 1989; Harborne, 1990; Jones and Firn, 1991; Murphy, 1993). Second, it assumes that the plant has the necessary resources, or that selection favors allocation of resources, to manufacture allelochemicals in an energetically cost-effective manner.

Although most plants contain the biochemical pathways (e.g., shikimate, isoprene) necessary to manufacture untold quantities of secondary metabolites, it has been argued that the energetic costs exceed the benefits in most cases, hence biological activity (including allelopathy) should be rare (Jones and Firn, 1991). It has been proposed that longer-lived perennials, such as *Phleum pratense*, should postpone sexual reproduction if they are to produce allelochemicals (Waller, 1988). It is then possible that there is a trade-off in resource allocation between allelochemicals and production of pollen, ovules, and seeds. This prediction was based on root allelochemicals (Waller, 1988) and it is unclear if it applies to pollen allelopathy. It should not, since pollen allelopathy implicitly depends upon some combination of allocation of resources to both secondary compounds and pollen.

Although resource allocation to both pollen and secondary compounds might seem prohibitively energy-expensive, there is much overlap in biosynthetic pathways through matrices, grids, and combined pathways that represent reductions in costs of production of secondary compounds (Jones and Firn, 1991). However, this may reduce the opportunity for selection to act on production (quantities, types) of allelochemicals. Nonetheless, the observed diversity in the types of chemicals produced by pollen may suggest that there is ample opportunity for such selection (Murphy, 1993). On a general scale, most of the key components of biosynthetic pathways are rather invariable; however, the highly branched nature of such pathways increases the realized variation (Jones and Firn, 1991). Selection is likely to operate on such diversity and specific portions of the branches, hence explaining the greater diversity of compounds than might otherwise be expected. Further, fine control of many of the enzymes of biosynthetic pathways of secondary compounds may allow increased opportunity for selection in that genotypes may have differential abilities to control these enzymes.

Changes in biological activity that result from differences in pathway branching and/or control of enzymes have been detected, indicating that selection does indeed operate (Murphy, 1993); however, the importance of these changes in pollen is unclear. It is known, however, that many pollen genes are highly regulated and enzymes and compounds may appear unexpectedly (i.e., versus that predicted by chemotaxonomy), perhaps in response to selection (Searcy and MacNair, 1990; Hormoza and Herrero, 1992). Genes that are involved in production of "sloppy" enzymes (ones that show low substrate affinity) may exhibit greater variability in their regulation, with respect to differences between genotypes (Jones and Firn, 1991). Of these sloppy enzymes, most relevant are those involved in terpenoid and phenol biosynthesis, that is, production of most allelochemicals (Harborne, 1984; Rice, 1984). In terms of energetic costs, phenolics and terpenoids are much more economical than alkaloids in that the latter requires investment of nitrogen (which is often most limiting) and/or pathways to recycle (salvage) nitrogen from alkaloid pathways (Harborne, 1990).

It may be that biosynthesis of one allelochemical should reduce production of a different one. For example, both the shikimate (phenolic) pathway and isoprene (terpenoid) pathway require the same precursors, thus necessitating regulation and lower activity of one of the pathways (Harborne, 1990). As a consequence, it might be expected that the allelochemicals would differ between pollen-allelopathic species. On a taxonomic basis, greater production and use of terpenoids is typical of Asteraceae, in particular, the tribe Lactuceae, which includes *Hieracium* (Gonzalez, 1977; Mabry and Bohlmann, 1977; Harborne, 1990). In Poaceae, phenolic compounds are more common and may be more likely to have biological (allelopathic) activity in most cases (Ortega et al., 1988; Anaya et al., 1992a, b). Both terpenoids and phenolics are water-soluble and, usually, acidic; in my studies on *Phleum pratense* and *Hieracium* species, the ion-exchange chromatography I used extracted phenolics and terpenoids (Harborne, 1984). I could not conclude that the allelochemicals were phenolics in *Phleum pratense* and terpenoids in *Hieracium* species. The precedents for such a conclusion are much too limited. Alternatively, it may be argued that because the qualitative and quantitative effects, *in vitro,* of pollen allelochemicals were similar in *Phleum pratense* and *Hieracium* species, the pollen allelochemicals are similar across the taxa. However, even within one species, the allelochemicals can vary in their chemical structure. The allelochemicals in *Parthenium hysterophorus* were identified as both phenolics and terpenoids, though parthenin (a terpenoid) apparently had most of the biological activity (Sukhada and Jayachandra, 1980a, b).

9.6.2 ARE POLLEN ALLELOCHEMICALS ACTUALLY SYNTHESIZED IN POLLEN?

It is often assumed that any pollen allelochemicals are within some portion of the cytoplasm of the pollen; this may be reasonable given the absence of evidence for abundant material in the chambers of the exine layer of pollen, tapetally derived or otherwise (Heslop-Harrison and Heslop-Harrison, 1985; Murphy, 1992). While the sporophyte does load some of the enzymes and chemicals found in pollen, the premise that the sporophyte loads allelochemicals (or its enzymes) is questionable (Murphy, 1992). The term "loading" is confusing as it can refer to direct transfer of compounds via the tapetum, or it may simply refer to genetic and biochemical overlap between pollen and the sporophyte. Although some have claimed that there are few tapetally derived compounds borne on the surface of pollen in the Poaceae, others claim that many enzymes are located on the exterior of the pollen, that is, in the outer exine or near the pore (Heslop-Harrison and Heslop-Harrison, 1985; Baraniuk et al., 1990). It has been difficult to determine the relative importance of such exine-borne compounds. It does not appear that they function in incompatibility responses nor are they antigenic in nature as these proteins are derived from the intine (Murphy, 1993). Therefore, if allelochemicals do originate from the intine, then they may be produced, *de novo,* by genes and enzymes unique to the pollen.

9.7 WHAT ARE THE MECHANISM(S) MEDIATING POLLEN ALLELOPATHY?

9.7.1 WHY DO POLLEN ALLELOCHEMICALS AFFECT SO MANY SPECIES?

To date, studies on pollen allelochemicals suggest that they have both broad effects and high biological activity (Char, 1977; Sukhada and Jayachandra, 1980a; Thomson et al., 1982; Jimenez et al., 1983; Ortega et al., 1988; Anaya et al., 1992a, b; Murphy and Aarssen, 1989, 1995a, b, c, d, 1996; Murphy, 1992, 1993). From an ecological perspective, this seems logical as these plants are exotic to the locations in which pollen allelopathic interactions have been found (it remains to be seen if these interactions exist in their regions of origin). As a result, the plants have short (less than 250 years) co-evolutionary histories with most of their target species; selection might favor broad spectrum compounds potentially affecting individuals of most any species that would be encountered (Murphy, 1992; Murphy and Aarssen, 1989). Nonetheless, there is some question as to whether allelochemicals, pollen or otherwise, should be broad-acting. There are opposing opinions on the breadth and intensity of activity of secondary compounds. High biological activity, especially if it is to be broad-acting, may be rare because most organisms do not have the specific site receptors (Jones and Firn, 1991). This argument, however, was developed mainly for compounds involved in defense against herbivores (Jones and Firn, 1991) and it is unclear if this is true of plant–plant allelopathic interactions. The alternative, that there are many broad-spectrum compounds with high biological activity has been supported, but, again, this was concerned mainly with chemicals toxic to animals (Downum et al., 1991).

9.7.2   TRANSFER OF ALLELOCHEMICALS FROM HETEROSPECIFIC POLLEN

9.7.2.1   The Role of Adherence of Heterospecific Pollen to Stigmas

It is unlikely that any transfer of allelochemicals is pollen grain to pollen grain, despite the possibility that even heterospecific grains can be associated closely (<10 mm) on stigmas (Heslop-Harrison, 1979b; Heslop-Harrison et al., 1984; Murphy, 1992). It seems that, *in situ*, pollen must adhere to the stigma if pollen allelopathic effects on pollen germination, ovule development, or seed set are to occur.

Studies of pollen allelopathy in *Phleum pratense* and *Hieracium* species provide a useful example. The stigmas of *Phleum pratense* and *Hieracium* species are termed "dry", that is, they exude proteinaceous compounds that form a thin pellicle (Heslop-Harrison and Heslop-Harrison, 1985). The importance of this to pollen allelopathy is related to its role in hydration and transport of any chemicals in the surface secretion. In Poaceae, pollen is released in a desiccated form, containing about 35 percent water (weight/volume) and the pollen must be rehydrated through contact with the stigma (Heslop-Harrison and Heslop-Harrison, 1985). In terms of pollen allelopathy, this requires that the foreign pollen be able to attach itself to the pellicle of a stigma (possibly through electrostatic attraction), form an aqueous meniscus (through establishment of capillary pathways), and create a free-flow of water from stigma to pollen, as the pellicle disperses and the stigma proteins fuse with the pollen exudate (Heslop-Harrison and Heslop-Harrison, 1985). In many cases, foreign pollen cannot form the initial capillary pathways; hence, it will not form a meniscus and rehydration will not occur (Heslop-Harrison and Heslop-Harrison, 1985) In addition, there may be tapetally derived glycoproteins that provide an additional cross-incompatibility barrier, although there is little evidence that this operates in Poaceae (Heslop-Harrison and Heslop-Harrison, 1985).

If pollen does not attach and rehydrate successfully on heterospecific stigmas, then it is unlikely that pollen allelochemicals could be transferred to the stigma or to any other tissues (e.g., heterospecific pollen on the stigma) since there would be no medium available for transport. There is evidence, however, that in Poaceae (among certain other families), foreign pollen attaches, rehydrates, germinates on stigmas (even those of different tribes), and that the pollen tube is able to penetrate the style (Heslop-Harrison et al., 1984; Hoekstra and Barten, 1986). I have also observed pollen of *Phleum pratense* rehydrating on stigma of all other species of Poaceae that I have studied to date (Murphy, 1992, 1993).

9.7.2.2   How Are Pollen Allelochemicals Transmitted from Pollen to Heterospecific Tissues?

Assuming that foreign pollen does rehydrate successfully on stigmas, then the opportunity for transfer of pollen allelochemicals may be great. The free-flow of water between stigma and pollen is not unidirectional, hence the allelochemicals may flow into the stigma (Heslop-Harrison and Heslop-Harrison, 1985). It has been suggested that some of the solutes exuded from pollen could form a concentrated and viscous solution on the stigma or with any stigmatic exudate (Baraniuk et al., 1990). If this occurs and if this solute fraction contains the pollen allelochemicals, then transport of allelochemicals would be expedited. It is cautioned, however, that much of this is concerned mainly with the higher molecular weight compounds, for example, proteins, that may be used in incompatibility reactions (Baraniuk et al., 1990). Nonetheless, pollen, during its rehydration phase, is able to mobilize and exude many compounds as the membranes do not act as barriers and the cell walls are breached in preparation for pollen germination (Heslop-Harrison, 1979b). Most importantly, the normal mechanism of pollen rehydration and germination provides a plausible means of exudation of intine-held pollen allelochemicals.

9.7.2.3 Which Tissues Are Affected by Pollen Allelochemicals?

Sukhada and Jayachandra (1980a, b) suggested that pollen allelochemicals act to induce necrosis in stigmas of target species. They did not, however, comment on the reason for such necrosis and, to date, there has been little research into this aspect of pollen allelopathy. It is unknown if the effect would be localized to the specific papillae or trichomes or if the chemicals could be transmitted throughout the stigma. Further, it is unclear if the effect would always be to cause localized or general stigma necrosis or if the allelochemicals might be transported into other pollen grains upon their adhesion to the stigma. In this manner, pollen allelochemicals could affect both the stigma and pollen of the heterospecific target plant. This may be probable as necrosis, while observable, was not always apparent in studies on pollen allelopathy in *Phleum pratense* (Murphy and Aarssen, 1989, 1995a, b, 1996). Further, since there are effects on a diversity of plant tissues, it is possible that allelochemicals act upon more than one tissue.

9.7.2.4 Possible Biochemical Mechanism(s) of Pollen Allelopathy

It is possible that allelochemicals may interfere with the ATPase that promotes proper adhesion of pollen to stigmatic papillae, hence the mechanism may be analogous to that involved in cross- or self-incompatibility (Heslop-Harrison and Heslop-Harrison, 1985). If this is true, it should involve something other than the S-alleles of incompatibility; experiments using small-bead resins in ion-exchange chromatography to isolate the biologically active pollen chemicals precluded a role of proteinaceous compounds (Murphy and Aarssen, 1995a, b, c). Further, pollen-allelopathic interactions do not necessarily require interaction with the stigma or style, based on the pollen–pollen studies *in vitro* (Murphy and Aarssen, 1989, 1995c).

Allelochemicals from pollen of *Zea mays* interfere with mitosis and/or uncouple oxidative phosphorylation in mitochondria, at a site between coenzyme Q and cytochrome *c* (Ortega et al., 1988). The effect on oxidative phosphorylation may be analogous to the deleterious effect on chlorophyll function observed by Sukhada and Jayachandra (1980a, b). It is unclear, however, if these are the actual targets of allelochemicals and/or if pollen would be affected in a similar manner. It also has been postulated that allelochemicals may increase permeability of membranes (Ortega et al., 1988). Given the need for rehydration and conservation of membrane integrity in order to germinate pollen of Poaceae, this may be as plausible a mechanism as any, at least in terms of *in vitro* studies. However, this does not account for the lack of toxicity of pollen allelochemicals to conspecifics (see next section).

9.7.2.5 Why Aren't Pollen Allelochemicals Toxic to Conspecifics in
              Phleum pratense?

Pollen allelochemicals are not toxic to conspecifics, at least in *Phleum pratense*. This seems logical from the perspective of natural selection unless there is an advantage to eliminating sexual reproduction in unrelated conspecifics and the detoxification mechanism was extremely specific. At present, we do not understand the reason for the lack of pollen-allelopathic effects from *Phleum pratense* on conspecifics or contribals (e.g., *Agrostis stolonifera*) and cannot predict if there is variation in detoxification at the smaller scale of close genetic relatives. In general, the mechanism of detoxification of allelochemicals is unknown, although it has been established that allelochemicals produced by many plants are not always autotoxic, unless modified by soil microbes (Rice, 1984). This latter factor may be less relevant to pollen allelopathy; however, microorganisms (e.g., yeasts) in the flower could alter the effectiveness and/or targets of pollen allelochemicals (Eisikowitch et al., 1990a, b). Plants also can avoid autotoxicity by sequestering allelochemicals in different organs (e.g., trichomes), specialized vacuoles, or in exportable substances such as wax (Harborne, 1990). In the case of pollen allelopathy, the role of any of these sequestering mechanisms is questionable because of the small scale of pollen and stigmas. Determining the reason for a lack of conspecific and congeneric toxicity probably will be difficult until the actual mechanism(s) of pollen allelopathy are specified.

9.8 EVOLUTIONARY ECOLOGY OF POLLEN ALLELOPATHY

9.8.1 COMPETITION AND COEXISTENCE

Pollen allelopathy must be examined in the broader context of evolutionary ecology and, more specifically, competition and coexistence. Competitive abilities may be best conceptualized as nonlinear hierarchies (Aarssen, 1989). Pollen allelopathy then can be defined as one of the many secondary traits of competitive ability (Aarssen, 1989). This means that pollen allelopathy mediates the proximate effect of interference with pollination of neighbors by denying the neighbor access to a limiting resource (viable pollen and/or stigmas) (Aarssen, 1989). Pollen allelopathy is one character of many that confers some degree of relative competitive advantage, at least in the short-term.

9.8.2 COMPETITIVE COMBINING ABILITY

Pollen allelopathy can therefore help test hypotheses such as the competitive combining ability hypothesis (Aarssen, 1983, 1989; Murphy and Aarssen, 1989; Taylor and Aarssen, 1990). In earlier versions, competitive combining ability was defined in terms of the reciprocal effect of continued selection such that difference in relative competitive abilities tend to be reduced over time and the competitors eventually coexist (Aarssen, 1983). Rather than an arms race, this resembles a perpetual tug-of-war in that coexistence belies the continual competition and selection (Aarssen, 1983). Essentially, superiority and inferiority of relative competitive ability become transitory as, barring any severe disturbance, the plants are locked in cycles of action and reaction, in terms of selection (Aarssen, 1983).

The competitive combining ability hypothesis has been expanded to include a mechanism involving intransitivity (Aarssen, 1989, 1992). Traditionally, it has been postulated that intransitivity networks of competitive ability may occur when no one species in a community can outcompete all others and each species is both a better and poorer competitor, relative to at least one other species (Taylor and Aarssen, 1990). In contrast, in terms of competitive combining ability, it is predicted that species are not the relevant unit of observation; rather, intransitivity occurs at the genotypic level (Taylor and Aarssen, 1990). No one genotype is competitively superior because the genotypes may each be relatively superior in one or several of a suite of characters that determine relative competitive ability in toto (Aarssen, 1989; Taylor and Aarssen, 1990). As a consequence, competitive exclusion at the level of species is unlikely to occur, simply because no one species has a monopoly on genotypes that are competitively superior (Aarssen, 1989; Taylor and Aarssen, 1990).

The hypothesis of coexistence as mediated by competitive combining ability is testable but information about the available genetic variability, the role of stochastic events, possible combinations of plant traits that confer competitive ability, relative roles of competitive aspects (denying resources to neighbors, tolerating denial of resources and allocating resources to fecundity), and even the importance of intransitivity itself is needed (Aarssen, 1983, 1992). If pollen allelopathy is studied as one component of several known competitive attributes of genotypes (individuals), then this may help explain the complexity of competitive relationships at the level of species (Aarssen, 1983, 1989). One approach may be to study pollen allelopathy in the context of other competitive phenomena, both intra- and inter-specific, that may be associated with pollination. Examples may include pollen competition, cryptic self-compatibility (and incompatibility), and outcrossing depression.

Alternatively, all competitive attributes associated with pollination (including pollen allelopathy) may be studied in relation to competitive ability. For example, genotypes of *Phleum pratense* may vary in their pollen-allelopathic effectiveness, while genotypes of other species may vary in their ability to intercept light or compete (interspecifically) for nutrients. Further, all genotypes of all species may vary in their relative ability to compete for all resources. The result may be intransitivity in competitive relations, which would be detected as coexistence at the species level. This also may explain why *Phleum pratense* is described as both an inferior and superior competitor (Taylor and Aarssen, 1990). Ultimately, all possible combinations and interactions of plant traits need to be examined, in terms of how they define relative competitive abilities. This ambitious approach is laudable, though rather unwieldy, given the myriad numbers of traits (hence combinations and interactions) that can exist. At least, however, this research paradigm suggests searching for general competitive mechanisms.

9.8.3   Ecological Combining Ability

Ecological combining ability, where there is selection for differences in niche requirements, can also lead to coexistence (Aarssen, 1983, 1989). A useful way of differentiating ecological combining ability from competitive combining ability is to use an example based on pollen allelopathic interactions between *Phleum pratense* and *Elytrigia repens*.

Elytrigia repens may have two main alternative responses to losses in fecundity caused by the allelopathic pollen of *Phleum pratense*. First, it may adjust its relative competitive ability through selection for plants that can detoxify allelochemicals or by reallocating resources to rhizomes, thereby competing for nutrient resources, instead of viable stigmas. As this would be consistent with the hypothesis of competitive combining ability, it may be expected that some individuals of *Phleum pratense* would respond, perhaps by reallocating resources to roots or increased height. In this case, there may or may not be intransitive relationships at the genotype, hence species, level but this would be testable.

Second, selection could favor individuals of *Elytrigia repens* that flower later in the day (afternoon), that is, distinct from flowering in *Phleum pratense*. Such divergence in diurnal flowering phenologies would be consistent with the hypothesis of ecological combining ability. Individuals of *Phleum pratense* may gain greater benefits (e.g., realized seed set) if they occupy the optimal window for pollination, which occurs in the early hours of the morning, hence avoid flowering at times less conducive to pollination (Raynor et al., 1972). Thus, it is unclear if selection would favor expansion of the diurnal phenology of *Phleum pratense* into the afternoon hours now occupied by *Elytrigia repens,* thereby allowing continued effectiveness of pollen allelopathy in reducing seed set in *Elytrigia repens.* Such a scenario is dependent upon the degree of genetic variation, intensity of selection, influence of stochastic factors (weather), and the costs and benefits of flowering (to individuals of all species) at certain times of the day (Ågren and Fagerstrom, 1980; Murphy, 1992). In this respect, it reflects the concerns raised for competitive (rather than ecological) combining ability but these concerns are common to both hypotheses (Aarssen, 1983, 1989, 1992).

9.8.4 PHENOTYPIC PLASTICITY

It is possible that selection may not act at all and *Elytrigia repens* may be phenotypically plastic in its response to the pollen allelochemicals of *Phleum pratense*. Although phenotypic plasticity in *Elytrigia repens* has been investigated (Taylor and Aarssen, 1990), there has yet to be any experiments in reference to the response to pollen allelopathy. There is evidence, however, that selection does operate to favor heterospecific genotypes that can respond (e.g., detoxify) to (soil-mediated) allelochemicals of *Ailanthus altissima* (Simaroubaceae) and that this response is not a result of phenotypic plasticity (Lawrence et al., 1991). As in the case of the hypotheses of combining ability, there have been few tests for phenotypic plasticity in the response to pollen allelopathy and further studies are needed. The example of pollen allelopathic interactions between *Phleum pratense* and *Elytrigia repens* may offer the opportunity to investigate phenotypic plasticity.

9.9 APPLYING POLLEN ALLELOPATHY TO AGRICULTURE

Allelochemicals may be used as models for new herbicides or directly as bioherbicides (Duke and Lydon, 1988; Einhellig and Leather, 1988; Putnam, 1988a, b; Lovett, 1991; Swanton and Weise, 1991; Swanton and Murphy, 1996). As bioherbicides, pollen allelochemicals are attractive because of their often brief persistence in soil; however, commercial applications depend on their specificity, active concentrations, and possibility of synthesis on an industrial scale (Duke and Lydon, 1988; Heisey, 1990; Murphy, 1993). To date, most research on using allelopathy in agriculture has focused on suppression of weed germination or emergence and the effect of weed allelochemicals on crop yields (Fischer et al., 1989; Hedge and Miller, 1990; Hegazy et al., 1990; Rizvi and Rizvi, 1992; Rice, 1995). The most sophisticated research program has been conducted in Mexico and was based on indigenous knowledge of farmers of the *chinampas* agroecosystems (Jimenez et al., 1983; Ortega et al., 1988; Anaya et al., 1987, 1992a, b). Farmers and researchers recognized that both crops, for example, watermelon (*Citrullus lanatus* [Thunb.] Matsum. and Nakai. cv. 'Peacock'), and weeds, for example, *Amaranthus leucocarpus* Wats. (Amaranthaceae), *Bidens pilosa* L. (Asteraceae), *Cassia jalapensis* L. (Fabaceae), *Echinochloa crus-galli* (L.) Beauv. (Poaceae), and *Rumex crispus* L. (Polygonaceae), suffered leaf damage from pollen of *Zea mays* var. *chalquiñocónico* (Jimenez et al., 1983; Ortega et al., 1988; Anaya et al., 1987, 1992a, b). This illustrates one problem with *in situ* use of pollen-allelopathic species: the pollen often affects desirable plants as well as weeds.

Implicitly this may be why no one has suggested the use of pollen allelochemicals from the noxious weed, *Parthenium hysterophorus*, as a bioherbicide; in fact, the presence of pollen allelopathy in such a prolific (and allergenic) weed has been viewed as a problem rather than an opportunity (Sukhada and Jayachandra, 1980a, b). Allelopathic pollen from the agamospermous *Hieracium* species does get transferred by bees to other weeds, albeit in amounts too low to cause pollen-allelopathic effects (McLernon et al., 1996; Murphy and Aarssen, 1995a). Since *Hieracium* species can be troublesome weeds and will transfer pollen to dicotyledonous crops (e.g., canola, *Brassica rapa* L. [Brassicaceae]), it is unlikely that farmers would welcome using live *Hieracium* as sources of bioherbicidal pollen (Murphy, 1993; Murphy and Aarssen, 1995a).

The wind-pollinated (and also allergenic), pollen-allelopathic *Phleum pratense* is a potential hazard to open-pollinated crops, especially anemophilous crops like *Zea mays*. However, current agricultural practice in North America (at least) does not encourage the use of *Phleum pratense* for forage and it generally has been relegated to abandoned pastures and roadsides, though it is widespread. Since these habitats are also reservoirs for many weeds that eventually can disperse seeds that can invade agricultural fields, the bioherbicidal action of pollen-allelopathic *Phleum pratense* might be beneficial. Again, this assumes the weeds are sympatric, flower concurrently with *Phleum pratense,* and that pollen allelopathy does not cause resistance or other selection pressures that promotes other weedy traits (e.g., vegetative apomixis via rhizomes). Anticipating a lack of resistance is probably wishful thinking. Following the discussion in Section 9.7 (evolutionary ecology of pollen allelopathy), I expect weeds exposed repeatedly to allelopathic pollen from live plants to develop some form of detoxification resistance (assuming that allelopathic pollen is a sufficiently strong selection pressure) just as they would to any mass-produced herbicide (bioherbicide or otherwise).

The resistance might be indirect in the sense that detoxification mechanisms may not be needed to escape pollen allelopathy (see Section 9.7). In an agricultural context, pollen allelopathy in living individuals of *Phleum pratense* might be effective at reducing seed set in annual, wind-pollinated weeds, as long as these weeds were physiologically constrained to flowering concurrently with *Phleum pratense*. Many weeds can easily reproduce earlier or later in the season and, from an agricultural perspective, anything that causes weeds to develop at a rate incompatible with current early season weed management programs (when it is most crucial for crop development) would not be desirable. If pollen allelopathy causes perennial weeds that are vegetative apomicts (e.g., *Elytrigia repens*) to produce more ramets (e.g., rhizomatous shoots), short-term weed management becomes more expensive regardless of any loss of genetic variation from reduced seed set.

The only species that might not pose a long-term threat in response to pollen allelopathy are partially cleistogamous weeds like *Danthonia compressa*. In these species, more self-fertilized cleistogenes would be produced. As long as these weeds are not also vegetatively apomictic, this could be an advantage because the loss of genetic variation means reduced adaptability to changing farming systems management techniques.

Pollen allelochemicals, whether they are used as commercial-scale bioherbicides or allowed to spread from live plants, may be natural but they cannot be used cavalierly or used as the sole method of weed management. They are merely a potential part of a larger integrated weed management/farming system (Swanton and Weise, 1991; Swanton and Murphy, 1996).

9.10 POLLEN ALLELOPATHY: RESEARCH BEYOND THE "EUREKA" STAGE

Feinsinger (1987) asked the question that turned out to be the foundation for my own research on pollen allelopathy: "Under what circumstances are pollen allelopathy or other effects of heterospecific pollen grains important factors in seed output? Only a systematic study will tell." More than 10 years later, the studies on *Hieracium* species, *Phleum pratense,* and *Zea mays* have begun to address pollen allelopathy in a systematic manner. Nonetheless, the evolutionary and agricultural implications are not yet well understood. It is clear that in terms of ecological interactions, allelopathic pollen can reduce seed set; it is less clear as to whether the pollen allelochemicals in all species discussed are similar and where and why pollen allelopathy originated.

Answering the question of where pollen allelopathy originated may be most crucial, but only as a first step in the continuing systematic study of pollen allelopathy. Asking "did it arise accidentally in *Phleum pratense* and Mexican varieties of *Zea mays* during the long history of agricultural breeding programs?" leads to other questions. Has pollen allelopathy persisted in these species simply because there is no fitness disadvantage? Since agricultural breeding accidents cannot explain pollen allelopathy in *Hieracium* species, why does it appear a genus unrelated to the Poaceae? Since species contribal with *Phleum pratense* are not susceptible to pollen allelochemicals, does this mean the detoxification mechanism is specific to this tribe? Does it mean that the mechanism is chemical? Will other susceptible species develop detoxification and/or avoidance of pollen allelopathy? Is pollen allelopathy a strong selection pressure or does it occur in species that occupy habitats too transient to make pollen allelopathy a selection pressure of any consequence?

These types of questions are not unique to pollen allelopathy or allelopathy in general. The true value of studying pollen allelopathy is probably not in the "eureka" reaction of finding an odd phenomenon or in the potential agricultural/commercial applications. Rather, to study pollen allelopathy is to try to answer the wider questions of evolutionary ecology. This is what makes it a topic of interest to more than just a small scientific community of allelopathy specialists.

ACKNOWLEDGMENTS

I thank Ana Luisa Anaya, Steve Gleissman, Kathleen Keating, Inderjit, and Darrell Miller for their constructive criticism and comments on previous drafts of this manuscript. Lonnie Aarssen played a major role in developing the theory of the evolutionary ecology of pollen allelopathy.

REFERENCES

Aarssen, L. W., Ecological combining ability and competitive combining ability in plants: towards a general evolutionary theory of coexistence in systems of competition, *Amer. Nat.*, 122, 707, 1983.

Aarssen, L. W., Competitive ability and species coexistence: a "plant's-eye" view, *Oikos*, 56, 386, 1989.

Aarssen, L. W., Causes and consequences of variation in competition ability in plant communities, *J. Veg. Sci.*, 3, 165, 1992.

Ågren, J. and Fagerstrom, T., Increased or decreased separation of flowering time? The joint effect of competition for space and pollination in plants, *Oikos*, 35, 161, 1980.

Anaya, A. L., Ramos, L., Hernandez, J., and Ortega, R. C., Allelopathy in Mexico, in *Allelochemicals: Role in Agriculture and Forestry (ACS Symposium Series 330)*, Waller, G. R., Ed., American Chemical Society Press, Washington, D.C., 1987, 89.

Anaya, A. L., Hernandez-Bautista, B. E., Jimenez-Estrada, M., and Velasco-Ibarra, L., Phenylacetic acid as a phytotoxic compound of corn pollen, *J. Chem. Ecol.*, 18, 897, 1992a.

Anaya, A. L., Ortega, R. C., and Rodriguez, V. N., Impact of allelopathy in the traditional management of agroecosystems in Mexico, *in Allelopathy: Basic and Applied Aspects*, Rizvi, S. J. H. and Rizvi, V., Eds., Chapman and Hall, London, UK, 1992b, 271.

Baraniuk, J. N., Bolick, M., and Buckely, C. E., III, Pollen grain chromatography: a novel method for separation of pollen wall solutes, *Ann. Bot.*, 66, 321, 1990.

Char, M. B. S., Pollen allelopathy, *Naturwiss.*, 64, 489, 1977.

Chatigny, M. A., Dimmick, R. L., and Mason, C. J., Atmospheric transport, *in Aerobiology: The Ecological Systems Approach*, Edmonds, R. L., Ed., Dowden, Hutchinson, & Ross, Stroudsburg, Germany, 1979, 85.

Dornbos, D. L., Jr. and Spencer, G. F., Natural products phytotoxicity: a bioassay suitable for small quantities of slightly water-soluble compounds, *J. Chem. Ecol.*, 16, 339, 1990.

Downum, K. R., Swain, L. A., and Faleiro, L. J., Influence of light on plant allelochemicals: a synergistic defense in higher plants, *Arch. Insect Biochem. Physiol.*, 17, 201, 1991.

Duke, S. O. and Lydon, J., Herbicides from natural compounds, *Weed Technol.*, 1, 122, 1988.

Einhellig, F. A. and Leather, G. R., Potentials for exploiting allelopathy to enhance crop production, *J. Chem. Ecol.*, 14, 1829, 1988.

Eisikowitch, D., Kevan, P. G., and Lachance, M-A., The nectar-inhibiting yeasts and their effect on pollen germination in common milkweed, *Asclepias syriaca, Isr. J. Bot.*, 39, 217, 1990a.

Eisikowitch, D., Lachance, M-A., Kevan, P. G., Wills, S., and Collins-Thompson, D. L., The effect of the natural assemblage of microorganisms and selected strains of the yeast *Metschinkowia reukauffi* in controlling the germination of pollen of the common milkweed, *Asclepias syriaca, Can. J. Bot.* 68, 1163, 1990b.

Evans, M. W., The flowering habits of timothy, *J. Amer. Soc. Agron.*, 8, 299, 1916.

Feinsinger, P., Effects of plant species on each other's pollination: is community structure influenced?, *Trends Ecol. Evol.*, 2, 123, 1987.

Fischer, N. H., Weidenhamer, J. D., and Bradow, J. M., Inhibition and promotion of germination by several sesquiterpenes, *J. Chem. Ecol.*, 15, 1785, 1989.

Ganeshaiah, K. N. and Shaanker, R. U., Regulation of seed number and female incitation of mate competition by a pH-dependent proteinaceous inhibitor of pollen in *Leucaena leucocephala, Oecologia*, 75, 110, 1988.

Gonzalez, A. G., Lactuceae - chemical review, in *The Biology and Chemistry of the Compositae*, Haywood, V. H., Harborne, J. B., and Turner, B. L., Eds., Academic Press, New York, 1977, 1081.

Harborne, J. B., *Phytochemical Methods: A Guide to Modern Techniques of Plant Analysis,* 2nd ed., Chapman and Hall, New York, 1984, 288.

Harborne, J. B., Constraints on the evolution of biochemical pathways, *Biol. J. Linn. Soc.*, 39, 135, 1990.

Hedge, R. S. and Miller, D. A., Allelopathy and autotoxicity in alfalfa: characterization and effects of preceding crops and residue incorporation, *Crop Sci.,* 30, 1255, 1990.

Hegazy, A. K., Mansour, K. S., and Abdel-Hady, N. F., Allelopathic and autotoxic effects of *Anastatica hierochuntica* L., *J. Chem. Ecol.,* 16, 2183, 1990.

Heisey, R. M., Allelopathic and herbicidal effects of extracts from tree of heaven (*Ailanthus altissima*), *Amer. J. Bot.,* 77, 662, 1990.

Heslop-Harrison, J., Pollen walls as adaptive systems, *Ann. Missouri Bot. Gard.,* 66, 813, 1979a.

Heslop-Harrison, J., Aspects of the structure, cytochemistry, and germination of the pollen of rye (*Secale cereale* L.), *Ann. Bot.* (Suppl. 1), 44, 1, 1979b.

Heslop-Harrison, J. and Heslop-Harrison, Y., Surfaces and secretions in the pollen-stigma interaction: a brief review, *J. Cell Sci.* (Suppl.) 2, 287, 1985.

Heslop-Harrison, J. and Heslop-Harrison, Y., Some permeability properties of angiosperm pollen grains, pollen tubes, and generative cells, *Sex. Plant Reprod.,* 1, 65, 1988.

Heslop-Harrison, J. and Heslop-Harrison, Y., Germination of monocolpate angiosperm pollen: effects of inhibitory factors and the Ca^{2+}-channel blocker, nifedipine, *Ann. Bot.,* 69, 395, 1992.

Heslop-Harrison, Y., Reger, B. J., and Heslop-Harrison, J., The pollen-stigma interaction in the grasses. 6. The stigma ("silk") of *Zea mays* L. as host to the pollens of *Sorghum bicolor* (L.) Moench. and *Pennisetum americanum* (L.) Leeke, *Acta Bot. Neer.,* 33, 205, 1984.

Hoekstra, F. A. and Barten, J. H. M., Do anti-oxidants and local anaesthetics extend pollen longevity during dry storage?, in *Biotechnology and Ecology of Pollen,* Mulcahy, D. L., Mulcahy, G. B., and Ottaviano, E., Eds., Springer-Verlag, New York, 1986, 339.

Hoekstra, F. A., Crowe, L. M., and Crowe, J. H., Differential desiccation sensitivity of corn and *Pennisetum* pollen linked to their sucrose content, *Plant Cell Environ.,* 12, 83, 1989.

Hoekstra, F. A., Crowe, J. H., and Crowe, L. M., Effect of sucrose on phase behavior of membranes in intact pollen of *Typha latifolia* L., as measured with Fourier transform infrared spectroscopy, *Plant Physiol.,* 97, 1073, 1991.

Hoekstra, F. A., Crowe, J. H., and Crowe, L. M., Germination and ion leakage are linked with phase transitions of membrane lipids during imbibition of *Typha latifolia* pollen, *Physiol. Plant,* 84, 29, 1992a.

Hoekstra, F. A., Crowe, J. H., Crowe, L. M., van Roekel, T., and Vermeer, E., Do phospholipids and sucrose determine membrane phase transitions in dehydrating pollen species?, *Plant Cell Environ.,* 15, 601, 1992b.

Hormoza, J. I. and Herrero, M., Pollen selection, *Theor. Appl. Genet.,* 83, 663, 1992.

Jackson, J. F. and Kamboj, R. K., Control of protein release from germinating pollen, in *Biotechnology and Ecology of Pollen,* Mulcahy, D. L., Mulcahy, G. B., and Ottaviano, E., Eds., Springer-Verlag, New York, 1986, 369.

Jimenez, J. J., Schultz, K., Anaya, A. L., Hernandez, J., and Espejo, O., Allelopathic potential of corn pollen, *J. Chem. Ecol.,* 9, 1011, 1983.

Jones, C. G. and Firn, R. D., On the evolution of plant secondary chemistry, *Phil. Trans. R. Soc. Lond., Ser. B,* 333, 273, 1991.

Kim, S. K., Lagerstedt, H. B., and Daley, L. S., Germination responses of filbert pollen to pH, temperature, glucose, fructose, and sucrose, *Hort. Sci.,* 20, 944, 1985.

Lawrence, J. G., Colwell, A., and Sexton, O. J., The ecological impact of allelopathy in *Ailanthus altissima* (Simaroubaceae), *Amer. J. Bot.,* 78, 948, 1991.

Leduc, N., Monnier, M., and Douglas, G. C., Germination of trinucleated pollen: formulation of a new medium for *Capsella bursa-pastoris, Sex. Plant Reprod.,* 3, 228, 1990.

Linskens, H.-F. and Schrauwen, J., The release of free amino acids from germinating pollen, *Acta Bot. Neer.,* 18, 605, 1969.

Lovett, J. V., Changing perceptions of allelopathy and biological control, *Biol. Agric. Hort.,* 8, 89, 1991.

Mabry, T. J. and Bohlmann, F., Summary of the chemistry of the Compositae, in *The Biology and Chemistry of the Compositae,* Haywood, V. H., Harborne, J. B., and Turner, B. L., Eds., Academic Press, New York, 1977, 1097.

McLernon, S. M., Murphy, S. D., and Aarssen, L. W., Heterospecific pollen transfer in sympatric grassland species, *Amer. J. Bot.,* 83, 1168, 1996.

Mulcahy, G. B. and Mulcahy, D. L., The effect of supplemented media on the growth *in vitro* of bi- and trinucleate pollen, *Plant Sci.,* 55, 213, 1988.

Murphy, S. D., The determination of the allelopathic potential of pollen and nectar, in *Modern Methods of Plant Analysis,* Linskens, H.-F. and Jackson, J. F., Eds., Springer-Verlag, New York, 1992, 333.

Murphy, S. D., *The Occurrence and Consequences of Pollen Allelopathy in Phleum and Hieracium.* Ph.D. Thesis, Queen's University, Kingston, Canada, 1993.

Murphy, S. D. and Aarssen, L. W., Pollen allelopathy among sympatric grassland species: *in vitro* evidence in *Phleum pratense* L., *New Phytol.,* 112, 295, 1989.

Murphy, S. D., and Aarssen, L. W., *In vitro* allelopathic effects of pollen from three *Hieracium* species (Asteraceae) and pollen transfer to sympatric Fabaceae, *Amer. J. Bot.,* 82, 37, 1995a.

Murphy, S. D. and Aarssen, L. W., Allelopathic pollen extract from *Phleum pratense* L. (Poaceae) reduces germination, *in vitro*, of pollen in sympatric species, *Int. J. Plant Sci.,* 156, 425, 1995b.

Murphy, S. D. and Aarssen, L. W., Allelopathic pollen extract from *Phleum pratense* L. (Poaceae) reduces seed set in sympatric species, *Int. J. Plant Sci.,* 156, 435, 1995c.

Murphy, S. D. and Aarssen, L. W., Allelopathic pollen of *Phleum pratense* reduces seed set in *Elytrigia repens* in the field, *Can. J. Bot.,* 73, 1417, 1995d.

Murphy, S. D. and Aarssen, L. W., Partial cleistogamy limits reduction in seed set in *Danthonia compressa* (Poaceae) by allelopathic pollen of *Phleum pratense* (Poaceae), *Écoscience,* 3, 205, 1996.

Ortega R. C., Anaya, A. L., and Ramos, L., Effects of allelopathic compounds of corn pollen on respiration and cell division of watermelon, *J. Chem. Ecol.,* 14, 71, 1988.

Paterniani, E. and Stort, A. C., Effective maize pollen dispersal in the field, *Euphytica,* 23, 129, 1974.

Pederson, G. A., White clover seed germination in agar containing tall fescue leaf extracts, *Crop Sci.,* 26, 1248, 1986.

Putnam, A. R., Allelochemicals from plants as herbicides, *Weed Technol.,* 1, 510, 1988a.

Putnam, A. R., Allelopathy: problems and opportunities in weed management, in *Weed Management in Agroecosystems: Ecological Approaches,* Altieri, M. A. and Liebman, M., Eds., CRC Press, Boca Raton, FL, 1988b, 77.

Raynor, G. S., Ogden, E. C., and Hayes, J. V., Dispersion and deposition of timothy pollen from experimental sources, *Agric. Meteorol.,* 9, 347, 1972.

Reger, B. J., Chaubal, R., and Pressey, R., Chemotropic responses by pearl millet pollen tubes, *Sex. Plant Reprod.,* 5, 47, 1992.

Rice, E. L., *Allelopathy,* 2nd ed., Academic Press, New York, 1984, 422.

Rice, E. L., *Biological Control of Weeds and Plant Diseases: Advances in Applied Allelopathy,* University of Oklahoma Press, Norman, OK, 1995, 439.

Rizvi, S. J. H. and Rizvi, V., *Allelopathy: Basic and Applied Aspects,* Chapman and Hall, London, UK, 1993, 480.

Searcy, K. B. and MacNair, M. R., Differential seed production in *Mimulus guttatus* in response to increasing concentrations of copper in the pistil by pollen from copper tolerant and sensitive sources, *Evolution,* 44, 1424, 1990.

Sedgley, M. and Blessing, M. A., Foreign pollination of watermelon (*Citrullus lanatus* [Thunb.] Matsum and Nakai), *Botan. Gaz.,* 143, 210, 1982.

Shivanna, K. R., Linskens, H. F., and Cresti, M., Pollen viability and pollen vigor, *Theoret. Appl. Genet.,* 81, 38, 1991.

Sidhu, R. J. K. and Malik, C. P., Metabolic role of boron in germinating pollen and growing pollen tubes, in *Biotechnology and Ecology of Pollen,* Mulcahy, D. L., Mulcahy, G. B., and Ottaviano, E., Eds., Springer-Verlag, New York, 1986, 373.

Sukhada, K. D. and Jayachandra, Pollen allelopathy—a new phenomenon, *New Phytol.,* 84, 739, 1980a.

Sukhada, K. D. and Jayachandra, Allelopathic effects of *Parthenium hysterophorus* L., Part IV, Identification of inhibitors, *Plant Soil,* 55, 67, 1980b.

Swanton, C. J. and Murphy, S. D., Weed science beyond the weeds: the role of integrated weed management (IWM) in agroecosystem health, *Weed Sci.,* 44, 437, 1996.

Swanton, C. J. and Weise, S. F., Integrated weed management: the rationale and approach, *Weed Technol.,* 5, 657, 1991.

Taylor, D. R. and Aarssen, L. W., Complex competitive relationships among genotypes of three perennial grasses: implications for species coexistence, *Amer. Nat.,* 136, 305, 1990.

Tepedino, V., Knapp, A. K., Eickwort, G. C., and Ferguson, D. C., Death camas (*Zigadenus nuttallii*) in Kansas: pollen collectors and a florivore, *J. Kansas Entomol. Soc.,* 62, 411, 1989.

Thomson, J. D., Andrews, B. J., and Plowright, R. C., The effect of a foreign pollen on ovule development in *Diervilla lonicera* (Caprifoliaceae), *New Phytol.,* 90, 777, 1982.

Van Aelst, A. C. and van Went, J. L., Effects of anoxia on pollen tube growth and tube wall formation of *Impatiens glandulifera, Sex. Plant Reprod.,* 2, 85, 1989.

Waller, D. M., Plant morphology and reproduction, in *Plant Reproductive Ecology: Patterns and Processes*, Lovett-Doust, J. and Lovett-Doust, L., Eds., Oxford University Press, Toronto, Canada, 1988, 203.

10 Allelopathic Bacteria

Oz Barazani and Jacob Friedman

CONTENTS

10.1 ABSTRACT

The impact of allelopathic, nonpathogenic bacteria on plant growth in natural and agricultural ecosystems has been discussed. In some natural ecosystems, evidence supports the view that in the vicinity of some allelopathically active perennials (e.g., *Adenostoma fasciculatum* Hook. & Arn. in California), in addition to allelochemicals leached from the shrub's canopy, accumulation of phytotoxic bacteria or other allelopathic microorganisms amplify retardation of annuals. In agricultural ecosystems where a crop is grown successively, the resulting yield decline cannot always be restored by application of minerals, but yield can often be improved by soil disinfestation. Transfer of soils from the areas where crop suppression had been recorded into an unaffected area, induced crop retardation without readily apparent symptoms of plant disease.

The allelopathic effect may occur directly, through release of allelochemicals by a bacterium that affects susceptible plant(s) or indirectly, through suppression of an essential symbiont. The process is affected by nutritional and other environmental conditions that control bacterial density and the rate of production of allelochemicals. Among these, water stress was suggested to govern the susceptibility of downy brome (*Bromus tectorum* L.) to deleterious rhizobacteria.

Allelopathic nonpathogenic bacteria are found in a wide range of genera and secrete a diverse group of plant growth-mediating allelochemicals. Although a limited number of plant growth-promoting bacterial allelochemicals have been identified, a considerable number of highly diversified growth inhibitors have been isolated and characterized. Efforts to use naturally produced allelochemicals as plant growth-regulating agents in agriculture have yielded two commercial herbicides, phosphinothricin, a product of *Streptomyces viridochromogenes,* and bialaphos from *S. hygroscopicus.*

Although many species of allelopathic bacteria are not plant specific, some do exhibit specificity. For example, dicotyledonous plants were more susceptible to *Pseudomonas putida* than were monocotyledons. Differential susceptibility of higher plants was noted also in much lower taxonomical categories, at the sub-species level, such as in different cultivars of wheat or of lettuce.

Therefore, when test plants are used to evaluate allelopathy, final evaluation must include those species that are suppressed in nature.

Release of allelochemicals from plant residues in plots of continuous crop cultivation or from aromatic shrubs may induce development of specific allelopathic bacteria. Both the rate by which a bacterium gains from its allelopathic activity through utilizing plant excretions, and the reasons for developing allelopathic bacteria where the same crop is grown repeatedly on the same soil, are important goals for further research.

10.2 INTRODUCTION

Interactions between plants and allelopathic nonpathogenic microorganisms are mainly concentrated in the root zone, on the root surface including root hairs (the rhizoplane), or within the layer of soil that is influenced by the root (the rhizosphere). Root colonizing bacteria are usually categorized as plant growth-promoting rhizobacteria (PGPR) (Kloepper and Schroth, 1978), or deleterious rhizobacteria (DRB) (Suslow and Schroth, 1982). Allelopathy is manifested when bacteria release allelochemicals that influence plant growth directly or indirectly by affecting growth of plant symbionts or parasites. Although a number of reports suggest that nonpathogenic bacteria affect the growth of higher plants, their allelopathic nature has not been elucidated.

Bacteria that release allelochemicals affecting higher plants may also compete with plants for essential nutrients. Direct and indirect effects of saprophytic, nonpathogenic bacteria, or of their allelochemicals on higher plants, and their ecological implications in natural or in agricultural ecosystems are reviewed.

10.3 ALLELOPATHIC BACTERIA IN NATURAL AND
AGRICULTURAL ECOSYSTEMS

Allelopathic effects of soil-borne microorganisms in a natural habitat were first studied by Kaminsky (1981). Inhibition of annuals growing close to *Adenostoma fasciculatum* (Rosaceae), an indigenous Californian shrub, has been attributed to soil-borne microorganisms and not to shoots leachates. When surface soil from under the canopy of the shrub was transferred to a control site in an open field, phytotoxicity was reproduced. This was not the case when soil from control sites was placed under the shrub. Fumigation of soil from under the canopy (UC) of *A. fasciculatum* reduced phytotoxicity, but the inhibitory effect was persisted after passing the UC soil extract through a 0.2 μm filter. The involvement of bacteria in the allelopathic phenomenon near some aromatic shrubs has also been studied in Israel, where suppression of annuals around *Coridothymus capitatus* L. (Labiatae) was observed on a sandstone formation (Katz et al., 1987). The density of annuals, adjacent to the shrub (Figure 10.1) decreased 16-fold compared with annuals 60 to 80 cm from the shrub canopy. Germination and development of planted annuals (e.g., *Plantago psyllium* L. and *Erucaria hispanica* (L.) Druce), were reduced 45 percent adjacent to the shrubs, compared with those planted near control plants. Isolates of some soil-borne actinomycetes recovered from annual-free sites and cultured on malt yeast-agar inhibited germination of test plants *Lactuca sativa* L. and *Anastatica hierochuntica* L. The addition of fresh shoots of *Coridothymus capitatus* to soil increased actinomycetes 9.5- and 36.2-fold in soils from nonallelopathic and allelopathic shrub, respectively (Katz et al., 1987). When essential oils of *Satureja thymbra* L. or of *Rosmarinus officinalis* L. were applied to soil, bacterial populations increased, suggesting that bacteria use essential oils as carbon and energy sources (Vokou et al., 1984). Similar data were obtained from the upper soil crust on north facing slopes, in the Negev Desert, Israel, near the aromatic shrub *Artemisia herba-alba* Asso. (white wormwood) (Chayen, 1991). Suppression of annuals around the canopies of aromatic shrubs may be a result of synergistic interactions of volatile essential oils and elevated densities of allelopathic, soil-borne bacteria.

FIGURE 10.1 Horse-shoe form of an annual-free belt, in the vicinity of an 'aggressive' shrub of *Coridothymus capitatus* L. (Labiatae) (after Katz et al., 1987).

Indirect allelopathic effect of actinomycetes suppressing mycorrhizal fungi in disinfested soil has been described (Krishna et al., 1982; Friedman et al., 1989). Inoculation of *Eleusine coracana* L. Gaertn. (Poaceae) with either *Streptomyces cinnamomeous,* or *Glomus fasciculatus* in a sterile phosphorus-deficient soil, improved the growth of the grass. When both species of microorganisms were applied simultaneously, or within a period of 2 weeks, these two microorganisms interacted antagonistically: *S. cinnamomeous* reduced infection and spore production by *G. fasciculatus,* while *G. fasciculatus* reduced the density of *S. cinnamomeous.* The authors suggested that inhibition of *G. fasciculatus* was induced by the allelochemicals cinnamycin and duramycin released by *S. cinnamomeous* (Krishna et al., 1982). A similar allelopathic effect was reported in clear-cut areas of Douglas fir (*Pseudotsuga menziesii* (Mirb.) Franco) in southern Oregon. Seedlings of Douglas fir planted after clear-cut on a coarse sandy skeletal soil failed to establish after a few repeated attempts of replanting. Field experiments with irrigation and addition of minerals did not improve the establishment (D. Perry, personal communication). It was assumed that actinomycetes were involved in the failure of Douglas fir regeneration. Four percent of the bacterial isolates from the clear-cut area inhibited growth of two common ectomycorrhizal fungi, *Laccaria laccata* and *Hebeloma crustuliniforme,* whereas only 2.6 percent of the isolates from the forest suppressed these fungi. In addition, isolates of actinomycetes from a clear-cut area suppressed germination of *Lactuca sativa* and *Anastatica hierochuntica.* Allelopathic actinomycetes were also recovered from the forest, but isolates from the clear-cut area had five times the phytotoxic effect as those from the forest. This suggests a combined effect of direct as well as of indirect allelopathy (Friedman et al., 1989).

Yield reductions following continuous cultivation of a single crop species on the same area have been recorded in agriculture, and have often been related to the accumulation of nonpathogenic, deleterious rhizobacteria (Schippers et al., 1987). Crop residues were suggested to support such microbial populations (Fredrickson et al., 1987). This was observed in several crops, for example, barley (*Hordeum vulgare* L.) (Alström, 1992; Olsson and Alström, 1996), corn (*Zea mays* L.) (Turco et al., 1990), and potatoes (*Solanum tuberosum* L.) (Bakker and Schippers, 1987), as well as in replanted deciduous apple orchards (Catska et al., 1982) and vines (Waschkies et al., 1994). When barley was grown continuously over about 30 years near Uppsala, Sweden, yields were reduced with time, compared with a control field with crop rotation (Alström, 1992). When infected by deleterious *Pseudomonas fluorescens,* the dry weight and the development of the second leaf of barley in crop rotation soil were reduced, but application of the bacterium to continuous crop soil did not exhibit any effect on barley. Tests to demonstrate the allelopathic effect of microbial populations were not always conclusive. By mixing crop rotation soil with continuous crop soil at a 20 percent level, Alström (1992) demonstrated that the inhibitory effect of the continuous crop soil was transferable. Nevertheless, inoculation of axenic barley seeds with the total microflora from soil of continuous barley or from soil of the control site, did not cause differences in root growth. Application of streptomycin sulphate (0 to 100 ppm) to the soil through the nutrient solution neutralized the phytotoxic effect in the continuously cultivated plots. Also, introducing microorganisms from plots of the continuous crop soil into the crop rotation soil (control) reproduced the phytotoxic effect, suggesting that suppression was maintained by deleterious rhizobacteria (Olsson and Alström, 1996). Yield decline was also monitored in continuously grown corn (for 12 successive years), on tile-drained Chalmers silty loam soil at the Purdue Agronomy farm, Indiana, United States. Since corn lacked any indications of plant disease, and fumigation of continuous corn plots resulted in yield increases, similar to the yield obtained in plots with corn/soybean rotation, the reduction was related to deleterious bacteria (Turco et al., 1990).

Potatoes grown in the Netherlands every second year for 12 years in the same area, yielded 30 percent less than potatoes grown every sixth year (Schippers et al., 1987). No known disease symptoms were observed. Since approximately 50 percent of the fluorescent pseudomonads isolated from the rhizosphere produced HCN *in vitro* (Table 10.1), yield reduction was assumed to be the result of HCN, although casual relationships were not studied (Bakker and Schippers, 1987; Schippers et al., 1987; Schippers et al., 1991).

In soil exhibiting apple replant disease, proliferation of fluorescent pseudomonads was enhanced after replanting of apple seedlings (Catska et al., 1982). Similarly, when roots of apple seedlings grown in replant disease soil were compared with those recovered in areas of unaffected apples, high densities of actinomycetes were found. This was also observed in clay loam, silt loam, and loam soils (Westcott and Beer, 1986). Since these findings were not associated with any known symptoms of plant disease, saprophytic nonpathogenic actinomycetes were assumed to be the reason for replant disease.

An indirect allelopathic effect was also reported in a replanted site in a nursery of cuttings of grape vine (*Vitis* sp.). Here, failure of establishment of the indigenous mycorrhizal fungus *Glomus mosseae* was related to the number of fluorescent pseudomonads, which increased over time. Since an increase in the population of fluorescent pseudomonads preceded the reduction of root and shoot weight, it was suggested that pseudomonads were responsible for the replant disease of grapevine (Waschkies et al., 1994).

In Israel, rose flower yield often declined when cultivated on stocks of *Rosa indica* L. under plastic cover on volcanic scoria (tuff) after replanting, or long cultivation. No evidence for the involvement of plant pathogens was observed; however, sterilization of the tuff improved flower yield. Nevertheless, in plants on control unused tuff, no such effect was observed. Densities of phytotoxic bacteria and actinomycetes in the replant disease tuff were 5.3 and 10 times higher compared with those populations in unused tuff (Stern, 1992).

The ecological advantage of allelopathic bacteria is perhaps obtained by the release of allelochemicals that enhance root exudation. The release of deleterious metabolites by bacteria may enhance the permeability of the plasmalemma, or damage plasmalemma protein function in the root. However, the release of root growth promoting substances enhances root exudation through stimulation of root growth/activity (Meharg and Killham, 1995), result in an increased root surface area for additional microbial colonization.

10.4 METHODOLOGY

When an allelopathic effect cannot be solely related to the presence of an allelopathically aggressive plant or plant residues, soil-borne microorganisms have been isolated and studied. Two major aspects have been stressed: (1) the impact of bacteria on development and establishment of plant communities (c.f., Kaminsky, 1981; Katz et al., 1987; Friedman et al., 1989), and (2) allelopathic bacteria as a source of allelochemicals (c.f., Mishra et al., 1987, 1988; Kennedy et al., 1991). Considerable effort has been focused on the isolation of allelochemicals with a potential for commercial use, mainly as weed killers. The search for allelopathic bacteria has often been guided by ecological parameters, for example, allelopathically active microorganisms have been isolated from areas where vegetation has been suppressed in natural or agricultural ecosystems.

In the search for allelopathic bacteria, many isolates have been tested under controlled conditions. To determine the allelopathic influence of a bacterium on the growth of a higher plant, the use of agar assays have been a common approach. Tests conducted with no contact between the bacteria and the tested plant have ensured that competition and parasitic or symbiotic inter-relationships are eliminated (Katz et al., 1987; Friedman et al., 1989). Other tests, with direct contact between the test microorganisms and the higher plant in Petri dishes or test tubes, have been reported (Elliott and Lynch, 1984; Fredrickson and Elliott, 1985; Turco et al., 1990; Kennedy et al., 1991; Gealy et al., 1996).

Young seedlings rely on their own storage materials, excluding water supply, for growth. Therefore, when allowed to interact with bacteria the possibility of competition for nutrients is nullified. Although screening tests on agar for evaluating the effects of allelopathic bacteria seem useful, they are limited to only short periods (48 to 72 h). Clearly, other methods to test bacterial effects over longer periods need to be used. Bacterial effects over long periods are usually tested in pots in vermiculite or soil. Bacteria are applied by inoculation of seeds, foliar spray, or addition to the soil. For the evaluation of the effect of bacteria on the development of sugar beet (*Beta vulgaris* L.), surface sterilized seeds were soaked 10 min in a bacterial suspension before placement in sterile growth pouches (Loper and Schroth, 1986). After 10 to 12 days, hypocotyl length, primary root length, branching pattern, and general root morphology were recorded. A similar method has been used for determining the effect of *Pseudomonas putida* GR12-2 on root elongation of canola (*Brassica campestris* L.) (Glick et al., 1994). Similarly, *Pseudomonas fluorescens* was applied to cuttings of sour cherry (*Prunus cerasus* L.) or black-currant (*Ribes nigrum* L.) (Dubeikovsky et al., 1993).

To fully evaluate the allelopathic effect, tests should be conducted under field conditions. However, under natural conditions, various environmental parameters such as adsorption, leaching, or oxygenation may hinder the allelopathic effect and its reproducibility. Plant growth may be affected by the method bacteria are employed. The application of deleterious bacteria directly to the soil suppressed downy brome (*Bromus tectorum* L.) (Kennedy et al., 1991). Seed inoculation with two isolates of *Pseudomonas cepacia* and two isolates of *P. putida* promoted grain yield of winter wheat (*Triticum aestivum* L.) by only about 10 percent. Nevertheless, when those bacteria were applied to the soil, yield of winter wheat increased by 17 to 40 percent (de Freitas and Germida, 1992). Under field conditions, seed inoculation does not always lead to significant effects on plant growth, probably due to the soil type, availability of nutrients, density of the bacterial inoculum, and the nature of the competing soil microflora (Kapulnik, 1991).

10.5 BACTERIAL ALLELOCHEMICALS

The increasing concern regarding synthetic chemicals in agriculture has promoted the interest in indigenous plant growth-mediating microorganisms. Thus, bacteria and their allelochemicals have been evaluated as biocontrol agents. The fact that natural products are often decomposed rapidly in nature (biodegradation) and do not accumulate in soil or contaminate the water table, has promoted

TABLE 10.1
Allelopathic Bacteria (Including Actinomycetes), and Allelochemicals and Susceptible Higher Plants

Allelopathic Bacteria	Allelochemical(s)***	Susceptible Plant Species		Dose/Conc.	Application** Alle.	Bac.	Effect(%)*	Reference
a. Deleterious bacteria								
Flavobacterium sp.; *Enterobacter taylorae*	IAA (I)	Sugar beet Bindweed	*Beta vulgaris* *Convolvulus arvensis*	1.3 µg/ml 72.2 µg/ml	p.d.	g.p. p.d.	-30 -77.5	Loper and Schroth, 1986; Sarwar and Kremer, 1995
Streptomyces hygroscopicus	Geldanamycin (II); Nigericin (III)	Garden cress	*Lepidium sativum*	1–2 ppm	p.d.		-50	Heisy and Putnam, 1986
Pseudomonas sp.	HCN	Potato	*Solanum tuberosum*	0.135 µg/ml		n.s.	-40	Bakker and Schippers, 1987
Streptomyces sp.	Cycloheximide (V)	Garden cress	*Lepidium sativum*	1 µg/ml	p.d		-50	Heisy et al., 1988
Streptomyces saganonensis	Herbicidin (VI)	Purslane Radish	*Portulaca oleracea* *Raphanus sativus* *Polygonum* spp. *Amaranthus* spp. *Commelina communis*	30–300 ppm	n.m.		n.m.	Cutler, 1988
Streptomyces hygroscopicus	Hydantocidin (IV)	Tomato Barnyardgrass Crabgrass Foxtail Cocklebur Wild mustard Nutsedge Johnsongrass Jimsonweed	*Lycopersicon esculentum* *Echinochloa crus-galli* *Digitaria ischaemum* *Setaria italica* *Xanthium pennsylvanicum* *Sinapis arvensis* *Cyperus rotundus* *Sorghum halepense* *Datura stramonium*	500 ppm	f.s.		-100	Nakajima et al., 1991
Bradyrhizobium japonicum	Rhizobitoxine (VII); IAA (I)			n.m	n.m.		n.m.	Minamisawa and Fukai, 1991
Thermoactinomycete sp.	5'-Deoxyguanosine (VIII)	Duckweed	*Lemna minor*	100 µg/ml	n.m.		n.m.	
Streptomyces sp.	Coaristeromycin (IX)	Barnyardgrass Johnsongrass	*Echinochloa crus-galli* *Sorghum halepense*	600 mg/m²	n.m.		n.m.	
Actinoplane sp.	Adenine 9-β-D--arabinofuranoside (Ara-A) (X)		*Arabidopsis thaliana*	25 µg/ml	n.m.		n.m.	Isaac et al., 1991
Streptomyces sp.	5'-Deoxytoyocamycin (XI)	Duckweed	*Lemna minor*	10 µg/ml	n.m.		bleaching	

Organism	Compound	Scientific name	Common name	Dose	Mode of application	Effect on plant growth	Reference
Unidentified	Coformycin (XII)	Sorghum halepense	Johnsongrass	600 mg/m²	n.m.	n.m.	
Streptomyces chromofuscus	Herboxidiene (XIV)	Echinochloa crus-galli Digitaria ischaemum Brassica juncea Zea mays Brassica napus Fagopyrum sagittatum	Barnyardgrass Crabgrass Indian mustard Maize Rape Buckwheat	6.9 mg/m²	f.s.	-100	Miller-Wideman et al., 1992
Streptomyces spp.	Blasticidin; 5-hydroxylmethyl- -blasticidin S (XIII)	Echinochloa crus-galli Bromus inermis Lolium multiflorum Vigna sinensis Stellaria media	Barnyardgrass Smooth broom Italian ryegrass Common cowpea Common chickweed	100 mg/m²	p.e.	-10 to -15	Scacchi et al., 1992
		Ipomea purpurea Convolvulus arvensis Veronica persica	Bush morning glory Field bindweed Persian speedwell	100 mg/m²	p.e.	-64 to -98	
	Phthoxazolin A (XV)	Raphanus sativus	Radish	25 µg/test tube	t.t.	-90	Tanaka et al., 1993
	4-chlorothreonine (XVI)	Raphanus sativus Sorghum bicolor	Radish Sorghum	<30 µg/tube	t.t.		Yoshida et al., 1994
	Phthoxazolin B (XVII)	Raphanus sativus Sorghum bicolor	Radish Sorghum	63 µg/test tube	t.t.	-90 -70	
	Phthoxazolin C	Raphanus sativus Sorghum bicolor	Radish Sorghum	63 µg/test tube	t.t.	-40 -40	Shiomi et al., 1995
	Phthoxazolin D	Raphanus sativus Sorghum bicolor	Radish Sorghum	250 µg/test tube	t.t.	-40 -40	

b. Plant growth promoting rhizobacteria

Organism	Compound	Scientific name	Common name	Dose	Mode of application	Effect on plant growth	Reference
Pseudomonas fluorescens	IAA (I)	Ribes nigrum	Black-currant	>30 µg/ml	p.e.	57	Dubeikovsky et al., 1993
Pseudomonas putida	Succinic acid Lactic acid	Asparagus officinalis	Garden asparagus	(1:1) 10 ppm	p.e.	30-40	Yoshikawa et al., 1993
	ACC deaminase	Brassica campestris	Chinese cabbage		g.p.		Glick et al., 1994

*The effect on plant growth (-represent inhibition)
**Mode of application: Alle.-Allelochemicals, Bac.-Bacteria; f.s.-Foliar spray, g.p.-Growth pouch,
n.m.-Not mentioned, p.d.-Petri dishes, p.e.-Tested in pots, t.t.-Test tubes
***For chemical structure see Figure 10.2.

FIGURE 10.2 Plant growth-mediating allelochemicals from bacterial origin (*c.f.*, Table 10.1).

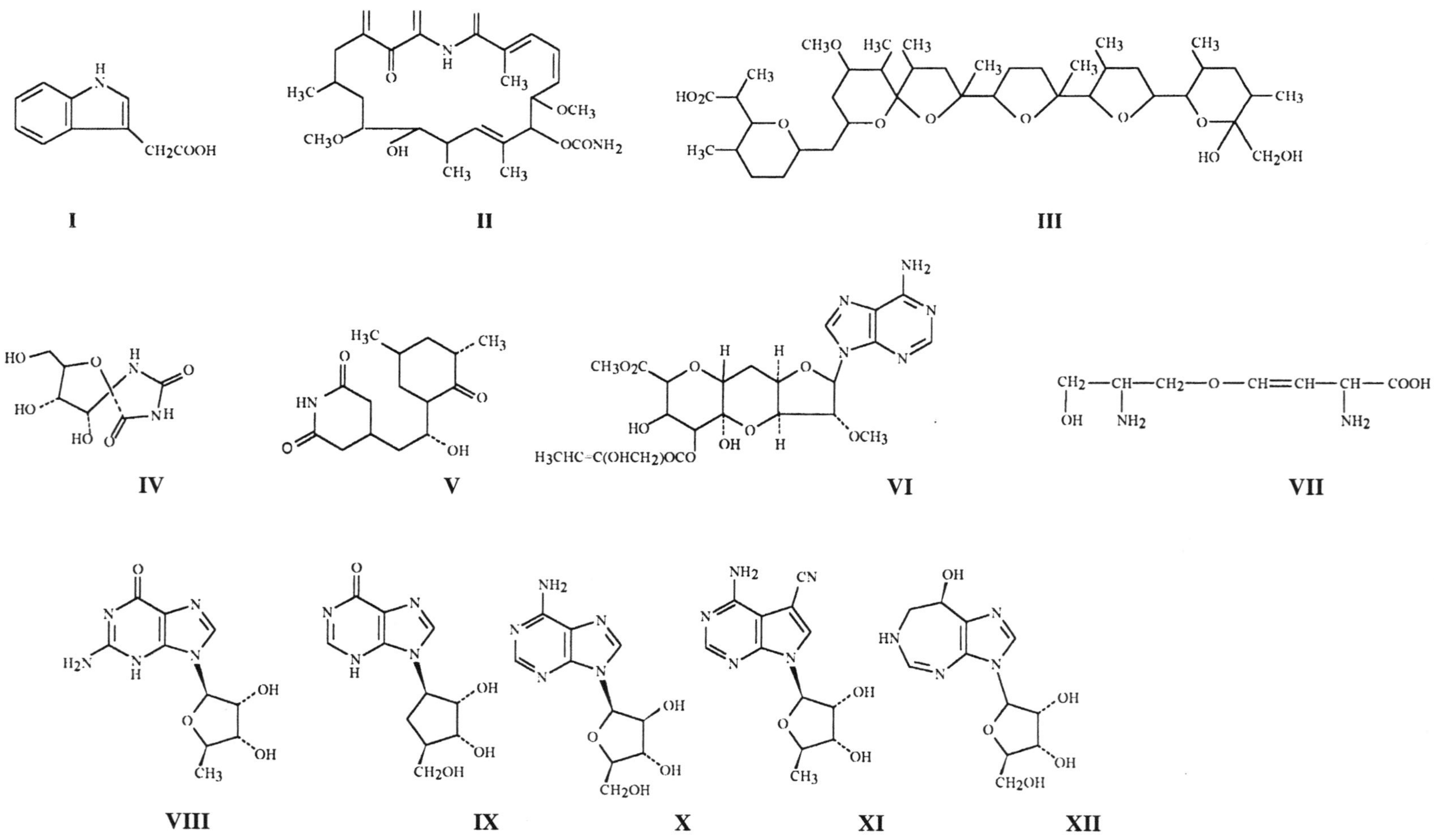

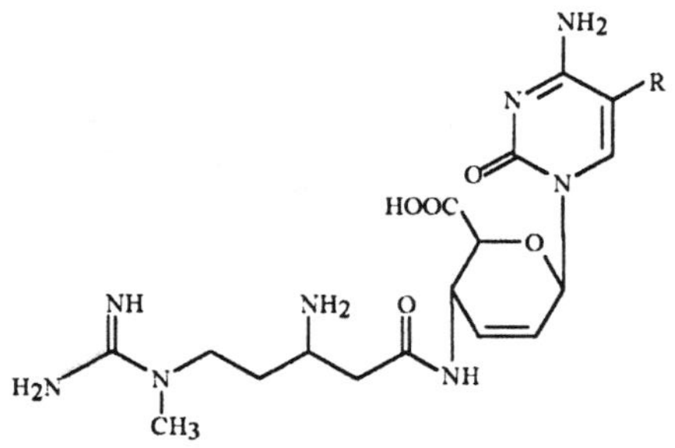

XV

Phthoxazolin D

XVII

XIV

Phthoxazolin B, C

XIII

R=H Blasticidin S
R=CH₂OH 5 hydroxylmethyl-blasticidin S

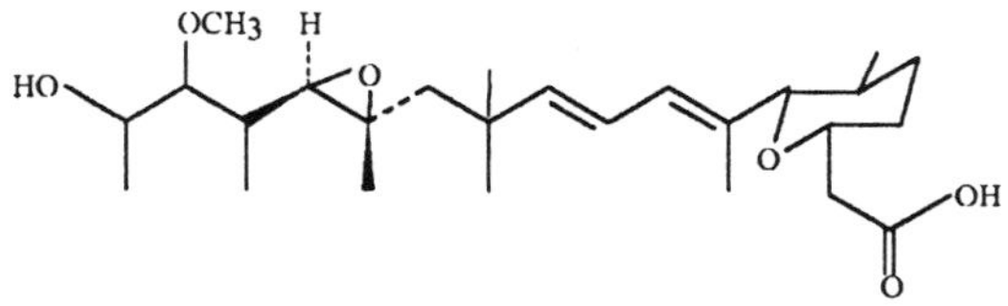

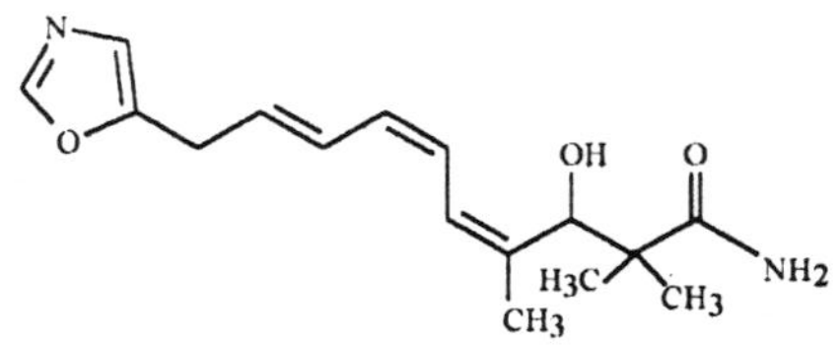

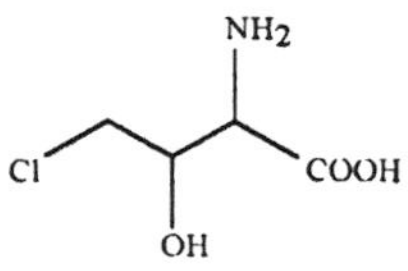

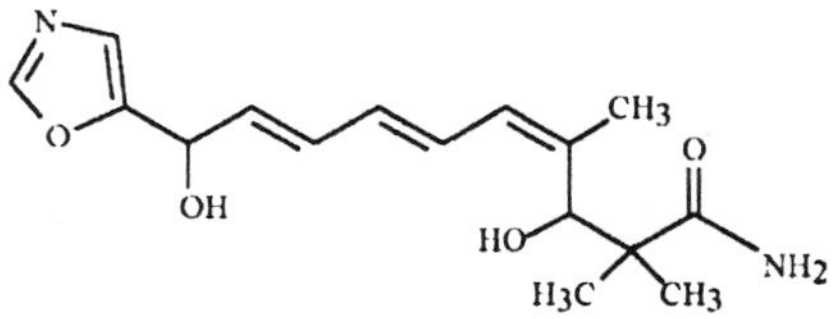

XVI

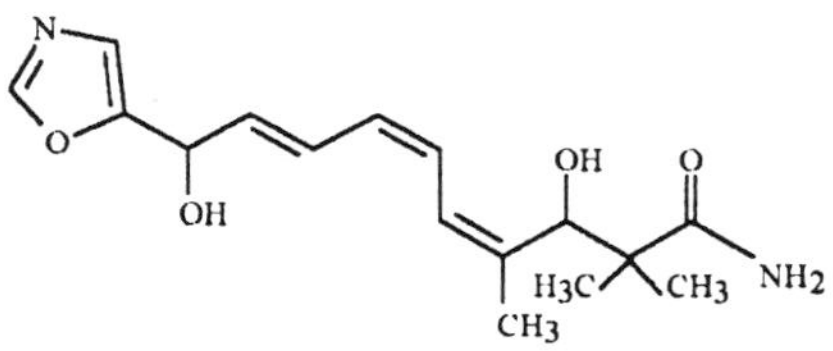

relevant research. As a result, a large number of allelochemicals have been isolated and character-ized from nonpathogenic soil-borne bacteria during the last decade (Table 10.1, Figure 10.2). In con-trast to the specific effect of allelochemicals that often characterize pathogenic microorganisms, allelochemicals released by saprophytic bacteria are often nonspecific, affecting many species of plants (Hoagland, 1990). For example, herbicidin (Figure 10.2, VI), from *Streptomyces saganonen-sis,* applied at 30 to 300 ppm (Table 10.1), inhibited several annuals and perennials of monocotyle-donous and dicotyledonous plants (Cutler, 1988). Nevertheless, certain allelochemicals exhibit some specificity. For example, blasticidin and 5-hydroxylmethyl-blasticidin S (Table 10.1), from the non-pathogenic *Streptomyces* sp., applied as foliar spray at 100 mg/m^2, were more phytotoxic to dicots than to monocots. When these compounds were applied to soil, dicot plants were inhibited by 98 and 64 percent, whereas monocots were almost unaffected (Scacchi et al., 1992).

Several allelochemicals have been isolated from a single species of bacteria, for example, gel-danamycin and nigericin (Table 10.1; Figure 10.2, II and III) from *Streptomyces hygroscopicus.* These compounds inhibited *Lepidium sativum* L. in Petri dishes at 1 to 2 ppm (Heisey and Putnam, 1986). Hydantocidin (Table 10.1; Figure 10.2, IV) isolated also from *Streptomyces hygroscopicus,* applied as a foliar spray at 500 ppm suppressed several annuals and perennials in soil (Nakajima et al., 1991). Such data reported by two different groups suggest occurrence of either two different bacterial chemotypes, or that the same species may produce several different effective allelo-che-micals. Nucleoside antibiotics 5'-deoxyguanosine (Table 10.1; Figure 10.2, VIII) produced by *Thermoactinomycete* sp. applied to the water substrate at 100 μg/ml, inhibited *Lemna minor* L. Coaristeromycin (Table 10.1; Figure 10.2, IX) produced by *Streptomyces sp.* inhibited barnyardgrass (*Echinochloa crus-galli* (L.) Beauv.) and johnsongrass (*Sorghum halepense* (L.) Pers) at a concen-tration of 600 mg/m^2 (Isaac et al., 1991). Among the bacterial allelochemicals evaluated as foliar spray, herboxidiene (Table 10.1; Figure 10.2, XIV), a polyketide from *S. chromofuscus,* inhibited rape (*Brassica napus* L.), buckwheat (*Fagopyrum sagittatum* Gilib.) , and maize (*Zea mays* L.) when applied to soil at only 6.9 mg/m^2 (Miller-Wideman et al., 1992).

Cycloheximide (Table 10.1; Figure 10.2, V) generated by *Streptomyces* sp. inhibited growth of garden cress in Petri dishes by 50 percent at 1 μg/ml (Heisy et al., 1988), suggesting that it may have potential use in the field. Although a wide range of allelochemicals have been isolated from plant growth mediating bacteria (Table 10.1), only two have been applied as commercial herbicides, phos-phinothricin (glufosinate when synthetic), a product of *S. viridochromogenes* and bialaphos, a tripeptide from *S. hygroscopicus* (Figure 10.2, XVIII and XIX) (Tomlin, 1994; Duke and Abbas, 1995).

An interesting group of bacteria are those secreting indole-3-acetic acid (IAA). Up to 80 per-cent of rhizosphere bacteria can synthesize IAA (Loper and Schroth, 1986). As a plant growth hor-mone, the effect of IAA is concentration dependent, that is, low concentrations of exogenous IAA can promote, whereas high concentrations can inhibit root growth (Arshad and Frankenberger, 1992). Secretion of IAA by various species of plant growth-promoting, or by plant growth-inhibit-ing bacteria have been reported. The inhibitory effect of some DRB through IAA secretion has been related to various bacterial species: *Enterobacter taylorae, Klebsiella planticola, Alcaligenes fae-calis, Xanthomonas maltophilia, Entrobacter* sp., *Agrobacterium radiobacter,* and *Flavobacterium* sp. (Sarwar and Kremer, 1995). Mutants of some PGPR that produced high levels of IAA inhibited root growth. For example, a mutant of *Pseudomonas putida* inhibited root growth of seedlings of canola (*Brassica campestris*) (Xie et al., 1996).

Among PGPR, the following secrete low amounts of IAA: *Alcaligenes faecalis, Enterobacter cloacae, Acetobacter diazotrophicus, Bradyrhizobium japonicum;* and also some species of *Azospirillum, Pseudomonas,* and *Xanthomonas* (Patten and Glick, 1996). In contrast to the diversi-fied group of plant growth-inhibiting allelochemicals produced by bacteria, a rather limited number of plant growth-promoting bacterial allelochemicals have been identified (Table 10.1). Succinic and lactic acids from *Pseudomonas putida* applied at 10 ppm promoted root growth of *Asparagus officinalis* by 30 to 40 percent (Yoshikawa et al., 1993). Glick et al. (1994) has shown that *Pseudomonas putida* utilized 1-aminocyclopropane-1-carboxylate (ACC) as a nitrogen source. The roots of *Brassica campestris,* produce ACC and eventually its uptake and metabolism by *Pseudomonas putida* accelerates exudation of ACC from roots. As a result, ethylene production within the roots is reduced and root elongation is promoted.

10.5.1 SPECIFICITY OF BACTERIAL ALLELOPATHY

Since allelopathy is an ecological phenomenon (c.f., Molisch, 1937), relevant research should be focused in natural or agricultural ecosystems. Nevertheless, due to the complexity of factors involved in nature and the interest in identifying natural allelochemicals for agriculture, considerable effort to evaluate the phytotoxic potential has been made under controlled or semi-controlled conditions. Provided the allelopathic effect is not specific, the use of fast germinating or growing seedlings has enabled efficient and a reproducible analysis. Seeds of domesticated species (less variable in germination or growth), have been preferred over indigenous plants. When the effect of soil toxicity on seed germination was examined in the field, seeds of *Avena fatua* L. and *Bromus diandrus* Roth. exhibited a similar inhibitory response (Kaminsky, 1981). When seedlings of *Lactuca sativa* (Asteraceae) and *Anastatica hierochuntica* (Brassicaceae) were exposed to different isolates of phytotoxic actinomycetes on agar in Petri dishes, they responded similarly, suggesting a nonspecific response (Katz et al., 1987; Friedman et al., 1989). Nevertheless, several other studies showed a rather specific, or partly specific response of higher plants to allelopathic bacteria. For example, among winter wheat (*Triticum aestivum*), spring barley (*Hordeum vulgare*), oats (*Avena sativa*), lentils (*Lens culinaris*), and peas (*Pisum sativum*), wheat was the most susceptible to the deleterious effects of culture filtrates of nonfluorescent pseudomonads (Fredrickson et al., 1987). Differences also were found in the susceptibility of cress (*Lepidium sativum*) and barnyardgrass (*Echinochloa crus-galli*) to metabolites of different bacterial isolates. Of 906 bacterial isolates, 8 percent inhibited cress seed germination, but only half of these (4 percent) also inhibited the germination of barnyardgrass (Mishra et al., 1988). Therefore, when test plants are used, evaluation of the allelopathic effect should be verified on those species that are suppressed in nature.

Since plant species differ in the concentration of their endogenous phytohormones, they may also differ in their response to the application of the same allelopathic bacteria. Thus, *Pseudomonas putida* GR12-2 promoted root growth by lowering endogenous levels of ethylene within three dicotyledonous plants (lettuce, tomato, and canola), but inhibited root elongation of wheat. Barley and oat were not affected by the bacterium, suggesting different levels of susceptibility of various plant species to ethylene (Hall et al., 1996).

When plant growth is tested in response to an allelopathic microorganism, genetic differences at the sub-species or cultivar levels should be examined. Specific allelopathic potential was revealed by different cultivars of cucumber (Putnam and Duke, 1974). Several wheat cultivars exhibited differential susceptibility when tested against deleterious pseudomonads (Elliott and Lynch, 1984). Two cultivars of both lettuce (*Lactuca sativa*) and wheat (*Triticum aestivum*) differed in their response to a deleterious *Pseudomonas fluorescens* (Aström, 1991).

Some studies have indicated that bacteria with antibiotic effect on microorganisms commonly exhibit growth regulation effects on plants. For example, deleterious pseudomonads that inhibited root growth of winter wheat also inhibited *Escherichia coli* (Fredrickson and Elliott, 1985). However, using the same method, bacterial isolates (non-fluorescent pseudomonads) that inhibited growth of several species, that is, morning-glory (*Ipomoea purpurea* Lam.), velvetleaf (*Abutilon theophrasti* Medicus), jimsonweed (*Datura stramonium* L.), and redroot pigweed (*Amaranthus retroflexus* L.), did not exhibit any inhibitory effect on *E. coli*. Several other bacteria, which exhibited antibiotic activity against *E. coli,* promoted growth of the above mentioned plant species (Kremer et al., 1990). Selection of a test plant, timing of exposure, and application methods are of crucial significance when the allelopathic effect of bacteria is evaluated. Of 700 different isolates of actinomycetes, only 60 inhibited chlorophyll content of the alga *Chlamydomonas reinhardtii* at levels greater than 30 percent. However, when 7- to 10-day-old seedlings growing in pots in clay soil were sprayed with these microbial extracts, only nine isolates (1.5 percent) decreased the dry weight of corn, soybeans, cucumbers, tomatoes, and sorghum (Mishra et al., 1987).

The significance of tests to evaluate bacterial allelopathic effects on higher plants in soil can be demonstrated by the following example. On evaluation of pseudomonads for differential inhibition of downy brome (*Bromus tectorum*) and lack of inhibition of winter wheat, Kennedy et al. (1991) screened 1000 isolates. When tested on agar, 8 percent inhibited root growth of downy brome, but did not affect root growth of winter wheat. However, when applied to the soil (10^8 c.f.u./ml) under nonsterile conditions only six isolates (~1 percent) inhibited growth of downy brome. In the field, when sprayed (10^8 c.f.u./m^2), two isolates (0.2 percent) suppressed downy brome by 31 to 53 percent and this increased winter wheat yield by 18 to 35 percent. This suggests that specificity does exist, and for an agricultural evaluation, tests in soil are essential.

10.6   ENVIRONMENTAL FACTORS THAT MODIFY THE ALLELOPATHIC EFFECT

The allelopathic effect is highly dependent on environmental conditions (Rice, 1984) such as water, nutrition, bacterial density, soil structure and texture. Drying and rewetting triggered soil toxicity, eventually reducing germination and root growth of lettuce (*Lactuca sativa*) (Kaminsky, 1981). Soil toxicity had been partly related to a flush of microbial activity due to the rewetting; however, drying may have produced some or all of the toxins through chemical alternation of the soil organic matter (Kaminsky, 1981). In studying inhibition of regeneration of Douglas fir by actinomycetes, it had been suggested that the upper soil layer of an exposed area dries out during short alternating periods of drought and humidity, thus increasing the density of actinomycetes (Friedman et al., 1989). In areas exposed to intermittent drought, drought tolerance of actinomycetes may account for their survival. Thus, water stress may alter microbial populations in the soil and/or affect the susceptibility of the higher plant to the allelopathic microorganisms. It was shown that the susceptibility of downy brome (*Bromus tectorum*) to deleterious rhizobacteria in the field was greatest at sites that received lower precipitation (Kennedy et al., 1991). Similarly, allelopathic actinomycetes in the vicinity of *Coridothymus capitatus,* were mostly found on the wind side only (Figure 10.1), restricted to some specific shrubs situated on those sites of sand-stone Kurkar, which were closer to parent rock material than to real soil (Katz et al., 1987). Similarly, the presence of allelopathic actinomycetes in clearcut areas of Douglas fir were observed on a coarse sandy skeletal soil rather than in areas covered by thick layers of organic material (Friedman et al., 1989). The restriction of the allelopathic effect to some specific localities in nature supports the view that high densities of allelopathic microorganisms that colonize sites where other microorganisms have been excluded account for the allelopathic effect.

Nutritional conditions may also control the allelopathic activity of bacteria. When grown on a medium containing nitrate, *Azotobacter chroococcum* suppressed germination of barley. Nevertheless, when grown on a nitrogen-free medium, this bacterium had either promoted germination of barley or had no effect (Lynch, 1978). At present, the effect of the level of nutrients on the allelopathic impact of a bacterium is not clear. When the fungistatic activity of two isolates of actinomycetes was tested on media containing different concentrations of a malt-yeast agar medium, the addition of malt-yeast extract increased the inhibitory effect of one isolate by 45 percent. However, reduction of the malt-yeast increased the inhibitory effect of another isolate by approximately 20 percent. It was suggested that one isolate inhibited the mycorrhizal fungi by competing with the fungus for nutrients, whereas the second isolate inhibited the mycorrhizal fungi by the release of allelochemicals (Friedman et al., 1989).

The relative concentration of microbial allelochemicals may result in a different response of higher plants. Therefore, density of allelopathic bacteria may be crucial in determining the allelopathic intensity.

Nonfluorescent pseudomonads applied at 10^6 and 10^7 c.f.u./ml on agar, reduced root growth of winter wheat, but when the inoculum was below 10^5 c.f.u./ml, root growth was not affected (Fredrickson and Elliott, 1985). Similar results were obtained when cuttings of sour cherry (*Prunus cerasus* L.) and black-currant (*Ribes nigrum* L.) were inoculated with a recombinant strain of *Pseudomonas fluorescens* that produced IAA. A high density of bacterial inoculum on the roots of cherry cuttings inhibited root growth, whereas lower densities on black-currant promoted growth (Dubeikovsky et al., 1993). Since the density of microflora may change considerably during the year, time of sampling may be crucial for evaluating the allelopathic potential of soil-borne microorganisms.

The inhibitory effect of two isolates of *Pseudomonas* on wheat root growth was exhibited at low inoculum densities ($< 10^3$ cells per plant), but required the addition of small amounts of nutrient broth. This suggested that the amount of substrate provided by the plant had been too low to induce measurable effects on plant growth (Aström and Gerhardson, 1989). However, filtrates of bacteria grown on root exudate, or on artificial medium, were equally inhibitory to root elongation of wheat (Aström et al., 1993). It was suggested that rhizobacterial phytotoxins may be produced, or become active, only in the rhizosphere, where plant root-derived materials serve as precursors for microbial metabolites and enhance their production (Schippers et al., 1987). It is therefore quite likely that after the establishment of deleterious bacteria around seeds or seedlings, inhibition of germination or root elongation may occur. Similarly, it is expected that in farm lands where a single crop species is grown continuously, or in areas adjacent to perennial shrubs that release allelchemicals, high densities of bacteria may develop. The question as to the factors that enhance the development of allelopathic bacteria in areas where a certain crop is continuously cultivated is a challenge for further research.

ACKNOWLEDGMENTS

The authors are most indebted to Prof. Eugene Rosenberg Tel Aviv University and Dr. John Cardina, Ohio State University, for critically reviewing the manuscript.

REFERENCES

Alström, S., Saprophytic soil microflora in relation to yield reductions in soil repeatedly cropped with barley (*Hordeum vulgare* L.), *Biol. Fertil. Soils,* 14, 145, 1992.

Arshad, M. and Frankenberger, W. T., Jr., Microbial production of plant growth regulators, *in Soil Microbial Ecology, Applications in Agricultural and Environmental Management,* Metting, F. B., Jr., Ed., Marcel Dekker, New York, 1992, 27.

Aström, B., Role of bacterial cyanide production in differential reaction of plant cultivars to deleterious rhizosphere pseudomonads, *Plant Soil,* 133, 93, 1991.

Aström, B. and Gerhardson, B., Wheat cultivar reactions to deleterious rhizosphere bacteria under gnotobiotic conditions, *Plant Soil,* 117, 157, 1989.

Aström, B., Gustafsson, A., and Gerhardson, B., Characteristics of a plant deleterious rhizosphere pseudomonad and its inhibitory metabolite(s), *J. Appl. Bacteriol.,* 74, 20, 1993.

Bakker, A. W. and Schippers, B., Microbial cyanide production in the rhizosphere in relation to potato yield reduction and *Pseudomonas spp.*-mediated plant growth-stimulation, *Soil Biol. Biochem.,* 19, 451, 1987.

Catska, V., Vancura, V., Hudska, G., and Prikryl, Z., Rhizosphere micro-organisms in relation to the apple replant problem, *Plant Soil,* 69, 187, 1982.

Chayen, S., Phytotoxic microorganisms and their impact on the allelopathic phenomenon near *Artemisia herba-alba* in the Negev desert, Thesis submitted toward M.Sc degree at Tel-Aviv University, Israel, 1991.

Cutler, H. G., Perspectives on discovery of microbial phytotoxins with herbicidal activity, *Weed Technol.,* 2, 525, 1988.

de Freitas, J. R. and Germida, J. J., Growth promotion of winter wheat by fluorescent pseudomonads under field conditions, *Soil Biol. Biochem.,* 24, 1137, 1992.

Dubeikovsky, A. N., Mordukhova, E. A., Kochetkov, V. V., Polikarpova, F. Y., and Boronin, A. M., Growth promotion of blackcurrant softwood cuttings by recombinant strain *Pseudomonas fluorescens* BSP53a synthesizing an increased amount of indole-3-acetic acid, *Soil Biol. Biochem.,* 25, 1277, 1993.

Duke, S. O. and Abbas, H. K., Natural products with potential use as herbicides, *in Allelopathy, Organisms, Processes and Applications,* Inderjit, Dakshini, K. M. M., and Einhellig, F. A., Eds., American Chemical Society, Washington, D.C., 1995, 348.

Elliott, L. F. and Lynch, J. M., Pseudomonads as a factor in growth of winter wheat (*Triticum aestivum* L.), *Soil Biol. Biochem.,* 16, 69, 1984.

Fredrickson, J. K. and Elliott, L. F., Effects of winter wheat seedling growth by toxin-producing rhizobacteria, *Plant Soil,* 83, 399, 1985.

Fredrickson, J. K., Elliott, L. F., and Engibous, J. C., Crops residues as substrates for host-specific inhibitory pseudomonads, *Soil Biol. Biochem.,* 19, 127, 1987.

Friedman, J., Hutchins, A., Li, C. Y., and Perry, D. A., Actinomycetes inducing phytotoxic fungistatic activity in a Douglas-fir forest and in an adjacent area of repeated regeneration failure in southwestern Oregon, *Biol. Plant.,* 31, 487, 1989.

Gealy, D. R., Gurusiddaiah, S., Ogg, A. G., Jr., and Kennedy, A. C., Metabolites from *Pseudomonas fluorescens* strain D7 inhibit downy brome (*Bromus tectorum*) seedling growth, *Weed Technol.,* 10, 282, 1996.

Glick, B. R., Jacobson, C. B., Schwarze, M. M. K., and Pasternak, J. J., 1-aminocyclopropane-1-carboxylic acid deaminase mutants of the plant growth promoting rhizobacterium *Pseudomonas putida* GR12-2 do not stimulate canola root elongation, *Can. J. Microbiol.,* 40, 911, 1994.

Hall, J. A., Peirson, D., Ghosh, S., and Glick, B. R., Root elongation in various agronomic crops by the plant growth promoting rhizobacterium *Pseudomonas putida* GR12-2, *Isr. J. Plant Sci.,* 44, 37, 1996.

Heisey, R. M., Mishra, S. K., Putnam, A. R., Miller, J. R., Whitenack, C. J., Keller, J. E., and Huang, J., Production of herbicidal and insecticidal metabolites by soil microorganisms, *in Biologically Active Natural Products, Potential Use in Agriculture,* Cutler, H. G., Ed., American Chemical Society, Washington, D.C., 1988, 65.

Heisey, R. M. and Putnam, A. R., Herbicidal effects of geldanamycin and nigericin, antibiotics from *Streptomyces hygroscopicus, J. Nat. Prod.,* 49, 859, 1986.

Hoagland, R. E., Microbes and microbial products as herbicides: an overview, *in Microbes and Microbial Products as Herbicides,* Hoagland, R. E., Ed., American Chemical Society, Washington, D.C., 1990, 2.

Isaac, B. G., Ayer, S. W., Letendre, L. J., and Stonard, R. J., Herbicidal nucleosides from microbial sources, *J. Antibiot.,* 44, 729, 1991.

Kaminsky, R., The microbial origin of the allelopathic potential of *Adenostoma fasciculatum* H&A, *Ecol. Monogr.,* 51, 365, 1981.

Kapulnik, Y., Plant-growth-promoting rhizobacteria, *in Plant Roots, the Hidden Half,* Waisel, Y., Eshel, A., and Kafkafi, U., Eds., Marcel Dekker, New York, 1991, 717.

Katz, D. A., Sneh, B., and Friedman, J., The allelopathic potential of *Coridothymus capitatus* L. (Labiatae). Preliminary studies on the roles of the shrub in the inhibition of annuals germination and/or to promote allelopathically active actinomycetes, *Plant Soil,* 98, 53, 1987.

Kennedy, A. C., Elliott, L. F., Young, F. L., and Douglas, C. L., Rhizobacteria suppressive to the weed downy brome, *Soil Sci. Soc. Am. J.,* 55, 722, 1991.

Kloepper, J. W. and Schroth, M. N., Plant growth-promoting rhizobacteria on radishes, in *Proceedings of the Fourth International Conference on Plant Pathogenic Bacteria,* Vol. 2, Angers, Station de Pathologie Vegetale et Phytobacteriology I.N.R.A., France, Ed., Gilbert-Clarey, Tours, 1978, 879.

Kremer, R. J., Begonia, M. F. T., Stanley, L., and Lanham, E. T., Characterization of rhizobacteria associated with weed seedlings, *Appl. Environ. Microbiol.,* 56, 1649, 1990.

Krishna, K. R., Balakrishna, A. N., and Bagyaraj, D. J., Interactions between a vesicular-arbuscular mycorrhizal fungus and *Streptomyces cinamomeous* and their effects on finger millet, *New Phytol.,* 92, 401, 1982.

Loper, J. E. and Schroth, M. N., Influence of bacterial source of indole-3-acetic acid on root elongation of sugar beet, *Phytopathology,* 76, 386, 1986.

Lynch, J. M., Microbial interactions around imbibed seeds, *Ann. Appl. Biol.,* 89, 165, 1978.

Meharg, A. A. and Killham, K., Loss of exudates from the roots of perennial ryegrass inoculated with a range of micro-organisms, *Plant Soil,* 170, 345, 1995.

Miller-Wideman, Makkar, N., Tran, M., Isaac, B., Biest, N., and Stonard, R., Herboxidiene, a new herbicidal substance from *Streptomyces chromofuscus* A 7847, *J. Antibiot.,* 45, 914, 1992.

Minamisawa, K. and Fukai, K., Production of indole-3-acetic acid by *Bradyrhizobium japonicum:* a correlation with genotype grouping and rhizobitoxine production, *Plant Cell Physiol.,* 32, 1, 1991.

Mishra, S. K., Taft, W. H., Putnam, A. R., and Ries, S. K., Plant growth regulatory metabolites from novel actinomycetes, *J. Plant Growth Regul.,* 6, 75, 1987.

Mishra, S. K., Whitenack, C. J., and Putnam, A. R., Herbicidal properties of metabolites from several genera of soil microorganisms, *Weed Sci.,* 36, 122, 1988.

Molisch, H., Der Einfluss einer Pflanze auf die andere- Allelopathie, Fischer, Jena, 1937.

Nakajima, M., Itoi, K., Takamatsu, Y., Kinoshita, T., Okazaki, T., Kawakubo, K., Shindo, M., Honma, T., Tohjigamori, M., and Haneishi, T., Hydantocidin: a new compound with herbicidal activity from *Streptomyces hygroscopicus, J. Antibiot.,* 44, 293, 1991.

Olsson, S. and Alström, S., Plant-affecting streptomycin-sensitive micro-organisms in barley monoculture soils, *New Phytol.,* 133, 245, 1996.

Patten, C. L. and Glick, B. R., Bacterial biosynthesis of indole-3-acetic acid, *Can. J. Microbiol.,* 42, 207, 1996.

Putnam, A. R. and Duke, W. B., Biological suppression of weeds: evidence for allelopathy in accessions of cucumber, *Science,* 185, 370, 1974.

Rice, E. L., *Allelopathy,* 2nd ed., Academic Press, Orlando, FL, 1984.

Sarwar, M. and Kremer, R. J., Enhanced suppression of plant growth through production of L-tryptophan-derived compounds by deleterious rhizobacteria, *Plant Soil,* 172, 261, 1995.

Scacchi, A., Bortolo, R., Cassani, G., Pirali, G., and Nielsen, E., Detection, characterization and phytotoxic activity of the nucleoside antibiotics, blasticidin S and 5-hydroxylmethyl-blasticidin S, *J. Plant Growth Regul.,* 11, 39, 1992.

Schippers, B., Bakker, A. W., and Bakker, P. A. H. M., Interactions of deleterious and beneficial rhizosphere microorganisms and the effect of cropping practices, *Annu. Rev. Phytopathol.,* 25, 339, 1987.

Schippers, B., Bakker, A. W., Bakker, P. A. H. M., and Van Peer, R., Beneficial and deleterious effects of HCN-producing pseudomonads on rhizosphere interactions, *in The Rhizosphere and Plant Growth,* Keister, D. L. and Cregan, P. B., Eds., Kluwer Academic Publishers, Dordrecht, The Netherlands, 1991, 211.

Shioimi, K., Arai, N., Shinose, M., Takahashi, Y., Yoshida, H., Iwabuchi, J., Tanaka, Y., and Omura, S., New antibiotics phthoxazolins B, C and D produced by *Streptomyces* sp. KO-7888, *J. Antibiot.,* 48, 714, 1995.

Stern, D., The involvement of phytotoxic microorganisms in the rose replant disease, Thesis submitted toward M.Sc degree at Tel-Aviv University, Israel, 1992.

Suslow, T. V. and Schroth, M. N., Role of deleterious rhizobacteria as minor pathogens in reducing crop growth, *Phytopatholgy,* 72, 111, 1982.

Tanaka, Y., Kanaya, I., Takahashi, Y., Shinose, M., Tanaka, H., and Omura, S., Phthoxazolin A, a specific inhibitor of cellulose biosynthesis from microbial origin, *J. Antibiot.,* 46, 1209, 1993.

Tomlin, C., *World Compendium, The Pesticide Manual, The Agro-Chemical Handbook,* 10th ed., British Crop Protection Council, Surrey, U.K., 1994.

Turco, R. F., Bischoff, M., Breakwell, D. P., and Griffith, D. R., Contribution of soil-borne bacteria to the rotation effect in corn, *Plant Soil,* 122, 115, 1990.

Vokou, D., Margaris, N. S., and Lynch, J. M., Effects of volatile oils from aromatic shrubs on soil microorganisms, *Soil Biol. Biochem.,* 5, 509, 1984.

Waschkies, C., Schropp, A., and Marschner, H., Relations between grapevine replant disease and root colonization of grapevine (*Vitis* sp.) by fluorescent pseudomonads and endomycorrhizal fungi, *Plant Soil,* 162, 219, 1994.

Westcott, S. W. and Beer, S. V., Infection of apple roots by actinomycetes associated with soil conducive to apple replant disease, *Plant Dis.,* 70, 1125, 1986.

Xie, H., Pasternak, J. J., and Glick, R. B., Isolation and characterization of mutants of the plant growth-promoting rhizobacterium *Pseudomonas putida* GR12-2 that overproduce indoleacetic acid, *Curr. Microbiol.,* 32, 67, 1996.

Yoshida, H., Arai, N., Sugoh, M., Iwabuchi, J., Shiomi, K., Shinose, M., Tanaka, Y., and Omura, S., 4-chlorothreonine, a herbicidal antimetabolite produced by *Streptomyces* sp. OH-5093, *J. Antibiot.,* 47, 1165, 1994.

Yoshikawa, M., Hirai, N., Wakabayashi, K., Sugizaki, H., and Iwamura, H., Succinic and lactic acids as plant growth promoting compounds produced by rhizospheric *Pseudomonas putida, Can. J. Microbiol.,* 39, 1150, 1993.

11 Allelochemistry in Plankton Communities

K. Irwin Keating

CONTENTS

11.1 ABSTRACT

With the benefit of water as a carrier, biologically active materials produced by planktonic organisms greatly influence community structure, subtly determining dominance and succession. In mesotrophic and eutrophic freshwater lakes, colors and odors resulting from excessive, or bloom, concentrations of single algal species periodically interfere with human activities, presenting both aesthetic and toxic insults. Such dominance is commonly obtained by phytoplankters capable of producing extracellular materials that enhance their competitive position. In marine waters, especially coastal waters, toxic red, brown, or green blooms devastate commercial fisheries, sickening fishermen and killing both finfish and bivalves. Although little corrective action has been possible to date, there are major research efforts currently aimed at prediction, control, and amelioration. Many see this extraordinary array of biologically active material to be of great promise in terms of ultimate biological control of nuisance blooms and in the pharmacological promise of antibiotics from the sea. However, a fuller understanding of the biosynthesis of the active molecules and of the environmental parameters that trigger and maintain their production is needed prior to commercial exploitation.

11.2 INTRODUCTION

11.2.1 CONTRASTS WITH TERRESTRIAL SYSTEMS

From the alkaloids of poison ivy to the pharmaceuticals of bread mold, the bio-active compounds produced by many terrestrial organisms play an intimate role in human activities. This anthropogenic involvement has brought allelochemistry in terrestrial systems to the attention of scientists and laymen alike. In contrast, *in situ* occurrence of aquatic allelochemistry is removed from daily contact and is often overlooked. Yet, both the physical and chemical characteristics of an aquatic medium and the structural and nutritional characteristics of aquatic organisms favor the role of chemical mediators in natural systems.

Water is often casually referred to as a universal solvent. While there is some hyperbole in this statement, water, as a carrier of bio-active material, is superior to air or soil. Receipt, retention, distribution, and delivery of the metabolic products of an aquatic organism are all facilitated by water whether fresh or marine. Additionally, the absence of desiccation protection, which in an air environment presents a physical barrier that isolates cell membranes from their environment, permits greater intimacy of the membrane with the surrounding medium and any bio-active molecules it might carry. This portends a considerable influence of such activity on *in situ* aquatic community structure. That, in turn, provides the evolutionary justification for the often quite high investment in biochemical pathways and molecular byproducts that do not contribute directly to either growth or reproduction. To date it has proven difficult to document this governing role. This reflects not only the state-of-the-art in aquatic culture, but also the lack of recognition due allelochemistry as a significant and pervasive phenomenon in aquatic systems.

11.2.2 HISTORIC PERSPECTIVE

Ancient references to what is most likely allelochemistry should be acknowledged: for example, the biblical story (Exodus 7:20–21) of the bloody Nile waters associated with Moses; and American Indian legends of the warnings offered by the glowing waters of the ocean (likely red tides). The earliest scientific references to allelochemicals associated with aquatic ecosystems, however, can be found in discussions of toxicity in the veterinary field, and on occasion in humans, in the medical literature of the 19th century. Some exceptionally entertaining reviews of these topics are offered by Schwimmer and Schwimmer (1955, 1964). Most reasonably documented examples involve phytoplankters, usually Prokaryotic blue-green algae. In the 20th century reports of probable allelochemical activity include both medical and ecological roles. It is this more recent dichotomy that this discussion will highlight. Several review, or overview, articles may be of interest. The following offer different perspectives on the state of the art through this century: Lucas, 1947; Pourriot, 1966; Fogg, 1971; Hellebust, 1974; Provasoli and Carlucci, 1974; Aubert, 1978; Maestrini and Bonin, 1981, Keating, 1981, 1987; Inderjit and Dakshini, 1994.

11.3 ECOLOGICAL CONSIDERATION—COMMUNITY STRUCTURE

11.3.1 TERMINOLOGY

The use of many terms to identify this single, and singular, phenomenon through the years has obscured the unusually broad spectrum of both *in situ* and *in vitro* allelochemical occurrences that have been documented in planktonic systems. This fractionalizing of information interferes with communication and semantically masks information from one research group that would be of paramount interest and value to another. Those pursuing possible pharmaceutical products would benefit from the ecological and evolutionary implications of allelochemical production since these would help to identify probable sources of biologically active materials. Those whose interests lie more in the realm of the roles biologically active metabolites might play in ecosystem structure would benefit from more precise identification of the biochemical mechanisms that are in play in each instance of allelochemistry. Finally, when interpretation of allelochemical phenomena are involved, researchers in marine and freshwater laboratories could benefit by elimination of the sometimes artificial distinction between the two environs.

Using a general definition of allelochemistry, that is, allelochemistry involves a response of one organism to contact with the biologically active metabolic products of another, as the bounding criterion is useful. This immediately exposes the extraordinary array of well-documented, or suspected, instances of significant allelochemical involvement in the structuring—dominance, resource allocation—of aquatic ecosystems. Mölisch's (1937) term "allelopathy" while commonly used introduces confusion, presenting the image of negative-only effects by using the suffix, "-pathy" when both positive and negative effects are discussed. The terms pro-biosis and antibiosis, which call to mind the fact that antibiotics involve a form of allelochemistry, were favored in our laboratory, but like other terms, Aubert's "telemediators" (Aubert, 1971a), Lefèvre's "auto-" and "hétéroantagonisme" (Lefèvre et al., 1952) have not been widely used. Fogg (1962) emphasized the distinction between extracellular metabolites released, "leaked," by healthy cells and those active materials available only on consumption of, or lysis of, the cells carrying them.

The distinction between the endotoxins that produce paralytic shellfish poisoning (*Gonyaulax*-based red tides, McFarren et al., 1957) and the more general mix of extracellular metabolic products (exotoxins) in the waters of red tides dominated by other dinoflagellates fit Fogg's consumption or lysis-released versus healthy cell-leaked distinction. In terms of the impact of biologically active materials on community structure, their presence and concentration would seem more important than the method of release, and we have adopted the more open interpretation of Lucas (1947), which includes both products leaked out of young, senescent or moribund cells, and products mainly emancipated when dying cells lyse. Ultimately, accurate communication must prevail. The term allelochemistry functioned well at several meetings in the last decade; it is used here in the hope that it will provide a unifying logic to the many studies of inter-species effects associated with biologically active metabolites that are separated more by the use of disparate terminology, than by differences in modes of action.

11.3.2 PRIMARY VERSUS SECONDARY ALLELOCHEMISTRY

Community structure, and the forces that design it, are of the utmost interest to limnologists and marine ecologists, alike. Apstein (1896) and Pütter (1908) are usually credited with the initial suggestions of allelochemical involvement in community structure; but, the logic of this role has escaped no one. The very high levels of production of extracellular materials, early documented by Watanabe (1951) and Fogg (1952; Fogg et al., 1965)—in a few cases almost 90 percent *in vitro* (Ignatiades and Fogg, 1973), 7 to 50 percent, occasionally above 90 percent, *in situ* (Fogg, 1966; Fogg et al., 1965)—require explanation in evolutionary terms. Such apparent inefficiency could not survive the rigors of fitness-based selection. In our Linsley Pond study (Keating, 1976, 1977, 1978, 1987) the demonstrated ubiquitous occurrence of allelochemical effects provided strong support for the conclusion that allelochemistry plays a major, *not solitary*, role in the structuring of the ecosystem's community. The extraordinary multi-year concurrence *in vitro* of all positive, negative, and neutral allelochemical effects of bloom dominant algal forms (blue-green Prokaryotes) with the actual *in situ* bloom sequence belies any other interpretation. This is additionally strengthened by the similar pattern of activity found in waters collected before, during, and after several of the blooms.

While the survival value of investing extracellular material in the elimination of current competing species is clear-cut, our studies showed considerable numbers of examples in which extracellular products of a bloom dominant improved the growth of its successor, or of some non–co-occurring species. Selection of the successor, or the favoring or disfavoring of non–co-occurring species offers no evolutionary value. Thus, our working assumption has always been that some other function, which does provide such value to the producer, must exist for the active substance. Nevertheless, the coincidental influence, the secondary role, is of considerable importance to the structure of the community as a whole. This is the concept of secondary allelochemistry (discussion, Keating, 1987). It calls to attention the likelihood that biologically active materials, once released, may have multiple roles in addition to that which fostered their development in the first place. The current hunt for pharmaceuticals in aquatic systems should benefit by the conscious identification of these distinct roles of primary and secondary allelochemistry. The secondary (antibiotic) activity they seek must be accompanied by primary activity. When this information is added to the autecological cues, vivid color, strong odor, that assure predator identification of undesirable prey, the pool of possible allelochemical producers can be limited.

11.4 FRESHWATER SYSTEMS

11.4.1 HISTORIC PERSPECTIVE

Akehurst (1931) offered the initial speculations concerning the roles of extracellular materials of phytoplankters in the structuring of *in situ* freshwater communities. And, the early work of Pratt and his coworkers (1940–1944) with *Chlorella vulgaris* established the existence of freshwater allelochemical activity *in vitro*. He named the complex substance he believed responsible for the activity, chlorellin. Several editions of the Merck Index (including the eighth, not the tenth) include chlorellin, labeling it an "antibacterial substance." Our Linsley work provided secure examples of the ubiquitous influence of allelochemical events in both the long-term changes in phytoplankton community dominance (1978) introduced by cultural eutrophication and the more immediate patterns of dominant phytoplankters in an annual bloom sequence (1977) in a freshwater lake. At that time, except for the observations that 1) the allelochemical activity (positive or negative) associated with any bloom could no longer be documented in natural waters after a week or two and that allelochemical activity was weaker and shorter-lived in bacterized cultures, and 2) that zooplankters were present only when the dominant allelochemical-producing blue-green, Prokaryotic, algae were in short supply (Keating, 1976), no intertrophic level work was reported. Nonetheless, the ubiquitous *in situ* presence of allelochemical material at concentrations sufficient to affect competitive encounters among phytoplankters strongly suggested that the same array of bio-active materials would affect all other trophic levels, and our work turned to that study objective.

More recent work has the advantage of partial, or full, identification of allelochemical materials (Watanabe et al., 1992; Pignatello et al., 1983), yet the questions remain the same. The certainty that the quantities *in situ* of active material are sufficient to assign them a role in community structure and that the materials produced by targeted organisms (commonly algae) *in vitro* are, for the most part, produced by those same organisms *in situ*.

11.4.2 INTERFERENCE OF ULTRA TRACE ELEMENT DEFICIENCIES *IN VITRO*

Warnings against making absolute judgments based on data from *in vitro* studies dependent upon bacterized cultures had always been taken into account in our work. One additional generalized interference had not. That was the array of unusually misleading possibilities associated with unknown and unsuspected nutritional requirements for ultra trace elements. In initial studies, an apparent negative allelochemical effect of a diatom on a variety of daphnid species proved to be evidence of an unexpected selenium deficiency for the diatom introducing an equally unexpected selenium deficiency for the zooplankter (Keating and Dagbusan, 1984). Soon thereafter we found that either a zinc or a copper deficiency introduced a secondary selenium deficiency (Keating and Caffrey, 1989). Clearly, heretofore unrecognized competitions for ultra trace nutrients readily masquerade as allelochemical phenomena and these distinct influences must be partitioned. Thereafter, as a basic quality control mechanism, we sought to partition nutritional interferences from initial studies of phytoplankton-zooplankton allelochemical relationships. Because it had not previously been possible to manipulate trace elements at the concentrations (commonly at, or below, 1 ppb) needed to demonstrate nutritional roles, little basic information existed. This forced us to focus away from the roles of trace organics (allelochemicals) and toward those of trace inorganics (nutrients).

The work of Conklin and Provasoli (1977, 1978) represented the state-of-the-art in zooplankton nutrition studies and, even there, little information concerning the specific requirements for inorganics, much less for ultra trace elements, was available. In our laboratory, with the advantage of Provasoli's continuing insights and critiques, efforts to overcome this deficit continued through the last decade of his life. At this time, a medium carrying all plausible required ultra trace elements is available for experimentation (currently unpublished). It requires additional adjustments; however, it will support unlimited generations of daphnids with relatively high reproduction (200+ progeny per female per generation). It is of particular interest that when the basic medium (Keating and Caffrey, 1989) is composed of 100 percent ACS (American Chemical Society) certified salts, it is sufficient to support permanent (20 continuous years) cultures of zooplankters. However, when the identical medium is composed of ultra pure or "spec pure" salts, it can support multigeneration cultures *only* if the additional ultra trace element solution, Table 11.1, is added.

To date completed studies indicate that 1) without the unusual ultra trace element solution animals cannot maintain reproduction; that is, they cannot survive, for multiple generations, and 2) preliminary studies indicate that the separate additions of each of the trace elements in the solution, at controlled concentrations, improves reproduction and vigor.

TABLE 11.1
Trace Nutrient Additions for Ultra Pure Medium

Element	Source Compound	Concentration in Complete MS Medium (ppb)
Ag	AgCl	0.2
Al	$AlCl_2$	3.0
As	Aldrich AA standard	0.05
Cd	$CdCl_2$	0.005
Cr	$CrCl_3.6H_2O$	0.01
F	NaF	0.5
Ge	Aldrich AA standard	0.1
Ni	$NiCl_2.6H_2O$	5.0
Pb	$PbCl_2$	0.3
Sb	Sb_2O_5	0.05
Sn	$SnCl_4$	0.04
Te	Aldrich AA standard	0.2
W	Aldrich AA standard	0.02

Note: These trace elements are suggested additions to MS media when media are assembled using very highly purified salts. This complete addition has proven essential to long-term maintenance of daphnids in ultra pure MS. Each chemical, at the listed concentration, has increased reproduction in a minimum of two consecutive generations when added, alone, to the basic MS medium. While much additional work is needed, this unusual information must be shared. (Basic MS media can be found in Keating and Caffrey, 1989.)

Although the roles of trace deficiency cannot be ruled out until all ultra trace element requirements of both the producer and sensitive species are known, intertrophic level studies that include isolation of the allelochemical material, with or without identification, continue to provide evidence for allelochemical influences in natural systems. The certain presence of the biologically active metabolite assures that the responses it elicits do not simply reflect trace element problems.

11.4.3 INTER TROPHIC LEVEL EFFECTS

In situ allelochemistry was first considered when in 1878 Francis reported a poisonous bloom of blue-greens in an Australian lake. It was not until 1948 (Lefèvre and Nisbit, 1948; Lefèvre and Jakob, 1949; Lefèvre, 1950), however, that Lefèvre offered convincing evidence of a parallel *in situ* and *in vitro* allelochemical effect.

After observing that the daphnid population in a freshwater system dropped dramatically when *Aphanizomenon gracile* dominated the phytoplankton, he demonstrated that cultured daphnids would actively eject *A. gracile* (Lefèvre et al., 1952). Thereafter, Ryther (1954) demonstrated that senescent *Chlorella vulgaris* were damaging to daphnids while actively growing cells were acceptable food. He also showed that critical differences in methodology can affect results by demonstrating differing negative effects on daphnid filtering rate depending on life stage and concentration of *C. vulgaris,* and on pre-feeding of animals. This was reminiscent of Pratt's (1940) observations that his *C. vulgaris* cultures responded differently to various culture conditions. More recent work examining the effects of algal metabolites on zooplankters (copepods and daphnids) includes that of Kirk and Gilbert (1992), DeMott and co-workers (1991), and Lampert (1982).

The earliest reports of effects of phytoplankton toxins on vertebrates, terrestrial or aquatic, are thoroughly reviewed by Gorham (1964; Gorham and Carmichael, 1979), and Schwimmer and Schwimmer (1964). Studies that focus on Eukaryotic aquatic organisms (discussion, E. Gross, this volume) are few.

Perhaps the most celebrated work implicating a zooplankter as producer is that of Gilbert (1975, 1980, 1981). He found *Asplanchna siebaldi* individuals presented various distinct morphs depending on diet. One of these morphs was triggered by the consumption of members of its own species, an example of allelochemistry in the broad sense favored by Lucas. Another example, more forcefully presented as allelochemical in nature, is that of *Epischura nevadensis* against *Diaptomus tyrrelli* offered by Folt and Goldman (1981). They offered a unique perspective on the evolutionary value of this interaction in which the sensitivity itself might play the role of selection. They suggest that the drop in filtering rate triggered by *E. nevadensis* could protect *D. tyrrelli* from predation because the absence of filtering activity would make *D. tyrrelli* less detectable by predators.

The detailed studies of Dodson and his coworkers from the 1980s to the present (Dodson, 1988, 1989; Parejko and Dodson, 1990; Dodson and Havel, 1988) have provided incontrovertible *in vitro* evidence of the occurrence of allelochemical interactions between zooplankters. Additionally, their efforts have shown that *Daphnia ambigua* (Hanazato, 1991), *D. middendorfiana* (Dodson, 1984), and *D. pulex* (Black and Dodson, 1990) suffer a general loss of vigor in the presence of the allelochemical, a kairomone (Parejko and S. Dodson, 1990), released by the predator, *Chaoborus americanus. D. ambigua* expresses the stress via an increased temperature sensitivity, *D. pulex* and *D. middendorfiana* via a loss of fecundity. Their current efforts (Dodson et al., 1995; Hanazato and Dodson, 1995) aimed at delineation of the role of zooplankter to zooplankter allelochemistry *in situ,* are focused on the synergistic effects of a series of chemical insults including pesticides and the kairomone of *Chaoborus* (Black and Dodson, 1990), with environmental stresses of the sort to be expected in natural systems. Their results continue to offer very meaningful information regarding the multiple forces forming the *in situ* community.

Fish and mollusks are known to respond to biologically active materials in aquatic systems (see especially the Gorham review, 1964, and the discussions of red and brown tides, *vide infra*), but they are rarely considered producers. Yet aficionados of Japanese cuisine are familiar with the toxins carried by certain fish organs, and tropical fish hobbyists are well-warned concerning the poisons released by lion fish. While the invertebrates of the oceans are suspected of the production of toxins, this author found no suggestions of freshwater mollusks that produced biologically active materials. This more likely reflects an absence of interest rather than an absence of biological activity. Machácek (1993), however, did report work with allelochemical-producing freshwater fish. As would be expected the affected organisms were prey.

11.5   MARINE SYSTEMS

11.5.1   Historic Perspective

Hardy in his theory of animal exclusion (Hardy, 1936; Hardy and Gunther, 1935) provided early observational support and ecological explanation for the phenomenon of allelochemistry in open ocean waters. But, it was Aubert's extensive studies in the 1960s and 1970s (1971a, b, 1978; Aubert et al., 1970a, 1970b) of telemediator involvement in the plankton community structure of the Mediterranean that focused attention on the probability of ubiquitous, intertrophic level, allelochemical influences in a marine community. By tying biologically active phytoplankter extracellular metabolites to bactericidal effects *in situ*, he established that these materials are present in an open marine system in sufficient concentrations to permit identifiable effects on natural populations. Carlucci and Bowes (1970a, b) established that marine algae not only use, but also produce, vitamins at least *in vitro*. These *in vitro* results parallel Carlucci's observations of increased vitamin presence in waters during blooms of suspected producers (discussion, Provasoli and Carlucci, 1974). Vitamin B_{12} studies (Droop, 1968) added an interesting twist to the story of vitamin production by algae. Producing species often release an accompanying B_{12}-binder that renders the B_{12} produced by these species useless to other phytoplankters, but available to the producer later in a bloom when the capacity to make B_{12} is no longer present. Daisley (1970) identified binders produced by *Euglena gracilis* as a glycoprotein.

11.5.2   Toxins and Toxic Blooms

Toxic blooms, red tides, brown tides (*vide infra*) and their like, have been reported for centuries. They are rarely seen as related to other reports of allelochemistry, yet they are. The toxic nature of these blooms sets them apart in societal terms because they often introduce significant public health issues and/or economic problems. Additionally, these blooms are usually obvious, repetitive events along coasts and are often associated with legends or folklore. Chronicles of the history of "red," "bloody," or strangely colored waters can be found in Hayes, Landau, and Austin (1951) and in Galstoff (1948). From Homer (Iliad) to Darwin (H. M. S. Beagle) these authors report episodes of discolored waters on the high seas. According to David Garrison (1995) of the University of California at Santa Cruz, Hitchcock's thriller, *The Birds*, reflected his intrigue with a 1961 California toxic tide episode. Collins (1978) reviewed both the occurrence and the chemistry of algal toxins known to that time.

11.5.3   Red Tides, Paralytic Shellfish Poisoning (PSP)

It has been the generally destructive results of red tides, however, and their damage in economic terms, that has, through the last half century, drawn most of the scientific attention and study of the *in situ* role of allelochemicals. The sudden appearance of huge populations (millions of cells per liter) often follows meteorological events that stir the sediments and bring very high concentrations of resting cysts into the water column to emerge as swimming forms. The terms reserved for allelochemistry (*vide supra*, II. Ecological Considerations) are not usually applied to red tides. Yet, the biological activities associated with these destructive phenomena are clearly allelochemical. Early reviews by Ballantine and Abbott (1957) and by McFarren and his coworkers (1957) focused on the effects of red tide organisms on marine animals, and the characteristics of the toxins, respectively. Perhaps the most important information the long history of red tides can provide is that it is indeed long. Red tides are not of recent origin. While human activity can be blamed for many environmental problems, occurrences of red tides are not among them. It would seem, however, that the duration and severity of such events must sometimes reflect the unnatural concentrations of nutrients carried into coastal waters by non-point source pollution.

Not all red tides are red, not all toxins associated with red tides are algal in origin, but those red tides that produce paralytic shellfish poisoning are mainly *Gonyaulax*-dominated, and the poison itself, saxitoxin, has been known for some time (Hutner and McLaughlin, 1958) as an acetylcholine inhibitor reminiscent of the actions of botulina, similar in effect to curare. It often kills within hours, with death being due to respiratory failure. Further study (Oshima et al., 1977) of the toxic materials produced by *Gonyaulax* sp. has revealed it to be a complex of related chemicals, all of which are among the most toxic substances known to humans. Studies since the 1950s have shown that those red tides that are less toxic and more generalized in their *in situ* damage are commonly dominated by other dinoflagellates; for example, *Prorocentrum* sp., and *Gymnodinium* sp. These are the infamous tides that introduce economic disasters along the gold coast of Florida, leaving behind miles of debris containing all manner of dead fish and invertebrates in a band several meters wide along the beaches.

11.5.4 BROWN TIDES

Unlike red tides, brown tides, may prove to be at least exacerbated by pollution. The occurrences of brown tides, particularly those occurring irregularly along the northeast coast of the United States and continually along the coast of Texas (DeYoe and Suttle, 1994), have introduced significant economic stresses. Much as red tides and toxic diatom blooms, their world-wide distribution represents a severe impact on fisheries and coastal commercial activities.

In the United States, the first severe brown tide occurred in Long Island Sound (NY) waters in 1985 (Nuzzi, 1988). Its occurrence disrupted near-shore communities, both human and marine, and devastated commercial clam and scallop fisheries. The effects of that incident on the lucrative shellfish businesses of Long Island Sound, and along the New England and Middle Atlantic coasts (Tracey, 1988), have focused both concern and funding on the phenomenon. Unfortunately, a considerable research investment since that time has not yet provided solutions to the problem.

The brown tides along the northeastern coast of North America are dominated by a single species, *Aureococcus anophagefferens,* an unusually small, nondescript, Chrysophyte. The most recalcitrant questions concern the circumstances that so favor its competitive position that although it always exists as one of many minor species, it occasionally produces an explosive bloom population. These blooms, however, occur only in waters that are partially confined (Anderson et al., 1993). Work has been generally hindered by the inability to develop robust cultures of *A. anophagefferens.* Only one secure culture is currently extant (Cosper, 1987). To further confound research efforts, success with clearing *A. anophagefferens* cultures of accompanying microbes has been even less forthcoming. In fact after ten years of effort, no microbe-free (axenic) cultures exist. More disturbing is the fact that those cultures that do exist are considered fragile (Anderson et al., 1993); that is, they are less healthy than naturally collected specimens. This strongly suggests that not even a suitable liquid medium has been identified for culturing this critical species.

The story of brown tides parallels that of red tides. Environmental triggers are assumed, but none are securely identified. Conflicting information exists. The bloom occurs during periods of low rainfall (Cosper et al., 1987), but does not appear to be the result of decreased nutrients due to decreased land runoff (Nixon et al., 1994). The bloom is not a response to high nitrogen or phosphorus concentrations (Nixon et al., 1994) although Cosper and her coworkers have observed that it responds to selected additions of trace elements (Milligan and Cosper, 1994). These *in situ* and *in vitro* observations would support the results of Keller and Rice's mesocosm experiments (1989), suggesting that one of the advantages of *A. anophagefferens* may be its ability to grow at relatively low macronutrient levels.

Its basic environmental requirements are not known (Anderson et al., 1993). Yentsch and his coworkers (1989) consider it an oceanic species. Yet, Sieburth and his coworkers (1988) have found that it can exist in salinities in the mid 20-parts-per-thousand range. Anderson and his coworkers (1993) find that it thrives in relatively high ambient temperatures. Yet, it successfully overwinters without encysting. Day length cannot provide a trigger since at the same longitude it began both in May (Barnaget Bay, New Jersey) and in July (Long Island Sound, NY) in 1995. (Cosper and her coworkers compiled the knowledge accumulated up to 1989 concerning the organism in a most valuable volume.)

At least the mechanism of its interference with bivalve survival is beginning to be understood. Clearly the animal's response is to cease filter-feeding. Gainey and Shumway (1991) suggest that on contact with the gill of sensitive filter-feeders a digestive enzyme is released from the gill. This digests the extracellular coat of the algae and releases a water soluble dopamine-mimetic compound that inhibits lateral cilia. This fits well with the conclusions of Ward and Targett (1989) that the filtering apparatus of *Mytilus edulis* can be shut down via pre-ingestion chemical cues from other marine microalgae. A final positive note concerning possible future control of brown tides—a virus that selectively infects *A. anophagefferens* has been found (Milligan and Cosper, 1994). With the investment of time and effort, it may ultimately be possible to develop a biological control of at least some brown tides by using this virus.

11.5.5 DEVASTATING NEWCOMER

The notoriety associated with its aerosols, which sicken fishermen and cause a disturbing array of physical and mental damages to laboratory personnel, and the apparently aggressive predatory capacity of the fish eating dinoflagellate, *Pfiesteria piscicida,* are fodder for science fiction writers (World Water and Environmental Engineering, 1996). This dinoflagellate, capable of damaging aquatic organisms from bacteria to fish, represents a threat, at once both new and novel, to fisheries along the southeastern coast of the United States. Because it also represents a public health risk, it has become a target not only of scientific and commercial interest, but also of political concern.

The pioneering work of Burkholder and her colleagues (Burkholder et al., 1995), aimed at a better understanding of this new phenomenon has proven costly in personal terms since she and several colleagues have been adversely affected by the aerosols produced by cultures of *P. piscicida* in their laboratories. Their studies, now accomplished under conditions of stringent isolation, are aimed at a better understanding of intra- and extracellular products of these extraordinary creatures, and at the environmental parameters and fisheries management practices that favor their dominance in plankton communities along the southeastern coast of the United States. To date this laboratory has produced solid data confirming the quite unexpected ferocity of this single-celled organism. Ultimately, their future success may determine the commercial survival of several major estuarine fisheries.

11.6 PHARMACEUTICAL CONSIDERATIONS

11.6.1 HISTORIC PERSPECTIVE

From De Giaxa's early reports (1889) of an anti-cholera substance produced by marine bacteria to the current listings on the NIH (National Institutes of Health, 1995) internet gopher server, aquatic organisms are acknowledged to be producers of "something" antibiotic in nature. Yet none have proven useful for general clinical applications.

Promise came early with Pratt's isolation of chlorellin (1940), Steemann-Nielsen's (1955) *in vitro* demonstration of inhibition of microbial respiration by antibiotics produced by either freshwater or marine algae, and Duff, Bruce, and Antia's (1966) demonstration of antibiotics from algae. Yet, in 1979 Faulkner reports a general disappointment that followed both the 1969 conference on Drugs from the Sea, and a 1978 conference on the same topic. Still, he enthusiastically reports dozens of promising results, and suggests a bright future for antibiotics from the sea. Austin (1989) tells us that the pharmaceutical compounds from marine bacteria are unique. Again, none are currently in general use.

As early as 1966 Baarn and his coworkers (1966a, b) concluded that there were more Gram-positive than Gram-negative marine producers of antibiotic materials. Of greater significance, in light of recent indications that hospital *Staphylococcus* infections are becoming increasingly more difficult to control (Begley, 1994), they also found that more Gram-positive bacteria (in particular, *Staphylococcus aureus*) were susceptible to the antibiotics from marine organisms. They isolated and reared the producing bacteria; thus, there can be no question as to the source of the active materials. In contrast many of the early reports of the antibacterial activity of seawater (Steeman-Nielsen, 1955) are open to challenges concerning the possible products of non-algal forms in their somewhat mixed study samples. In 1989 Kellam and Walker screened 132 marine algae for antibacterial activity. They found much promise in general. They also found, reminiscent of the results of Baarn and his coworkers, that Gram-positive forms were more sensitive than Gram-negative forms, and that *Staphylococcus aureus* was quite sensitive to the materials from these marine organisms—this time, from algae.

11.6.2 NEW ANTIBIOTICS

Current searches for allelochemical pharmaceuticals require an understanding of their roles in nature. Without this, the decisions concerning just where to look for allelochemicals will be random. That is, appreciation for the *in situ* value of investing resources in the production of an allelochemical, if known, can help in selecting sites for the hunt. Faulkner (1979), discussing the work of A. J. Weinheimer and his co-workers, notes that they found much higher rates of antibiotic production in marine than terrestrial organisms; that invertebrates more than algae are producers; that sessile forms are more likely sources than mobile forms; and that producers are more likely to be found in tropical than in temperate waters. All in all, his conclusion is that there is great value in using ecological principles in the hunt for antibiotics. Surely in agreement, David Newman, project officer for the Marine Collection Program of the National Cancer Institute, NIH, is quoted (Rouhi, 1995) as saying that any organism that is "fat, fleshy, slow moving, and brightly colored" is a candidate for antibiotic (antitumor) study since it needs defense as it moves through the water with a sign that says, "Eat me!"

Since there appears to be agreement that there are abundant antibiotic materials produced in aquatic systems, the current absence of approved drugs from marine sources requires consideration. First, one cannot find what is never sought. Aquatic systems have not been studied with the intensity of terrestrial systems. Also, there are fewer native medicines (used by "primitive" tribes) likely to have been taken in the past from inaccessible marine systems; thus, there have been fewer clues to pursue. But, it is not simply a matter of less attention to aquatic systems, it is also a problem of the state-of-the-art in aquatic culture.

There are no defined media in which marine organisms can be routinely grown. Luigi Provasoli (with his colleagues and students) invested a productive lifetime in the pursuit of complete and defined media-nutritional systems for maintenance of normal, healthy aquatic organisms. He developed many media for both autotrophs and heterotrophs (commonly phytoplankters and zooplankters), but was never satisfied that sufficient information was available concerning the trace nutrient requirements of aquatic organisms. The state-of-the-art in chemical manufacturing interfered with his examinations of the ultra trace element requirements of cultured organisms. Only in the last decade were sufficiently purified chemicals available for test media. While he enjoyed the promise of future incontrovertible proofs that aquatic organisms require a long list of ultra trace elements (Table 11.1), he is no longer with us to observe their development. This writing is dedicated to him, with gratitude, *in memoriam.*

ACKNOWLEDGMENTS

The editorial assistance of Dr. N. Adin is gratefully acknowledged. This work was supported by the New Jersey Agricultural Experiment Station and by additional state funds. NJAES number D-07130297.

REFERENCES

Akehurst, S., Observations on pond life, with special reference to the possible causation of swarming of phytoplankton, *J. R. Microscop. Soc.,* 51, 237, 1931.

Anderson, D., Keafer, B., Kulis, D., Waters, R., and Nuzzi, R., An immuno-fluorescent survey of the brown tide Chrysophyte *Aureococcus anophageʃferens* along the northeast coast of the United States, *J. Plankton Res.,* 15, 563, 1993.

Apstein, C., *Das Süsswasserplankton Methode und Resultäte der Quantitätiven Untersuchung,* University of Kiel, 1896, 200.

Aubert, M., Télémediateurs chimiques et équilibre biologique océanique. Prèmiere Partie. Theorie generale. Nature chemique de l'inhibiteur de la synthèse d'un antibiotique produit par un diatomée, *Revue Internat. Océanogr. Médical,* 21, 5, 1971a.

Aubert, M., Télémédiateurs chimiques et équilibre biologique océanique. Deuxieme Partie. Nature chemique de l'inhibiteur de la synthese d'un antibiotique produit par un diatomée, *Revue Internat. Océanogr. Médical,* 21, 17, 1971b.

Aubert, M., Télémédiateurs et rapports inter-espèces dans le domaine des micro-organismes marine, in *Actualités de biochimie marine,* Colloquium GABIM-CNRS, 1978, 179 (cited in Maestrini and Bonin, 1981).

Aubert, M., Pesando, D., and Gauthier, M., Phénomènes d'antibiose d'origine phytoplanktonique en milieu marine. Substances antibactériènne produites par un diatomée *Asterionella japonica* (Cleve), *Revue Internat. Océanogr. Médical,* 18, 69, 1970a.

Aubert, M., Pesando, D., and Pincemin, J., Médiateurs chimiques et relations inter-espèces. Mise en évidence d'un inhibiteur de synthèse métabolique d'une diatomée produit par un péridinien (étude *"in vitro"*), *Revue Internat. Océanogr. Médical,* 17, 5, 1970b.

Austin, B., Novel pharmaceutical compounds from marine bacteria, *J. Appl. Bacteriol.,* 67, 461, 1989.

Baarn, R., Gandhi, N., and Freitas, Y., Antibiotic activity of marine micro-organisms, *Helgo. Wissensch. Meeres.,* 13, 181, 1966a.

Baarn, R., Gandhi, N., and Freitas, Y., Antibiotic activity of marine micro-organisms: the anti-bacterial spectrum, *Helgoländer Wissenschaftliche Meeresuntersuchungen,* 13, 188, 1966b.

Black, A. and Dodson, S., Demographic costs of *Chaoborus*-induced phenotypic plasticity in *Daphnia pulex, Oecologia,* 83, 117, 1990.

Ballantine, D. and Abbott, B., Toxic marine flagellates; their occurrence and physiological effects on animals, *J. Gen. Microbiol.,* 16, 274, 1957.

Begley, S., The end of antibiotics, *Newsweek,* p. 46, March 28, 1994.

Burkholder, J., Glasgow, C., and Hobbs, C., Fish kills linked to a toxic Ambush-predator dinoflagellate: Distribution and environmental conditions, *Mar. Ecol. Prog. Ser.,* 124, 43, 1995.

Carlucci, A. and Bowes, P., Production of vitamin B12, thiamin, and biotin by phytoplankton, *J. Phycol.,* 6, 351, 1970a.

Carlucci, A. and Bowes, P., Vitamin production and utilization by phytoplankton in mixed culture, *J. Phycol.,* 6, 393, 1970b.

Collins, M., Algal toxins, *Microbiol. Rev.,* 42, 725, 1978.

Conklin, D. and Provasoli, L., Nutritional requirements of the water flea *Moina macrocopa, Biol. Bull.,* 152, 337, 1977.

Conklin, D. and Provasoli, L., Biphasic particulate media for the culture of filter-feeders, *Biol. Bull.,* 154, 47, 1978.

Cosper, E., Culturing the "Brown Tide" alga, *Appl. Phycol. Forum,* 4, 3, 1987.

Cosper, E., Carpenter, E., and Bricelj, V., Eds., *Novel Phytoplankton Blooms: Causes, and Impacts of Recurrent Brown Tides and Other Unusual Blooms,* Springer Verlag, Berlin, 1989.

Cosper, E., Dennison, W., Carpenter, E., Bricelj, V., Mitchell, J., Kuenstner, S., Colflesh, D., and Dewey, M., Recurrent and persistent brown tide blooms perturb coastal marine ecosystem, *Estuaries,* 10, 284, 1987.

Daisley, K., The occurrence and nature of *Euglena gracilis* proteins that bind vitamin B_{12}, *Internat. J. Biochem.,* 1, 561, 1970.

DeMott, W., Zhang, Q., and Carmichael, W., Effect of toxic cyanobacteria on the survival and feeding of a copepod and three species of *Daphnia, Limnol. Oceanogr.,* 36, 1346, 1991.

De Giaxa, Über das Verhalten einiger pathogener Mikro-organismen in Meerwasser, *Zentralb. Hyb. und Infekt.,* 6, 162, 1889.

DeYoe, H. and Suttle, C., The inability of the Texas "Brown Tide" alga to use nitrate and the role of nitrogen in the initiation of a persistent bloom of this organism, *J. Phycol.,* 30, 800, 1994.

Dodson, S., Predation of *Heterocope septentrionalis* on two species of *Daphnia:* morphological defenses and their cost, *Ecology,* 65, 1249, 1984.

Dodson, S., The ecological role of chemical stimuli for the zooplankton: predator-avoidance behavior in *Daphnia, Limnol. Oceanogr.,* 33, 1431, 1988.

Dodson, S., The ecological role of chemical stimuli for the zooplankton: predator-induced morphology in *Daphnia, Oecologia,* 78, 361, 1989.

Dodson, S., Hanazato, T., and Gorski, P., Behavioral responses of *Daphnia pulex* exposed to carbaryl and *Chaoborus* kairomone, *Environ. Toxicol. Chem.,* 14, 43, 1995.

Dodson, S. and Havel, J., Indirect effects: some morphological and life history responses of *Daphnia pulex* exposed to *Notonecta undulata, Limnol. Oceanogr.,* 33, 1274, 1988.

Droop, M., Vitamin B_{12} and marine ecology. IV. The kinetics of uptake, growth and inhibition in *Monochrysis lutheri, J. Mar. Biol. Assoc. UK,* 48, 689, 1968.

Duff, D., Bruce, D., and Antia, N., The antibacterial activity of marine planktonic algae, *Can. J. Microb.,* 12, 876, 1966.

Faulkner, D., The search for drugs from the sea, *Oceanus,* 22(2), 44, 1979.

Fogg, G., The production of extracellular nitrogenous substances by a blue-green algae, *Proc. Roy. Soc. Brit.,* 139, 372, 1952.

Fogg, G., Extracellular products, in *Physiology and Biochemistry of Algae,* Lewin, R., Ed., Academic Press, NY, 1962, 475.

Fogg, G., The extracellular products of algae, *Oceanogr. Mar. Biol. Ann. Rev.,* 4, 195, 1966.

Fogg, G., The extracellular products of Algae in fresh water, *Archiv für Hydrobiol.,* 5, 1, 1971.

Fogg, G., Nalewajko, C., and Watt, W., Extracellular products of phytoplankton photosynthesis, *Proc. Roy. Soc. Brit.,* 162, 517, 1965.

Folt, C. and Goldman, C., Allelopathy between zooplankton: a mechanism for interference competition, *Science,* 213, 1133, 1981.

Francis, G., Poisonous Australian lake, *Nature,* 18, 11, 1878.

Gainey, L. and Shumway, S., The physiological effects of *Aureococcus anophagefferens* ("brown tide") on the lateral cilia of bivalve mollusks, *Biol. Bull.,* 181, 298, 1991.

Galstoff, P., Red tide: progress report, *Fish and Wildlife Service, Special Scientific Report,* No. 46, 1948.

Garrison, D., (quoted in) Earth Almanac, *Natl. Geograph.* 188(3), 1995.

Gilbert, J., Polymorphism and sexuality in the rotifer *Asplanchna,* with special reference to the effects of prey-type and clonal variation, *Archiv für Hydrobiol.,* 75, 442, 1975.

Gilbert, J., Feeding in the rotifer *Asplanchna:* behavior, cannibalism, selectivity, prey defenses, and impact on rotifer communities, *Limnol. Oceanogr. Symp., 3,* 158, 1980.

Gilbert, J., Developmental polymorphism in the Rotifer *Asplanchna siebaldi., Amer. Scient.* 68, 636, 1981.

Gorham, P., Toxic algae, in *Algae and Man,* Jackson, D., Ed., Plenum Press, NY, 1964, 307.

Gorham, P. and Carmichael, W., Phycotoxins from blue-green algae, *Pure Appl. Chem.,* 52, 165, 1979.

Hanazato, T., Influence of food density on the effects of a *Chaoborus*-released chemical on *Daphnia ambigua*, *Freshw. Biol.*, 25, 477, 1991.

Hanazato, T. and Dodson, S., Synergistic effects of low oxygen concentration, predator kairomone, and a pesticide on the cladoceran *Daphnia pulex*, *Limnol. Oceanogr.*, 40, 700, 1995.

Hardy, A., Plankton ecology and the hypothesis of animal exclusion, *Proc. Linn. Soc. London* 148, 64, 1936.

Hardy, A. and Gunther, E., The plankton of the South Georgia whaling grounds and adjacent waters, 1926–1927, *Disc. Rep.*, 11, 1, 1935.

Hayes, L., Landau, H., and Austin, T., The distribution of discolored sea water, *Texas J. Sci.*, 4, 530, 1951.

Hellebust, J., Extracellular products, in *Algal Physiology and Biochemistry*, Stewart, W., Ed., Botanical Monographs 10, University of California Press, Berkeley, 1974, 838.

Hutner, S. and McLaughlin, J., Poisonous tides, *Scient. Amer.*, 199(8), 92, 1958.

Ignatiades, L. and Fogg, G., Studies on the factors affecting the release of organic matter by *Skeletonema costatum* (Greville) Cleve in culture, *J. Marine Biol. Assoc. UK*, 73, 1973.

Inderjit, and Dakshini, K. M. M., Algal allelopathy, *Bot. Rev.*, 60, 182, 1994.

Keating, K. I., *Algal Metabolite Influence on Bloom Sequence in Eutrophic Freshwater Ponds*, E.P.A. Ecological Research Monograph Series (EPA-600/3-76-081), Government Printing Office, Washington, D.C., 1976, 148.

Keating, K. I., Allelopathic influence on blue-green bloom sequence in a eutrophic lake, *Science*, 196, 885, 1977.

Keating, K. I., Blue-green algal inhibition of diatom growth: transition from mesotrophic to eutrophic community structure, *Science*, 199, 971, 1978.

Keating, K. I., Extracellular metabolite involvement in plankton community structure, in *Algal Management And Control*, Proceedings, USEPA/Army Corps. Engineering Conference, Monterey, 1981, 146.

Keating, K. I. and Dagbusan, B. C., The effect of selenium deficiency on cuticle integrity in the Cladocera (Crustacea), *Proc. Natl. Acad. Sci. USA*, 81, 3433, 1984.

Keating, K. I., Aquatic allelochemistry: challenges, in *Allelochemicals: Role in Agriculture and Forestry*, Waller, G., Ed., American Chemical Society Symposium Series 330, ACS, Washington, D.C., 1987, 136.

Keating, K. I. and Caffrey, P. B., A selenium deficiency induced by zinc deprivation, *Proc. Natl. Acad. Sci. USA*, 86, 6436, 1989.

Kellam, S. and Walker, J., Antibacterial activity from marine microalgae in laboratory culture, *Brit. Phycol. J.*, 24, 191, 1989.

Keller, A. and Rice, R., Effects of nutrient enrichment on natural populations of the brown tide phytoplankton *Aureococcus anophagefferens* (Chrysophyceae), *J. Phycol.*, 25, 632, 1989.

Kirk, K. and Gilbert, J., Variation in herbivore response to chemical defenses: zooplankton foraging on toxic cyanobacteria, *Ecology*, 73, 2208, 1992.

Lampert, W., Further studies on the effect of the toxic blue-green *Microcystis aeruginosa* on the filtering rate of zooplankton, *Archiv für Hydrobiol.*, 95, 207, 1982.

Lefèvre, M., *Aphanizomenon gracile* Lemm. Cyanophyte défavorable au zooplankton, *Ann. Sta. Centr. Hydro. App.*, 3, 205, 1950.

Lefèvre, M. and Jakob, H., Sur quelques propriétés des substances activers tirées des cultures d'algues d'eau douce, *Comptes Rendus Acad. Sci.*, 229, 234, 1949.

Lefèvre, M., Jakob, H., and Nisbet, M., Auto-et hétéroantagonisme chez les algues d'eau douce *in vitro* et dans les collections d'eau naturelles, *Ann. Sta. Centr. Hydro. App.*, 4, 1, 1952.

Lefèvre, M. and Nisbit, M., Sur la sécrétion par certaines espèces d'algues, de substances inhibitrices d'autres espèces d'algues, *Comptes Rendus Acad. Sci.*, 226, 107, 1948.

Lucas, C., The ecological effects of external metabolites. *Biol. Rev., Cambridge Philosophil. Soc.*, 22, 270, 1947.

Machácek, J., Comparison of the responses of *Daphnia galeata* and *Daphnia obtusa* to fish-produced chemical substance, *Limnol. Oceanogr.*, 38, 1544, 1993.

Maestrini, S. and Bonin, D., Allelopathic relationships between phytoplankton species, *Can. Bull. Fish. Aquat. Sci.*, 210, 323, 1981.

McFarren, E., Schafer, M., Cambell, J., Lewis, K., Jensen, E., and Schantz, E., Public health significance of paralytic shellfish poison: a review of literature and unpublished research, *Proc. Natl. Shellfish. Assoc.*, 47, 114, 1957.

Milligan, K. and Cosper, E., Isolation of virus capable of lysing the brown tide microalga, *Aureococcus anophagefferens*, *Science*, 266, 805, 1994.

Molisch, H., *Die Einfluss einer Pflanze auf die Andere—Allelopathie,* Fisher Verlag, Jena, Germany, 1937.

National Institutes of Health, USA, 1995.

Nixon, S., Granger, S., Taylor, D., Johnson, P., and Buckley, B., Subtidal volume fluxes, nutrient inputs and the brown tide—An alternative hypothesis, *Estuarine, Coastal, Shelf Sci.,* 39, 303, 1994.

Nuzzi, R., New York's brown tide, *The Conservationist,* Department of Environmental Conservation, NY, p30, Sept.—Oct., 1988.

Oshima, Y., Buckley, L., Alam, N., and Shimizu, Y., Heterogeneity of paralytic shellfish poisons. Three new toxins from cultured *Gonyaulax tamarensis* cells, *Mya arenaria* and *Saxidomus giganteus, Comp. Biochem. Physiol.,* 57, 31, 1977.

Parejko, K. and Dodson, S., Progress towards characterization of a predator/prey kairomone: *Daphnia pulex* and *Chaoborus americanus, Hydrobiologia,* 198, 51, 1990.

Pignatello, J., Porwall, J., Carlson, R., Xavier, A., Gleason, F., and Wood, J., Structure of the antibiotic cyanobacterin, a chlorine-containing Y-lactone from freshwater cyanobacterium, *Scytonema hofmanni, J. Org. Chem.,* 48, 4035, 1983.

Pourriot, R., Metabolites externes et interactions biochemiques chez les organisms aquatiques, *Année Biol.,* 7–8, 337, 1966.

Pratt, R., Studies on *Chlorella vulgaris.* I. Influence of the size of the inoculum on the growth of *Chlorella vulgaris* in freshly prepared culture medium, *Am. J. Bot.,* 27, 52, 1940.

Pratt, R., Studies on *Chlorella vulgaris.* V. Some properties of the growth-inhibitor formed by *Chlorella* cells, *Am. J. Bot.,* 29, 142, 1942.

Pratt, R., Studies on *Chlorella vulgaris.* VI. Retardation of photosynthesis by growth-inhibiting substance from *Chlorella vulgaris, Am. J. Bot.,* 30, 32, 1943.

Pratt, R. and Fong, J., Studies on *Chlorella vulgaris.* II. Further evidence that *Chlorella* cells form a growth-inhibiting substance, *Am. J. Bot.,* 27, 431, 1940.

Pratt, R. and Fong, J., Chlorellin, an antibacterial substance from *Chlorella, Science,* 99, 51, 1944.

Provasoli, L. and Carlucci, A., Vitamins and growth regulators, in *Algal Physiology and Biochemistry,* Stewart, W., Ed., Botanical Monographs 10, University of California Press, Berkeley, 1974, 741.

Pütter, A., Der Stoffhaushalt des Meeres. *Zeitsch. allge. Physiol.,* 7, 321, 1908.

Rouhi, A., Supply issues complicate trek of chemicals from sea to market, *Chem. Eng. News,* 73(47), 42, 1995.

Ryther, J., Inhibitory effects of phytoplankter on the feeding of *Daphnia magna* with reference to growth, reproduction, and survival, *Ecology,* 35, 522, 1954.

Schwimmer, M. and Schwimmer, D., *The Role of Algae and Plankton in Medicine,* Grune & Stratton, NY, 1955, 85.

Schwimmer, M. and Schwimmer, D., Algae and medicine, in *Algae and Man,* Jackson, D., Ed., Plenum Press, NY, 1964, 368.

Sieburth, J., Johnson, P., and Hargraves, P., Ultrastructure of *Aureococcus anophagefferens gen. et sp. nov.* (Chrysophyceae): the dominant picoplankter during a bloom in Narragansett Bay, Rhode Island, summer 1985, *J. Phycol.,* 24, 416, 1988.

Steemann-Nielsen, E., An effect of antibiotics produced by plankton algae, *Nature,* 176, 553, 1955.

Tracy, G., Feeding reduction, reproductive failure, and mortality in *Mytilus edulis* during the 1985 "brown tide" in Narragansett Bay, Rhode Island, *Mar. Ecol. Prog. Ser.,* 50, 73, 1988.

Ward, J. and Targett, N., Influence of marine microbial metabolites on feeding behavior of the blue mussel *Mytilus edulis, Mar. Biol.,* 101, 313, 1989.

Watanabe, A., Production in cultural solution of some amino acids by the atmospheric nitrogen fixing blue-green algae, *Arch. Biochem. Biophys.,* 34, 50, 1951.

Watanabe, M., Kaya, K., and Takanwia, N., Fate of the toxic cyclic heptapeptides, the microcystins, from blooms of *Microcystis* (cyanobacteria) in a hypertrophic lake, *J. Phycol.,* 28, 761, 1992.

Killer plankton responsible for millions of fish deaths, in *World Water and Environmental Engineering,* 1996, 6.

Yentsch, C., Phinney, D., and Shapiro, L., Absorption and fluorescent characteristics of the brown tide Chrysophyte—Its role in light reduction in coastal marine environments, in *Coastal and Estuarine Studies,* Cosper, E., Bricelj, V., and Carpenter, E., Eds., Springer-Verlag, Berlin, 1989, 77.

12 Allelopathy in Benthic and Littoral Areas: Case Studies on Allelochemicals from Benthic Cyanobacteria and Submersed Macrophytes

Elisabeth M. Gross

CONTENTS

12.1 ABSTRACT

Photosynthetic organisms in littoral and benthic habitats are in general situated in close proximity to their competitors. Benthic algae and cyanobacteria compete for space and grow adjacent to each other. Submersed macrophytes can be overgrown by epiphytic algae and are surrounded by other macrophytes and phytoplankton. The particular spatial setup in these habitats makes allelopathy a powerful strategy. Released compounds can more or less directly reach and act on target organisms. Allelopathic compounds may either be released into the water or transferred by direct cell–cell contact from donor to recipient.

This chapter is not intended as a review on allelopathy in benthic and littoral areas. Two case studies of allelopathic interactions are discussed: 1) photosynthesis inhibitors in the benthic cyanobacteria *Fischerella muscicola* and *F. ambigua*, and 2) algicidal hydrolyzable polyphenols from the submersed macrophyte *Myriophyllum spicatum.*

12.2 INTRODUCTION

Benthic cyanobacteria and submersed macrophytes face a similar challenge; they both have to compete for light and, therefore, for space with other photosynthetically active organisms. Direct competition for nutrients is generally less pronounced between submersed macrophytes and algae. Macrophytes obtain their major nutrients from the sediment, whereas algae obtain nutrients from the open water (Carignan and Kalff, 1980, 1982; Phillips et al., 1978). Nutrient competition can play a role in cyanobacterial mats (Fong et al., 1993). In both the benthic and the littoral habitat, the spatial distance between organisms is relatively small compared with the situation in the open water. Nevertheless, both systems have specific characteristics that influence the interaction with other competing primary producers. In the benthic system, cyanobacterial mats often consist of only one species. Light and oxygen gradients are very steep (Revsbech et al., 1989). Near-damaging light intensities at the surface of such mats are only millimeters away from zones of light limitation lower in the cyanobacterial mat, generating a highly stratified environment. Potential hazards arise from other cyanobacteria that would overgrow the mat and deprive it of sufficient light. Possible counterbalancing strategies would be to grow faster or to release growth inhibitors. Cell–cell contact between competing species is highly possible, thus facilitating the direct transfer of allelochemicals.

In the littoral system, thick epiphyte cover, scums of filamentous algae, and high phytoplankton densities can deprive submersed macrophytes of sufficient light. In addition, with increasing depth, light intensity decreases and the light spectrum changes. Counterbalancing strategies used by submersed macrophytes include the formation of finely dissected leaves, low light and CO_2-compensation points, fast growth, canopy building (Wetzel, 1983), and the excretion of algal growth inhibitors (Wium-Andersen, 1987; Gopal and Goel, 1993). Epiphytes covering submersed macrophytes would be directly subjected to allelochemicals released by the host macrophyte. But even phytoplankton in littoral zones can be affected since the water exchange in dense macrophyte beds is very limited (Losee and Wetzel, 1993). Because the competing species in both systems are in close physical proximity, potential allelochemicals can act more efficiently and are not diluted as in the open water. Therefore, allelopathy in these environments would be a forceful strategy.

Since Molisch (1937) coined the term, allelopathy has been used often in conflicting ways. Molisch (1937) used allelopathy only to describe biochemical interactions among higher plants and between higher plants and microorganisms. However, other authors use the term allelopathy for biochemical interactions between plants and animals and among animals (Seitz, 1984; Rizvi and Rizvi, 1992). Whittaker and Feeny (1971) proposed the use of *allelochemics*. They wrote: "We review here a class of interactions termed *allelochemic* (. . .) involving chemicals by which organisms of one species affect the growth, health, behavior, or population biology of organisms of another species (excluding substances used only as foods by the second species)." Thus, the term *allelopathy* should only be used for biochemical interactions among plants and between plants and microorganisms. *Allelochemistry* would be the overall concept and would include biochemical interactions of any two different species, including animals. The same plant-derived compound may affect other primary producers as well as animals, so that we should use the precise terms to describe each specific interaction.

The original definition of allelopathy by Molisch (1937) includes both stimulatory and inhibitory actions. However, the majority of allelopathic studies focuses on negative effects exerted by one species upon another. This is reflected in a too restrictive translation of the term allelopathy. Allelopathy is derived from the Greek words *allelon* meaning 'one another,' 'mutually,' or 'in turn' and *pathos* meaning 'suffering,' but also 'accident,' 'experience,' 'feeling.' Most definitions of allelopathy only use the translation 'suffering' for *pathos*. *Pathos,* as used for example in "sympathetic" or "pathetic" clearly translates not as 'suffering' but rather as 'feeling' or 'sensitive.' Rice (1974) concentrated exclusively on inhibitory allelopathic interactions in his first volume, but corrected this perception in the second volume (1984). It is now commonly accepted that certain allelochemicals may stimulate target organisms at very low concentrations but inhibit at higher concentrations (Rice, 1984).

All proposed new definitions for allelopathic interactions (see Willis, 1994) do not seem to improve the original meaning but only to cause more confusion. Staying closer to the original definition by Molisch (1937) will allow most of the processes observed in this field to be covered (see also discussion in Rice, 1974, 1984).

Whereas allelopathy still seems to be a mechanism that has not received its appropriate attention, more and more scientists show interest in chemical ecology. In the field of limnology this is reflected by recent reviews on planktonic (Larsson and Dodson, 1993) and benthic organisms (Dodson et al., 1993). Chemical ecology tries to close the gap between natural product chemistry and ecology. In the past, natural product chemists would often describe new compounds from (aquatic) organisms without paying attention to their ecological function. On the other side, ecologist and limnologists investigate chemical cues responsible for observed patterns without knowing about the molecular structure of the compounds involved. The increased use of analytical techniques in ecology and limnology and more frequent cooperations of ecologists and natural product chemists have greatly improved our understanding of allelochemicals.

Allelochemicals found in cyanobacteria and macrophytes are mainly secondary metabolites. The wide array of secondary metabolites in cyanobacteria attracts natural product chemists because many of these compounds show pharmacological activity. Few of these studies cover ecological aspects of these compounds. Still, we can hypothesize that these compounds are biosynthesized not for pharmaceutical reasons but for a purpose beneficial to the producing organism. Every year, new compounds and even new classes of compounds are elucidated from cyanobacteria (Patterson et al., 1994; Borowitzka, 1995). The situation is totally different for macrophytes. Submersed macrophytes are thought to be inferior in secondary metabolism (McClure, 1970) compared with emerged macrophytes or even terrestrial vegetation. Macrophytes similar to other angiosperms will most likely produce well-known compounds or derivatives of known compounds as allelochemicals. The most recent review on macrophyte secondary metabolites can be found in the volumes of "Chemotaxonomie der Pflanzen" (translation, Chemotaxonomy of Plants) (Hegnauer, 1962–1995). Only a couple of other surveys cover this field (Bate-Smith, 1962; McClure, 1970; Su and Staba, 1973; Hutchinson, 1975; Ostrofsky and Zettler, 1986; Pip, 1992). However, without more convincing data the conclusion that (submersed) macrophytes are less active producers of secondary metabolites seems premature. The increasing number of publications (Della Greca et al., 1995; Yoshikawa et al., 1993) rather indicates that this has been due to a lack of research in this field.

Two examples of allelochemical interactions will be presented here that focus on the benthic cyanobacterium *Fischerella* and the submersed macrophyte *Myriophyllum*. Both studies involve a threefold approach: 1) to isolate and identify the major allelochemical(s), 2) to study possible release mechanisms for the allelochemicals in order to evaluate how they reach target organisms, and 3) to reveal the mode of action of the active compounds to its target organism. Identifying the active compounds present in the producing organisms is essential for investigating whether or not this compound is released into the environment. Controlled studies can provide insight into mechanisms ruling allelopathy. Both studies would have been impossible without axenic cultures. Non-axenic cultures always have accompanying organisms that interfere with the identification of released compounds. It is nearly impossible to separate the organic compounds excreted by accompanying organisms from those of the allelopathically active organism. Additionally, the accompanying organisms can metabolize allelochemicals and change their original structure. In both cases presented here, the structure of the algicides hints at the possible mode of action, or *vice versa*. Overall, this threefold approach covers a wide range of possible interactions between producer and target organism and permits the evaluation of ecological implications, even when substantial methodical problems prevent field studies.

12.3 LIPOPHILIC ALLELOCHEMICALS FROM BENTHIC CYANOBACTERIA: PHOTOSYNTHETIC INHIBITORS FROM *FISCHERELLA*

Allelopathic interactions of algae, although not referred to as such, were observed as early as 1917 by Harder. He found that aging cultures of the cyanobacterium *Nostoc* showed a decreased growthrate and explained this by the accumulation of autotoxic organic compounds in the culture medium. Biochemical interactions as a cause for phytoplankton succession were first proposed by Akehurst (1931). Further information was gained with cultures of algae and cyanobacteria by Rice, 1954; Proctor, 1957; Pratt et al., 1944; Pratt, 1966; and Harris, 1971. None of these studies allowed an estimation of the importance of allelopathy under natural conditions. Reviews on algal allelopathy can be found in articles by Rice (1984), and Inderjit and Dakshini (1994). The most convincing experiments for the involvement of allelopathy in phytoplankton succession were conducted by Keating (1977, 1978). Two thirds of her mostly axenic algal isolates from Linslay Pond exhibited allelopathic activity. The direction of the allelopathic effects followed the natural phytoplankton succession: the culture filtrate of one algae or a certain cyanobacterium showed neutral or inhibitory activity toward the predecessor but neutral or stimulatory activity towards the following algae.

Allelochemicals released by planktonic algae and cyanobacteria must be effective at very low concentrations considering high dilution effects. For this reason, Lewis (1986) questioned if allelopathy is an important factor in phytoplankton succession, since individual cells or colonies are separated from each other by large distances—dozens to hundreds of cell diameters—even in dense populations. However, in benthic habitats allelopathy should be a powerful competitive strategy because of the proximity of adjacent species. In order to avoid light limitation, benthic cyanobacteria have to compete for space with other cyanobacteria and benthic algae. Cyanobacterial mats consist very often of only one species, suggesting that this species has out-competed others, either by fast growth or due to allelopathically active compounds. Strong evidence for the involvement of allelochemistry in this habitat is provided by the fact that most of the yet chemically characterized allelochemicals from cyanobacteria have been isolated from benthic species. For example, *Scytonema hoffmanni* produces the low molecular lipophilic compound cyanobacterin (Pignatello et al., 1983) that inhibits photosynthetic electron flow in other algae (Gleason and Paulson, 1984), but also in chloroplasts isolated from angiosperms (Gleason and Case, 1986). Algicidal indolederivatives have been isolated from *Hapalosiphon fontinalis* (Moore et al., 1984).

Fischerella muscicola turned out to be the most active species in a screening of 65 filamentous, nitrogen-fixing cyanobacteria for the production of cyanophages or cyanobactericidal compounds (Flores and Wolk, 1986). None of the investigated strains contained cyanophages but seven produced cyanobactericidal compounds. Our studies used a combined effort to isolate and identify the main allelochemicals from *Fischerella* and consequently study possible release mechanisms and the mode of action (Gross et al., 1991, 1994; Papke et al., 1997). The following sections focus especially on particular features of the *Fischerella* allelochemicals: 1) the structure elucidation, 2) the transfer to target organisms, 3) their herbicidal action, and 4) environmental factors influencing their production.

12.3.1   STRUCTURE ELUCIDATION OF *FISCHERELLA* ALLELOCHEMICALS

Fischerella muscicola and *F. ambigua* produce allelopathically active compounds against other cyanobacteria, chlorophytes, and diatoms. The allelopathic activity mainly is based on one major allelochemical; some minor, less active compounds are also involved (Gross et al., 1991; Papke et al., 1997). The major allelochemical is lipophilic, protease-insensitive, and of low molecular weight (Gross et al., 1991). The isolation of this compound, which we named fischerellin (Gross et al., 1991) was guided by a special agar diffusion assay (Flores and Wolk, 1986; Gross et al., 1991). C18-HPLC and chemical derivatizations were used to isolate and characterize this allelochemical (Gross, 1990; Gross et al., 1991). The structure of the main allelochemical, now renamed more precisely fischerellin A (Figure 12.1), has been elucidated recently (Hagmann and Jüttner, 1996). The isolation yield is reported with approximately 0.005 percent of the dry weight.

Photodiode-array analysis of crude extracts from *F. muscicola* and *F. ambigua* show at least three to four compounds with ultraviolet (UV)-spectra very similar to fischerellin A (maxima at 240, 252, 267, and 283 nm). We isolated and identified one of these compounds (Papke et al., 1997) and named it fischerellin B. Fischerellin B inhibits cyanobacteria in the agar diffusion assay (Papke et al., 1997) and inhibits photosynthesis of *Trichormus variabilis* by 40 percent at a concentration of 10 μM (Gross, unpublished results). Purified fischerellin B (0.55 mg) was obtained from 70 g of freeze-dried cyanobacteria, giving a yield of less than 0.0008 percent of the dry weight. The structure was elucidated by UV-, NMR-, and mass-spectroscopy. In this case, a derivatization of the natural product and stereocontrolled synthesis of the derivative allowed the determination of the absolute configuration by means of chiral gas-chromatography. The compound was identified as (3*R*,5*S*)-3-methyl-5-((5*E*)-pentadec-5-ene-7,9-diynyl)-pyrrolidin-2-one (Figure 12.1). To our knowledge, fischerellins A and B are the first fully characterized enediyne metabolites from cyanobacteria (Hagmann and Jüttner, 1996; Papke et al., 1997).

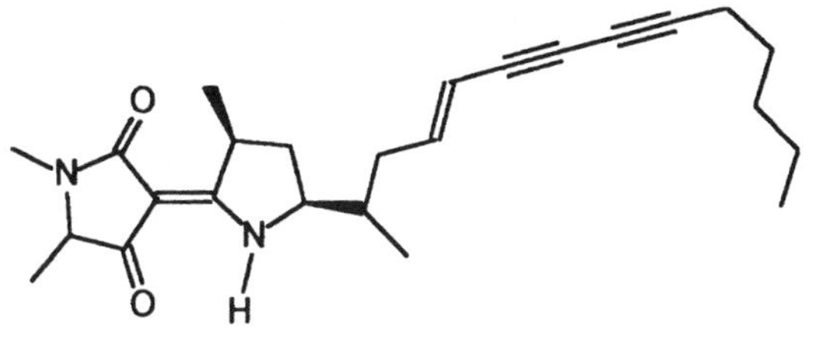

a) Fischerellin A from *Fischerella muscicola* (Hagmann and Jüttner, 1996)

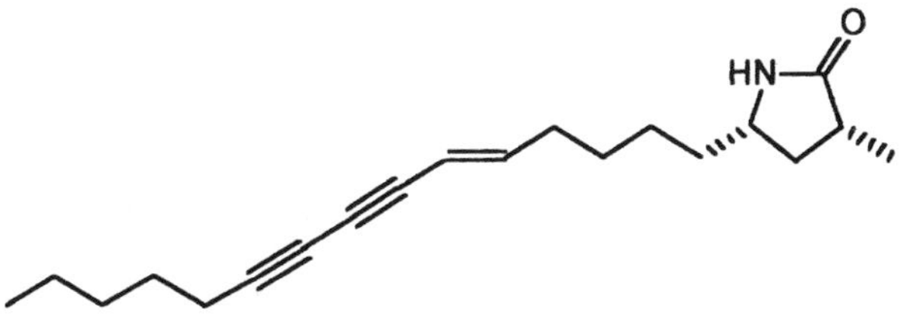

b) Fischerellin B from *Fischerella ambigua* and *F. muscicola* (Papke et al., 1997)

FIGURE 12.1   Structures of fischerellins A and B from *Fischerella muscicola* and *F. ambigua*.

These two enediyne allelochemicals seem also to be the first cyanobacterial metabolites in which the ecological function was described before their potential pharmaceutical value. Fischerellins A and B contain an enediyne sidechain; other compounds with similar structural features are used as antibiotic and anticancer drugs (Nicolaou and Dai, 1991; Smith and Nicolaou, 1996). Furthermore, structural features of the N-containing heterocyclus of fischerellins A and B indicate their activity as potent inhibitors of photosystem II electron transport (see Section 12.3.3).

12.3.2 TRANSFER OF ALLELOCHEMICALS IN BENTHIC CYANOBACTERIA

How do *Fischerella* allelochemicals come in contact with target algae and cyanobacteria? Intact cell filaments of *Fischerella* cause clearing areas in the agar diffusion assay; thus, it is obvious that somehow the allelochemical reaches other cyanobacteria and algae. Consequently, we looked for fischerellin A in the culture medium. The *Fischerella* culture medium was subjected to solid phase extraction (SPE) with C18-cartridges. The C18-adsorbent would trap lipophilic compounds released into the mineral medium. But the HPLC separation of the SPE-eluate did not exhibit any trace of fischerellin A (Gross, 1990). Furthermore, this fraction did not show allelopathic activity in our bioassay. In these experiments we found no evidence for an inhibition of indicator cells when undiluted, cell-free culture supernatant fluids from *Fischerella musicola* UTEX 1829 were used, as has been reported in the original work by Flores and Wolk (1986). The authors do not report how dense the producing strain was grown to cause such an effect. We rather hypothesize that the active compound(s) are transmitted by direct contact of cells. For this purpose the lipophilic nature and low molecular weight of fischerellins A and B would be very suitable. Lipophilic compounds of low molecular weight (MW < 600 Da) can easily pass cell membranes.

To mimic direct cell contact as it would happen in adjacently growing cyanobacteria, we used XAD beads in coculture with a *Fischerella* culture (Gross et al., 1994). The XAD-16 beads had a mean diameter of 1.6 mm and a lipophilic surface. All cyanobacteria adherent to cell surfaces, benthic as well as epilithic and symbiotic organisms, have a hydrophobic cell surface; in contrast, all planktonic cyanobacteria have a cell surface that is highly hydrophilic (Shilo, 1989). The experiments were run for one or six weeks. Fischerellin A could be detected in the methanolic extract of the XAD-beads but not in the SPE-eluate of the culture medium. The allelochemical must have bound to the surface of the XAD beads. This proved our hypothesis that *Fischerella* allelochemicals are not released into the culture medium but directly transferred to target organisms by cell–cell contact.

Further support for this hypothesis comes from the work by Flores and Wolk (1986). When producing cells were placed on a piece of dialysis membrane and actively grown for three to four days, the inhibitor diffused through the dialysis membrane and into the agar. After the membrane and cells were removed, the agar was overlaid with indicator cells in soft agar. A clearing zone could be observed where the *Fischerella* allelochemicals had passed the dialysis membrane. However, this test does not prove that fischerellins or other allelochemicals from *Fischerella* can pass dialysis membranes if they are placed into culture medium.

12.3.3 FISCHERELLINS, BIOLOGICAL HERBICIDES

Fischerella cells and isolated fischerellins coming in close contact with other cyanobacteria and algae lead to cell death in these organisms indicating that vital functions in these photosynthetic organisms are interrupted. In contrast, various Eubacteria are not inhibited by *Fischerella* or isolated fischerellin (Gross et al., 1991). The respiratory electron transport of the chlorophyte *Nannochloris* is not affected by the addition of fischerellin A, but the photosynthetic electron transport is severely inhibited. Further studies with the filamentous cyanobacterium *Trichormus (Anabaena) variabilis* ATCC 29413, strain P-9 (for renaming see Komarek and Anagnostidis, 1989) elucidated photosystem II (PS II) as the target site of fischerellin A, similar to many synthetic herbicides such as atrazine or DCMU. Structural features of the fischerellins meet those of urea and triazine herbicides. All members of the serine family of herbicides contain a lipophilic group in close association with an sp^2 hybrid and an essential positive charge (Trebst et al., 1984). Further experiments are warranted to reveal the exact mode of action of fischerellins on PS II. Photosynthesis in *Fischerella muscicola* itself was not affected by fischerellin A (Gross, 1990; Gross et al., 1991). It is not clear yet how *Fischerella* protects itself from the deleterious effect of fischerellins. Since neither whole cells nor cell fragments, which were obtained by french-press treatment, responded to the addition of fischerellin A, a structural rather than a metabolic protection is likely.

Inhibition of photosynthesis seems to be a widespread mode of action for aquatic allelochemicals, especially from benthic cyanobacteria. The cyclic sulfur compounds dithiane and trithiane from *Chara* sp. (Wium-Anderson et al., 1982) inhibit photosynthesis of diatoms and samples of natural phytoplankton; the target site of these biological herbicides is not given. Similar to fischerellin A, cyanobacterin from *Scytonema hofmanni* (Gleason and Paulson, 1984) and an allelochemical from *Oscillatoria* (Bagchi, 1995) inhibit PS II electron transport. All these cyanobacterial PS II inhibitors are lipophilic and of low molecular weight. It is striking that all these different species produce chemically different compounds with the same biological activity. Since light is the most important factor for the survival of benthic cyanobacteria, inhibition of photosynthesis of competing species would be especially effective. The widespread use of this defense strategy indicates a convergent evolution of this pattern among benthic cyanobacteria.

12.3.4 ENVIRONMENTAL FACTORS AFFECTING THE PRODUCTION OF ALLELOCHEMICALS IN *FISCHERELLA*

Allelochemicals should be especially effective and useful to the producing organism if certain essential resources are limiting. On the other hand, the biosynthesis of allelochemicals certainly involves metabolic costs, so that under limiting conditions, their production could be decreased. Examples from literature provide evidence for both cases. The content of secondary metabolites in cyanobacteria depends on the culture conditions, but is generally highest under nitrogen (N) or phosphorus (P) limitation (Moore et al., 1988). The content of the cyclic didepsipeptide cryptophycin from the cyanobacterium *Nostoc* ATCC 53789 increases under P-limitation (Schwartz et al., 1990). The same response happens with the intracellular toxin content in the dinoflagellate *Prymnesium parvum*, but toxin content stays constant during N-, thiamin, and vitamin B_{12} limitation (Shilo, 1971). P-limitation also increases toxin content in the dinoflagellate *Protogonyaulax tamarensis,* but in this case toxin levels decrease under N-limitation (Boyer et al., 1987). *Trichormus doliolum* excretes 30 times more of an allelochemical under P-limitation than in full strength medium (von Elert, 1994; von Elert and Jüttner, 1997). Since nitrogen, phosphorus, and light limitation may affect the production of allelochemicals, their impact on fischerellin production in *Fischerella muscicola* was investigated (Gross et al., 1994).

Phosphorus is often the most limiting resource for freshwater cyanobacteria (Schindler, 1977; Sommer, 1989). Assuming that the release of allelochemicals is of competitive advantage to the producing organism, this should be especially effective during peak population growth in summer when phosphorus is limiting. But P-limitation (1 μM instead of 40 μM P-PO_4) of *F. muscicola* did not change the content of fischerellin A per unit biomass. P-limited cultures reached only 40 percent of the biomass of control cultures and had 60 percent less chlorophyll content per unit biomass. This indicates that *Fischerella* invests the same amount of energy per unit biomass for the production of fischerellin, irrespective of P-availability.

Fischerella is a nitrogen fixing species, so nitrogen deprivation should not necessarily influence the production of fischerellins. Nevertheless, nitrogen depletion (1/100 of normal N-NO_3 supply, 10 μM instead of 1 M) significantly decreased fischerellin content by 75 percent. In the planktonic environment, this pattern would not be a disadvantage for *Fischerella*, since chlorophytes, diatoms, and other algae are not capable of N-fixation and would be already inhibited by N-depletion. But in benthic areas, the main competitors are other N-fixing cyanobacteria. Cyanobacteria are adapted to low light conditions. Since competition for light is a major factor for benthic cyanobacteria, it would be advantageous for *Fischerella* to maintain or even increase the production of fischerellin A under low light conditions. However, light limitation (10 compared with 60 μmol photons PAR m^{-2} sec^{-1} under normal light conditions) led to a significant decrease of fischerellin A production by more than 90 percent. It appears that both nitrogen and light depletion cause an energy shortage in the cyanobacterial cells, which then are no longer capable of synthesizing the same amount of this allelochemical. N-fixation is very costly in terms of ATP for cyanobacteria. Furthermore, fischerellin is an N-containing metabolite, and under nitrogen shortage, N rather might be used for protein biosynthesis. Light shortage reduces the ATP yield gained by photophosphorylation. These results provide evidence for the metabolic or energetic costs of the production of such allelochemicals. However, considering the high algicidal activity of fischerellins, even light- and nitrogen-depleted cells may produce sufficient quantities to inhibit other competing species.

12.4   ALLELOPATHY AND INTERFERENCE IN SUBMERSED MACROPHYTES: INSIGHTS FROM *MYRIOPHYLLUM SPICATUM*

Shallow eutrophic lakes can have two alternative equilibria, either turbid, dominated by algal blooms or clear with a dominance of submersed macrophytes (Phillips et al., 1978; Blindow et al., 1993; Scheffer et al., 1993). Shifts between one state and the other occur rather quickly without prolonged intermittent phases. Observations in shallow eutrophic lakes show that a dense cover of submersed macrophytes can keep the water clear despite high enough nutrient concentrations to support phytoplankton growth (Blindow et al., 1992; Ozimek et al., 1993; Scheffer et al., 1993). Many ecological mechanisms are probably involved in this process. The release of allelochemicals by submersed macrophytes to suppress algal growth seems to be one strategy (Phillips et al., 1978; Wium-Anderson, 1987; Scheffer et al., 1993). Further evidence for allelopathic interactions of macrophytes comes from observations of monospecific stands of some macrophytes (*Chara*: Wium-Anderson et al., 1982; *Myriophyllum spicatum*: Grace and Wetzel, 1978; Smith and Barko, 1990) and low epiphyte densities on certain macrophytes (*Chara*: Wium-Anderson et al., 1982; *Ceratophyllum demersum*: Gough and Woelkerling, 1976). A certain degree of host-specificity of epiphytes (Gough and Woelkerling, 1976; Eminson and Moss, 1980; Burkholder and Wetzel, 1990) indicates biochemical interactions between epiphytes and their macrophyte.

Proving allelopathy in aquatic systems under field conditions is very difficult, therefore, most studies on macrophyte allelopathy have been performed under controlled laboratory conditions. This may be considered a drawback. However, this work with *Myriophyllum* shows that laboratory studies can provide insights in ecological mechanisms that would be nearly impossible to obtain from field observations alone. This section focuses on five issues that arose from work on allelopathy of submersed macrophytes, especially members of the Haloragaceae as follows:

1. A high percentage of submersed macrophytes from northern German lakes produce allelochemicals. Although the focus will be on inhibitory biochemical interactions, possible stimulatory allelopathic relationships will be discussed.
2. Phenolic compounds, especially hydrolyzable polyphenols, appear to be common allelochemicals in the family of the Haloragaceae. Such allelochemicals may have certain advantages for aquatic plants with respect to concentration, solubility in the aquatic medium, and cost of biosynthesis and turn-over.

3. Algicidal hydrolyzable polyphenols released by *Myriophyllum* will end up in the large pool of humic-like compounds in lake water. Possible release mechanisms, metabolic features, and effects on target organisms will be given.

4. Structure-activity relationships define the main algicidal hydrolyzable polyphenol from *M. spicatum* not only as a very potent inhibitor of algal exoenzymes, but also of other modes of action.

5. According to the original definition, allelopathy does not imply competition (Molisch 1937, Rice 1984). The hypothesis will be tested that interference, meaning coupled allelopathic and competitive interactions (Muller, 1969), may be an effective strategy for some submersed macrophytes.

12.4.1 WIDESPREAD OCCURRENCE OF ALLELOPATHIC INTERACTIONS IN SUBMERSED MACROPHYTES

In agreement with Molisch's (1937) definition of allelopathy, both stimulatory and inhibitory biochemical interactions will be considered. However, negative effects may be predominant because submersed macrophytes face a severe competition especially for light with epiphytes and phytoplankton, and also with other macrophytes for space and nutrients. Inhibitory allelochemicals seem to be rather broad in their specificity toward certain target algae or cyanobacteria (Gross et al., 1991; Gross, 1995). This might be of advantage for the producing organism in order to affect most epiphytes and phytoplankton with only one or a few inhibitors. On the other side, stimulating allelopathic interactions should be very specific. Positive effects on all algae and excessive epiphyte or phytoplankton growth could easily threaten the survival of the macrophyte. It might be this difference in specificity of action that has biased research on allelopathy toward inhibitory interactions, overlooking stimulatory effects.

12.4.1.1 Stimulatory Allelopathic Interactions

Observations on freshwater (Wetzel, 1983) and marine macrophytes (Harlin, 1973) indicate that they rely in part on organic compounds obtained from accompanying microorganisms and algae. Some macrophytes may no longer be capable of a sufficient synthesis of certain vitamins, phytohormones, or enzymes (Godmaire and Nalewajko, 1989) and, therefore, obtain those, perhaps in exchange for extracellular organic compounds (EOC), from bacteria or microalgae. A couple of studies with axenic cultures of macrophytes and algae provide further evidence for stimulatory biochemical interactions. It seems that macrophytes are not at all a neutral substrate for epiphytes (Cattaneo and Amireault, 1992).

Axenic cultures of macrophytes grow better when vitamins were added to the culture medium (Vitamin B_{12} to *Chara*: Wetzel and McGregor, 1968; Thiamin-HCl to *Myriophyllum spicatum*: Gross, 1995). In carbonate-poor culture medium, some macrophytes, capable of using bicarbonate (Grace and Wetzel, 1978; Wetzel, 1983), grow better when bacteria are present compared with axenic cultures (*Potamogeton pectinatus*: Ailstock et al., 1991; *Myriophyllum spicatum*: Godmaire and Nalewajko, 1989; Gross, 1995). It might be easier for those macrophytes to use respiratory CO_2 from the bacteria than to rely on their own carboanhydrase activity. Such biochemical interactions between submersed macrophytes and their epiphytes can be described as stimulatory allelopathic relationships. No final conclusions can be drawn at this point since vigorous studies on isolated stimulatory allelochemicals are still unavailable. Some studies (Gough and Woelkerling, 1976; Eminson and Moss, 1980; Burkholder and Wetzel, 1990) indicate a certain host-specificity of epiphytes on macrophytes, suggesting the involvment of stimulatory allelopathic interactions. There are indications that certain epiphytes are only found on certain submersed macrophyte species; but no comprehensive study has been published thus far.

12.4.1.2   Inhibitory Allelopathic Interactions

The allelopathic potential of submersed macrophytes seems to be high, as described above. This is reflected in the reviews by Wium-Andersen (1987), Gopal and Goel (1993), and Elakovich and Wooten (1995). The identified allelochemicals belong to rather different chemical classes such as sulfur compounds, polyacetylenes, and oxygenated fatty acids (see Figure 12.2). A screening of 17 different, mainly submersed, macrophytes of northern German lakes revealed that 8 species show a significant cyanobactericidal and algicidal activity of the crude methanolic extract (Gross, 1995). Very active macrophyte species are *Ceratophyllum* sp., *Myriophyllum* sp., and *Hottonia palustris* as well as the aquatic moss *Fontinalis antipyretica*. No published data exist thus far on the allelopathic activity of *Fontinalis* and *Hottonia*, but allelopathy has been described for *Ceratophyllum demersum* (Kogan and Chinnova, 1972; Wium-Anderson et al., 1983), *Myriophyllum spicatum* (Fitzgerald, 1969; Planas et al., 1981; Agami and Waisel, 1985), and *M. verticillatum* (Aliotta et al., 1992). Wium-Anderson et al. (1983) postulated elemental sulfur as the active allelochemical in

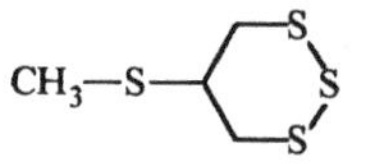

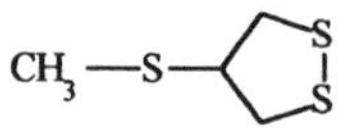

Trithiane   Dithiane

a) Algicidal cyclic sulfur compounds from *Chara globularis* (Wium-Andersen et al., 1982)

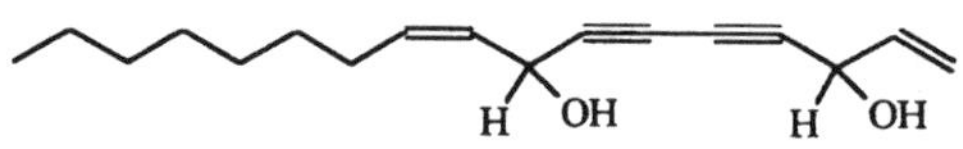

Falcarindiol

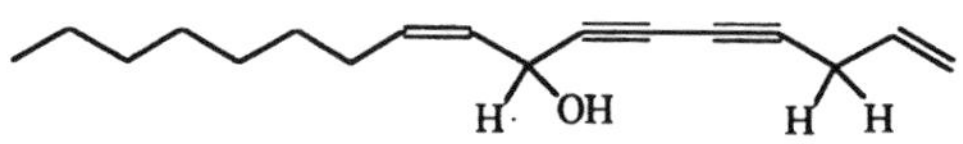

Falcarindol

b) Algicidal polyacetylenes from *Berula erecta (Sium erectum)* (Wium-Andersen et al., 1987)

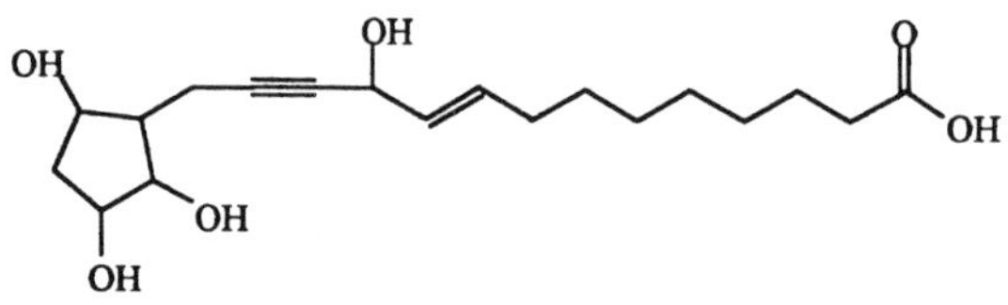

Trihydroxycyclopentenyl-fatty acid

c) Algicidal fatty acid derivative from *Eleocharis microcarpa* (van Aller et al., 1983)

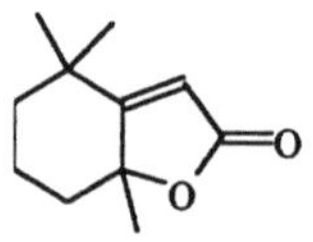

Dihydroactinidiolid

d) Allelochemical against macrophytes from *Eleocharis coloradensis* (Stevens and Merril, 1980)

FIGURE 12.2   Structures of allelochemicals in macrophytes.

Ceratophyllum demersum. Gas-chromatography/Mass spectroscopy (GC/MS) studies of volatile organic compounds released into the incubation water by *C. demersum* indicate that low molecular weight sulfur compounds might be responsible for the release of algicidal elemental sulfur (Gross, 1995). *C. demersum* also seems to have some allelopthic activity against other macrophytes (Elakovich and Wooten, 1995). Two further species of the Haloragaceae from different locations also exhibited a strong algicidal activity, namely *Myriophyllum heterophyllum*, a neophyte from North America that invaded a small lake in western Germany (Heider Bergsee) and *Proserpinaca palustris* from the Talladega wetland, Alabama, USA. This screening shows that there is an overall high probability of finding inhibitory allelopathic activity of submersed macrophytes.

12.4.2 HYDROLYZABLE POLYPHENOLS: A NEW CLASS OF ALLELOCHEMICALS
IN HALORAGACEAE

Hydrolyzable polyphenols (synonymous for tannins) are known as herbivore deterrents or as antibiotics (Haslam, 1989). Only recently it became obvious that they also have potent algicidal properties (Saito et al., 1989; Aliotta et al., 1992; Gross et al., 1996). Phenolic compounds and especially hydrolyzable polyphenols are well suited as allelochemicals in the littoral zone, since their molecular structure has lipophilic and hydrophilic sites, allowing both attachment to plant surfaces and dissolution in water.

A simple assay can reveal the presence of phenolic allelochemicals. Treating a crude extract of any allelochemical-producing macrophyte with insoluble polyvinylpyrrolidone (PVP) removes over 95 percent of the phenolic compounds (Loomis and Battaile, 1966; Gross et al., 1996). Phenolic allelochemicals are presumed responsible for the allelopathic activity when an extract is active prior to PVP-treatment and is inactive afterward. Most of the crude extracts of macrophytes used in the above-mentioned screening were subjected to the PVP-assay. *Fontinalis antipyretica* as well as all members of the Haloragaceae produce phenolic compounds as allelochemicals. For more precise information on the structure of the phenolic compounds, enzymatic and acid hydrolysis of all

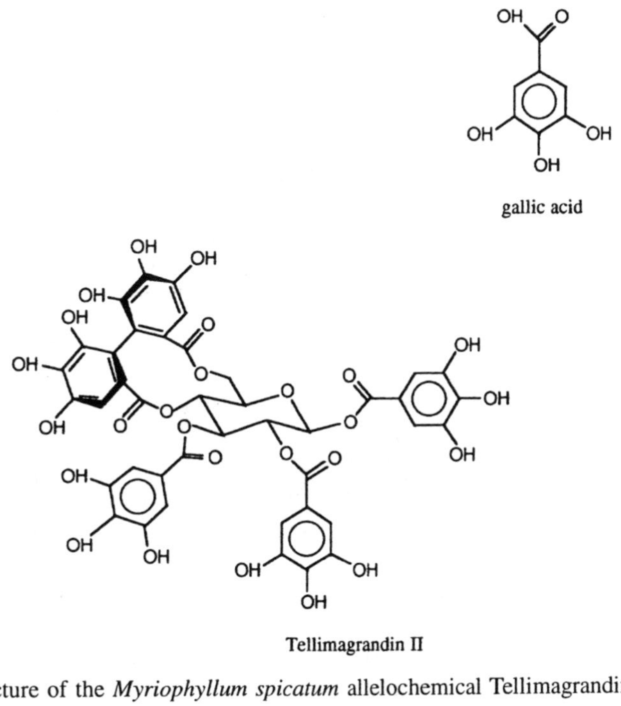

gallic acid

Tellimagrandin II

FIGURE 12.3 Structure of the *Myriophyllum spicatum* allelochemical Tellimagrandin II and its precursor gallic acid.

haloragacean crude extracts were performed. Both tests revealed that all haloragacean species produce hydrolyzable polyphenols since the enzymatic as well as the acid hydrolysis showed the presence of gallic and ellagic acid (Gross, 1995; Gross et al., 1996). All members of the Haloragaceae contain polyphenols, according to a literature survey (Mole, 1993). Tellimagrandin II (Figure 12.3) is the main allelochemical in *Myriophyllum spicatum* (Gross et al., 1996). None of the other haloragacean species produces this allelochemical. Their biological activity is based on other, yet unidentified, hydrolyzable polyphenols.

The polyphenolic content in all haloragacean species, measured with the Folin-Ciocalteau assay (Box, 1983; Gross et al., 1996), ranges between 6 and 15 percent of the dry weight, which is much higher than in other submersed macrophytes (less than 1 percent of the dry weight; Gross, unpublished results). Other macrophytes producing hydrolyzable polyphenols are *Trapa japonica* (Nonaka et al., 1981), *Trapa bicornis* (Yoshida et al., 1989), *Nuphar japonica* (Ishimatsu et al., 1989), *Nuphar variegatum* (Nishizawa et al., 1990), and *Nymphaea tetragona* (Kurihara et al., 1993). All of these species have floating leaves; nothing is known about algicidal activities of their polyphenols. However, *Trapa bicornis* contains tellimagrandin I and II (Yoshida et al., 1989). A thorough screening of (submersed) macrophytes for the production of (poly)phenolic allelochemicals could provide valuable information.

Contents of polyphenols in angiosperms are generally higher than other secondary metabolites, such as alkaloids, terpenoids, and cyanogenic compounds. Nonphenolic macrophyte-derived allelochemicals, as shown in Figure 12.2, are present in very low concentrations (Stevens and Merril, 1980; Wium-Anderson et al., 1982, 1983) with extraction yields usually much lower than 1 percent of the dry weight. Only very few nitrogen-containing allelochemicals (e.g., alkaloids) are known from submerged macrophytes. *Chara globularis,* for example, produces the antibiotically active charamin, a quarternary amine (Anthoni et al., 1987). Nitrogen-containing secondary metabolites are generally considered to be costly for the producing plant (Bryant et al., 1983). Therefore, the production of phenolic compounds for defense against other organisms may be a less costly method and allows use of superfluous assimilated carbon.

12.4.3 EXCRETION OF ALGICIDAL HYDROLYZABLE POLYPHENOLS

The study of the release of allelochemicals by aquatic organisms faces extreme difficulties due to abiotic and biotic interferences. Nevertheless, this study is necessary to verify the release of allelochemicals and their effect on target organisms. *Myriophyllum spicatum* (Eurasian watermilfoil, or short milfoil) may release polyphenols by several mechanisms, two of which are exudation from intact, living tissue and leaching from decaying shoots. Both processes happen simultaneously in a macrophyte bed (Wetzel, 1983). Further leaching can occur due to injury by herbivory, mechanical damage (wave action) or during autofragmentation of shoots. Milfoil has secretory trichomes (Mayr, 1915; Godmaire and Nalewajko, 1990), which contain polyphenol-like substances, according to histochemical studies (Tunman, 1913; Janson, 1918). Tannins (polyphenols) have been considered responsible for the low epiphyte density on the green algae *Spirogyra* (Pankow, 1961). These algicidal compounds must be actively released, since only living cells showed this pattern and dead *Spirogyra* filaments were covered with epiphytes. Hydrolyzable polyphenols have been identified from *Spirogyra* sp., which contains 2.0 to 6.4 percent of the dry weight as hydrolyzable polyphenols, among them tetra- to undeca-galloylglucose derivatives (Nishizawa et al., 1985). *Spirogyra varians* produces the α-glucosidase-inhibitor 1,2,6-tri-*O*-galloyl-3-digalloyl-β-glucose (Cannel et al., 1988).

Axenic cultures are the method of choice for undisturbed studies on the release of allelochemicals (Keating, 1977; 1978; Gross et al., 1991; Gross, 1995). Some scientists may argue that non-axenic specimens reflect the more natural state. However, using such material makes it much more difficult to distinguish compounds originating from the producing organism itself from those released by accompanying algae and microorganisms. For example, macrophyte-derived compounds are metabolized by epiphytes. Furthermore, many bacterial contaminants in lab cultures probably arrived long after isolation of the donor species and did not play any role in the original field relationship. An axenic culture of *M. spicatum* has been established (Gross, 1995; Gross et al., 1996) to study the release of polyphenolic allelochemicals under controlled conditions. In the incubation medium of axenic cultures, single phenolic compounds have been identified. Among them are traces of the main allelochemical tellimagrandin II as well as ellagic acid and several other, yet unidentified low molecular hydrolyzable polyphenols (Gross and Sütfeld, 1994; Gross, 1995). Non-axenic shoots exhibit different patterns of released compounds indicating that they are rapidly metabolized by bacteria. Only short-term incubations of non-axenic milfoil shoots in sterile mineral medium reveal similar release patterns to axenic cultures. Considering the high phenolic content (approximately 10 percent of the dry weight) of milfoil and the often dense stands of this macrophyte, an estimate of the amount of released phenolic allelochemicals can be made. Given a maximum biomass of milfoil of 280 to 1150 g m^{-2} (Grace and Wetzel, 1978) and the content of Tellimagrandin II ranging from 10 to 15 µg mg^{-1} dry weight of *M. spicatum*, dense stands could contain 1 to 6 mg/l (1.1 to 6.4 µM) of this highly algicidal hydrolyzable polyphenol. Further, given that 5 to 50 µg of tellimagrandin II led to a distinct inhibition of all tested algae and cyanobacteria in the agar diffusion assay, a release of only 1 percent of the main inhibitor would be sufficient to severely affect epiphytic or phytoplanktic organisms. Other hydrolyzable polyphenols found in and released by milfoil exert further but less inhibitory activity. Substantial improvements of methods are necessary to confirm such interactions *in situ*. It is not impossible that microbially degraded hydrolyzable polyphenols from milfoil are also allelopathically active. Phenolic compounds may be active before and after microbial degradation (Inderjit, 1996). A couple of studies indicate that milfoil releases allelochemicals. Agami and Waisel (1985) have shown that culture water from milfoil inhibits the growth of *Najas* sp. This appears to be caused by allelopathic interactions since nutrient effects could be ruled out. Organic compounds released by milfoil inhibited phytoplankton growth in a western German lake (G. Friedrich, unpublished results; Gross, 1995).

12.4.4 HYDROLYZABLE POLYPHENOLS: INHIBITORS OF ALGAL EXOENZYMES AND MORE

A predominant biological action of polyphenols is their strong interaction with proteins (Haslam, 1989; Appel, 1993). Polyphenols may attenuate digestive enzymes of herbivores (Feeny, 1976; Appel, 1993). In aquatic systems, humic material includes allochthonous (leaf-litter) and autochthonous (macrophyte- and algal-derived) polyphenols. Allochthonous polyphenols decrease phytoplankton densities and change community structure of the phytoplankton in freshwater systems in the Doñana-National Park, southern Spain (Serrano and Guisande, 1990). Wetzel (1990, 1991, 1992, 1993) showed that humic compounds and simple phenolic acids inhibit algal exoenzymes, such as alkaline phosphatase activity (APA).

A special bioassay system has been developed to test whether tellimagrandin II and other milfoil polyphenols inhibit APA. Inhibition of APA is measured fluorescence-spectroscopically with MUF-P (Methylumbelliferyl-Phosphate) as a substrate (Gross, 1995; Gross et al., 1996). Polyphenols released by milfoil into the surrounding medium, as well as polyphenols extracted from plant tissue, strongly inhibit APA of selected cyanobacteria, chlorophytes, diatoms, and a sample of natural epiphytes (Gross, 1995; Gross et al., 1996). Polyphenols extracted from the culture medium by means of solid phase extraction (Gross and Sütfeld, 1994; Gross et al., 1996) led to a 60 percent inhibition of APA at concentrations as low as 1.6 mg l^{-1}. Higher concentrations in the range of 5 to 15 mg l^{-1} of humic acid or monophenols are needed for a comparable degree of inhibition (Wetzel, 1991, 1993). An exponential relationship exists between the concentration of tellimagrandin II used and the effective inhibition of APA in the cyanobacterium *Trichormus variabilis* P-9 (Gross et al., 1996). As little as 0.2 mg l^{-1} (0.2µM) tellimagrandin lead to a 10 percent inhibition, 3.4 mg l^{-1}

(3.6 μM) to a 40 percent inhibition. Hydrolyzable polyphenols are unspecific inhibitors of enzymes; the type of inhibition should therefore be noncompetitive (Haslam, 1989). Preliminary data suggest that the interaction between milfoil polyphenols and APA follows the kinetics of a noncompetitive inhibition (Gross, 1995).

Previous studies on hydrolyzable polyphenols have shown that the biological activity is correlated with the size of the molecule, the degree of oxidation, and the number of hydroxyl groups and aromatic systems (Zucker, 1983; Haslam, 1989; Appel, 1993). For example, β-1,2,3,4,6-penta-O-galloyl-D-glucose exerts a tenfold stronger inhibition on β-glucosidase than β-1,2,3-Tri-O-galloyl-D-glucose (Haslam, 1989). In agreement with this, the large and complex hydrolyzable polyphenol tellimagrandin II is a stronger inhibitor of APA of the cyanobacterium *Trichormus variabilis* P-9 and the chlorophyte *Scenedesmus falcatus* than the simple phenolic compound gallic acid even when the latter is used in a fivefold concentration to provide the same amount of hydroxy groups and aromatic systems than tellimagrandin II (Gross, 1995; Gross et al., 1996).

The strong algicidal activity observed with milfoil polyphenols cannot only be explained by interactions with membrane-bound and extracellular enzymes of algae, but also is probably a result of further, yet unknown, interference with essential metabolic processes of target cells. Polyphenols with a molecular weight of 800 to 3000 Da are considered small enough to pass through bacterial cell membranes (Field et al., 1990). Most of the milfoil polyphenols have molecular weights lower than 1000 Da (Gross, 1995; Gross et al., 1996) and would fall into this category. Lemke et al. (1995) have shown that humic acids can change membrane permeability in bacteria with a hydrophilic surface structure. Tellimagrandin II and other milfoil polyphenols probably can pass through both algal and bacterial cell membranes and exert further deleterious action on metabolic processes inside the target cell.

12.4.5 ALLELOPATHY AND INTERFERENCE

Complex interactions in lakes do not permit easy distinction between allelopathic and competitive interactions. Evidence for coupled allelopathic and competitive interactions have been described from many shallow eutrophic lakes (Phillips et al., 1978; Scheffer et al., 1993; van Donk et al., 1993). Competition for resources implies the uptake or removal of resources (exploitative competition), whereas allelopathy involves the addition of biochemically active compounds (interference competition) (e.g., Willis, 1994). Muller (1969) used interference in a slightly different way to describe combined allelopathic and competitive interactions. The hypothesis is raised that allelopathy is especially effective under high competition for nutrients and light. This hypothesis was tested with Eurasian watermilfoil. The influence of phosphorus, nitrogen, and light on the production of algicidal hydrolyzable polyphenols in milfoil will be discussed.

Milfoil derives most of its phosphorus from the sediment (Carignan and Kalff, 1980). It never seems to be limited by this element (Gerloff and Krombholz, 1966; Anderson and Kalff, 1986). On the other hand, tellimagrandin II and other released or plant-bound polyphenols inhibit alkaline phosphatase of target organisms as shown above. Milfoil biomass peaks in mid-summer when phytoplankton in freshwater lakes is severely P-limited (Schindler, 1977; Sommer, 1989). Epiphytes may also be affected since they obtain most of their phosphorus from the water and not from their milfoil host (Carignan and Kalff, 1982). Inhibiting algal extracellular alkaline phosphatase would then be especially deleterious.

Nitrogen and light can both influence the biosynthesis of phenolic compounds (Haslam, 1986). The activity of phenylalanin-ammoniumlyase, a crucial enzyme in the biosynthesis of phenolic compounds, is regulated by light (Hahlbrock et al., 1976; Kuhn et al., 1984). Low light decreased the total phenolic content in milfoil, but the tellimagrandin II content stayed constant (Gross, unpublished results). Nitrogen limitation has frequently been shown to increase the phenolic content in

Myriophyllum spicatum in shallow water

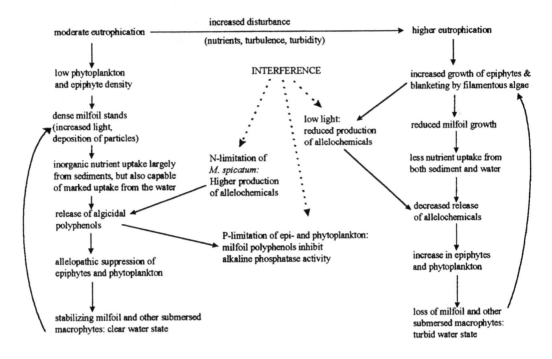

FIGURE 12.4 Coupled interaction of allelopathy and competition (interference) in *Myriophyllum spicatum* stands.

angiosperms (Hall et al., 1982; Haslam, 1986; Estiarte et al., 1994; Arnold et al., 1995). Milfoil growth is often N-limited (Gerloff and Krombholz, 1966; Sytsma and Anderson, 1993), which could influence the polyphenol content. Non-axenic milfoil shoots from Schöhsee that were obviously N-limited (tissue C:N ratio of 24:1, see Duarte, 1992), had significantly higher total amounts of total phenolics and tellimagrandin II than axenic milfoil grown in N-rich culture medium (Gross, 1995; Gross et al., 1996). Nitrogen limitation led to a higher content of tellimagrandin II in milfoil but did not affect the total phenolic content (Gross, unpublished results). This indicates a strong interaction between resource availability and the production of phenolic allelochemicals. Moreover, the main allelochemical found in milfoil exhibits different regulation patterns than the total phenolic compounds.

Figure 12.4 illustrates possible interactions of competition and allelopathy in the relationship between milfoil and algae. It describes how some major nutrients affect the production and effectiveness of milfoil allelochemicals and how this may influence the dominance of either milfoil or algae. Phillips et al. (1978) have described alternating stable states of submersed macrophytes and phytoplankton and considered the excretion of organic suppressors of algal growth by macrophytes to be an important mechanism in this relationship. Figure 12.4 extends the model of Phillips and coworkers (1978) and adjusts it for the interactions associated with milfoil. Overall, milfoil has been shown to be a highly competitive species (Grace and Wetzel, 1978; Smith and Barko, 1990). Furthermore, according to this and other studies (Fitzgerald, 1969; Planas et al., 1981; Agami and Waisel, 1985; Gross and Sütfeld, 1994; Gross et al., 1996) allelopathic interactions have to be included as a potent defensive strategy.

12.5 CONCLUSIONS

Limited spatial distance between competing organisms in benthic and littoral habitats makes allelopathy a powerful strategy. However, only detailed investigations can reveal how and in what form allelochemicals reach target organisms. Lipophilic allelochemicals of low molecular weight from the benthic cyanobacteria *Fischerella* sp. are transmitted by direct cell–cell contact. More polar hydrolyzable polyphenols from *Myriophyllum spicatum* and other Haloragaceae are released into the surrounding water, where they are rapidly metabolized. Still, also metabolized phenolic compounds exert a deleterious effect on membrane-bound and extracellular enzymes of algae. Valuable information about the ecological impact of allelochemicals was gained in these two systems by applying a threefold approach: 1) isolating active compounds, 2) studying possible release mechanisms, and 3) evaluating the mode of action in target organisms. Additionally, both systems revealed that the production and potency of allelochemicals is influenced by the availability of certain resources, such as phosphorus, nitrogen, and light. The production of allelochemicals can incur metabolic costs for the producing organism as has been shown when *Fischerella muscicola* was grown under nutrient and light-limiting conditions. Interference, meaning coupled competitive and allelopathic interactions, seems to be a powerful strategy in littoral areas, and may well be the driving force in the successful establishment of *M. spicatum*.

ACKNOWLEDGMENTS

Appreciation goes to Colleen Kearns and Kathleen I. Keating for improved spelling, grammar, and syntax of the manuscript. I thank K. I. Keating, Inderjit, and two anonymous reviewers for helpful comments and discussion. This work was supported in part by the Volkswagen Foundation Germany and the Max-Planck Society Germany.

REFERENCES

Agami, M. and Waisel, Y., Inter-relationship between *Najas marina* L. and three other species of aquatic macrophytes, *Hydrobiologia*, 126, 169, 1985.

Ailstock, M. S., Fleming, W. J., and Cooke, T. J., The characterization of axenic culture systems suitable for plant-propagation and experimental studies of the submersed aquatic angiosperm *Potamogeton pectinatus* (Sago pondweed), *Estuaries*, 14, 57, 1991.

Akehurst, S. C., Observations on pond life, *J. R. Microsc. Soc. London*, 51, 236, 1931.

Aliotta, G., Molinaro, A., Monaco, P., Pinto, G., and Previtera, L., Three biologically-active phenylpropanoid glucosides from *Myriophyllum verticillatum*, *Phytochemistry*, 31, 109, 1992.

Anderson, M. R. and Kalff, J., Nutrient limitation of *Myriophyllum spicatum* growth *in situ*, *Freshw. Biol.*, 16, 735, 1986.

Anthoni, U., Nielsen, P. H., Smith-Hansen, L., Wium-Andersen, S., and Christophersen, C., Charamin, a quaternery ammonium ion antibiotic from the green alga *Chara globularis*, *J. Org. Chem.*, 52, 694, 1987.

Appel, H. M., Phenolics in ecological interactions: the importance of oxidation, *J. Chem. Ecol.*, 19, 1521, 1993.

Arnold, T. M., Tanner, C. E., and Hatch, W. I., Phenotypic variation in polyphenolic content of the tropical brown alga *Lobophora variegata* as a function of nitrogen availability, *Mar. Ecol. Progr. Ser.*, 123, 177, 1995.

Bagchi, S. N., Structure and site of action of an algicide from a cyanobacterium, *Oscillatoria late-virens*, *J. Plant Physiol.*, 146, 372, 1995.

Bate-Smith, E. C., The phenolic constituents of plants and their taxonomic significance, I. Dicotyledones, *J. Linn. Soc. London (Bot.)*, 58, 95, 1962.

Blindow, I., Long-term and short-term dynamics of submerged macrophytes in two shallow eutrophic lakes, *Freshw. Biol.*, 28, 15, 1992.

Blindow, I., Andersson, G., Hargeby, A., and Johansson, S., Long-term pattern of alternative stable states in two shallow eutrophic lakes, *Freshw. Biol.*, 30, 159, 1993.

Borowitzka, M. A., Microalgae as sources of pharmaceuticals and other biologically active compounds, *J. Appl. Phycol.,* 7, 3, 1995.

Box, J. D., Investigation of the Folin-Ciocalteu phenol reagent for the determination of polyphenolic substances in natural waters, *Water Res.,* 17, 511, 1983.

Boyer, G. L., Sullivan, J. J., Andersen, R. J., and Taylor, J. R., Effects of nutrient limitation on toxin production and composition in the marine dinoflagellate *Protogonyaulax tamarensis, Mar. Biol.,* 96, 123, 1987.

Bryant, J. P., Chapin, F. S., III, and Klein, D. R., Carbon/nutrient balance of boreal plants in relation to vertebrate herbivory, *Oikos,* 40, 357, 1983.

Burkholder, J. M. and Wetzel, R. G., Epiphytic alkaline-phosphatase on natural and artificial plants in an oligotrophic lake—reevaluation of the role of macrophytes as a phosphorus source for epiphytes, *Limnol. Oceanogr.,* 35, 736, 1990.

Cannell, R. J. P., Farmer, P., and Walker, J. M., Purification and characterization of pentagalloylglucose, an a-glucosidase inhibitor antibiotic from the freshwater green alga *Spirogyra varians, Biochem. J.,* 255, 937, 1988.

Carignan, R. and Kalff, J., Phosphorus sources for aquatic weeds: water or sediments?, *Science,* 207, 987, 1980.

Carignan, R. and Kalff, J., Phosphorus release by submerged macrophytes: significance to epiphyton and phytoplankton, *Limnol. Oceanogr.,* 27, 419, 1982.

Cattaneo, A. and Amireault, M. C., How artificial are artificial substrata for periphyton, *J. North Am. Benthol. Soc.,* 11, 244, 1992.

Della Greca, M., Fiorentino, A., Monaco, P., Previtera, L., and Zarelli, A., Effusides I–V: 9,10-dihydrophenanthrene glucosides from *Juncus effusus, Phytochemistry,* 40, 533, 1995.

Dodson, S. I., Crowl, T. A., Peckarsky, B. L., Kats, L. B., Covich, A. P., and Culp, J. M., Non-visual communication in freshwater benthos: an overview, *J. North Am. Benthol. Soc.,* 13, 268, 1994.

Duarte, C. M., Nutrient concentration of aquatic plants—patterns across species, *Limnol. Oceanogr.,* 37, 882, 1992.

Elakovich, S. D. and Wooten, J. W., Allelopathic, herbaceous, vascular hydrophytes, in *Allelopathy: Organisms, Processes, and Applications,* Inderjit, Dakshini, K. M. M., and Einhellig, F. A., Eds., American Chemical Society, Washington, D.C., 1995, 58.

Eminson, D. and Moss, B., The composition and ecology of periphyton communities in freshwaters. I. The influence of host type and external environment on community composition, *Brit. Phycol. J.,* 15, 429, 1980.

Estiarte, M., Filella, I., Serra, J., and Penuelas, J., Effects of nutrient and water-stress on leaf phenolic content of peppers and susceptibility to generalist herbivore *Helicoverpa armigera* (Hubner), *Oecologia,* 99, 387, 1994.

Feeny, P. P., Plant apparency and chemical defense, *Rec. Adv. Phytochem.,* 10, 1, 1976.

Field, J. A. and Lettinga, G., Biodegradation of tannins, *Metal Ions Biol. Sys.,* 28, 61, 1992.

Field, J. A., Lettinga, G., and Habets, L. H. A., Measurement of low molecular weight tannins: indicators of methanogenic toxic tannins, *J. Ferment. Bioeng.,* 69, 148, 1990.

Fitzgerald, G. P., Some factors in the competition or antagonism among bacteria, algae and aquatic weeds, *J. Phycol.,* 5, 351, 1969.

Flores, E. and Wolk, C. P., Production, by filamentous, nitrogen-fixing cyanobacteria, of a bacteriocin and of other antibiotics that kill related strains, *Arch. Microbiol.,* 145, 215, 1986.

Fong, P., Donohoe, R. M., and Zedler, J. B., Competition with macroalgae and benthic cyanobacterial mats limits phytoplankton abundance in experimental microcosms, *Mar. Ecol. Progr. Ser.,* 100, 98, 1993.

Gerloff, G. C. and Krombholz, P. H., Tissue analysis as a measure of nutrient availability for the growth of angiosperm aquatic plants, *Limnol. Oceanogr.,* 11, 529, 1966.

Gleason, F. K. and Case, D. E., Activity of the natural algicide cyanobacterin on angiosperms, *Plant Physiol.,* 80, 834, 1986.

Gleason, F. K. and Paulson, J. L., Site of action of the natural algicide, cyanobacterin in the blue-green alga, *Synechococcus* sp., *Arch. Microbiol.,* 138, 273, 1984.

Godmaire, H. and Nalewajko, C., Growth, photosynthesis, and extracellular organic release in colonized and axenic *Myriophyllum spicatum, Can. J. Bot.,* 67, 3429, 1989.

Godmaire, H. and Nalewajko, C., Structure and development of secretory trichomes on *Myriophyllum spicatum* L., *Aquat. Bot.,* 37, 99, 1990.

Gopal, B. and Goel, U., Competition and allelopathy in aquatic plant-communities, *Bot. Rev.,* 59, 155, 1993.

Gough, S. B. and Woelkerling, W. J., Wisconsin Desmids. II. Aufwuchs and plankton communities of selected soft water lakes, hard water lakes and calcareous spring ponds, *Hydrobiologia,* 49, 3, 1976.

Grace, J. B. and Wetzel, R. G., The production biology of Eurasian watermilfoil (*Myriophyllum spicatum* L.): a review, *J. Aquat. Plant Manage.*, 16, 1, 1978.

Gross, E. M., Isolation, chemical and physiological characterization of cyanobacteriocidal compounds from *Fischerella* (Cyanobacteria) (in german), Master thesis, Eberhard-Karls-Universität, Tübingen, Germany, 1990.

Gross, E. M., Allelopathische Interaktionen zwischen Makrophyten und Epiphyten: Die Rolle hydrolysierbarer Polyphenole aus *Myriophyllum spicatum*, Dissertation, Christian-Albrechts-Universität, Kiel, Germany, 1995.

Gross, E. M., Wolk, C. P., and Jüttner, F., Fischerellin, a new allelochemical from the fresh-water cyanobacterium *Fischerella muscicola*, *J. Phycol.*, 27, 686, 1991.

Gross, E. M. and Sütfeld, R., Polyphenols with algicidal activity in the submerged macrophyte *Myriophyllum spicatum* L., *Acta Hortic.*, 381, 710, 1994.

Gross, E. M., von Elert, E., and Jüttner, F., Production of allelochemicals in *Fischerella muscicola* under different environmental conditions, *Verh. Internat. Verein. Limnol.*, 25, 2231, 1994.

Gross, E. M., Meyer, H., and Schilling, G., Release and ecological impact of algicidal hydrolyzable polyphenols in *Myriophyllum spicatum*, *Phytochemistry*, 41, 133, 1996.

Hagmann, L. and Jüttner, F., Fischerellin A, a novel photosystem-II inhibiting allelochemical of the cyanobacterium *Fischerella muscicola* with antifungal and herbicidal activity, *Tetrahed. Lett.*, 37, 6539, 1996.

Hahlbrock, K., Knobloch, K. H., Kreuzaler, F., Potts, J. R. M., and Wellmann, E., Coordinated induction and subsequent activity changes of two groups of metabolically interrelated enzymes, *Eur. J. Biochem.*, 61, 199, 1976.

Hall, A. B., Blum, U., and Fites, R. C., Stress modification of allelopathy of *Helianthus annuus* L. debris on seed germination, *Am. J. Bot.*, 69, 776, 1982.

Harder, R., Ernaehrungsphysiologische Untersuchungen an Cyanophyceen, hauptsaechlich dem endophytischen *Nostoc punctiforme*, *Zeitschr. Bot.*, 9, 145, 1917.

Harlin, M. M., Transfer of products between epiphytic marine algae and host plants, *J. Phycol.*, 9, 243, 1973.

Harris, D. O., Growth inhibitors produced by the green algae (Volvocaceae), *Arch. Microbiol.*, 76, 47, 1971.

Haslam, E., Secondary metabolism—fact and fiction, *Nat. Prod. Rep.*, 3, 217, 1986.

Haslam, E., Plant Polyphenols: Vegetable Tannins Revisited, Cambridge University Press, Cambridge, NY, 1989.

Hegnauer, R., Chemotaxonomie der Pflanzen, Birkhäuser Verlag, Basel, 1962–1995.

Hutchinson, G. E., A Treatise in Limnology, Vol. III, Limnological Botany, John Wiley & Sons, New York, 1975.

Inderjit, Plant phenolics in allelopathy, *Bot. Rev.*, 62, 186, 1996.

Inderjit and Dakshini, K. M. M., Algal allelopathy, *Bot. Rev.*, 60, 182, 1994.

Ishimatsu, M., Tanaka, T., Nonaka, G. I., Nishioka, I., Nishizawa, M., and Yamagishi, T., Tannins and related compounds. LXXXV. Isolation and characterization of novel diastereomeric ellagitannins, nupharins *a* and *b*, and their homologues from *Nuphar japonicum* DC, *Chem. Pharm. Bull.*, 37, 129, 1989.

Janson, E., Über die Inhaltskörper der *Myriophyllum*-Trichome, *Flora (Germ.)*, 110, 265, 1918.

Kane, M. E. and Gilman, E. F., *In vitro* propagation and bioassay systems for evaluating growth regulator effects on *Myriophyllum* species, *J. Aquat. Plant Manage.*, 29, 29, 1991.

Keating, K. I., Allelopathic influence on blue-green bloom sequence in a eutrophic lake, *Science*, 196, 885, 1977.

Keating, K. I., Blue-green algal inhibition of diatom growth: transition from mesotrophic to eutrophic community structure, *Science*, 199, 971, 1978.

Kogan, S. I. and Chinnova, G. A., Relations between *Ceratophyllum demersum* (L.) and some blue-green algae, *Gidrobiol. Zh.*, 8, 21, 1972.

Komarek, J. and Anagnostidis, K., Modern approach to the classification system of Cyanophytes 4—Nostocales, *Arch. Hydrobiol.*, 82, 247, 1989.

Kuhn, D. N., Chappel, J., Boudet, A., and Hahlbrock, K., Induction of phenylalanin-ammonia-lyase and 4-coumarat: CoA ligase mRNAs in cultured plant cells by UV light or fungal elicitors, *Proc. Natl. Acad. Sci. USA*, 81, 1102, 1984.

Kurihara, H., Kawabata, J., and Hatano, M., Geraniin, a hydrolyzable tannin from *Nymphaea tetragona* Georgi (Nymphaeaceae), *Biosci. Biotech. Biochem.*, 57, 1570, 1993.

Larsson, P. and Dodson, S., Invited review—Chemical communication in planktonic animals, *Arch. Hydrobiol.*, 129, 129, 1993.

Lemke, M. J., Churchill, P. F., and Wetzel, R. G., Effect of substrate and cell-surface hydrophobicity on phosphate utilization in bacteria, *Appl. Env. Microbiol.*, 61, 913, 1995.

Lewis, W. M. and Lewis, J., Jr., Evolutionary interpretation of allelochemical interactions in phytoplankton algae, *Am. Nat.,* 127, 184, 1986.

Loomis, W. D. and Battaile, J., Plant phenolic compounds and the isolation of plant enzymes, *Phytochemistry,* 5, 423, 1966.

Losee, R. F. and Wetzel, R. G., Littoral flow-rates within and around submersed macrophyte communities, *Freshw. Biol.,* 29, 7, 1993.

Mayr, F., Hydropoten an Wasser-und Sumpfpflanzen, *Beih. Bot. Centralbl. Abt.* I, 32, 278, 1915.

McClure, J. W., Secondary constituents of aquatic angiosperms, *in Phytochemical Phylogeny,* Harborne, J. B., Ed., Academic Press, London, 1970, 233.

Mole, S., The systematic distribution of tannins in the leaves of angiosperms—A tool for ecological-studies, *Biochem. Syst. Ecol.,* 21, 833, 1993.

Molisch, H., Der Einfluß einer Pflanze auf die Andere—Allelopathie, Fischer, Jena, Germany, 1937.

Moore, R. E., Cheuk, C., and Patterson, G. M. L., Hapalindoles: new alkaloids from the blue-green algae *Hapalosiphon fontinalis, J. Am. Chem. Soc.,* 106, 6456, 1984.

Moore, R. E., Patterson, G. M. L., and Carmichael, W. W., New pharmaceuticals from cultured blue-green algae, *in Biomedical Importance of Marine Organisms,* Fautin, D. G., Ed., 1988, 143.

Muller, C. H., Allelopathy as a factor in ecological process, *Vegetatio,* 18, 348, 1969.

Nicolaou, K. C. and Dai, W. M., Chemistry and biology of the enedyine anticancer antibiotics, *Angew. Chem.* (Int. ed.), 30, 1387, 1991.

Nishizawa, M., Yamagishi, T., Nonaka, G. I., Nishioka, I., and Ragan, M. A., Tannins and related compounds, Part 34: Gallotannins of the freshwater green algae *Spirogyra* sp, *Phytochemistry,* 24, 2411, 1985.

Nishizawa, K., Nakata, I., Kishida, A., Ayer, W. A., and Browne, L. M., Some biologically active tannins of *Nuphar variegatum, Phytochemistry,* 29, 2491, 1990.

Nonaka, G., Harada, M., and Nishioka, I., Eugeniin, a new ellagitannin from cloves, *Chem. Pharm. Bull.,* 28, 685, 1980.

Nonaka, G., Matsumoto, Y., and Nishioka, I., Trapain, a new hydrolyzable tannin from *Trapa japonica* Flerov, *Chem. Pharm. Bull.,* 29, 1184, 1981.

Ostrofsky, M. L. and Zettler, E. R., Chemical defences in aquatic plants, *J. Ecol.,* 74, 279, 1986.

Ozimek, T., Van Donk, E., and Gulati, R. D., Growth and nutrient-uptake by two species of *Elodea* in experimental conditions and their role in nutrient accumulation in a macrophyte-dominated lake, *Hydrobiologia,* 251, 13, 1993.

Pankow, H., Über die Ursachen des Fehlens von Epiphyten auf Zygnemalen, *Arch. Protistenk.,* 105, 417, 1961.

Papke, U., Gross, E. M., and Francke, W., Isolation, identification and determination of the absolute configuration of fischerellin B. A new algicide from the freshwater cyanobacterium *Fischerella muscicola* (Thuret.), *Tetrah. Lett.,* 38, 379, 1997.

Patterson, G. M. L., Larsen, L. K., and Moore, R. E., Bioactive natural products from blue-green algae, *J. Appl. Phycol.,* 6, 151, 1994.

Phillips, G. L., Eminson, D., and Moss, B., A mechanism to account for macrophyte decline in progressively eutrophicated freshwaters, *Aquat. Bot.,* 4, 103, 1978.

Pignatello, J. J., Porwoll, J., Carlson, R. E., Xavier, A., Gleason, F. C., and Wood, J. M., Structure of the antibiotic cyanobacterin, a chlorine-containing γ-lactone from the freshwater cyanobacterium *Scytonema hofmanni, J. Org. Chem.,* 48, 4035, 1983.

Pip, E., Phenolic compounds in macrophytes from the lower Nelson River System, Canada, *Aquat. Bot.,* 42, 273, 1992.

Planas, D., Sarhan, F., Dube, L., Godmaire, H., and Cadieux, C., Ecological significance of phenolic compounds of *Myriophyllum spicatum,* Verh. Internat. Verein. Limnol., 21, 1492, 1981.

Pratt, D. M., Competition between *Skeletonema costatum* and *Olistodiscus luteus* in Narragansett Bay and in culture, *Limnol. Oceanogr.,* 11, 447, 1966.

Pratt, R., Daniels, R. C., Eiler, J. J., Gunnison, J. B., and Kumler, W. D., Chlorellin, an antibacterial substance from *Chlorella, Science,* 99, 351, 1944.

Proctor, V. W., Studies on algal antibiosis using *Haematococcus* and *Chlamydomonas, Limnol. Oceanogr.,* 2, 125, 1957.

Revsbech, N. P., Christensen, P. B., and Nielsen, L. P., Microelectrode analysis of photosynthetic and respiratory processes in microbial mats, *in Microbial Mats—Physiological Ecology of Benthic Microbial Communities,* Cohen, Y. and Rosenberg, E., Eds., American Society for Microbiology, Washington, D.C., 1989, 153.

Rice, T. R., Biotic influences affecting growth of planktonic algae, *Fish. Bull. USA*, 54, 227, 1954.

Rice, E. L., Allelopathy, 1st ed., Academic Press, New York, 1974.

Rice, E. L., Allelopathy, 2nd ed., Academic Press, Orlando, FL, 1984.

Rizvi, S. J. H. and Rizvi, V., Allelopathy—Basic and Applied Aspects, Chapman and Hall, London, 1992.

Saito, K., Matsumoto, M., Sekine, T., Murakoshi, I., Morisaki, N., and Iwasaki, S., Inhibitory substances from *Myriophyllum brasiliense* on growth of blue-green algae, *J. Nat. Prod.*, 52, 1221, 1989.

Scheffer, M., Hosper, S. H., Meijer, M. L., Moss, B., and Jeppesen, E., Alternative equilibria in shallow lakes, *TREE*, 8, 275, 1993.

Schindler, D. W., Evolution of phosphorus limitation in lakes, *Science*, 195, 260, 1977.

Schwartz, R. E., Hirsch, C. F., Sesin, D. F., Flor, J. E., Chartrain, M., Fromtling, R. E., Harris, G. H., Salvatore, M. J., Liesch, J. M., and Yudin, K., Pharmaceuticals from cultured algae, *J. Ind. Microbiol.*, 5, 113, 1990.

Seitz, A., Are there allelopathic interactions in zooplankton? Laboratory experiments with *Daphnia*, *Oecologia*, 62, 94, 1984.

Serrano, L. and Guisande, C., Effects of polyphenolic compounds on phytoplankton, *Verh. Intern. Verein. Limnol.*, 24, 282, 1990.

Shilo, M., Toxins of Chrysophyceae, in *Microbial Toxins*, Kadis, S., Ciegler, A., and Ajl, S. J., Eds., Academic Press, New York, 1971, 67.

Shilo, M., The unique characteristics of benthic cyanobacteria, in *Microbial Mats—Physiological Ecology of Benthic Microbial Communities*, Cohen, Y. and Rosenberg, E., Eds., American Society for Microbiology, Washington, D.C., 1989, 207.

Smith, A. L. and Nicolaou, K. C., The enedyine antibiotics, *J. Med. Chem.*, 39, 2103, 1996.

Smith, C. S. and Barko, J. W., Ecology of Eurasian watermilfoil, *J. Aquat. Plant Manage.*, 28, 55, 1990.

Sommer, U., The role of competition for resources in phytoplankton succession, in *Plankton Ecology, Succession in Plankton Communities*, Sommer, U., Ed., Springer Verlag, Berlin, 1989, 57.

Stevens, K. L. and Merril, G. B., Growth inhibitors from Spikerush, *J. Agric. Food Chem.*, 28, 644, 1980.

Su, K. L. and Staba, E. J., Preliminary chemical studies on aquatic plants form Minnesota, *LLoydia*, 36, 72, 1973.

Sytsma, M. D. and Anderson, L. W. J., Nutrient limitation in *Myriophyllum aquaticum*, *J. Freshw. Ecol.*, 8, 165, 1993.

Trebst, A., Donner, W., and Draber, W., Structure activity correlation of herbicides affecting plastoquinone reduction by photosystem II: electron density distribution in inhibitors and plastoquinone species, *Z. Naturf.*, 39C, 405, 1984.

Tunmann, O., *Pflanzenmikrochemie*. Bornträger, Berlin, Germany, 1913.

van Aller, R. T., Pessoney, G. F., Rogers, V. A., Watkins, E. J., and Leggett, H. G., Oxygenated fatty acids: a class of allelochemicals from aquatic plants, in *The Chemistry of Allelopathy*, Thompson, A. C., Ed., American Chemical Society, Washington, D.C., 1985, 387.

van Donk, E., Gulati, R. D., Iedema, A., and Meulemans, J. T., Macrophyte-related shifts in the nitrogen and phosphorus contents of the different trophic levels in a biomanipulated shallow lake, *Hydrobiologia*, 251, 19, 1993.

von Elert, E., Analytical investigation of allelopathic interactions between cyanobacteria and phytoplankton in eutrophic Ostholsteinische Seen (in German), Dissertation thesis, Eberhard-Karls-Universität, Tübingen, Germany, 1994.

von Elert, E. and Jüttner, F., Factors influencing the allelopathic activity of the planktonic cyanobacterium *Trichormus doliolum*, *Phycologia*, 35, 68, 1997.

Wetzel, R. G., Limnology, 2nd ed., W. B. Saunders, Fort Worth, TX, 1983.

Wetzel, R. G., Edgardo Baldi Memorial Lecture. Land-water interphases, metabolic and limnological regulators, *Verh. Internat. Verein. Limnol.*, 24, 6, 1990.

Wetzel, R. G., Extracellular enzymatic interactions: storage, redistribution and interspecific communication, in *Microbial Enzymes in Aquatic Environments*, Chrost, R. J., Ed., Springer Verlag, New York, 1991, 6.

Wetzel, R. G., Gradient-dominated ecosystems—sources and regulatory functions of dissolved organic matter in fresh-water ecosystems, *Hydrobiologia*, 229, 181, 1992.

Wetzel, R. G., Humic compounds from wetlands: complexation, inactivation, and reactivation of surface-bound and extracellular enzymes, *Verh. Internat. Verein. Limnol.*, 25, 122, 1993.

Wetzel, R. G. and McGregor, D. L., Axenic culture and nutritional studies of aquatic macrophytes, *Am. Midl. Nat.*, 80, 52, 1968.

Whittaker, R. H. and Feeny, P. P., Allelochemics: chemical interactions between species, *Science,* 171, 757, 1971.

Willis, R. J., Terminology and trends in allelopathy, *Allelop. J.,* 1, 6, 1994.

Wium-Andersen, S., Allelopathy among aquatic plants, *Arch. Hydrobiol., Beih. Ergebn. Limnol.,* 27, 167, 1987.

Wium-Andersen, S., Anthoni, U., and Houen, G., Elemental sulphur, a possible allelopathic compound from *Ceratophyllum demersum, Phytochemistry,* 22, 2613, 1983.

Wium-Andersen, S., Anthoni, U., Christophersen, C., and Houen, G., Allelopathic effects on phytoplankton by substances isolated from aquatic macrophytes (*Charales*), *Oikos,* 39, 187, 1982.

Wium-Andersen, S., Jorgensen, K. H., Christophersen, C., and Anthoni, U., Algal growth inhibitors in *Sium erectum* Huds, *Arch. Hydrobiol.,* 111, 317, 1987.

Yoshida, T., Yazaki, K., Usman Memon, M., Maruyama, I., Kurokowa, K., and Okuda, T., Bicornin, a new hydrolyzable tannin from *Trapa bicornis* and revised structure of alnusiin, *Heterocycles,* 29, 86, 1989.

Yoshikawa, M., Hatakeyama, S., Tanaka, N., Matsuoka, T., Yamahara, J., and Murakami, N., Crude drugs from aquatic plants. 2. On the constituents of the rhizome of *Alisma orientale* Juzep originating from Japan, Taiwan, and China—absolute stereostructures of 11-deoxyalisols-b and b-23-acetate, *Chem. Pharm. Bull.,* 41, 2109, 1993.

Zucker, W. V., Tannins: does structure determine function? An ecological perspective. *Am. Nat.* 121, 335, 1983.

13 Australian Studies on Allelopathy in *Eucalyptus:* A Review

Rick Willis

CONTENTS

13.1   ABSTRACT

Allelopathy has been the subject of much speculation in vegetation dominated by *Eucalyptus* in Australia. Direct allelopathy seldom has been demonstrated in native eucalypt woodlands or forests, although many species can be shown to have allelopathic potential. There are many reports of eucalypt seedling suppression associated with mature tall open-forests, especially those dominated by species of the subgenus *Monocalyptus*. Most studies indicate that progressive change in the microbiology of the forest soil can result in an environment antagonistic to seedling growth, but that this may be alleviated by fire. Direct allelopathy is most likely to occur in vegetation dominated by eucalypts where the understory species have not had a substantial period of co-evolution, for example with introduced plants beneath a native eucalypt canopy or in regions where introduced eucalypts may affect local understory species.

0-8493-2116-6/99/$0.00+$.50
© 1999 by CRC Press LLC

13.2   INTRODUCTION

Shortly before his tragic death in a shooting accident, David Burton, sponsored by Joseph Banks to collect plants and seeds in the infant colony of New South Wales, wrote to Governor Arthur Phillip on February 24, 1792:

> "I have attended to the land at and around Parramatta, ...the land is excellent. It is black rich light soil, ...I beg leave to observe here that where different species of red gum-trees grow the earth has a great portion of oils mixed with it, and unless the ground is properly worked and turned over to meliorate and dissolve those oils, the first crop will come to little account." (Britton, 1892)

Herein lies one of the earliest observations and formulations relating to the concept of allelopathy, from a young man unlikely familiar with the writings of Pliny, or the more contemporary work of the 18th century Dutch physicians, H. Boerhaave and S. J. Brugmans, which offered early accounts of the chemical interactions of plants (Willis, 1985).

Allelopathy concerns the ecological effects that plant metabolites may have on other plants once they leave the plant. More often, it is seen as the harmful effect that such compounds have on other plants and microorganisms, although most compounds may act as promoters or inhibitors, according to the concentrations present. While there have been dozens of overseas studies on allelopathy in *Eucalyptus*, particularly in India where the planting of eucalypts is greatly controversial (Willis 1991), the subject of allelopathy in Australian native forests has only been reviewed in passing (Trenbath, 1978; Poore and Fries, 1985; Lovett, 1986, 1989; Stoneman, 1994). It is the intent of this chapter to present an overview of Australian studies on *Eucalyptu*, a summary of these is presented in Table 13.1.

TABLE 13.1
Summary of Australian Studies Concerning Allelopathy in *Eucalyptus*

Eucalyptus sp.	Allelochemical Source	Target Species	Basis	Reference
E. astringens	Grasses		No data	Reid and Wilson 1985
E. baxteri	Roots, leaves, litter, soil	*Triticum, E. viminalis, Allocasuarina, Leptospermum*	Bioassay	del Moral et al. 1978
	Leaves	*Allocasuarina*	Field trial	Frood 1979
E. blakelyi	Litter	*Lemna*	Bioassay	May and Ash 1990
E. calophylla	Dead leaves	Cyanobacteria	Chemistry	Wrigley and Cowan 1995
E. camaldulensis	Grasses		No data	Cremer 1990
E. crebra		Various sp.	No data	Story 1967
E. dawsonii		Various sp.	No data	Story 1967
E. delegatensis		Autotoxic?	Pattern	Ellis et al. 1980
	Roots, leaves, litter	Autotoxic	Pot trial	Bowman and Kirkpatrick 1986
	Soil	Autotoxic	Field trial	Lovett 1986
	Soil	Autotoxic	Microbial	Ellis and Pennington 1992
E. diversicolor		Autotoxic?	Pattern	Rotheram 1983
E. dives		Autotoxic?	No data	Costin 1954
E. drepanophylla		Autotoxic?	Pattern	Henry and Florence 1966
E. elata	Litter	*Lemna*	Bioassay	May and Ash 1990
E. fraxinoides		Autotoxic?	Pattern	Prober 1992
E. globulus ssp. *bicostata*	Grasses		No data	Reid and Wilson 1985
	Litter, stemflow, root leachate, oils	*Lolium, Lemna*	Bioassay	May and Ash 1990

	Leaves	*Sinapis, Trifolium,*	Bioassay	Trenbath and Fox 1976, 1977
		Festuca	Pot trial	Trenbath and Silander 1978, 1980
E. leucoxylon	Litter	*Allocasuarina*	Bioassay	Withers 1978
E. macrorhynca		Autotoxic?	No data	Jarrett and Petrie 1929
		Autotoxic	No data	Purdie 1977
	Litter, stemflow	*Lolium, Lemna*	Bioassay	May and Ash 1990
E. maculata		Autotoxic?	Pattern	Henry and Florence 1966
	Litter	*Lemna*	Bioassay	May and Ash 1990
E. mannifera		Autotoxic	No data	Purdie 1977
	Litter	*Lemna*	Bioassay	May and Ash 1990
E. marginata		Autotoxic?	Pattern	Florence 1967
	Dead leaves	Cyanobacteria	Chemistry	Wrigley and Cowan 1995
E. melliodora		Various sp.	No data	Story 1967
	Litter	*Lemna*	Bioassay	May and Ash 1990
E. microcarpa		*Gonocarpus*	Pattern	Lange and Reynolds 1981
E. miniata		Various sp.	Inference	Stocker 1969
E. moluccana		Various sp.	No data	Story 1967
E. obliqua		Autotoxic	No data	Mount 1964, 1969
	Leaves, litter	*Triticum, E. viminalis, Allocasuarina*	Bioassay	del Moral et al. 1978
E. ovata	Litter	*Allocasuarina*	Bioassay	Withers 1978
E. paliformis		Autotoxic?	Pattern	Prober 1992
E. pilularis	Soil	Autotoxic?	Pot trial	Florence and Crocker 1962
E. polyanthemos	Litter	*Lemna*	Bioassay	May and Ash 1990
E. radiata		Autotoxic	No data	Costin 1954
	Oils	*Lemna*	Bioassay	May and Ash 1990
E. regnans		Autotoxic		Ashton 1981
	Root exudate	Autotoxic	Pot trial	Ashton and Willis 1982
	Soil	Autotoxic	Bioassay	Willis 1980
E. rossii		Autotoxic	No data	Purdie 1977
	Litter, stemflow	*Lolium, Lemna*	Bioassay	May and Ash 1990
E. rubida	Litter, oils	*Lolium, Lemna, Acacia, E. globulus*	Bioassay	May and Ash 1990
E. siderophloia		Autotoxic?	Pattern	Henry and Florence 1966
E. sideroxylon	Grasses		No data	Reid and Wilson 1985
E. sieberi		Autotoxic?	Pattern	Incoll 1979
E. tetradonta		Various sp.	Inference	Stocker 1969
E. todtiana		Various sp.	Inconclusive	Turner et al. 1985
E. wandoo	Leaves, soil	*Hakea, E. wandoo, Gastrolobium*	Pot trial	Lamont 1985
	Leaves	*Avena*	Bioassay	Hobbs and Atkins 1991

It is not surprising that the precocious observation by Burton concerns the genus *Eucalyptus*.[1] Of all the genera of trees that dominate the world's forests and woodlands, none dominate regionally to the extent that *Eucalyptus* does, and allelopathy offers a ready explanation of eucalypt dominance (van Steenis, 1971). Few genera would have a better chemical armoury than *Eucalyptus*. The chemical constituents of eucalypt foliage have been the subject of many studies. All species of *Eucalyptus* have foliar oil glands that are rich in essential oils, principally terpenoids; typically 1 to 5 percent of the fresh weight is essential oil (Baker and Smith, 1920; Guenther, 1950). The leaves also contain diverse phenolic compounds (Hillis, 1966a,b, 1967a,b,c,d; Hillis and Brown, 1978). It has long been recognized that many eucalypt extracts have strong antibiotic properties (Kinney,

[1]Nomenclature of *Eucalyptus* spp. follows Chippendale (1988).

1895; Hall, 1904; Penfold and Grant, 1923; Atkinson and Rainsford, 1946), and there have even been warnings about using eucalypt sawdust as a potting mix ingredient (Nicholls and Yazaki, 1979) although composted eucalypt bark in potting mix may help in preventing fungal root diseases (e.g., Hardy and Sivasithamparam, 1991). Recently it has been reported that leachates of decomposing leaves of *Eucalyptus marginata* Donn ex Smith and *E. calophylla* R. Br. may contribute to the suppression of cyanobacterial blooms in certain areas of the Swan Coastal Plain in Western Australia (Wrigley and Cowan, 1995).

Although there are many early accounts of "die-back" in natural forests of eucalypts (e.g., Norton, 1886; MacPherson, 1885), the explanations given do not include any suggestion of inhibition or allelopathic mechanisms.

The first modern suggestion that inhibition may be a factor in eucalypt ecology originated with Jarrett and Petrie (1929), who observed the profuse regeneration of eucalypt species, such as *E. macrorhyncha* F. Muell following the 1926 fires in the Blacks' Spur region, north of Healesville, Victoria:

"It seems, then, that, for those seeds which lie dormant on the ground and escape destruction, and which are normally permeable to water like *Eucalyptus*, either the fire or the deposition of ash has some action which is a stimulus to germination, or else removes some previously operating inhibiting agent."

This line of reasoning recurs frequently in the Australian ecological literature, although direct evidence is sorely lacking and authors such as Gilbert (1959) did not concur. The main protagonist of this theory has been Mount (1964, 1969) who suggested that the copious litter of species such as *Eucalyptus obliqua* Labill. (messmate stringybark) allowed the build-up of inhibitory substances in the soil. Similar ideas were expressed by Purdie (1977) in reference to dry sclerophyll forest in the Australian Capital Territory (A.C.T.) dominated by *E. macrorhynca* F. Muell., *E. rossii* R. T. Baker & H. G. Smith, and *E. mannifera* Mudil. sp. *maculosa* (R. Baker) L. Johnson, and by Ashton (1981) in reference to tall open-forests dominated by *E. regnans* F. Muell. Westman and Anderson (1970) suggested that aggregates of eucalypt individuals may be partly in response to the patchy removal of inhibitors by fire.

Following the lead of Muller (1966), Mount (1968) in a paper presented to the Institute of Foresters of Australia Conference in Perth, endeavored to develop the more general theme that plants, in the process of metabolism, produce waste byproducts, which they must eliminate. As plants are obviously sedentary organisms, "waste" elimination is somewhat problematic but is achieved through the shedding of disused plant parts (litter) and the release of volatile and water-soluble metabolites, many of which are potentially phytotoxic. The progressive accumulation of this material in the soil can lead to conditions unfavorable to plant growth; however, fire may act to remove or detoxify harmful substances. The waste theory in allelopathy is certainly not new, dating back to the 18th century, but has not attracted much interest in recent times (Willis, 1985).

13.3   EUCALYPTS OF OPEN-FORESTS AND WOODLANDS

The low open-woodland known as mallee in Australia has long been a source of interest for ecologists. As seedlings of mallee eucalypts (or mallees) are rarely seen, it is widely believed that the seeds do not readily germinate under natural conditions. Prescott and Piper (1932) suggested that eucalypt oils may prevent seed germination, as certain soils have an aromatic odor reminiscent of sandalwood (*Santalum album* L.). The presence of terpenes in eucalypt soils also has been reported by del Moral et al. (1978) and Lovett (1985). Many soils are water repellent, although this phenomenon is generally believed to be linked to fungal activity (Hubble et al., 1983). Prescott (1941) concluded that the usual paucity of mallee seed germination is a reflection of harsh conditions and that germination occurs readily under favorable conditions.

Costin (1954), having read the work of Bonner concerning inhibitors produced by foliage of *Encelia farinosa* A. Gray ex Torrey in California, reckoned that the peppermints, *Eucalyptus dives* Schau and *E. radiata* Sieb ex DC, of the Monaro region in New South Wales may exert an analogous influence on surrounding vegetation.

It has often been reported that there is a ring of sparse vegetation beneath single eucalypt trees, which is particularly obvious in aerial photographs. Story (1967) suggested that such vegetational patterning beneath *Eucalyptus crebra* F. Muell., *E. dawsonii* R. T. Baker, *E. melliodora* A. Cunn. ex Schaner, and *E. moluccana* Roxb. in the Hunter Valley region of New South Wales was not explainable by grazing or differences in soil moisture, and was possibly due to the action of associated microorganisms or exudates. Similar observations were made by Lange and Reynolds (1981) who reported the selective suppression of the herb *Gonocarpus elatius* (A. Cunn. ex Fenzi) Orchard (Haloragaceae) beneath canopies of *Eucalyptus microcarpa* (Maiden) Maiden. Cremer (1990) reports that *E. camaldulensis* Dehnh. may exert a negative effect on the grazing yield of pasture in Western Australia, and Reid and Wilson (1985) suggest that *E. astringens* Maiden, *E. globulus* Labill., and *E. sideroxylon* Benth. may inhibit grass growth through allelopathy. Turner et al. (1985) investigated the effects of the Western Australian endemic *E. todtiana* F. Muell. on associated vegetation, but results concerning allelopathy were inconclusive. Withers (1978) found that the litter of *E. ovata* Labill. and *E. leucoxylon* F. Muell. at Ocean Grove, Victoria was inhibitory to *Allocasuarina littoralis* (Salisb.) L. Johnson (*Casuarina littoralis*) germination, but not to eucalypt germination.

There is often a pattern of suppressed tree seedlings or young trees around older, veteran trees. While the effect of the canopy of such trees has been found to extend up to four times the crown radius in the case of *Eucalyptus sieberi* L. Johnson (Incoll, 1979) and twice the crown radius in *E. diversicolor* F. Muell. (Rotheram, 1983), the causes were not clear. In a study of forest in southeastern Queensland dominated by *Eucalyptus maculata* Hook., *E. drepanophylla* F. Muell. ex Benth., and *E. siderophloia* Benth., Henry and Florence (1966) could only conclude that the suppression of seedlings in the lignotuberous stage was due to some unknown factor associated with the larger trees.

On Melville Island in the Northern Territory, Stocker (1969) investigated the apparent difference in fertility in soils collected from monsoon forest dominated by *Myristica insipida* R. Br., *Bombax malabaricum* DC., *Alstonia actinophylla* (A. Cunn.) Schumann, and *Vitex acuminata* R. Br., and from nearby woodland supporting *Eucalyptus miniata* A. Cunn. ex Schauer and *E. tetrodonta* F. Muell. Stocker found that, although there were differences in the concentrations of nutrients in the soils, the growth of test species, especially in the poorer eucalypt soils could not be substantially improved through the addition of fertilizers or, in the case of *B. malabaricum*, through soil sterilization, and he concluded that soil inhibitors may be a cause. More recently, Bowman and Panton (1993) examined the causes of the absence of the monsoon forest species *Bombax ceiba* L. and *Sterculia quadrifida* R. Br. in savanna dominated by *E. tetrodonta*, *E. bleeseri* Blakely, and *E. miniata* near Darwin. Eucalypt litter was not inhibitory but microbiological factors such as poor mycorrhizal development may be involved.

Recently, the allelopathic potential of various plant parts of eucalypts found commonly in the A.C.T., either as natives or as introductions, including *E. globulus* ssp. *bicostata*, *E. maculata*, *E. macrorhynca*, *E. rossii*, *E. rubida*, *E. mannifera*, *E. blakelyi*, *E. polyanthemos* Schauer, *E. melliodora*, *E. elata* Dehnh., and *E. radiata* have been studied in some detail (May and Ash, 1990). Generally, fresh foliar and root leachates prepared at concentrations simulating natural conditions showed no activity in bioassay with *Lolium* and *Lemna*. However, leaf litter leachates of *E. macrorhynca* and *E. rubida* were inhibitory in bioassay, as were bark litter leachates of *E. rossii* and *E. rubida*. The most potent source of allelochemicals was stemflow, especially in *E. globulus* and *E. macrorhynca,* and this may be significant in explaining suppression zones near trees (see also, Ashton and Willis, 1982). Most of the extracts when artificially concentrated became strongly inhibitory and, again, this may have significance in the field where successive wetting and evaporation can cause effective concentration of allelochemicals in the soil. The authors found differences in soil response to allelochemicals as eucalypt soil treated with leachate showed 24 percent germination of *Lolium* seed, whereas adjacent grassland soil showed only 5 percent germination. There was an estimated 8 percent per day decline in allelopathic activity of leachates. The authors concluded that while the production of allelochemicals may be greater in wetter climates, allelopathy may be of greater significance in drier climates where competition for moisture and the concentration of allelochemicals are more likely to cause suppression.

The regeneration of eucalypts is often poor, especially on undisturbed sites, and bracken (*Pteridium esculentum* (G. Forst.) Cockayne), which is a common understory element in many eucalypt open-forests, has been investigated recently as a potential agent in inhibiting eucalypt seedling growth. Allelopathy in bracken is well known overseas (Gliessman, 1976; Nava et al., 1987). Taylor and Thomson (1990), working in New South Wales, found that the natural leachate from juvenile fronds was inhibitory to the radicle extension of *E. haemastoma* Smith. Mature fronds were not inhibitory and the degree of phytotoxicity of the young fronds varied according to the time since the last rainfall. Tolhurst and Turvey (1992) also investigated the effect of bracken in open-forests of *E. obliqua* Labill., *E. radiata*, and *E. rubida* in west-central Victoria, but they found that frond leachates had no effects on eucalypt seed germination, although they stated that they were working with senescing fronds and that there had been much recent rainfall.

13.3.1 *Eucalyptus baxteri*

In 1977 Roger del Moral, well known for his work concerning allelopathy in *Eucalyptus globulus* and *E. camaldulensis* in California (del Moral and Muller, 1969, 1970), worked at the University of Melbourne. He revisited the domain of allelopathy and studied eucalypts in their native habitat. Student work at Wilson's Promontory, Victoria had shown a distinctive zone of plant suppression beneath canopies of stunted coastal trees of *E. baxteri* (brown stringybark), a tree of widely varying stature found in open-forest and woodland situations in Victoria and eastern South Australia (del Moral et al., 1978).

Shrub species such as *Allocasuarina pusilla* (*Casuarina pusilla*) and *Leptospermum myrsinoides*, which are normally dominant in the coastal heath of the study area, were absent or severely suppressed beneath the canopies of *E. baxteri*, as were several other shrubby species. A similar pattern was noted for stunted coastal individuals of *E. obliqua*. Studies of soil moisture content, xylem water potential of shoots, nutrient status of soil, and light intensities failed to adequately explain the phenomenon. Consequently, root and foliar leachates were bioassayed to assess if there were potential phytotoxic substances emanating from *E. baxteri* and *E. obliqua*, although most work concerned the former.

Foliar leachates of both eucalypt species proved inhibitory to seedling growth in bioassay using *E. viminalis*, and *E. obliqua* also inhibited seedling growth of *Allocasuarina pusilla* (Mackin) L. Johnson. Litter extracts of both eucalypts were inhibitory in bioassays with *E. viminalis* Labill. and wheat (*Triticum aestivum* L.), whereas that of *E. nitida* Hook. f. was not. Extracts of soils collected from beneath trees and shrubs also proved inhibitory; and the most inhibitory was that of *E. baxteri*. A number of potential phytotoxins, including several phenolic acids, were identified in foliar and litter leachates of both *E. baxteri* and *E. obliqua*.

The work of del Moral et al. (1978) was followed by that of Frood (1979). Frood conducted two sets of experiments relating to investigating allelopathy in *E. baxteri*. First, he bioassayed foliar leachates from a range of woody species to assess the presence of phytotoxic substances and, again, the foliar leachate of *E. baxteri* showed inhibitory activity.

More importantly, Frood designed a novel experiment in an attempt to test the allelopathic effects of *E. baxteri* leachates in the field. Seedlings of test species, including *Allocasuarina pusilla*, were planted in specially designed pots with collecting covers that either fed natural leachate to the pot or fed it away from the pot. Pots were placed in fenced enclosures beneath the canopy of *E. baxteri*, in heath, or in the open. All pots were watered, such that lack of water was not a variable.

Unfortunately, the experiment could be followed for only a relatively short period, 106 days. Plants of *A. pusilla* receiving leachate from *E. baxteri* showed distinctly less growth (as measured by plant height) in comparison with protected plants, although the probability of the difference ($P = .06$) just failed to reach the normally accepted level of statistical significance.

13.3.2 *EUCALYPTUS WANDOO*

Eucalyptus wandoo (wandoo) occurs commonly as an open-forest or woodland tree in southwestern Western Australia. As with many eucalypts there is a discernible pattern of reduced undergrowth associated with the canopies of *E. wandoo*, particularly where present as individual trees or small clumps (Lamont, 1985). A detailed study was conducted at Wongamine Reserve, approximately 80 km northeast of Perth, and the suppression zones were commonly found to extend beyond the tree canopies. As well as examining soil moisture and nutrients, Lamont included an assessment of allelopathy in his study. Soil analyses indicated that soil phenolics were higher in concentration near a *E. wandoo* tree, but that the boundary of the suppression zone did not correspond to a sudden change in phenolic content. *E. wandoo* foliar leachate (2.25 kg dry weight equivalent branches in 110 L water for 20 h at 5°C) was applied to pots filled with reconstituted soil from the study site, and planted with seeds of *Hakea trifurcata* (Smith) R. Br. (Proteaceae), *Gastrolobium spinosum* Lindley (Fabaceae), and wandoo. Germination was not greatly affected by leachate treatment but was affected by soil type; soils collected near the tree generally showed poor wettability. Similarly, the growth of shrub seedlings over ten weeks showed little suppressive effect of leachates, and seedlings of *E. wandoo* showed enhanced growth.

In a separate study, Hobbs and Atkins (1991) examined species interactions in Durokoppin Reserve, approximately 200 km east of Perth. Preliminary studies of allelopathy were conducted using foliar leachates prepared from 50 g fresh weight leaves soaked in 250 ml distilled water for 24 h. The leachates of six species including *E. wandoo* were bioassayed using seeds and subsequent seedlings of wild oat (*Avena fatua* L.); no germination and subsequently no growth at all was recorded with *E. wandoo* leachate, which contrasts markedly with the results of Lamont (1985). However, the difference in extraction methods, bioassay, and site were all likely important factors.

13.4 EUCALYPTS OF TALL OPEN-FORESTS

The best studied species of eucalypts in Australia with regard to allelopathy are those associated with what is termed tall open-forests; here, the dominant trees are in excess of 30 m in height and the projective canopy cover ranges between 30 and 70 percent. The reasons for interest in these forests undoubtedly are their great commercial value and concern regarding their successful regeneration. It is interesting that the species that have attracted most attention, namely *E. regnans* (mountain ash), *E. delegatensis* R. T. Baker (alpine ash), and *E. pilularis* (blackbutt), lack lignotubers and all fall within the informal subgenus *Monocalyptus* (Pryor and Johnson, 1971). Florence (1967) regarded the poor regrowth of these species, as well as another *Monocalyptus* species, *E. marginata* Donn ex Smith (jarrah), as Australian examples of soil sickness. Prober (1992) stated that the poor growth of two other *Monocalyptus* species, *E. fraxinoides* Deane & Maiden and *E. paliformis* L. Johnson & D. Blaxell, on unburnt mature forest soil in southeast New South Wales could be partly due to inhibition. Finally, it is ironic that *E. regnans* and *E. delegatensis* are considered good agroforestry species because of their alleged lack of allelopathic activity (Reid and Wilson, 1985).

13.4.1 *EUCALYPTUS REGNANS*

In the 1950s, David Ashton, at the University of Melbourne, was working on *E. regnans* (mountain ash) at Wallaby Creek, a catchment reserve 50 km north of Melbourne, Victoria. He became intrigued by the lack of regeneration beneath mature *E. regnans,* the most important native hardwood species in Victoria and Tasmania, and the world's tallest angiosperm. Inexplicably poor regeneration of *E. regnans* after logging is well known to foresters (Powles, 1940; Cunningham, 1960) and it is commonly observed that seedlings rarely survive past the cotyledonary stage (Gilbert, 1959).

This very problem was to trouble Ashton and his students for the next 30 years (Ashton and Willis, 1982). The results of some of Ashton's early experiments on allelopathy were first presented in 1962 to the Third Conference of the Institute of Foresters of Australia in Melbourne in a paper entitled "Some aspects of root competition in *E. regnans*." A manuscript of this address is lodged with the University of Melbourne library. Some of these results and those from further experiments were eventually published by Ashton and Willis (1982).

In the 1962 unpublished paper Ashton detailed results from competition experiments concerning the relative survival of seedlings of the common understory shrub *Pomaderris aspera* Sieber ex DC. and *E. regnans* when grown at various light intensities in the field. After 12 years, only 2 of 4214 *E. regnans* seedlings originally recorded in monitored plots were still alive, and these were deemed unlikely to survive. Soil moisture, generally, remained above wilting point and experimental plots were fertilized, which suggested that competition for water and nutrients were not greatly important. Observations, such as that seedlings nearest established trees were usually the first to succumb, led Ashton to hypothesize that inhibitory substances may be released from the roots of large trees.

The idea that inhibitors may play a role in the poor regeneration of mountain ash was first suggested to Ashton by F. W. Went who visited Australia during 1955 (Ashton and Willis, 1982).

In 1957 Ashton commenced glasshouse experiments to investigate the nature of root interactions. *E. regnans* seedlings, grown in contact with a larger plant of either *E. regnans* or *Pomaderris aspera,* after 9 months were severely retarded compared with a control, even when fertilized. In a subsequent experiment, there were marked differences in the growth of *E. regnans* seedlings, depending on their degree of contact with the roots of the large seedling, particularly after 60 days, which suggested that the roots or mycorrhizae of the established *E. regnans* seedlings were interfering with the growth of the small seedlings, possibly through some inhibitory exudate.

In 1962 Ashton, again, planted more *E. regnans* seedlings in the field, within fenced plots. Four-leaf seedlings were planted under full *E. regnans* canopy or half-canopy conditions; half of the seedlings were planted out and half remained in countersunk bitumen-lined pots. All seedlings were fertilized and were watered weekly. After about two months of growth, seedlings in the pots, grown under the full canopy of mature *E. regnans,* showed significantly better growth, whereas there was no such difference in plots further away from large trees. Similar experiments in 1964 and 1967 showed that very young seedlings showed less response than older seedlings.

Ashton collected soil leachate from well-watered 1.8 m seedlings of *E. regnans* growing in buckets. This leachate, a control leachate from forest soil and concentrated ($\times$10) leachate, were then applied every 2 days to young seedlings of *E. regnans* maintained in pots. After two months there was significantly less growth with respect to seedling dry weight and seedling height when treated associated with the *E. regnans* leachates, although their efficacy and importance in the field remained to be established.

In a manner reminiscent of the *E. pilularis* story (*q.v.*), Ashton and Willis (1982) reported that the fungus *Cylindrocarpon destructans* (Zins.) Scholten (*C. radicicola*) was present in mature *E. regnans* soil and bioassays demonstrated that the fungus produced a phytotoxin that caused symptoms similar to those observed in the field. The same fungus has also been recorded in another *Monocalyptus* species, *E. obliqua* (Jehne, 1976), and Bowling and McLeod (1968) have associated dieback of both *E. regnans* and *E. obliqua* in Tasmania with the fungus *Armillaria*.

Largely following the impetus of del Moral and Muller (1969, 1970), the author, Ashton's Ph.D. student, investigated the role of allelopathy in the regeneration problem of *E. regnans* at Toolangi, Victoria (Willis, 1980, 1982). Foliar and litter leachates only inhibited seedling growth if concentrated at least tenfold. Further studies revealed potent terpenoid inhibitors in leaves of *E. regnans* but whether these are ever released to the soil remains doubtful. A search for soil toxins also proved elusive. Soil phenolics showed great variability from season to season and site to site. The forest soil was found to contain relatively large amounts of lipid material (3.5 percent dry weight) in comparison with neighboring agricultural soils, and the soil lipid fraction demonstrated great phytotoxicity in bioassays in which seedling radicles were in contact with the lipid material (Willis and Ashton, 1978; Willis, 1980, 1982). Other studies investigated the nature of the nitrogen economy of *E. regnans* seedlings in relation to that of larger individuals and this is discussed elsewhere in this chapter.

The understory of the tall open-forests dominated by *E. regnans* is characteristically rich in shrub species and some of these may have allelopathic properties. Gilbert (1959) observed that the aromatic shrub *Atherosperma moschatum* Labill. (Monimiaceae) may be allelopathic. Willis (1980) bioassayed the major understorey species found in *E. regnans* forests and found that, in particular, foliar leachates of the shrub *Cassinia aculeata* (Labill.) R. Br. (Asteraceae) were inhibitory to *E. regnans* germination and seedling growth. Recent work by McDowell (1993) suggested that foliar leachates of *Correa lawrenciana* Hook. (Rutaceae) have strong allelopathic potential.

13.4.2 *EUCALYPTUS PILULARIS*

Contemporary with Ashton's work was that of Ross Florence, at the University of Sydney, who completed his doctoral thesis in 1961, entitled "Studies in the ecology of blackbutt (*Eucalyptus pilularis* Sm.)." *E. pilularis* is the most important hardwood of the coastal regions of New South Wales and southern Queensland, and as with *E. regnans,* seedling growth is often poor (Burgess, 1975).

The key results from Florence's thesis were published in 1962 (Florence and Crocker, 1962). During trials in which seedlings of *E. pilularis* were raised on soils originating from different forest types, Florence found that the seedlings were most severely inhibited when grown in soils from mature *E. pilularis* stands. Seedlings were, typically, stunted and remained a purplish color. Florence conducted a number of experiments to elucidate the nature of the inhibition. Trials with soil from mature virgin *E. pilularis* forests, amended with various nutrients, indicated that the inhibition could be partly overcome with heat sterilization or with the addition of nitrogen and phosphorus; however, the addition of litter to the soil greatly reduced growth. Soil that had been air-dried and stored for $4^1/_2$ months lost much of the inhibitory effect and when supplemented with nutrients, yielded seedlings with growth double that in the fertilized fresh soil.

Florence attempted to discover if there were any seasonal patterns to the inhibition. With soils collected from March to October, he found that growth was less inhibited on soil collected in May but, otherwise, soil was uniformly deleterious to growth and even when amended with phosphorus and nitrogen, the soils produced seedlings with distinct yellowing of the leaves.

Florence postulated that nutrients, especially nitrogen, may become immobilized during the processes of litter degradation and hence cause a nutrient imbalance in the soil. Seedlings were provided with varying dosages of nitrogen and phosphorus on both fresh and stored *E. pilularis* soils. While there was a range of growth responses, as measured by leaf area, in all combinations the eucalypt seedlings remained manifestly retarded and abnormal in appearance in the fresh soil. In stored soil, growth responses accorded more with expectation, which suggested that there is some labile factor in the fresh soil that can inhibit seedling growth.

Florence further investigated the effect of heat on the soil and was surprised to find that, over the range of temperatures assayed from ambient to 160°C, by far the best growth was attained in soil heated at just 35°C. Seedlings grown in soil heated to this low temperature looked healthy and responded as expected to nutrients.

The enhanced response on the moderately heated soil suggested that populations of nitrifying bacteria may have been stimulated and caused the flush in growth. Florence heated fresh soil samples at various temperatures ranging from 25°C to 105°C for 48 hours and then incubated the soils at 25°C for 65 days. The levels of ammonium and nitrate were monitored periodically and as expected, there was a very marked surge in nitrification as a result of the 35°C treatment, whereas initial ammonification correlated with the temperature. Florence also collected soil from both virgin forest and a mature regrowth site, rapidly air-dried the soils, remoistened them, and followed the course of ammonification and nitrification for 70 days. Unlike the previous experiment, there was a marked initial flush of ammonification in both soils, in both the fresh and remoistened states, although the remoistened soils did reach higher levels. The most interesting aspect of this experiment relates to nitrification. In the remoistened soils, as expected, there was a distinct rise in nitrification following the flush of ammonification. However, in the fresh soils, this did not occur and there was a considerable lag of 26 to 60 days before nitrification increased. Florence did not regard these results as the total answer to the problem of seedling inhibition, especially as *E. pilularis* seedlings during glasshouse trials remained suppressed well after flushes of ammonification and nitrification should have had some impact.

Florence continued to investigate the role of microbiological effects of the fresh soil. While heat treatments had produced different patterns of growth, the treatment itself likely physically and chemically altered the soil. To avert this problem, Florence exposed fresh soil to gamma ray irradiation in order to effect heatless sterilization. Even in comparison with seedlings grown on heat-treated soils, seedlings in irradiated soils were markedly healthy. An examination of the root systems demonstrated that control seedlings were depauperate in root hair development, whereas seedlings in irradiated soil had vigorous root hair development.

While Florence regarded this work as exploratory, he published no further reports on the issue. Florence and Crocker (1962) concluded that there is a soil-litter complex that progressively can become antagonistic to the establishment of blackbutt seedlings, if microorganisms inhibitory to *E. pilularis* growth are not suppressed. Evans et al. (1967) reported that isolates of the noninvasive pathogen *Cylindrocarpon radicicola* collected from roots of *E. pilularis* were shown to be capable of producing a phytotoxin, which caused symptoms in *E. pilularis* seedlings consistent with Florence and Crocker's work and which these authors named nectrolide.

13.4.3 *EUCALYPTUS DELEGATENSIS*

Eucalyptus delegatensis (alpine ash) is closely related to *E. regnans,* and is found in cool, humid mountain regions in southeastern Australia. As with *E. regnans,* it is reported that regrowth of *E. delegatensis* may prove difficult on unburnt, mature forest soils (Grose, 1960; Keenan and Candy, 1983; O'Dowd and Gill, 1984). Also, particularly in Tasmania, there are numerous reports of dieback in stands of *E. delegatensis,* which putatively has been linked with microbial immobilization of nitrogen due to heavy litter production (Ellis, 1964) or the build-up of phytotoxins in the soil (Ellis et al., 1980). Recently, dieback in some Victorian *Eucalyptus* spp. has been tied more certainly to imbalances in the nitrogen cycle and concomitant increased levels of insect herbivory (Granger et al., 1994; Marsh and Adams, 1995).

The tussock grass *Poa labillardieri* Steud. is often associated with poor seedling growth of *E. delegatensis* in northeastern Tasmania (Keenan and Candy, 1983; Ellis and Pennington, 1992). Fensham and Kirkpatrick (1992) reported results of inhibition of *E. rodwayi* R. T. Baker & H. G. Smith by *Poa*-dominated swards consistent with allelopathy, but Webb et al. (1983) working with *E. delegatensis* found no effects that were attributable to allelochemicals, although litter and soil extracts were generally growth-promoting.

Bowman and Kirkpatrick (1986), in the course of a broad study on the growth of *E. delegatensis* in Tasmania, included a few experiments to assess the role of allelopathy. Germination and seedling growth of *E. delegatensis* were monitored following treatment with various extracts of leaves, roots, and litter of *E. delegatensis*. Results were inconclusive, largely because of the high variation within treatments; however, there was significant inhibition of germination and hypocotyl growth with the 1:10 (w/w) leaf extract prepared by soaking fresh (assumedly macerated) leaves for 24 hours. It is interesting that a leachate prepared by spraying fresh intact leaves with water (approximately 1:2.2 basis) reduced growth but not significantly. In another experiment, seedlings were grown outdoors in pots with and without shredded litter for three months but, again, there was reduced growth that did not prove to be statistically significant.

Ellis and Pennington (1989) investigated the nitrification status of soils associated with *E. delegatensis* forests and related vegetation in northeast Tasmania, and found that rates of nitrification were generally very low in fire-stabilized eucalypt forests, possibly due to the presence of nonlabile, water-insoluble inhibitors; however, no such compounds were actually isolated.

Lovett (1986) reported some preliminary work by Goodwin, in *E. delegatensis* forests of Tasmania, which pointed to autotoxicity. Ground beneath a single mature *E. delegatensis* tree was divided into four sectors and given four treatments: 1) no treatment (control), 2) burning, 3) trenching, and 4) burning and trenching. *E. delegatensis* seedlings were planted and allowed to grow for 22 months. In particular, seedlings in the burned/trenched plot showed growth up to eight times that of the control. Soil leachates from the four plots were bioassayed with *Linum usitatissimum* L. seedlings, but soil from trenched or burned plots was surprisingly stimulatory, which suggested that biologically active compounds were present only at low concentrations in the soil leachates.

In the most recent study relating to *E. delegatensis,* Ellis and Pennington (1992) bioassayed soils from several secondary succession sites, from old *Poa* grassland, and from under individual trees growing in old grassland. In most cases, the growth of *E. delegatensis* seedlings in potted soil reflected the condition of the trees in the field and poor growth was only partially overcome by the addition of nitrogen and phosphorus, or by soil sterilization. The most interesting result is that inhibition in mature forest or grassland soils was successfully overcome by the addition of 10 to 20 percent soil from a healthy eucalypt forest, which strongly suggests that changes in soil microbiology can be major causes of both die-back and seedling decline in these forests.

13.5   INDIRECT ALLELOPATHY

It has been recognized that indirect allelopathy may occur (Grodzinskii, 1978; Lovett and Jackson, 1980) in which, for example, fungi and bacteria feeding on plant parts may cause or accelerate the release of phytotoxic metabolites.

13.5.1   *EUCALYPTUS GLOBULUS SSP. BICOSTATA*

Trenbath and Fox (1976, 1977) introduced the novel idea that eucalypts may chemically influence the growth of plants beneath the canopy by means of an insect vector. This work and subsequent reports (Trenbath and Silander, 1978, 1980) formed the basis of a more accessible account by Silander et al. (1983).

First, Trenbath and Fox (1976, 1977) observed that bare zones commonly develop under certain individual mature trees of *E. globulus* spp. *bicostata* (*E. bicostata*) in parks in Canberra, A.C.T.; this species is not native to the A.C.T., but is native to Victoria and New South Wales. Not all trees showed this effect, and it was suggested that the pattern was associated with the presence of a folivorous chrysomelid beetle, *Chrysophtharta m-fuscum* and its rain of frass. Aqueous extracts of the foliage showed little allelopathic activity, whereas the beetle frass severely inhibited germination of mustard (*Sinapis alba* L.) seed in bioassay. Thus, the beetles provide a means of release and transport of phytotoxins from the foliage of *E. globulus* spp. *bicostata.*

In subsequent trials, Trenbath and Silander (1978) investigated the phytotoxicity of frass from another chrysomelid beetle, *Poropsis atomaria,* feeding on leaves of *E. bicostata.* Frasses originating from both adult and juvenile foliage were bioassayed, but showed no significant difference. Consequently, amounts of frass from juvenile foliage were placed in Petri dishes such that they were equivalent to a range of concentrations, from 40 to 1800 kg/ha. These were then tested with seeds of mustard (*Sinapis alba* L.), clover (*Trifolium repens* L.), and Chewing's fescue (*Festuca rubra* Thuill var. *fallax*). Mustard and fescue proved most sensitive and radicle extension was almost completely inhibited at frass concentrations equivalent to 120 to 360 kg/ha.

Natural rates of frass fall were estimated to range from 20 to 2500 kg/ha/y (Silander et al., 1983) and the monitoring of one tree of *E. globulus* spp. *bicostata* with a light insect infestation revealed that frass fall averaged 328 kg/ha over 6 months, with most of the fall occurring in a 4-week period. Jacobs (1955) recorded that, during a particularly severe infestation, the rate of frass fall was 6700 kg/h in one week, which achieved a maximal rate of 62 kg/ha. Trenbath and Silander (1978) estimated that for each gram of eucalypt foliage produced, 0.4 g becomes insect frass, 0.4 g eventually becomes litter, and 0.2 g is metabolized by the insect. Silander et al. (1983) undertook preliminary investigations on the nature of the inhibitors in frass but, following chromatography and bioassay, could only conclude that there were both similarities and differences in phytotoxic components of the foliage and insect frass.

Trenbath and Silander (1978) observed that, in the field, clover and the native wallaby grass (*Themeda triandra* Forssk. [*T. australis*]) were rarely found in the bare zone beneath *E. globulus* spp. *bicostata,* whereas catsear or flatweed (*Hypochoeris radicata* L.) and speargrass (*Stipa scabra* ssp. *falcata* (Hughes) Vickary, S. W. L. Jacobs & J. Everett [*S. falcata*]) were commonly present. Consequently, cores of grassland soil were taken and dressed with the equivalent of 500 kg/h frass and were seeded with the four species. Seedlings of clover and *Themeda* were moribund, whereas seedlings of *Hypochoeris* and *Stipa* became established, although growth was slow. Trenbath and Silander (1980) report that in the same or a similar experiment when the plants were harvested after 54 days of growth, there was no significant decrease in the dry weight of the shoots of *Hypochoeris* and *Stipa* in frass amended cores, whereas that of clover and *Themeda* was significantly reduced.

In another experiment, Trenbath and Silander (1980) applied frass at a rate of 1000 kg/ha to swards of grassy vegetation. After three months, it was found that the clover:grass ratio had increased by 73 percent, which suggests that frass can have a significant effect on the patterning of understorey species.

The work of Trenbath and his coworkers attracted considerable notice at the time and Ohmart (1985), in particular, criticized numerous aspects of the work, including the reality of the rates of frass fall used in the experiments. Ohmart stated that the rate of frass fall is more realistically given by a value of 100 to 150 kg/ha/y and that the methodology of Silander et al. (1983) lacks relevance in that frass does not normally fall in a single massive dose. Silander et al. (1985) answered the criticisms of Ohmart at length and in reiterating the validity of their work, pointed out that inhibition was demonstrated even with a frass dosage equivalent to 40 kg/ha. The only other subsequent work that has been stimulated by these original studies is that of Carlberg (1988) who suggested that frass-mediated allelopathy in eucalypts has been an important component in the evolution of the life history of the phasmatid *Extatosoma tiaratum*. If the frass of the insect inhibits the growth of vegetation beneath the canopy, Carlberg postulated that there has been selection for insects that can throw their eggs outside the bare zone, such that they will be less exposed to predation.

13.6   INHIBITION OF NITRIFICATION

In 1972 the American ecologist E. L. Rice and his doctoral student S. K. Pancholy published a theory, supported with data from grassland communities, which stated that nitrification should be progressively inhibited in seral stages of succession. The key to this theory is the notion that nitrate is leached from soils, whereas ammonium is attracted to clay micelles and, hence, efficient ecosystems should have mechanisms to conserve nitrogen and promote plant nitrogen acquisition through ammonium uptake. This theory has attracted considerable attention in Australia, particularly with regard to vegetation dominated by *Eucalyptus.*

In summary, results from Australia do not support Rice and Pancholy (1972). A number of investigators have found that nitrification is suppressed (Florence and Crocker, 1962) and/or nitrifiers are often scarce in soils dominated by eucalypts (Jones and Richards, 1977a; Hopmans et al., 1980; Adams and Attiwill, 1986). Jones and Richards (1977b) examined nitrifiers at sites with native vegetation dominated by *E. racemosa* Cav. and sites reforested with *Pinus elliottii* var. *ellittii* Little & Dorman, but the nitrifier population density could not be related to forest type. Lamb (1980) tested Rice and Pancholy's hypothesis in a successional series of secondary rainforest plots in Queensland and, while Australian rainforest does not generally contain *Eucalyptus,* the study concluded that nitrification actually increased overall with site age since disturbance and nitrogen mineralization was related primarily to site fertility. Ellis and Pennington (1989) investigated nitrification in soils of differing successional status from Tasmania, which included sites dominated by *E. delegatensis,* and nitrification increased with age of the eucalypts. Willis (1980), in attempting to elucidate the causes of seedling failure in forests of *E. regnans* in Victoria, examined soil nitrogen and nitrifying bacteria over a 12-month period and concluded that nitrification is limited by the availability of substrate, not by inhibitors. This is in accord with subsequent work by Adams and Attiwill (1982, 1986) who, working at a different *E. regnans* site, at Mt. Disappointment, Victoria concluded that the rate of nitrification was controlled by the rate of mineralization and is linked to the C:N ratio and that there was no evidence to support the Rice-Pancholy theory in eucalypt forest vegetation in Australia. This contrasts, again, with results with *Eucalyptus* grown as an exotic (Vargues, 1954). For example, Dyck et al. (1983) indicate that *E. saligna* Smith foliar leachates may be inhibitory to nitrifiers in New Zealand (contrary to a claim of no allelopathic activity by Reid and Wilson [1985]).

In view of the recent theories of White (1991, 1994) concerning the effects of terpenes on nitrification, it is interesting to speculate whether these compounds play any such role in eucalypt ecosystems, particularly as eucalypts on agricultural land, with high levels of nitrification, are prone to dieback disorders.

13.7   EFFECTS OF FIRE

The ecology of eucalypts in Australia is intimately bound with the effects of fire and many eucalypt species show enhanced seedling growth following fire, although the causes remain uncertain (Kimber and Schuster, 1979; Attiwill, 1994). Recent work by Chambers and Attiwill (1994) indicates that enhanced growth following heating or sterilization of *E. regnans* soil is associated with the release of nutrients from otherwise unavailable substrates. The effects of fire on allelopathy are very poorly understood. It has long been recognized by Australian foresters that the burning of waste wood and debris in windrows often results in greatly enhanced growth of the trees planted in the heat-affected soil (Hatch, 1960; Cremer 1962; Pryor, 1963; Cromer, 1967) although, in the United States, burning of slash has been reported to increase both the hydrophobicity and the allelopathic potential of sand below (Everett et al., 1995). The "ashbed effect" is the term given in Australia to enhanced plant growth on soils that have been heated to temperatures usually in excess of 150°C. This phenomenon has also been observed commonly overseas and was the object of much interest in the early part of this century in Europe, England, and the United States.

While the ashbed effect is well known to foresters, it has not been thoroughly studied and the results are often paradoxical. Pryor (1960) heated soil from a stand of *E. fastigata* Deane & Maiden and *E. viminalis* Labill. to 150°C and found that seedling growth improved 100 percent compared with a 25 percent increase through fertilization with wood ash. Leaching of control soil reduced plant growth, but subsequent leaching of the baked soil resulted in improved plant growth, which suggested that toxic compounds may be formed in heated soils, although no toxins could be isolated. Attiwill reported enhanced growth of *E. obliqua* Labill. seedlings on heat-treated soils at the Third Conference of the Institute of Foresters of Australia in 1962 but this work was not published. Work by Renbuss et al. (1972) suggested that the increase in plant growth may have resulted from reduced competition for nutrients by soil microorganisms and/or the removal of antagonistic organisms but, as Attiwill (1985) indicated, there has been very little progress in our understanding of the effect of fire on soil microbiology. Numerous investigators have hypothesized that the improved germination and/or growth of plants on burnt soil may be due to the removal of inhibitors from the litter or soil (Jarret and Petrie, 1929; Florence and Crocker, 1962; Cremer and Mount, 1965; Mount, 1968, 1969; Ashton, 1970; Smith, 1990), but hard evidence is, plainly, lacking.

13.8 CONCLUSIONS

Despite many attempts and many tantalizing leads, there has been no unambiguous demonstration of allelopathy in native eucalypt forests in Australia. This is in marked contrast to situations overseas where eucalypts are well known to affect the growth of understory species or in Australia where introduced species may grow poorly beneath eucalypts.

The above accords well with the ideas of Rabotnov (1977, 1982), who was of the opinion that allelopathy should be of greatest significance in plant communities containing a mixture of native and introduced plants, where the species have had only a brief, if any, history of co-evolution. Clearly, in the eucalypt forests of Australia, the present species have mostly co-existed for millions of years and, in this environment, have seemingly developed mechanisms of tolerance to allelochemicals, especially from the dominant eucalypts. Those species that are not tolerant have long been excluded. Thus, allelopathy in native vegetation is better seen as a subtle filter that helps govern the species composition of the community. Direct allelopathy, in which eucalypts inhibit other species in order to promote their own chances of survival, is likely unusual as a major factor in Australian eucalypt forests, although it may occur on poor sites, for example in *E. baxteri* (del Moral et al., 1978). Many of the studies linked to allelopathy in Australian eucalypts involved the apparent inhibition of seedling growth, or autotoxicity, and this is difficult to justify from an evolutionary point of view (Newman, 1978; Wilson and Agnew, 1992).

Allelochemicals are continuously present in the environment and simply because they cannot be directly related to vegetation patterning, this does not mean that they have no effect. The ecology of eucalypts in Australia is intimately connected to fire, which allows for periodic and radical changes in soil conditions. The regimen of soil microorganisms that ensues after fire must come under the influence of the allelochemicals of the establishing vegetation. Most allelopathic studies of eucalypt vegetation indicate that an understanding of the changes in the soil microbiology is essential. In particular, the copious amounts of eucalypt litter with its high C:N ratio, accompanied by sundry allelochemicals, profoundly influence the balance of soil microorganisms, including ammonifiers and nitrifiers, mycorrhizae, and pathogens, which ultimately affect the status and health of eucalypt forests. Similarly, leaf and litter leachates may be of significance not because of their allelochemicals but because of their inputs of carbon (Michelsen et al., 1995).

The allelopathic relationships of species in Australian native forests and woodlands are complex and reflect a lengthy history of species co-evolutionary interaction. The use of overseas studies on *Eucalyptus* as models for understanding the allelopathic impact of eucalypts in Australian ecosystems and vice versa should, thus, be regarded with great caution.

REFERENCES

Adams, M. A. and Attiwill, P. M., Nitrogen mineralization and nitrate reduction in forests, *Soil Biol. Biochem.*, 14, 197, 1982.

Adams, M. A. and Attiwill, P. M., Nutrient cycling and nitrogen mineralization in eucalypt forests of south-eastern Australia. II. Indices of nitrogen mineralization. *Plant Soil,* 92, 341, 1986.

Ashton, D. H., The effects of fire on vegetation, *in Second Fire Ecology Symposium,* Monash University, Melbourne, 1970, 1.

Ashton, D. H., Fire in tall open forests (wet sclerophyll forests), *in Fire and the Australian Biota,* Gill, A. M., Groves, R. H., and Noble, I. R., Eds., Australian Academy of Science, Canberra, 1981, 339.

Ashton, D. H. and Willis, E. J., Antagonisms in the regeneration of *Eucalyptus regnans* in the mature forest, *in The Plant Community as a Working Mechanism,* Newman, E. I., Ed., Blackwell Scientific, Oxford, 1982, 113.

Atkinson, N. and Rainsford, K. M., Antibacterial substances produced by flowering plants. 1. Preliminary survey, *Aust. J. Exp. Biol. Med. Sci.,* 24, 49, 1946.

Attiwill, P. M., Effects of fire on forest ecosystems, *in Research for Forest Management,* Landsberg, J. J. and Parsons, W., Eds., Commonwealth Scientific and Industrial Research Organization, Melbourne, 1985, 249.

Attiwill, P. M., Ecological disturbance and the conservative management of eucalypt forests in Australia, *For. Ecol. Manage.,* 63, 301, 1994.

Baker, R. T. and Smith, H. G., *A Research on the Eucalypts and Their Essential Oils,* 2nd ed., Government Printer, Sydney, 1920, 472.

Bowling, P. J. and McLeod, D. E., A note on the presence of *Armillaria* in second growth eucalypt stands in southern Tasmania, *Aust. For. Res.,* 3, 38, 1968.

Bowman, D. M. J. S. and Kirkpatrick, J. B., Establishment, suppression and growth of *Eucalyptus delegatensis* R. T. Baker in multiaged forests. III. Intraspecific allelopathy, competition between adult and juvenile for moisture and nutrients, and frost damage to seedlings, *Aust. J. Bot.,* 34, 81, 1986.

Bowman, D. M. J. S. and Panton, W. J., Factors that control monsoon-rainforest seedling establishment and growth in north Australian *Eucalyptus* savanna, *J. Ecol.,* 81, 297, 1993.

Britton, A., Letter to Gov. A. Phillip, dated February 24, 1792 from D. Burton, *in Historical Records of New South Wales,* Vol. 1 (Part 2), Phillip. C. Potter, Sydney, 1892, 599.

Burgess, I. P., A provenance trial with blackbutt: 9-year results, *Aust. For. Res.,* 7, 1, 1975.

Carlberg, U., Aspects of evolution in relation to defecation and oviposition behaviour of *Extatosoma tiaratum* Macleay Insecta Phasmida, *Biol. Zentralbl.,* 107, 541, 1988.

Chambers, D. P. and Attiwill, P. M., The ash-bed effect in *Eucalyptus regnans* forest: chemical, physical and microbiological changes in soil after heating or partial sterilisation, *Aust. J. Bot.,* 42, 739, 1994.

Chippendale, G. M., *Eucalyptus, in Flora of Australia,* Vol. 19, *Myrtaceae, Eucalyptus, Angophora,* Australian Government Publishing Service, Canberra, 1988, 1.

Costin, A. B., *A Study of the Ecosystems of the Monaro Region of New South Wales with Special Reference to Soil Erosion,* Government Printer, Sydney, 1954, 860.

Cremer, K. W., Effects of burnt soil on the growth of eucalypt seedling, *Inst. For. of Aust. Newslett.,* 3, 2, 1962.

Cremer, K. W., *Trees for Rural Australia,* Inkata Press, Melbourne, 1990, 455.

Cremer, K. W. and Mount, A. B., Early stages of plant succession following the complete felling and burning of *Eucalyptus regnans* forest in the Florentine Valley, Tasmania, *Aust. J. Bot.,* 13, 303, 1965.

Cromer, R. N., The significance of the "ashbed effect" in *Pinus radiata* plantations, *APPITA,* 20, 104, 1967.

Cunningham, T. M., *The Natural Regeneration of* Eucalyptus regnans, Bulletin No. 1 of the School of Forestry, University of Melbourne, Australia, 1960, 158.

del Moral, R. and Muller, C. H., Fog drip: a mechanism of toxin transport from *Eucalyptus globulus, Bull. Torrey Bot. Club,* 96, 467, 1969.

del Moral, R. and Muller, C. H., The allelopathic effects of *Eucalyptus camaldulensis, Amer. Midl. Nat.,* 83, 254, 1970.

del Moral, R., Willis, R. J., and Ashton, D. H., Suppression of coastal heath vegetation by *Eucalyptus baxteri, Aust. J. Bot.,* 26, 203, 1978.

Dyck, W. J., Gosz, J. R., and Hodgkiss, P. D., Nitrate losses from disturbed ecosystems in New Zealand: a comparative analysis, *New Zealand J. For. Sci.,* 13, 14, 1983.

Ellis, R. C., Dieback of alpine ash in north eastern Tasmania, *Aust. For.,* 28, 75, 1964.

Ellis, R. C., Mount, A. B., and Mattay, J. P., Recovery of *Eucalyptus delegatensis* from high altitude dieback after felling and burning the understorey, *Aust. For.,* 43, 29, 1980.

Ellis, R. C. and Pennington, P. I., Nitrification in soils of secondary vegetational successions from *Eucalyptus* forest and grassland to cool temperate rainforest in Tasmania, Australia, *Plant Soil,* 115, 59, 1989.

Ellis, R. C. and Pennington, P. I., Factors affecting the growth of *Eucalyptus delegatensis* seedlings in inhibitory forest and grassland soils, *Plant Soil,* 145, 93, 1992.

Evans, G., Cartwright, J. B., and White, N. H., Nectrolide, a phytotoxic compound produced by some root surface isolates of *Cylindrocarpon radicicola,* Wr., *Plant Soil,* 26, 253, 1967.

Everett, R. L., Javasharpe, B. J., Scherer, G. R., Wilt, F. M., and Ottmar, R. D., Co-occurrence of hydrophobic-

Fensham, R. J. and Kirkpatrick, J. B., The eucalypt forest-grassland/grassy woodland boundary in central Tasmania, *Aust. J. Bot.,* 40, 123, 1992.

Florence, R. G., Factors that may have a bearing upon the decline of productivity under forest monoculture, *Aust. For.,* 31, 50, 1967.

Florence, R. G. and Crocker, R. L., Analysis of blackbutt (*E. pilularis*) seedling growth in a blackbutt forest soil, *Ecology,* 43, 670, 1962.

Frood, D., *Dynamics of a Post-1951 Fire Heathland, Tidal Overlook, Wilson's Promontory, Victoria.* B.Sc. (Hons.) thesis, University of Melbourne, Australia, 1979.

Gilbert, J. M., Forest succession in the Florentine Valley, Tasmania, *Proc. R. Soc. Tasmania,* 93, 129, 1959.

Gliessman, S. R., Allelopathy in a broad spectrum of environments as illustrated by bracken, *Bot. J. Linn. Soc.,* 73, 95, 1976.

Granger, L., Kasel, S., and Adams, M. A., Tree decline in south-eastern Australia: nitrate reductase activity and indications of unbalanced nutrition in *Eucalyptus ovata* (Labill.) and *E. camphora* (R.T. Baker) communities at Yellingbo, Vic., *Oecologia,* 98, 221, 1994.

Grodzinskii, A. M., Study of indirect allelopathy. I. Statement of the problem [Russian], *in Problemy Allelopatii,* Grodzinskii, A. M., Ed., Naukova Dumka, Kiev, 1978, 5.

Grose, R. J., Effective seed supply for the natural regeneration of *Eucalyptus delegatensis* R.T. Baker, syn. *Eucalyptus gigantea* Hook. f., *APPITA,* 13, 141, 1960.

Guenther, E., *The Essential Oils,* Vol. 4, *Individual Essential Oils of the Plant Families Gramineae, Lauraceae, Burseraceae, Myrtaceae, Umbelliferae and Geraniaceae,* Van Nostrand Reinhold, New York, 1950.

Hall, C., *On Eucalyptus Oils, Especially in Relation to their Bactericidal Power,* Little & Co., Parramatta, Australia, 1904, 38.

Hardy, G. E. S. and Sivasithamparam, K., Suppression of *Phytophthora* root rot by a composted *Eucalyptus* bark, *Aust. J. Bot.,* 39, 153, 1991.

Hatch, A. B., Ashbed effects in Western Australian forest soils, *West. Aust. For. Dep. Bull.,* 64, 1, 1960.

Henry, N. B. and Florence, R. G., Establishment and development of regeneration in spotted gum–ironbark forests (*Eucalyptus* spp.), *Aust. For.,* 30, 304, 1966.

Hillis, W. E., Variation in polyphenol composition within species of *Eucalyptus* l'Hérit, *Phytochemistry,* 5, 541, 1966a.

Hillis, W. E., Polyphenols in the leaves of *Eucalyptus* l'Hérit.: a chemotaxonomic survey. I. Introduction and a study of the series Globulares, *Phytochemistry,* 5, 1075, 1966b.

Hillis, W. E., Polyphenols in the leaves of *Eucalyptus* l'Hérit.: a chemotaxonomic survey. II. The sections Renantheridae and Renantherae, *Phytochemistry,* 6, 259, 1967a.

Hillis, W. E., Polyphenols in the leaves of *Eucalyptus* l'Hérit.: a chemotaxonomic survey. III. The series Transversae, Exsertae, Subexsertae, Microcarpae, Semiunicolores, Viminales, Argyrophyllae and Paniculatae of the section Macrantherae, *Phytochemistry,* 6, 275, 1967b.

Hillis, W. E., Polyphenols in the leaves of *Eucalyptus* l'Hérit.: a chemotaxonomic survey. IV. The sections Porantheroideae and Terminales, *Phytochemistry,* 6, 373, 1967c.

Hillis, W. E., Polyphenols in the leaves of *Eucalyptus* l'Hérit.: a chemotaxonomic survey. V. The series Cornutae and Subcornutae of the section Macrantherae and the section Platyantherae, *Phytochemistry,* 6, 845, 1967d.

Hillis, W. E. and Brown, A. G., *Eucalypts for Wood Production,* Commonwealth Scientific & Industrial Research Organization, Melbourne, 1978, 434.

Hobbs, R. J. and Atkins, L., Interactions between annuals and woody perennials in a Western Australian nature reserve, *J. Vege. Sci.,* 2, 643, 1991.

Hopmans, P., Flinn, D. W., and Farrell, P. W., Nitrogen mineralisation in a sandy soil under native eucalypt forest and exotic pine plantations in relation to moisture content, *Comm. Soil Sci. Plant Anal.,* 11, 71, 1980.

Hubble, G. D., Isbell, R. F., and Northcote, K. H., Features of Australian soils, *in Soils: an Australian Viewpoint,* Commonwealth Scientific & Industrial Research Organization, Melbourne, 1983, 17.

Incoll, W. D., Effect of overwood trees on growth of young stands of *Eucalyptus sieberi, Aust. For.,* 42, 110, 1979.

Jacobs, M. R., *Growth Habits of the Eucalypts,* Forestry and Timber Bureau, Canberra, Australia, 1955, 262.

Jarret, P. H. and Petrie, A. H. K., The vegetation of the Blacks' Spur region. A study in the ecology of some Australian mountain *Eucalyptus* forests. II. Pyric succession, *J. Ecol.,* 17, 249, 1929.

Jehne, W., Phytotoxins in *Cylindrocarpon* root-rot of *Eucalyptus obliqua* regrowth trees, *Aust. Plant Pathol. Soc. Newslett.,* 5 (Suppl), 217, 1976.

Jones, J. M. and Richards, B. N., Changes in the microbiology of eucalypt forest soils following reafforestation with exotic pines, *Aust. For. Res.,* 7, 229, 1977a.

Jones, J. M. and Richards, B. N., Effect of reforestation on turnover of ^{15}N-labeled nitrate and ammonium in relation to changes in soil microflora, *Soil Biol. Biochem.,* 9, 383, 1977b.

Keenan, R. J. and Candy, S., Growth of young *Eucalyptus delegatensis* in relation to variation in site factors, *Aust. For. Res.,* 13, 197, 1983.

Kimber, P. C. and Schuster, C. J., Fire and the regeneration of karri (*Eucalyptus diversicolor* F. Muell.), *in Abstracts of Symposium on the Biology of Native Australian Plants, Perth, Western Australia, August 6–10, 1979,* Dept. of Botany, University of Western Australia, Perth, 1979, 19.

Kinney, A., *Eucalyptus,* B. R. Baumgardt & Co., Los Angeles, 1895, 298.

Lamb, D., Soil nitrogen mineralisation in a secondary rainforest succession, *Oecologia,* 47, 257, 1980.

Lamont, B., Gradient and zonal analysis of understorey suppression by *Eucalyptus wandoo, Vegetatio,* 63, 49, 1985.

Lange, R. T. and Reynolds, T., Halo-effects in native vegetation, *Trans. R. Soc. South Aust.,* 105, 213, 1981.

Lovett, J. V., Defensive stratagems of plants, with special reference to allelopathy, *Papers Proc. R. Soc. Tasmania,* 119, 31, 1985.

Lovett, J. V., Allelopathy: the Australian experience, *in The Science of Allelopathy,* Putnam, A. R., and Tang, C.-S., Eds., Wiley, New York, 1986, 75.

Lovett, J. V., Allelopathy research in Australia: an update, *in Phytochemical Ecology: Allelochemicals, Mycotoxins, and Insect Pheromones and Allomones,* Chou, C.-H. and Waller, G. R., Eds., Academia Sinica Monograph Series 9, Inst. of Botany, Academia Sinica, Taipei, 1989, 49.

Lovett, J. V. and Jackson, H. F., Allelopathic activity of *Camelina sativa* (L.) Crantz in relation to its phyllosphere bacteria, *New Phytol.,* 86, 273, 1980.

MacPherson, P., Some causes of the decay of Australian forests, *J. Proc. R. Soc. New South Wales,* 19, 83, 1885.

Marsh, N. R. and Adams, M. A., Decline of *Eucalyptus tereticornis* near Bairnsdale, Victoria: insect herbivory and nitrogen fractions in sap and foliage, *Aust. J. Bot.,* 43, 39, 1995.

May, F. E. and Ash, A. E., An assessment of the allelopathic potential of *Eucalyptus, Aust. J. Bot.,* 38, 245, 1990.

McDowell, A., *Victoria's Wet Sclerophyll Forest Species—their Allelopathic Potential,* B.Sc. (Hons.) thesis, University of Melbourne, Australia, 1993.

Michelsen, A., Schmidt, I. K., Jonasson, S., Dighton, J., Jones, H. E., and Callaghan, T. V., Inhibition of growth, and effects on nutrient uptake of arctic graminoids by leaf extracts— allelopathy or resource competition between plants and microbes?, *Oecologia,* 103, 407, 1995.

Mount, A. B., The interdependence of eucalypts and forest fires in southern Australia, *Aust. For.,* 28, 166, 1964.

Mount, A. B., The effect of plant wastes on forest productivity, Unpublished paper presented to Inst. For. Aust. Conf., Perth, 1968.

Mount, A. B., Eucalypt ecology as related to fire, *Proc. Ann. Tall Timbers Fire Ecol. Conf.,* 9, 75, 1969.

Muller, C. H., The role of chemical inhibition (allelopathy) in vegetational composition, *Bull. Torrey Bot. Club,* 93, 332, 1966.

Nava, R.V., Fernandez, E., and Del Amo, R. S., Allelopathic effects of green fronds of *Pteridium aquilinum* on cultivated plants, weeds, phytopathogenic fungi and bacteria, *Agri. Ecosyst. Environ.,* 18, 357, 1987.

Newman, E. I., Allelopathy: adaptation or accident?, *in Biochemical Aspects of Plant and Animal Coevolution,* Harborne, J. B., Ed., Academic Press, London, 1978, 327.

Nichols, D., and Yazaki, Y., Phytotoxicity of *Eucalyptus camaldulensis* and *E. regnans* sawdusts, *Aust. For. Res.,* 9, 35, 1979.

Norton, A., On the decadence of Australian forests, *Proc. R. Soc. Queensland,* 3, 15, 1886.

O'Dowd, D. J. and Gill, A. M., Predator satiation and site alteration following fire: mass reproduction of alpine ash (*Eucalyptus delegatensis*) in southeastern Australia, *Ecology,* 65, 1052, 1984.

Ohmart, C. P., Chemical interference among plants mediated by grazing insects: a reassessment, *Oecologia,* 65, 456, 1985.

Penfold, A. R. and Grant, R., The germicidal values of the principal commercial Eucalyptus oils and their pure constituents, with observations on the value of concentrated disinfectants, *J. Proc. R. Soc. New South Wales,* 57, 80, 1923.

Poore, M. E. D. and Fries, C., *The Ecological Effects of Eucalyptus,* FAO Forestry Paper no. 59, FAO, Rome, 1985, 88.

Powles, R., Artificial regeneration of mountain ash (*E. regnans*), *Aust. For.,* 5, 26, 1940.

Prescott, E. E., Germination of the seeds of mallee eucalypts, *Victorian Nat.*, 58, 8, 1941.

Prescott, J. A. and Piper, C. S., The soils of the South Australian mallee, *Trans. R. Soc. South Aust.*, 56, 118, 1932.

Prober, S. M., Environmental influences on the distribution of the rare *Eucalyptus paliformis* and the common *E. fraxinoides, Aust. J. Ecol.*, 17, 51, 1992.

Pryor, L. D., The "ash-bed" effect in eucalyptus ecology, *Inst. For. Aust. Newslett.*, 2, 23, 1960.

Pryor, L. D., Ash bed growth response as a key to plantation establishment on poor sites, *Aust. For.*, 27, 48, 1963.

Pryor, L. D. and Pryor, L. A. S., *A Classification of the Eucalypts*, Australian National University, Canberra, 1971, 102.

Purdie, R. W., Early stages of regeneration after burning in dry sclerophyll vegetation. II. Regeneration by seed germination, *Aust. J. Bot.*, 25, 35, 1977.

Rabotnov, T. A., The significance of the coevolution of organisms for the formation of phytocoenoses [Russian], *Byull. Mosk. O-ra, Ispyt. Prir., Otd. Biol.*, 82, 91, 1977.

Rabotnov, T. A., Importance of the evolutionary approach to the study of allelopathy, *Soviet J. Ecol.*, 12, 127, 1982.

Reid, R. and Wilson, G., *Agroforestry in Australia and New Zealand*, Goddard & Dobson, Box Hill, Victoria, 1985, 223.

Renbuss, M. A., Chilvers, G. A., and Pryor, L. D., Microbiology of an ashbed, *Proc. Linn. Soc. New South Wales*, 97, 302, 1972.

Rice, E. L. and Pancholy, S. K., Inhibition of nitrification by climax ecosystems, *Amer. J. Bot.*, 59, 1033, 1972.

Rotheram, I. R., Suppression of growth of surrounding regeneration by veteran trees of karri (*Eucalyptus diversicolor*), *Aust. For.*, 46, 8, 1983.

Silander, J. A., Fox, L. R., and Trenbath, B. R., The ecological importance of insect frass: allelopathy in eucalypts, *Oecologia*, 67, 118, 1985.

Silander, J. A., Trenbath, B. R., and Fox, L. R., Chemical interference among plants mediated by grazing insects, *Oecologia*, 58, 415, 1983.

Smith, D., *Continent in Crisis: a Natural History of Australia*, Penguin Books, Ringwood, Victoria, 1990, 201.

Stocker, G. C., Fertility differences between the surface soils of monsoon and eucalypt forests in the Northern Territory, *Aust. For. Res.*, 4, 31, 1969.

Stoneman, G. L., Ecology and physiology of establishment of eucalypt seedlings from seed: a review, *Aust. For.*, 57, 11, 1994.

Story, R., Pasture patterns and associated soil water in partially cleared woodland, *Aust. J. Bot.*, 15, 175, 1967.

Taylor, J. E. and Thomson, J. A., Allelopathic activity of frond run-off from *Pteridium esculentum, in Bracken Biology and Management*, Thomson, J. A. and Smith, R. T., Eds., AIAS Occasional Publication No. 40, Australian Institute of Agricultural Science, Sydney, 1990, 203.

Tolhurst, K. G. and Turvey, N. D., Effects of bracken (*Pteridium esculentum* (Forst. f.) Cockayne) on eucalypt regeneration in west-central Victoria, *For. Ecol. Manage.*, 54, 45, 1992.

Trenbath, B. R. and Fox, L. R., Insect frass and leaves from *Eucalyptus bicostata* as germination inhibitors, *Aust. Seed Sci. Newslett.*, 2, 34, 1976.

Trenbath, B. R. and Fox, L. R., Suppression of vegetation beneath trees of Tasmanian blue gum (*Eucalyptus bicostata*), *Aust. For. Res. Newslett.*, 4, 57, 1977.

Trenbath, B. R. and Silander, J. A., Bare zones under eucalypts and the toxic effects of frass of insects feeding on the leaves, *Aust. Seed Sci. Newslett.*, 4, 13, 1978.

Trenbath, B. R. and Silander, J. A., The effects of insect frass falling from *Eucalyptus globulus* ssp. *bicostata* on associated vegetation, *Aust. Seed Sci. Newslett.*, 6, 13, 1980.

Turner, J. M., Fox, J. D. E., and Lamont, B. B., Notes on *Eucalyptus todtiana* F. Muell., *Mulga Res. Centre J.*, 8, 101, 1985.

van Steenis, C. G. G. J., *Nothofagus,* key genus of plant geography in time and space, living and fossil, ecology and phylogeny, *Blumea*, 19, 65, 1971.

Vargues, H., Étude de quelques activités microbiennes dans les sols plantés d'*Eucalyptus, Bull. Soc. Hist. Nat. Afr. Nord.*, 45, 323, 1954.

Webb, D. P., Ellis, R. C., and Hallam, P. M., *Growth Check of* Eucalyptus delegatensis *(R.T. Baker) Regeneration at High Altitudes in Northeastern Tasmania,* Information Report O-X-348, Great Lakes Forest Research Centre, Canadian Forestry Service, Sault Ste. Marie, Ontario, 1983, 55.

Westman, W. E. and Anderson, D. J., Pattern analysis of sclerophyll trees aggregated to different degrees, *Aust. J. Bot.*, 18, 237, 1970.

White, C. S., The role of monoterpenes in soil nutrient cycling processes in ponderosa pine, *Biogeochemistry,* 12, 43, 1991.

White, C. S., Monoterpenes: their effect on ecosystem nutrient cycling, *J. Chem. Ecol.,* 20, 1381, 1994.

Willis, E. J., *Allelopathy and its Role in Forests of* Eucalyptus regnans *F. Muell.,* Ph.D. thesis, University of Melbourne, Australia, 1980.

Willis, E. J., Allelopathy and its role in forests of *Eucalyptus regnans F. Muell., Aust. J. Ecol.,* 7, 315, 1982.

Willis, R. J., The historical bases of the concept of allelopathy, *J. Hist. Biol.,* 18, 71, 1985.

Willis, R. J., Research on allelopathy on *Eucalyptus* in India and Pakistan, *Commonwealth For. Rev.,* 70, 279, 1991.

Willis, R. J. and Ashton, D. H., The possible role of soil lipids in the regeneration problem of *Eucalyptus regnans* (F. Muell.), *Bull. Ecol. Soc. Aust.,* 8, 11, 1978.

Wilson, J. B. and Agnew, A. D. Q., Positive-feedback switches in plant communities, *Adv. Ecol. Res.,* 23, 263, 1992.

Withers, J. R., Studies on the status of unburnt Eucalyptus woodland at Ocean Grove, Victoria. II. The differential seedling establishment of *Eucalyptus ovata* Labill. and *Casuarina littoralis* Salisb., *Aust. J. Bot.,* 26, 465, 1978.

Wrigley, T. J. and Cowan, M., Octanol partition coefficients for wetland humus, *Water Res.,* 29, 11, 1995.

14 The Research of Allelopathy in Australia: 1988–1993

Bong-Seop Kil and John V. Lovett

CONTENTS

14.1   ABSTRACT

Although the number of active workers in Australia is small, the range of species associated with allelopathic activity is at least as large as that reported elsewhere in the world. In this chapter, work on native and introduced plants is discussed under the general heading of conventional allelopathy, that is, interactions between plants that are chemically mediated. Work in which compounds associated with allelopathy in the conventional sense, are biologically active in different context is included under unconventional allelopathy. Examples involve microorganisms, corals, and other marine species.

14.2   INTRODUCTION

Australia in the 1990s has been described in the media as a "megabiodiverse" continent. Certainly, in terms of its flora, Australia enjoys not only a rich array of native plants, but also crop, pasture, and weed species familiar throughout the "westernized" world.

While Australia's native forests continue to provide the basis of a sometimes controversial industry, the remainder of the native flora has been under-exploited. Given the unique and highly variable climate and ancient, impoverished soils of Australia, this is a matter of regret to many plant scientists. Adaptation to these difficult conditions by the native flora, it is argued, should afford advantages that may outweigh the apparent superiority of introduced species of crop and pasture plants in terms of yield potential.

What is beyond dispute is that the Australian flora is rich in secondary metabolites. Some, like the oils of *Eucalyptus* spp. or the alkaloids of *Duboisia* form the basis of small industries. The oil of the tea tree, *Melaleuca* spp. has pharmaceutical properties that have only recently been thoroughly explored. Since the compounds found in these species belong to families that have long been associated with allelopathy, it is understandable that Australian native plants have attracted the attention of scientists interested in this discipline. Introduced plants, especially weeds but also some crops, notably, barley (*Hordeum vulgare* L.), have also received attention, as have their residues as sources of phytotoxins. Aquatic environments, too, have provided examples of allelopathy in recent times.

This chapter provides an update to studies of allelopathy in Australia during the period of 1988 to 1993 with some reference made to earlier works when this is necessary in providing a context.

14.3 CONVENTIONAL ALLELOPATHY

14.3.1 *Eucalyptus*

Several researchers such as Igboanugo (1986, 1987), Vicherkova and Polova (1986), Bowman and Kirkpatrick (1986a,b,c), Andrew (1986), and Lovett (1989) have contributed to studies of allelopathy in *Eucalyptus* species. From an experiment in plantation mixed stands of *Eucalyptus citriodora* Hook., *E. camaldulensis* Dehnh, and *E. grandifolia* R. Br. in Nigeria, Igboanugo (1988) found that beans can be incompatible with eucalypts, while maize and sorghum may be compatible with eucalypts for agrosilvicultural practices, and that fertilization can offset, to an extent, the depressive effects of eucalypts on crops.

An Australian researcher, May (1989), pointed out that various *Eucalyptus* species can yield allelopathic chemicals that may be effective in suppressing understory vegetation. Furthermore, May and Ash (1990) used techniques resembling more natural ecological processes, that is, extraction mimicked natural rainfall rates, root leachates, stemflow, soil leachates, and volatiles from leaves. The results of these experiments demonstrated that fresh intact leaves caused little growth suppression, in contrast to coarsely chopped leaves and extracted leaf essential oils, which were both highly suppressive. Whole leaf litter, shed bark, and, especially, stemflow yielded suppressive leachates. Evaporative concentrations of leachates in soil showed an increase in their inhibitory effects. Decay was shown to reduce the allelopathic effects of leaf and bark litter leachates, but some inhibitory chemicals persisted after five months. Therefore, allelopathy must be considered in relation to rainfall and the soil water balance, and is likely to be a cause of understory suppression by *Eucalyptus* species, especially in drier climates. Essential oils from *Eucalyptus globulus* Labill. and *E. citriodora* Hook. were extracted by steam distillation and the rich components, cineole and limonene were fractionated. Two experiments were carried out to study 1) the effect of eucalypt oil vapors and their pure components absorbed on the soil alone or in combination with a vapor column on germination behavior of *Phaseolus aureus* Roxb., and 2) the effect of eucalypt oil vapors on growth of young *P. aureus* plants. Both experiments showed lower germination percentage compared with the controls.

Singh and Kohli (1992) have studied the impact of *Eucalyptus tereticornis* Sm. shelterbelts on some crops in India. They concluded that the poor performance of crops in the eucalypt-sheltered area was related to an allelopathic effect of *Eucalyptus*.

According to Lisanework and Michelsen (1993), aqueous leaf extracts of *Eucalyptus globulus*, *E. camaldulensis* Dehnh., and *E. saligna* Sm. significantly reduced both germination and radicle growth of tested crops such as chickpea (*Cicer arietinum* L.), maize (*Zea mays* L.), pea (*Pisum sativum* L.), and teff (*Eragrostis tef* Trotter) mostly starting from a concentration of 1 or 2.5 percent (Figure 14.1).

Chemical compounds from eucalypts were isolated by Wang and Fujimoto (1993). They identified nine compounds: triterpene esters named tereticornate A and B, ursolic acid lactone, ursolic acid, betulonic acid, undulatoside, sideroxylin, 8-demethylsideroxylin, and 1-triacontananol from the dried leaves of *Eucalyptus tereticornis* Sm. A study on antioxidant activity of an extract from *Eucalyptus rostrata* Cav. has been performed in Japan (Okamura et al., 1993). From the leaves of *E. rostrata* Cav. nine active compounds were isolated: 1,2,6-tri-*O*-gallolyl-β-D-glucose and tellima-grandin 1 as tannins; spiraeoside, hyperoside, isoquercetin, and myricetin 3-*O*-*a*-L-arabinoside as flavonol glycoside; quercetin 3-*O*-*a*-arabinopyranoside-2"-gallate, kaempferol 3-*O*-*a*-arabinopyra-noside-2"-gallate, and quercetin 4'-*O*-β-D-glucopyranoside-6"-gallate as acylated flavonol glyco-sides. Antioxidant activity of the tannins and acylated flavonol glycosides, all with galloyl groups, was much higher than that of a synthetic antioxidant. Spiraeoside and quercetin 4'-glucosylgallate showed especially high superoxide dismantase activity.

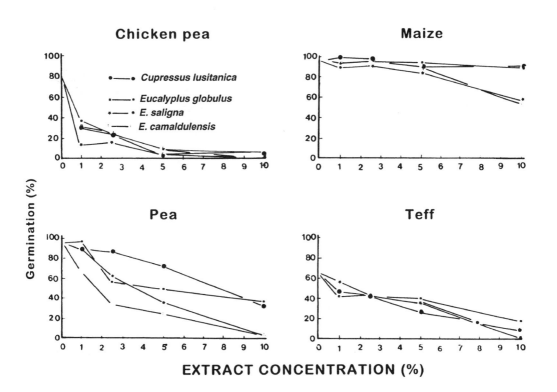

EXTRACT CONCENTRATION (%)

FIGURE 14.1 The effect of aqueous extracts of leaves of *Cupressus lusitanica* Mill., *Eucalyptus globulus* Labill, *E. camaldulensis* Dehnh, and *E. saligna* Sm. on the seed germination of four crops. N = 5 per extract type and concentration. Results of the statistical analyses are not presented. The most important significant dif-ferences are mentioned in the results, and a table with statistics is available upon request (Reproduced from Lisanework and Michelsen, Agroforestry System 21, 1993, 67 [Figure 1]. With permission from Kluwer Academic Publishers).

14.3.2 WEEDS

Some 30,000 species have been classified as weeds worldwide (Chandler, 1985). The Australian weed flora includes many species introduced from foreign countries. For grasses alone, more than 250 members of the Poaceae are regarded as weeds in standard Australian weed texts and lists (Wilding et al., 1986; Auld and Medd, 1987). Of these, 170 species, including most of the important weeds, are of exotic origin. Combellack (1987) has estimated the total cost of all weeds to Australia as between $2.50 and $2.75 billion per annum, that is, about the gross value of the Australian wheat crop in recent years (Australian Bureau of Statistics, 1987).

Allelopathy has been documented in both exotic and indigenous Australian weeds (Lovett, 1986, 1989). Work undertaken by Purvis and Jessop (1985) and Purvis (1990) with *Avena sterilis* spp. *ludoviciana* (Durieu) Nyman has shown that wild oat germination and growth are regulated by the biochemical characteristics of soil cropped to wheat, rendering the weed a "specialist" in this crop. Experimental evidence indicated that the presence of germinating wheat seeds can stimulate wild oats to emerge at the same time as the crop, whereas emergence into an already established crop is somewhat inhibited.

The majority of pasture in southeastern Australia contains either subterranean clover (*Trifolium subterraneum* L.) or white clover (*Trifolium repens* L.) and the most commonly sown grass is perennial ryegrass (*Lolium perenne* L.). Radicle elongation in *Trifolium* species was significantly reduced when grown in the presence of senescent, endophyte colonized, ryegrass vegetation, relative to endophyte-essential vegetation. And phytotoxicity of the vegetation remained despite weathering of the vegetation in the field for several months (McFarlane et al., 1992).

New research focusing on the use of allelopathic plants and their allelochemicals as natural herbicides for the biological control of leafy spurge (*Euphorbia esula* L.) has been reported. The challenge was to target the perennating organs of leafy spurge as the site of action for phytotoxic compounds (Leather and Lovett, 1992).

14.3.3 RESIDUES

Phytotoxic substances from plant residues are leached out by rainfall or irrigation, or may be released into the rhizosphere. Phytotoxins contribute to the complex interactions between residues, the environment, and microorganisms. Allelopathy sometimes results and is one of the many stress factors that operates in the crop environment. At times allelopathy may be dominant, at others it is of minor significance (Lovett and Hurney, 1992).

In previous reviews (Lovett, 1986, 1989), allelopathy has been associated with the residues of a diverse range of Australian crop species. Purvis and Jones (1990), from results of glasshouse experiments, showed that wheat seedlings emerging in sunflower stubble treatments exhibited a threefold increase in stubble quantity, from 0.5 to 1.5 percent w+/w+ soil (Figure 14.2).

14.3.4 ALLELOCHEMICALS

There are many thousands of secondary metabolites, but only a limited number of them have been identified as being involved in allelopathy (Rice, 1984). Whittaker and Feeny (1971) stated that, with few exceptions, the secondary compounds could be classified into five major categories: phenylpropanes, acetogenins, terpenoids, steroids, and alkaloids.

14.3.4.1 Phenolic Compounds

Phenolic compounds have been identified as allelopathic agents more commonly than any other substance (Stowe and Kil, 1983). In Australia they have been identified as allelochemicals in both native and introduced plants. Thus in central New South Wales, Australia, phenolic content among ever-

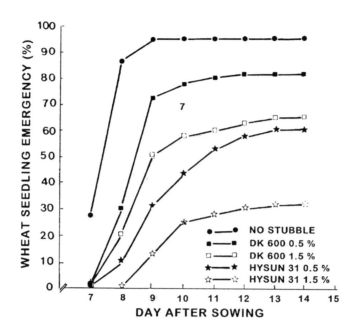

FIGURE 14.2   Wheat emergence in the presence of 0, 0.5 and 1.5 percent w+/w+ Dekalb 600 and Hysun 31 sunflower stubbles in chocolate soil. Treatments with common letters at the end-points of the lines are not significantly different at day 14 (P = .05) (Reproduced from Purvis and Jones, *Aust. J. Agric. Res.*, 41, 247, 1990 [Figure 5]. With permission from CSIRO Publishing).

green canopy trees was investigated. Polyphenols increased as leaves aged and corresponding insect grazing decreased (Lowman and Box, 1983). Phenolic compounds secreted from the roots of white clover plants either had stimulatory or inhibitory effects upon the induction of the nodulation (nod) gene expression in *Rhizobium trifolii* Dangeard (Djordjevic et al., 1987). Vanillin and isovanillin were identified from extracts of wheat seedlings and shown to have interactions with the gene nod D1 from *Rhizobium* strain to induce the expression of other genes (Le Strange et al., 1990).

Lovelock et al. (1992) investigated phenolic compounds in mangroves. Phenolic compounds were present as a band in the epidermal layers of the upper surface of the sun leaf. Sun leaves had greater contents of phenolic compounds than shade leaves. Over all tested species and sites, it was found that soluble phenolics were accumulated as a constant proportion of dry weight per leaf area.

Phenolic compounds seem to have several different functions in plants, in addition to their possible allelopathic activity. Christen and Lovett (1993) found that 1.81 mM *p*-hydroxybenzoic acid significantly reduced the radicle length of barley, whereas coleoptile elongation was less sensitive (Table 14.1). The higher tiller categories in general showed a greater sensitivity toward an application of *p*-hydroxybenzoic acid and, therefore, could not compensate for the yield decrease in the main tiller.

14.3.4.2   Terpenoids

Terpenoids and their relative compounds are not as widespread in the plant kingdom as phenolic compounds, but they have a high potential as allelopathic agents because they volatilize readily from intact leaves and they can be phytotoxic at concentrations as low as 1 to 3 10^{-6} M (Asplund, 1968). The basic types of terpenoids are the monoterpenoids (C10), sesquiterpenoids (C15), diterpenoids (C20), triterpenoids (C30), and tertraterpenoids (C40) (Rice, 1984).

TABLE 14.1

Effects of *p*-Hydroxybenzoic Acid (HBA) and pH Adjustment on Coleoptile and Radicle Elongation (mm) of Barley Seeds

Treatment	Concentration (mM)			
	Control	1.81	3.62	7.24
Coleoptile				
HBA (pH adj. to 5.5)	33.7a	33.6a	30.9abc	29.0bc
HBA (pH according to HBA concentration)	32.3a	31.6ab	28.0bc	14.8d
Radicle				
HBA (pH adj. to 5.5)	57.3a	46.0b	39.0c	32.9d
HBA (pH according to HBA concentration)	57.9a	48.6b	30.9d	8.5c

Values followed by the same letter within the same feature are not significantly different at $P < .05$ level.

Source: From Christen and Lovett, Plant and Soil 151, Kluwer Academic Publishers, Dordrecht, The Netherlands, 1993, 282. With permission.

Approximately 46 volatile terpenes have been identified in grape (Hardie and O'Brien, 1988). These compounds are characteristic of *Eucalyptus* spp. (Wang and Fujimoto, 1993) but occur also in other Australian plants. For example, a significant weed, *Prostanthera rotundifolia* R. Br. (round-leaved mint-bush), in northwestern New South Wales and southeastern Queensland, bears trichomes on its leaves from which terpenoid chemicals are liberated. These chemicals can toxify seedlings of crop plants (Lovett, 1985).

14.3.4.3 Flavonoids

There is a huge variety of flavonoids and they are widespread in seed plants (Harborne and Simmonds, 1964). In spite of the large number and wide distribution, only a few have been implicated in allelopathy. This may be due, in part, to the difficulties involved in identifying many of the flavonoids and their numerous glycosides (Rice, 1984).

Allelochemicals may influence gene expression (see Section 14.3.4.1). In Australian work, Redmond et al. (1986) reported that flavones such as authentic 7,4'-dihydroxyflavone found in washings of undamaged clover roots induced nod gene expression. Rolfe (1988) reviewed flavones and isoflavones as inducing substances of legume nodulation. In interactions between plants and bacteria, rhizobia have adapted to the use of flavonoid compounds, released by the plant root, as part of a regulatory system to initiate the transcription of their infection (nod) genes. Stimulatory compounds have been isolated from clovers, 7, 4-dihydroxyflavone; from alfalfa, luteolin; from peas, apigenin; and soybeans, the isoflavones daidzein and genistein. These plant-derived compounds are responsible for the activation of the nod genes. These hydroxylated flavonoid compounds are derived from the phenylpropanoid biosynthetic pathways (Figure 14.3).

Curir et al. (1990) found that flavonoid (identified as quercetin glycosides) accumulation is correlated with adventitious root forming in *Eucalyptus gunnii* Hook. micropropagated through axillary bud stimulation.

14.3.4.4 Tannins

The only logical reason for including the hydrolyzable and condensed tannins in one category is that both types possess an astringent taste and tan leather. The former contain ester linkages, which can be hydrolyzed by boiling with dilute mineral acid. Condensed tannins are only partially broken down

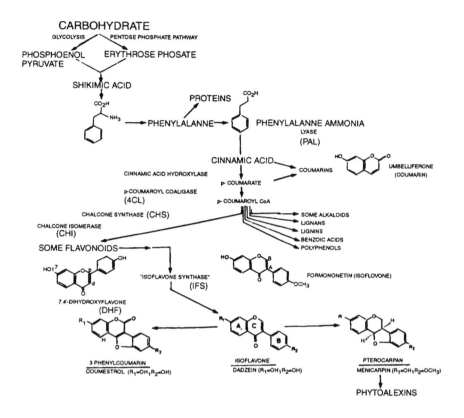

FIGURE 14.3 Proposed phenylpropanoid biosynthetic pathways for shikimic acid and the phenolic plant compounds flavonoids, isoflavonoids, and phytoalexins. The flavonoids induce the fast-growing rhizobia *Rhizobium trifolii* Dangeard, *R. meliloti* Dangeard, *R. leguminosarum* Frank, while the isoflavonoids induce the fast-growing strain NGR234 and the slow-growing Bradyrhizobia. The proposed synthesis of the isoflavones is via a flavone to an isoflavone, which involves a 2,3 aryl migration of the b-ring linked to the C-ring (Hahlbrock, 1981; Ebel et al., 1986). (Reproduced from Rolfe, *BioFactors* 1, 1988, 6. With permission from IOS Press, Amsterdam).

by rather drastic heating with concentrated acid to release cyanidin chloride, which has a bright red color, and some red-brown polymers often termed phlobaphenes (Robinson, 1983).

There are few reports of the involvement of condensed tannins in allelopathy in Australia. Duarsa et al. (1993) investigated soil moisture and temperature effects on condensed tannin concentration in *Lotus corniculatus* L. and *L. pedunculatus* Carv. The potential for elevated condensed tannin levels in *L. pedunculatus* appeared to be enhanced when growth rates were reduced by moisture stress and temperatures were high. Condensed tannin concentrations were elevated, primarily, in proportion to the size of the reduction in growth rate.

In terms of the distribution and content of tannins in *Lotus* from two experiments, it can be concluded that young plant parts as well as young plants contained higher tannins than old ones, and that leaves accumulated higher tannins than stems (Duarsa et al., 1992).

14.3.4.5 Alkaloids

The alkaloids are a structually diverse group of secondary chemical compounds that are widely distributed in plants. Alkaloids are distinguished generally from most other plant components by being basic. They usually occur in plants as the salts of various organic acids (Robinson, 1983). Alkaloids have long been used by humans in pharmacy, food preparation, and as poisons, but they play a wider role as chemical messengers between the plants that produce them and an array of organisms (Levitt and Lovett, 1985).

Alkaloids of barley, gramine, and hordenine affect radicle elongation of test species in the typical fashion of biologically active compounds (Liu and Lovett, 1987). Effects on white mustard (*Brassica alba* [L.] Rabenh.) by gramine and hordenine included reduction of radicle length and an apparent reduction in health and vigor of radicle tips (Liu and Lovett, 1990). The ability to produce one of the barley metabolites, gramine, may be a heritable characteristic (Lovett and Hoult, 1992). Hordenine was released from the roots of barley in a hydroponic system for up to 60 days. The amount reached a maximum, 2 μg per plant per day, at 36 days, then declined. Transmission electron microscopic examination of white mustard radicle tips exposed to hordenine and gramine showed damage to cell walls, increase in both size and number of vacuole, autophagy, and disorganization of organelles. These biologically active secondary metabolites of barley may play a significant role in self-defense by the crop (Liu and Lovett, 1993). Hoult and Lovett (1993) reported on the isolation and quantification of gramine and hordenine from barley material using Sep-Pak C18 cartridges to prepare the samples, followed by reversed-phase liquid chromatography. The method was faster and easier to use than methods hitherto reported.

Pyrrolizidine alkaloids were analyzed from a weed, *Heliotropium europaeum* L. (Boraginacae). The alkaloids varied significantly between sites and sampling dates but no marked seasonal trend was detected. The three major pyrrolizidine alkaloids were europine, lasiocarpine, and heliotrine (O'Dowd and Edgar, 1989) (Figure 14.4).

Duboisia species, Australian native plants, have long been harvested for their alkaloid content. Gritsanapan and Griffin (1992) detected alkaloids from a *Duboisia* hybrid, that is, scopolamine, 6-hydroxyhyoscyamine and hyoscyamine as major alkaloids, and butropine, valtropine, norvale-

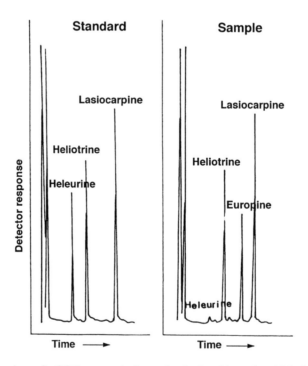

FIGURE 14.4 Comparison of a GC from a typical sample of n-butyl boronic acid derivatives of PA extracted from *Heliotropium europaeum* L. with that of authentic standards. A pure standard of europine was not available so europine was quantified from the lasiocarpine peak. Supinine is not shown because its n-butyl boronic acid derivative co-chromatographs with heliotrine (Reproduced from O'Dowd and Edgar, *Aust. J. Ecol.*, 14, 98, 1989 [Figure 2]. With permission from Blackwell Science).

roidine, valeroidine, tropine, acetyltropine, norscopolamine, hygrine, nornicotine, tetramethylpu-trescine, anabasine, and aposcopolamine as minor alkaloids. In addition, protoanemonin has been isolated from the Australian "Headache Vine" *Clematis glycinoises* DC. by Southwell and Tucker (1993).

The tropane alkaloids, scopolamine and hysocyamine, are also liberated from seeds of thornap-ple (*Datura stramonium* L.), an important weed, and interfere with the growth of seedling roots of some crop plants, including sunflower. TEM of damaged sunflower root tip cells exposed to thorn-apple allelochemicals at 0.5 and 0.05 percent concentrations showed cell division apparently dis-rupted at metaphase and anaphase. There was evidence also of damage to mitochondria, increasing with concentration of allelochemicals (Lovett and Ryuntyu, 1988).

14.4 UNCONVENTIONAL ALLELOPATHY

14.4.1 MICROORGANISMS

Interrelationships between allelopathy and microorganisms in Australia have been reported by sev-eral researchers (Lovett, 1987; Djordjevic et al., 1987; Gabriel et al., 1988). Bacteria play an impor-tant but often ill-defined role as ameliorants in allelopathy. The genera so far identified are frequently cosmopolitan and capable of activity in both the phyllosphere and rhizosphere. It seems likely that allelopathy occurs when environmental conditions retard the sequence of breakdown and permit accumulation of concentrations of phytotoxic compounds (Lovett, 1987).

In the clover-Rhizobia symbiosis, a distinct cocktail of phenolic compounds (flavonoids) has been recognized. Flavonoids of the correct structure induce the expression of several bacterial nod and other genes required for plant infection. Flavonoids of the incorrect, but related, structure can antagonize nod gene induction (Djordjevic and Weinman, 1991) (Figure 14.5).

McGee et al. (1991) showed that two isolates of *Acremonium strictum* significantly inhibited, *in vitro,* the rate of growth of five fungi commonly associated with grasses. Extracts from cultures of the isolates also inhibited the rate of hyphal elongation.

14.4.2 CORAL AND OTHER MARINE SPECIES

Conventionally, studies of allelopathy have focussed on terrestrial environments. However, there is a significant amount of Australian literature on marine coral algae. Earlier work is reviewed in Lovett (1989). Coll and Sammarco (1988) have pointed out that the chemistry of soft corals reveals the presence of a variety of terpenoid metabolites, often in unexpectedly high quantities. These rich terpenes in soft corals relate to roles in their defense, for example, concerning toxicity, feeding deter-rence, physical defense, and allelopathic interactions between soft corals and hard corals.

The specialized predator *Chaetodon melannotus,* which feeds on highly toxic and sometimes allelopathic octocorals, initially locates its prey using visual rather than chemical cues (Alino et al., 1992). Extraction of the soft coral, *Sinularia flexibilis* (Coelenterata, Octocorallia) and quantitative chemical analysis for the three major diterpene components, flexibilide, dihydroflexibilide, and sin-ulariolide, afforded average ratios of 4:3:1, respectively. Possible biosynthetic pathways were pre-sented (Maida et al., 1993) (Figure 14.6).

Shallow and deep marine sponges were classified into six toxicity groups, falling into three broad categories: 100 percent lethal to a test fish, *Gambusia affinis* (Baird et Birard). Over 73 per-cent of the sponges ranged from harmful to toxic, with deep sponges being generally more lethal (La Barre et al., 1988).

Antonelli et al. (1988) have applied mathematical modelling of allelopathic interactions in sessile communities to the cost of terpene production to colonizing soft corals and their highly toxic effects on scleractinian corals. The model predicted that any factor that reduced soft coral

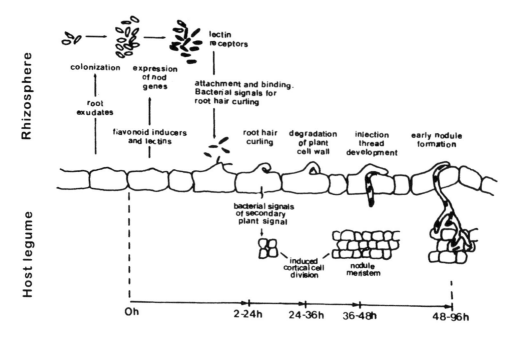

FIGURE 14.5   Summary of the early infection events evident during Rhizobium legume interactions. A diagrammatic representation is presented of the events leading to and culminating in legume root infection by Rhizobium over a 96-h period. Flavonoids, lectins, and nutritional factors affect the rate of colonization and responsiveness of *Rhizobium* during the early infection process. Immature root hairs are targeted for infection and evidence of root hair curling, infection, and colonization are already clearly evident after 24 h. Evidence of completed cycles of cortical cell division is clearly seen after 24 to 36 h (Ridge and Rolfe, 1986). Infection threads grow through and between cells without causing cell death. Rhizobia actively colonize the confines of the infection thread and are eventually released into the cytoplasm of the infected plant cells before the onset of nitrogen fixation (Reproduced from Djordjevic and Weinman), *Aust. J. Plant Physiol.*, 18, 544, 1991 [Figure 1]. With permission from CSIRO Publishing).

encroachment (which includes direct toxic effects) had survival value. The allelochemic strategies of corals would be passive.

Secondary compounds, particularly terpenes, in alcyonacean soft corals (Coelenterata, Octocorallia) may play a role in predator defense, competitor defense, anti-fouling/anti-biosis (Sammarco, 1988).

Two *bis*-prenylated phenols have been isolated from brown algae, *Encyothalia cliftonii*. The major metabolite, 2,4-bis(3-methylbut-2-enyl) phenol showed significant feeding deterrence toward the herbivorous sea urchin (Roussis et al., 1993).

Extracts from four species of bryozoans, found in Tasmanian coastal waters, have been demonstrated to exhibit selective antibacterial activity (Walls et al., 1993) (Table 14.2).

14.4.4   BREAK CROPS

According to Kirkegaard et al. (1994), alternative crops to wheat such as winter oilseeds and grain legumes are being increasingly grown in the cropping phase of farms in the traditional wheat/sheep zone of southeastern Australia (Mead, 1992). This trend reflects the need for producers to diversity in the face of uncertain commodity prices, and the growing awareness of the benefits to wheat yield arising from sound crop rotations. These break crops can increase the yield of subsequent wheat crops by depriving soil-borne wheat pathogens of a host and reducing infection of the subsequent crop (Kollmorgen et al., 1983).

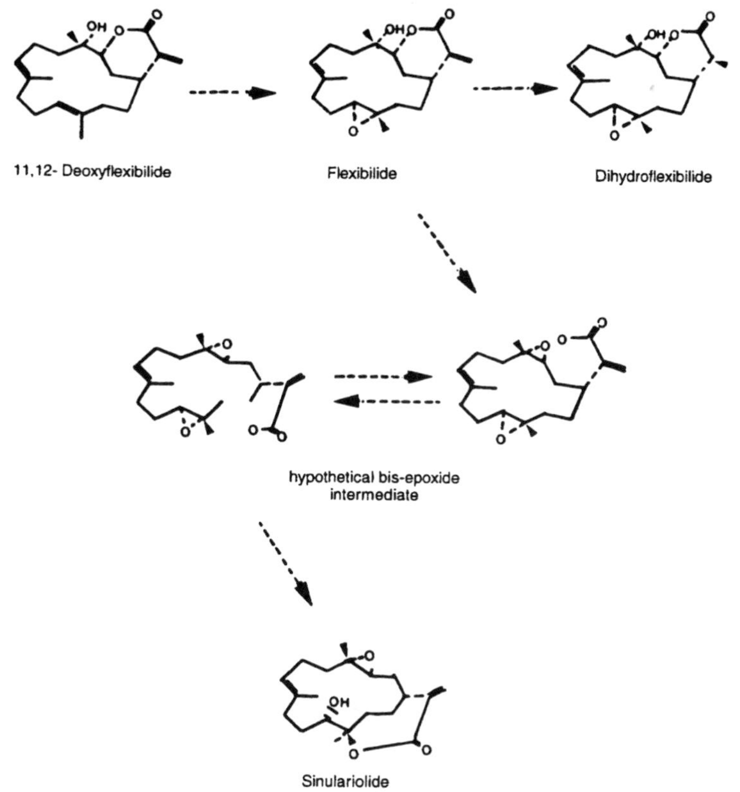

11,12- Deoxyflexibilide   Flexibilide   Dihydroflexibilide

hypothetical bis-epoxide
intermediate

Sinulariolide

FIGURE 14.6   Cembranolide diterpene metabolites derived from *Sinularia flexibilis* and possible biosynthetic pathways linking them (Reproduced from Maida et al., *J. Chem. Ecol.*, 19, 2293, 1993 [Figure 3]. With permission from Plenum Publishing, NY).

Kirkegaard et al. (1993) demonstrated the effect of the *Brassica* crops, canola and Indian mustard, on the growth and yield of subsequent wheat crops at four sites in southern New South Wales. The effect of break crops on grain yield was influenced by water availability after anthesis. At one site, where significant rainfall occurred after anthesis, the early improvements in growth persisted to maturity, and yield was significantly improved following the break crops. At the other three sites, dry conditions occurred after anthesis and the greater biomass of wheat following break crops resulted in more rapid depletion of soil water.

14.5   CONCLUSION

Mukerji and Garg (1988) stated, "Almost any process, occurring naturally or done artificially, which affects the relationship between organisms in such a way that the natural biological balance is restored, can be regarded as biocontrol." Allelopathy is encompassed by this definition and deployment of allelopathy in agriculture, aquaculture, horticulture, or silviculture may best be regarded in this context.

Studies of allelopathy in Australia include all of these milieux. However, reports of allelopathic potential in Australia outnumber instances of the application of allelopathy and allelochemicals as components of management systems for weeds or other pest organisms. In this respect, the Australian experience mirrors that elsewhere in the world.

Direct use of allelopathy, on a field scale, by plant to plant activity has been documented in India by Joshi and Mahadevappa (1986). These investigators found that the leguminous plant, *Cassia sericea*, could exert effective control of *Parthenium hysterophorus* L. in the field through

TABLE 14.2
Inhibition of Bacterial Growth by Extracts from Bryozoans

	2A	2B	4A	6A	6B	8A	8B	10A	10B
Am DEM	++	++	++	++	-	++	++	++	++
Am MeOH	+++	-	+	+	-	+	+	++	-
Ortho DCM	+++	++++	+	++++	+++	++++	++++	++++	-
Ortho MeOH	++++	+	++++	+++	++++	+	+	++++	-
Cel DCM	-	-	-	-	-	+	-	+	-
Cel MeOH	-	-	-	-	-	-	-	-	-
Bug DCM	+	+	-	+	-	-	-	+	-
Bug MeOH	-	-	-	-	-	-	-	-	-
Gram stain	-	+	-	-	-	-	-	+	-
Morphology	rod	rod	curved rod	rod	rod	curved rod	rod	rod	rod
Size	var	var	var	var	small	var	var	small	small
Motility	m	m	nm	m	m	m	m	nm	m

Size of inhibition zone: +, 0–2 mm; ++, 2–4 mm; +++, 4–6 mm; ++++, 6–8 mm. Am, *A. wilsoni*; Ortho, *O. ventricosa*; Cel, *C. pilosa*; Bug, *B. dissimilis*. DCM, dichloromethane; MeOH, methanol. 2A–10B indicates plates incubated in seawater two to ten days. A and B indicate two different bacterial strains tested at interval (at the 4-day interval strain, B failed to grow). var, variable size; small, 5–10 μm; m, motile; nm, non-motile.

Source: From Walls et al., *J. Exp. Mar. Biol. Ecol.*, 169, 5, 1993. With permission from Elsevier Science.

allelopathic activity. Parthenium weed is becoming widespread in Australia, where other species of *Cassia* occur, suggesting a potential to emulate the Indian work. But the significance of the work of Joshi and Mahadevappa (1986) lies not only in the application of allelopathy to solving a weed problem but also in the fact that Joshi (1990) attempted a benefit cost analysis in respect of his findings.

At the beginning of the 1990s decade, Australia's Commonwealth Scientific and Industrial Research Organisation (CSIRO) compared an investment of $2,330,000 in biocontrol programs with estimated benefits of $261,200,000—an average return of $112 per $1 invested (Lovett et al., 1990). Returns of such magnitude to allelopathy programs would finally confirm the place of the discipline as a positive contributor to food and fiber production systems.

REFERENCES

Alino, P. M., Coll, J. C., and Sammarco, P. W., Toxic prey discrimination in a highly specialized predator *Chaetodon melannotus* (Block et Schneider): visual vs. chemical cues, *J. Exp. Mar. Biol. Ecol.*, 164, 209, 1992.

Andrew, M. H., Population dynamics of the tropical annual grass *Sorghum intrans* in relation to local patchiness in its abundance, *Aust. J. Ecol.*, 11, 209, 1986.

Antonelli, P. L., Sammarco, P. W., and Coll, J. C., A model of allelochemical interactions between soft and scleractinian corals on the great barrier reef, *J. Biol. Syst.*, 1, 1, 1998.

Asplund, R. O., Monoterpenes: relationships between structure and inhibition of germination, *Phytochemistry*, 7, 1995, 1968.

Auld, B. A. and Medd, R. W., Weeds: an illustrated botanical guide to the weeds of Australia, Inkata Press, Melbourne, 1987, 255.

Australian Bureau of Statistics, Crops and Pastures Australia 1984–85, Commonwealth Government Printer, Canberra, Australia, 1987.

Bowman, D. M. J. S. and Kirkpatrick, J. B., Establishment, suppression and growth of *Eucalyptus delegatensis* R. T. Baker in multiaged forests. I. The effect of fire on mortality and seedling establishment, *Aust. J. Bot.*, 34, 63, 1986a.

Bowman, D. M. J. S. and Kirkpatrick, J. B., Establishment, suppression and growth of *Eucalyptus delegatensis* R. T. Baker in multiaged forests. II. Sampling growth and its environmental correlates, *Aust. J. Bot.,* 34, 73, 1986b.

Bowman, D. M. J. S. and Kirkpatrick, J. B., Establishment, suppression and growth of *Eucalyptus delegatensis* R. T. Baker in multiaged forests. III. Intraspecific allelopathy, competition between adult and juvenile for moisture and nutrients, and frost damage to seedlings, *Aust. J. Bot.,* 34, 81, 1986c.

Chandler, J. M., Economics of weed contriol in crops, *in The Chemistry of Allelopathy: Biochemical Interactions Among Plants,* Thompson, A. C., Ed., American Chemical Society Symposium Series 268, Washington, D. C., 1985, 9.

Christen, O. and Lovett, J. V., Effects of a short-term *p*-hydroxybenzoic acid application on grain yield and yield components in different tiller categories of spring barley, *Plant Soil,* 151, 279, 1993.

Coll, J. C. and Sammarco, P. W., The role of secondary metabolites in the chemical ecology of marine invertebrates: a meeting ground for biologists and chemists, 6th Int. Coral Reef Congress, Townsville, Qld., 1988, 2.

Combellack, J. H., Invited Editorial, *Plant Protect. Quart.,* 2, 2, 1987.

Curir, P., Van Sumere, C. F., Termini, A., Barthe, P., Marchesini, A., and Dolci, M., Flavonoid accumulation is correlated with adventitious roots formation in *Eucalyptus gunnii* Hook. micropropagated through auxiliary bud stimulation, *Plant Physiol.,* 92, 1148, 1990.

Djordjevic, M. A., and Weinman, J. J., Factors determining host recognition in the clover. *Rhizobium* symbiosis, *Aust. J. Plant Physiol.,* 18, 543, 1991.

Djordjevic, M. A., Remond, J. W., Batley, M., and Rolfe, B. G., Clovers secrete specific phenolic compounds which either stimulate or repress nod gene expression in *Rhizobium trifolii, EMBO J.* 6, 1173, 1987.

Duarsa, M. A. P., Hill, M. J., and Lovett, J. V., The distribution of tannins in Lotus spp.: variations within plant and stage of maturity, Proceedings of the 6th Australian Agronomy Conference, University of New England, Armidale, 1992, 446.

Duarsa, M. A. P., Hill, M. J., and Lovett, J. V., Soil moisture and temperature affect condensed tannin concentrations and growth in *Lotus corniculatus* and *Lotus pedunculatus, Aust. J. Agric. Res.,* 44, 1667, 1993.

Gabriel, D. W., Loschke, D. C., Rolfe, B. G., Gene-for-gene recognition: the ion channel defense model, *in Molecular Plant-Microbe Interaction,* Verma, D. P. S. and Palacios, R., Eds., Proc. 4th Int. Symp., APS Press, St. Paul, MN, 1988, 3.

Gritsanapan, W. and Griffin, W. J., Alkaloids and metabolism in a *Duboisia* hybrid, *Phytochemistry,* 3, 471, 1992.

Harborne, J. B. and Simmonds, N. W., The natural distribution of the phenolic aglycones, *in Biochemistry of Phenolic Compounds,* Harborne, J. B., Ed., Academic Press, New York, 1964, 77.

Hardie, W. J. and O'Brien, T. P., Considerations of the biological significance of some volatile constituents of grape (*Vitis* spp.), *Aust. J. Bot.,* 36, 107, 1988.

Hoult, A. H. C. and Lovett, J. V., Biologically active secondary metabolites of barley. III. A method for identification and quantification of hordenine and gramine in barley by high-performance liquid chromatography, *J. Chem. Ecol.,* 19, 2245, 1993.

Igboanugo, A. B. I., Phytotoxic effects of some eucalypts on food crops, particularly on germination and radicle extension, *Trop. Sci.,* 26, 19, 1986.

Igboanugo, A. B. I., Effects of some eucalypts on growth and yield of *Amaranthus caudatus* and *Abelmoschus esculentus, Agric. Ecosyst. Environ.,* 18, 243, 1987.

Igboanugo, A. B. I., Effects of some eucalypts on yield of *Vigna unguiculata* L. Walp., *Zea mays* L. and *Sorghum bicolor* L., *Agric. Ecosys. Envir.,* 24, 453, 1988.

Joshi, S., An economic evaluation of control methods for *Parthenium hysterophorus* Linn., *Biol. Agric. Horticult.,* 6, 285, 1990.

Joshi, S. and Mahadevappa, M., *Cassia sericea* Sw. to fight *Parthenium hysterophorus* Linn., *Curr. Sci.,* 55, 261, 1986.

Kirkegaard, J. A., Gardner, P. A., Angus, J. F., and Koetz, E., Effect of *Brassica* break crops on the growth and yield of wheat, *Aust. J. Agric. Res.,* 45, 529, 1994.

Kollmorgen, J. F., Griffiths, J. B., and Walsgott, D. N., The effects of various crops on the survival and carry over of the wheat take-all fungus *Gaeumannomyces graminis* var. *tritici, Plant Pathol.,* 32, 73, 1983.

La Barre, S., Laurent, D., Sammarco, P., Williams, W. T., and Coll, J. C., Comparative ichthyotoxicity study of shallow and deep water sponges of New Caledonia, 6th Int. Coral Reef Congress, Townsville, Qld., 1988.

Leather, G. R. and Lovett, J. V., *Euphorbia esula*: a challenge for non-herbicidal weed control in north America, Proceedings of the 6th Australian Agronomy Conference, University of New England, Armidale, 1992, 430.

Le Strange, K. K., Bender, G. L., Djordjevic, M. A., Rolfe, B. G., and Redmond, J. W., The *Rhizobium* strain NGR234 *nod* D_1 gene product responds to activation by the simple phenolic compounds vanillin and isovanillin present in wheat seedling extracts, *Mol. Plant-Microb. Interact.*, 3, 214, 1990.

Levitt, J. and Lovett, J. V., Alkaloids, antagonisms and allelopathy, *Biol. Agric. Hort.*, 2, 289, 1985.

Lisanework, N. and Michelsen, A., Allelopathy in agroforestry systems: the effects of leaf extracts of *Eucalyptus lusitanica* and three *Eucalyptus* spp. on four Ethiopian crops, *Agrofor. Systems*, 21, 63, 1993.

Liu, D. L. and Lovett, J. V., Crop allelochemical interference with weed seed germination, Proceedings of the Weed Seed Biology Workshop, Orange, New South Wales, 1987, 116.

Liu, D. L. and Lovett, J. V., Allelopathy in barley: potential for biological suppression of weeds, *in Alternatives to the chemical control of weeds*, Bassett, C., Whitehouse, L. J., Zabkiewicz, J. A., Eds., Proceedings of the International Conference, Rotorua, New Zealand, July 1989. Ministry of Forestry, FRI Bulletin 155, 1990, 85.

Liu, D. L. and Lovett, J. V., Biologically active secondary metabolites of barley. II. Phytotoxicity of barley allelochemicals, *J. Chem. Ecol.*, 19, 2231, 1993.

Lovelock, C. E., Clough, B. F., and Woodrow, I. E., Distribution and accumulation of ultraviolet-radiation-absorbing compounds in leaves of tropical mangroves, *Planta*, 188, 143, 1992.

Lovett, J. V., Defensive stratagems of plants with special reference to allelopathy, *Paper Proc. R. Soc. Tasmania*, 119, 31, 1985.

Lovett, J. V., Allelopathy: the Australian experience, *in The Science of Allelopathy*, Putnam, A. R. and Tang, C. S., Eds., John Wiley, New York, 1986, 75.

Lovett, J. V., On communities and communication in agriculture, An Inaugural Public Lecture, Armidale, New South Wales, 1987, 27.

Lovett, J. V., Allelopathy research in Australia: an update, *in Phytochemical Ecology: Allelochemicals, Mycotoxins and Insect Pheromones and Allomones*, Chou, C. H. and Waller, G. R., Eds., Institute of Botany, Academia Sinica Monograph Series No. 9., Taipei, ROC, 1989.

Lovett, J. V. and Hoult, A. H. C., Gramine: the occurrence of a self defence chemical in barley, *Hordeum vulgare* L., Proceedings of the 6th Australian Agronomy Conference, University of New England, Armidale, 1992, 426.

Lovett, J. V. and Hurney, A. P., Allelopathy: a possible contributor to yield decline in sugar cane, *Plant Prot. Quart.*, 17, 180, 1992.

Lovett, J. V. and Ryuntyu, M. Y., Vacuolar activity and phagocytosis as responses to allelopathic stress, Inst. Phys. Conf. Ser. No. 93, Vol. 3, EUREM 88, York, England, 1988, 103.

Lovett, J. V. Parbery, I. H., and Guest, R. J., Plant production systems, *in Australian Agriculture*, Lovett, J. V., Ed., A series of six texts, University of New England, Armidale, 1990, 56.

Lowman, M. D. and Box, J. D., Variation in leaf toughness and phenolic content among five species of Australian rain forest trees, *Aust. J. Ecol.*, 8, 17, 1983.

Maida, M., Carroll, A. R., and Coll, J. C., Variability of terpene content in the soft coral *Sinularia flexiblis* (Coelenterata: Octocorallia), and its ecological implications, *J. Chem. Ecol.*, 19, 2285, 1993.

May, F. E., The allelopathic potential of *Eucalyptus*, B.Sc. (Hons.) thesis, Australian National University, 1989.

May, F. E. and Ash, J. E., An assessment of the allelopathic potential of *Eucalyptus*, *Aust. J. Bot.*, 38, 245, 1990.

McFarlane, N. M., Jowett, B., Schroder, A., and Quigley, P. E., Potential harmful effects of high endophyte content in perennial ryegrass on companion legumes, Proceedings of the 6th Australian Agronomy Conference, University of New England, Armidale, 1992, 434.

McGee, P. A., Hincksman, M. A., and White, C. S., Inhibition of growth of fungi isolated from plants by *Acremonium strictum*, *Aust. J. Agric. Res.*, 42, 1187, 1991.

Mead, J. A., Rotations and farming systems-the current situation, *in Rotations and Farming Systems for Southern and Central New South Wales*, Murray, G. M. and Heenan, D. P., Eds., NSW Agriculture: Wagga NSW, 1992, 5.

Mukerji, K. G. and Garg, K. L., Eds., Biocontrol of plant diseases, Vol. 1, CRC Press, Boca Raton, 1988, 211.

O'Dowd, D. J. and Edgar, J. A., Seasonal dynamics in the pyrrolizidine alkaloids of *Heliotropium europaeum*, *Aust. J. Ecol.*, 14, 95, 1989.

Okamura, H., Mimura, A., Yakou, Y., Niwano, M., and Takahara, Y., Antioxidant activity of tannins and flavonoids in *Eucalyptus rostrata*, *Phytochemistry*, 33, 557, 1993.

Purvis, C. E., Allelopathy: a new direction in weed control, *Plant Protect. Quart.,* 5, 55, 1990.

Purvis, C. E. and Jessop, R. S., Biochemical regulation of wild oat germination and growth by wheat and wheat crop residues, 1985 British Crop Protection Conference—Weeds, 1985, 661.

Purvis, C. E. and Jones, G. P. D., Differential response of wheat to retained crop stubbles. II. Other factors influencing allelopathic potential; intraspecific variation, soil type and stubble quantity, *Aust. J. Agric. Res.,* 41, 243, 1990.

Redmond, J. W., Batley, M., Djordjevic, M. A., Innes, R. W., Kuempel, P. L., and Rolfe, B. G., Flavones induce expression of nodulation genes in *Rhizobium, Nature,* 323, 632, 1986.

Rice, E. L., *Allelopathy,* Academic Press, New York, 1984.

Robinson, T., *The Organic Constituents of Higher Plants,* 5th ed., Cordus Press, North Amherst, MA, 1983.

Rolfe, B. G., Flavones and isoflavones as inducing substances of legume nodulation, *BioFactors,* 1, 3, 1988.

Roussis, V., King, R. L., and Fenical, W., Secondary metabolite chemistry of the Australian brown alga *Encyothalia cliftonii*: evidence for herbivore chemical defence, *Phytochemistry,* 34, 107, 1993.

Sammarco, P. W., The multiple functions of secondary metabolites in soft corals and their interaction with functionally related adaptations, 5th Annual Meeting, Int. Soc. Chem. Ecol. University of Georgia, Athens, 1988.

Singh, D. and Kohli, R. K., Impact of *Eucalyptus tereticornis* Sm. shelterbelts on crops, *Agrofor. Syst.,* 20, 253, 1992.

Southwell, I. A. and Tucker, D. J., Protonemonin in Australian *Clematis, Phytochemistry,* 33, 1099, 1993.

Stowe, L. G. and Kil, B. S., The role of toxins in plant–plant interactions, *in Handbook of Natural Toxins, Vol. 1, Plant and Fungal Toxins,* Keeler, R. F. and Tu, A. T., Eds., Marcel Dekker, New York, 1983.

Vicherkova, M. and Polova, M., Effect of essential oil vapors of different concentration upon leaf transpiration of bean *Vicia faba* cultivar *Prerovsky* and sunflower *Helianthus annuus* cultivar Slovenska-Siva, *Fac. Sci. Nat. Univ. Purkynianae Brun.,* 16, 109, 1986.

Walls, J. T., Ritz, D. A., and Blackman, A. J., Fouling, surface bacteria and antibacterial agents of four bryzoan species found in Tasmania, Australia, *J. Exp. Mar. Biol. Ecol.,* 169, 1, 1993.

Wang, H. and Fujimoto, Y., Triterpene esters from *Eucalyptus tereticornis, Phytochemistry,* 33, 151, 1993.

Whittaker, R. H. and Feeny, P. P., Allelochemics: chemical interactions between species, *Science,* 171, 757, 1971.

Wilding, J. L., Barnett, A. G., and Amor, R. L., Crop weeds, Inkata Press, Melbourne, Australia, 1986, 153.

15 Historical Review and Current Models of Forest Succession and Interference

Daniel W. Gilmore

CONTENTS

15.1 ABSTRACT

Difficulties in separating the effects of allelopathy from other forest interferences under field conditions are well established. In this chapter, compounding factors including successional and competitive influences on forest ecosystems are presented to provide a holistic framework for the discussion of allelopathy. The natural herbicidal effect of one species on another is not only ecologically interesting, but with testing and experience it may be possible to use these and other allelopathic relationships to reduce competition in forests managed for fiber production.

15.2   INTRODUCTION

During the latter half of the 20th century, allelopathy has become established as an independent branch of physiological plant ecology (Rice, 1984; Inderjit et al., 1995). Authors of ecological text books have included their respective discussions of allelopathy in the context of nutrient cycling (Spurr and Barnes, 1973), disturbance (Waring and Schlesinger, 1985), competition (Barbour et al., 1987), plant succession (Kozlowski et al., 1991; Perry 1994), the plant environment (Larcher, 1995), and community ecology (Kimmins, 1997).

Allelopathy is defined as "a competitive strategy of plants in which there is the production of chemical compounds (allelochemicals) by such plants that interfere with the germination, growth, or development of another plant" (Dunster and Dunster, 1996). Allelopathic compounds, or allelo-chemicals, are produced from roots, as leachates from stems and leaves, through the decomposition of plant parts, or from the volatilization of substances such as terpenes into the air. Allelopathy includes both promoting and inhibitory activities and is a concentration-dependent phenomenon. Allelochcmicals may also operate via the effects of mycorrhizal symbionts, or decomposer organ-isms. Inderjit (1996) discussed the necessary experimental steps in order to determine allelopathy. In brief they are 1) demonstrate correlation between the allelopathic species and the plant affected; 2) experiment to demonstrate cause and effect, usually in laboratory conditions; and 3) return to the field situation to determine if the results obtained in the laboratory occur in the ecosystem. In a more recent paper, Inderjit and del Moral (1997) discussed the experimental design difficulties in making a conclusive determination of allelopathy under field conditions.

Allelopathy is difficult to separate from other forest interferences under field conditions because of our incomplete understanding of ecological processes and our inability to control these processes in an experimental framework. This chapter focuses on two ecological processes, succession and competition, in order to provide an ecological perspective for other chapters in this book that focus exclusively on allelopathy.

Forest interferences may lead to the suppression of one species by another, or to the establish-ment of a single species at higher or lower relative population densities. It is not uncommon for the competitive relationship among plants in forest communities to change during ecological succes-sion. The process of ecological succession, however, is also dependent on factors that are external to the forest community such as climate change or human-caused disturbances. Abiotic factors affect-ing ecological succession include solar radiation, temperature, wind, soil, water, and fire. Biotic fac-tors affecting ecological succession include herbivory, the introduction of exotic species, intraspecific competition, and human influences. Many of these abiotic and biotic factors affect allelopathic vectors (Inderjit and Dakshini, this volume).

15.3   SUCCESSION

15.3.1   Ecological Succession

An ecosystem is defined as an ecological system composed of living organisms and their nonliving environment. Ecosystems are characterized by five major attributes: structure, function, complexity, interaction of the components, and change over time. The last attribute, change over time is the process by which a series of different plant communities and associated animals and microbes suc-cessively occupy and replace each other over time in a particular ecosystem or landscape location following a disturbance. A thorough review of ecological succession is beyond the scope of this review. Interested readers are referred to the pioneer works of Clements (1916), Gleason (1926), Tansley (1935), and thorough reviews of the topic by West et al., (1981), Barbour et al., (1987), and Kimmins (1997).

Clements (1916) proposed the monoclimax theory of ecological succession that equated ecosystems to "supraorganisms," which would always, given enough time, "grow to be" the same ecosystem. The monoclimax theory maintains that climate is the dominant community-forming factor and discounts the importance of soil, topographic relief, biota, time, and disturbance. Tansley (1935) proposed the polyclimax theory of succession, which recognizes that successional processes are disturbance-dependent and that frequent disturbances in many cases preclude the existence of a climax. Gleason (1926) proposed the climax pattern hypothesis, which maintains that plant communities were not individuals in themselves but more or less chance aggregations of various species. The climax pattern hypothesis proposes a complex pattern of integrating communities rather than a mosaic of distinct communities. This hypothesis is the basis for current models of successional pathways or trajectories that recognize the importance of the type and intensity of forest disturbance (Oliver, 1981; West et al., 1981; Seymour, 1992; Kimmins, 1997).

Applied ecologists now recognize that while its usage will undoubtedly continue into the foreseeable future, the term succession is less appropriate than in the past because it implies a single pathway for community development. The alternative term of vegetation dynamics has been suggested in several text books of applied forest ecology (Spurr and Barnes, 1973; Oliver and Larson, 1990; Kimmins, 1997).

15.3.2 TYPES OF SUCCESSION

There are three types of primary succession that occur on lands not previously vegetated: xerarch, mesarch, and hydrarch. Primary xerarch succession occurs on dry rock surfaces or on the exposed solum. Primary mesarch succession occurs on deep, medium-textured soils of moderate fertility or on soil parent material. Landslides, the deposition of alluvial materials, or severe human-caused disturbances would precede this type of primary succession. Primary hydrarch succession occurs on wet substrates, generally in the wake of retreating glaciers, during the aggradation of water courses, and during the progressive process that occurs as lakes accumulate sediment and a terrestrial ecosystem develops along the former shoreline.

Secondary succession is the invasion of land that has been previously vegetated and destroyed by natural or human disturbances. Old-field succession is secondary succession that occurs on abandoned fields that were at one time cleared for crop production or pasture.

Allogenic succession is the result of major environmental changes beyond the control of the organisms indigenous to the ecosystem. Examples of allogenic disturbances would include wildfire, severe windstorms, floods, and mass wasting or landslides. Autogenic succession is the result of changes in the living community and in the soil and microclimatic characteristics of an ecosystem caused by the living community itself. The major mechanisms of autogenic succession are invasion and colonization, local environmental alteration through single tree senescence, and species exclusion.

Progressive succession refers to the development of communities having greater complexity or diversity. Regressive succession refers to the process in which ecosystems become simpler or depauperate. Progressive and regressive successional patterns can be either allogenic or autogenic. An interesting example of a retrogressive, autogenic successional pattern commonly occurs in late successional forests at their northern range limit (Spurr and Barnes, 1973; Barbour et al., 1987; Perry, 1994; Kimmins, 1997). Low temperatures result in the slow decomposition of dead organic matter, which accumulates in the forest floor. This slow decomposition has a negative effect on tree growth because nutrients are not mineralized and hence are unavailable. Additionally, as the forest floor thickness increases, its insulating factor causes either permafrost to develop or the permafrost level to rise. The development of permafrost close to the soil surface further reduces decomposition and soil drainage. Cooler soils further decrease decomposition rates, which results in a greater accumulation of organic matter; this in turn, causes organic acids to accumulate, which decreases the soil and forest floor pH. Reduced nutrients, increased acidity, and a higher water table favor the survival and growth of mosses, which further contributes to the decline of the forest. The end result of this retrogressive successional process is a treeless muskeg or bog. These late successional forests at the northern extension of their range cannot perpetuate themselves except in the presence of disturbance caused by fire or windthrow, which removes the insulating duff layer to warm the soil and prepare a favorable germination site for conifer seeds.

15.3.3 STAGES OF SUCCESSIONAL DEVELOPMENT

Clements (1916) proposed that succession was a six-stage process:

1. nudation, the exposure of a new surface in primary succession or the clearing away of previous vegetation in secondary succession;
2. migration of seeds, spores, or vegetative propagules from adjacent areas, though in secondary succession these may be present in the soil;
3. ecesis, germination, early growth, and establishment of plants;
4. competition among the established plants;
5. reaction, the autogenic effects of plants on the habitat; and
6. stabilization, the climax.

Egler (1954) studied the colonization of abandoned fields in southern New England and presented two models of old-field succession (Figure 15.1). Egler (1954) modified the Clementsian model through his observations and models of relay floristics and initial floristics that involve first, field abandonment, then the development of a herbaceous weed community followed by a grass community, then a shrub community, and ending in a forest. The relay floristic model is likely to be more important in primary succession and best fits the Clementsian model in that each seral stage appears to prepare the site for the next stage. The initial floristics model allows for the presence of dormant seeds and the presence of plants from an earlier successional stage model is likely to be more important in secondary succession. For example, tree species may be present as seedlings immediately after field abandonment but may take years to germinate or become dominant over the herbaceous, grass, and shrub communities.

Oliver (1981) and Oliver and Larson (1990) proposed a four-stage model of forest stand development: 1) stand initiation stage, 2) stem exclusion stage, 3) understory reinitiation stage, and 4) old-growth stage (Figure 15.2). This model is analogous to the four-stage forest development model presented by Bormann and Likens (1979) and Peet (1992): 1) establishment phase, 2) thinning phase, 3) transition phase, and 4) steady-state phase. Both of these models describe secondary succession of forests and have been found to be applicable across North America (Oliver, 1981; Seymour, 1992), and, with slight modification, in the Mexican tropics (Gómez-Pompa and Vázquez-Yanes, 1981).

15.3.4 GENERAL MECHANISMS AND MODELS OF SUCCESSION

Most early studies of succession lacked scientific rigor because they neglected to consider mechanisms responsible for succession or consider alternative successional models (Peet, 1992). In an influential paper, Conell and Slatyer (1977) proposed a three-pathway model that depicted succession as following one of three pathways: facilitation, tolerance, or inhibition (Figure 15.3).

According to the facilitation pathway, only certain early successional species can invade a site, after a disturbance. These species then modify and ameliorate the site, thereby making it suitable for invasion and growth of later successional species. Their concept of facilitation was similar to Egler's (1954) relay floristic pathway and was intended to apply to primary succession or old-field succession. Following the tolerance pathway, early-colonizing plants have little effect on future-colonizing plants, and the sequence of various species is determined predominantly by their life history characteristics. The tolerance pathway closely resembles Gleason's (1926) climax pattern hypothesis.

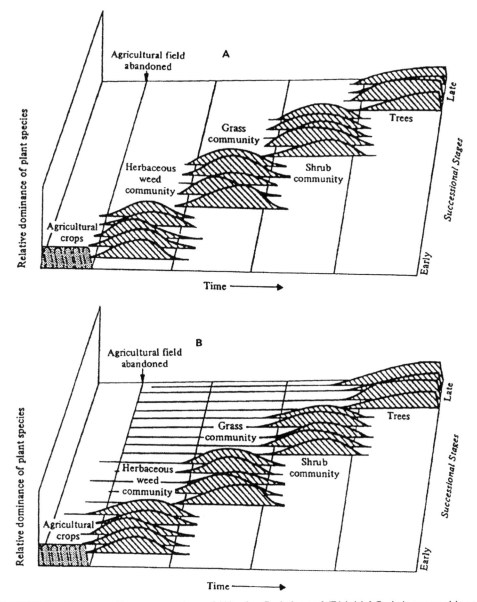

FIGURE 15.1   Diagrammatic representations of (A) relay floristics and (B) initial floristic composition types of old-field succession as proposed by Egler (1954). (*Source:* From Egler, F. E., Vegetation science concepts. I. Initial floristic composition—a factor in old-field vegetation development, *Vegetatio,* 4, 412, 1954. With permission of Kluwer Academic Publishers.

According to the inhibition pathway, early-colonizing plants secure space and resources and inhibit future invasion of other species, or suppress those already present.

Conell and Slatyer (1977) provided a valuable framework for examining the role of interactions between species during succession, particularly in the context of forest interferences. Ecological succession is far too complex, however, to be so easily described. In reality, most successional sequences involve a mixture of different pathway mechanisms (Egler, 1954). For example, facilitation may occur during the early stages of primary or secondary succession and may be replaced by either tolerance or inhibition mechanisms depending on the life histories of the particular species involved. Furthermore, inhibition of one species by another in an early successional stage may

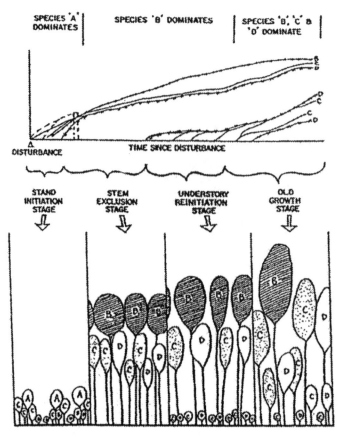

FIGURE 15.2 Schematic diagram of stages of stand development following major disturbances as proposed by Oliver (1981). All trees comprising the forest start soon after the disturbance; however, dominant tree type changes as stem number decreases and vertical stratification of species progresses. Height attained and time lapsed during each stage would vary with species, disturbances during community development, and site quality. Letters designate different species. (*Source:* From Oliver, C. D., Forest development in North America following major disturbances, *For. Ecol. Manage.*, 3, 153, 1981. With permission of Elsevier Science).

benefit another species. The introduction of nonindigenous and exotic plant and animal species to forest ecosystems coupled with human-caused disturbances have altered successional pathways to create many ecosystems for which there exists no natural analog (White, 1985, 1991; Seymour, 1992; Stelfox, 1995). Because of this mixture of pathway mechanisms, all models of ecological succession should be considered in their proper context. In spite of shortcomings due to its oversimplicity, the Conell and Slayter three-pathway successional model provides an excellent setting for a discussion of forest interferences in the context of community ecology.

15.4 COMPETITION

15.4.1 COMPETITIVE STRATEGIES BETWEEN PLANTS

Stands are the basic forest management unit used by resource managers (Smith et al., 1997). Competition between plants in stands is important because stands are not homogeneous in regards to species composition. The discussion in this section focuses on intraspecific competition and interspecific competition. No discussion of competition is complete, however, without reviewing the seminal works of MacArthur and Wilson (1967), Grime (1977), and Tilman (1988) to obtain their perspectives on life history strategies and resource competition.

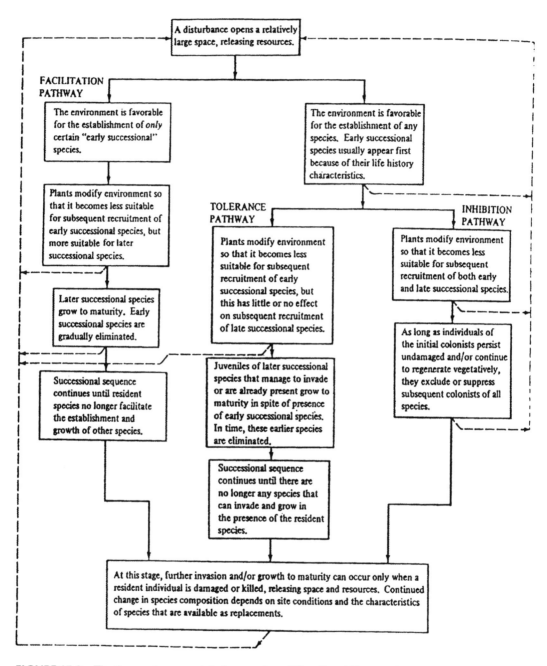

FIGURE 15.3 The three-pathway model of succession of Conell and Slatyer (1977). (*Source:* From Connell, J. H. and Slatyer, R. O., Mechanisms of succession in natural communities and their role in community stability and organization, *Amer. Nat.,* 111:1119, 1977. With permission of The University of Chicago Press).

15.4.2 LIFE HISTORY STRATEGIES AND RESOURCE COMPETITION

MacArthur and Wilson (1967) suggested that the life history patterns of all organisms could be placed somewhere on a continuum between the extremes of reproductive (r) or conservative (K) survival strategies. The concept of r- and K-selection life history strategies was further detailed by Pianka (1970) as summarized in Table 15.1.

TABLE 15.1
Climate and Life History Attributes Associated with r- and K-Selection

	r-Selection	K-Selection
Climate	Variable or uncertain	Constant or predictable
Mortality	Density-independent	Density-dependent
Population size	Variable over time, usually below carrying capacity, frequent recolonization	Constant over time, at or near carrying capacity, minimal recolonization
Intra- and interspecific competition	Variable, often little	usually intense
Selection favors	1. Rapid development	1. Slower development, greater competitive ability
	2. High rate of reproduction	2. Lower resource thresholds
	3. Early reproduction	3. Delayed reproduction
	4. Small body size	4. Larger body size
	5. Single reproduction	5. Repeated reproductions
Length of life	Short, usually less than 1 year	Longer, usually more than 1 year
Leads to	Productivity	Efficiency

Source: From Pianka, E. R., On r- and K-selection, *Amer. Nat.,* 104, 592, 1970. With permission of The University of Chicago Press.

The r and K concept of life history strategies was further expanded by Grime (1977, 1979) to include competitive (C), stress tolerant (S), and ruderal (R) life history strategies (Table 15.2). C-selection plant characteristics maximize vegetative growth in productive, relatively undisturbed environments. S-selection adaptations allow for establishment on continuously unproductive environments that have either, or both, severe environmental stresses and severe resource depletion at the expense of vegetative and reproductive vigor. R-selection traits include a short lifespan and high seed production on severely disturbed but potentially productive environments.

Through various combinations of life history strategies (e.g., stress-tolerant ruderals, competitive ruderals, stress-tolerant competitors, and C-S-R strategists), this model emphasizes that the species with the greatest capacity for acquiring resources under a given set of conditions will be the superior competitor. A fundamental assumption of this model is that a species that is superior in competing for one resource will also be superior in competing for other resources. For example, according to Grime's model the better competitor above ground is also the better competitor below ground.

Tilman (1982, 1985, 1988), on the other hand, proposed the resource-ratio hypothesis, which emphasizes resource tradeoffs within a plant. In other words, unlike Grime's model, a superior above ground competitor is not by default the superior below ground competitor. Tilman (1987) compared his model with Grime's and observed that traits that Grime associated with superior competitive ability were the same traits that would maximize the ability of a plant to compete in a highly disturbed nutrient-rich habitat. Traits that the Grime model associated with poor competitive ability were those that the Tilman model associated with the maximization of competitive ability in undisturbed habitats.

15.4.2.1 Intraspecific Competition

Conceptually, intraspecific competition, competition between individuals of the same species, is considered the most severe form of competition because of identical resource demands (Darwin, 1859). On the other hand, intraspecific competition between plants should be no more important than interspecific competition, competition between different species. Most plants require the same handful of basic resources in similar amounts, and although plant size typically confers superiority, size depends on age or growth form rather than on species identity.

TABLE 15.2

Life History Characteristics of Competitive, Stress-Tolerant, and Ruderal Plants

	Competitive	Stress-Tolerant	Ruderal
Morphology of shoot	High dense canopy of leaves	Wide range of growth forms	Small stature, limited lateral spread
Leaf form	Robust, often mesomorphic	Small leathery, or needle-like	Various, often mesomorphic
Litter	Copious, often persistent	Sparse, sometimes persistent	Sparse, not usually persistent
Maximum relative growth rate	Rapid	Slow	Rapid
Life forms	Perennial herbs, shrubs, trees	Lichens, perennial herbs, shrubs, trees	Annual herbs
Leaf longevity	Relatively short	Long	Short
Phenology of leaf production	Well-defined peaks of leaf production	Evergreens with various patterns of leaf production	Short period of leaf production during periods of high potential productivity
Phenology of flowering	Flowers produced after leaves	No general relationships	Flowers produced at the end of favorable periods
Proportion of annual production devoted to reproduction	Small	Small	Large

Source: From Grime, J. P., Evidence for the existence of three primary strategies in plants and its relevance to ecological and evolutionary theory, *Amer. Nat.,* 111, 1169, 1977. With permission of The University of Chicago Press and the author J. P. Grime.

Some ecologists maintain that intraspecific competition does not affect the composition of the forest type and therefore has no effect on forest succession (Spurr and Barnes, 1973; Kozlowski et al., 1991). This statement is not always valid, however, because intraspecific competition will have an effect on population characteristics, such as density, which may affect forest health (Kimmins, 1997), which in turn could alter both the rate and trajectory of ecological succession.

15.4.2.2   Interspecific Competition

Interspecific competition between plant species in the forest takes place in the forest canopy, understory, and below ground. The larger vascular plants, which include trees, large shrubs, and lianas compete for sunlight and growing space in the forest canopy. Trees in certain stages of their life cycle (seedling and sapling stage), shrubs, forbs, and grasses compete for sunlight and growing space in the forest understory. Depending on the density of the forest canopy, light can be the most limiting factor for plant growth in the forest understory. All plants, with the notable exception of epiphytes, compete for below-ground moisture, nutrients, and growing space.

15.4.2.3   Interferences in the Forest Canopy

Photosynthesis is the physiological process that produces all of the oxygen that organisms breath, all of the plant food consumed by humans and other organisms, and all of the fixed carbon in fossil fuels used by humans (van Overbeek, 1976). In the context of forestry, leaf area is used as a surrogate of photosynthesis in ecosystem-based models of forest productivity (Kimmins, 1996). A strong relationship between foliar leaf area and crop yield has been demonstrated in agriculture (e.g., Watson, 1952; Rees, 1963) and in forest tree species (e.g., Waring, 1983; Gilmore and Seymour, 1996; Stoneman and Whitford, 1996).

The competitive status of trees in a single-cohort stand is associated with their crown position. The most widely used crown classification system in forestry classifies trees as dominant, codominant, intermediate, and overtopped (Smith et al., 1997). Dominant trees have crowns extending above the general level of the canopy and receive full sunlight from above and partial sunlight from the side. Codominant trees have crowns forming the general level of the canopy and receive full sunlight from above but comparatively little from the sides. Intermediate trees are shorter than the main canopy, receive little direct sunlight from above and none from the sides. Overtopped trees, also called suppressed trees, have crowns entirely below the main canopy and receive no direct sunlight. As a general rule, shade-intolerant tree species are found in single-cohort stands. Shade-tolerant trees can be found in both single-cohort and multi-cohort stands.

Studies of the crown architecture of forest trees have shown that tree crowns are highly plastic in regards to changes in resource availability, particularly sunlight and growing space (e.g., Mitchell, 1975; Maguire, 1983; Barthelemy et al., 1991; Seymour, 1992; Gilmore and Seymour, 1997). Crown plasticity is more pronounced in tree species that are shade-tolerant (e.g., McClendon and McMillen, 1982; Canham, 1988; Clark and Clark, 1992; Takahashi, 1996) essentially because tree species that are not shade tolerant would not be able to survive in the understory. Crown plasticity, likely supplemented by root plasticity, has been an important factor in the success of advance regeneration through adaptations in crown morphology. Advance regeneration, in turn, has had a profound influence on forest stand development following a forest disturbance (e.g., Lorimer, 1983; Cogbill, 1985; Kelty, 1986; Deal et al., 1991; Clark and Clark, 1992; Seymour, 1992; Fajvan and Seymour, 1993; Palik and Prezitzer, 1993).

15.5 INTERFERENCES IN THE FOREST UNDERSTORY AND THE FOREST FLOOR

Forest interferences in the understory involve a greater number of plant species and plant forms than in the overstory. Interferences at the forest floor level would also include tree seedlings, forbs, grasses, pteridophytes, and bryophytes. Competition in the understory and at the forest floor level is usually greatest during the stand initiation stage of stand development (Oliver, 1981; Oliver and Larson, 1990). Allelopathy has been invoked on numerous occasions as one agent of tree regeneration failure (del Moral and Cates, 1971; Gabriel, 1975; Gant and Clebsch, 1975; Horsley, 1976a,b; Fisher, 1980). The effects of allelopathy are more pronounced during reforestation or regeneration (Fisher, 1980).

Conclusive evidence of allelopathy under natural conditions, however, is difficult to obtain (Inderjit, 1996; Inderjit and del Moral, 1997). While investigating the interference potential of *Kalmia angustifolia* L. to black spruce [*Picea mariana* (Mill.) B.S.P.], Mallik (1987) was unable to demonstrate an allelopathic vector responsible for observed decreases in *P. mariana* seedling growth. Later Inderjit and Mallik (1996) found significantly higher phenolic content in soils amended with leaf litter of *K. angustifolia*. In addition, significant nutrient alteration was observed in amended soils. These authors concluded that *K. angustifolia* has the potential to cause nutrient interference to seedling growth of black spruce, and did not rule out the possible allelopathic effects of *K. angustifolia* to black spruce. Inderjit and Mallik (1997) investigated the effects of *Ledum groenlandicum* Oeder on soil charactertistics and black spruce growth. They reported that phenolics from *L. groenlandicum* and changes in nutrient availability after amendment are the likely cause of *L. groenlandicum* interference to black spruce. In a separate study of forest interferences, Zackrisson et al. (1997), suggested that a three-part interacting system of feathermosses, mycorrhizae, and ericacaeous shrubs may inhibit tree regeneration and immobilize nutrients in late successional boreal forests.

15.5.1 SUCCESSION, COMPETITION, AND ALLELOPATHY

Competition occurs when two species compete for resources that are in limited supply. In allelopathy, chemical compounds are contributed to the environment by the donor plant, which influences growth, establishment, and survival of its associated species. There are instances when allelopathy evolved due to competition, and also vice-versa (Williamson, 1990). Higher production of allelo-chemicals in plants growing in nutrient-poor soil has often been suggested (Rice, 1984). Furthermore, allelochemicals may also influence the soil nutrient dynamics (Inderjit and Dakshini, this volume; Northup et al., this volume). Rice and his coworkers (see Rice, 1984) published several papers on the role of allelopathy in old-field succession in Oklahoma. They reported a low rate of nitrification as succession proceeds toward climax in many ecosystems. Phenolic allelochemicals released from the certain grasses, tolerant to low nitrogen levels, inhibit rate of nitrification, and this, in turn, influences the invasion of nitrophilous species (Rice, 1984).

While studying the allelopathy in the first stages of secondary succession on the Piedmont of New Jersey, Jackson and Willemson (1976) reported that ragweed (*Ambrosia artemisiifolia* L.), a dominant species of the first year of old-field succession, rarely exists for more than two years. They found that the ragweed and wild radish (*Raphanus raphanistrum* L.) could not re-establish in plots dominated by *Aster pilosus* Willd. Field soil collected from the first successional stage had no affect on plant growth, but field soil from the second successional stage inhibited the growth of ragweed and wild radish. These authors implicated the partial role of allelopathy in vegetational changes from the first to second stage of succession. In natural systems, different mechanisms of interference such as resource competition and allelopathy operate simultaneously and/or sequentially, and their sepa-ration is extraordinarily difficult (Inderjit and del Moral, 1997). This illustrates how competition, allelopathy, and succession are inter-linked in nature. More research, however, is needed for a better and broad understanding of such allelochemical interactions in plant ecology (Dakshini et al., 1998).

15.6　INTERFERENCES BELOW-GROUND AND ABIOTIC INFLUENCES

The understanding of allelopathic compounds in natural systems is difficult because they are inter-woven with other stressing agents in the environment. Many allelopathic compounds enter the soil, but only a few are destined to be absorbed by plants. Abiotic factors that influence allelopathic inter-actions include moisture, temperature, water-holding capacity, pH, and aeration and are thoroughly discussed by Inderjit and Dakshini (this volume).

Evidence exists to demonstrate that allelopathic interactions and fine-root turnover or root life-span are site specific. Fisher (1978) reported that juglone, an allelopathic compound produced by black walnut, damaged and sometimes killed red and white pine on wet sites, while on dry sites it had little effect. Schmidt (1988) suggested that the rapid degradation of juglone and other suspected allelochemical by soil bacteria negates their role as allelopathic compounds under natural condi-tions. Keyes and Grier (1981) presented evidence that demonstrated that fine-root turnover on a lower quality site was greater than that on higher quality site for Douglas-fir. In contrast, Nadelhoffer et al., (1985) reported that fine-root turnover rates were higher in nitrogen-rich soils. Shaver and Billings (1975) showed that fine-root turnover rates can differ considerably between species. The biomass of roots at the ecosystem scale indicates the magnitude of below-ground production, and below-ground interferences influence plant survival and reproductive success. Observation of roots will always be more difficult than observation of leaves. Eissenstat and Yanai (1997) suggested a modeling approach to further develop theories regarding the ecology of root lifespans and to develop testable hypotheses. Future opportunities exist for the inclusion of allelopathic influences in such a below-ground modeling framework.

15.7　APPLICATIONS TO RESOURCE MANAGERS

It is apparent from this review that as the resolution of the ecological scale regarding forest interfer-ences and allelopathy increases, there is a corresponding decrease in our understanding of allelopa-thy. At the community level, allelopathy has been suggested to influence ecological succession. At

TABLE 15.3
Some Allelopathic Plants Important in Forestry, the Chemicals they Produce, and the Plants They Are Reported to Affect

Allelopathic Species	Class of Chemical Produced	Affected Species
Trees		
sugar maple	Phenolics	yellow birch
hackberry	Coumarins	herbs, grasses
eucalyptus	Phenolics, terpenes	shrubs, herbs, grasses
walnut	Quinone	trees, shrubs, herbs
juniper	Phenolics	grasses
sycamore	Coumarins	grasses
black cherry	Cyanogenic glycosides	red maple
oaks	Coumarins, phenolics	herbs, grasses
sassafras	Terpenoids	elm, maple
Shrubs		
laurel	Phenolics	black spruce
manzanita	Coumarins, phenolics	herbs, grasses
bearberry	Phenolics	pine, spruce
sumac	Phenolics, terpenoids	Douglas-fir
rhododendron	Phenolics	Douglas-fir
elderberry	Phenolics	Douglas-fir
Other		
aster	Phenolics, terpenoids	sugar maple, black cherry
goldenrod	Phenolics, terpenoids	sugar maple, black cherry
New York fern	Phenolics	black cherry
bracken fern	Phenolics	Douglas-fir
fescue	Phenolics	sweetgum
short husk grass	Phenolics	black cherry
clubmoss	Phenolics	black cherry
reindeer lichen	Phenolics	jack pine, white spruce

Source: From Fisher, R. F., Allelopathy: potential cause of regeneration failure, *J. For.,* 78, 346, 1980. With permission of The Society of American Foresters and the author R. F. Fisher. Not for further reproduction.

the population level, the role of allelopathy becomes less clear due to the compounding effects of multiple species and site-specific factors. Despite these difficulties in determining the exact relationship between allelopathic and afflicted plants, Fisher (1980) provided a thorough summary of allelopathic plants of interest to foresters (Table 15.3). The natural herbicidal effect of one species on another is not only ecologically interesting, but with testing and experience it may be possible to use these (Table 15.3) and other allelopathic relationships to reduce competition in forests managed for fiber production. This could have the net effect of reducing the amount of commercial herbicides used in competition control.

ACKNOWLEDGMENTS

Appreciation is extended to Dr. Inderjit for his encouragement to contribute to this volume and his thorough and thoughtful review of an early draft of this chapter.

REFERENCES

Barbour, M. G., Burk, J. H., and Pitts, W. D., Terrestrial Plant Ecology, 2nd ed., Benjamin/Cummings Publishing, Menlo Park, CA, 1987.

Barthelemy, D., Edelin, C., and Hall, F., Canopy architecture, *in Physiology of Trees,* Raghavendra, A. S., Ed., John Wiley & Sons, New York, 1991, 1.

Borman, F. H. and Likens, G. E., Patterns and Processes in a Forested Ecosystem, Springer-Verlag, New York, 1979.

Canham, C. D., Growth and canopy architecture of shade-tolerant trees: response to canopy gaps, *Ecology,* 69, 786, 1988.

Clark, D. H. and Clark, D. B., Life history diversity of canopy and emergent trees in a neotropical rain forest, *Ecol. Monogr.,* 62, 314, 1992.

Clements, F. E., Plant Succession. An Analysis of the Development of Vegetation, Carnegie Institute, Washington, D.C., 1916.

Cogbill, C. V., Dynamics of the boreal forests of the Laurentian Highlands, Canada. *Can. J. For. Res.,* 15, 52, 1985.

Conell, J. H. and Slatyer, R. O., Mechanisms of succession in natural communities and their role in community stability and organization, *Amer. Nat.,* 111, 1119, 1977.

Dakshini, K. M. M., Foy, C. L., and Inderjit, Allelopathy: one component in a multifaceted approach to ecology, 1999 (this volume), Chapter 1.

Darwin, C. R., On the Origin of the Species, J. Murray, London, 1859.

Deal, R. L., Oliver, C. D., and Bormann, B. T., Reconstruction of mixed hemlock-spruce stands in coastal southeast Alaska, *Can. J. For. Res.,* 21, 643, 1991.

del Moral, R. and Cates, R. G., Allelopathic potential of the dominant vegetation of western Washington, *Ecology,* 52, 1030, 1971.

Dunster, J. and Dunster, K., Dictionary of Natural Resource Management, University of British Columbia Press, Vancouver, 1996, 363.

Egler, F. E., Vegetation science concepts. I. Initial floristic composition—a factor in old-field vegetation development, *Vegetatio,* 4, 412, 1954.

Eissenstat, D. M. and Yanai, R. D., The ecology of root lifespan, *Adv. Ecolog. Res.,* 27, 1, 1997.

Fajvan, M. A. and Seymour, R. S., Canopy stratification, age structure, and development of multicohort stands of eastern white pine, eastern hemlock, and red spruce, *Can. J. For. Res.,* 23, 1799, 1993.

Fisher, R. F., Juglone inhibits pine under certain moisture regimes, *Soil Soc. Amer. J.,* 42, 801, 1978.

Fisher, R. F., Allelopathy: a potential cause of regeneration failure, *J. For.,* 78, 346, 1980.

Gabriel, W. J., Allelopathic effects of black walnut on white birches, *J. For.,* 73, 234, 1975.

Gant, R. E. and Clebsch, E. E. C., The allelopathic influences of *Sassafras albidum* in old-field succession in Tennessee, *Ecology,* 56, 604, 1975.

Gilmore, D. W. and Seymour, R. S., Alternative measures of stem growth efficiency applied to Abies balsamea from four canopy positions in Central Maine, USA, *For. Ecolog. Manage.,* 84, 209, 1996.

Gilmore, D. W. and Seymour, R. S., Crown architecture of *Abies balsamea* from four canopy positions, *Tree Physiol.,* 17, 71, 1997.

Gleason, H. A., The individualistic concept of the plant association, *Bull. Torrey Bot. Club,* 53, 7, 1926.

Gómez-Pompa, A. and Vázquez-Yanes, C., Successional studies of a rain forest in Mexico, *in Forest Succession, Concepts and Applications,* West, D. C., Shugart, H. H., and Botkin, D. B., Eds., Springer-Verlag, New York, 1981, 246.

Greenwood, M. S., Juvenility and maturation in conifers: current concepts, *Tree Physiol.,* 15, 433, 1995.

Grime, J. P., Evidence for the existence of three primary strategies in plants and its relevance to ecological and evolutionary theory, *Amer. Nat.,* 111, 1169, 1977.

Grime, J. P., *Plant Strategies and Vegetation Processes,* John Wiley and Sons, Chichester, England, 1979, 222.

Horsley, S. B., Allelopathic inhibition of black cherry by fern, grass, goldenrod and aster, *Can. J. For. Res.,* 7, 205, 1977a.

Horsley, S. B., Allelopathic inhibition of black cherry. II. Inhibition by woodland grass, fern and clubmoss, *Can. J. For. Res.,* 7, 515, 1977b.

Inderjit, Plant phenolics in allelopathy, *Bot. Rev.,* 62, 182, 1996.

Inderjit and Dakshini, K. M. M., Bioassays for allelopathy: interactions of soil organic and inorganic constituents, 1999 (this volume), Chapter 4.

Inderjit and Mallik, A. U., The nature of interference potential of *Kalmia angustifolia, Can. J. For. Res.,* 26,
1899, 1996.

Inderjit and del Moral, R., 1997, Is separating resource competition from allelopathy realistic?, *Bot. Rev.,* 63,
221, 1997.

Inderjit and Mallik, A. U., Effects of *Ledum groenlandicum* amendments on soil characteristics and black
spruce seedling growth, *Plant Ecol.,* 133, 29, 1997.

Inderjit, Dakshini, K. M. M., and Einhellig, F. A., Eds., *Allelopathy: Organisms, Processes and Applications,*
American Chemical Society, Washington, D.C., 1995.

Jackson, J. R. and Willemsen, R. W., Allelopathy in the first stages of secondary succession on the piedmont of
New Jersey, *Am. J. Bot.,* 63, 1015, 1976.

Kelty, M. J., Development patterns in two hemlock-hardwood stands in southern New England, *Can. J. For.
Res.,* 16, 885, 1986.

Keyes, M. R. and Grier, C. C., Above- and below-ground net production in 40-year-old Douglas-fir stands on
low and high productivity sites, *Can. J. For. Res.,* 11, 599, 1981.

Kimmins, J. P., Importance of soil and role of ecosystem disturbance for sustained productivity of cool temperate and boreal forests, *Soil Sci. Am. J.,* 60, 1643, 1996.

Kimmins, J. P., *Forest Ecology, A Foundation for Sustainable Development,* 2nd ed., Prentice-Hall, Upper
Saddle River, NJ, 1997, 596.

Kozlowski, T. T., Kramer, P. J., and Pallardy, S. G., *The Physiological Ecology of Woody Plants,* Academic
Press, San Diego, 1991, 657.

Larcher, W., *Physiological Plant Ecology,* 3rd ed., Springer-Verlag, New York, 1995, 506.

Lorimer, C. G., Eighty-year development of northern red oak after partial cutting in a mixed-species Wisconsin
forest, *For. Sci.,* 29, 371, 1983.

MacArthur, R. H. and Wilson, E. O., *The Theory of Island Biogeography,* Princeton University Press, Princeton,
NJ, 1967, 203.

Maguire, D. A., Suppressed crown expansion and increased bud density after precommercial thinning in
California Douglas-fir, *Can. J. For. Res.,* 13, 1246, 1983.

Mallik, A. U., Allelopathic potential of *Kalmia angustifolia* to black spruce (*Picea mariana*), *For. Ecol.
Manage.,* 20, 43, 1987.

McClendon, J. H. and McMillen, G. G., The control of leaf morphology and the tolerance of shade by woody
plants, *Bot. Gaz.,* 143, 79, 1988.

Mitchell, K. J., Dynamics of simulated yield of Douglas-fir, For. Sci. Monogr. 17, Society of American
Foresters, Bethesda, MD, 1975, 39.

Nadelhoffer, K. J., Aber, J. D., and Melilo, J. M., Fine roots, net primary production, and soil nitrogen availability: a new hypothesis, *Ecology,* 66, 1377, 1985.

Northup, R. R., Dahlgren, R. A., Aide, T. M., and Zimmerman, J. K., Effect of plant polyphenols on nutrient
cycling and implications for community structure, 1999 (this volume), Chapter 22.

Oliver, C. D., Forest development in North America following major disturbances, *For. Ecol. Manage.,* 3, 153,
1981.

Oliver, C. D. and Larson, B. C., *Forest Stand Dynamics,* McGraw-Hill, New York, 1990, 467.

Palik, B. J. and Pregitzer, K. S., The vertical development of early successional forests in northern Michigan,
USA, *J. Ecol.,* 81, 271, 1993.

Peet, R. K., Community structure and ecosystem function, *in Plant Succession: Theory and Prediction,* Glenn-
Lewin, D. C., Peet, R. K., and Veblen, T. T., Eds., Chapman and Hall, London, 1992, 103.

Perry, D. A., *Forest Ecosystems,* John Hopkins University Press, Baltimore, 1994, 649.

Pianka, E. R., On r- and K-selection, *Amer. Nat.,* 104, 592, 1970.

Rees, A. R., Relationship between crop growth and leaf index to the oil palm, *Nature,* 197, 63, 1963.

Rice, E., *Allelopathy,* 2nd ed., Academic Press, San Diego, 1984.

Schmidt, S. K., Degradation of juglone by soil bacteria, *J. Chem. Ecol.,* 14, 1561, 1988.

Seymour, R. S., The red spruce—balsam fir forest of Maine: Evolution of silvicultural practice in response to
stand development patterns and disturbances, *in The Ecology and Silviculture of Mixed-species Forests,*
Kelty, M. J., Larson, B. C., and Oliver, C. D., Eds., Kluwer, Norwell, MA, 1992, 217.

Shaver, G. R. and Billings, W. D., Root production and root turnover in a wet tundra ecosystem, Barrow, Alaska, *Ecology,* 56, 401, 1975.

Smith, D. M., Larson, B. C., Kelty, M. J., and Ashton, P. M. S., *The Practice of Silviculture—Applied Forest Ecology,* 9th ed., John Wiley and Sons, New York, 1997, 537.

Spurr, S. H. and Barnes, B. V., *Forest Ecology,* 2nd ed., Ronald Press, New York, 1973, 571.

Stelfox, J. B., Ed., Relationships between stand age, stand structure and biodiversity in aspen mixedwood forests in Alberta, Jointly published by Alberta Environmental Center, Vegreville and Canadian Forest Service, Edmonton, 1995, 308.

Stoneman, G. L. and Whitford, K., Analysis of the concept of growth efficiency in *Eucaplyptus marginata* (jarrah) in relation to thinning, fertilising and tree characteristics, *For. Ecol. Manage.,* 76, 47, 1996.

Takahashi, K., Plastic response of crown architecture to crowding in understory tree of two co-dominating conifers, *Ann. Bot.,* 77, 159, 1996.

Tansley, A. G., The use and abuse of vegetational concepts and terms, *Ecology,* 16, 284, 1935.

Tilman, D., *Resource Competition and Community Structure,* Princeton University Press, Princeton, NJ, 1982, 296.

Tilman, D., The resource ratio hypothesis of plant succession, *Amer. Nat.,* 125, 827, 1985.

Tilman, D., On the meaning of competition and the mechanisms of competitive superiority, *Func. Ecol.,* 1, 304, 1987.

Tilman, D., *Plant Strategies and the Dynamics and Structure of Plant Communities,* Princeton University Press, Princeton, NJ, 1988, 360.

van Overbeek, J., Plant physiology and the human ecosystem, *Ann. Rev. Plant Physiol.,* 27, 1, 1976.

Waring, R. H., Estimating forest growth and efficiency in relation to canopy leaf area, *Adv. Ecol. Res.,* 13, 327, 1983.

Waring, R. H. and Schlesinger, W. H., *Forest Ecosystems, Concepts and Management,* Academic Press, San Diego, 340.

Watson, D. J., The physiological basis of variation in yield, *Adv. Agron.,* 4, 101, 1952.

West, D. C., Shugart, H. H., and Botkin, D. B., Eds., *Forest Succession, Concepts and Applications,* Springer-Verlag, New York, 1981, 517.

White, A. S., Presettlement regeneration patterns in a southwestern ponderosa pine stand, *Ecology,* 66, 589, 1985.

White, A. S., The importance of different forms of regeneration to secondary succession in a Maine hardwood forest, *Bull. Torrey Bot. Club,* 118, 303, 1991.

Williamson, G. B., Allelopathy, Koch's postulates and the neck riddle, *in Perspectives in Plant Competition,* Grace, J. B. and Tilman, D., Eds., Academic Press, New York, 1990, 143.

Zackrisson, O., Nilsson, M. C., Dahlberg, A., and Jäderlund, A., Interference mechanisms in conifer-Ericaceae-feathermoss communities, *Oikos,* 78, 209, 1997.

Section III

Ecological Aspects

16 Plant Phenolics and Terpenoids: Transformation, Degradation, and Potential for Allelopathic Interactions

Inderjit, H. H. Cheng, and H. Nishimura

CONTENTS

16.1 ABSTRACT

In natural systems, transformation and degradation of chemicals can alter innocuous compounds into toxic byproducts as well as reduce toxic compounds into innocuous byproducts. Although more and more allelopathic research is giving attention to the interaction between allelochemicals and soil in terms of their sorption onto soil particles and their microbial degradation in the soil environment, there is still little attention given to the role of soil inorganic components in abiotic transformation of these chemicals. This chapter discusses the transformation and degradation of some secondary plant metabolites, which could potentially be involved in allelopathy, and discusses how abiotic and biotic factors affect the importance of these chemicals and their transformation products in allelopathic interactions in natural and managed ecosystems.

16.2 INTRODUCTION

Many different secondary plant metabolites such as phenolics, terpenoids, alkaloids, polyacetylenes, fatty acids, and steroids provide enormous potential for allelopathy (Rice, 1984; Waller, 1987;

Langenheim, 1994; Macias, 1995; Inderjit, 1996). Allelopathic compounds are released into the environment through leaching or emission from living plant parts, root exudation, volatilization, and residue decomposition (Rice, 1984; Putnam and Tang, 1986; Waller, 1987; Inderjit et al., 1995). Allelopathic expression largely depends upon the presence of chemicals at biologically active concentrations and their persistence and fate in the soil. In addition to allelopathic compounds *per se,* allelopathy may also arise from degraded and transformed products of other chemicals important in natural systems. On entry into soil, chemical compounds may undergo processes such as retention, transport, and/or transformation (Cheng, 1995). Degradation/transformation products of innocuous compounds from the donor plant may be toxic to its immediate surroundings (Chang et al., 1969; Turner and Rice, 1975; Rice, 1984; Tanrisever et al., 1987). Research on the factors controlling the rate of biodegradation of toxic compounds demonstrated the significance of mass transfer constraints in the soil matrix, microbial population dynamics, and other limiting factors (Scow and Hutson, 1992; Hess and Schmidt, 1995; Hess et al., 1996). Various abiotic and biotic factors are responsible for degradation and transformation of chemical compounds in the soil. Abiotic factors mainly include physical and chemical factors such as light, heat, soil texture, soil inorganic components, and organic matter (Blum et al., 1987; Cheng, 1995; Dalton, 1989; Dalton et al., 1983; Dao, 1987; Einhellig, 1995; Huang et al., 1977; Inderjit and Dakshini, 1994; Rice, 1984; Wang et al., 1978). Biotic agents responsible for degradation/transformation of organic molecules are mainly soil microbes such as bacteria and fungi (Gibson, 1968; Rice, 1984; Blum and Shafer, 1988). Plant phenolics and terpenoids are widely implicated in allelopathy (Inderjit 1996; Inderjit et al., 1995). This chapter focuses on the degradation and transformation products of these compounds in natural and managed ecosystems and their potential for allelochemical interactions.

16.3 SIGNIFICANCE OF DEGRADATION AND TRANSFORMATION

In natural systems, degradation and/or transformation of chemical compounds may play a key role of ecological importance. Transformation can alter innocuous chemicals into toxic products as well as reduce toxic chemicals into innocuous products. The significance of degradation and transformation in allelopathy can be illustrated by the following related studies. Fischer et al. (1994) studied two communities in Florida: 1) a sandhill with longleaf pine, *Pinus palustris* Mill., and slash pine, *P. elliotii* Engelm., co-occurring with dense growth of native grasses and herbs; and 2) a sand pine scrub dominated by sandpine, *P. clausa* (Chapm. ex Engelm.) Vasey ex Sarg., almost devoid of any grasses or herbs. In the sandhill community, surface fire occurs every three to eight years; however, in the fire-sensitive sand pine community, surface fire occurs at intervals of 20 to 50 years. Richardson and Williamson (1988) hypothesized that the longer fire cycle in the sand pine scrub community may be a result of terpenoids released by the shrubs, which inhibit the growth of grasses and herbs. Fischer et al. (1994) suggested that terpenoids are probably involved in allelopathic interactions in this Florida plant community. These terpenoids may also undergo degradation and/or transformation. For instance, wild rosemery *Ceratiola ericoides* A. Gray (Heller), a perennial shrub of Florida plant community (the predominant pine species in the scrub is sand pine, *Pinus clausa*), produces an inactive dihydrochalcone ceratiolin that, due to light, heat, and acidic soil conditions, undergoes transformation to produce the toxic compound hydrocinnamic acid (Tanrisever et al., 1987). Hydrocinnamic acid further undergoes microbial degradation to form acetophenone (Fischer et al., 1994). Both the transformed (hydrocinnamic acid) and degraded (acetophenone) products are inhibitory to germination and growth of *Schizachyrium scoparium* (Michx.) Nash, a fire fuel grass of sand pine scrub. Rather than compounds directly contributed by plants, in many instances it is the degraded products that are involved in the allelopathic effects. Degradation and transformation processes could also play a key role in the processes of minimizing autotoxicity (Williamson, 1990). Ceratiolin, released from *Ceratiola ericoides*, is an inactive compound prone to degradation when exposed to light, heat, and acidic conditions particular to that habitat (Tanrisever et al., 1987;

Williamson et al., 1992). Plants therefore appear to produce inactive compounds to avoid autotoxicity, which may subsequently degrade or transform to yield toxic compounds.

Sparling et al. (1981) amended soils from fallow or carried crops of potatoes, peas, or barley with phenolic acids (*p*-hydroxybenzoic, ferulic, caffeic, and vanillic acids). They recorded data on changes in phenolic acid concentration, the soil biomass, the respiration rate, and soil amylase activity. From the results, Sparling et al. (1981) concluded that soil biomass can utilize phenolic acid as a substrate. Their study suggests that some crops, for example, potatoes, stimulate phenolic acid-degrading microorganisms, and these changes may be of importance in crop rotation. Several phytopathogenic fungi have potential to degrade the defense compounds of their hosts. Willeke et al. (1983) reported that two isolates of *Nectria haematococca*, mating population of VI (*Fusarium solani*) have ability to degrade the isoflavonoid biochanin A. The pathway of catabolism reported by these authors was: biochanin A → dihydrobiochanin A → 3-(*p*-methoxyphenyl)-6-hydroxy-γ-pyrone → *p*-methoxyphenylacetic acid → *p*-hydroxyphenylacetic acid → 3,4-dihydroxyphenylacetic acid.

Another example is the activities of two anthraquniones, emodin and physcion, which have been implicated in allelochemical interactions involving *Polygonum sachalinense* F. Schmidt ex Maxim. (Inoue et al., 1992). Inderjit and Nishimura (unpublished data) found that the degradation products of emodin and physcion influence the nutrient availability in soils amended with these two anthraquinones, which may have a significant role in *Polygonum* allelopathy in natural systems.

Many dicotyledonous plants (e.g., Brassicaceae) produce glucosinolates. Enzymatic degradation products of glucosinolates are biologically active against herbivorous insects and many fungi (Bialy et al., 1990; Schung and Ceynowa, 1990; Borek et al., 1994). Borek et al. (1994) reported that sinigrin degradation by endogenous enzyme myrosinase yielded allylnitrile and allyl isothiocynate. While Fe^{2+} addition to buffer solution enhanced the formation of allylnitrile, Fe^{3+} inhibited the enzymatic decomposition of sinigrin. Phenolic detoxification by degradation was suggested by Levy and Carmeli (1995). Chemical compounds such as phenolics and terpenoids are present in all plants and soils. It is the rate of their production, their residence time, and their persistence and fate in the soil that determine their phytotoxicity in natural systems. Degradation of toxic chemical compounds to inactive compounds is nature's own means of detoxifying the environment. If they are not degraded, the environment would be saturated with toxic chemical compounds. For example, phloroglucinol is common to many soils, and reported to be a toxic compound in many instances. The bacterium *Rhodococcus* sp. can utilize phloroglucinol as a sole carbon source (Levy and Carmeli, 1995). Levy and Carmeli (1995) suggested that *Rhodococcus* sp. can be an advantage in detoxifying environments containing toxic concentrations of phloroglucinol. Below, we will discuss how degradation and transformation can affect the interpretation of allelopathy.

16.4   NON-MICROBIAL TRANSFORMATION AND DEGRADATION

16.4.1   PLANT PHENOLICS

Non-volatile compounds are released into the soil environment prior to their direct/indirect involvement in allelopathic interactions. Thus, an understanding of the behavior and fate of allelopathic compounds in soils is of utmost importance. Many allelopathic studies have focused on microbial transformation and sorption of allelopathic compounds onto soil particles (Rice, 1984; Blum, 1995; Cheng, 1995). Less attention has been given to the involvement of soil inorganic components in the abiotic transformation of chemical compounds. Shindo et al. (1978) reported that the concentrations of phenolic acids with no chelating ability such as *p*-hydroxybenzoic, vanillic, syringic, *p*-coumaric, and ferulic acids were related to pH and organic content of soil. The concentrations of phenolic acids with chelating ability such as protocatechuic and salicylic acids, however, were not related to pH and organic carbon in the soil. Phenolic compounds occur in soil in three different forms: free, reversibly bound, and bound forms (Rice, 1984). Katase (1981a) reported that the concentrations of free, reversibly bound, and bound forms in peat soils are 2.4, 5, and 92.7 percent, respectively. Wang and Huang (1989) studied the interaction of pyrogallol, often present in soil, with oxides of Mn, Fe, Al, and Si in systems free of microbial activities. They reported that abiotic ring cleavage of polyphenols such as pyrogallol was caused by soil inorganic components. Wang et al. (1983) reported that the amount of humic substances transformed from catechol was higher in the presence of aluminum oxide. The chemical speciation of Al of soils is even more important than their content of Al in the transformation of catechol. Forest mineral soils associated with ericaceous plants, for example, *Kalmia angustifolia* L., are reported to have iron pans in lower soil horizons (Damman, 1971). Vance et al. (1986) reported that iron and aluminium chelate complexes with protocatechuic acid and may play an important role in the formation of the sodic horizon. Aristovskaya and Zavarin (1971) reported that microorganisms decomposed organic metal complexes by degrading organic acids resulting in precipitation of aluminium and iron. Any allelopathic studies on ericaceous plants should take this point into account while investigating the fate of phenolic compounds in the soil. Shindo and Huang (1982, 1984) discussed the role of Mn(IV), Fe(III), Al, and Si oxides on the formation of phenolic polymers and humic substances in the environment. Phenolic acids can be readily oxidized abiotically in the presence of Mn and Fe oxides (Lehmann et al., 1987; Lehmann and Cheng, 1988). Huang and his coworkers (this volume) discuss the catalytic transformation of phenolic compounds in soils. Cheng (1989) concluded from earlier studies that abiotic transformation of phenolic compounds in soil was related to the extent of exposure of organic molecules to manganese and iron surfaces, and not related to the total amount of these minerals in the soil. Soil organic matter can coat the surface of phenolic acids and therefore prevent their oxidation. There is a need to understand how soil chemical composition and structure influence the degradation process. For example, Cheng (1989) reported that the degree of phenolic acid oxidation increased with increased methoxy substitution on the benzene ring and increased length of alkyl side chain. Lehman et al. (1987) studied the reactivity of *p*-hydroxybenzoic, vanillic, and syringic acids, and their longer side-chain analogs, *p*-coumaric, ferulic, and sinapic acids, in the presence of the Palouse soil. They found that after 30 minutes, recovery of these phenolic acids was 85, 81, and 56 percent; and 85, 64, and 5 percent, respectively. After 72 hours, recovery of *p*-hydroxybenzoic, vanillic, and *p*-coumaric acids was 85, 49, and 62 percent, respectively, while syringic, ferulic, and sinapic acids could no longer be recovered in the parent form in the soil. The position of phenolic hydroxyls in the structure should also be considered in interpreting the data. Many isoflavonoids such as formononetin and biochanin A were reported from red clover; however, red clover soil sickness was due to degradation products of isoflavonoids such as *p*-methoxybenzoic, *p*-hydroxybenzoic, 2,4-dihydroxybenzoic, and salicyclic acids (Chang et al., 1969). In aqueous 0.1 M NaOH for 12 weeks, biochanin A degraded into phloroglucinol, *p*-methoxyphenylacetic acids and several unidentified compounds. Degradation products of daidzein were *p*-hydroxybenzoic, *p*-hydroxyphenylacetic, 2,4-dihydroxybenzoic acids, and α-resorcinol (Rice, 1984). Katase (1981b) reported the degradation products of *p*-coumaric, and ferulic acids. They found that *trans*-forms of these phenolic acids transformed into *cis*-forms by exposure to fluorescent light.

Catechol has been reported from leaf and needle litter of several deciduous (e.g., beech, birch, oak, hazelnut, maple, willow, poplar) and coniferous trees (e.g., spruce-fir, Douglas-fir, and larch) (Kuiters and Sarnik, 1986; Kuiters and Denneman, 1987; Gallet and Lebreton, 1995). Catechol has also been reported to form during degradation of many naturally occurring aromatic substances (Snook and Fortson, 1979). Cheng et al. (1983) reported degradation of catechol in soil under laboratory conditions using ^{14}C-tracer technique. They found 30 percent catechol was degraded to CO_2 after six months. Catechol was found to be more stable than chlorocatechols and less extractable from soil. The presence of catechol in plant material is therefore not sufficient evidence to confirm its role in allelopathy. It is important to isolate and quantify catechol from soils using appropriate aqueous solutions as extractants. The amount of catechol extracted with organic solvents does not likely correspond to amounts of catechol available in soil in the free form. Rapid polymerization of catechol in soil could be caused by enzymatic reaction or autooxidation (Martin et al., 1979). Rapid polymerization of catechol can be promoted by abiotic catalysts, for example, Mn oxides (Professor P. M. Huang, personal communication). Binding of xenobiotics to organic matter has been reported (Bollag and Loll, 1983) to detoxify toxic substances (Berry and Boyd, 1985; Bollag, 1987). Dec and Bollag (1988) reported that microbial release of bound catechol from synthetic organic matter was limited. Therefore, the significance of catechol in allelopathic interactions may be questioned. It should be of interest to study recovery of catechol in different soils with different inorganic ion (e.g., Cu, Zn, Mg, Ca, K, Na, Al, Fe, Ba, Mn, etc.) concentrations to determine its bioavailability.

16.4.2 PLANT TERPENOIDS

Compared to phenolic compounds, terpenoids have not been investigated as often for their non-microbial degradation and transformation. Terai et al. (1994) reported the transformation of grayan-otoxin, a diterpenoid found in leaves of many Ericaceae species. They found grayanotoxin II transformed, using Pb(II) acetate as an oxidizing agent, into 1(R)-spiro-3, 6(s), 14, 16-tetra-hydroxy-5-keto derivative. They also reported the transformation of grayanotoxin-II-tetracetate, using Ti(III) acetate in HOAc or benzene, to 1, 5-seco-GTX-1(s) derivatives. However, these transformation studies were not conducted in natural soil environments. Therefore, the potential impact of their presence in soil on allelopathy still needs to be studied.

16.5 MICROBIAL TRANSFORMATION AND DEGRADATION

16.5.1 PLANT PHENOLICS

Phenolic acids are constituents of plant lignins. Lignin biodegradation in soil is the source of many phenolic acids. Soil microorganisms have the potential to toxify and detoxify allelopathic compounds. Although transformation or degradation of chemical compounds is common in soil, it does not necessarily mean detoxification (Liebl and Worsham, 1983; Kaminsky, 1981; Williamson and Weidenhamer, 1990). Dao (1987) suggested that fungal degradation of lignin is mainly an oxidative process. Monomeric phenolic compounds released from lignin degradation, might be benzoic acid or cinnamic acid derivatives (Haider and Martin, 1975). Phenolic acids may degrade aerobically and/or anaerobically (Dao, 1987). Colberg and Young (1985) reported the anaerobic degradation of the soluble form of [^{14}C-lignin]lignocellulose in enriched cultures. Their results indicated that the degree of degradation to gaseous end products was inversely related to the size of the molecular fraction, that is, the smaller the size of the molecular fraction, the faster the degradation. Phenolic acids may undergo different degradation under anaerobic conditions, such as in flooded soil with poor drainage (Dao, 1987). This aspect should be considered in laboratory bioassay studies in which soils are amended with plant debris. Artificially flooded conditions could bring qualitative and quantitative differences in phenolic profiles compared with that in natural systems.

Gibson (1968) reviewed the metabolic pathways for the microbial degradation of aromatic compounds. Catechol and protocatechuic acid are intermediates in microbial degradation of many aromatic compounds. For example, benzoic acid, salicyclic acid, phenol, naphthalene, phenanthrene, anthracene, mandelic acid, o-cresol, and benzene were metabolized via catechol, and m-cresol, p-cresol, p-hydroxybenzoic acid, p-hydroxymandelic acid, p-aminobenzoic acid, and phthalic acid were microbially metabolized via protocatechuic acid. Many phenolic acids disappeared very rapidly in soils. To demonstrate their involvement in allelopathy, it will be necessary to consider their residence time and factors influencing their concentration in the soil.

Many soil microbes use phenolic acids as carbon sources (Blum and Shafer, 1988). Kunc (1971) found that soil enriched with vanillin as a carbon source had bacterial counts tripled three days after inoculation. Treatments with added vanillin and increased numbers of soil microbes resulted in a significantly higher decomposition of vanillin. Degradation of ferulic, vanillic, cinnamic, and *p*-hydroxybenzoic acids by *Rhodotorula rubra* and *Cephalosporium curtipes* was reported to be pH-dependent (Turner and Rice, 1975; Rice, 1984). However, utilization of ferulic acid by *C. curtipes* was rapid. Not all species of a genus are equally capable of decomposing phenolic acids. For example, out of four species of *Cephalosporium* (*C. furcatum*, *C. khandalense*, *C. nordinii*, and *C. roseum*), only *C. furcatum* was capable of decomposing ferulic acid (Rice, 1984). Many bacteria such as *Pseudomonas*, *Cellulomonas*, and *Achromobactor* are capable of decomposing vanillin. Various flavonoids such as quercetin, rutin, kaempferol, rhamnetin, fisetin, and galagin are known to be degraded by soil fungi (Rice, 1984). Rutin was decomposed by *Pullularia fermenmtans* into phloroglucinol, protocatechuic, and 2-protocatechuoylphloroglucinol carboxylic acids (Hattori and Noguchi, 1959). Species of aspergilli such as *Aspergillus flavus* and *A. niger* are known to degrade flavonoids. Microbial degradation of phlorizin to phloretin, *p*-hydroxyhydrocinnamic acid, phloroglucinol, *p*-hydroxybenzoic acid was reported by Rice (1984). The bacterium *Pseudomonas putida* was isolated from soils beneath walnut trees, and this bacterium could convert juglone to 2-hydroxymuconic acid (Rettenmaier, 1983; Schmidt, 1988). Schmidt (1988) reported that the bacterium could easily use juglone (5 hydroxy-1,4-napthoquinone) as a carbon source. Schmidt (1990) further argued that juglone is susceptible to biotic and abiotic degradation in soil and the likelihood that juglone persists at phytotoxic concentrations is very remote. Sutherland et al. (1983) reported the metabolism of cinnamic, *p*-coumaric, and ferulic acids by *Streptomyces setonii* strain 75Vi2 in liquid yeast extract media at 45°C. They found cinnamic acid was degraded via benzaldehyde, benzoic acid, and catechol; *p*-coumaric acid was catabolized via *p*-hydroxybenzaldehyde, *p*-hydroxybenzoic and protocatechuic acids; ferulic acid was degraded via vanillin, vanillic acid, and protocatechuic acid. However, these studies were conducted in the soil environment and to comment on significance of above degradation, it is necessary to study degradation of these phenolic compounds in soil with *Streptomyces setonii*.

Soil microbes have been known to transform substituted aniline compounds to azo compounds (Bartha and Pramer, 1972). Rye allelochemicals such as benzoxazinones are prone to microbial degradation after entering the soil environment. Nair et al. (1990) speculated that benzoxazinones could be involved in rye allelopathy as either these compounds are resistant to microbial degradation, or their degradation products are allelopathically active. Patrick and Koch (1958), however, reported that toxic substances were detected from soils only after decomposition of rye had occurred. Microbial degradation thus plays a key role in rye allelopathy. Barnes et al. (1987) reported that allelopathic activity of rye mulch was largely due to 2,4-dihydroxy-1,4(2H)-benzoxazin-3-one (DIBOA) and its degradation product 2(3H)-benzoxazolone (BOA). Nair et al. (1990) reported that the transformation product of BOA from soil was 2,2'-oxo-1,1'-azobenzene, a chemical involved in rye allelopathy. Chase et al. (1991) found that *Acinetobacter calcoaceticus* was responsible for microbial transformation of 2,3-benzoxazolinone. Gagliardo and Chilton (1992) reported the non-sterile soil transformation of 2(3H)-benzoxazolone (BOA) from rye into 2-amino-3H-phenoazin-3-one. They reported that BOA probably transformed first into o-aminophenol followed by its oxidation to aminophenoxazinone; this microbially transformed compound may play a significant role in rye allelopathy. Kumar et al. (1993) reported that in non-sterile soils, 7-methoxy-2,4-dihydroxy-1,4(2H)-benzoxazin-3-one (DIMBOA) from wheat and 7,8-dimethoxy-2,4-dihydroxy-1,4(2H)-benzoxazin-3-one (DIM$_2$BOA) of corn were transformed to 2-amino-7-methoxy-3H-phenoxazin-3-one and 2-amino-4,6,7-trimethoxy-3H-phenoxazin-3-one, respectively. The transformation was via 6-methoxy-2(3H)-benzoxazolone (MBOA) and 6,7-dimethoxy-2(3H)-bezoxazolone (M$_2$BOA) in wheat and corn, respectively. Keeping the significance of degraded products in mind, the importance of studying the fate of allelopathic compounds in the rhizosphere becomes evident. Mere isolation and identification of compounds from plants is not enough to confirm allelopathic relationships. The soil rhizosphere is very important in studying allelopathic compounds; the soil rhizosphere is the "bottle neck" for interactions of allelopathic biochemicals with plant. Katase and Bollag (1991) reported that *trans*-4-hydroxycinnamic acid when incubated with extracellular laccase from the fungus *Trametes versicolor* transformed into *trans*-4-hydroxy-3-(*trans*-4'-cinnamyloxy) cinnamic acid. The enzyme laccases is produced by several species of basidiomycetes and few ascomycetes and deuteromycetes, and is involved in the formation of humus, the degradation of lignin, and the detoxification of phenolic compounds (Sjoblad and Bollag, 1981; Katase and Bollag, 1991). Physical, chemical, and biological soil components have been known to influence qualitative and quantitative availability of allelopathic compounds in the rhizosphere.

Hashidoko et al. (1993) reported that two gram-negative bacteria, *Klebsiella oxytoca* and *Erwinia uredovora,* converted hydroxycinnamic acids into hydroxystyrenes by decarboxylation. These two bacteria constituted epiphytic microflora on yacon (*Polymnia sonchifolia*) leaves. Their studies, however, were not conducted from an allelochemical perspective. Phloroglucinol and its derivatives have often been implicated in allelopathy. Phloroglucinol has also been reported to be a degradation product of many phenolic compounds (Rice, 1984). Armstrong and Patel (1993) reported the biodegradation of phloroglucinol by the bacterium *Rhodococcus* sp. BPG-8, isolated from oil-rich soils of Newfoundland, Canada. Their proposed pathway of phloroglucinol degradation was: phloroglucinol → 1,2,3,5-tetrahydroxybenzene → (2,4-dihydroxy-2,4-hexadiene-1,6-dioic acid) → (2,4-dihydroxy-6-oxo-2,4-hexadienoic acid) → acetopyruvate + formate → acetate + pyruvate → tricarboxylic acid. It would be of interest to make a comparative study of soils with and without *Rhodococcus* sp. BPG-8, and find out which of these soils are more inhibitory. It is important to study the fate of phloroglucinol in the soil environment because it is prone to degradation. It is also important to isolate its degraded products from soils and collect data on their residence time and allelopathic potential.

Leaf leachate and acid sand bioassays were used in many studies to evaluate allelopathic responses (Rice, 1984; Inderjit and Dakshini, 1995). However, there might be little correlation between these bioassays and field interactions. Care should be taken to keep physical (e.g., texture), chemical (e.g., inorganic components), and biological (e.g., microbes) factors of soil to be used in allelopathic bioassays similar to that of natural soils of the allelopathic plant (Inderjit and Dakshini, 1994, 1995).

16.5.2 PLANT TERPENOIDS

Heterotrophic soil bacteria, such as *Umbellularia californica* have been shown to metabolize terpenoids (Langenheim, 1994). Compared with plant phenolics, not many studies were carried out on the implication of plant terpenoid degradation products in allelopathy. However, there are many reports of microbial degradation of plant terpenoids without any implication in allelopathy (Miyazawa et al., 1991, 1993a,b, 1994, 1995a,b,c,d; Hoffman and Fraga, 1993). Nishimura et al., (1982) reported the microbial transformation of 1,8-cineole to 3-exo-, and 3-endo-hydroxycineole by *Aspergillus niger.* Biotransformation of (−)-*cis*-carveol to (+)-bottrospicatol by *Streptomyces bottropensis* isolated from soil has been reported by Nishimura et al., (1983). They found significant inhibitory activity of bottrospicatol on lettuce seed germination. However, such bioassays to determine bioactivity of bottrospicatol has little ecological relevance. This is because lettuce is not growing in association with *Eucalyptus* in the natural environment. To understand the bioactivity of bottrospicatol, it is important to select test plant species growing in association with *Eucalyptus* under natural conditions (Inderjit and Dakshini, 1995). Noma et al. (1992a) reported biotransformation of (+),-(−)- and (±)-limonenes by *Aspergillus cellulosae.* Major transformation products of (+)-limonene are (+)-isopiperitenone, (+)-limonene-1,2-*trans*-diol, (+)-*cis*-carveol, and (+) perillyl alcohol; and transformation products of (−)-limonene are (−)-perillyl alcohol, (−) -limonene-1, 2-*trans*-diol, and (+)-neodihydrocarveol. Microbial transformation of the sesquiterpenoid cedrol by *Cephalosporium aphidicola, Aspergillus niger,* and *Beauveria sulfurescens* was reported (Hanson and Nasir, 1993; Gand et al., 1995).

Biotransformation of terpenes by microbes other than bacteria and fungi is also reported. Noma et al. (1991a,b) reported the biotransformation of monoterpenes and aromatic aldehydes by *Euglena gracilis*. Noma et al. (1992b) reported the biotransformation of different classes of monoterpene aldehydes and aromatic aldehydes by a unicellular green alga, *Dunaliella tertiolecta*. They reported the biotransformation of mono- and sesquiterpenes such as *d*-limonene, *p*-methane, α and β-pinene, caryophyllene, and longifollene, and monoterpene aldehydes such as myrtenol, *l*-phellandrol, citrol, *d*-citronellel, *l*-citronellel, *dl*-citronellel, etc. It was reported that *Dunaliella tertiolecta* reduced all saturated and unsaturated terpene aldehydes and many aromatic and related aldehydes to corresponding primary alcohols.

The above biotransformation studies on terpenoids were not conducted directly in relation to allelopathy. However, many of these terpenoids such as cedrol, bornyl acetate, geraniol, geranyl acetate, and 1, 8-cineole, etc., have been implicated in allelopathy (Fischer, 1986; Nishimura and Mizutani, 1995; Weidenhamer et al., 1993). Nishimura and Noma (1996) studied the microbial transformation of monoterpenes, (−)-carvone and 1,8-cineole in *Eucalyptus* species from a flavor and biological activity standpoint.

16.6 RESEARCH NEEDS

There is a great need for establishing the cause–effect relationship between the fate of a chemical in soil and its allelopathic potential. To understand the probable role of a chemical in allelopathy, it should be isolated, identified, and quantified from the plant material as well as from the soil environment. Identification and isolation of allelochemical in the rhizosphere soil is especially important. The amount of compound in the plant is not very important in terms of allelopathic effects. The amount released and actually present in the rhizosphere in an available form is more important. It is also important to know whether compounds are present in original form or degraded form. Furthermore, all compounds from plants, irrespective of their toxicity, should be studied for their behavior in soil. This is because nontoxic chemicals in the plant could degrade into toxic chemicals in the rhizosphere (Tanrisever et al., 1987). If chemicals in plants that are thought to be allelopathic are not detected in the soil environment, it becomes necessary to study their potential degradation/transformation products to detect allelochemical activity. The ecological significance of a particular compound can only be fully realized when there is sufficient information available on possible transformation/degradation products and their allelopathic potential.

ACKNOWLEDGMENTS

We sincerely thank Professors B. R. Dalton, P. M. Huang, and Steve Schmidt for their valuable comments to improve the quality of this chapter.

REFERENCES

Aristovskaya, T. V. and Zavarin, G. A., Biochemistry of iron in soil, in *Soil Biochemistry,* McLaren, A. D. and Skujins, J., Eds., Marcel Dekker, New York, 1971, 385.

Armstrong, S. and Patel, T. R., 1,3,5-trihydroxybenzene biodegradation by *Rhodococcus* sp. BPG-8, *Can. J. Microbiol.,* 39, 175, 1993.

Barnes, J. P., Putnam, A. R., Burke, B. A., and Aasen, A. J., Isolation and characterization of allelochemicals in rye herbage, *Phytochemistry,* 26, 1385, 1987.

Bartha, R. and Pramer, D., Biochemical transformation of herbicide-induced anilines: requirements of molecular configuration, *Can. J. Microbiol.*, 161, 1617, 1972.

Berry, D. F. and Boyd, S. A., Decontamination of soil through enhanced formation of bound residues, *Environ. Sci. Technol.*, 9, 1132, 1985.

Bialy, Z., Oleszek, W., Lewis, J., and Fenwick, G. R., Allelopathic potential of glucosinolates (mustard oil glycosides) and their degradation products against wheat, *Plant Soil*, 129, 277, 1990.

Blum, U., The value of model plant-microbe-soil systems for understanding processes associated with allelopathic interaction: one example, in *Allelopathy: Organisms, Processes, and Applications*, Inderjit, Dakshini, K. M. M., and Einhellig, F. A., Eds., American Chemical Society, Washington, D.C., 1995, 127.

Blum, U. and Shafer, S. R., Microbial populations and phenolic acids in soil, *Soil Biol. Biochem.*, 20, 793, 1988.

Blum, U., Weed, S. B., and Dalton, B. R., Influence of various soil factors on the effects of ferulic acid on leaf expansion of cucumber seedlings, *Plant Soil*, 98, 111, 1987.

Bollag, J. M., Decontamination of the environment through binding of pollutants to humic materials, in *Extended Abstracts, New Orleans, LA*, Amer. Chem. Soc., Div. of Environmental Chemistry, Washington, D.C., 1987, 342.

Bollag, J. M. and Loll, M. J., Incorporation of xenobiotics into soil humus, *Experientia*, 39, 1221, 1983.

Borek, V., Morra, M. J., Brown, P. D., and McCaffrey, J. P., Allelochemicals produced during sinigrin decomposition in soil, *J. Agric. Food Chem.*, 42, 1030, 1994.

Chang, C. F., Suzuki, A., Kumai, S., and Tamura, S., Chemical studies on 'clover sickness.' Part II. Biological functions of isoflavonoids and their related compounds, *Agric. Biol. Chem.*, 33, 398, 1969.

Chase, W. R., Nair, M. G., Putnam, A. R., and Mishra, S. K., 2, 2′-oxo-1,1′-azobenzene: microbial transformation of rye (*Secale cereale* L.) allelochemicals in field soils by *Acinetobacter calcoaceticus*: III, *J. Chem. Ecol.*, 17, 1575, 1991.

Cheng, H. H., Assessment of the fate and transport of allelochemicals in the soil, in *Phytochemical Ecology: Allelochemicals, Mycotoxins, and Insect Pheromones and Allomones*, Chou, C. H. and Waller, G. R., Eds., Institute of Botany, Academia Sinica, ROC, 1989, 209.

Cheng, H. H., Characterization of the mechanisms of allelopathy: modeling and experimental approaches, in *Allelopathy: Organisms, Processes and Applications*, Inderjit, Dakshini, K. M. M., and Einhellig, F. A., Eds., ACS Symposium Series 582, American Chemical Society, Washington, D.C., 1995, 132.

Cheng, H. H., Haider, K., and Harper, S. S., Catechol and chlorophenols in soil: degradation and extractability, *Soil Biol. Biochem.*, 15, 311, 1983.

Colberg, P. J. and Young, L. Y., Anaerobic degradation of soluble fractions of [^{14}C-lignin]lignocellulose, *Appl. Environ. Microbiol.*, 49, 345, 1985.

Dalton, B. R., Physiochemical and biological processes affected the recovery of exogenously applied ferulic acid from tropical forest soils, *Plant Soil*, 115, 13, 1989.

Dalton, B. R., Blum, U. and Weed, S. B., Allelopathic substances in ecosystems: effectiveness of sterile soil component in altering recovery of ferulic acid, *J. Chem. Ecol.*, 9, 1185, 1983.

Damman, A. W. H., Effect of vegetation changes on the fertility of a Newfoundland forest site, *Ecol. Monogr.*, 41, 253, 1971.

Dao, T. H., Sorption and mineralization of plant phenolic acids in soil, in *Allelochemicals: Role in Agriculture and Forestry*, Waller, G. R., Ed., American Chemical Society, Washington, D.C., 1987, 358.

Dec, J. and Bollag, J. M., Microbial release and degradation of catechol and chlorophenols bound to synthetic humic acid, *Soil Sci. Soc. Am. J.*, 52, 1366, 1988.

Einhellig, F. A., Allelopathy: current status and future goals, in *Allelopathy: Organisms, Processes, and Applications*, Inderjit, Dakshini, K. M. M., and Einhellig, F. A., Eds., American Chemical Society, Washington, D.C., 1995, 1.

Fischer, N. H., The function of mono- and sesquiterpenes as plant germination and growth regulators, in *The Science of Allelopathy*, Putnam, A. R. and Tang, C. S., Eds., John Wiley & Sons, New York, 1986, 203.

Fischer, N. H., Williamson, G. B., Weidenhamer, J. D., and Richardson, D. R., In search of allelopathy in Florida scrub: the role of allelopathy, *J. Chem. Ecol.*, 20, 1355, 1994.

Gagliardo, R. W. and Chilton, W. S., Soil transformation of 2(3H)-benzoxazolone of rye into phytotoxic 2-amino-3H-phenoxazin-3-one, *J. Chem. Ecol.*, 18, 1683, 1992.

Gallet, C. and Lebreton, P., Evolution of phenolic patterns in plants and associated litters and humus of a mountain forest ecosystem, *Soil Biol. Biochem.*, 27, 157, 1995.

Gand, E., Hanson, J. R., and Nasir, H., The biotransformation of 8-epicedrol and some relatives by *Cephalosporium aphicola*, *Phytochemistry*, 39, 1081, 1995.

Gibson, D. T., Microbial degradation of aromatic compounds: metabolic pathways reveal a general formula for the degradation of several compounds, *Science,* 161, 1093, 1968.

Haider, K. and Martin, J. P., Decomposition of specifically carbon-14 labeled benzoic and cinnamicacid derivatives in soil, *Soil Sci. Soc. Am. Proc.,* 39, 657, 1975.

Hanson, J. R. and Nasir, H., Biotransformation of sesquiterpenoid, cedrol, by *Cephalosporium aphidicola, Phytochemistry,* 33, 835, 1993.

Hashidoko, Y., Urashima, M., Yoshida, T., and Mizutani, J., Decarboxylative conversion of hydrocinnamic acids by *Klebsiella oxytoca* and *Erwinia uredovora,* epiphytic bacteria of *Polymnia sonchifolia* leaf, possibly associated with formation of microflora on the damaged leaves, *Biosci. Biotech. Biochem.,* 57, 215, 1993.

Hattori, S. and Noguchi, I., Microbial degradation of rutin, *Nature,* 184, 1145, 1959.

Hess, T. F. and Schmidt, S. K., Improved procedure for obtaining statistically valid parameters estimates from soil respiration data, *Soil Biol. Biochem.,* 27, 1, 1995.

Hess, T. F., Schmidt, S. K., and Colores, G. M., Maintenance energy model for microbial degradation of toxic chemicals in soil, *Soil Biol. Biochem.,* 28, 907, 1996.

Hoffman, J. J. and Fraga, B. M., Microbial transformation of diterpenes: hydroxylation of 17-acetoxy-kolavenol acetone by *Mucor plumbeus, Phytochemistry,* 33, 827, 1993.

Huang, P. M., Wang, T. S. C., Wang, M. K., Wu, M. H., and Hsu, N. W., Retention of phenolic acids by noncrystalline hydroxy-aluminum and -iron compounds and clay mineral of soil, *Soil Sci.,* 123, 213, 1977.

Inderjit, Plant phenolics in allelopathy, *Bot. Rev.,* 62, 186, 1996.

Inderjit and Dakshini, K. M. M., Allelopathic effect of *Pluchea lanceolata* (Asteraceae) on characteristics of four soils and tomato and mustard growth, *Am. J. Bot.,* 81, 799, 1994.

Inderjit and Dakshini, K. M. M., On laboratory bioassays in allelopathy, *Bot. Rev.* 61, 28, 1995.

Inderjit, Dakshini, K. M. M., and Einhellig, F. A., Eds., *Allelopathy: Organisms, Processes and Applications,* ACS Symposium Series 582, American Chemical Society, Washington, D.C., 1995.

Inoue, M., Nishimura, H., Li, H. H., and Mizutani, J., Allelochemicals from *Polygonum sachalinense* Fr. Schm. (Polygonaceae), *J. Chem. Ecol.,* 18, 1833, 1992.

Kaminsky, R., The microbial origin of the allelopathic potential of *Adenostoma fasciculatum* H & A, *Ecol. Monogr.,* 51, 365, 1981.

Katase, T., The different forms in which *p*-coumaric acid exists in peat soil, *Soil Sci.,* 131, 271, 1981a.

Katase, T., Stereomerization of *p*-coumaric and ferulic acids during their incubation in peat soil extract solution by exposure to fluorescent light, *Soil Sci. Plant Nutr.,* 27, 421, 1981b.

Katase, T. and Bollag, J.-M., Transformation of *trans*-4-hydrocinnamic acid by a laccase of the fungus *Trametes versicolor*: its significance in humification, *Soil Sci.,* 151, 291, 1991.

Kuiters, A. T. and Denneman, C. A. J., Water-soluble phenolic substances in soils under several coniferous and deciduous tree species, *Soil Biol. Biochem.,* 19, 765, 1987.

Kuiters, A. T. and Sarink, H. M., Leaching of phenolic compounds from leaf and needle litter of several deciduous and coniferous trees, *Soil Biol. Biochem.,* 18, 475, 1986.

Kumar, P., Gagliardo, R. W., and Chilton, W. S., Soil transformation of wheat and corn metabolites MBOA and DIM$_2$BOA into aminophenoxazinones, *J. Chem. Ecol.,* 19, 2453, 1993.

Kunc, F., Decomposition of vanillin by soil microorganisms, *Folia Microbiol.,* 16, 41, 1971.

Langenheim, J. H., Higher plant terpenoids: a phytocentric overview of their ecological roles, *J. Chem. Ecol.,* 20, 1223, 1994.

Lehmann, R. G. and Cheng, H. H., Reactivity of phenolic acids in soil and formation of oxidation products, *Soil Sci. Soc. Am. J.,* 52, 1304, 1988.

Lehmann, R. G., Cheng, H. H., and Harsh, J. B., Oxidation of phenolic acids by iron and manganese oxides, *Soil Sci. Soc. Am. J.,* 51, 352, 1987.

Levy, E. and Carmeli, S., Biological control of plant pathogen by antibiotic-producing bacteria, in *Allelopathy: Organisms, Processes and Applications,* Inderjit, Dakshini, K. M. M., and Einhellig, F. A., Eds., ACS Symposium Series 582, American Chemical Society, Washington, D.C., 1995, 300.

Liebl, R. and Worsham, A. D., Inhibition of pitted morning glory (*Ipomea lacunosa* L.) and other weed species by phytotoxic components of wheat (*Triticum aestivum* L.) straw, *J. Chem. Ecol.,* 9, 256, 1983.

Martin, J. P., Haider, K., and Linhares, L. F., Decomposition and stabilization of ring-^{14}C labeled catechol in soil, *Soil Sci. Soc. Am. J.,* 43, 100, 1979.

Miyazawa, M., Nakaoka, H., Hyakumachi, M., and Kameoka, H., Biotransformation of 1,8-cineole to (+)-2-endo-hydroxy-1,8-cineole by *Glomerella cingulata, Chem. Express,* 6, 667, 1991.

Miyazawa, M., Uemura, T. and Kameoka, H., Biotransformation of sesquiterpenoids, (−)-globulol and (+)-ledol by *Glomerella cingulata, Phytochemistry*, 37, 1027, 1994.

Miyazawa, M., Uemura, T., and Kameoka, H., Biotransformation of (−)-a-bisabolol by plant pathogenic fungus, *Glomerella cingulata, Phytochemistry*, 39, 1077, 1995a.

Miyazawa, M., Yokote, K., and Kameoka, H., Biotransformation of the monoterpenoid, rose oxide, by *Aspergillus niger, Phytochemistry*, 39, 85, 1995b.

Miyazawa, M., Suzuki, Y., and Kameoka, H., Biotransformation of (−)-nopol by *Glomerella cingulata, Phytochemistry*, 39, 337, 1995c.

Nair, M. G., Whiteneck, C. J., and Putnam, A. R., 2, 2′-oxo-1,1′-azobenzene, a microbially transformed allelochemical from 2, 3-benzoxazolinone: I, *J. Chem. Ecol.*, 16, 353, 1990.

Nishimura, H. and Mizutani, J., Identification of allelochemicals in *Eucalyptus citriodora* and *Polygonum sachalinense*, in *Allelopathy: Organisms, Processes and Applications*, Inderjit, Dakshini, K. M. M., and Einhelli, F. A., Eds., ACS Symposium Series 582, American Chemical Society, Washington, D.C., 1995, 310.

Nishimura, H. and Noma, Y., Microbial transformation of monoterpenes: flavor and biological activity, in *Biotechnology for Improved Foods and Flavors*, Takeoka, G. R., Teranishi, R., Williams, P. J., and Kobayashi, A., Eds., ACS Symposium Series 637, American Chemical Society, Washington, D.C., 1996, 173.

Nishimura, H., Noma, Y., and Mizutani, J., *Eucalyptus* as biomass. novel compounds from microbial conversion of 1,8-cineole, *Agric. Biol. Chem.*, 46, 2601, 1982.

Nishimura, H., Hiramoto, S., Mizutani, J., Noma, Y., Furusaki, A., and Matsumoto, T., Structure and biological activity of bottrospicatol, a novel monoterpene produced by microbial transformation of (−)-*cis*-carveol, *Agric. Biol. Chem.*, 47, 2697, 1983.

Noma, Y., Takahashi, H., and Asakawa, Y., Biotransformation of terpene aldehyde by *Euglena gracilis* Z, *Phytochemistry*, 30, 1147, 1991a.

Noma, Y., Okajima, Y., Takahashi, H., and Asakawa, Y., Biotransformation of aromatic aldehydes and related compounds by *Euglena gracilis* Z, *Phytochemistry*, 30, 2969, 1991b.

Noma, Y., Yamasachi, S., and Asakawa, Y., Biotransformation of limonene and related compounds by *Aspergillus cellulosae, Phytochemistry*, 31, 2725, 1992a.

Noma, Y., Akehi, E., Miki, N., and Asakawa, Y., Biotransformation of terpene aldehydes, aromatic aldehydes and related compounds by *Dunaliella tertiolecta, Phytochemistry*, 31, 515, 1992b.

Nukina, M., Otsuki, T., and Kuniyasu, N., Microbial transformation of 1- and 2-phenyl-1-propenes by the fungus *Pyricularia oryzae* Cavara, *Biosci. Biotech. Biochem.*, 58, 2293, 1994.

Patrick, Z. A. and Koch, L. W., Inhibition of respiration, germination, and growth by substances arising during the decomposing of certain plant residues in soil, *Can. J. Bot.*, 36, 621, 1958.

Putnam, A. R., Allelopathic chemicals: nature's herbicides in action, *Chem. Eng. News*, 61, 34, 1983.

Putnam, A. R. and Tang, C. S., Allelopathy: state of the science, in *The Science of Allelopathy*, Putnam, A. R. and Tang, C. S., Eds., John Wiley & Sons, New York, 1986, 1.

Rettenmaier, H., Kupas, U., and Lingens, F., Degradation of juglone by *Pseudomonas putida* J1, *FEMS Microbiol., Lett.*, 19, 193, 1983.

Rice, E. L., *Allelopathy*, Academic Press, Orlando, FL, 1984.

Rice, E. L., *Biological Control of Weeds and Plant Diseases: Advances in Applied Allelopathy*, University of Oklahoma Press, Norman, OK, 1995, 448.

Richardson, D. R. and Williamson, G. B., Allelopathic effects of shrubs of the sand pine scurb on pines and grasses of the sandhills, *For. Sci.*, 34, 592, 1988.

Schmidt, S. K., Degradation of juglone by soil bacteria, *J. Chem. Ecol.*, 14, 1561, 1988.

Schmidt, S. K., Ecological implication of destruction of juglone (5-hydroxy-1, 4-napthoquinone) by soil bacteria, *J. Chem. Ecol.*, 16, 3547, 1990.

Schung, E. and Ceynowa, J., Phytopathological aspects of glucosinolates in oilseed rape, *J. Agron. Crop Sci.*, 165, 319, 1990.

Scow, K. M. and Hutson, J., Effect of diffusion and sorption on the kinetics of degradation: theoretical considerations, *Soil Sci. Soc. Am. J.*, 56, 119, 1992.

Shindo, H. and Huang, P. M., Role of Mn(IV) oxide in abiotic formation of humic substances in the environment, *Nature*, 298, 363, 1982.

Shindo, H. and Huang, P. M., Significance of Mn(IV) oxide in abiotic formation of organic nitrogen complexes in natural environment, *Nature,* 308, 87, 1984.

Shindo, H. and Kuwatsuka, S., Behaviour of phenolic substances in the decaying process of plants. III. Degradation pathway of phenolic acids, *Soil Sci. Plant Nutr.,* 21, 227, 1975.

Shindo, H., Ohta, S., and Kuwatsuka, S., Behaviour of phenolic substances in the decaying process of plants. IX. Distribution of phenolic acids in soil of paddy fields and forests, *Soil Sci. Plant Nutr.,* 24, 233, 1978.

Sjoblad, R. D. and Bollag, J.-M., Oxidative coupling of aromatic compounds by enzymes from soil microorganisms, in *Soil Biochemistry,* Vol. 5, Paul, E. A. and Ladd, J. N., Eds., Marcel Dekker, New York, 1981, 113.

Snook, M. E. and Fortson, P. J., Gel chromatographic isolation of catechols and hydroquinones, *Anal. Chem.,* 51, 1814, 1979.

Sparling, G. P., Ord, B. G., and Vaughan, D., Changes in microbial biomass and activity in soils amended with phenolic acids, *Soil Biol. Biochem.,* 13, 455, 1981.

Sutherland, J. B., Crawford, D. L., and Pometto, A. L., III, Metabolism of cinnamic, *p*-coumaric and ferulic acids by *Streptomyces setonii, Can. J. Microbiol.,* 29, 1253, 1983.

Tanrisever, N., Fronczek, F. R., Fischer, N. H., and Williamson, G. B., Ceratiolin and other flavonoids from *Ceratiola ericoides, Phytochemistry,* 26, 175, 1987.

Terai, T., Nishioku, Y., and Goto, K., Transformation of grayanotoxin-II/grayanotoxin-II tetraacetate to 1,5-seco grayanotoxin derivatives, *Biosci. Biotech. Biochem.,* 58, 2251, 1994.

Turner, J. A. and Rice, E. L., Microbial decomposition of ferulic acid in soil, *J. Chem. Ecol.,* 1, 41, 1975.

Vance, G. F., Mokma, D. L., and Boyd, S. A., Phenolic compounds in soils of hydrosequences and developmental sequences of sodzols, *Soil Sci. Soc. Am. J.,* 50, 992, 1986.

Waller, G. R., Ed., *Allelochemicals: Role in Agriculture and Forestry,* ACS Symposium Services 330, American Chemical Society, Washington, D.C., 1987.

Wang, T. S. C., Li, S. W., and Ferng, Y. L., Catalytic polymerization of phenolic compounds by clay minerals, *Soil Sci.,* 126, 15, 1978.

Wang, M. C. and Huang, P. M., Pyrogallol transformation as catalyzed by oxides of Mn, Fe, Al and Si, in *Phytochemical Ecology: Allelochemicals, Mycotoxins, and Insect Pheromones and Allomones,* Chou, C. H. and Waller, G. R., Eds., Institute of Botany, Academia Sinica, ROC, 1989, 195.

Wang, T. S. C., Wang, M. C., and Huang, P. M., Catalytic synthesis of humic substances by using aluminas as catalysts, *Soil Sci.,* 136, 226, 1983.

Weidenhamer, J. D., Macias, F. A., Fischer, N. H., and Williamson, G. B., Just how insoluble are monoterpenes?, *J. Chem. Ecol.,* 19, 1827, 1993.

Willeke, U., Weltring, K. M., Barz, W., and VanEtten, H. D., Degradation of the isoflavone biochanin A by isolates of *Nectria haematococca (Fusarium solani), Phytochemistry,* 22, 1539, 1983.

Williamson, G. B., Allelopathy, Koch's postulates and the neck riddle, in *Perspectives in Plant Competition,* Grace, J. B. and Tilman, D., Eds., Academic Press, New York, 1990, 143.

Williamson, G. B. and Weidenhamer, J. D., Bacterial degradation of juglone: evidence against allelopathy?, *J. Chem. Ecol.,* 16, 1739, 1990.

Williamson, G. B., Richardson, D. R., and Fischer, N. H., Allelopathic mechanism in fire-prone communities, in *Allelopathy: Basic and Applied Aspects,* Rizvi, S. J. H. and Rizvi, V., Eds., Chapman and Hall, London, 1992, 59.

17 Variation of Flavonoid Synthesis Induced by Ecological Factors

N. Chaves and J. C. Escudero

CONTENTS

17.1 ABSTRACT

Conditions of environmental stress may restrict plant growth and photsynthetic rate. Such conditions lead to the synthesis of compounds deriving from secondary metabolism, with the most important group being the flavonoids, of which some 4000 structures are known. These secondary metabolites play a major role in protecting the plant against microorganism or animal attack, or in giving a competitive advantage to a certain species over others in a given habitat. Flavonoid build-up may occur in plants as constitutive of the plant itself with no apparent external cause, or may be induced by some external physical signal or as the response to some type of cellular damage associated with microbial infections.

Flavonoid accumulation is highly tissue specific, and can vary qualitatively and quantitatively from one organ to another, according to the stage of development of the organ, the age of the plant, and the conditions of growth, and it can vary from one season to another. A number of studies have shown that there is a greater degree of induction during certain seasons. Since the main parameters differentiating one season from another are climatological, it is natural to think that these seasonal changes might be primarily responsible for the variation in flavonoid secretion. It has been clearly demonstrated that flavonoid build-up is specifically induced by ultraviolet light and that the dependence is linear. Different plant species may use different physiological routes to trigger the production of some flavonoid or may respond to the same stimulus by producing different chemical structures. The responses of plants to increased levels of ultraviolet light varied markedly between individuals of the same species and also between species, with each species responding differently in the synthesis of these compounds.

0-8493-2116-6/99/$0.00+$.50
© 1999 by CRC Press LLC

Another point worthy of note is the response of the plant to the interaction of various stressors. A plant's response to an adverse situation will depend on the degree of exposure to the stress, and the result in regard to flavonoid synthesis can be cumulative. The conjoint action of hydric stress and ultraviolet irradiation has been shown to be additive in inducing these compounds, with more stress leading to more synthesis. While ultraviolet irradiation clearly has a quantitative effect, hydric stress triggers also a qualitative change, augmenting the synthesis of position-4′-methylated flavonoids. Temperature is another climatological factor with a qualitative effect; specifically, high temperatures strongly induce the activity of the enzyme 7-methyltransferase, with an increase in the accumulation of flavonoids that are methylated at this position.

Biological stress such as herbivore attack provokes the production of dissuasory chemical compounds deriving from secondary metabolism. These include terpenoids, alkaloids, and flavonoids, which reduce the growth and survival of the attackers. Many studies have looked at the question of whether herbivore defense is constitutive of the plant or is an attack-induced response. Their results indicate that the information content of the environment has the potential to determine whether constitutive or induced resistance is involved. If the herbivore is completely unpredictable, then plants might also evolve a level of constitutive defense that is optimal on average. In other words, induced resistence will be favored only if the plant is likely to be attacked by herbivores in the future.

17.2   INTRODUCTION

There is no reason to expect that plant biochemistry, like plant physiology or morphology, is not subject to selective pressures exerted by the environment. During the course of evolution, many secondary products have been synthesized by different plant species, and when one species in particular has conferred some selective advantage, it has been maintained, leading to an enhancement in survival of the species and of that specific product. The random synthesis of new compounds has provided plants with the great variability that is a prerequisite for the operation of natural selection and the evolution of all living organisms (Bell, 1980).

Stressful environmental conditions can restrict plant growth and photosynthesis rate. Nonstructural carbohydrates then tend to accumulate and thus one can explain the increased synthesis of carbon-based defensive substances (compounds deriving from secondary metabolism). A confirmation of this carbon/nutrient balance is that species that grow in media with low nutrient or water availability produce high levels of tannins and phenols (Bryant et al., 1983; Gershenzon, 1984; Waring et al., 1985; Nicolai, 1988). These secondary metabolites are of great importance in protecting the plant against microorganisms or animal attack, or in enhancing the ability of a certain species to compete with others in a given habitat.

One of the basic problems of plant biology research is to establish the physiological role played by secondary compounds in plants due to the multitude of structures that are encountered. However, a small number of these compounds stand out as being extensively distributed, and of these the most important are the flavonoids, of which some 4000 structures are known (Harborne, 1988).

One can describe three routes by which flavonoids accumulate in plants (Hartmann, 1985):

1. A basal route, without any apparent environmental stimulus. The flavonoids are preformed, endogenous compounds constitutive of the plant itself.
2. Induced flavonoids, which are produced in the plant after the reception of an external signal. They are usually induced by some physical or chemical stress. It has been found that the expression of the enzymes of the flavonoid biosynthetic pathway depends on the plant's response to environmental stimuli (Chappell and Hahlbrock, 1984; Bolwell et al., 1985; Hungria et al., 1991; Jordan et al., 1994).

3. Flavonoids produced in response to some kind of cell damage and associated with infections produced by microbial pathogens (Darwill and Albersheim, 1984; Porter et al., 1986; Rao, 1990; Klepzig et al., 1995).

In the last 20 years there have been many studies of the complex mixture of phenolic compounds found in the different organs of most plant species. The physiology of flavonoid accumulation is highly tissue specific and can be channeled and regulated in different ways in each tissue type. The mix can vary quantitatively and qualitatively according to the specific organ involved, the stage of development, the age of the plant, the season, and the conditions for growth (Harborne et al., 1975; Stewart et al., 1979; Schüte, 1985; Vogt et al., 1988; Curir et al., 1990; Graham, 1991; Lamaison and Carnat, 1991; Chaves et al., 1993, 1997).

17.3 FLAVONOID DISTRIBUTION IN DIFFERENT TISSUES OF PLANTS: QUALITATIVE AND QUANTITATIVE VARIATIONS

Variations have been detected in the flavonoid content of plants according to the different stages of development of the organs. Stewart et al. (1979) studied the anthocyanin content in *Poinsettia* and found that it varied according to the maturity of the leaves, flowering season, irradiation quality, and the part of the bract studied. The specific distribution of these molecules is highly characteristic of a given organ and is strongly affected by its developmental age (Graham, 1991).

Studies of *Citrus aurantium* (Castillo et al., 1992) showed the dependence of the naringenin and neohesperidin concentrations on the development of the leaf. These flavonoids reach their highest concentrations in the first stages of leaf development and decline gradually with the growth of the leaves. The flowers present the same pattern as the leaves, but in the fruit the neohesperidin content reaches the highest levels in the first stages of development, whereas naringenin remains practically constant at all times, only declining finally in the phase of senescence. The biosynthesis of these compounds is therefore shown to be more intense during the stage of cellular differentiation, and declining during the growth period and subsequent ripening. It must be noted that regulation of the formation of these compounds is different in each organ of the plant. Also, not only are there quantitative variations that depend on the stage of development, but shifts in the biosynthetic pathway may occur. For instance, in *Matthiola incana* (L.) R. Br., flavonol synthesis is strictly limited to the period of shoot development, the formation of these compounds being inhibited by maturation when the anthocyanins are activated (Schüte, 1985).

Not all plants show the same behaviour. In *Betula pubescens* Ehrh., *Aesculus hippocastanum* L., *Fraxinus excelsior* L., and *Cistus ladanifer* L. no differences are found in flavonoid content between different stages of development. The "flavonic composition" of young and adult leaves is the same both qualitatively and quantitatively (Tissut and Egger, 1972; Chaves, 1994).

17.4 SEASONAL VARIATION IN FLAVONOID SECRETION

Variations in the type and content of phenolic compounds depend on season and the stage of development of the organ. Such changes were detected in *Pteridium aquilinum,* which presented two peaks in flavonoid production, one at the end of June and the other in August (Cooper-Driver et al., 1977).

Studies by Pacheco et al. (1985) of the flavonoids of seven species of *Robinsonia* found only one of the species, *R. evenia,* to show clear qualitative seasonal flavonoid dependence, with different compounds being detected at different seasons of the year. Apigenin 7-O-diglycoside appeared only during November, and quercetin was detected in February. In *Menzesia ferruginea* Smith, the greatest changes in the composition of flavonoids appeared during the course of just a single season (summer). These changes did not seem to be correlated with any particular stage of development, but rather with the activation of the formation of certain flavonoids and the inhibition of the production of others (Bohm et al., 1984). Likewise, in the Mediterranean scrub species, *Calluna vulgaris* Salisb., the composition and quantity of phenolic compounds, as with the species described above, varies considerably over the course of the year. There are quantitative and qualitative variations (Jalal et al., 1982). The amounts of simple phenols and flavonoids vary according to the season, and certain compounds are formed only in specific months.

It has been shown that the levels of aglycone flavonoids in the exudate from leaves and photosynthetic shoots in *Cistus ladanifer* have a seasonal variation (Chaves, 1991; Chaves et al., 1993, 1997). In summer, these compounds are four to five times more abundant than in spring. Both photosynthetic shoots and leaves are able to synthesize flavonoids and to vary the rate of biosynthesis of these compounds over the course of the summer (Figure 17.1).

In *C. ladanifer*, the qualitative and quantitative seasonal variation is not correlated with any specific organ or with the stage of development. The leaves emerging in August have four times the amount (22 mg/g dry mass) of flavonoids of those formed in March (5 mg/g dry mass). Furthermore, the relative percentage composition in apigenins and kaempferols of leaves formed in spring is different from those of summer (apigenin and kaempferol-3,7-di(O) methyl, 11 and 7 percent, respectively, for leaves formed in spring, and 3 and 22 percent, respectively, for leaves formed in summer). This implies that the exudate analyzed in summer does not depend on a particular stage of development, nor is it due to a progressive build-up. Instead it appears to have been induced by external factors during the summer.

In *C. ladanifer*, all of the flavonoids in the exudate increase in absolute terms, but not all equally. The contribution from each varies from one season to another (Table 17.1). In general, concentrations of apigenins decrease from spring to summer, and increase in the level of kaempferol synthesis. The most abundant apigenins in spring are apigenin and apigenin-4'-(O)methyl, with their percentage falling as summer progresses. The greatest summer increase in percentage composition among the kaempferols is kaempferol-3,7-di(O) methyl (Chaves et al., 1993). From the study of *C. ladanifer*, therefore, one sees that not only is kaempferol synthesis being induced to a greater degree than the apigenins, but also there are shifts from apigenin and apigenin-4'-(O)methyl toward apigenin-7-(O)methyl and apigenin-7,4'-di(O)methyl, and from kaempferol-3-(O)methyl to predominantly kaempferol-3,7-di(O)methyl. There is a clear tendency toward more methylated compounds, and in particular to methylation at position 7.

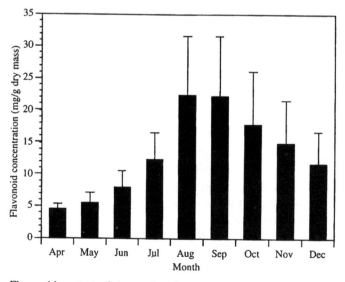

FIGURE 17.1 Flavonoid content of the exudate from leaves of *Cistus ladanifer* gathered from April to December. Vertical lines at the top of the bars show ± standard deviation.

TABLE 17.1
Relative Seasonal Variation of Each of the Flavonoids in the Exudate from the Leaves of
Cistus ladanifer

Flavonoid	Relative Seasonal Variation (%)				
	March	May	August	October	December
Apigenin	14	6	4	3	4
Kaempferol-3-(O)methyl	57	57	47	43	48
Apigenin-4'-(O)methyl	11	8	4	4	5
Apigenin-7-(O)methyl	7	9	10	9	11
Kaempferol-3,4'-di(O)methyl	3	5	4	5	4
Kaempferol-3,7-di(O)methyl	4	10	22	26	21
Apigenin-7,4'-di(O)methyl	1	3	3	3	3
Kaempferol-3,7,4'-tri(O)methyl	2	2	3	2	2
Total apigenins	33	25	22	18	23
Total kaempferols	68	75	77	86	78

17.5 ECOLOGICAL FACTORS INDUCING FLAVONOID SYNTHESIS

The response of plants in terms of quantitative and/or qualitative variation of the flavonoids, independently of the stage of development of the organs, can be assumed to be an adaptive response induced by some physical, chemical, or biological factor. Such induction has greatest relevance in certain seasons, such as summer in the case of *C. ladanifer*. The variation in the levels of these compounds is determined by their rate of synthesis, which in turn is modulated by the production of enzymes responsible for their formation. Studies on *Robinia pseudoacacia* L. have shown that the variation in activity of the enzymes PAL (phenylalanine ammonia lyase) and CHS (chalcone synthase) clearly depends on the season. Their greatest activity is detected at two times of the year, April and September, with trace amounts of these enzymes appearing in January and November (Magel et al., 1991).

The results obtained from this study of *C. ladanifer* indicate that there has to exist an induction of the activity of enzymes involved in the synthesis of flavonoids beginning in spring and increasing as summer progresses. But apart from the greater degree of induction during summer, there exists a different regulation of the production of group II enzymes—flavone synthase, flavonol synthase, and flavonol 3-hydroxylase (Figure 17.2). The increase in kaempferols relative to apigenins implies an enhanced induction of the genes, and transcription of the mRNA, that code for flavonol synthase and flavonol 3-hydroxylase, leading to a greater build-up of kaempferols as summer progresses. This quantitative and qualitative variation of the flavonoids must be derived from induction by environmental factors to which the plant is subject (Nuñez-Olivera et al., 1996).

Factors that might induce enzymes of the flavonoid pathway may be divided into four groups:

Climatic factors—temperature, light intensity, and humidity.

Animals—herbivores are particularly hostile to plants, and defensive adaptations against them are known (Shaver and Lukefahr, 1969; Feeny, 1970; Mooney, 1972; Rhoades and Cates, 1976; Bryant and Kuropat, 1981; Harborne, 1985; Howe and Westly, 1988; Karban and Myers, 1989; Krischik et al., 1991; Karban, 1993a).

Edaphic factors—the soil is the source of mineral nutrients for plants that may suffer biochemical stress due to mineral deficiencies in the soil.

Interference from other plants—this would include both competition for resources and allelopathy, inhibiting the germination and/or growth of other species.

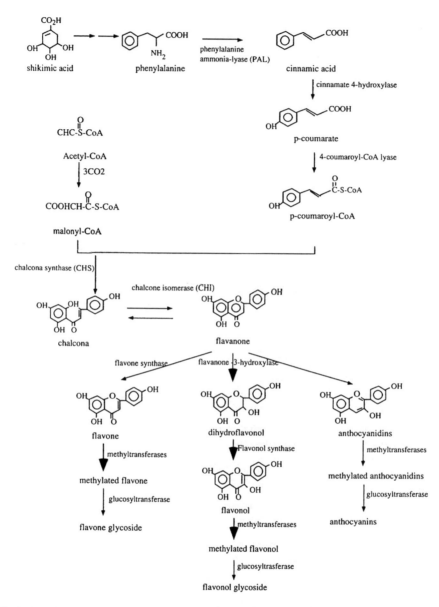

FIGURE 17.2 Biosynthetic pathway of flavonoids. Bold arrows indicate enzymes induced during the summer.

Since the main parameters differentiating one season from another are climatological, it is reasonable to think that they will be primarily responsible for the seasonal variation of flavonoid secretion. Thus, studies on *Cistus ladanifer* showed the secretion to be clearly enhanced during summer, which is characterized in its area (the Mediterranean) by high temperatures, high intensities of light, and marked hydric stress. The importance of these factors is also manifested in the difference in behavior between jarales (*C. ladanifer* scrub) located on south- or on north-facing slopes (Figure 17.3). The accumulation of flavonoids in the latter is less than in the former. Other investigators have also found differences in flavonoid levels in plants that have grown on south-facing or north-facing slopes (Cooper-Driver et al., 1977; McDougal and Parks, 1986; Lovelock et al., 1992). This implies that the differences between these two environmental situations determine the greater or lesser secretion of flavonoids in these plants.

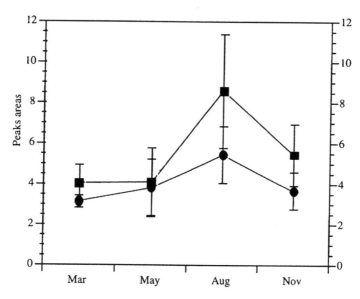

FIGURE 17.3 Seasonal variation of total flavonoids in exudate extracted from leaves of *C. ladanifer* plants growing in open areas (■) and shaded areas (●). The data are expressed as the sum of peak areas of flavonoids obtained from HPLC chromatograms. Vertical lines at the top of the bars show ± standard deviation.

17.6 EFFECT OF ULTRAVIOLET LIGHT, HYDRIC STRESS, AND TEMPERATURE ON FLAVONOID INDUCTION

Numerous studies have shown that hydric stress and ultraviolet radiation stand out among the most relevant factors in the induction of flavonoid synthesis. It has been demonstrated clearly that the build-up of flavonoids is induced specifically by ultraviolet light and that the dependence is linear (Wellmann, 1985). Also, the visible radiation flux is particulary influential. Plants that grow under high light intensities contain greater quantities of these substances (Cen and Bornman, 1990) and exibit decreased susceptibility to UVB-induced damage. A wide range of species show this factor to be determinant in their flavonoid synthesis. For example, the study performed with *Prasiola crispa* (an Antarctic alga) shows the importance of the quantity of light received by each algae in the synthesis of these compounds, indicating that ultraviolet-light-absorbing pigments depend on the degree of exposure to that light (Post and Larkum, 1993). Selective screening of UVB radiation in the epidermis appears to be the dominant defense mechanism of these pigments in protecting the algae (Caldwell et al., 1983); in *Prasiola crispa* photosynthesis is at its peak when the ultraviolet-absorbing pigment content is maximum.

The phylogenetically close algae *Prasiola crispa* and *Enteromorpha bulbosa* are clearly different in content of these compounds, with very low levels in *Enteromorpha* (Post and Larkum, 1993). This is due to the quantity of light these two algae receive in natural conditions. *Prasiola* is exposed directly to sunlight while *Enteromorpha* grows in the shade of other algae. Reaffirmation of these results is found in *Prasiola* gathered from zones with less light intensity, for example, in Japan (Sivalingam, 1974), where the amount of pigment is found to be four times less than that analyzed during the Antarctic summer.

More detailed studies verify the importance of ultraviolet light in phenol synthesis and show that there is a direct relationship between the inducing factor and the function that these compounds may have. Thus in *Brassica napus* L. irradiated with ultraviolet light, there is practically a 20 percent increase in ultraviolet screening compounds, of which 30 percent are located in the first 40 μm of the leaf, which includes the adaxial epidermis, forming a true filter against these wavelengths of radiation (Cen and Bornman, 1993). Levels of these compounds vary considerably between plant species and between plant life forms. In some sclerophyllous species more than 90 percent of the incident UVB radiation is attenuated before reaching the mesophyllous tissues (Day et al., 1992; DeLucia et al., 1992); with other species the situation is generally similar (Caldwell et al., 1983).

It must be remarked that different plant species may use different physiological routes to trigger the production of same given flavonoid, or may respond to the same stimulus by producing different flavonoids. The response of plants to increased ultraviolet levels varies notably both intra- and interspecifically. For instance, light induces cyanidin 3-glucoside to build-up in *Sorghum,* but it is produced in *Brassica* cells cultured under light only after exposure to low temperatures. Phytochrome activation by light, which induces production of anthocyanins in some plants and flavonol glycosides in others, is another clear example of this behavior (Smith, 1972; McClure, 1975). A similar case is that of *Cistus laurifolius* and *Cistus ladanifer* which, though belonging to the same genus, respond differently to ultraviolet light in their flavonoid synthesis. In experiments performed at the beginning of the vegetative period (Vogt et al., 1991), with *Cistus laurifolius* leaves covered with plexiglass boxes to screen out ultraviolet light, less intracellular flavonoid glycosides were found (50 percent) than in leaves left exposed to the sun. In contrast, the aglycone flavonoids did not vary significantly under the influence of ultraviolet radiation. In similar experiments with *Cistus ladanifer,* the plants exposed to ultraviolet light presented three times more aglycone flavonoids than those that were protected from ultraviolet light, where the flavonoid secretion was practically constant (Chaves, 1994; Chaves et al., 1997). The light-induced flavonoids in *C. ladanifer* are therefore the aglycones, whereas in *C. laurifolius* they are the glycosides, a clearly different response of the two species in the synthesis of these compounds.

These examples show that ultraviolet light is a physical factor that induces the phenolic biosynthetic pathway in the plant, but that one or another type of flavonoid can be accumulated according to the species.

Another notable point is the plant's response when various environmental factors interact. The stress provoked, for example, by the joint action of ultraviolet radiation and water deficit adversely affects plant physiology. Sullivan and Teramura (1990) reported that interaction between solar UVB radiation and drought resulted in a reduction in the apparent quantum efficiency of photosynthesis. The response of plants to adverse conditions will depend on the degree of exposure, and can be cumulative in the case of flavonoid synthesis. One deduces that the amount of flavonoids synthesized will depend on the stress to which the plant is exposed.

The effect of the interaction of UV irradiation and hydric stress were studied in *C. ladanifer* under controlled conditions in three types of experiment: 1) field-grown *C. ladanifer* plants under plastic housing; 2) glasshouse studies; and 3) culture room studies.

17.6.1 FIELD-GROWN *CISTUS LADANIFER*

Large plastic boxes were constructed to cover selected groups of 15 plants. Two types of plastic were used for the boxes. Two boxes were made of plexiglass (to cover a total of ten plants), a plastic that filters out ultraviolet light below 380 nm (Vogt et al., 1991), one box was made of a greenhouse plastic with little ability to filter out ultraviolet light (five plants), and ten plants were used as controls with no cover. To avoid excessive heating inside the boxes, several ventilation slits were made at the bottom and near the top of the plastic walls without allowing sunlight to penetrate the box without passing through the plastic. The relative humidity and maximum and minimum temperatures inside and outside the boxes were measured weekly.

17.6.2 GLASSHOUSE STUDIES

Small *C. ladanifer* plants (each treatmentent, n=10) of approximately 1 year in age were collected in November and re-planted in a glasshouse and grown under temperature and humidity conditions close to those attained in summer (e.g., maximum and minimum temperatures of 37.5°C and 15°C at midday and midnight, respectively, and with 25 and 50 percent relative humidity during daylight and during the night, respectively). Day/night cycles were simulated by illumination for 16 hours with the lamps indicated below, followed by 8 hours of darkness. Daylight was simulated by combining visible light Philips 36W/56 and ultraviolet light Philips TL/40W 09 and Mazoafluor TFWN 18 lamps. Hydric stress was simulated by controlled irrigation of the plants divided into two groups. The plants irrigated every day were taken as the control group without hydric stress, and the other group had hydric stress provoked by moderate irrigation every 15 days.

17.6.3 CULTURE ROOM STUDIES

In these experiments plants were grown in the absence of ultraviolet light and the effect of hydric stress was studied. Plants of *C. ladanifer* like those used for the glasshouse trials were placed in the culture room, separated into two groups (ten plants per group); they were irrigated daily (control unstressed group) or every 15 days (hydric stress group). The culture room was kept at 30°C with a relative atmospheric humidity of 5 percent, and 16 hours irradiation with a daylight lamp (Sylvania Gro-lux F30W/Gro-T8), followed by an 8-hour darkness period.

In addition, seven other sampling sites were chosen in the southwest of Spain to give the highest possible variation in the following climatic factors: average maximum temperatures; average minimum temperatures; and annual rainfall (ranging between 16.8°C and 23.8°C; 3.5°C and 10.7°C, and 455 and 933 mm, respectively). At each site, samples were collected from both open and shaded areas.

For plants covered by plastic boxes and controls, additional samples were taken immediately before covering the plants with the plastic (mid-June), and then every 15 days until the end of December. For glasshouse and culture room trials, plants were allowed to grow for 2 to 3 months in the glasshouse or culture room, and then samples of leaves (of approximately 0.3 g, 3 to 4 leaves) born during this period were collected as indicated above.

The exudate was extracted from the samples by following the protocols indicated in Vogt and Gülz (1991); the leaves were dipped several times into 2 ml of chloroform. The chloroform was evaporated and the extract was dissolved in hot methanol, then cooled to -20°C and after 12 hours the precipitate containing waxes and other hydrocarbons was removed. Flavonoids and terpenoids remain soluble in methanol and were separated by chromatography on a column of Sephadex LH-20, using methanol as eluent. Terpenoids elute first and are well separated from flavonoids as shown in Vogt and Gülz (1991). After extraction of the exudate the leaves were weighed, oven-dried at 60°C for 12 hours, and weighed again to measure their dry biomass. From these measurements, we determined the water content of the leaves, which can be used as an index of the hydric stress suffered by the plant (Balakumar et al., 1993). Quantitative analysis of the flavonoids in all the samples was performed by HPLC, under the following conditions: 20 µl of the extract (eight-fold dilution) was injected into a Nucleosil 5µ C-18 (150 x 4 mm) column, and eluted with water: methanol: acetonitrile: tetrahydrofuran (56:16:6:22) at a flow rate of 0.7 ml/min (Chaves et al., 1993).

Experiments with *Cistus ladanifer* under controlled conditions simulating hydric stress and with ultraviolet irradiation showed that the accumulated action of the two types of stress had an additive effect on flavonoid induction (Chaves, 1994; Chaves et al., 1997). The plants with the least flavonoid content were those that had not been subjected to any stress, were maintained in daylight, and with no hydric deficit. The level was 4 mg/g dry weight of flavonoids, corresponding to the basal level of flavonoids secreted by these plants in winter and early spring. Plants subjected to a single stressor, whether hydric or ultraviolet, presented intermediate levels (7-8 mg/g dry weight), while those subjected to both possessed 11 mg/g dry weight, the maximum of these compounds. The same experiment was performed with another species, *Vigna unguiculata* (L.) Walp., where the total phenol content rose by 19 percent under hydric stress, 58 percent under ultraviolet stress, and 63 percent under both (Balakamar et al., 1993).

It has to be remarked that the studies carried out over this period of years on different plant species demonstrated the cited relationship between flavonoid build-up and the amount of ultraviolet light received. There have been fewer studies of the effect of hydric stress, but as was shown with *Cistus ladanifer* and *Vigna unguiculata,* there is similarly a direct relationship with this physical factor, which furthermore causes a greater flavonoid build-up when accumulated with ultraviolet light. Hydric deficit affects not only the quantitative variations in flavonoid synthesis, but is also involved in the qualitative variation of these compounds. This is clear from studies with *Cistus ladanifer* as we shall now describe.

In *C. ladanifer* there exist qualitative variations in the composition of the exudate as well as the clear increase in flavonoid secretion in summer. Total apigenins (apigenin, apigenin-4'-(*O*)methyl, apigenin-7-(*O*)methyl, and apigenin-7,4'-di(*O*)methyl) are clearly abundant in spring, and decline considerably in summer, mainly due to a fall in apigenin and apigenin-4'-(*O*)methyl levels. This behavior is not uniform in all of the jarales studied, either along a latitudinal gradient crossing the Iberian Peninsula, or in local comparisons between south- and north-facing slopes (Chaves et al., 1997). With respect to jarales on south-facing slopes, there is not the same variation in apigenin and apigenin-4'-(*O*)methyl relative total apigenins in all localities (Table 17.2). Thus, in Monesterio, Llerena, and Jerez de los Caballeros the decrease in apigenin-4'-(*O*)methyl is 20 to 22 percent and in apigenin is 10 to 14 percent. In Monterrubio and Alburquerque the two percentages of decline are equal, and in Jaraiz de la Vera the greatest decline is for apigenin (17 percent) as against apigenin-4'-(*O*)methyl (only 4 percent). There are also differences in apigenin-4'-(*O*)methyl levels between south-facing and north-facing jarales, with greater amounts in the latter. Another difference between these two orientations of jaral slopes is the greater amount of kaempferol-3,4'-di(*O*) methyl in north-facing jarales.

A positive correlation (r = 0.975) is found between the amount of apigenin- 4'-(*O*)-methyl, which declines from spring to summer, and the rainfall received at the different localities in the months of June to September (summer months characterized by severe hydric stress) (Figure 17.4,A). The results therefore indicate that the greater or lesser decline in apigenin-4'-(*O*)methyl is

TABLE 17.2

Seasonal Changes of the Apigenin Distribution in the Exudate of *C. ladanifer* Leaves from March to August in Different Sampling Areas

	SAMPLING SITE [a]									
	Ap4'/Apt Open Area						Ap4'/Apt Shade Area			
	M	LLE	JE	MO	AL	JA	M	JE	AL	JA
March	0.47	0.54	0.47	0.48	0.33	0.32	0.54	0.4	0.35	0.44
August	0.25	0.34	0.23	0.31	0.19	0.28	0.36	0.33	0.29	0.28
Decrease (%)	22%	20%	24%	17%	14%	4%	18%	7%	6%	19%
	Ap7/Apt Open Area						Ap7/Apt Shade Area			
March	0.12	0.12	0.13	0.11	0.2	0.18	0.1	0.15	0.22	0.13
August	0.38	0.32	0.4	0.32	0.47	0.32	0.27	0.28	0.34	0.35
Increase (%)	26%	20%	27%	21%	27%	14%	17%	13%	12%	22%

Expressed as relative ratios of apigenin-4'-(O)methyl (Ap4') and apigenin-7-(O)methyl (Ap7) content over total a apigenins (Apt).

[a]Field sites: M, Monesterio; LLE, Llerena; JE, Jerez de los Caballeros; MO, Monterrubio; AL, Alburquerque; JA, Jaraiz.

Source: Chaves, N., et al., Role of ecological variables in the seasonal variation of flavonoid content of *Cistus ladanifer* exudate, *J. Chem. Ecol.,* 23, 2577, 1997. With permission.

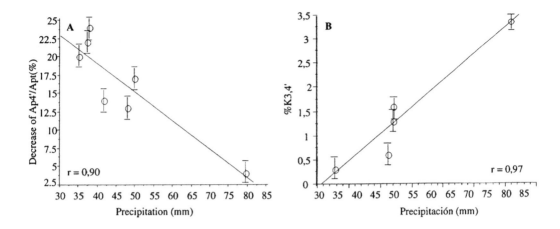

FIGURE 17.4 A. Effect of rainfall from June to September on the seasonal decrease of the content of apigenin -4'-(O)methyl (Ap4') in the exudate, expressed as percent of total apigenins (Apt). B. Effect of rainfall from June to September on the seasonal increase of the content of kaempferol-3,4'di(O)methyl (K3,4') in the exudate from Chaves et al., 1997.

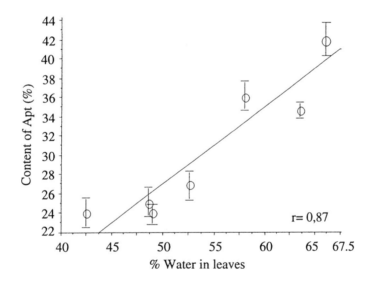

FIGURE 17.5 Decrease of total apigenin content (Apt) in the exudate of leaves of *C. ladanifer* as a function of the water content of the leaves.

determined by the rainfall received during these months. There is a similar positive correlation (r = 0.97) between the rainfall of those months and the kaempferol-3,4'-di(O)methyl content of the exudate (Figure 17.4,B).

In *C. ladanifer* plants studied under controlled conditions, the greatest amount of total apigenins were present in plants that underwent no stress. There is a significant positive correlation between the amount of water in the leaves and the accumulated total apigenin content (r = 0.87) (Figure 17.5). These results indicate that summer rainfall and leaf water content are correlated with the level of apigenins, and in particular with apigenin-4'-(O)methyl and kaempferol-3,4'-di(O) methyl. This suggests that the activity of 4'-methyltransferase is inversely related to hydric stress, and that the fall of apigenins relative to kaempferols is due to the hydric stress that occurs in summer.

Temperature is another factor that clearly differentiates the seasons as is evident comparing the north and south facing orientations. There have been very few studies of the relationship of temperature with flavonoid synthesis. The results obtained with *Cistus ladanifer* show that it is not this variable that is mainly responsible for the summer rise in flavonoid secretion. There was no induction of flavonoid secretion in experiments on *C. ladanifer* plants subjected to high temperatures with no other stressor. Low temperatures, however, did provoke inhibition of the synthesis of these compounds, as was found in trials with the plants subjected to hydric stress, ultraviolet irradiation, and temperatures between 0°C and 10°C (Chaves, 1994).

Although they do not affect the quantity of flavonoids secreted by *Cistus ladanifer,* the high summer temperatures (40 to 50°C) where this species grows do influence the qualitative composition of the secretion. In summer, the *C. ladanifer* exudate is especially enriched in apigenin and kaempferol derivatives with methylation at position 7. At the sampling sites located on south-facing slopes the temperature is a few degrees above the north-facing slopes, and there are greater levels of position-7 methylated apigenins and kaempferols in the former than in the latter. It seems reasonable to think that the shift toward this group of compounds is due to the temperature difference. There is a positive correlation (Figure 17.6) between the level of kaempferol-3,7-di(*O*)methyl and temperature. On the contrary, there is no correlation between temperature and position-7 methylated apigenins. This may be because the more stimulated biosynthetic pathway in summer is that of the kaempferols. Position-7 methylation of the flavones is stimulated, however, by high temperatures. This was shown by analysis of the exudate from plants studied under controlled conditions of high temperatures with no other stressor. In these plants, 59 percent of the apigenins correspond to apigenin-7-(*O*)methyl. Under normal conditions, this flavone does not surpass 15 percent.

One deduces from these studies that high temperatures strongly induce the activity of the 7-methyltransferase enzyme, and that the shift toward the kaempferols and position-7 methylated apigenins is due to this physical factor.

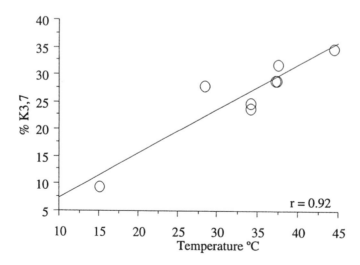

FIGURE 17.6   Content of kaempferol-3,7-di(*O*)methyl (k,3,7) in the exudate of *C. ladanifer* plants as a function of the maximum daily temperature during the sampling period.

The climatic factors analyzed here are closely related, of course. In Mediterranean climates, the periods of greatest hydric stress coincide with the hottest months when the light intensities are also the highest. It is, therefore, difficult to decide which parameter or parameters have the most influence on the qualitative variations and to determine the minimum or maximum thresholds for methyltransferase induction to occur. For example, in the studies with plants under controlled conditions, the individuals with no hydric stress secreted a greater amount of total apigenins, but this does not correspond with greater amounts either of apigenin-4'-(O)-methyl or of apigenin. It is rather that the high temperatures shift production toward apigenin-7-(O)methyl, so that in this case temperature is a more important factor than hydric stress. The situation is the same with kaempferols, since the plants subjected to either ultraviolet or hydric stress synthesized very little kaempferol-3-(O)methyl; the synthesis was practically toward kaempferol-3,7-di(O)methyl (Table 17.3).

In conclusion, one observes from the studies discussed in the present section that climatic factors are primarily responsible for quantitative (ultraviolet and hydric stress) and qualitative (temperature and hydric stress) variations in flavonoid synthesis. Variations in these physical parameters coincide with seasonal changes, with the result being a seasonal change in the composition of flavonoids possessed by the species investigated.

17.7 INFLUENCE OF HERBIVORES

Plants defend themselves against animals on the basis of both physical and chemical factors. Physical defenses are patently clear—thorns, spikes, stinging hairs, and increased thickness of the epidermis. Chemical defenses are the production of substances that can be toxic and/or the development of repellants.

The evolutonary response of forest species (especially perennials) to the biological stress of herbivore attack has been to produce secondary chemical compounds that dissuade the herbivores; these compounds are often quite toxic (Feeny, 1970; Bryant and Kuropat, 1981). The alkaloids, such as nicotine, the pyretrines, and rotenoids (Harborne, 1985; Krischik et al., 1991), are characterized as truly toxic substances. But it also has been shown that one of the ecological functions of the flavonoids is the protection of leaves against herbivores (Mooney, 1972; Rhoades and Cates, 1976), and some flavonol glycosides such as rutin, quercitrin, and isoquercitrin are very toxic to certain insect species (Shaver and Lukefahr, 1969).

Compounds to dissuade browsers may appear as terpenoids, alkaloids, and flavonoids. The last group includes the most important barrier of angiosperms against herbivore ingestion, the tannins, which appear at high concentrations in the leaves of certain woody plants. Tannins make plant

---

TABLE 17.3
Percentage of Each of Flavonoids in the Exudate from Leaves of
Cistus ladanifer **Without Stress**

Flavonoid	Percentage
Apigenin	4
Kaempferol-3-(O)methyl	17
Apigenin-4'-(O)methyl	4
Apigenin-7-(O)methyl	25
Kaempferol-3,4'-di(O)methyl	3
Kaempferol-3,7-di(O)methyl	33
Apigenin-7,4'-di(O)methyl	10
Kaempferol-3,7,4'-tri(O)methyl	4
Total apigenins	43
Total kaempferols	57

tissues unpalatable and indigestible to animals. Plants with lower tannin concentration showed higher levels of contact with herbivores than the plants containing more tannins (Bryant, 1987; Clausen et al., 1987; Nichols-Orians, 1991). They act by attaching to proteins, often irreversibly, blocking the proteins from being attacked by trypsin and other digestive enzymes, or interfering with proteinase activity in the intestine (Howe and Westly, 1988) with the resulting loss in nutritional value of the leaves. It is shown that flavonoids contained in *Diplacus aurantius* resin (Lincoln, 1980), such as diplacone and diplacol, are negatively correlated to the growth and survival of *Euphydras chalcedona* larvae (primary herbivore of *D. aurantius*) (Lincoln et al., 1982; Lincoln, 1985) possibly in a manner similar to that of tannins and other phenols in their capacity to form complexes with proteins.

The production and storage of secondary compounds requires a supply of energy derived from primary metabolism. There has been debate about the "metabolic cost" to the plant in producing toxins as a defense against herbivores (Rhoades, 1979; Harvell, 1986; Karban and Myer, 1989; Baldwin et al., 1990; Karban, 1993b). This cost can be reduced if the toxins are synthesized and stored only when really needed. For instance, to be a direct response to avoid being eaten, their synthesis would increase immediately after damage occurred to the plant. (Adler and Karban, 1994).

It has been demonstrated that a herbivore attack can change the secondary chemistry of a plant and reduce its nutritive value in 24 hours (Rhoades, 1979, 1983). These changes are called the short-term induced defense (STID) triggered in the plant by insect attack. It has been demonstrated in *Populus tremuloides* Michx. leaves after simulated insect attack causes increase in phenols such as salicortin and tremulacin. The levels of these phenols rose following leaf damage within approximately 24 hours by 3 to 13 mg phenol glycoside/g leaf tissue. This plant is therefore able to selectively alter the levels of phenol glycosides in its leaves after tissue damage (Clausen et al., 1989).

In addition, the grinding of the leaf tissue that may occur when it is eaten by insects leads to a complete conversion of salicortin and tremulacin into salicin and tremuloiden, respectively. This second change does not occur in healthy leaves, which points to its being enzymatically mediated. It was found that these compounds have an effect on the pupal development of the insect that attacks this plant (Clausen et al., 1989). These observations suggest that during the time the insect spends eating, the food quality of the leaf declines sufficiently to be rejected, and the increase in salicortin and tremulacin provides a defense for the leaves, which under normal circumstances have very low levels of phenol glycosides. Figure 17.7 reflects a proposed model of STID in *Populus tremuloides* leaves. After the initial attack, the levels of salicortin and tremulacin in the damaged leaves rises over the first 24 hours. After ingestion, the salicortin and tremulacin are converted to 6-hydroxy-2-cyclohexenone (6-HCN) by esterases of the leaf. In the herbivore gut, 6-HCN is converted to toxins such as catechol or phenol depending on the conditions in the gut, or it is trapped by nucleophils.

While *Populus tremuloides* clearly responds to herbivore attack immediately after being attacked, other studies show chemical protection against herbivores to be developed throughout the life of the plant as an evolutionary response. This is the case for plants that grow in nutrient-poor environments. It contrasts with the case of plants that grow in rich media and that grow rapidly. These plants do not possess abundant chemical defenses because the damaged or eaten part is rapidly replaced or they induce a short-term response in the synthesis of secondary compounds.

A chemical barrier like the production of tannins is a constitutive defense rather than an immediately induced response following attack. Studies with *Pteridium aquilinum* (Cooper-Driver et al., 1977) have shown the production of tannins to vary seasonally, determining the palatability of the leaves to insects and mammalian herbivores. Nevertheless, the change in total flavonoid content seems to be of little importance in this aspect, although the variation in individual compounds may be significant as toxic agents. Another example is the study by Feeny (1970) with *Quercus robur* L. It was found that tannins rise from April to September from 0.66 to 5.55 percent dry weight. It is during the summer that the plants are subject to more intense stresses from the high temperatures, hydric deficit, and ultraviolet light, leaving them more defenseless against herbivores. They are

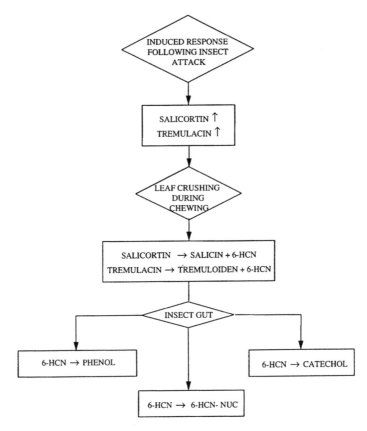

FIGURE 17.7 Chemical model for short-term induction (STID) in *Populus tremuloides,* from Clausen et al., 1989.

therefore more capable of withstanding herbivore attack by increasing the tannin concentration. Previous models have indicated that the information content of the environment has the potential to determine whether constitutive or induced resistance is favored (Riessen, 1992; Frank, 1993; Adler and Karban, 1994). If the herbivore is completely unpredictable, then plants might also evolve a level of constitutive defense that is best on average. In other words, induced resistance will be favored only if the plant is likely to be attacked by herbivores in the future.

It must be emphasized that when plants respond to herbivore damage by inducing the synthesis of compounds, it is unclear what signal triggers the synthesis of compounds that are toxic to the herbivore. In studies of *Sorghum bicolor* (L.) Moench. (Woodhead 1981), it was shown that simulation of damage could be done to leaves by puncturing them. Partial herbivory by insects, corrosion, or oxidation do not affect the phenolic content of these crops. The mean values in plants with obvious signs of damage do not differ significantly from those of healthy, undamaged plants. Darkening due to oxidation of phenolic compounds was observed around the damage site, but the phenolic content of the adjacent area of the leaf or in other leaves of the plant did not vary. However, in experiments where the larvae were introduced into the verticils of the plant and allowed to feed, there was an increase in phenolic compounds. This implies that the damage caused by the insect is obviously different from purely mechanical damage and that it is perhaps some substance in the digestive tract of the insect that provokes the synthesis of these phenols. The same behavior is observed when insect-caused mechanical damage was simulated in *Cistus ladanifer.* After damaging leaves of a single plant grown under controlled conditions with no type of stress, it was found that mechanical damage to the leaf neither induced the synthesis of aglycone flavonoids that are the constituents of the exudate, nor caused their composition to vary (Chaves, 1994). Although the flavonoids constituting the exudate of this species are induced by climatic factors rather than by herbivore-associated mechanical damage, this does not mean that they do not serve to dissuade herbivores. To confirm or reject this hypothesis, further studies are required in this respect, including an analysis of the toxicity of the aglycone flavonoids in the exudates of *Cistus ladanifer* to herbivores that attack or may potentially attack this species.

REFERENCES

Adler, F. R. and Karban, R., Defended fortresses or moving targets? Another model of inducible defenses inspired by military metaphors, *Am. Nat.*, 144, 813, 1994.

Balakumar, T., Vicent, H. B., and Paliwal, K., On the interaction of UV-B radiation (280-315 nm) with water stress in crop plants, *Physiol. Plant.*, 87, 217, 1993.

Baldwin, I. T., Sims, C. L., and Kean, S. E., The reproductive consequences associated with inducible responses in wild tobacco, *Ecology*, 71, 252, 1990.

Bell, E. A., The possible significance of secondary compounds in plants, *in Secundary plant products*, Bell, E. A. and Charlwood, B. V., Eds., Springer-Verlag, New York, 1980, 11.

Bohm, B. A., Banek, H. M., and Maze, J. R., Flavonoid variation in North American *Menziesia* (Ericaceae), *Syst. Bot.*, 9, 324, 1984.

Bolwell, G. P., Bell, J. N., Cramer, C. L., Schuch, W., Lamb, C. J., and Dixon, R. A., PAL from *Phaseolus vulgaris:* Characterisation and differential induction of multiple forms from elicitor-treated cell suspension cultures, *Eur. J. Biochem.*, 146, 411, 1985.

Bryant, J. P., Feltleaf willow-snowshoe hare interactions: plant carbon/nutrient balance and floodplain succession, *Ecology*, 68, 1319, 1987.

Bryant, J. P. and Kuropat, P. J., Selection of winter forage by suartic browsing vertebrates: the role of plant chemistry, *Ann. Rev. Ecol. Syst.*, 11, 261, 1981.

Bryant, J. P., Chapin, F. S., and Klein, D. R., Carbon/nutrient balance of boreal plants in relation to vertebrate herbivory, *Oikos*, 40, 357, 1983.

Caldwell, M. M., Robberecht, R., and Flint, S. D., Internal filters: prospect for UV-acclimation in higher plants, *Physiol. Plant.*, 58, 445, 1983.

Castillo, J., Benavente, O., and del Rio, J. A., Naringin and Neohesperidin levels during development of leaves, flower buds and fruits of *Citrus aurantium*, *Plant Physiol.*, 99, 67, 1992.

Cen, Y. P. and Bornman, J. F., The response of bean plants to UV-B radiation under different irradiances of background visible light, *J. Exp. Bot.*, 41, 1489, 1990.

Cen, Y. P. and Bornman, J. F., The effect of exposure to enhanced UV-B radiation on the penetration of monochromatic and polychromatic UV-B radiation in leaves of *Brassica napus*, *Physiol. Plant.*, 87, 249, 1993.

Curir, P., Vansumere, C. F., Termini, A., Barthe, P., Marchesini, A., and Dolci, M., Flavonoid accumulation is correlated with adventitious roots formation in *Eucalyptus gunnii* Hook. micropropagated through axillary bud stimulation, *Plant Physiol.*, 92, 1148, 1990.

Chapell, J. and Hahlbrock, K., Transcription of plant defence genes in response to UV light or fungal elicitor, *Nature*, 311, 76, 1984.

Chaves, N., Estudio sobre la variación estacional de la composición del ládano en *Cistus ladanifer* L, Master thesis, Facultad de Ciencias, Universidad de Extremadura, Badajoz, Spain, 1991.

Chaves, N., Variación cualitativa y cuantitativa de los flavonoides del exudado de *Cistus ladanifer* L. como respuesta a diferentes factores ecológicos, PhD dissertation, Universidad de Extremadura, Badajoz, Spain, 1994.

Chaves, N., Escudero, J. C., and Gutierrez-Merino, C., Seasonal variation of exudate of *Cistus ladanifer, J. Chem. Ecol.*, 19, 2577, 1993.

Chaves, N., Escudero, J. C., and Gutierrez-Merino, C., Role of ecological variables in the seasonal variation of flavonoid content of *Cistus ladanifer* exudate, *J. Chem. Ecol.*, 23, 2577, 1997.

Chappell, J. and Hahlbrock, K., Transcription of plant defence genes in response to UV light or fungal elicitor, *Nature*, 311, 76, 1984.

Clausen, T. P., Reichardt, P. B., McCarthy, M. C., and Werner, R. A., The effect of nitrogen fertilization upon the secondary chemistry and nutritional value of quaking aspen (*Populus tremuloides* Michx.) over the large aspen tortix (*Choristoneura conflictana* (Walker)), *Oecologia*, 73, 513, 1987.

Clausen, T. P., Reichardt, P. B., Bryant, J. P., Werner, R. A., Post, K., and Frisby, K., Chemical model for short-term induction in quaking aspen (*Populus tremuloides*) foliage against herbivores, *J. Chem. Ecol.*, 15, 2335, 1989.

Cooper-Driver G., Finch, S., and Swain, T., Seasonal variation in secondary plant compounds in relation to the palatability of *Pteridium aquilinum, Biochem. Syst. Ecol.*, 5, 177, 1977.

DeLucia, E. H., Day, T. A., and Volgelmann, T. C., Ultraviolet B and visible light penetration into needles of two species of subalpine conifers during foliar development, *Plant Cell Environ.*, 15, 921, 1992.

Day, T. A., Volgelmann, T. C., and DeLucia, E. H., Are some plant life forms more effective than others in screening out ultraviolet-B radiation?, *Oecologia*, 92, 513, 1992.

Darwill, A. G. and Albersheim, P., Phytoalexins and their elicitors. A defence against microbial infection in plants, *Ann. Rev. Plant Physiol.*, 35, 243, 1984.

Feeny, P., Seasonal changes in oak leaf tannins and nutrients as a cause of spring feeding by winter moth caterpillars, *Ecology*, 51, 565, 1970.

Frank, S. A., A model of inducible defense, *Evolution*, 47, 325, 1993.

Gershenzon, J., Changes in the levels of plant secondary metabolites under water and nutrient stress, *in Phytochemical Adaptations to Stress, Recent Advances in Phytochemistry*, Timmermann, B. N., Steelink, C., and Loewus, F. A., Eds., Plenum Press, New York, 1984, 273.

Graham, L. T., Flavonoid and isoflavonoid distribution in developing soybean seedling tissues and in seed and root exudates, *Plant Physiol.*, 95, 594, 1991.

Harborne, J. B., *Introducción a la bioquímica ecológica*, Alhambra, Madrid, 1985.

Harborne, J. B., *Flavonoids: Advances and Research*, Chapman and Hall, London, 1988.

Harborne, J. B, Mabry, T. J., and Mabry, H., *The Flavonoids*, Chapman and Hall, London, 1975.

Hartmann, T., Prinzipien des pflanzlinchen sekundärstoffwechsels, *Plant Syst. Evol.*, 150, 15, 1985.

Harvell, C. D., The ecology and evolution of inducible defenses in a marine bryozoan: cues, costs, and consequences, *Am. Nat.*, 128, 810, 1986.

Howe, H. F. and Westley, L. C., *Ecological Relationships of Plants and Animals*, Oxford University Press, New York, 1988.

Hungria, M., Joseph, C. M., and Phillips, D. A., Anthocyanidins and flavonols, major nod gene inducers from seeds of a black-seeded common bean (*Phaseolus vulgaris* L.), *Plant Physiol.*, 97, 751, 1991.

Jalal, M. A. F., Read, D. J., and Haslam, E., Phenolic composition and its seasonal variation in *Calluna vulgaris, Phytochemistry*, 21, 1397, 1982.

Jordan, B. R., James, P. E., Strid, A., and Anthony, R. G., The effect of ultraviolet-B radiation on gene expression and pigment composition in etiolated and green pea leaf tissues: UVB induced changes are gene specific and dependent upon the developmental stage, *Plant Cell Environ.*, 17, 45, 1994.

Karban, R. and Myers, J. M., Induced plant responses to herbivory, *Ann. Rev. Ecol. Syst.*, 20, 331, 1989.

Karban, R., Induced resitence and plant density of a native shrub, *Gossypium thurberi*, effect its herbivores, *Ecology*, 74, 1, 1993a.

Karban, R., Costs and benefits of induced resistance and plant density of a native shrub, *Gossypium thurberi, Ecology*, 74, 9, 1993b.

Klepzig, K. D., Kruger, E. L., Smalley, E. B., and Raffa, K. F., Effects of biotic and abiotic stress on induced accumulation of terpenes and phenolics in red pines inoculated with bark beetle-vectores fungus, *J. Chem. Ecol.*, 21, 601, 1995.

Krischik, V. A., Goth, R. W., and Barbosa, P., Generalized plant defense: effects on multiple species, *Oecologia*, 562, 1991.

Lamaison, J. L. and Carnart, A., Teneurs en pricipaux flavonoïdes des fleurs et des feuilles de *Crataegus monogyna* et de *C. Laevigata*(Poiret) DC. en fonction de la période de végétation, *Plant. Méd. Phytothér.* XXV, 12, 1991.

Lincoln, D. E., Leaf resin flavonoids of *Diplacus aurantiacus, Biochem. Syst. Ecol.*, 8, 397, 1980.

Lincoln, D. E., Host plant protein and phenolic resin effects on larval growth and survival of a butterfly, *J. Chem. Ecol.*, 11, 1459, 1985.

Lincoln, D. E., Newton, F. S., Ehrlich, P. R., and Williams, K. S., Coevolution of the checkerspot butterfly *Euphydras chalcedona* and its larval food plant *Diplacus aurantiacus:* larval response to protein and leaf resin, *Oecologia,* 52, 216, 1982.

Lovelock, C. E., Clough, B. F., and Woodrow, I. E., Distribution and accumulation of ultraviolet-radiation-absorbing compounds in leaves of tropical mangroves, *Planta,* 188, 143, 1992.

Magel, E. A., Drouet, A., Claudot, A. C., and Ziegler, H., Formation of heartwood substances in the stem of *Robinia pseudoacacia* L. I. Distribution of PAL and CHS across the trunk, *Tress,* 5, 203, 1991.

McClure, J. M., Physiology and functions of flavonoids, *in The Flavonoids,* Harborne, J. B., Mabry, T. J., and Mabry, H., Eds., Academic Press, London, 1975, 970.

McDougal, K. M. and Parks, C. R., Environmental and genetic components of flavonoid variation in red oak, *Quercus rubra, Biochem. Syst. Ecol.,* 14, 291, 1986.

Mooney, H. A., The carbon balance of plants, *Ann. Rev. Ecol. Syst.,* 3, 315, 1972.

Nichols-Orians, C. M., Environmentally induced differences in plant traits: consequences for susceptibility to a leaf-cutter ant, *Ecology,* 72, 1609, 1991.

Nicolai, V., Phenolic and mineral content of leaves influences decomposition in European forest ecosystems, *Oecologia,* 75, 575, 1988.

Nuñez-Olivera, E., Martinez-Abaigar, J., and Escudero, J. C., Adaptability of leaves of *Cistus ladanifer* L. to widely varying environmental conditions, *Func. Ecol.,* 10, 636, 1996.

Pacheco, P., Crawford, D. J., Stuessy, T. F., and Silva, M., Flavonoid evolution in Robinsonia (Compositae) of the Juan Fernandez islands, *Amer. J. Bot.,* 72, 989, 1985.

Porter, P. M., Banwant, W. L., and Hassett, J. J., Phenolic acids and flavonoids in soybean root and leaf extracts, *Exp. Bot.,* 26, 65, 1986.

Post, A. and Larkum, A. W. D., UV-absorbing pigments, photosynthesis and UV exposure in Antarctica: comparison of terrestrial and marine algae, *Aquatic Bot.,* 45, 231, 1993.

Rao, A. S., Root flavonoids, *Bot. Rev.,* 56, 1, 1990.

Rhoades, D. F. and Cates, R. G., Toward a general theory of plant antiherbivore chemistry, *in Biochemical Interaction Between Plants and Insects, Recent Advances in Phytochemistry,* Wallace, J. W., and Mansell, R. L., Eds., Plenum Press, New York, 1976, 168.

Rhoades, D. F., Evolution of plant chemical defences against herbivores, *in Herbivores: Their Interaction with Secondary Plant Metabolites,* Rosenthal, G. A., and Janzen, D. H., Eds., Academic Press, New York, 1979, 4.

Rhoades, D. F., Herbivore population dynamics and plant chemistry, *in Variable Plants and Herbivores in Natural and Managed Systems,* Denno, R. F., and McClure, M. S., Eds., Academic Press, New York, 1983, 155.

Riessen, H. P., Cost-benefit model for the induction of an antipredator defense, *Am. Nat.,* 140, 349, 1992.

Schüte, H. R., Secondary plant substances special topics of the flavonoid metabolism, *Prog. Bot.,* 47, 118, 1985.

Shaver, T. N. and Lukefahr, M. J., Effect of flavonoid pigments and gossypol on growth and development of the bollworm, tobacco budworm, and pink bollworm, *J. Econ. Entomol.,* 62, 643, 1969.

Sivalingam, P. M., Ikawa, T., Yokohama, Y., and Nisizawa, K., Distribution of a 334 UV-absorbing-substance in algae, with special regard of its possible physiological roles, *Bot. Mar.,* 17, 23, 1974.

Smith, H., The photocontrol of flavonoid biosynthesis, *in Phytochrome,* Mitrakos, K. and Shropshire, W., Jr., Eds., Academic Press, New York, 1972, 43.

Stewart, R. N., Asen, S., Massie, D. R., and Norris, K. H., The identification of *Poinsettia* cultivars by high performance liquid chromatographic analysis of their anthocyanin content, *Biochem. Syst. Ecol.,* 7, 281, 1979.

Sullivan, T. H. and Teramura, A. H., Field study of the interaction between solar ultraviolet B radiation and drought on photosynthesis and growth in soybean, *Plant Physiol.,* 92, 141, 1990.

Tissut, M. and Egger, K., Les glycosides flavoniques foliaires de quelques arbres, au cours du cycle vegetatif, *Phytochemistry,* 11, 631, 1972.

Vogt, T., Gülz, P. G., and Wray, V., Epicuticular 5-o-methyl flavonols from *Cistus laurifolius, Phytochemistry,* 27, 3712, 1988.

Vogt, T. and Gülz, P. G., Isocratic column liquid chromatographic separation of a complex mixture of epicuticular flavonoid aglycones and intracellelar flavonol glycosides from *Cistus laurifolius* L., *J. Chromatogr.,* 537, 453, 1991.

Vogt, T., Gülz, P. G., and Reznik, H., UV radiation dependent flavonoid accumulation of *Cistus laurifolius* L., *Z. Naturforsch.,* 46, 37, 1991.

Waring, R. H., Mcdonald, A. J. S., Larsson, S., Ericsson, T., Wiren, A., Arwidsson, E., Ericsson, A., and Lohammar, T., Differences in chemical composition of plants grown at constant relative growth rates with stable mineral nutrition, *Oecologia,* 66, 157, 1985.

Wellmann, E., UV-B-signal/response-Beziehungen unter natürlichen und artifiziellen Liychtbedingungen, *Ber. Deutsch. Bot. Ges.,* 98, 99, 1985.

Woodhead, S., Environmental and biotic factors affecting the phenolic content of different cultivars of *Sorghum bicolor, J. Chem. Ecol.,* 7, 1035, 1981.

18 Catalytic Transformation of Phenolic Compounds in the Soils

P. M. Huang, M. C. Wang, and M. K. Wang

CONTENTS

18.1 ABSTRACT

Many phenolic compounds have been implicated as allelopathic biochemicals in soil environments. The transformation of phenolic compounds to phenolic polymers can proceed through abiotic and biotic catalytic processes and their interactions. Abiotic catalysts, which include a series of soil inorganic constituents, promote the transformation of phenolic compounds to humic macromolecules through their oxidative polymerization, ring cleavage, decarboxylation, and/or dealkylation. Oxidoreductases, namely, polyphenol oxidases and peroxidases, catalyze the polymerization of phenolic compounds by oxidative coupling during the process of humus formation. Limited research data indicate that abiotic and biotic catalysts differ in their capacity to mediate the polymerization of phenolic compounds. The ability of abiotic and biotic catalysts to catalyze the transformation should vary with their structural configuration and surface chemistry and the structure and functionality of the phenolic compounds concerned. Abiotic catalysts constantly interact with biotic catalysts in the

0-8493-2116-6/99/$0.00+$.50
© 1999 by CRC Press LLC

soil. The formation of enzyme and the activity of immobilized and desorbed enzymes can be affected by abiotic catalysts. The transformation of phenolic compounds to phenolic polymers by interactions of abiotic and biotic catalytic reactions in influencing the allelopathy in terrestrial ecosystems deserves increasing attention. Further, the potentiality of the use of biotic and/or abiotic catalysts in coping with allelopathic problems remains to be uncovered.

18.2   INTRODUCTION

Phenolic compounds play many roles in the environment (Swain, 1978). Many phenolic compounds in soils have been considered to be allelopathic biochemicals (Wang et al, 1967; Chou and Young, 1975; Rice, 1984; Young and Chou, 1985; Waller et al., 1987; Chou, 1995; Einhellig, 1995). Phytotoxic phenolic compounds are widely distributed in plants and soils (Whitehead, 1964; Rice, 1984; Einhellig, 1995). Concentrations of some free soil phenolic compounds have been reported to be in the order of 10^{-5} M depending on soil type, vegetation, and seasonal variation (Whitehead, 1964). Free phenolic compounds may accumulate in rhizosphere soils and reach levels sufficient to cause phytotoxicity (Young et al., 1989). Whitehead et al. (1983) found that more than 50 percent of water-soluble phenolic compounds are in the bound form. Phytotoxicities of many phenolic compounds in allelopathy greatly depend on whether they are free or bound forms. Free phenolic compounds evidently play a phytotoxic role in allelopathy in soil environments. Therefore, the transformation and fate of these phenolic compounds should profoundly influence their phytotoxic role in allelopathy.

Phenolic compounds are important precursors of soil humus (Martin and Haider, 1971). The polymer formation from phenols appears to proceed through auto-oxidation to some extent. Phenols, at neutral or alkaline pH, react with atmospheric oxygen to form reactive quinones and/or radicals that can polymerize with other phenols. The oxidative coupling (polymerization) plays a central role in the formation of phenolic polymers from phenols. The oxidative polymerization of phenol can be substantially accelerated by enzymatic (Mayaudon et al., 1973; Sulflita and Bollag, 1981; Bollag, 1992) and abiotic (Scheffer and Ulrich, 1960; Wang et al., 1978a; Shindo and Huang, 1982; Huang, 1990) processes.

Besides polymerization of the same monomers, polycondensation of phenols with a variety of organic components such as nitrogen- and sulfur-bearing substances (Wang and Huang, 1987, 1991; Wang, 1995) and amino acids with carbohydrates (Maillard, 1916) results in the formation of humic-like macromolecules. Relatively little is known about the incorporation of protein and saccharide (Mayaudon, 1968; Liu et al., 1985) and aliphatic (Dinel et al., 1990) moieties into humus. Polycondensation could prevail over simple polymerization in the transformation of phenolic compounds to humus (Kononova, 1966).

Although both biotic and abiotic catalysts can oxidatively polymerize phenolic compounds, these agents may differ in their capacity and kinetics in mediating oxidative coupling (Shindo and Huang, 1992; Pal et al., 1994). From a practical point of view, a catalyst is a substance that changes the rate of a reaction, regardless of the fate of the catalyst itself (Moore and Pearson, 1988). Biotic and abiotic catalysts may interact with each other to modify the transformation kinetics and mechanisms of phenolic compounds. Further, the influence of environmental factors on the catalytic transformation merits close attention in understanding the dynamics of phenolic compounds.

The objective of this chapter is to integrate the existing information, especially the more recent findings on abiotic and biotic catalytic reactions in the transformation of phenolic compounds and the comparison and interactions of these two groups of catalytic reactions. This information is essential for understanding the fate of phenolic compounds and the impact on allelopathy in soil environments.

18.3 ABIOTIC CATALYSTS IN THE TRANSFORMATION OF PHENOLIC COMPOUNDS

18.3.1 PRIMARY MINERALS

Polyphenols are common in the decomposition products of plant and animal materials and microbial metabolites (Henderson, 1955; Martin and Haider, 1980b; Wang et al., 1986). These phenolics are regarded as important precursors to form humic substances in soils (Flaig et al., 1975; Hayes and Swift, 1978). Hydroquinone is a phenolic compound and a well-defined precursor for the formation of humic substances.

Primary minerals differ in their ability to catalyze the abiotic polymerization of hydroquinone, which is used as a model phenolic compound (Shindo and Huang, 1985a). The sequence of the catalytic power of the primary minerals is tephroite > actinolite > hornblende > fayalite > augite > biotite > muscovite ≅ albite ≅ orthoclase ≅ microcline ≅ quartz. Tephroite is a Mn-bearing olivine. Within a seven-day reaction period, the conversion of hydroquinone to humic acid (HA) by tephroite reaches 21 percent, whereas in the systems of the other primary minerals, the yields of HA are much smaller (actinolite, 2.2 percent; hornblende, 1.1 percent; augite and biotite, trace; and microcline and quartz, not detectable).

Except for the tephroite, the reaction products of abiotic polymerization of hydroquinone catalyzed by primary minerals are largely low-molecular-weight phenolic polymers (Shindo and Huang, 1985a). The infrared (IR) spectrum of the hydroquinone polymers (Shindo and Huang, 1985a) is similar to that of humic substances as reported by Schnitzer and Khan (1972) and Schnitzer (1978). The spectrum indicates great molecular complexity of the humic polymers. The scanning electron micrographs of the hydroquinone polymers formed in the presence of tephroite show various surface features (Shindo and Huang, 1985a). The smallest discrete particles detected are spheroids with diameters of 0.1 to 0.2 μm. The reaction products also show the presence of aggregates consisting of a number of individual spheroids. In the case of the small aggregates, they appear like moss. In contrast, the large aggregates indicate the nodule-like shapes with diameters of 1 to 5 μm and doughnut-like shapes with diameters of 6 to 8 μm. The surface morphology of these humic polymers is similar to that of humic acids (HAs) and fulvic acids (FAs) of soil reported by Stevenson and Schnitzer (1982).

18.3.2 LAYER SILICATES

The clay-size layer silicates have the ability to catalyze the oxidative polymerization of phenolic compounds and the subsequent formation of humic substances. Kumada and Kato (1970) were the pioneers in the study of browning of pyrogallol as affected by clay-size layer silicates. Wang et al. (1971) found the polymerization of ^{14}C-labeled p-coumaric and ferulic acids to HAs in soils. Wang and his co-workers reported that layer silicates can catalyze the abiotic formation of model humic substances through the oxidative polymerization of many phenolic compounds common in soils, plants, and microbial metabolites (Wang and Li, 1977; Wang et al., 1978a,b, 1980). The formation of aromatic radical cations can be catalyzed by intracrystal surfaces of transition metal saturated layer silicates (Pinnavia et al., 1974). Mortland and Halloran (1976) reported that aromatic molecules may donate electron to the metal ions such as Cu(II) or Fe(III) on the cation exchange complex of smectite clay. Filip et al. (1977) stated that certain layer silicates, such as muscovite, can catalyze the oxidation of some phenolic substances important in humification.

During the 1980s and mid-1990, Huang and his co-workers investigated the sequence of catalytic power of layer silicates and their reaction sites in the polymerization of phenolic compounds and the subsequent formation of humic substances. The promoting effect of 2:1 layer silicates is higher than that of 1:1 layer silicates (Shindo and Huang, 1985b). The oxygen molecules or radicals adsorbed on the surface of smectite (Thompson and Moll, 1973) may partially contribute to oxidation of hydroquinone by smectites. The larger the specific surface and lattice imperfection of the layer silicates, the easier the adsorption of oxygen molecules or radicals. The specific surface of 2:1 layer silicates is larger than that of 1:1 layer silicates (Jackson, 1979) and the possibility of lattice imperfections is higher in the former (Thomas and Thomas, 1967). This explains the differences in the catalytic power of 1:1 layer silicates and 2:1 layer silicates. The edges of kaolinite are virtually the only catalytic sites for the formation of hydroquinone-derived humic macromolecules. The edges of nontronite have a very important role as catalytic sites in the formation of hydroquinone-derived polymers (Wang and Huang, 1988). The internal and external planar surfaces of nontronite are also substantially involved in promoting the catalytic reaction. The polymerization of hydroquinone is greatly enhanced by the synergistic effect of Ca-nontronite and oxygen adsorbed on the silicate surface. Further, nontronite is an $Fe(III)$-bearing smectite. The $Fe(III)$ in the octahedral sheet of nontronite apparently serves as a Lewis acid site to accept electron from phenolic compounds to catalyze their oxidative polymerization. One of the well identified precursors for the formation of humic substances, hydroquinone (Flaig et al., 1975) can be transformed to humic macromolecules at pH 6.5 and deposited in the interlayers of nontronite saturated with Ca, which is the predominant exchangeable cation in soils (Wang and Huang, 1986). Most of the macromolecules deposited in the interlayers of nontronite are humin-type materials, because they are highly resistant to alkali extraction. Aging effects on the stability of the interlayer humic macromolecules in soil environments remain obscure.

Nontronite also has the ability to cleave the ring of pyrogallol, catechol, and hydroquinone (Wang and Huang, 1994). These polyphenols are very common in soils (Flaig et al., 1975; Martin and Haider, 1980a). They are often selected as precursors to synthesize humic-like polymers (Shindo and Huang, 1982, 1984a,b, 1985a,b; Wang et al., 1986; Wang and Huang, 1986). The amount of CO_2 released from the pyrogallol-, catechol-, and hydroquinone-nontronite systems are 4.9, 2.6, and 2.3 times higher, respectively, than that from the corresponding polyphenol system in the absence of nontronite (Table 18.1). The sequence of the amounts of CO_2 released from the reaction system was pyrogallol-nontronite $>>$ catechol-nontronite $>$ pyrogallol $>$ hydroquinone-nontronite $>$ catechol $>$ hydroquinone. The extent of ring cleavage of the polyphenols to release CO_2 varies with their structure and nontronite greatly promotes their ring cleavages.

The absorbances at 400 and 600 nm (Kumada, 1955), 465 and 665 nm (Kononova, 1966; Schnitzer and Khan, 1972), or at 472 and 664 nm (Welte, 1956) of the solutions of humic materials are frequently used as indices of the extent of humification as well as molecular weight status (Hayes and Himes, 1986) of soil organic matter. The presence of nontronite in the solution of the polyphe-

TABLE 18.1
Release of Carbon Dioxide in the Nontronite-Polyphenol Systems at the End of a 90-h Reaction Period

Reaction Condition		
Nontronite	Polyphenol	CO_2 Release (μmol)[a]
+	Pyrogallol	263
−	Pyrogallol	54
+	Catechol	88
−	Catechol	34
+	Hydroquinone	49
−	Hydroquinone	21

Note: + indicates in the presence; − indicates in the absence.

[a] The amounts of CO_2 released in the systems that contained one gram of Ca-nontronite (0.2 to 2 μm), 5 mmol of pyrogallol, catechol, or hydroquinone in 30 ml of aqueous solution adjusted to pH 6.0.

Source: From Wang and Huang, 1994. With permission.

nols greatly enhances absorbance of the supernatants at both 472 and 664 nm. The sequence for the absorbances of the supernatants at both 472 and 664 nm in the system is: pyrogallol-nontronite >> catechol-nontronite >> pyrogallol $\geq$ hydroquinone-nontronite > hydroquinone > catechol. The sequence for the yields of the total humic polymers formed in the systems is basically the same as that for absorbance of the supernatants at both 472 and 664 nm. The IR spectra of the FA formed in the reaction systems show that the sequence for the intensity of the IR absorption at ~ 1715 cm^{-1} is: pyrogallol-nontronite > catechol-nontronite > hydroquinone-nontronite (Figure 18.1). The ~ 1715 cm^{-1} band is mainly due to carboxyl groups as confirmed by the titration data. The solid state CPMAS ^{13}C NMR data (not shown) is in accord with the IR evidence and the data on the CO$_2$ release, revealing the catalytic power of nontronite in the ring cleavage of pyrogallol and the associated formation of aliphatic fragments. Therefore, part of the reaction process may proceed as shown below:

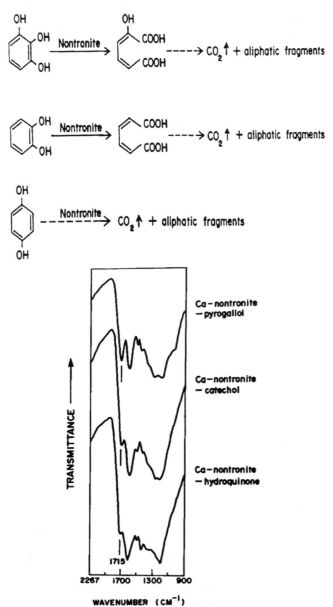

FIGURE 18.1　Infrared spectra recorded from 900 to 2267 cm^1 of the FA (MW >1000) formed in the Ca-nontronite-polyphenol systems (Wang and Huang, 1994).

Catechol, which has two hydroxyls in *ortho* positions is evidently more easily cleaved than hydro-quinone, which has two hydroxyls in *para* positions. Furthermore, pyrogallol, which has three hydroxyls in consecutive positions is even more easily cleaved than catechol. The resultant carboxyl group-containing intermediates are further oxidized to form CO_2 and aliphatic fragments. In the reaction systems, intermediate products and aliphatic fragments may form polycondensates which in turn enhance the absorbance of the supernatants at 472 and 664 nm, and increase the yields of humic polymers. The structure and functionality of polyphenols, thus, have an important role in determining the extent of the ring cleavage of polyphenols, release of CO_2, formation of aliphatic fragments, contents of carboxyl group, and yields of humic polymers formed by catalysis of nontronite. Nontronite is one of the smectites that are among the most common and important clay minerals in soils and sediments (Borchardt, 1989). Its structural Fe(III) is regarded as Lewis acid (an electron acceptor) in oxidation-reduction reactions (Solomon, 1968; Theng, 1971, 1974). Pyrogallol, catechol, and hydroquinone are very common polyphenols in soils (Flaig et al., 1975; Martin and Haider, 1980b). Therefore, the data obtained implicate the significance of nontronite in catalyzing the transformation of polyphenols with different structure and functionality in soil environments.

Calcium-saturated illite, which is a 2:1 layer silicate, catalyzes the formation of N-containing HAs in phenolic compounds–amino acids systems at neutral pH. The yields and nitrogen contents of the synthesized HAs vary with the kind of amino acids (Wang et al., 1985). The formation of free radicals from phenolic compounds by catalysis of Ca-illite and the subsequent reaction of the free radicals with amino acids appears to be the primary mechanism for the formation of N-containing humic polycondensates.

In view of the complexity of structural and surface chemistry of a series of layer silicates and hydroxy-interlayered minerals (Barnhisel and Bertsch, 1989; Chiang et al., 1994), their ability to catalyze the polycondensation of phenolic compounds with amino acids and peptides merits in-depth research.

18.3.3 METAL OXIDES, HYDROXIDES, AND OXYHYDROXIDES

Catalytic power of Mn(IV), Fe(III), and Al oxides in the abiotic formation of humic polymers is greatly influenced by the nature of the oxides. Scheffer et al. (1959) reported that hydroquinone can be oxidized and polymerized in the presence of iron oxides and oxyhydroxides at pH 3 to 7. The catalytic power of the iron oxides and oxyhydroxides follows the sequence: hydrohematite > maghemite > lepidocrocite > hematite. In soil environments, iron oxides are commonly present in the fine colloidal forms. These oxides may form surface coatings on layer silicates and humic substances and thus catalyze the oxidative transformation of phenolic compounds in the soil. Shindo and Huang (1984b) reported that the importance of hydrous oxides of Fe in catalyzing the oxidative polymerization of phenolic compounds varies with the structure and functionality of phenolic compounds.

Hydrous oxides of Al may catalyze the oxidative polymerization of phenolic compounds (Wang et al., 1983). The yields of humic substances formed from the catechol and pyrogallol are significantly greater in the presence of Al oxides than in its absence. Al^{3+} complexes with phenolic compounds and semiquinone radicals (Eaton, 1964). By displacing protons from the phenolic groups, Al^{3+} appears to promote delocalization of electrons from phenolic oxygens into the π orbital system, which may then be more susceptible to oxidation. Displacement of the highly electronegative proton allows electron density to be delocalized from oxygen atoms into the aromatic ring. In addition to promoting oxidation at low pH, Al^{3+} facilitates the formation of charge-transfer complexes. These complexes can be stabilized by H-bonding and/or coulombic attraction. Aluminum has been reported to increase radical content in melanins (Felix et al., 1978). This effect is also attributed to metal complexation in promoting the pairing of oxidized molecules and creating radical characters by charge-transfer processes. The ability of various Al hydroxides and their surface-coating on soil particles in promoting oxidative polymerization of phenolic compounds remains obscure.

Manganese oxides (birnessite, cryptomelane, and pyrolusite) are commonly present in soils (McKenzie, 1989). They are very reactive in promoting the polymerization of phenolic compounds and the subsequent formation of HAs (Shindo and Huang, 1982). Shindo and Huang (1984b) reported that the sequence for promoting the polymerization of polyphenols and the subsequent formation of HA is Mn oxide $\gg$ Fe oxide $>$ Al oxide $>$ Si oxide. Manganese oxides are thus by far the most powerful catalysts among metal oxides, hydroxides, and oxyhydroxides in catalyzing the transformation of phenolic compounds. Manganese oxides act as Lewis acids, which accept electrons form diphenols, leading to their oxidative polymerization and the subsequent formation of humic macromolecules. The rate determining steps in the formation of HAs from phenolic compounds by oxidative polymerization seems to be the formation of a semiquinone radical (Schnitzer, 1982). Semiquinones will couple with each other to form stable HA macromolecules. In contrast to electron transfer reactions, the coupling of radicals requires little heat of activation (Chang and Allan, 1971). Coupling of semiquinones rather than the formation of quinones should, thus, be kinetically the preferred reaction path.

Manganese oxides (birnessite) also greatly promotes the abiotic formation of NH_3-N and nitrogeneous polymers in hydroquinone-glycine systems in the common pH range of soils (Shindo and Huang, 1984a). Wang and Huang (1987) reported that the deamination of glycine from the birnessite-glycine-pyrogallol systems at initial pHs of 7.00 and 5.00 are 95.5 and 68.9 times, respectively, higher than that from the glycine-pyrogallol systems at the corresponding pHs of 7.00 and 5.00 (Table 18.2). Appreciable amount of NH_3 is released from the deamination of glycine by the catalysis of birnessite. The formation of N-containing humic polymers in the glycine-pyrogallol systems at pH values of 7.00 and 5.00 is also greatly increased by the presence of birnessite. The data clearly indicate that the polycondensation of glycine and pyrogallol is strongly catalyzed by birnessite. Birnessite is poorly crystalline and thus should have extensive exposed edge surfaces. These edges contain manganese of high oxidation numbers, for example, Mn(III) and Mn(IV) (McKenzie, 1989), which can adsorb and polarize oxygen molecules, leading to catalysis of the ring cleavage of pyrogallol, the formation of aliphatic fragments, and the decarboxylation and dealkylation of glycine as shown below:

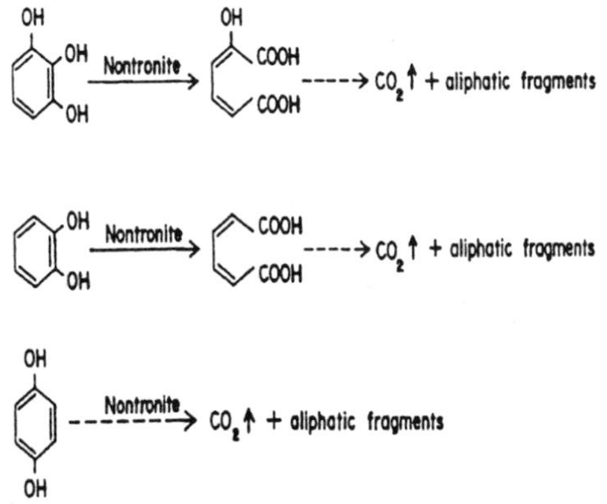

TABLE 18.2

Distribution of the Converted Glycine-N to NH$_4^+$, NH$_3$, and N-Polymers in the Birnessite–Glycine–Pyrogallol Systems at the End of 90-h Reaction Period

Reaction Condition[a]							N in Humic Polymers (μmol)				Total Converted Glycine-N
									Humic Polymer Adsorbed on		
Birnessite	Glycine	Pyrogallol	Initial pH	NH$_4$	NH$_3$	Sum	HA	FA[b]	Birnessite	Sum	
+	+	+	7.0	823	17	840	19	75	2	96	936
+	+	+	5.0	249	6	255	9	31	5	45	300
+	+	−	7.0	68	4	72	NA	NA	NA	NA	72
+	+	−	5.0	54	ND	54	NA	NA	NA	NA	54
−	+	+	7.0	8	1	9	7	26	NA	33	42
−	+	+	5.0	4	ND	4	2	7	NA	9	13
−	+	−	7.0	ND	ND	ND	ND	ND	ND	ND	ND
−	+	−	5.0	ND	ND	ND	ND	ND	ND	ND	ND

[a]One hundred mg of birnessite was suspended in 30 ml of aqueous solution which contained 1 mmol of glycine and 0.5 mmol of pyrogallol.

[b]Including N in FA (MW > 1000) and FA (MW < 1000).

Note: NA, not applicable; ND, not detectable.

Source: From Wang and Huang, 1987. With permission.

The ESR data of humic macromolecules derived from pyrogallol (Table 18.3) and the ESR evidence of the presence of the Mn(II) in the supernatants of the reaction systems (Wang, 1987) indicate that Mn(IV) is reduced to Mn(II) in the oxidation of pyrogallol to form semiquinone free radicals (Wang and Huang, 1987). The semiquinone could react with glycine; these free radicals could then react with the aliphatic fragments derived from the ring cleavage of pyrogallol and decarboxylation and dealkylation of glycine. These reactions apparently have contributed to the polycondensation of glycine and pyrogallol by the catalysis of birnessite.

The significance of Mn in the oxidation of phenolic compounds is also illustrated in the strong catalytic power of tephroite, a Mn-bearing olivine (Shindo and Huang, 1985a). The role of Mn oxides in oxidative polymerization of phenolic compounds through reduction of Mn(IV) of Mn oxides to Mn(II) (Shindo and Huang, 1982, 1984a; Wang and Huang, 1987) is supported by the findings of Stone and Morgan (1984), Stone (1987), and Lehmann and Cheng (1988). Redox reactions are one of the catalytic mechanisms (Moore and Pearson, 1988). Dissolved Mn(II) in the amino acid-pyrogallol-birnessite systems, which apparently can be adsorbed on the surface of birnessite (Wang and Lin, 1993), should be subject to oxidation by O$_2$ in the atmosphere (Coughlin and Matsui, 1976). Therefore, O$_2$ molecules present in the atmosphere play an important role in enhancing the catalytic reactions.

18.3.4 POORLY CRYSTALLINE ALUMINOSILICATES

The ability of poorly crystalline aluminosilicates such as allophane to catalyze the polymerization of polyphenols has been shown (Kyuma and Kawaguchi, 1964). Wang et al. (1983) reported that the infrared spectra of humified polyphenols catalyzed by silicoalumina resemble those of natural humic substances. There are a series of other poorly crystalline aluminosilicates and solution and adsorbed hydroxyaluminosilicates in the soils (Huang, 1991). These include imogolite, proto-imogolite allophane, proto-imogolite, and hydroxyaluminosilicates intercalated in smectites and vermiculites. Their catalytic effects on the humification of phenolic compounds remain to be uncovered.

TABLE 18.3
ESR Spectroscopic Properties of Humic Polymers Formed in Birnessite–Glycine–Pyrogallol Systems at an Initial pH of 7.0

Reaction Conditions[a]			HA (MW > 1000)		FA (MW > 1000)	
Birnessite	Glycine	Pyrogallol	g-Value	Breadth	g-Value	Breadth
				—G—		—G—
+	+	+	2.0032	3.8	2.0035	3.5
+	−	+	2.0031	4.0	2.0032	3.7
−	+	+	2.0035	3.5	2.0034	3.5
−	−	+	2.0036	3.3	2.0038	3.2

[a]One hundred milligrams of birnessite was suspended in 30 ml of aqueous solution, which contained 1.0 mmol glycine and 0.5 mmol of pyrogallol.

Source: From Wang and Huang, 1987. With permission.

18.4 BIOTIC CATALYSTS IN THE TRANSFORMATION OF PHENOLIC COMPOUNDS

Biotic catalysts (enzymes) are vital in the transformation of phenolic compounds. Phenolic compounds may be polymerized during the process of humus formation by oxidative coupling reactions. This is one of the most important reactions in the binding of phenolic derivatives to humus. In this function, oxidoreductases play a very important role (Sjoblad and Bollag, 1981).

Oxidoreductases are classified as either polyphenol oxidases or peroxidases (Bollag, 1992). Polyphenol oxidases are divided into two subclasses, namely, laccases and tyrosinases. Both enzyme groups require bimolecular oxygen, but no coenzyme, for activity. These enzymes differ in the mechanism by which they oxidize the parent compounds. Tyrosinases oxidize the parent compounds and subsequently release an oxidized, usually highly reactive, o-quinone (Sjoblad and Bollag, 1981). In an alkaline environment, the quinone products slowly polymerize through auto-oxidative processes. In contrast, laccases oxidize phenolic compounds to form their corresponding anionic free radicals. Laccases may prove to be the most useful of the oxidoreductases because they produce these very reactive radicals. Further, unlike peroxidases, they do not require the presence of hydrogen peroxide as stated below.

Peroxidases are produced by plants and microorganisms. They catalyze a wide variety of reactions (Bollag, 1992). All peroxidases contain an iron porphyrin ring and require the presence of peroxides, for example hydrogen peroxide, for activity. In particular, horseradish peroxidase catalyzes the polymerization of a wide range of phenolic compounds.

Phenolic compounds such as catechol, pyrogallol, orcinol, ferulic acid, and syringic acid are deemed to be the most important substrates for the polymerization reactions. Many researchers have reported the formation of humic-like substances by enzymatic catalysis of polymerization of one or more phenol-derived compounds (Ladd and Butler, 1975; Mathur and Schnitzer, 1978; Martin and Haider, 1980b; Dec and Bollag, 1988; Shindo and Huang, 1992). The humic polymers produced are quite similar to natural humic substances in elemental analysis, cation exchange capacity, total acidity, and resistance to microbial degradation in soil.

In the first step of oxidative coupling reactions, susceptible phenolic compounds are oxidized to form unstable free radicals. The free radicals then proceed to react with nearby molecules to form polymers. This reaction leads to the formation of C-C and C-O bonds between phenolic species. The mechanism of coupling requires the removal of a proton and an electron from the hydroxyl group. This reaction results in the formation of free radicals or reactive quinones (Bollag, 1983). These intermediates then couple at positions *ortho* and *para* to the hydroxyl groups to form a dimer. The *meta* position is not reactive in the coupling reactions. Phenolic dimers are further oxidized to form polymers. If the potential of the oxidizing agent is high enough, C-C coupled dimers are oxidized to form extended quinones. Such a reaction pathway was demonstrated with guaiacol (Simmons et al., 1988).

Soils are complex and heterogeneous systems. Therefore, earlier studies focused on studying the polymerization of a single phenolic compound in reactions catalyzed by isolated oxidoreductase. Liu et al. (1981) used a simple model involving one of the components of humus, that is, syringic acid to demonstrate the formation of various oligomers, which range from dimers to hexamers, by a laccase isolated from the fungus *Rhizoctonia praticola*. Simmons et al. (1988) used mass spectroscopy and NMR techniques to identify the products of oligomerization of guaiacol catalyzed by horseradish peroxidase. The polymers of syringic acid and guaiacol are similar to the structure presented by Stevenson (1994), especially with regard to the oxygen bridges and the carbon–carbon linkages between aromatic rings. It has also been demonstrated that a peptide unit similar to that depicted in the structure proposed by Stevenson can be incorporated into a model humic acid (Liu et al., 1985). Compounds incorporated into humus are stabilized, resistant to microbial degradation (Batistic and Mayaudon, 1970; Verma et al., 1975), and may remain intact over hundreds of thousands of years (Bollag et al., 1998).

A variety of phenolic compounds exist in soil environments. Polymerization is, thus, most likely to occur between different types of phenolic molecules. Further, natural phenolic compounds may react with xenobiotics to form coupled hybrid phenolic compounds (Bollag et al., 1980). More research is needed in this area to elucidate the mechanisms of the copolymerization and the nature and toxicity of the copolymerization products of allelopathic phenolic compounds and other biochemicals.

18.5 COMPARISON OF ABIOTIC AND BIOTIC CATALYTIC TRANSFORMATIONS

Although both abiotic and biotic catalysts can accelerate the transformation of phenolic compounds, relatively limited studies have been conducted to compare the catalytic activity of biotic and abiotic catalysts in the transformation of phenolic substances. The goal of this section is to discuss the similarities and differences of the products in abiotic and biotic catalytic transformations of phenolic compounds.

Simmons et al. (1988) investigated the oxidative coupling and polymerization of guaiacol catalyzed by oxidoreductases and abiotic catalysts. They reported that, in the initially formed low-molecular-weight oligomers, the guaiacol-derived product mixture formed by peroxidase catalysis is composed of five dimers (compounds B to F in Figure 18.2) and two trimers (compounds G and H in Figure 18.2) with quinoid and phenolic structural features. The same reaction products are formed in the systems catalyzed by other biotic (tyrosinase and laccases) and abiotic (manganese dioxide and bentonite clay) catalysts. These abiotic and biotic catalysts each cause qualitatively equivalent transformations of guaiacol. However, the products obtained in the clay-mediated oxidation of guaiacol do not include compound B, a dimer. It is likely that this dimer, which is an orange-colored quinone, is adsorbed on the clay surface as indicated by the orange color of the clay residue. Further, some of the reaction products might also not be detectable in the reaction solutions due to their adsorption on the abiotic catalysts. Therefore, to critically compare the nature of the reaction products in abiotic and biotic catalytic reactions, the reaction products adsorbed on the surfaces of abiotic catalysts are of concern.

More recently, research has been conducted to further our understanding of the role of abiotic and biotic catalysts in the transformations of phenolic compounds. The catalytic effects of Mn(IV) oxide and tyrosinase on the oxidative polymerization of diphenols (hydroquinone, catechol, and resorcinol) in the pH range of 4 to 8 were compared by measuring the degree of darkening and the formation of humic acids (Shindo and Huang, 1992). Manganese oxide influences the darkening of

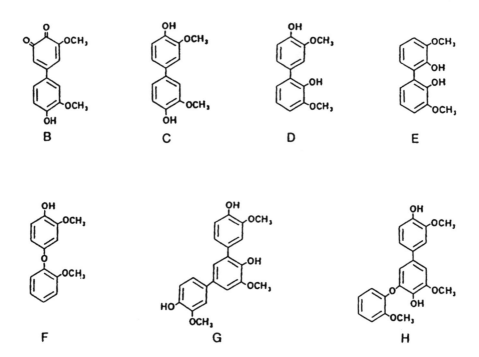

FIGURE 18.2 Structures of the oligomers formed following the one-electron oxidation of guaiacol (Simmons et al., 1988).

hydroquinone and resorcinol to a larger extent than does tyrosinase, while the reverse is true for catechol (Table 18.4). In the tyrosinase system, the darkening of catechol is more efficient compared with hydroquinone and resorcinol. In both the Mn oxide and tyrosinase systems, the degree of darkening of resorcinol is generally much lower than that for the other phenols studied. The yields of humic acids are also significantly influenced by the kind of catalysts and diphenols (Table 18.5). In the Mn oxide system, the yield of humic acid is in the order: hydroquinone > catechol > resorcinol. In the tyrosinase system, catechol produces the highest yield of humic acid, followed by hydroquinone and resorcinol. These findings indicate that the relative catalytic effects of Mn oxides and tyrosinases in promoting the formation of diphenol-derived humic substances would vary with the type of diphenols in natural systems.

Pal et al. (1994) reported that the rate of transformation of 2,6 dimethoxyphenol (2,6 DMP) is considerably higher in reactions catalyzed by enzymes as compared with birnessite-mediated reactions. In subsequent studies, they determined the effect of continuous additions of substrate catechol. During the first eight hours of incubation, birnessite oxidizes catechol, but it is unable to transform the two subsequent catechol additions. In contrast, both laccase and tyrosinase retained the ability to transform with repeated addition of the substrate. The results indicate that both abiotic and biotic catalysts are capable of oxidizing phenolic compounds, but these catalysts differ in their capacity and kinetics in mediating oxidative coupling reactions.

Minerals are much more abundant than enzymes in soil environments. Further, in contrast to enzymes, the catalytic ability of minerals can be sustained at warm temperatures common in subtropical and tropical regions (P. M. Huang, unpublished data, 1997). Therefore, abiotic catalysis of soil minerals in transformation of phenolic compounds and the impact on allelopathy in nature merit increasing attention.

TABLE 18.4

Effects of Mn Oxide and Tyrosinase on the Darkening of Hydroquinone, Resorcinol, and Catechol Solutions at Different pH Values at the End of 24 h

Diphenol[a]	Catalyst[b]	Final pH	Initial pH 4.0 Absorbance[c]		Final pH	Initial pH 6.0 Absorbance		Final pH	Initial pH 7.8 Absorbance	
			400 nm	600 nm		400 nm	600 nm		400 nm	600 nm
Hy	—	4.0	Colorless[d]	Colorless	6.0	Colorless	Colorless	7.1	0.85	0.18
Hy	Ty	4.0	0.08	0.05	6.0	2.06	0.55	6.7	4.20	1.13
Hy	Mn	4.0	0.76	0.23	6.9	6.98	1.58	7.6	8.00	1.74
Re	—	4.0	Colorless	Colorless	6.0	Colorless	Colorless	7.1	Colorless	Colorless
Re	Ty	4.0	0.07	0.04	6.0	0.21	0.09	7.5	0.09	<0.01
Re	Mn	4.0	0.25	0.02	6.9	0.95	0.15	7.9	4.15	0.65
Ca	—	4.0	Colorless	Colorless	6.0	Colorless	Colorless	6.8	0.18	0.04
Ca	Ty	4.0	1.67	0.46	5.9	5.81	1.91	6.3	14.96	4.84
Ca	Mn	4.1	1.06	0.26	6.7	2.01	0.73	7.6	2.32	0.78

[a]Hy, Re, and Ca represent 0.01 M solutions in acetic acid/acetate buffer of hydroquinone (1,4-dihydroxybenzene), resorcinol (1,3-dihydroxybenzene), and catechol (1,2-dihydroxybenzene) of certified reagent grade, respectively.

[b]Ty and Mn represent tyrosinase and Mn (IV) oxide (birnessite), respectively. Ten grams of each catalyst was used in this experiment.

[c]The absorbances of supernatant in the system. The absorbances at 600 nm in the tyrosinase systems without the diphenols were 0.04, 0.09, and 0.11 at the initial pH of 4.0, 6.0, and 7.8, respectively; while the absorbances at 400 nm were 0.23, 0.44, and 0.54. These values were substracted in tyrosinase system in the presence of diphenols. In the Mn oxide system without the diphenols, the supernatants were colorless under the conditions studies.

[d]The supernatant was visibly colorless and its absorbance was < 0.01.

Source: From Shindo and Huang, 1992. With permission.

TABLE 18.5

Effects of Mn Oxide and Tyrosinase on the Synthesis of Diphenol-Derived Humic Acid at the Initial pH of 6.0 at the End of 24 h

Diphenol[a]	Catalyst[b]	Humic acid[c] (mg carbon)
Hy	—	ND
Hy	Ty	2.52
Hy	Mn	3.02
Re	—	ND
Re	Ty	0.11
Re	Mn	0.30
Ca	—	ND
Ca	Ty	11.85
Ca	Mn	1.00

[a]Hy, Re, and Ca represent hydroquinone, resorcinol, and catechol, respectively, at a concentration of 0.01 mol/L in acetic acid/acetate buffer solution.

[b]Ty and Mn represent tyrosinase and Mn (IV) oxide (birnessite), respectively. Ten grams of each catalyst was used in this experiment.

[c]The amount of carbon in the tyrosinase system without the diphenols was 0.43 mg and this value was subtracted in the tyrosinase system in the presence of diphenols. In the Mn oxide system without the diphenols, carbon was not detectable. ND, not detectable.

Source: From Shindo and Huang, 1992. With permission.

18.6 INTERACTIONS OF ABIOTIC AND BIOTIC CATALYTIC REACTIONS IN THE TRANSFORMATION OF PHENOLIC COMPOUNDS

Abiotic and biotic catalysts constantly interact with each other in soil environments. Abiotic catalysts are inorganic soil components that include metal oxides, layer silicates, poorly crystalline aluminosilicates, and certain primary minerals (Wang et al., 1986; Huang, 1990). These soil components can influence microbial activity (Stotzky, 1986) and enzymatic activity (Burns, 1986). Further, these inorganic components also affect microbial formation of enzymes such as phenoloxidases (Filip and Claus, 1995).

Clay additions to the culture of soil fungi and actinomycetes markedly accelerate growth, especially in well-aerated cultures (Haider et al., 1970; Filip et al., 1972; Martin et al., 1976). The influence of clays and other solids (1 percent wt/vol) in batch cultures of selected microorganisms is shown in Table 18.6. The addition of kaolinite results in a slight decrease of the total protein production. However, the enzyme activities are sometimes enhanced, especially the laccase from *P. versicolor,* which increases up to 50 percent. The presence of different bentonites and of a bentonite-humus complex in the culture media results in a strong reduction of phenoloxidase activity. For example, in the *Streptomyces michiganesis* cultures, protein contents are reduced by 20 to 34 percent and tyrosinase activity ceases almost completely. The bentonite-humus complex also inhibits microbial growth and enzyme activity of the laccases from *Pleuroptus ostreatus* and *Polyporus versicolor*. Quartz sand has no significant effect on the behavior of *S. michiganesis* and *P. ostreatus*, but it enhances the production of laccase in *P. versicolor*. Porous glass has little effect on the microbial growth and the activity of laccase. In contrast, porous glass substantially enhances the tyrosinase activity of *S. michiganesis*. With increasing concentrations of bentonite from 0.01 to 0.5 percent (wt/vol), the microbial growth (protein content) remains almost unaffected in cultures of *Streptomyces eurythermus*, but there is a continuous decrease in both the bulk and specific activity of tyrosinase.

TABLE 18.6
Production of Protein and Activity of Phenoloxidases in the Presence of Solids (as % of control)

Subject	K1	K2	B1	B2	B3	B1+HA	QS	PG
				Solids[a]				
Streptomyces michiganensis								
Protein content	85	88	66	69	80	76	92	73
Tyrosinase activity	71	102	0.3	0.6	2.2	0.6	94	149
Pleurotus ostreatus								
Protein content	87	79	75	82	ND	62	116	93
Laccase activity	102	78	34	42	ND	56	81	91
Polyporus versicolor								
Protein content	96	96	78	77	97	60	119	100
Laccase activity	147	109	6.0	10	54	6.0	131	83

[a]K1: Kaolinite (Merck, Germany), K2: Kaolinite (KGa-1, Ga, clay Minerals Society, USA), B1: Bentonite (Sigma, Germany), B2: Bentonite (Südchemie, Germany), B3: Bentonite (similar to B2 but minimum contents of montmorillonite 60 percent), HA: Humic acid (commercial product, Roth, Germany), QS: Quartz sand (from a groundwater aquifer near Hamburg, Germany), PG: Porous Glass (Schott, Germany).

Source: From Claus and Filip, 1990. With permission.

Homoionic kaolinites (1 percent wt/vol) have different effects on the growth of microorganisms and the activity of enzymes (Table 18.7). The H^+-, Na^+-, Ca^{2+}- and Cu^{2+}-saturated clays inhibit the growth of *S. michiganesis* by approximately 20 percent and the Al^{3+}-saturated kaolinite inhibits the growth by 30 percent. The activity of tyrosinase also decreases substantially by the presence of these clays. In contrast, the growth of *P. versicolor* is less affected and the production of laccase is enhanced. In sand cultures of *P. ostreatus*, the growth of the indigenous microorganisms appears to completely inhibit the production of laccase (Claus and Filip, 1990). Leonowicz and Bollag (1987) also made similar observations for laccase producers in soil cultures.

Gianfreda and Bollag (1994) investigated the behavior of a laccase (from the fungus *Trametes versicolor*) and a peroxidase (from horseradish [*Armoracia rustricana* P. Gaertner, Meyer & Scherb.]) in the presence of a montmorillonite, a kaolinite, or a silt loam soil. The various supports show different enzyme immobilization capabilities (Table 18.8). Montmorillonite has the highest binding capacity, immobilizing 71 and 43 percent of the laccase and peroxidase, respectively, added to the immobilization mixtures. The smallest amounts of laccase and peroxidase are immobilized on glass beads. There is considerable variation in the retained activities of the two enzymes on the four supports, as well as the same enzymes on the four supports. The residual specific activities (calculated as percentages of the specific activity of the free enzyme) of laccase and peroxidase immobilized on all supports are high. Further, laccase immobilized on montmorillonite shows specific activities higher than that of the free enzyme, indicating that some sort of reactions take place during the immobilization process. The data also indicate that immobilized enzymes in the soil environment have catalytic activity, but their performance is affected by soil constituents.

Many soil abiotic catalysts not only influence the performance of immobilized enzymes, but also the activity of the desorbed enzymes. The specific activity of desorbed phenoloxidases is distinctly decreased by soil abiotic catalysts as compared with the controls (Claus and Filip, 1988). The ability of abiotic soil catalysts in influencing the activity of immobilized enzymes and the desorbed enzymes should vary with their structural configuration and surface reactivity and the nature of the enzymes. More research is needed in this exciting area of science. The findings expected in

---

TABLE 18.7

Production of Protein and Activity of Phenoloxidases in the Presence of Homoionic Kaolinite (as % of control)

Subject	Kaolinite Homoionic to				
	H^+	Na^+	Ca^{2+}	Cu^{2+}	Al^{3+}
Streptomyces michiganensis					
Protein content	79	78	80	82	68
Tyrosinase activity	32	71	76	78	60
Streptomyces eurythermus					
Protein content	86	92	93	91	70
Laccase activity	69	109	68	161	72
Pleurotus ostreatus					
Protein content	74	88	82	158	82
Laccase activity	112	119	80	93	91
Polyporus versicolor					
Protein content	71	86	97	100	81
Laccase activity	130	128	122	155	133

Source: From Claus and Filip, 1990. With permission.

TABLE 18.8

Immobilization of a Laccase (from *Trametes versicolor*) and a Peroxide (from horseradish) on Different Supports

Enzyme and Support	Protein Adsorbed[a] (mg/%)	Enzymatic Activity		
		Units Adsorbed[b]	Specific Activity[c]	Residual Specific Activity[d](%)
Laccase				
Glass beads	0.452/56	28.8	63.7	236
Montmorillonite	0.622/71	19.8	31.8	118
Kaolinite	0.566/64	13.1	23.1	85.5
Soil	0.644/73	15.7	24.4	90.4
Peroxidase				
Glass beads	0.092/17	8.4	91.6	93.8
Montmorillonite	0.224/43	23	102.8	105.2
Kaolinite	0.120/23	9.5	78.9	80.7
Soil	0.162/31	15	92.6	94.8

[a]Difference between proteins initially added to 200 mg of support (0.88 mg laccase and 0.52 mg of peroxidase) and those recovered in the supernatant and washings.

[b]Expressed as μmol O_2 consumed min^{-1} for laccase and μmol guaiacol transformed min^{-1} for peroxidase.

[c]Units adsorbed/protein adsorbed.

[d]Calculated as percentage of the specific activity (sa) of the free enzyme (laccase, sa = 27 μmol^{-1} min^{-1} mg^{-1}; peroxidase, sa = 97.7 μmol min^{-1} mg^{-1}).

Source: From Gianfreda and Bollag, 1994. With permission.

this area of research would lead to further understanding of catalytic transformation of phenolic compounds in the soil and the possible use of phenoloxidases, peroxidases, and/or abiotic catalysts to transform phenolic compounds in soil environments.

18.7 SUMMARY AND CONCLUSIONS

Phenolic compounds are of allelopathic concern in soil environments. The transformation of phenolic compounds can proceed through abiotic and biotic catalytic processes and their interactions. These reactions lead to the formation of phenolic polymers from phenolic compounds and should, thus, be of concern in coping with allelopathic problems in terrestrial ecosystems.

Abiotic catalysts include primary minerals, layer silicates, metal oxides, hydroxides, oxyhydroxides, and poorly crystalline aluminosilicates. They promote the formation of humic macromolecules from phenolic compounds especially in the presence of amino acids through oxidative polymerization, ring cleavage, decarboxylation, and/or dealkylation. The ability of inorganic soil constituents to catalyze the transformation substantially varies with their structural configuration and surface chemistry and the structure and functionality of the biochemicals involved.

Oxidoreductases are vital biotic catalysts in the transformation of phenolic compounds. These enzymes promote the polymerization of phenolic compounds by oxidative coupling during the process of humus formation. Limited research data indicate that although both oxidoreductases and abiotic catalysts oxidize phenolic compounds, they differ in their capacity and kinetics in mediating oxidative coupling reactions.

Abiotic and biotic catalysts coexist in soil environments. Abiotic catalysts can influence microbial formation of enzymes and enzymatic activity. Further, many soil abiotic catalysts also influence the activity of the desorbed enzyme. The influence of interactions of abiotic and biotic catalysts on the transformation and toxicity of phenolic compounds and the impact on allelopathy and terrestrial ecosystem health is, thus, an issue that will be of intense interest for years to come.

ACKNOWLEDGMENTS

This study was supported by research Grant No. GP 2383-Huang of the Natural Sciences and Engineering Research Council of Canada and Grants No. NSC#85-2621-B002-017, 85-2811-B002-019, 85-2321-B005-041, and 85-2321-B005-127-A12 of the National Science Council of Republic of China, Taiwan.

REFERENCES

Barnhisel, R. I. and Bertsch, P. M., Chlorites and hydroxy-interlayered vermiculite and smectite, in *Minerals in Soil Environments*, Dixon, J. B. and Weed, S. B., Eds., Soil Science Society of America, Madison, WI, 1989, 729.

Batistic, L. and Mayaudon, J., Stabilisation biologique dans le sol de l'acide ferulique ^{14}C, de l'acide vanillique ^{14}C et de l'acide *p*-coumaric ^{14}C, *Ann. Inst. Pasteur,* 188, 199, 1970.

Bollag, J.-M., Cross-coupling of humus constituents and xenobiotic substances, in *Aquatic and Terrestrial Humic Materials,* Christman, F. R. and Gjessing, E. T., Eds., Ann Arbor Press, Ann Arbor, MI, 1983.

Bollag, J.-M., Decontaminating soil with enzymes, *Environ. Sci. Technol.,* 26, 1876, 1992.

Bollag, J.-M., Dec, J., and Huang, P. M., Formation mechanisms of complex organic structures in soil habitats, *Adv. Agron.,* 63, 237, 1998.

Bollag, J.-M., Liu, S.-Y., and Minard, R. D., Cross-coupling of phenolic humus constituents and 2,4-dichlorophenol, *Soil Sci. Soc. Am. J.,* 44, 52, 1980.

Borchardt, G. A., Montmorillonite and other smectite minerals, in *Minerals in Soil Environments,* Dixon, J. B. and Weed, S. B., Eds., Soil Science Society of America, Madison, WI, 1989, 675.

Burns, R. G., Interactions of enzymes with soil mineral and organic colloids, in *Interactions of soil Minerals with Natural Organics and Microbes*, Huang, P. M. and Schnitzer, M., Eds., Soil Science Society of America, Madison, WI, 1986, 429.

Chang, H. M. and Allan, G. G., Oxidation, in *Lignins*, Sarkanen, K. V. and Ludwig, C. H., Eds., Wiley Interscience, New York, 1971, 433.

Chiang, H. C., Cheng, Y. W., Yang, J. H., Horng, F. W., and Wang, M. K., Iron oxides in placic horizons of alpine forest soils, *J. Chin. Agri. Chem. Soc.,* 32, 666, 1994.

Chou, C.-H., Allelopathy and sustainable agriculture, in *Allelopathy, Processes, and Applications*, Inderjit, Dakshini, K. M. M., and Einhellig, F. A., Eds., ACS Symposium Series 582, American Chemical Society, Washington, D.C., 1995, 211.

Chou, C. H. and Young, C. C., Phytotoxic substances in twelve subtropical grasses, *J. Chem. Ecol.,* 1, 183, 1975.

Claus, H. and Filip, Z. K., Behavior of phenoloxidases in the presence of clays and other soil-related adsorbents, *Appl. Microbiol. Biotechnol.,* 28, 506, 1988.

Claus, H. and Filip, Z. K., Effects of clays and other solids on the activity of phenoloxidases produced by some fungi and actinomycetes, *Soil Biol. Biochem.,* 22, 483, 1990.

Coughlin, R. W. and Matsui, I., Catalytic oxidation of aqueous Mn(II), *J. Catalysis,* 41, 108, 1976.

Dec, J. and Bollag, J.-M., Microbial release and degradation of catechol and chlorophenols bound to synthetic humic acid, *Soil Sci. Soc. Am. J.,* 52, 1366, 1988.

Dinel, H., Schnitzer, M., and Mehuys, G. R., Soil lipids: origin, nature, content, decomposition, and effect on soil physical properties, in *Soil Biochemistry,* Vol. 6, Bollag, J.-M. and Stotzky, G., Eds., Marcel Dekker, New York, 1990, 397.

Eaton, D. G., Complexing of metal ions with semiquinones. An electron spin resonance study, *Inorg. Chem.,* 3, 1268, 1964.

Einhellig, F. A., Allelopathy: current status and future goals, in *Allelopathy: Organisms, Processes, and Applications*, Inderjit, Dakshini, K. M. M., and Einhellig, F. A., Eds., ACS Symposium Series 582, American Chemical Society, Washington, D.C., 1995, 1.

Felix, C. C., Hyde, J. S., Sarna, T., and Sealy, R. C., Interactions of melanins with metal ions. Electron spin resonance evidence for chelate samples of metal ions with free radicals, *J. Am. Chem. Soc.*, 100, 3922, 1978.

Filip, Z. K. and Claus, H., Effect of soil minerals on the microbial formation of enzymes and their possible use in remediation of chemically polluted sites, in *Environmental Impact of Soil Component Interactions*, Vol. I, *Natural and Anthropogenic Organics*, Huang, P. M., Berthelin, J., Bollag, J.-M., McGill, W. B., and Page, A. L., Eds., CRC Press/Lewis, Boca Raton, FL, 1995, 409.

Filip, Z. K., Flaig, W., and Rietz, E., Oxidation of some phenolic substances as influenced by clay minerals, in *Isot. Radiat. Soil Org. Matter Studies*, II. International Atomic Energy Agency Bulletin, IAEA, Vienna, 1977, 91.

Filip, Z. K., Haider, K., and Martin, J. P., Influence of clay minerals on growth and metabolic activity of *Epicoccum nigrum* and *Stachybotrys chartarum, Soil Biol. Biochem.*, 4, 135, 1972.

Flaig, W., Beutelspacher, H., and Rietz, E., Chemical composition and physical properties of humic substances, in *Soil Components*, Vol. 1, *Organic Components*, Gieseking, J. E., Ed., Springer-Verlag, New York, 1975, 1.

Gianfreda, L. and Bollag, J.-M., Effect of soils on the behavior of immobilized enzymes, *Soil Sci. Soc. Am. J.*, 58, 1672, 1994.

Haider, K., Filip, Z. K., and Martin, J. P., Einfluss von Montmorillonit auf die Bildung von Biomasse und Stoffwechselprodukten durch einige Mikroorganismen, *Arch. Microbial.*, 73, 201, 1970.

Hayes, M. H. B. and Himes, F. L., Nature and properties of humus-mineral complexes, in *Interactions of Soil Minerals with Natural Organics and Microbes*, Huang, P. M. and Schnitzer, M., Eds., Soil Science Society of America, Madison, WI, 1986, 103.

Hayes, M. H. B. and Swift, R. S., The chemistry of soil organic colloids, in *The Chemistry of Soil Constituents*, Greenland, D. J. and Hayes, M. H. B., Eds., John Wiley & Sons, Chichester, 1978, 179.

Henderson, M. E. K., Release of aromatic compounds from birch and spruce sawdusts during decomposition by white-rot fungi, *Nature* (London), 175, 634, 1955.

Huang, P. M., Role of soil minerals in transformation of natural organics and xenobiotics in soil, in *Soil Biochemistry*, Vol. 6, Bollag, J.-M. and Stotzky, G., Eds., Marcel Dekker, New York, 1990, 29.

Huang, P. M., Ionic factors affecting the formation of short-range ordered aluminosilicates, *Soil Sci. Soc. Am. J.*, 55, 1172, 1991.

Jackson, M. L., *Soil Chemical Analysis—Advanced Course*, published by the author, Department of Soil Science, University of Wisconsin, Madison, WI, 1979.

Kononova, M. M., *Soil Organic Matter*, 2nd ed., Pergamon Press, New York, 1966.

Kumada, K., Absorption spectra of humic acids, *Soil Plant Food*, 1, 29, 1955.

Kumada, K. and Kato, H., Browning of pyrogallol as affected by clay minerals, *Soil Sci. Plant Nutr.* (Tokyo), 16, 195, 1970.

Kyuma, K. and Kawaguchi, K., Oxidative changes of polyphenols as influenced by allophane, *Soil Sci. Soc. Am. Proc.*, 28, 371, 1964.

Ladd, J. N. and Butler, J. H. A., Humus-enzyme systems and synthetic, organic polymer-enzyme analogs, in *Soil Biochemistry*, Vol. 4, Paul, E. A. and McLaren, A. D., Eds., Marcel Dekker, New York, 1975, 143.

Lehmann, B. G. and Cheng, H. H., Reactivity of phenolic acids in soil and formation of oxidation products, *Soil Sci. Soc. Am. J.*, 52, 1304, 1988.

Leonowicz, A. and Bollag, J.-M., Laccases in soil and the feasibility of their extraction, *Soil Biol. Biochem.*, 19, 237, 1987.

Liu, S.-Y., Minard, R. D., and Bollag, J.-M., Oligomerization of syringic acid, a lignin derivative, by a phenoloxidase, *Soil Sci. Soc. Am. J.*, 45, 1100, 1981.

Liu, S.-Y., Freyer, A. J., Minard, R. D., and Bollag, J.-M., Enzyme-catalyzed complex-formation of amino acid esters and phenolic humus constituents, *Soil Sci. Soc. Am. J.*, 49, 337, 1985.

Maillard, L. C., Synthèse des matières humiques par action des acides aminès sur les sucres reducteurs, *Ann. Chim. Phys.*, 9(5), 258, 1916.

Martin, J. P. and Haider, K., Microbial activity in relation to soil humus formation, *Soil Sci.*, 111, 54, 1971.

Martin, J. P. and Haider, K., A comparison of the use of phenolase and peroxidase for the synthesis of model humic acid-type polymers, *Soil Sci. Soc. Am. J.*, 44, 983, 1980a.

Martin, J. P. and Haider, K., Microbial degradation and stabilization of ^{14}C-labeled lignins, phenols, and phenolic polymers in relation to soil humus formation, in *Lignin Biodegradation: Microbiology, Chemistry, and Potential Applications*, Vol. 1, Kirk, T. K., Higuchi, T., and Cheng, H., Eds., CRC Press, Boca Raton, FL, 1980b, 77.

Martin, J. P., Filip, Z. K., and Haider, K., Effect of montmorillonite and hematite on growth and metabolic activity of some actinomycetes, *Soil Biol. Biochem.,* 8, 409, 1976.

Mathur, S. P. and Schnitzer, M., A chemical and spectroscopic characterization of some synthetic analogues of humic acid, *Soil Sci. Soc. Am. J.,* 42, 591, 1978.

Mayaudon, J., Stabilization biologique des protèins 14-C dans le sol, in *Isotopes and Radiation in Soil Organic Matter Studies,* IAEA, Vienna, 1968, 177.

Mayaudon, J., El Halfawi, M., and Chalvignac, M. A., Propertes des diphenol oxydases exraits des sols, *Soil Biol. Biochem.,* 5, 369, 1973.

McKenzie, R. M., Manganese oxides and hydroxides, in *Minerals in Soil Environments*, Dixon, J. B. and Weed, S. B., Eds., Soil Science Society of America, Madison, WI, 1989, 439.

Moore, J. W. and Pearson, R. G., *Kinetics and Mechanisms,* 3rd ed., John Wiley & Sons, New York, 1988.

Mortland, M. M. and Halloran, L. J., Polymerization of aromatic molecules on smectites, *Soil Sci. Soc. Am. J.,* 40, 367, 1976.

Pal, S., Bollag, J.-M., and Huang, P. M., Role of abiotic and biotic catalysts in the transformation of phenolic compounds through oxidative coupling reactions, *Soil Biol. Biochem.,* 26, 813, 1994.

Pinnavaia, T. J., Hall, P. L., Cady, S. S., and Mortland, M. M., Aromatic radical cation formation on the intracrystal surfaces of transition metal layer lattice silicates, *J. Phys. Chem.,* 78, 994, 1974.

Rice, E. L., *Allelopathy,* 2nd ed., Academic Press, New York, 1984.

Scheffer, F., Meyer, B., and Niederbudde, E. A., Huminstoffbildung unter katalyischer Einwirkung nathrlich vorkommender Eisenverbindungen im Modellversuch, *Z. Pflanzenernaehr. Bodenkd.,* 87, 26, 1959.

Scheffer, F. and Ulrich, B., *Humus und Humusduengung,* Ferdinand Enke Verlag, Stuttgart, Germany, 1960.

Schnitzer, M., Humic substances: chemistry and reactions, in *Soil Organic Matter*, Schnitzer, M. and Khan, S. U., Eds., Elsevier, Amsterdam, 1978, 1.

Schnitzer, M., Quo vadis soil organic matter research. Panel discussion paper, in *Whither soil research,* Publications of the 12th Int. Congr. Soil Sci., New Delhi 5, 67, 1982.

Schnitzer, M. and Khan, S. U., *Humic Substances in the Environment,* Marcel Dekker, New York, 1972.

Shindo, H. and Huang, P. M., Role of Mn(IV) oxide in abiotic formation of humic substances in the environment, *Nature* (London), 298, 363, 1982.

Shindo, H. and Huang, P. M., Significance of Mn(IV) oxide in abiotic formation of organic nitrogen complexes in natural environments, *Nature* (London), 308, 57, 1984a.

Shindo, H. and Huang, P. M., Catalytic effects of manganese (IV), iron (III), aluminum, and silicon oxides on the formation of phenolic polymers, *Soil Sci. Soc. Am. J.,* 48, 927, 1984b.

Shindo, H. and Huang, P. M., Catalytic polymerization of hydroquinone by primary minerals, *Soil Sci.,* 39, 505, 1985a.

Shindo, H. and Huang, P. M., The catalytic power of inorganic components in the abiotic synthesis of hydroquinone-derived humic polymers, *Appl. Clay Sci.,* 1, 71, 1985b.

Shindo, H. and Huang, P. M., Comparison of the influence of Mn(IV) oxide and tyrosinase on the formation of humic substances in the environment, *Sci. Total Environ.,* 117/118, 103, 1992..

Simmons, K. E., Minard, R. D., and Bollag, J.-M., Oxidative coupling and polymerization of guaiacol, a lignin derivative, *Soil Sci. Soc. Am. J.,* 52, 1356, 1988.

Sjoblad, R. D. and Bollag, J.-M., Oxidative coupling of aromatic compounds by enzymes from microorganisms, in *Soil Biochemistry*, Vol. 5, Paul, E. A. and Ladd, J. N., Eds., Marcel Dekker, New York, 1981, 113.

Solomon, D. H., Clay minerals as electron acceptors and/or electron donors in organic reactions, *Clays Clay Miner.,* 16, 31, 1968.

Stevenson, F. J., *Humus Chemistry: Genesis Composition, Reactions,* Wiley-Interscience, New York, 1994.

Stevenson, I. L. and Schnitzer, M., Transmission electron microscopy of extracted fulvic and humic acids, *Soil Sci.,* 133, 179, 1982.

Stone, A. T., Reduction dissolution of manganese (III/IV) oxides by substituted phenols, *Environ. Sci. Technol.,* 21, 979, 1987.

Stone, A. T. and Morgan, J. J., Reduction and dissolution of manganese(III) and manganese(IV) oxides by organics: 2. Survey of the reactivity of organics, *Environ. Sci. Technol.,* 18, 617, 1984.

Stotzky, G., Influence of soil mineral colloids on metabolic processes, growth, adhesion, and ecology of microbes and viruses, in *Interactions of Soil Minerals with Natural Organics and Microbes,* Huang, P. M. and Schnitzer, M., Eds., Soil Science Society of America, Madison, WI, 1986, 305.

Suflita, J. M. and Bollag, J.-M., Polymerization of phenolic compounds by a soil-enzyme complex, *Soil Sci. Soc. Am. J.,* 45, 297, 1981.

Swain, T., Phenolics in the environment, in *Rec. Adv. Phytochem.,* 12, 617, 1978.

Theng, B. K. G., Mechanisms of formation of colored clay-organic complexes. A review, *Clays Clay Miner.,* 19, 383, 1971.

Theng, B. K. G., The Chemistry of Clay-Polymer Complexes, Elsevier, Amsterdam, 1974, 283.

Thomas, J. M. and Thomas, W. J., *Introduction to the Principles of Heterogeneous Catalysis,* Academic Press, New York, 1967.

Thompson, T. D. and Moll, W. F., Jr., Oxidative power of smectites measured by hydroquinone, *Clays Clay Miner.,* 21, 337, 1973.

Verma, L., Martin, J. P., and Haider, K., Decomposition of carbon-14-labeled proteins, peptides, and amino acids: Free and complexed with humic polymers, *Soil Sci. Soc. Am. Proc.,* 39, 279, 1975.

Waller, G. R., Krenzer, E. G., McPherson, J. K., Jr., and McGown, S. R., Allelophatic compounds in soil from no tillage vs. conventional tillage in wheat production, *Plant Soil,* 98, 5, 1987.

Wang, M. C., Catalytic role of selected soil minerals in the abiotic formation of humic substances and the associated reactions, Ph.D. thesis, University of Saskatchewan, Saskatoon, Canada, 1987.

Wang, M. C., Influence of pyrogallol on the catalytic action of iron and manganese oxides in amino acid transformation, in *Environmental Impact of Soil Component Interactions,* Vol. I, *Natural and Anthropogenic Organics,* Huang, P. M., Berthelin, J., Bollag, J.-M., McGill, W. B., and Page, A. L., Eds., CRC Press/Lewis Publishers, Boca Raton, FL, 1995, 169.

Wang, M. C. and Huang, P. M., Humic macromolecule interlayering in nontronite through interaction with phenol monomers, *Nature* (London), 323, 529, 1986.

Wang, M. C. and Huang, P. M., Polycondensation of pyrogallol and glycine and the associated reactions as catalyzed by birnessite, *Sci. Total Environ.,* 62, 435, 1987.

Wang, M. C. and Huang, P. M., Catalytic power of nontronite, kaolinite, and quartz and their reaction sites in the formation of hydroquinone-derived polymers, *Appl. Clay Sci.,* 4, 43, 1988.

Wang, M. C. and Huang, P. M., Nontronite catalysis in polycondensation of pyrogallol and glycine and the associated reactions, *Soil Sci. Soc. Am. J.,* 55, 1156, 1991.

Wang, M. C. and Huang, P. M., Structural role of polyphenols in influencing the ring cleavage and related chemical reactions as catalyzed by nontronite, in *Humic Substances in the Global Environment and Implications on Human Health,* Senesi, N. and Miano, T. M., Eds., Elsevier, Amsterdam, 1994, 173.

Wang, M. C. and Lin, C. H., Enhanced mineralization of amino acids by birnessite as influenced by pyrogallol, *Soil Sci. Soc. Am. J.,* 57, 88, 1993.

Wang, T. S. C., Chen, J.-H., and Hsiang, W.-M., Catalytic synthesis of humic acids containing various amino acids and dipeptides, *Soil Sci.,* 140, 3, 1985.

Wang, T. S. C., Huang, P. M., Chou, C.-H., and Chen, J.-H., The role of soil minerals in the abiotic polymerization of phenolic compounds and formation of humic substances, in *Interactions of Soil Minerals with Natural Organics and Microbes,* Huang, P. M. and Schnitzer, M., Eds., Soil Science Society of America, Madison, WI, 1986, 251.

Wang, T. S. C., Kao, M.-M., and Li, S. W., A new proposed mechanism of formation of soil humic substance, in *Studies and Essays in Commemoration of the Golden Jubilee of Academia Sinica,* Academia Sinica, Taipei, Taiwan, 1978b.

Wang, T. S. C., Kao, M.-M., and Huang, P. M., The effect of pH on the catalytic synthesis of humic substances by illite, *Soil Sci.,* 129, 333, 1980.

Wang, T. S. C. and Li, S. W., Clay minerals as heterogeneous catalysts in preparation of model humic substances, *Z. Pflanzenernaehr. Bodenkd.,* 140, 669, 1977.

Wang, T. S. C., Li, S. W., and Ferng, Y. L., Catalytic polymerization of phenolic compounds by clay minerals, *Soil Sci.,* 126, 15, 1978a.

Wang, T. S. C., Wang, M. C., and Huang, P. M., Catalytic synthesis of humic substances by using aluminas as catalysts, *Soil Sci.,* 136, 226, 1983.

Wang, T. S. C., Yang, T., and Chuang, T., Soil phenolic acids as plant growth inhibitors, *Soil Sci.,* 103, 239, 1967.

Wang, T. S. C., Yeh, K. L., Cheng, S. Y., and Yang, T. K., Behavior of soil phenolic acids, in *Biochemical Interaction Among Plants,* National Academy of Sciences, Washington, D.C., 1971, 113.

Welte, E., Zur Konzentrationsmessung von Huminsauren, *Z. Pflanzenernaehr. Dhng. Bodenk.,* 74, 219, 1956.

Whitehead, D. C., Identification of *p*-hydroxybenzoic, vanillic, *p*-coumaric and ferulic acids in soils, *Nature* (London), 202, 417, 1964.

Whitehead, D. C., Dibb, H., and Hartley, R. D., Bound phenolic compounds in water extracts of soils, plant roots and leaf litter, *Soil Biol. Biochem.,* 15, 133, 1983.

Young, C. C. and Chou, T. C., Autointoxication of residues of *Asparagus officinalis* L., *Plant Soil,* 85, 385, 1985.

Young, C. C., Tsai, C. S., and Chen, S. H., Allelochemicals in rhizosphere soils of *Dendrocalamus latiflorus* Munro and *Asparagus officinalis* L., in *Phytochemical Ecology: Allelochemicals, Mycotoxins and Insect Pheromones and Allomones,* Chou, C. H. and Waller, G. R., Eds., Institute of Botany, Academia Sinica, Monograph 9, Taipei, ROC, 1989, 227.

19 The Role of Flavan-3-ols and Proanthocyanidins in Plant Defense

W. Feucht and D. Treutter

CONTENTS

0-8493-2116-6/99/$0.00+$.50
© 1999 by CRC Press LLC

19.1   ABSTRACT

Flavanols occur as monomers or undergo structural complexing. They are intimately involved in the complex interactions between plants and their environment. In considerations of plant defense, major emphasis is laid on the large array of responses evolved by plants using the diversity of flavanols. Each process of defense is the result of a combination not only of biochemical aspects but also of anatomical mechanisms. Histological reactions, including the build-up of physical barriers, participate in the complex regulation of resistance. Even at the cellular level the amount and location of flavanols might determine resistance or susceptibiliy of the host. Some of these aspects have been discussed with special reference to host-pathogen interactions and fruit trees exposed to long periods of abiotic and biotic stresses.

19.2   INTRODUCTION

Among the huge group of flavonoids the monomeric flavan-3-ols together with their oligomeric derivatives, the proanthocyanidins, posses some special properties that give them a great significance in plant defense. They are widely distributed in the plant kingdom, particularly in the woody species, and the oligomers belong to that group of polyphenols that are commonly called "tannins." Among these they are distinguished from the derivatives of gallic acid, which are described as gallotannins or "hydrolysible tannins." In the literature the oligomeric flavan-3-ol–derived proanthocyanidins are often named "condensed tannins." This older nomenclature is based on the analytical methods formerly used to separate these two groups. Since both the chemical nature of the compounds and their main biosynthetic pathways have been worked out in detail, presently therefore, the current structural names are being used in place of earlier terms.

The structural diversity within the flavan-3-ols does not justify a simplifying nomenclature, since small differences in structure, such as the degrees of oxygenation and polymerization or type of interflavan-bonding, change their chemical properties, thus leading to various biological activities (Treutter, 1996). It must be expected that both similarity and diversity within the flavan-3-ols account for their physiological importance.

In this chapter the term flavan-3-ol is used both for the monomers and their oligomeric derivatives, the proanthocyanidins. In our opinion, this is justified by the fact that the proanthocyanidins possess at least one flavan-3-ol unit, which is also regarded to be a prerequisite for the very sensitive color reaction with the diagnostic aldehyde reagents (Treutter, 1998a) used for post-column derivatization combined with HPLC and for tissue localization in microscopic studies.

The functions of flavan-3-ols in plant physiology range from scavenging of free radicals to growth regulating activity, from protein precipitation to DNA protection, and from hydration to impregnation of macromolecules of the cell wall. Comprising these properties the role of flavan-3-ols in plant defense is described and discussed in this chapter.

19.3   STRUCTURE AND NOMENCLATURE

The most common monomeric flavan-3-ol is catechin (3,5,7,3′,4′-pentahydroxyphenyl-benzopyran) with 2,3-trans configuration (Figure 19.1). The corresponding epimer with 2,3-*cis* configuration is characterized by the epi-prefix and is called epicatechin. According to the nomenclature proposed

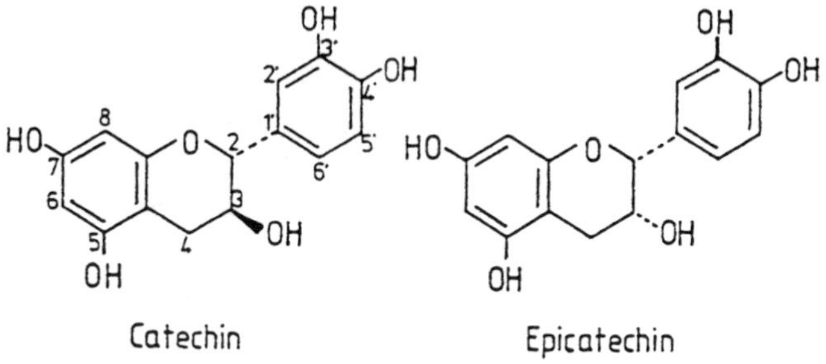

FIGURE 19.1 Structures of catechin and epicatechin.

by Porter (1988), the enantiomers are named ent-catechin (2,3-*cis*) and ent-epicatechin (2,3-*trans*), respectively. These basic structures can be modified by the number of hydroxyl-groups mainly in the B-ring, resulting in the formation of gallocatechin and epigallocatechin (3', 4',5'-triOH) or afzelechin and epiafzelechin (4'-OH). The monomers can be condensed to oligomeric and polymeric flavanols via C-C and/or C-O-C-bonding (Figure 19.2). These condensed products are called proanthocyanidins (formerly leucoanthocyanidins) since they release anthocyanidins after being boiled with strong mineral acid. However, the term leucoanthocyanidin is now reserved for the flavan-3,4-diols.

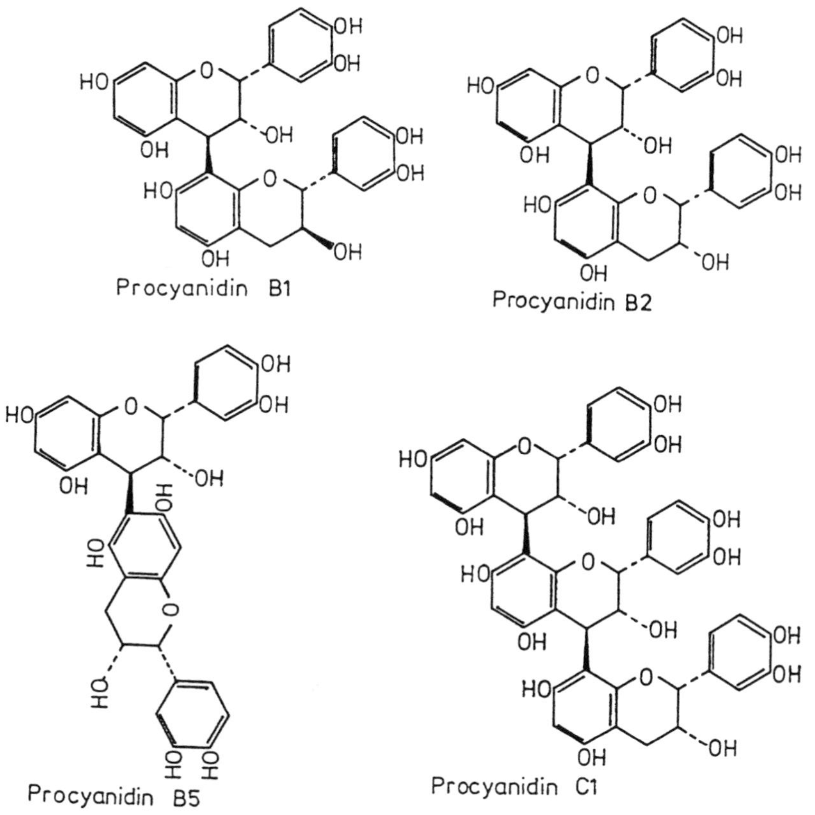

FIGURE 19.2 Structures of procyanidins.

A crude classification is given in Table 19.1, which, however, does not cover the possibility of condensation between members of different classes resulting in an additional number of structures by the mixed-type proanthocyanidins. At the time the first proanthocyanidins were structurally elucidated, a rather simple (plain) nomenclature was suggested (Weinges et al., 1968; Thompson et al., 1972) as is listed in Table 19.2. This nomenclature is commonly used when dealing with procyanidins consisting of one single type of monomeric unit, and can also describe other classes of proanthocyanidins when the word procyanidin is replaced by prodelphinidin or propelargonidin, for instance. Further structural variability originates from esterification with phenolic acids, from glycosylation, C-substitution, or from condensation with other flavonoids as reviewed by Porter (1994).

19.3.1   BIOSYNTHESIS

Flavan-3-ols are derived from leucoanthocyanidin (flavan-3,4-diols) via a reaction catalyzed by a NADPH-depending reductase (Stafford and Lester, 1984, 1985). The resulting product is a

---

TABLE 19.1

Classification of Proanthocyanidins and Corresponding Monomeric Anthocyanidins, Flavan-3-ols and Leucoanthocyanidins

Proanthocyanidin Class	Pattern of	Anthocyanidin	Flavan-3-ol	Flavan-3,4-diol
Propelargonidin	3, 4', 5,7	pelargonidin	afzelechin (pelargonidol)	leucopelargonidin
Procyanidin	3,3',4', 5,7	cyanidin catechin	(cyanidanol)	leucocyanidin
Prodelphinidin	3,3',4',5',5,7	delphinidin gallocatechin	(delphinidol)	leucodelphinidin
Proguibourtinidin	3,4',7		guibourticacidol	guibourtacacidin
Profisetinidin	3,3',4',7	fisetinidin	fisetinidol	leucofisetinidin
Prorobinetinidin	3,3',4',5',7		robinetinidol	leucorobinetinidin
Proteracacidin	3,4',7,8			teracacidin
Promelacacidin	3,3',4',7,8		mesquitol	melacacidin
Proapigeninidin	4',5,7	apigeninidin		
Proluteolinidin	3',4',5,7	luteolinidin		

Based on Porter, 1994.

TABLE 19.2

Common Nomenclature of Procyanidins According to Weinges et al. (1968) and Thompson et al. (1972)

Common Name	Structural Name
Procyanidin B1	Epicatechin-(4β 8)-catechin
Procyanidin B2	Epicatechin-(4β 8)-epicatechin
Procyanidin B3	Catechin-(4α 8)-catechin
Procyanidin B4	Catechin-(4α 8)-epicatechin
Procyanidin B5	Epicatechin-(4β 6)-epicatechin
Procyanidin B6	Catechin-(4α 6)-catechin
Procyanidin B7	Epicatechin-(4β 6)-catechin
Procyanidin B8	Catechin-(4α 6)-epicatechin
Procyanidin A1	Epicatechin-(2β O 7, 4β 8)-catechin
Procyanidin A2	Epicatechin-(2β O 7, 4β 8)-epicatechin
Procyanidin C1	Epicatechin-(4β 8)-epicatechin-(4β 8)-epicatechin
Procyanidin C2	Catechin-(4α 8)-catechin-(4α 8)-catechin

flavan-3-ol unit, such as catechin or gallocatechin. There is no information on how the epimers and the enantiomers are synthesized. It is also unclear by which mechanism the proanthocyanidins are formed. It is widely accepted that the flavan-3-ols function as an initial unit that is expanded on by condensation with a flavan-3,4-diol.

19.3.2 SECONDARY ALTERATIONS TO THE FLAVANOL STRUCTURES

The metabolic fate of flavan-3-ols is often characterized as an oxidative coupling reaction that occurs spontaneously or is catalysed by oxidases and peroxidases. The reaction products that are initially colorless can subsequently be oxidised to form yellow to brown polymers, especially in senescing cells (McGraw, 1989; Laks, 1989). Intermediate products that have been identified are the dicatechin (Weinges and Ebert, 1968) and catechinic acid (Kennedy et al., 1984). The most prominent oxidation products of flavan-3-ols are the theaflavins and the thearubigens in black tea as well as the phlobaphenes. The latter show a reddish-brown color and are insoluble in water. They also include glucose and other phenolic compounds in their polymeric structure (Foo and Karchesy, 1989; Steenkamp et al., 1985). It is assumed that similar compounds occur in senescing tissues and during autolysis of plant cells.

19.4 ANALYTICAL METHODS AND HISTOCHEMICAL LOCALIZATION

For the analytical determination of flavan-3-ols in plant extracts, principally two colorimetric techniques are used. The total amount of flavanols including monomers and proanthocyanidins can be estimated by using aldehyde-containing reagents such as vanillin and p-dimethylaminocinnamaldehyde (DMACA). The photometrically measured intensity of the colored product (500 nm or 640 nm, respectively) gives a quantitative result. The concentration of soluble oligomeric proanthocyanidins is determined after boiling the extract in strong mineralic acid, yielding the colored anthocyanidins that are quantified photometrically. Apart from these old techniques, which are extensively reviewed by Ribereau-Gayon (1972), Treutter (1989a,b), and Treutter et al. (1994a), several HPLC separation methods have been developed in the last two decades (Porter, 1988; Karchesy et al., 1989). Instead of the normally used UV-detection, the combination of HPLC and post-column derivatization using the DMACA-reagent (Figure 19.3) can be applied to selectively detect flavan-3-ols (monomers and oligomers) in crude extracts (Treutter, 1989; Treutter et al., 1994b; Treutter and Santos-Buelga, 1995).

The color reactions described above are also used for localized stains of both catechins and proanthocyanidins histochemically at the tissue and cellular level (Feucht and Schmid, 1983; Gutmann and Feucht, 1991; Gutmann, 1993). Other less sensitive and unspecific techniques are reviewed by Chalker-Scott and Krahmer (1989). Toluidine blue O gives a differential staining for tissues. Various phenols attain a greenish coloration. Cell walls stain bluish and pectins reddish-violet (Gutmann, 1995). The quantification of wall-bound proanthocyanidins is described by Mathews et al. (1996) who solubilized intact flavan-3-ol units from the cell wall matrix by treatment with toluol-thiol.

Positioning and arrangement of the flavanols within the different organs and tissues is of outstanding importance for our understanding of host/pathogen interrelationships (Adaskaveg, 1992).

The objectives of the histological investigations are, therefore, to determine at the cellular level the sites of damage and the formation of biochemical flavanol barriers. In mandarins, the newly formed wound phelloderm may consist of only two cell layers or of up to seven layers of cells depending on the cultivar (Achor et al., 1991).

Studies that measure differences in the extension of wound healing zones would finally allow more correct sampling and more realistic conditions in determining analytically special defense compounds.

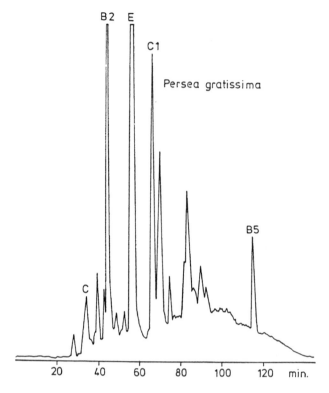

FIGURE 19.3 Separation of flavan-3-ols from a crude extract of avocado fruit skin by rp-HPLC with chemical reaction detection using *p*-dimethylaminocinnamaldehyde (DMACA) as a selective reagent; detection wavelength: 640 nm.C, (+)-catechin; E, epicatechin; procyanidins B2, B5, C1.

19.5 CHEMICAL REACTIONS OF FLAVAN-3-OLS

19.5.1 MOLECULAR INTERACTION WITH PROTEINS

Irreversible complexes of proanthocyanidins with proteins are formed via o-quinone-intermediates resulting in stable covalent bondings (Pierpoint, 1983; Beart et al., 1985). In senescing or damaged plant tissues o-quinones originate from the action of phenoloxidases after stress-induced loss of cellular compartmentation (Mayer and Havel, 1981). In tannin-containing food the digestibility of proteins is generay reduced (Griffiths, 1991).

Since the irreversible binding of polyphenols to proteins that leads to precipitation of the latter is well-known, the reversible interactions (reviewed by Spencer et al., 1988) are often ignored. According to Hagerman and Butler (1981) and Asquith and Butler (1986), the affinity of polyphenols to proteins does not depend on the structure of the polyphenol but on the nature of the protein. High portions of hydrophobic amino acids, in particular prolin, favor the binding of a protein to proanthocyanidins. It was also found that even monomeric catechins as well as other phenolics can be bound to proteins (Arora et al., 1988).

19.5.2 INHIBITION OF ENZYME ACTIVITIES

Many reports have described the inhibition of lytic enzymes, such as pectinases, cellulases, and proteases, by proanthocyanidins (Etchells et al., 1958; Loomis and Bataile, 1966; Hathway and

Seakins, 1958; Bell and Etchells, 1958; Pollard et al., 1958; Bell et al., 1962; Porter et al., 1961; Reese, 1963; Griffiths and Jones, 1977; Strobel and Sinclair, 1991). In sorghum grains the activity of amylases is inhibited by procyanidins (Strumeyer and Malin, 1970). Much more is known about the interaction between flavan-3-ols and mammalian enzymes (Regnault-Roger, 1988). The inhibition of enzyme activities by proanthocyanidins has not been studied in detail; however, it is possible that the above-mentioned affinity to proteins may contribute to the regulation of enzyme activities.

The inhibition of the activity of a membrane bound H^+-ATPase by catechin was reported by Erdei et al. (1994). They suggest that the influence is not attributed to bonding to the protein, but rather to the function of the phenol as a redox component.

19.5.3 COMPLEXATION WITH POLYSACCHARIDES

The reversible complexation of polyphenols with polysaccharides was studied by Gaffney et al. (1986). The molecular interaction is based on Van der Waal forces, hydrogen bridges, and hydrophobic interactions (Haslam, 1989). The affinity of polyphenols to polysaccharides largely depends on the structure of the latter, that is, on molecular size, on the degree of branching, on the occurrence of cavities, and on the flexibility of the molecule (Ya et al., 1989). The complexation of proanthocyanidins with cell wall components is responsible for the drought resistance of several plants growing under arid climates (Pizzi and Cameron, 1986). The oligomeric flavan-3-ols stabilize the hemicellulose-fibers when they lose water.

19.5.4 ANTIOXIDATIVE AND RADICAL-SCAVENGING ACTIVITY

Free radicals are often formed in stressed plant cells. These very reactive molecules may cause damage to DNA, to membranes, and to SH-proteins; they may also destroy nucleotides and/or initiate lipid-peroxidation. These lead to severe damage of tissues and to the necrotic symptoms that often occur during pathogenesis. Many researchers have described the antioxidative and radical-scavenging activities of flavan-3-ols, which are found to be even better than those of vitamins, ascorbic acid, and α-tocopherol, as well as those of other phenolic compounds (Baumann et al., 1980; Ariga et al., 1988; Uchida et al., 1988; Perchellet et al., 1994; Rice-Evans, 1995; Teissedre et al., 1996). Flavan-3-ols and other phenolics were reported to protect unsaturated fatty acids against free radicals (Affany et al., 1987; Salvayre et al., 1988). The flavanols operate as reducing agents in the maintenance of membrane integrity (Mukherjee and Choudhouri, 1983). Flavanols were qualified as small antioxidative first aid molecules, with or without the chance of being metabolically repaired themselves (Elstner et al., 1994). Extracts of the yellowing beech leaves showed a greater antioxidative potential than that of green leaves (Feucht et al., 1997). An improved stability of ascorbic acid by the presence of proanthocyanidins in biological systems was described by Clemetson and Andersen (1966).

19.5.5 LOSS OF CELLULAR INTEGRITY BY PARAQUAT-INDUCED OXYGEN RADICALS AND THE ROLE OF FLAVANOLS

Leaves of sweet cherry, when treated with paraquat (Feucht et al., 1996c), indicated a loss of compartmentation (Figure 19.4). The plasmic material was dislocated to the cellular periphery, similar to the xylem rays of the incompatible graft union. This effect was largely overcome by the addition of catechin. The chloroplasts were observed to stain for phenols, indicating a change in membrane properties.

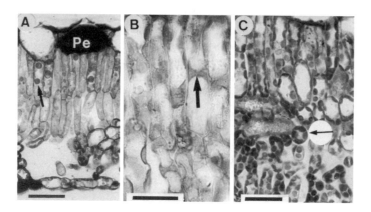

FIGURE 19.4   Toluidine blue-stained (A–C) cross sections of leaves of *Prunus avium*. (A) Untreated cherry leaf with some flavanol inclusions (arrow) in the palisade cells (Pe, subepidermal cell filled with pectins). (B) Palisade leaf section treated with paraquat showing dissolution of the flavanol globules and displacement of the cytoplasm (arrow). (C) Combined treatment with paraquat and catechin showing recovery of normal cellular structures and the chloroplasts attaining a greenish coloration with toluidine blue (arrow). Bars = 50 μm (A and C), and 20 μm (B).

19.5.6   INTERACTION WITH DNA

Haslam (1989) pointed to the evidence that the stereochemistry of proanthocyanidins fits well into the groove of the DNA double helix. He speculated that this would permit hydrogen bonds between the two molecules. Haslam further directed to the possible biological significance involved, since there are similarities to the complexation between DNA and histones in eucaryotic organisms.

19.6   ALLELOPATHIC EFFECTS BY GROWTH-REGULATING ACTIVITY

The general rule that the monophenols show growth-inhibiting effects, whereas o-diphenols are growth promoting (Nitsch and Nitsch, 1962; Kefeli and Kutacek, 1977; Grambow, 1986; Krylov et al., 1994; Volpert et al., 1995), can also be applied to flavan-3-ols possessing one or two OH-groups in ring B, respectively. Lavee et al. (1994) found that the growth of olive callus was promoted by chlorogenic acid even in the absence of auxin. This influence on plant growth is mainly attributed to the regulation of IAA-oxidases. However, on the other hand, the interaction with other macromolecules could also interfere with the effect on auxin turnover. It was found that dimeric propelargonidin and the corresponding monomer afzelechin inhibit growth of rice seedlings and could be responsible for restricted growth of peach roots (Ohigashi et al., 1982). This negative effect of the secreted propelargonidin, which is a monophenol, may contribute to the replanting problem of *Prunus persica* (L.) Batsch.

Among the many observations of auxin-protection only a few examples will be cited here. Procyanidins were identified as growth-promoting components of cocos and *Aesculus* extracts (Shantz and Steward, 1955). Catechin was found to stimulate growth of *in vitro* cultures (Feucht and Nachit, 1977; Feucht and Schmid, 1980). It has been shown that catechin protected IAA in the nutrient solution from peroxidative attack (Feucht and Treutter, 1995). Evidence was obtained in our laboratory that catechin applied in the 0.34 to 1.0 μM range was capable of stimulating callus growth to twice the values of controls in the following fruit crops: *Prunus avium* (L.) L., *Prunus domestica* L., *Prunus cerasus* L. (Feucht and Treutter, 1995; Feucht et al., 1993, 1996a), *Vitis vinifera* L.,

Sambucus nigra L., and *Rosa canina* L. (unpublished). Growth promotion by catechin reacting synergistically with auxin was found in tissues of beech cultivated *in vitro* (Feucht et al., 1997a,b). Callus grown *in vitro* on catechin forms isotropic undifferentiated cells and has an increased ratio of fresh weight to dry weight. Those cells need a longer time to accept morphogenetic competence compared with meristems (Lyndon, 1990). Rapid wound healing is best achieved by the callusing cell type. Supposedly, catechin inhibits the effects of auxin to perform polar diffusion from cell to cell and rapid determination of a special cell type. Catechin seems to be required for attaining the parenchymatic status.

Flavonoids are also considered to be a class of endogenous auxin-transport regulators. Catechin-treated tissues showed slightly increased IAA uptake (Faulkner and Rubery, 1992). Some flavonoids were found to hold auxin at special receptors at the plasma membrane in cells near a wound by inhibiting special auxin efflux carriers (Jacobs and Rubery, 1988).

19.7   GENERAL ASPECTS OF FLAVANOL DISTRIBUTION IN TREES

Based on ample histological studies in our laboratory there is convincing evidence for a whole-tree strategy in the distribution of stored flavanol. In apple trees, flavanol concentrations are maximal in the youngest shoots and young fruitlets with the concentration rapidly falling off during maturation (Mayr et al., 1995, 1997). This distribution pattern, which provides chemical protection for the expanding thin-walled tissues, is common for cultivated fruit trees.

This is not to say that large species-dependent variations do not exist. Rosaceous species can well be grouped in high-flavanol synthesizers and low-flavanol synthesizers. For instance, the flavanol content in leaves of sour cherry (*Prunus cerasus*) is higher than in leaves of sweet cherry (*P. avium*).

A threshold level of carbohydrate and ATP is needed as starting material for an increased carbon flux through the flavonoid pathway. Notably, there is no such clear relationship between the carbohydrates produced in a tree canopy and the pool size of phenols as that existing between leaf nitrogen and photosynthetic capacity in natural ecosystems (Field and Mooney, 1986). In whatever way the carbohydrates are up- and downregulated, light exposure and photosynthesis are doubtless of substantial importance to the flavanol production (see also Creasy, 1968). In grape vine callus cultures, 3 percent sucrose solution yielded several-fold more flavanols than 1 percent sucrose (Feucht et al., 1996b). However, 6 percent sucrose concentration or addition of phenylalanine into the medium resulted in less flavanol accumulation.

To survive in a hostile environment, groups of cells of a plant are specialized for defense. However, phenols that are formed in special cells are dangerous to physiological processes in the cytoplasm. Sequestering of those phenols is, therefore, indispensable. This may occur in large vacuoles or in localized subcellular sites.

The distribution pattern of phenol cells in shoots, leaves, fruits, and roots is coordinated genetically and modulated by information from the environment. Precise coordination of the functions of individual cells including the spatial distribution of the defense cells is a prerequisite of plant development and stress tolerance.

19.8   BORDERING AND EXPOSED TISSUES—THE FRONT LINE
         OF DEFENSE ACTION

Flavanols develop rapidly in the growing shoot tips (Figure 19.5, A). This, however, decreases with the thickening of cell walls (at about 10 cm from tip) (Figure 19.5, C). Upon maturing of the shoot tissue the periderm is formed (Figure 19.5, B). In a number of fruit crop plants the thick-walled epidermis and the phellem contain flavanols as absorbing pigments. The periderm constitutes a barrier against biotic and abiotic stress. Upon wounding of the periderm, two basic strategies come

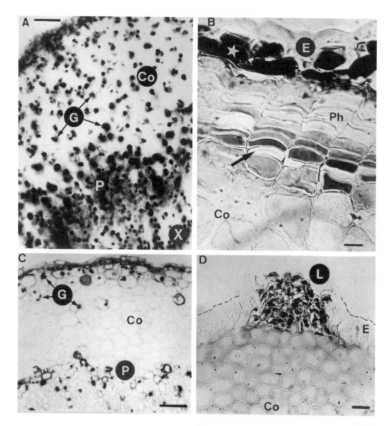

FIGURE 19.5 DMACA-stained cross sections of young shoots (A–D). (A) *Pyrus domestica.* Cortex (Co), phloem (P), and young xylem (X) of elongating internodes of *Pyrus domestica.* Cells near the shoot periphery and in the phloem contain abundant flavanols, often filling the entire central vacuoles. Throughout the inner cortex, smaller flavanol globules (G) are frequently found in the central vacuoles. (B) *Prunus avium.* The thick walled epidermis (E) is rich in dark-staining flavanols (asterisk). Outer phellem (Ph) layers lacking flavanols. Inner phellem layers with flavanols. Phellogen cells do not contain flavanols (arrow). The adjacent cells of the Co lack flavanols. (C) *Pyrus domestica.* Co and P of mature non-elongating internodes. Only the outer protective layers of the shoot periphery and the phloem contain numerous flavanol G. (D) *Rosa canina.* The lenticels (L) of young shoots contain dark-staining flavanol cells. Bars = 50 μm (A, C, D), and 10 μm (B).

into play: 1) renewed cell division, and 2) additional phenol synthesis. Lenticels constitute natural, programmed wounds. The cell layers of the undersurface are heavily loaded with flavanols (Figure 19.5, D). This pattern is found in Rosaceae (several species), *Juglans regia* L., *Sambucus nigra,* and *Robinia pseudoacacia* L.

The epidermis of anthers of the cultivated *Prunus* species contains abundant flavanols, thus attenuating a large portion of incoming UV-B radiation. The pollen itself, however, is free from flavanols. Pistils show increased accumulation of flavanols, except in the stigma, which stains faintly in all cultivated *Prunus* species. Trichomes located on shoots, leaves, or fruits have protective functions. They are often so numerous as to densely cover the epidermis. Thus, even fungal spores are prevented from contacting the epidermis. Their lumen was found to be frequently filled with densely staining flavanols. Most prominent in this respect are conifers, walnut, *Juglans nigra* L., and *Rosa canina.* The thin-walled glandular trichomes were rapidly destroyed by mechanical touching and this leads to the spread of the flavonols on the surface causing the damage to the insects. In the dormant buds of *Aesculus hippocastanum* L., the elongated hairs are heavily loaded with flavanols (Figure 19.6, A–D). The youngest root tips are extremely exposed to harsh environments. Root tips stain intensely for flavanols. This was studied in our laboratory for *Vitis vinifera, Prunus* species, and some conifers (not shown). Forrest and Bendall (1969) found the young roots of the tea plant to contain flavanols.

Lees et al. (1995) studied the distribution of tannins in the peel and pulp of apple fruits. For mature fruits of 24 fruit crops, a histological semiquantitative survey on the presence of flavanols is given by Feucht et al. (1994a,b). The flavanols were most concentrated in the epidermis and subepidermal layers of the fruits. Sour cherry has rather low values of flavanols (1 to 3 mg per 100 g dry weight). Fruits of *Sorbus domestica* L. tree have the highest flavanol contents (1.800 to 2.200 mg per 100 g dry weight). Thus, the difference is approximately 1:1,000. Ranking in amounts of flavanols is as follows: low in amounts are prune (*Prunus domestica*), strawberry (*Fragaria x ananassa*), sweet cherry (*Prunus avium*), sour cherry (*P. cerasus*), blueberry (*Vaccinium australe* Small), wild apple (*Malus baccata*), cv.'Landsberger' (*Malus domestica*), cv.'Jonagold' (*Malus domestica* Borkh), and gooseberry (*Ribes rubrum* L.). Medium amounts are contained in 'Sir Price' (*Malus domestica*), Kiwi (*Actinidia arguta* Sieb. et Zncc.),'Packhams Triumph' (*Pyrus domestica*), and the fruit skin of orange (*Citrus sinensis* [Linn.] Osbeck) and mandarin (*Citrus reticulata* Blanco). Higher amounts were found in mesplar (*Mespilus germanica*), dog rose (*Rosa canina*), pear (*Pyrus communis* L.), mountain ash (*Sorbus aucuparia* L., *var. moravica*), cherimola (*Annona cherimola* Mill.), walnut fruit skin (*Juglans regia* L.), banana skin (*Musa accuminata* Colla), sloe (*Prunus spinosa* L.) and service tree (*Sorbus domestica*).

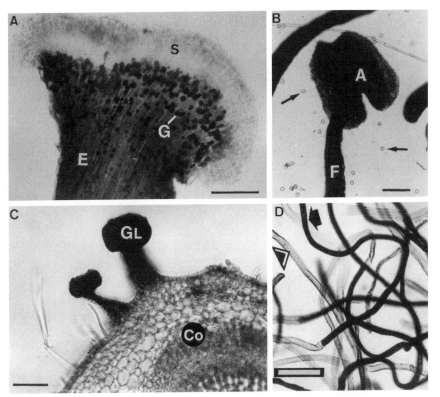

FIGURE 19.6 DMACA-stained tissue (A–D). (A) Stigma and upper style of *Prunus avium.*. Diffuse lightly stained flavanols on the stigma (S) and densely packed flavanol globules (G) in the outer layers of the style. Much of the dark-staining style can be attributed to flavanols lining the entire tonoplasts of the epidermal cells (E). (B) Anther (A) and filament (F) of *Prunus cerasus* showing heavy flavanol loading. Pollen grains (arrows) do not stain. (C) Cross section of a fruit stalk of *Rosa canina*. The large multicellular glandular trichomes (Gl) are rich in flavanols. The cortical tissue (Co) stains moderately for flavanols. (D) Elongated hair trichomes from dormant buds of *Aesculus hippocastanum* being filled with flavanols (arrow). Some of them lacking flavanols (arrowhead). Bars = 50 μm (D), 200 μm (A, B), and 100 μm (C).

Leaves need protection against an array of potential aggressors. In addition, flavonoids have a protecting role against UV radiation (Lois and Buchanan, 1994). The scab (*Venturia inaequalis*) resistant genotype Coop 50 shows the mesophyll enriched with flavanol vacuoles. However, the apple cultivar 'Summerred' exhibits flavanols in a very low range (Figure 19.7, A,B). Youngest leaves of *Prunus tomentosa* achieve full capacity to store high amounts of flavanols throughout the upper and lower epidermis (Figure 19.7, C and D). These epidermal depositions were found to be diminished in mature leaves. The subsidiary cells of the guard cell complex of *Prunus* species contain small flavanol globules. But in guard cells flavanols were only found sporadically. The elongated palisade cells of *Prunus* species frequently contain one flavanol vacuole in both the upper and lower half, each occupying a large portion of the cell area. In the spongy parenchyma cells the flavanol deposits are less prominent.

In *Prunus* species the bundle sheaths circumventing xylem and phloem were most pronounced in their capacity to store flavanols. There are often two to four medium-sized inclusions within the central vacuole. The larger a vein is developed, the more distinguished are the bundle sheaths in storing flavanols. Assimilate transport in the phloem offers a great nutritive potential for insects and pathogens. At the level of light microscopy, sieve tubes themselves are devoid of flavanol deposits. Occasionally, very small flavanol vacuoles were present in the companion cells of *Prunus* species. However, most of the parenchyma cells near the sieve tubes are heavily loaded with flavanols (Figure 19.8, A and B). Transverse sections of young shoot internodes of *Actinidia* revealed a particularly high density of phenol cells in the phloem region. (Figure 19.9). The phloem flavanols can be qualified as an inner circular defense barrier, in addition to the outer periderm ring.

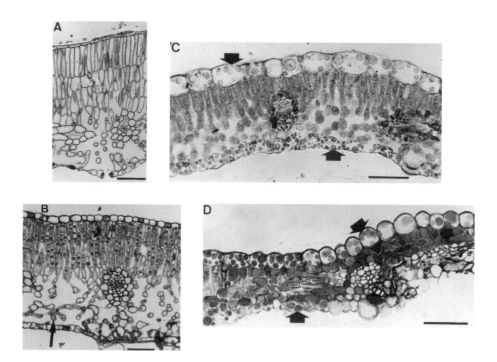

FIGURE 19.7   Flavanols in leaves (A–D, toluidine blue stained). (A) Cultivar 'Summerred' (*Malus domestica*) lacking flavanol globules in the mesophyll. (B) Coop 50 (Malus hybrid) showing numerous round globules in the palisade cells. Note the small size of these globules in the spongy parenchyma (arrow). (C) Young leaf of *Prunus tomentosa* with numerous globules in the upper and lower epidermal layers (arrows). (D) Young leaflets (*P. tomentosa*) from the early spring flush showing deep flavanol staining of globules (arrows) from the upper epidermis as well as in most mesophyll cells. Bars = 50 μm (A–D).

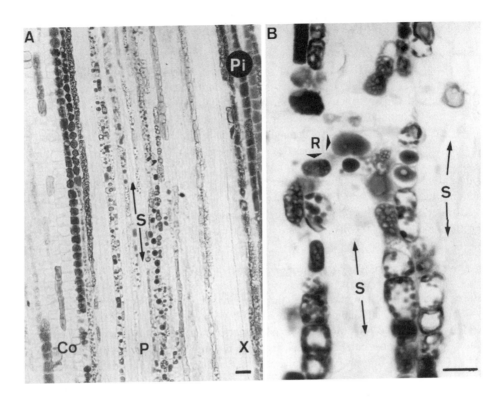

FIGURE 19.8 DMACA-stained flavanols in parenchymatous cells of the phloem. (A) Longitudinal section of a shoot of *Prunus avium* with numerous flavanol globules (X, xylem; P, phloem; Co, cortex). Sieve tubes (S) are devoid of flavanols. Note the heavy flavanol loading of the quadrangular cells bordering the pith (Pi). (B) Longitudinal section of a shoot of *Ulmus americana*. The cells contain either flavanols in globular deposits or uniformly distributed in the large central vacuole. The sieve tubes lack flavanols. Ray cells (R) traversing radially the phloem show a high flavanol stainability. Bars = 100 μm (A), and 50 μm (B).

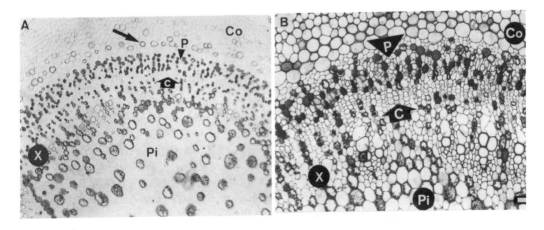

FIGURE 19.9 Cross sections (A, DMACA; B, toluidine blue) of young shoots of *Actinidia arguta*. Both staining procedures reveal flavanols located in the cortex (Co), phloem (P), young xylem (X), and pith (Pi). Note the lack of flavanols in the cambium (C). DMACA-staining permits localize faintly staining flavanols lining the inside tonoplast in cells of the cortex (arrow). With the toluidine reaction (B), all cell walls are well recognizable, giving a bluish coloration. The flavanols, particularly those in the phloem and in the rays traversing the cambium stain darkly green. Bars = 100 μm (A and B).

19.8.1   Is There a Strategic Positioning of Flavanol Cells?

The distribution pattern of flavanols within the tree system strengthens the view of a strategic positioning of the phenol cells. This is quite convincing when dealing with the bordering plant tissues as the primary sites of defense. Flavanol deposition in the inner shoot regions such as phloem and pith could be expected to play a less important role in plant responses to environmental and biotic stresses. The positioning pattern of particular cell types within a tissue of a plant species is ruled by hormonal regulation, which in turn depends on gene-regulated hormonal competence. Lyndon (1990) reported that a given cell becomes determined physiologically by virtue of its position within the tissue. It may suggest that the positioning of phenol cells is self-adjusting and in some way is under the influence of the neighboring cells. Thus, one phenol cell may be the site of an inhibitor production preventing the induction of the same cell type nearby. Hence, only at a certain distance away, a further phenol cell is allowed to develop. Summing up, the gene-controlled spacing of flavanol cells is modified by the plant's developmental state and its environment.

19.9   PATHOGEN-PLANT CONFRONTATION

The strong interaction of polymeric proanthocyanidins with proteins is responsible for the astringency of many unripe fruits and other plant parts. This protects many plants from herbivoral attack. These kinds of defensive compounds are usually referred to as constitutive or preformed (cf. review of Schlösser, 1994). Those compounds that are produced only during pathogenesis are named "phytoalexins" (Müller and Börger, 1941). However, there are a number of reports dealing with induced biosynthesis following biotic or abiotic stress. Therefore, the suggestion of Stafford (1997) should be accepted and the term "constitutive" be replaced by "under endogenous control."

Some plants accumulate polymeric proanthocyanidins in their leaves after an insect attack, leading to protection against further herbivoral damage (Feeny, 1976; Schultz and Baldwin, 1982). The deterrent effects of tannin-containing higher plants are summarized by Bernays (1981). It is not yet clear whether protein precipitation is the only reason for the deterrent effect. Schultz (1989) pointed to other possible mechanisms of biological activity such as membrane binding, selective inhibition of enzymes, formation of toxic quinones, or other adverse metabolites.

It should be noted that the uptake of polymeric flavan-3-ols does not always affect herbivores adversely. It was found that the uptake of tannin-rich legumes by ruminants prevents bloat, which is caused by proanthocyanidin-free forage species (Jones et al., 1976).

19.9.1   Antibacterial Effects

Not much is reported about the specific activity of flavan-3-ols against bacteria. The growth of some non-phytopathogenic bacteria, such as *Staphylococcus aureus*, *Lactobacillus casei*, and *Escherichia coli*, is inhibited in the presence of procyanidins and prodelphinidins isolated from grapes (Somaatmadja et al., 1965). Epicatechin-3-O-gallate and epigallocatechin-3-O-gallate affect the growth of *Streptococcus mutans* (Sakanaka et al., 1990), and catechin and epicatechin were reported to impair *Pseudomonas maltophilia* and *Enterobacter cloacae* (Waage et al., 1984). A long list of bacteria that inhibit the *in vitro* development was given by Hara and Watanabe (1989a,b). The structural prerequisites for other flavonoids and their antimicrobial activities have been discussed but still remain unclear (Mori et al., 1987; Weidenbörner and Jha, 1994).

19.9.2   Virus-Diseases

The question as to whether tannins are capable of protecting plants from virus infection was discussed by Cadman (1960), who noticed that tannin-rich plant extracts hindered the inoculation of healthy plants with tobacco rattle virus. As for the mode of action, the inhibition of polymerases or

TABLE 19.3
Fungitoxicity of Flavan-3-ols

	Microorganism	Reference
Catechin	*Rhizoctonia solani*	Hunter, 1978
Catechin	*Trichoderma viride*	Malterud et al., 1985
Catechin	*Fomes annosus*	Alcubilla et al., 1971
Procyanidin	*Rhizoctonia solani*	Rao and Rao, 1986
Fraction from *Pinus sylvestris*		
Fraction from *Anona squamosa*		
Propelargonidin	*Rhizoctonia solani*	Rao and Rao, 1986
Fraction from *Cassia javanice*		
Fraction from *Peltophorum pterocarpum*		
Prorobinetinidin	*Rhizoctonia solani*	Rao and Rao, 1986
Fraction from *Acacia leucophloea*		

direct interaction with nucleic acids has been proposed leading to a reduced multiplication of the viruses (Okada et al., 1977; Selway, 1986; Middleton et al., 1986). The inhibition of the reproduction of influenza and herpes viruses was also reported (Serkedjieva et al., 1992).

19.9.3 ANTIFUNGAL ACTIVITY

The antifungal activity of flavan-3-ols is often assumed but only rarely demonstrated (Table 19.3). Agar diffusion tests sometimes show a brown-colored ring around the inhibition zone indicating the involvement of oxidative turnover of flavanols most likely via quinonoid intermediates (Hunter, 1978). It is speculated that the intermediates are the actual active compounds, not the original flavan-3-ols.

19.9.4 FUNGAL DISEASES OF FRUIT CROPS

A number of fungi and bacteria cause leaf spotting symptoms, common to fruit crops, particularly in stone fruits. At earlier stages, the leaf spots consist of small purple lesions. Later, a shot hole effect can be produced by the excision and dropping out of the necrotic tissue. At the very beginning of the infection, the leaves respond at the site of attempted invasion with localized cell death. As the disease develops, the affected area becomes circular in outline and is characterized by a hypersensitive reaction. Resistance is now often characterized by an oxidative burst; the superoxidanion radical O_2^- and H_2O_2 react with one another, yielding the highly toxic OH radical in what is known as the Haber-Weiss reaction. These radicals may diffuse into the circumventing green "halo." In a response to this, the reaction zone around the necrotic center could accumulate flavanols up to several times as compared with control (Feucht et al., 1993, 1996a,b). The mesophyll cells around the dying center become proliferative. Cell divisions occur in all planes. The dead area becomes encapsulated by new cells.

The accumulated flavanols (Figure 19.10) participate in various defense reactions. For example,

- Callus cultures grown on a catechin medium principally show an increased fresh weight to dry weight ratio compared with controls (Feucht and Treutter, 1995). Notably, ascorbic acid improved the water potential of previously water-stressed tissues (Mukherjee and Choudhuri, 1983). Both ascorbic acid and flavanols have reducing properties. Raising the turgor pressure of a cell might increase H_2O_2 levels (Yahraus et al., 1995), which is highly diffusive and is known to operate as a signal for defense (Dempsey and Klessig, 1995).
- Close cellular contact between turgid cells provides better transmission of cellular signals.

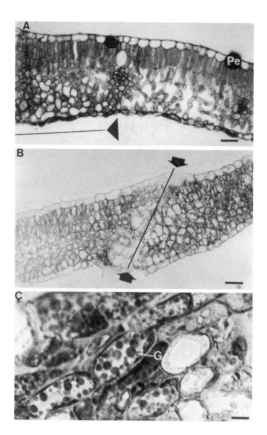

FIGURE 19.10   (A, stained with toluidine blue; B and C, stained with DMACA). (A) Histological changes in shot hole diseased leaves of *Prunus domestica.* To the left from the arrowhead, intense cell division occurring especially in the spongy parenchyma. The healthy sector showing large intercellular spaces. Some of the epidermal cells stain violet with toluidine blue, which is indicative for pectins (Pe). (B) Intense cell division, loss of intercellular spaces in the spongy parenchyma, degradation of the epidermis, and enlarging cells (along the line between both arrows) indicate the development of an abscission zone. Deep flavanol staining coincides with the regions of preferential cell division. (C) Enlarged view of cells close the infection site. Flavanols are deposited as globules (G) or peripherally attached toward the cell wall. Bars = 50 μm (A and B), and 10 μm (C).

- Activation of ATPase occurs (Feucht and Treutter, 1995), thereby regulating membrane transport of ions and metabolites related with cell division, protein synthesis, and finally with defense reactions against pathogenic invaders.
- Cell division in a catechin-modulated environment (Feucht and Treutter, 1995) yields new undetermined cells that are more competent at adapting to new gene-controlled strategies for coping with a given stress situation. Flavanols are involved in shifting the tissue into reductive state undergoing cell division as was noted 25 years ago by Stonier and Yang (1973). Down-regulation of auxin after wounding or infection must be prevented to allow wound healing through cell division (Antonelli and Daly, 1966; Thornburg and Li, 1991).

Similar to prune (*P. domestica*), the mesophyll of sweet cherry is replaced by dividing parenchyma cells filling the entire leaf sector after fungal invasion (Figure 19.11, A). The affected palisade cells are enriched with flavanols that are oxidized during necrotization. The entire phloem is bordered by a "defense girdle" of heavily loaded flavanol cells. At the border line between healthy and dying tissues, there are cell walls that stain for flavanols (Figure 19.11, B). The enlarged parenchyma cells repeatedly show some flavanol inclusions.

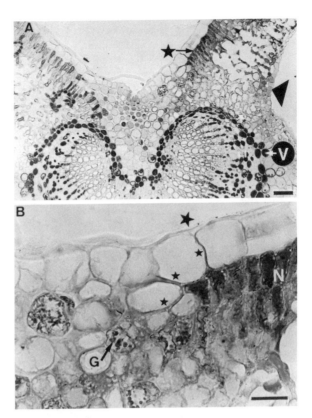

FIGURE 19.11 DMACA-stained leaf sections of *Prunus avium* showing shot hole injury. (A) Note the necrotic infection site with oxidized palisade layers and gradually shriveled spongy parenchyma cells (asterisk) to the right from the leaf vein. Multiple cell divisions are apparent between the necrotic area and the leaf vein (arrowhead). The vacuoles (V) of the bundle sheath cells surrounding the vascular bundle are rich in flavanols. (B) Enlarged view of the transition zone (large asterisk) from necrotic (N) to dividing tissue. Some cells contain flavanol globules (G). Some cell walls stain blue with DMACA, indicating an affinity for flavanols (small asterisks). Bars = 100 μm (A), and 50 μm (B).

19.9.5 TWO SPECIES, TWO TYPES OF FUNGI, AND TWO REACTION PATTERNS

Plum leaves (*P. domestica*) infected by *Cylindrosporium* show disintegrated cell walls and distorted masses of cytoplasm at the infestation site. Some phellem-like cells are formed by periclinal divisions. They are densely filled with flavanols and the spherical globules in some cells reacted in the same way (Figure 19.12, A). While the leaf area in the plum is greatly eroded by the fungus, the pear (*Pyrus domestica*) leaves become swollen on infection with *Gymnosporangium sabinae*. The enlarged mesophyll cells in the pre-necrotic state are heavily loaded with flavanols (Figure 19.12, B).

19.9.6 ROLE OF FLAVAN-3-OLS IN THE PATHOGENESIS OF APPLE SCAB— INTERRELATION BETWEEN CONSTITUTION AND INDUCTION

It has been documented that the lesions produced by apple tissues following an attack by the fungus *Venturia inaequalis* are surrounded by some cell layers rich in flavan-3-ols (Figure 19.13; Treutter and Feucht, 1990a). The accumulation of epicatechin and procyanidins is induced by the action of the fungus and may be attributed to an unspecific wound response since the fungus produces pectinases and cellulases (Valsangiacomo et al., 1992; Kollar, 1994), which destroy the cell wall. As a consequence of this, chemical destruction elicitors might be released stimulating the phenyl-propanoid

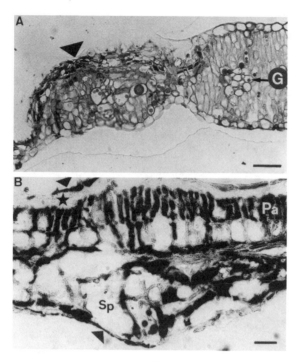

FIGURE 19.12   Different disease symptoms upon fungal invasion (A, toluidine blue; and B, DMACA staining). (A) Invasive fungi had begun to degrade (arrowhead) much of the infected leaf site of *P. domestica*. Greenish toluidine reactions were observed in the affected area. Globules (G) stain dark green. (B) The sector of a pear leaf (*Pyrus domestica*) invaded by *Gymnosporangium sabinae* is thickened, cells are enlarged and stain heavily for flavanols. Spongy parenchyma (Sp) is severely affected. The disintegrating epidermal cells stain only partially for flavanols (arrowhead). They are separated from the palisade cells (Pa) by nonstaining mucous material (asterisk). Bars = 50 μm (A and B).

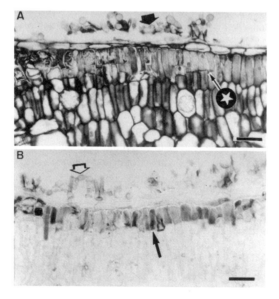

FIGURE 19.13   (A) Toluidine blue-stained scab infected leaf sector (hyphae below arrow) of *Malus domestica*. Just below the hyphae the palisade cells begin to divide periclinally (asterisk). (B) Same infection site with hyphae (black-white arrow) stained with DMACA. The upper halves of the newly divided palisade cells stain for flavanols (arrow); however, the lower halves are unresponsive. Bars = 50 μm (A and B).

pathway. It may be speculated that the flavan-3-ols represent a part of a barrier restricting fungal growth. The mechanism by which the phenols may inhibit still remains unclear; however, an inhibition of fungal enzymes is being considered (A. Kollar, B. Bühler, and D. Treutter, unpublished results). It was found that scab-resistant genotypes possess a high constitutive level of flavan-3-ols under field conditions (Treutter and Feucht, 1990b). This led to the assumption that the preformed phenolic compounds play a role in the defense mechanism.

Recent studies have confirmed the importance of flavan-3-ols in the defense of apple against scab. The activity of the phenylalanine-ammonia-lyase (PAL), the key enzyme of the phenylpropanoid pathway, was reduced by a competitive inhibitor just before inoculation of a resistant apple genotype with conidia of *Venturia inaequalis* (Mayr et al., 1997). In contrast to the control with its unaffected flavan-3-ol synthesis, the inhibitor-treated leaves showed severe symptoms of the disease. It can be deduced from this observation that the resistance of apple to scab is a function of a rapid biosynthesis induced by the action of the invader.

It must also be noted that environmental conditions modify the phenol metabolism. This could also be observed for susceptible apple cultivars, which produced high levels of flavan-3-ols under certain environmental conditions and thus were found to be field-resistant (Mayr et al., 1995, 1997). Apart from climate factors, the nutritional status of plants influences the production of secondary metabolites to a great extent (Graham, 1983). It is generally accepted that excessive nitrogen supply favors the primary metabolism and the vegetative growth at the cost of the synthesis of flavan-3-ols (Pankhurst and Jones, 1979; Hakulinen et al., 1995).

The defensive potential of a plant tissue is also influenced by other cultivation techniques such as spray-treatments with pesticides or growth regulators. It was found that surfactants may induce stress and are able to stimulate the production of flavan-3-ols as unspecific stress metabolites (Mayr et al., 1994, 1995). Growth regulators, in particular cytokinins, are also found to stimulate the phenol synthesis (Treutter and Feucht, 1986) and may alter disease resistance. In general, the effect of phytohormones on the metabolism of flavan-3-ols is very complex and closely connected to their growth-regulating activity (Zaprometov, 1988).

19.10   ROLE OF FLAVAN-3-OLS IN WOUND REACTIONS

Occasionally, young shoots of sweet cherry (*P. avium*) are attacked by insects, which ruptures the recently formed cells. The wounds produced are then enclosed by excessive flavanol deposition. These cells then disintegrate (Figure 19.14, A and B). Adjacent to these wounded cells, the flavanols were seen as 1) large inclusions within the central vacuole, or as 2) thick ribbons lining the cellular periphery or 3) attached to a cell wall site (Figure 19.14, C and D).

Fungal invasion, viewed on a microscale, can be considered as a wounding phenomenon similar to a physical incision or rupture of tissues. Many fungal diseases cause an increase in the levels of auxin, often to very high concentrations (Isaac, 1992). Conversely, around the fungal wound, a limited number of host cells may be induced to synthesize more IAA. On a molecular level, the different interactions between a distinct isoenzyme and its substrates are very complicated, mainly due to the polyfunctionality of peroxidases. Peroxidases are capable of generating H_2O_2 from which various oxygen radicals arise. Most importantly, catechin, when oxidized by peroxidases, also yields H_2O_2 (Jiang and Miles, 1993), which in turn "orchestrates" the hypersensitive response in disease resistance (Tenhaken et al., 1995).

19.10.1   RELEASE OF FLAVANOLS: KEY FUNCTION IN DEFENSE

Pathogenic fungi producing wounded tissues were described long ago as being able to cause membranes to become extremely permeable and to trigger the decompartmentalization of plant phenolic storage pools (Beckman, 1966). Leaching of fungistatic flavanols into intercellular spaces provides a toxic liquid environment against intercellularly spreading pathogens. Experiments in our laboratory

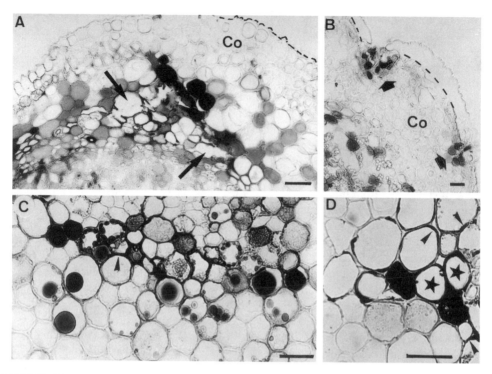

FIGURE 19.14   DMACA staining of flavanols (A–D) in cross sections of cherry shoots (*Prunus avium*) damaged by insect feeding. (A) The ruptured cells (arrows) in the cortex (Co) are surrounded by cells filled with flavanols (stippled line, epidermis). (B) Small lesions at the shoot epidermis (stippled line) induce few neighboring cells in the cortex (Co) to accumulate flavanols (arrows). (C and D). Arrowheads point out to cell wall portions with flavanols. In some cells the complete cell wall contains flavanols (asterisks). Bars = 50 μm (A–D).

proved that phenols are releasable from cells in different amounts. Abscisic acid is capable of inducing release of flavanols into the extracellular space (Feucht et al., 1997). In tissues invaded by pathogens, elevated levels of abscisic acid should be expected, which, in turn, affect membrane permeability (Owen, 1988). Both, the pathogen and the host tissue were found to synthesize abscisic acid (Isaac, 1992). Hair trichomes of *Fagus sylvatica* or *Aesculus hippocastanum* were thick-walled, allowing easy recognition of the location of flavanols. Imbibing the trichomes with ABA at 40 μM for several hours caused the flavanols to be partially released from the lumen and to enter into the walls (Figure 19.15, A and B). Callus tissues from shoot segments of sweet cherry that have been established in vitro when cultivated with ABA, show flavanol globules that stain dark blue; additional flavanols are observed at the cellular periphery, that is, inside the tonoplast practically lining the cell wall (Figure 19.15, C and D). Dislocation of flavanols toward the cell wall might reflect metabolic disturbances.

Natural cell wall flavanols are shown by DMACA staining in the inner layers of the exocarp of oak (*Quercus robur* L.) (Figure 19.16, A). External flavanols are also capable of binding to cell walls, as shown by the addition of a mixture of flavanols to tissue sections of dog rose (Figure 19.16, B).

In the calyx of *Prunus tomentosa,* the flavanols are found to be unevenly distributed due to minute lesions produced either by pathogens or insects (Figure 19.16, C). Some cell walls underlying the epidermis clearly contain flavanols. An enlarged view of the stressed epidermal cells indicates clearly that the outer cell wall stains only when the cytoplasm contains flavanols (Figure 19.16, D and E).

Severe lesions on fruit skins of *Malus domestica* as caused by a hail storm produce extensive flavanol deposition without formation of a special wound periderm (Figure 19.17).

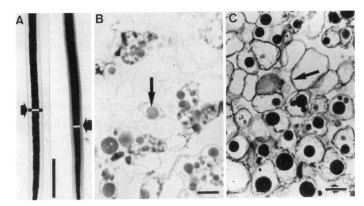

FIGURE 19.15 (A) DMACA-stained hair trichomes of beech galls without (left) and with ABA treatment (right). Imbibition of trichomes with ABA caused release of flavanols into the thick cell wall (arrow; white line, cellular lumen). Water controls were without effect. The flavanols remained in their original position in the cellular lumen (arrow, white line; cell wall, black line). (B) *In vitro* cultivated callus tissue (without ABA) of *Prunus avium* with large intravacuolar less intensely stained flavanol globules (arrow). The cellular periphery stains poorly or not at all. (C) Callus tissue cultivated on ABA showing heavy flavanol loading of the globules and additionally an intense flavanol deposition at the cellular periphery (arrow). Bars = 50 μm (A–C).

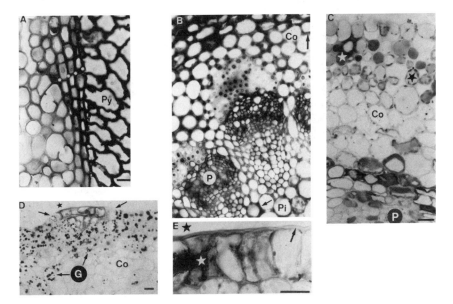

FIGURE 19.16 DMACA staining of flavanols (cross sections A–E) associated with cell walls. (A) Thick-walled cells of the fruit pericarp of *Quercus robur* stain for flavanols. A solid strand consisting of two or three small-sized cells gives intense flavanol coloration of the cell walls (4 asterisks). Enlarged folded and very thin-walled parenchyma cells (Py) of the outer pericarp showing broad flavanol ribbons inside the peripherally placed tonoplasts. (B.) Cortex (Co), phloem (P), and pith (Pi) of the petiole of dog rose (*Rosa canina*) fruit. Cell wall bound flavanols are found in the cortex (arrow) and in the pith (arrows). (C) One subepidermal cell (above white asterisk) of a damaged flower calyx of *Prunus tomentosa* showing flavanols associated with the cell wall. Cells of the Co showing chloroplasts stained with DMACA (black asterisk; P, phloem). The region is rich in flavanols. (D) Wounded epidermal cells of *P. tomentosa* showing flavanols bound to cell walls (asterisk). Note the lack of flavanols in the nonstaining cell walls of the healthy epidermis (arrows). The underlying healthy cells retain their flavanols in the intact globules (G). (E) Enlarged view of epidermal cells of *P. tomentosa*. Note stainability for flavanols in the outer epidermal cell wall (black asterisk) in response to injury. Healthy cell walls do not stain (arrow). Patches of disintegrating flavanol-positive material within the affected cells (white asterisk). Bars = 25 μm (A–E).

19.11   ADAPTATION OF PATHOGENS TO DEFENSIVE FLAVAN-3-OLS

As plants developed the use of flavan-3-ols for defense, several organisms adapted to these compounds and developed their own defensive systems. It was found that some *Rhizobium* and *Fusarium* species are able to cleave the aromatic ring (Gajendiran and Mahadevan, 1988; Barz et al., 1976). Some species of the *Colletotrichum* produce special proteins that inactivate the phenols by binding them, and thus increasing the virulence of the pathogen (Nicholsen et al., 1986). Walkinshaw (1989) showed that the rust fungus *Cronartium quercium* is able to use the proanthocyanidins as a carbon source. The reverse of a role in defense has been described for *Ulmus americana* L. A catechin-glycoside synthesized by the elm turned out to act as a feeding stimulant for the beetle *Scolytus multistriatus* (Doskotch et al., 1973).

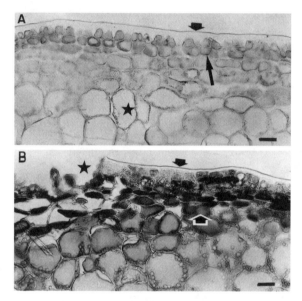

FIGURE 19.17   Cross sections of apple fruit skin stained with DMACA. (A) Healthy fruit skin of *Malus floribunda*. Thick cuticle does not stain (arrow). Some flavanol staining in the epidermal and subepidermal layers (large arrow). The large parenchyma cells show a thin flavanol layer inside the tonoplast (asterisk). (B) Wounded skin (asterisk) of *Malus floribunda* showing deep flavanol staining of hypodermal cells (black-white arrow). The black arrow points out shrinkage of the cuticle. Bars = 50 μm (A and B).

19.12   ENVIRONMENTAL STRESS—PLANT CONFRONTATION

19.12.1   AIR POLLUTION

Air pollutants cause serious impacts on biomass production of sensitive trees. Global ozone concentrations are steadily increasing, and the rates of photosynthesis are being reduced in a number of tree species. The damaging effects of H_2O_2-containing acidic fog is described by Masuch et al. (1986). Overall, there are many interactive effects of several concurrent stresses from air pollution. Ecophysiological studies are therefore complex in their final interpretations. Beech trees (*Fagus sylvatica* L.) from a heavily air-polluted site in the Black Forest (Germany) show a green-yellowish leaf appearance after prolonged periods of hot and dry weather (Feucht et al., 1994, 1997). The conversion of chloroplasts into chromoplasts is as apparent as the visible damage. But it is of physiological significance that carotinoids, in addition to vitamins C and E (Kunert and Ederer, 1985), play a pivotal role as free radical scavengers delaying the destruction of the green pigments. A similar role is ascribed to the flavanols. Analysis of the yellowing beech leaves for flavanol levels revealed a several-fold increase over that found in the green controls (Feucht et al., 1994, 1997). Promotion of flavanol production consumes some of the energy normally used for growth.

The drastic changes in the flavanols from healthy to yellowing leaves are shown in Figures 19.18 and 19.19. The dissolution of the flavanol globules in response to the stress is noteworthy. At lower flavanol concentrations a varied intracellular distribution of the leaching flavanols is apparent.

19.12.2 CLIMATIC STRESS—*PRUNUS AVIUM*

Sweet cherry trees were freeze-damaged at temperatures of $-25°C$ in the orchard. Investigation in following spring revealed that the shoots had suffered from these extreme conditions. Cambial division and spring flush of the buds were somewhat retarded. The young xylem as produced prior to the frost period was analyzed the following spring, and showed batches with intense blue flavanol coloration (Figure 19.20, A and B). The walls of the tracheids are linked with flavanols. Chilling of these cell walls apparently results in structural changes allowing the entry of prestored flavanols. Alternatively, during the following spring, higher amounts of flavanols might have been synthesized in response to the freezing stress. Evidently, the flavanols were not oxidized in the xylem to structures unresponsive to DMACA. Auto-oxidation, chemical oxidation, or peroxidase-catalyzed oxidation leading to destruction of flavanols (Guyot et al., 1996) is expected to occur, particularly in the cell wall.

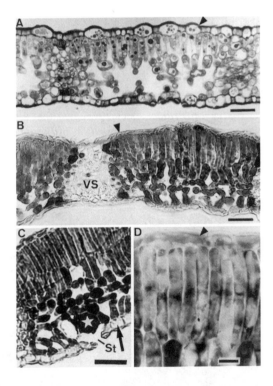

FIGURE 19.18 Flavanols in leaves of *Fagus sylvatica* (A, toluidine blue staining; B, C, and D, DMACA staining). (A) Dark green flavanol globules in the upper (arrowhead) and lower epidermis and in the mesophyll of healthy leaves. (B) All cells of the stressed yellowed beech leaves being filled with intensely staining flavanols, except lower epidermis and vascular strand (VS). The upper epidermis is shrunken (arrowhead). (C) Maximal flavanol deposition in a stressed leaf sector. Enlarged spongy parenchyma cell (white asterisk) with abundant flavanols (St, stomatal cavity). The lower epidermis still does not stain for flavanols (arrow). (D) Moderate flavanol staining throughout the palisade cells as well as in the upper epidermis (arrowhead). Bar = 50 μm (A, B, and C), and 10 μm (D).

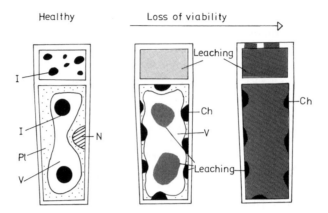

FIGURE 19.19   Schematic illustration of the flavanol deposition as evidenced in Figure 19.18. Flavanol inclusions (I) located in the central vacuole (V) and stress-induced flavanol release outside the inclusions. Leaching flavanols can be deposited on chloroplasts (Ch) and cytoplasm (Pl). With increasing loss of viability the cells stain darkly for flavanols.

19.13   PLANT–PLANT CONFRONTATION

19.13.1   GRAFT INCOMPATIBILITY—A SPECIAL CASE OF ALLELOPATHIC INTERACTION

The incompatibility of interspecific graftings is characterized by a necrotic line just at the union (Tanrisever and Feucht, 1978; Poessel et al., 1980). This brown to black cell layer can be interpreted as a defense reaction since the metabolisms of the two partners are incongruent and do not form a durable connection. Several studies on *Prunus* species revealed that a dramatic accumulation of flavan-3-ols in living cells precedes the definitive separation of the diverging metabolic systems (Treutter et al., 1986a; Feucht and Treutter, 1991; Errea et al., 1992a,b). It should be emphasized, however, that the accumulation of flavan-3-ols occurs at a very late stage when incompatibility between the two partners is completely manifested. In the case of delayed incompatibility of *Prunus-*

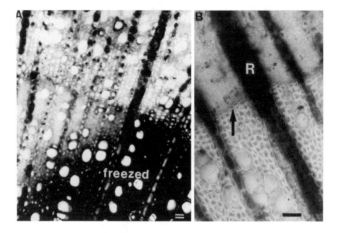

FIGURE 19.20   Chilling stress and flavanol deposition. (A) DMACA staining of xylem of *Prunus avium* as affected by freezing damage. The injured xylem cell walls are heavily loaded with flavanols. (B) Non-freezed controls show the typical flavanol staining of the xylem rays (R). Arrow indicates cambium. Bars = 50 μm (A and B).

heterografts, the first stage is characterized by an enforced synthesis of monophenols with *p*-coumaroyl-glucoside, the flavanone naringenin 7-glucoside, and the isoflavone genistein 7-glucoside (Treutter et al., 1986a,b; Treutter, 1989). The accumulation of soluble flavan-3-ols further coincides with the formation of a border zone at the union including physical barriers (Feucht and Treutter, 1991; Treutter and Feucht, 1991; Feucht et al., 1983a,b), thus amplifying their efficiency by adding a chemical defense. The flavan-3-ols may play a role as part of an unspecific defensive reaction at the graft union.

Longitudinal sections were made through the graft union of an incompatible *P. avium/P. cerasus* combination (Figure 19.21). It was suggested that an increased availability of cytokinins transported from the roots is affected by grafting. Cytokinin entering the union disturb physiological processes in the cell layers of *Prunus avium*. Tissues located directly along the line of the union show a more disorganized growth. Strongly staining, irregular clumps of flavanols were found to occur more frequently.

The formation of a barrier zone between the rootstock and the scion results in a restricted acropetal transport of water and nutrients as well as in a diminished basipetal flux of assimilates (Schmid and Feucht, 1986a,b; Moing et al., 1987). As a consequence, the metabolism of the whole tree is affected. This is apparent from the occurrence of yellowing and epinastic leaves with a tendency to accumulate carbohydrates, and a significant rise of flavan-3-ols (Bauer et al., 1989). The beneficial effects of flavan-3-ols may be the reason that the plant accumulates these compounds.

19.14 FLAVANOLS AND DEFENSE—A UNIFYING CONCEPT

Any condition in which cellular redox homeostasis is disrupted can be defined as oxidative stress (Alscher et al., 1997). The reactive oxygen species are capable of operating destructively at different sites in the cell. Even the nucleus is exposed to attacks by toxic radicals (Britt, 1996). Summing up, integrated defense processes against biotic or abiotic stress include the following flavanol functions:

1. Flavanols participate in the so-called front line defense action. Exposed and bordering cells/tissues as well as wounds or infections are the preferred sites for flavanol deposition.
2. Dimeric, trimeric, and tetrameric proanthocyanidins were found to be prominent protein kinase inhibitors mediating defense functions related to herbivory and fungal invasion (Polya and Foo, 1994).
3. As a response to stress, ions and organic compounds freely pass through membranes. Even if it is difficult to quantify, it appears the released flavanol molecules traversing the cytosol are capable of modulating enzymes.
4. Stimulation of ATPases by catechin (Feucht and Treutter, 1995), in a mode similar to auxin, is an indirect proof that flavanols act in modulating electron transport. Auxin affects microviscosity of membranes, and flavanols might participate synergistically.
5. Reactivation by flavanols of mitotically dormant mesophyll cells to proliferating undifferentiated cell clusters is a method of repairing wounded tissue.
6. Catechin is capable of promoting the biosynthesis of proteins (Feucht and Schmid, 1980) and hence may be involved in growth processes.
7. Flavanols may regulate the IAA peroxidase pathway.
8. The superior antioxidant status of catechin is shown by its effectiveness in overcoming lethal processes such as that induced by paraquat. Flavanols participate as antioxidants in breaking radical-induced chain reactions.
9. Besides those influences on plant physiology, the direct antimicrobial and toxic effects complete the possible roles of flavan-3-ols in plant defense.

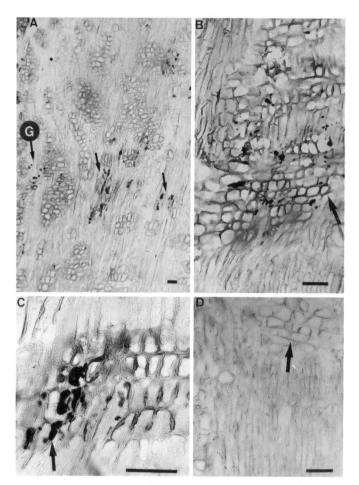

FIGURE 19.21   DMACA-stained longitudinal sections (A–D) of the phloem in an incompatible graft union of *P. avium/P.cerasus*. (A) High frequency of flavanol deposits (arrows) just above the union (about 2 cm). (B) In the transition region of the two species, flavanols appear to be closely attached to the inner side of the cell walls of the phloem rays (arrow). (C) Enlarged view from the transition zone with severe disintegration of ray cells and dense flavanol depositions (arrow). (D) Phloem area just below the union (about 2 cm). Note the absence of distinct flavanol depositions, except some staining occurring in the ray cells (area above the arrow). Bars = 50 μm (A–D).

REFERENCES

Achor, D. S., Albrigo, L. G., and McCoy, C. W., Developmental anatomy of lesions on 'Sunburst' mandarin leaves initiated by citrus rust mite feeding, *Amer. Soc. Hort. Sci.,* 116, 663, 1991.

Adaskaveg, J. E., Defense mechanisms in leaves and fruit of trees to fungal infection, in *Defense Mechanisms of Woody Plants Against Fungi,* Blanchette, R. A., and Biggs, A. R., Eds., Springer-Verlag, Berlin, 1992, 207.

Affany, A., Salvayre, R., and Douste-Blazy, L., Comparison of the protective effects of various flavonoids against lipid peroxidation of erythrocyte membranes (induced by cumene hydroperoxide), *Fundam. Clin. Pharm.,* 1, 451, 1987.

Alcubilla, M., Diaz-Palacio, P., Kreutzer, K., Laatsch, W., Rehfuess, K. E., and Wenzel, G., Beziehungen zwischen dem Ernahrungsustand der Fischte (*Picea abies* Karst.), ihrem Kernfaulebefall und der Pilzhemmung ihres Basts, *Eur. J. Forest Path.,* 2, 100, 1971.

Alscher, R. G., Donahue, J. L., and Cramer, C. L., Reactive oxygen species and antioxidants: relationships in green cells, *Physiol. Plant.,* 100, 224, 1997.

Antonelli, E. and Daly, J. M., Decarboxylation of indoleacetic acid by near-isogenic lines of wheat resistant or susceptible to *Puccinia graminis* f. sp. *tritici, Phytopathology,* 56, 610, 1966.

Ariga, T., Koshiyama, I., and Fukushima, D., Antioxidative properties of procyanidins B-1 and B-3 from Azuki beans in aqueous systems, *Agr. Biol. Chem.,* 52, 2717, 1988.

Arora, J. P. S., Pal, C., and Jain, P. B., pH-metric studies on the binding of catechin to proteins, *Studia Biophysica,* 126, 61, 1988.

Asquith, T. N. and Butler, L. G., Interactions of condensed tannins with selected proteins, *Phytochemistry,* 25, 1591, 1986.

Barz, W., Schlepphorst, R., and Laimer, J., Uber den Abbau von Polyphenolen durch Pilze der Gatfung *Fusarium, Phytochemistry,* 15, 87, 1976.

Bauer, H., Treutter, D., Schmid, P. P. S., Schmitt, E., and Feucht, W., Specific accumulation of 0-diphenols in stressed leaves of *Prunus avium, Phytochemistry,* 28, 1363, 1989.

Baumann, J., Wurm, G., and Bruchhausen, V., Hemmung der Prostaglandin Synthetase durch Flavonoide und Phenolderivate in Vergleich mit deren O_2 Radikalfangereigenschaften, *Arch. Pharm.,* 313, 330, 1980.

Beart, J. E., Lilley, T. H., and Haslam, E., Polyphenol interactions. Part 2. Covalent binding of procyanidins to proteins during acid-catalysed decomposition: observation on some polymeric proanthocyanidins, *J. Chem. Soc.,* 1439, 1985.

Beckman, C. H., Cell irritability and location of vascular infections in plants, *Phytopathology,* 56, 821, 1966.

Bell, T. A. and Etchells, J. L., Pectinase inhibitor in grape leaves, *Bot. Gaz.,* 119, 192, 1958.

Bell, T. A., Etchells, J. L., Williams, C. F., and Porter, W. L., Inhibition of pectinase and cellulase by certain plants, *Bot. Gaz.,* 123, 220, 1962.

Bernays, E. A., Plant tannins and insect herbivores: an appraisal, *Ecol. Entomol.,* 6, 353, 1981.

Britt, A. B., DNA damage and repair in plants, *Ann. Rev. Plant Physiol. Plant Mol. Biol.,* 47, 75, 1996.

Cadman, C. H., Inhibition of plant virus infection by tannins, *in Phenolics in Plants in Health and Disease,* Pridham, J. B., Ed., Pergamon Press, Oxford, 1960, 101.

Chalker-Scott, L. and Krahmer, R. L., Microscopic studies of tannin formation and distribution in plant tissues, in *Chemistry and Significance of Condensed Tannins,* Hemingway, R. W. and Karchesy, J. J., Eds., Plenum Press, New York, 1989, 345.

Clemetson, C. A. B. and Andersen, L., Plant polyphenolics as antioxidants for ascorbic acid, *Ann. New York Acad. Sci.,* 341, 1966.

Creasy, L. L., The increase in phenylalanine ammonia-lyse activity in strawberry leaf disks and its correlation with flavonoid synthesis, *Phytochemistry,* 7441, 1968.

Dempsey, D. A. and Klessig, D. F., Signals in plant disease resistance, *Bull. Inst. Pasteur,* 93, 167, 1995.

Doskotch, R. W., Mikhail, A. A., and Chatterjii, S. J., Structure of water-soluble feeding stimulant for *Scolyptus multistriatics, Phytochemistry,* 12, 1153, 1973.

Elstner, E. F., Osswald, W., Volpert, R., and Schempp, H., Phenolic compounds, *in* Geibel, M., Treutter, D., Feucht, W., Eds., International Symposium on Natural Phenol in Plant Resistance, Weihenstephan. *Acta Hort.* (Suppl.) 381, 304, 1994.

Erdei, L., Szabo-Nagy, A., and Laszlavik, M., Effects of tannin and phenolics on H^+-ATPase activity in plant plasma membrane, *J. Plant Physiol.,* 144, 49, 1994.

Errea, P., Treutter, D., and Feucht, W., Scion-rootstock effects on the content of flavan-3-ols in the union of heterografts consisting of apricots and diverse *Prunus* rootstocks, *Gartenbauwiss,* 57, 131, 1992a.

Errea, P., Treutter, D., and Feucht, W., Specificity of individual flavan-3-ols interfering with the grafting stress with apricots, *Angew. Bot.,* 66, 21, 1992b.

Etchells, J. L., Bell, T. A., and Williams, C. F., Inhibition of pectinolytic and cellulolytic enzymes in cucumber by Scuppernong grape leaves, *Food Technol.,* 12, 204, 1958.

Faulkner, I. J. and Rubery, P. H., Flavonoid and flavonoid sulphates as probes of auxins-transport regulation in *Cucurbita pepo* hypocotyl and vesicles, *Planta,* 186, 618, 1992.

Feeny, P. P., Plant apparency and chemical defence, *Phytochemistry,* 10, 1, 1976.

Feucht, W. and Nachit, M., Flavolans and growth-promoting catechins in young shoot tips of *Prunus* species and hybrids, *Physiol. Plant.,* 40, 230, 1977.

Feucht, W. and Schmid, P. P. S., Effect of ortho-dihydroxyphenols on growth and protein pattern of callus cultures from *Prunus avium, Physiol. Plant.,* 50, 309, 1980.

Feucht, W. and Schmid, P. P. S., Selektiver histochemischer Nachweis von Flavanen (Catechinen) mit p-dimethylminozimtaldehyd in Sprossen einiger Obstgehölze, *Gartenbauwiss,* 48, 119, 1983.

Feucht, W. and Treutter, D., Phenol gradient in opposing cells of *Prunus* heterografts, *Adv. Hort. Sci.,* 3, 107, 1991.

Feucht, W. and Treutter, D., Catechin effects on growth related processes in cultivated calli of *Prunus avium, Gartenbauwiss,* 60, 7, 1995.

Feucht, W., Treutter, D., and Christ, E., Cell division as a response to shoot hole infection in *Prunus domestica*: involvement of flavanols, *J. Plant Dis. Protect.,* 100, 488, 1993.

Feucht, W., Treutter, D., and Christ, E., Accumulation of flavanols in yellowing beech leaves from forest decline, *Tree Physiol.,* 14, 403, 1994a.

Feucht, W. D., Christ, E., and Treutter, D., Flavanols as defence barriers of the fruit surface, *Angrew Bot.,* 68, 122, 1994b.

Feucht, W., Treutter, D., and Christ, E., Shoot hole disease in sour cherry: defensive flavanol barrier, *J. Plant Dis. Prot.,* 103, 279, 1996a.

Feucht, W., Treutter, D., and Christ, E., Flavanols in grape vine: *in vitro* accumulation and defence reactions in shoots, *Vitis,* 35, 113, 1996b.

Feucht, W., Treutter, D., Santos-Buelga, C., and Christ, E., Catechin as a radical scavenger in paraquat-treated *Prunus avium, Angew. Bot.,* 70, 119, 1996c.

Feucht, W., Treutter, D., and Christ, E., Role of flavanols in yellowing beech trees of the Black Forest, *Tree Physiol.,* 17, 335, 1997a.

Field, C. and Mooney, H. A., The photosynthesis-nitrogen relationship in wild plants, in *On the Economy of Plant Form and Function,* Givnish, T. J., Ed., Cambridge University Press, Cambridge, 1986, 25.

Foo, L. Y. and Karchesy, J. J., Chemical nature of phlobaphenes, in *Chemistry and Significance of Condensed Tannins,* Hemingway, R. W. and Karchesy, J. J., Eds., Plenum Press, New York, 1989, 109.

Forrest, G. I. and Bendall, D. S., The distribution of polyphenols in the tea plant (*Camellia sinensis* L.), *Biochem. J.,* 113, 741, 1969.

Gaffney, S. H., Martin, R., Lilley, T. H., Magnolato, E., and Haslam, D., The association of polyphenols with caffeine and *a* and *b* cyclodextrin in aqueous media, *J. Chem. Soc. Chem. Commun.,* 107, 1986.

Gajendiran, N. and Mahadevan, A., Utilization of catechin by *Rhizobium* sp., *Plant Soil,* 108, 263, 1988.

Graham, R. D., Effects of nutrient stress on susceptibility of plants to disease with particular reference to the trace elements, *Adv. Bot. Res.,* 10, 221, 1983.

Grambow, H. J., Pathway and mechanism of the peroxidase-catalyzed degradation of indole-3-acetic acid, in *Molecular and Physiological Aspects of Plant Peroxidases,* Greppin, H., Penel, C., and Gaspar, T., Eds., University of Geneva, Switzerland, 1986, 31.

Griffiths, D. W. and Jones, D. I. H., Cellulose inhibition by tannins in the tests of field beans (*Vicia faba*), *J. Sci. Food Agric.,* 28, 983, 1977.

Griffiths, W., Condensed tannins, in *Toxic Substances in Crop,* D'Mello, J. P. F., Duffus, C. M., and Duffus, J. H., Eds., Cambridge University Press, Cambridge, 1991, 181.

Gutmann, M., Localization of proanthocyanidins using *in situ*-hydrolysis with sufuric acid, *Biotech. Histochem.,* 68, 161, 1993.

Gutmann, M. and Feucht, W., A new method for selective localization of flavan-3-ols in plant tissues involving glycolmethacrylate embedding and microwave irradiation, *Histochemistry,* 96, 83, 1991.

Guyot, S., Vercauteren, J., and Cheynier, V., Structural determination of colourless and yellow dimers resulting from (+)-catechin coupling catalysed by grape phenoloxidase, *Phytochemistry,* 42, 1279, 1996.

Hagerman, A. E. and Butler, L. G., The specificity of proanthocyanidin-protein interactions, *J. Biol. Chem.,* 256, 4494, 1981.

Hakulinen, J., Julkunen-Tiitto, R., and Tahvanainen, J., Does nitrogen fertilization have an impact on the trade-off between willow growth and defensive secondary metabolism?, *Trees,* 9, 235, 1995.

Hara, Y. and Watanabe, M., Antibacterial activity of tea polyphenols against *Clostridium botulinum, J. Jap. Soc. Food Sci. Technol.,* 36, 951, 1989.

Haslam, E., *Plant Polyphenols,* Cambridge University Press, Cambridge, 1989.

Hathway, D. E. and Seakins, J. W. T., The influence of tannins in the degradation of pectin by pectinase enzymes, *Biochem. J.,* 70, 158, 1958.

Hunter, R. E., Effects of catechin in culture and in cotton seedlings on growth and polygalacturonase activity of *Rhizoctonia solani, Phytopathology,* 68, 1032, 1978.

Isaac, S., *Fungal-Plant Interactions,* Chapman and Hall, London, 1992.

Jacobs, M. and Rubery, P. H., Naturally-occurring auxin trasport regulators, *Science,* 241, 346, 1988.

Jiang, Y. and Miles, P. W., Generation of H_2O_2 during enzymatic oxidation of catechin, *Phytochemistry,* 33, 29, 1993.

Jones, W. T., Broadhurst, R. B., and Lyttleton, J. W., The condensed tannins of pasture legume species, *Phytochemistry,* 15, 1407, 1976.

Karchesy, J. J., Bae, Y., Chalker-Scott, L., Helm, R. H., and Foo, L. Y., Chromatorgraphy of proanthocyanidins, in *Chemistry and Significance of Condensed Tannins,* Hemingway, R. W. and Karchesy, J. J., Eds., Plenum Press, New York, 1989, 139.

Kefeli, V. I. and Kutacek, M., Phenolic substances and their role in plant growth regulation, in *Plant Growth Regulation,* Pilet, P. E., Ed., Springer-Verlag, 1977, 181.

Kennedy, J. A., Munro, M. H. G., Powell, H. K. J., Porter, L. J., and Foo, L. Y., The protonation reactions of actechin, epicatechin and related compounds, *Aust. J. Chem.,* 37, 885, 1984.

Kollar, A., Characterization of specific induction, activity and isoenzyme polymorphism of extracellular cellulases from *Venturia inaequalis* detected *in vitro* and on the host plant, *Mol. Plant-Microbe Interact.,* 7, 603, 1994.

Krylov, S. N., Krylova, S. M., Chebotarev, I. G., and Chebotareva, A. B., Inhibition of enzymatic indole-acetic acid oxidation by phenols, *Phytochemistry,* 36, 263, 1994.

Kunert, K. J. and Ederer, M., Leaf aging and lipid peroxidation: the role of antioxidants vitamins C and E, *Physiol. Plant.,* 65, 85, 1985.

Laks, P. E., Condensed tannins as a source of novel biocides, in *Chemistry and Significance of Condensed Tannins,* Hemingway, R. W. and Karchesy, J. J., Eds., Plenum Press, London, 1989, 503.

Lavee, S., Avidan, N., and Pierik, R. L. M., Chlorogenic acid—an independent morphogensis regulator or a cofactor, *Acta Horticult.,* 381, 405, 1994.

Lees, G. L., Wall, K. M., Beveridge, T. H., and Suttill, N. H., Localization of condensed tannins in apple fruit peel, and seeds, *Can. J. Bot.,* 73, 1897, 1995.

Lois, R. and Buchanan, B. B., Severe sensitivity to ultraviolet radiation in an *Arabidopsis* mutant deficient in flavonoid accumulation, *Planta,* 194, 504, 1994.

Loomis, W. D. and Battaile, J., Plant phenolic compounds and the isolation of plant enzymes, *Phytochemistry,* 5, 423, 1966.

Lyndon, R. F., *Plant Development. The Cellular Basis,* Unwin Hyman, London, 1990.

Malterud, K. E., Bremnes, T. E., Faegr, A., and Moe, T., Flavonoids from the wood of *Salix caprea* as inhibitors of wood-destroying fungi, *J. Nat. Prod.,* 48, 559, 1985.

Masuch, G., Kettrup, A., Mallant, R. K., and Slanina, J., Effects of H_2O_2-containing acidic fog on young trees, *Intern. J. Environ. Anal. Chem.,* 27, 183, 1986.

Mathews, S., Mila, I., Scalbert, A., and Donnelly, D. M. X., Insoluble proanthocyanidins in barks, *Polyphenols Comm.,* 96, 229, 1996.

Mayer, A. M. and Havel, E., Polyphenol oxidases in fruits—changes during ripening, in *Recent Advances in the Biochemistry of Fruits and Vegetables,* Friend, J. and Rhodes, M. J. C., Eds., Academic Press, London, 1981, 161.

Mayr, U., Batzdorfer, R., Treutter, D., and Feucht, W., Surfactant-induced changes in phenol content of apple leaves after wounding, *Acta Horticult.,* 381, 479, 1994.

Mayr, U., Fünfgelder, S., Treutter, D., and Feucht, W., Induction of phenol accumulation by pesticides and the control of environmental factors, *Proc. Eur. Found. Plant Pathol.,* 399, 1995a.

Mayr, U., Treutter, D., Santos-Buelga, C., Bauer, H., and Feucht, W., Developmental changes in the phenol concentration of 'Golden Delicious' apple fruit and leaves, *Phytochemistry,* 38, 1151, 1995b.

Mayr, U., Michalek, S., Treutter, D., and Feucht, W., Phenolic compounds of apple and their relationship to scab resistance, *J. Phytopathol.,* 145, 69, 1997.

McGraw, G. W., Reactions at the A-ring of proanthocyanidins, in *Chemistry and Significance of Condensed Tannins,* Hemingway, R. W., and Karchesy, J. J., Eds., Plenum Press, New York, 1989, 227.

Middleton, E., Faden, H., Drzewiecki, G., and Perrissoud, D., Correlation of antiviral and histamine release-inhibitory activity of several synthetic flavonoids, in *Plant Flavonoids in Biology and Medicine. Biochemical, Pharmacological, and Structure-Activity Relationships,* Cody, V., Middleton, E., and Harborne, J. B., Eds., Alan R. Liss, New York, 1986, 541.

Moing, A., Salesses, G., and Saglio, H., Growth and the composition and transport of carbohydrate in compatible and incompatible peach/plum grafts, *Tree Physiol.,* 3, 345, 1987.

Mori, A., Nishino, C., Enoki, N., and Tawata, S., Antibacterial activity and mode of action of plant flavanoids against *Proteus vulgaris* and *Stephylococcus aureus, Phytochemistry,* 26, 2231, 1987.

Mukherjee, S. P. and Choudhuri, M. A., Implications of water stress-induced changes in the levels of endogenous ascorbic acid and hydrogen peroxide in *Vigna* seedlings, *Physiol. Plant.*, 58, 166, 1983.

Müller, K. O. and Börger, H., Experimentelle Untersuchungen über die Phytophthora Resistenz der Kartoffel,
Arb. Bio. Anst. Reichsanst., 23, 189, 1941.

Nicholson, R. L., Butler, L. G., and Asquith, T. N., Glycoproteins from *Colletotrichum graminicola* that bind phenols: implications for survival and virulance of phytopathogenic fungi, *Phytopathology*, 76, 1315, 1986.

Nitsch, J. P. and Nitsch, C., Composés phenoliques et croissance végétale, *Annal. Physiol. Vég.*, 4, 211, 1962.

Ohigashi, H., Minami, S., Fukui, H., Koshimizu, K., Mizutani, F., Sugiura, A., and Tomana, T., Flavanols as
plant growth inhibitors from the roots of peech *Prunus persica* 'Hakuto,' *Agric. Biol. Chem.*, 46, 2555,
1982.

Okada, F., Takeo, T., Okada, S., and Tamemasa, O., Antiviral effect of theaflavins on tobacco mosaic virus,
Agric. Biol. Chem., 41, 791, 1977.

Owen, J. H., Role of ABA in a Ca^{2+} second messenger system, *Physiol. Plant.*, 72, 637, 1988.

Pankhurst, C. E. and Jones, W. T., Effectiveness of *Lotus* root nodules. III. Effect of combined nitrogen on nod-
ule effectiveness and flavanol synthesis in plant roots, *J. Exp. Bot.*, 30, 1109, 1979.

Perchellet, J. P., Gao, X. M., Perchellet, E. M., Gali, H. O., Rodriguez, L., and Hemingway, R. W., Anti-tumor
promoting activity of lobolly pine bark condensed tannin in mouse epidermis *in vivo, Polyphenols*,
Brouillard, R., Yay, M., and Scalbert, A., eds., INRA, 94, 407, 1994.

Pierpoint, W. S., Reaction of phenolic compounds with proteins, and their relevance to the production of leaf
protein, in *Leaf Protein Concentrates*, Telek, L. and Graham, H. D., Eds., Avi Publishing, Westport, 1983,
235.

Pizzi, A. and Cameron, F. A., Flavonoid tannins—structural wood components for drought-resistant mechanisms of plants, *Wood Sci. Technol.*, 20, 119, 1986.

Poëssel, J.-L., Martinez, J., Macheix, J.-J., and Jonard, R., Variation saisonniéres de l'aptitude au greffage in
vitro d'apex de Pêcher (*Prunus persica* Batsch). Reactions avee les teneurs en composés phenoliques
endogenes et les activities peroxoidases et polyphenolxydasique, *Physiol. Vég.*, 18, 665, 1980.

Pollard, A., Kieser, M. E., and Sissons, D. J., Inactivation of pectic enzymes by fruit phenolics, *Chem. Ind.*, 952,
1958.

Polya, G. M. and Foo, L. Y., Inhibition of eukaryotic signal-regulated protein kinases by plant-derived catechin-
related compounds, *Phytochemistry*, 35, 1399, 1994.

Porter, L. J., Flavans and proanthocyanidins, in *The Flavonoids*, Harborne, J. B., Ed., Chapman and Hall,
London, 1988, 21.

Porter, L. J., Flavans and proanthocyanidins, in *The Flavonoids*, Harborne, J. B., Ed., Chapman and Hall,
Cambridge, 1994, 23.

Porter, W. L., Schwartz, J. H., Bell, T. A., and Etchells, J. L., Probable identity of the pectinase inhibitor in grape
leaves, *J. Food Sci.*, 26, 600, 1961.

Rao, S. S. R. and Rao, K. V. N., Fungitoxic activity of proanthocyanidins, *Ind. J. Plant Physiol.*, 19, 278, 1986.

Reese, E. T., *Advances in enzymic hydrolysis of cellulose and related materials*, Pergamon Press, London, 1963.

Regnault-Roger, C. R., The nutritional incidence of flavonoid: some physiological and metabolic considerations, *Experientia*, 44, 725, 1988.

Ribereau-Gayon, P., *Plant Phenolics*, Oliver & Boyd, Edinburgh, 1972.

Rice-Evans, C., Plant polyphenols: free radical scavengers or chain-breaking antioxidants?, *Biochem. Soc.
Symp.*, 61, 103, 1995.

Sakanaka, S., Sato, T., Kim, M., and Yamamoto, T., Inhibitory effects of green tea polyphenols on glucan synthesis and cellular adherence of cariogenic sterptococci, *Agric. Biol. Chem.*, 54, 1925, 1990.

Salvayre, R., Nègre, A., Affany, A., Lenoble, M., and Douste-Blazy, L., Protective effect of plant flavonoids,
analogs and vitamin E against lipid peroxidation of membranes, in *Plant Flavonoids in Biology and
Medicine II Biochemical, Cellular, and Medicinal Properties*, Cody, V., Middleton, E., Harborne, J. B., and
Beretz, A., Eds., Alan R. Liss, New York, 1988, 313.

Schlösser, E., Performed phenols as resistance factors, *Acta Hortic.*, 381, 615, 1994.

Schmid, P. P. S. and Feucht, W., Carbohydrates in the phloem of *Prunus avium/Prunus cerasus* graftings and of
homospecific controls, *Angew. Botanik*, 60, 201, 1986a.

Schmid, P. P. S. and Feucht, W., Kohlehydrate, Chlorophyll, Protein, Polyphenole in Blättern von
Kirchkombinationen (*Prunus avium* L. auf *P. cerasus* L.) mit unterschiedlichen Streβsymptomen, *Angew.
Botanik*, 60, 365, 1986b.

Schultz, J. C., Tannin-insect interactions, in *Chemistry and Significance of Condensed Tannins,* Hemingway, R. W. and Karchesy, J. J., Eds., Plenum Press, New York, 1989, 417.

Schultz, J. C. and Baldwin, T. T., Oak leaf quality declines in response to defoliation by gypsy moth larve, *Science,* 217, 149, 1982.

Selway, J. W. T., Antiviral activity of flavones and flavans, in *Plant Flavonoids in Biology and Medicine, Biochemical, Pharmacological, and Structure-activity Relationships,* Cody, V., Middleton, E., and Harborne, J. B., Eds., Alan R. Liss, New York, 1986, 521.

Serkedjeva, S., Manolova, N., and Bankova, V., Anti-influenza virus effect of some propolis constituents and their analogues, *J. Nat. Prod. Lloydia,* 55, 294, 1992.

Shantz, E. M. and Steward, F. C., The general nature of some nitrogen free growth promoting substances from *Aesculus* and *Cocos, Plant Physiol.* (Suppl), 30, 35, 1955.

Somaatmadja, D., Powers, J. J., and Wheeler, R., Action of leucoanthocyanins of cabernet grapes on reproduction and respiration of certain bacteria, *Am. J. Enol.,* 16, 54, 1965.

Spencer, C. M., Cai, Y., Martin, R., Gaffney, S. H., Goulding, P. N., Magnolato, D., Lilley, T. H., and Haslam, E., Polyphenol complexation—some thoughts and observations, *Phytochemistry,* 27, 2397, 1988.

Stafford, H. A., Roles of flavonoids in symbiotic and defense functions in legume roots, *Bot. Rev.,* 63, 27, 1997.

Stafford, H. A. and Lester, H. H., Flavan-3-ol biosynthesis. The conversion of (+)-dihydroquercetin and flavan-3,4-cis-diol (leucocyanidin) to (+)-catechin by reductases extracted from cell suspension cultures of Douglas fir, *Plant Physiol.,* 76, 184, 1984.

Stafford, H. A. and Lester, H. H., The conversion of (+)-dihydromyricetin to its flavan-3,4-diol (leucodelphinidin) and to (+)-gallocatechin by reductases extracted from tissue cultures of *Ginkgo biloba* and *Pseudotsuga menziesii, Plant Physiol.,* 78, 791, 1985.

Steenkamp, J. A., Steynberg, J. P., Brandt, E. V., Ferreira, D., and Roux, D. G., Phlobatannins, a novel class of ring isomerized condensed tannins, *J. Chem. Soc. Chem. Commun.,* 1, 671, 1985.

Stonier, T. and Yang, H., Studies on auxins protectors. XI. Inhibition of peroxidase catalzed oxidation of glutathione by auxin protectors and o-dihydroxyphenols, *Plant Physiol.,* 51, 391, 1973.

Strobel, N. E. and Sinclair, W. A., Role of flavonolic wall infusion in the resistance induced by *Laccaria bicolor* to Fusarium oxysporum in primary roots of Douglas fir, *Phytopathology,* 81, 420, 1991.

Strumeyer, D. H. and Malin, M. J., Resistance of extracellular yeast invertase and other glycoproteins to denaturation by tannins, *Biochem. J.,* 118 (1970).

Tanrisever, A. and Feucht, W., Strutur des Phloems bei veschieden stark wachsenden kirschartigen Prunusgehölzen, *Gartenbauwiss.,* 43, 59, 1978.

Teissedre, P. C., Frankel, E. N., Waterhouse, A. L., Peleg, H., and German, J. B., Inhibition of *in vitro* human LDL oxidation by phenolic antioxidants from grapes and wines, *J. Sci. Food Agric.,* 70, 55, 1996.

Tenhaken, R., Levine, A., Brisson, L. F., Dixon, R. A., and Lamb, C., Function of the oxidative burst in hypersensitive disease resistance, *Proc. Natl. Acad. Sci. USA,* 92, 4158, 1995.

Thompson, R. S., Jacques, D., Haslam, E., and Tanner, R. J. N., Plant proanthocyanidins. Part I. Introduction: the isolation, structure and distribution in nature in plant procyanidins, *J. Chem. Soc. Perkin Trans.,* 1, 1387, 1972.

Thornburg, R. N. and Li, X., Wounding *Nicotiana* leaves causes a decline of endogenous indoleacetic acid, *Plant Physiol.,* 96, 802, 1991.

Treutter, D., Chemical reaction detection of catechins and proanthocyanidins with 4-dimethylaminocinnamaldehyde, *J. Chromatogr.,* 467, 185, 1989a.

Treutter, D., Polyphenole des Phloems in Beziehung zur Inkompatibilität von interspezifischen *Prunus*-Veredlungen (*Prunus avium* L., *Prunus cerasus* L.). II. Akkumulation von p-Cumaroylglucose über der Veredlungsstelle, *Gartenbauwiss.,* 54, 261, 1989b.

Treutter, D., Flavane in pflanzlichen Lebensmitteln, in *Deutsche Gesellschaft für Qualitätsforshung; XXXI, Vortragstagung,*Nahrungsmittel, P., Ed., 31, 81, 1996.

Treutter, D. and Feucht, W., Taxonomically relevant flavonoid glycosides of *Prunus*-phloem and their response according to various physiological conditions, *Groupe Polyphenols: Bulletin de Liaison,* 13, 45, 1986.

Treutter, D. and Feucht, W., The pattern of flavan-3-ols in relation to scab resistance of apple cultivars, *J. Hort. Sci.,* 65, 511, 1990a.

Treutter, D. and Feucht, W., Accumulation of flavan-3-ols in fungus-infected leaves of Rosaceae, *J. Plant Dis. Protect.,* 97, 634, 1990b.

Treutter, D. and Feucht, W., Accumulation of phenolic compounds above the graft union of cherry trees, *Gartenbauwiss.,* 56, 134, 1991.

Treutter, D., Feucht, W., and Schmid, P. P. S., Polyphenole des Phloems in Beziehung zur Inkompatibilität von interspezifischen *Prunus*-Veredlungen (*Prunus avium* L., *Prunus cerasus* L.). I. Flavanone und flavanole über der Veredlungsstelle, *Gartenbauwiss.*, 52, 77, 1986.

Treutter, D. and Santos-Buelga, C., Sensitive detection and identification of actechins and proanthocyanidins by HPLC and post-column-derivatization, in *Flavonoids and Bioflavonoids 1995*, Antus, S., Gábor, M., and Vetschera, K., Eds., Budapest, Hungary, 1995, 303.

Treutter, D., Santos-Buelga, C., and Feucht, W., Determination of catechins and procyanidins in plant extracts—a comparison of methods, *Acta Hort.*, 381, 789, 1994a.

Treutter, D., Santos-Buelga, C., Gutmann, M., and Kolodziej, H., Identification of flavan-3-ols and procyanidins by HPLC and chemical reaction detection, *J. Chromatogr. A*, 667, 290, 1994b.

Uchida, S., Ohta, H., Edamatsu, R., Hiramatsu, M., Mori, A., Nonaka, G., Nishioka, I., Niwa, M., Akashi, T., and Ozaki, M., Active oxygen free radicals are scavenged by condensed tannins, in *Plant Flavonoids in Biology and Medicine II. Biochemical, Cellular, and Medicinal Properties*, Cody, V., Middleton, E., Harborne, J. B., and Beretz, A., Eds., Alan R. Liss, New York, 1988, 135.

Valsangiacomo, C., Ruckstuhl, M., and Gessler, C., *In vitro* degradation of cell walls of apple leaves by pectinolytic enzymes of the scab fungus, *Venturia inaequalis*, and by commercial pectinolytic and cellulytic enzyme preparations, *J. Phytopathol.*, 135, 20, 1992.

Volpert, R., Osswald, W., and Elstner, E. F., Effects of cinnamic acid derivatives on indole acetic acid oxidation by peroxidase, *Phytochemistry*, 38, 19, 1995.

Waage, S. K., Hedin, P. A., and Grimley, E., A biologically active procynidin from *Machaerium floribundum*, *Phytochemistry*, 23, 2785, 1984.

Walkinshaw, C. H., Are tannins resistance factors against rust fungi?, in *Chemistry and Significance of Condensed Tannins*, Hemingway, R. W. and Karchesy, J. J., Eds., Plenum Press, New York, 1989, 435.

Weidenbörner, M. and Jha, H. C., Antifungal activity of flavonoids in relation to degree of hydroxylation, methoxylation and glycosidation, *Acta Hort.*, 381, 702, 1994.

Weinges, K. and Ebert, W., Isolierung eines kristallisierten Dehydrierunggsdimeren aus (+)-Catechin, *Phytochemistry*, 7, 153, 1968.

Weinges, K., Kaltenhauser, W., Marx, H.-D., Nader, E., Nader, F., Perner, J., and Seiler, D., Procyanidine aus Früchten, *Liebigs Annalen*, 711, 184, 1968.

Ya, C., Gaffney, S. H., Lilley, T. H., and Haslam, E., in *Chemistry and Significance of Condensed Tannins*, Hemingway, R. W. and Karchesy, J. J., Eds., Plenum Press, New York, 1989, 307.

Yahraus, T., Chandra, S., Legendre, L., and Low, P. S., Evidence for a mechanically induced oxidative burst, *Plant Physiol.*, 109, 1259, 1995.

Zaprometov, M. N., Proanthocyanidins and catechins, in *Cell Culture and Somatic Cell Genetics of Plants. Vol. 5. Phytochemicals in Plant Cell Cultures*, Constabel, F. and Vasil, I. K., Eds., Academic Press, London, 1988, 77.

20 Microbial Competition and Soil Structure Limit the Expression of Allelochemicals in Nature

S. K. Schmidt and R. E. Ley

CONTENTS

20.1 ABSTRACT

In this chapter we review the literature on the fate of allelochemicals in soil with special emphasis on the microbial mineralization of such compounds by soil microorganisms. It is our thesis that most purported alleochemicals could not build up to phytotoxic levels under natural conditions. Recent literature indicates that even in the absence of microbial metabolism, soil renders compounds much less toxic than they are in laboratory solutions. This lessened toxicity is due to slowed diffusion rates in soil and to various complexation and sorptive reactions (discussed below and by Huang et al., this volume). Added to these abiotic checks on toxicity is the potential for microbial destruction of allelochemicals in soil. Most soil microbes are carbon limited and many soil organisms can very rapidly mineralize aromatic compounds. On a per weight basis, aromatic compounds are more energy rich than simple sugars and thus some soil bacteria prefer these purported allelochemicals to sugars. In addition, many soil bacteria are chemotactically attracted to phenolic compounds and can therefore intercept them at their source. We conclude our chapter by discussing possible scenarios in which allelopathy might occur in nature and what kinds of chemicals could stand up to the gauntlet of soil and cause allelopathy.

20.2   INTRODUCTION

In seeking to establish if allelochemicals can be important mediators of plant–plant dynamics, the interactions of allelochemicals with soil cannot be overlooked. Whether they originate in the rhizosphere and therefore directly in soil, or on the soil surface in decomposing plant material, chemicals must traverse the soil matrix if they are to have allelopathic effects on other plants. Many phenolic compounds that are often invoked as being allelopathic have been identified in the rhizosphere and in litter, yet these same compounds often are not found in the surrounding soil (Siqueira et al., 1991a). In this chapter we discuss how the structural complexity of soil acts as a barrier to chemical dispersion: a barrier hoisted higher by the activity of soil microorganisms.

Many studies have demonstrated the ability of soils to lessen the toxic impact of chemicals compared with toxicity in laboratory bioassays (Bonner, 1946; Buchanan et al., 1978; Heisey and Delwiche, 1985). For example, Heisey and Delwiche (1985) showed that leaf extracts from *Trichostema lanceolatum* Benth. were very toxic in Petri dish bioassays, but the same levels in soil had very little effect. This phenomenon has also been observed with regard to very toxic human-made chemicals. For instance, one of the most acutely toxic chemicals known is 2,4-dinitrophenol (DNP), which kills even those microorganisms that can metabolize it (and are likely to have some resistance to it) at concentrations as low as 20 parts per million (ppm) in liquid culture (Hess et al., 1990). When DNP is introduced into various soils, however, it takes a dose of up to 30 times higher than in liquid culture (Schmidt and Gier, 1989) to inhibit the same organisms. Explanations for the decreased toxicity of chemicals in soil vary but they can be broken down into two categories: 1) decreased bioavailability due to interactions with the soil matrix (including complexation reactions and slowed diffusion rates in soil), and 2) microbial uptake and metabolism. These important processes, and possible strategies plants may use to overcome them, are discussed in the following sections.

20.3   ABIOTIC ATTENUATION OF TOXICITY

Before an understanding of the direct effect of allelochemicals on target organisms can be achieved, researchers must address the question of whether purported allelochemicals are available to these organisms under natural conditions. The effects of abiotic processes on the bioavailability of organic molecules has received much recent attention by ecotoxicologists and soil microbiologists (Alexander, 1994; Donker et al., 1994; Hoffman et al.,1995) and this information should be of great utility to those studying allelopathy. In addition, many empirical studies have demonstrated the rapid decrease in availability of purported allelochemicals after addition to soil (e.g., Dalton et al., 1989; De Scisciolo et al., 1990; Sparling et al., 1981). For example, Dalton et al. (1989) showed that cinnamic acid (0.1 mg/g) could not be extracted from sterile soil after only two days of incubation. Furthermore, Sparling et al. (1981) demonstrated that even high concentrations (5 mg/g) of phenolic acids such as ferulic and *p*-hydroxybenzoic acids were rendered 75 to 97 percent unavailable even during the first day of incubation. Another good recent example of this is the work of De Scisciolo et al. (1990), who found that even minutes after the introduction of juglone into soil, most of it was not extractable using water and therefore probably not available to plant roots. Although some of the non-extractable juglone is probably on exchange sites and may therefore become available later, it still cannot be toxic to plants if it is not in the soil solution. Dao (1987) summed up research in this area as follows: "It has been difficult to detect unbound phenolic acids in the soil solution and the compounds do not seem to accumulate in appreciable amounts under aerobic conditions." We suggest that soil may commonly protect plants against the allelochemicals released by allegedly antagonistic plants. This decreased bioavailability of chemicals in soil can be due to several factors including: 1) slow rates of chemical diffusion in soil, 2) sorption of chemicals to soil particles, 3) processes such as sequestration into soil organic matter or micropores, and 4) microbial enzymatic effects.

20.3.1 SLOWED DIFFUSION

One way the protective effect of soil can be explained is in terms of the complexity of the soil matrix and its effect on the rate of movement of chemicals in soil. In order for a compound to get from point A to point B, it not only has to diffuse through the soil solution, but it also has to diffuse around and through the complex three-dimensional structure of soil that holds the soil solution. This tortuosity greatly slows the movement of chemicals from their point of release to potential target organisms. The size, structure, charge, and polarity of the allelochemical are all likely to affect this rate of diffusion. Another consequence of the tortuosity of soil is that chemicals interact with many surfaces and micropores as they move through soil. This increases the chances of the chemical being sorbed or sequestered before it reaches potential target organisms.

20.3.2 SORPTION

Many purported allelochemicals can be classified as phenolic compounds and there has been some research on the availability and movement of these types of chemicals in soil (reviewed by Dao, 1987). In general, such compounds are subject to sorption into or onto both mineral and organic matter fractions of the soil. Soils with high levels of organic matter or high clay contents generally retain phenolic compounds more than sandy soils (Dalton et al., 1989; Huang et al., 1977). The types of clay and aluminum and iron hydroxides in soil can influence the ability of soil to adsorb phenolic molecules (Wang et al., 1978). Because many phenolic acids have relatively low pKas (<5), they are usually ionized at the pH values of most soils (Dalton et al., 1989) and as such are more likely to interact with positively charged sites on clay minerals.

20.3.3 ABIOTIC COMPLEXATION

Sorption is usually onto exchange sites, but permanent complexation of allelochemicals can also occur, decreasing their bioavailability in soil. Although the exact mechanisms of these reactions are not completely understood, they do seem to be of some importance in certain soils. In general, such reactions lead to strong (probably covalent) linkages with the humic fraction of soil, rendering the chemical unavailable to both microorganisms and plants (Dao, 1987; Alexander, 1994). These humic acid complexes can also be formed abiotically by the catalysis by metal oxides of phenolic acids adsorbed on colloidal surfaces (Dao, 1987; Wang et al., 1978); or biotically during microbiological production of humic and fulvic acids (see below). A more thorough exploration of abiotic complexation is provided by Huang et al. in an earlier work (1997) and in this volume.

20.3.4 ENZYMATIC COMPLEXATION

Microbial enzymes can act in concert with soil abiotic factors in immobilizing allelochemicals. Isolated soil microorganisms can transform allelochemicals into more recalcitrant and less toxic forms, which contribute to the humic fraction of soils. A number of microorganisms have been shown to polymerize phenolic compounds using enzymes such as polyphenoloxidases and laccases (Haider et al., 1975; Tatsumi et al., 1994). Polymers produced by these fungi are usually dark colored (Rahouti et al., 1989; Saiz-Jimenez et al., 1975) and can appear rapidly in response to additions to soil of suitable substrates such as ferulic acid (Rahouti et al., 1989). There is also a body of literature that indicates that phenolic compounds can be co-polymerized with compounds containing amino groups to yield humic-like polymers (Saiz-Jimenez et al., 1975).

In addition to studies with isolated microorganisms, soil incubation studies have also demonstrated the incorporation of compounds such as anisic, benzoic, ferulic, p-coumaric, p-hydroxycinnamic, and vanillic acids into soil humic fractions (Batistic and Mayaudon, 1970; Kassim et al., 1981). A good review of some of the possible pathways of humic formation from phenolic compounds is presented by Dao (1987).

20.4 COMPETITION BETWEEN PLANTS AND MICROBES FOR ALLELOCHEMICALS

There has been much recent interest by ecologists in the relative competitive abilities of microbes and plants for inorganic nutrients (Jackson et al., 1989; Michelsen et al., 1995; Schimel et al., 1989; Verhagen et al., 1995; Zak et al., 1990). In situations in which both microbes and plants are nitrogen-limited, microbes can usually out-compete plants for inorganic forms of nitrogen. The superior competitive ability of soil microorganisms is usually attributed to their higher surface-to-volume ratio as compared with plant roots. This advantage in taking up solute molecules is even more pronounced with regard to organic molecules. This is because microbes are often carbon limited and have very high affinity active transport systems for energy-rich molecules like simple phenolic compounds. In contrast, plants have little reason to take up such compounds from the soil and high affinity systems for plant uptake of phenolic compounds have not been demonstrated. Little research has been carried out, however, on the relative abilities of plants and microbes to take up allelochemicals. Four fundamental differences between soil microorganisms and plant roots are important in considering the fate of allelochemicals in soil: 1) the greater uptake potential of microorganisms compared with plant roots, 2) the distribution of microbes in soil, 3) the motility and chemotactic abilities of many soil bacteria, and 4) the degradation potential of microorganisms.

20.4.1 UPTAKE

It can generally be stated that many soil microbes have evolved to take up phenolic compounds, whereas this cannot be said of plant roots. The surface to volume ratio (S:V) of soil microbes (especially bacteria) is far greater than that of plant roots. In addition, soil microbes are not only concentrated in the rhizosphere but also have a more extensive distribution in the soil than do plant roots. This combination of uptake potential and distribution is such that microorganisms are much more likely to encounter and take up allelochemicals in the soil than are plants. Even with equal exposure to allelochemicals, differences between plants and microbes in the affinity of their membrane-bound receptors for allelochemicals could determine the outcome of competition for these chemicals. Table 20.1 lists the half-saturation constants (Km) of several taxa of plants, fungi, and bacteria for phenolic compounds. Although the data for plants are quite limited, the examples given show that plants have Km values 100 to 1000 times higher than bacteria indicating that the bacteria have an affinity for phenolics up to 1000 times greater than plants do. Further research into the affinity of plant roots for phenolics is needed to corroborate this trend. Nonetheless, the high affinity of microbes for phenolic compounds is expected in light of the uses many bacteria have for phenolics as carbon and energy sources as well as chemical cues (see below). The low plant root affinities for phenolics is not surprising given that it is unclear what use plants may have for phenolics and their uptake is probably largely a result of passive diffusion. Although there is some evidence that plants can use absorbed phenolics for the production of lignin (Shann and Blum, 1987), the low affinity (high Km) values of plants for these compounds make it unlikely that plant uptake of these compounds is important in nature.

TABLE 20.1
A Comparison of Uptake Affinities for Phenolic Compounds of Plants, Bacteria, and Fungi

Organism	Compound	μmax[a] (h-1)	Km[b] (μM or mM)	Reference
Bacteria				
Pseudomonas sp.	Phenol	0.38	1.1 μM	Schmidt et al. (1987)
P. cepacia	Phenol	ND	8.5 μM	Folsom et al. (1990)
P. putida	Phenol	0.50	20.1 μM	Yang and Humphrey (1975)
P. putida	Phenol	0.53	<11μM	Hill and Robinson (1975)
P. putida	Juglone	0.56	5.5 μM	Schmidt (1988)
Fungi				
Trichosporon cutaneum	Phenol	0.44	16.2 μM	Yang and Humphrey (1975)
Aspergillus niger	Tyrosine	ND	8 μM	Cain (1988)
A. niger	p-hydroxy-benzoate	ND	0.27mM	Cain (1988)
Plants				
Raphanus sativas	Guaiacol	NA	8 mM[c]	Lee and Kim (1990)
Hordeum vulgare	Arbutin	NA	5 mM	Glass and Bohm (1971)
Eichhornia crassipes	Phenol	NA	0.58 mM	Nor (1994)
Zea mays	Glucose	NA	0.8 mM	Jones and Darrah (1996)

Values for some nonphenolic substrates are included because a lack of values for plant uptake of phenolics.

Abbreviations: ND, not determined; NA, not applicable.

[a] μmax = the maximum specific growth rate (ln 2/generation time).

[b] Km = the half-saturation constant for uptake.

[c] Enzyme Km

20.4.2 DISTRIBUTION

In addition to running the gauntlet of soil's tortuosity and sorptive abilities discussed in the previous section, allelochemicals probably encounter many billions of active soil microbes while in transit through the soil. Allelochemicals released in root exudates must first diffuse through the microbe-dense rhizosphere of the plant of origin, then through the soil between plants, and again through the rhizosphere of the target plant. Plant rhizospheres support large numbers of microbes, many of which are capable of degrading allelochemicals. For example, fluorescent pseudomonads are very common rhizosphere inhabitants and are among the most metabolically versatile organisms known. The fluorescent pseudomonad *Pseudomonas putida* can metabolize over 100 different organic compounds (Palleroni, 1984; Schmidt, 1988; Stanier et al., 1966) including purported allelochemicals such as juglone, ferulic acid, chlorogenic acid, and tannic acids (Schmidt, 1988). The rhizosphere of most plants is also inhabited by mycorrhizal fungi. These fungi have not yet been shown to metabolize allelochemicals but ectomycorrhizal fungi can act as an additional barrier to be crossed by ions in the soil solution (Ashford et al., 1989). For allelochemicals to be active against ectomycorrhizal plant roots they would have to pass through the symplast of the fungal sheath around the roots. This scenario seems very unlikely, but more research is needed to determine the role mycorrhizal fungi may play in protecting plant roots from allelochemicals in the soil solution.

20.4.3 MOTILITY

Soil microorganisms are often motile and many have been shown to be chemotactically attracted to a wide variety of organic compounds. Aromatic compounds are chemoattractants for a diversity of bacterial genera including soil dwellers such as *Azospirillum* spp. (Lopez de Victoria and Lovell, 1993), *Pseudomonas* spp. (Scher et al., 1985; Harwood et al., 1984), *Rhizobium trifoli* (Parke et al.,

1985) and *Agrobacterium* spp. (Caetano-Anolles et al., 1988). Soil microbes can move toward the source of chemicals in soil and in effect intercept them before they can reach plant roots. Indeed, many purported allelochemicals induce taxis at very low concentrations. Threshold concentrations needed to attract *Bradyrhizobium japonicum* were found to increase with chemical complexity from 3×10^{-9} M to 5×10^{-6} M for cinnamic, *p*-coumaric, caffeic, ferulic, and sinapic acids (Kape et al., 1991). Thus soil bacteria would be attracted to the source of potential alleochemicals long before they built up to phytotoxic levels.

There is also evidence that chemotaxic responses within bacterial genera can vary, possibly because of the selective pressures that the presence of certain compounds exert in a habitat. That different *Pseudomonas fluorescens* strains show varying degrees of chemotaxis to the same compounds might reflect their different soil origins (Reinhold et al., 1985). Compounds such as *p*-coumarate and *p*-hydroxybenzoate not only attract bacteria at low concentrations, but support growth of the same bacterium as well (Parke et al., 1985). In addition, several compounds implicated in plant–plant allelopathic interactions are known to mediate plant–microbe interactions. Plant phenolics have been shown to attract and also induce virulence genes in *Agrobacterium tumefaciens* (Parke et al., 1987). Flavonoids and isoflavonoids induce mycorrhizal root infections (Siqueira et al., 1991b). Flavonoid compounds that induce the symbiotic nod genes in *Rhizobium* spp. (Redmond et al., 1986) also elicit a chemotactic response in *R. meliloti* (Armitage et al., 1988; Aguilar et al., 1988). Phytogenic phenolic compounds can act similarly; for example, luteolin attracts *A. tumefaciens* and *R. meliloti* in addition to inducing the nod gene in *R. meliloti* (Caetano-Anolles et al., 1988). A study of behavioral mutants of *R. meliloti* provides evidence that the metabolic pathways involved in chemotaxis for nutrients and chemotaxis for signals are distinct (Bergman et al., 1988). This level of sophistication in chemotaxis indicates the complexity of chemical interactions between plants and microbes, and underscores the fact that bacteria and fungi living in plant rhizospheres are well adapted to responding to root exudates to meet their needs. These complex interactions also make it very unlikely that the compounds involved are also involved in plant–plant interactions. It would make no sense for compounds that stimulate plant–microbe interactions to move too far or persist too long in soil.

20.4.4 MICROBIAL METABOLISM

Given that soil microorganisms have a higher probability of encountering allelochemicals than do plant roots, effective allelochemicals should be somewhat resistant to microbial breakdown. Compounds that do not meet this criterion have to be suspect as active agents of allelopathy. In this regard, one of the least likely chemical classes to be allelochemicals are simple aromatic compounds such as ferulic, caffeic, vanillic, or *p*-coumaric acids. These compounds are energy rich (Linton and Stephenson, 1978), and are rapidly metabolized by a wide variety of bacteria and fungi (Donnelly et al., 1994; Henderson and Farmer, 1955; Jones et al., 1993; Palleroni, 1984, Parke and Ornston, 1984; Schmidt, 1988; Schmidt et al., 1987; Scow et al., 1990a; Vaughan et al., 1983). In fact, early workers in this field even claimed that compounds such as *p*-hydroxybenzoic, *m*-hydroxybenzoic, quinic, uric, and benzoic acids were "universal substrates" for growth of large groups of soil bacteria, especially the pseudomonads (den Dooren de Jong, 1926 as cited in Stanier et al., 1966). In addition, some pseudomonads prefer phenolic compounds to simple sugars such as glucose. For example, *Pseudomonas acidovorans* does not even have the enzymatic ability to mineralize glucose (Wettermark et al., 1979). As a result of all of the above, standard enrichment and isolation procedures for many soil pseudomonads involve using relatively high concentrations of phenolic compounds to encourage their growth.

Soil incubation studies also give strong evidence that many phenolic allelochemicals would not build up in soil. For example, Blum and Shafer (1988) showed that high concentrations of ferulic acid (approximately 0.1 mg/g) were readily degraded in soil. This level of ferulic acid is higher than is usually found in soil and is a concentration that could not realistically be generated via root exudation. Nonetheless, even repeated addition of this high level of ferulic acid did not lead to a build up of ferulic acid or a metabolic product (vanillic acid) until it had been added every two days over a six-day period. Even then build-up of vanillic acid was minimal. The results of Blum and Shafer (1988) therefore should be interpreted as strong evidence that phytotoxic levels of ferulic and vanillic acid would not build up in aerobic soil. It should also be stressed, however, that even low concentrations of phenolics are rapidly removed by soil microorganisms. Scow et al. (1986) showed that compounds such as phenol, aniline, and benzylamine were rapidly mineralized at concentrations ranging from 0.3 to 100 parts per billion in natural soil samples. The results of Scow et al. (1986) and Blum and Shafer (1988) are consistent with the idea that simple aromatic compounds are too desirable as carbon sources for soil microorganisms to pose a serious threat to plants in natural soils (Schmidt, 1990).

An indication of the ability of soil microbes to respond to and destroy allelochemicals is to measure their maximum growth rate on such chemicals. This is because heterotrophic microbes in soil are usually carbon limited (Lockwood and Filonow, 1981) and therefore can respond quite rapidly to inputs of easily mineralized compounds like simple phenolics. Schmidt (1988) showed that a strain of *Pseudomonas putida* found under black Walnut trees in both the United States and Germany (Rettenmaier et al., 1983) could grow at a maximum specific growth rate (μmax) of 0.56 h-1 when supplied with a low concentration of juglone (10 μg/ml) as its carbon and energy source. This is equivalent to a generation time (G) of 1.2 hours, a very fast G for a soil bacterium growing at an environmentally relevant temperature (23°C). In fact this organism grew faster on juglone and other allelochemicals than it did on simple sugars such as glucose and fructose (Schmidt, 1988). Other growth rates for microorganisms that mineralize phenolics are given in Table 20.1.

Not all microbial transformations may act to alleviate soil toxicity, however. Nair et al., (1990) found that addition of 2,3-benzoxazolinone (BOA) to soil resulted in the microbial production of 2,2'-oxo-1,1'-azobenzene (AZOB), which is far more phytotoxic (but did not build up in soil). Whether such transformations occur in soils under natural conditions is unknown. It is possible that the more toxic transformed chemical may again be metabolized or complexed before it can build up to toxic concentrations.

A more likely negative effect of microbial transformations is via the immobilization of plant nutrients caused by adding carbon compounds (i.e., allelochemicals) to soil (Azam et al., 1989; Harper, 1977; Heisey, 1990). This sort of inhibition can be easily demonstrated by adding sugar or plant litter with a high carbon to nitrogen ratio to soil. The mechanism of this type of inhibition is not normally called allelopathy because it results from the fact that heterotrophic microorganisms in soil are usually carbon limited and when they are supplied with excess carbon they can easily outcompete plants (Michelsen et al., 1995) or nitrifying bacteria (Bremner and McCarty, 1988) for available nutrients. This increased microbial immobilization of nutrients has also been demonstrated in field studies (e.g., Hunt et al., 1988). Few studies of allelopathy have included controls for microbial immobilization of inorganic nutrients (Heisey, 1990; Inderjit and Dakshini, 1995; Schmidt and Reeves, 1989).

20.5 OVERCOMING SOIL AND MICROBIAL CONSTRAINTS

As described above, the combined biotic and abiotic constraints on the movement and persistence of allelochemicals results in their low levels of bioavailability in soil. Several strategies of allelochemical release might overcome these hurdles. These strategies involve the release of allelochemicals that are recalcitrant to microbial attack, and variation in space and time of allelochemicals release. These possibilities are discussed in the following sections.

20.5.1 MASS RELEASE

A strategy that potentially could be used to avoid microbial destruction of allelochemicals would be for the chemicals to be released only occasionally but in large doses. This would be similar to the phenomenon of "mast" years of seed production (Harper, 1977), which allows certain plants to avoid complete destruction of their seed crop by producing a big crop of seeds after many years of producing few or none. This strategy does not lead to the build up of high populations of seed predators, thus allowing for a higher probability of survival in years in which seeds are produced. In a similar fashion, if plants only produced allelochemicals in spurts they could swamp the demand of the existing microbial populations and cause harm to surrounding plants before microbial populations could grow. However, in a study that addressed the effect of phenolic acids on the microbial populations in the rhizosphere of cucumber, Shafer and Blum (1991) found that chronic additions had the same effect as one-time additions. In both treatments, high populations of fast-growing bacteria and fungi were seen relative to controls. These results underscore the rapidity with which rhizosphere flora can respond to carbon amendments under the right environmental conditions.

20.5.2 GREATER RECALCITRANCE

Given the metabolic diversity and rapid response capabilities of soil microorganisms, it would be reasonable to expect that effective allelochemicals would be compounds that are resistant to microbial attack. Clues to the types of compounds to look for can be gleaned from the literature on microbe–microbe interactions. For instance, most effective antibiotics produced by soil microbes are fairly novel compounds with complex structures that are resistant to microbial breakdown. These compounds often contain heterocyclic rings and usually have higher molecular weights than simple phenolic compounds. Even most antibiotics, however, are susceptible to slow microbial attack, but do seem to persist long enough to be effective in soil (Anderson and Domsch, 1975; Schmidt and Gier, 1990). Figure 20.1 shows the structure of some common antibiotics that are effective in the soil

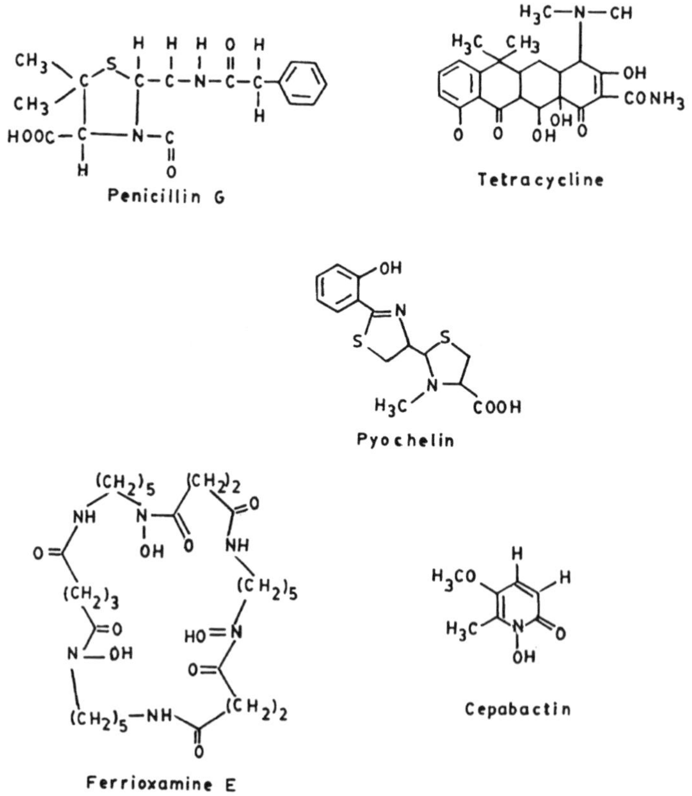

FIGURE 20.1 Structure of some antibiotics and microbial siderophores produced by soil microorganisms and probably functional in the soil environment. Note the chemical complexity and/or occurrence of heterocyclic rings in these compounds. Structures of cepabactin, ferroxamine E, and pyochelin are from Höfte (1993).

environment. Another class of compounds that are probably resistant to microbial attack and function in the soil environment are microbial siderophores (Hofte, 1993). Some representative structures of these compounds are also shown in Figure 20.1.

It should be kept in mind, however, that microbes do evolve rapidly and can adapt to degrade even complex compounds that are regularly released into the soil. Thus, if plants that produce new types of potentially allelopathic compounds are selected for, the advantage conferred might be short-lived because of the concurrent selection pressure for allelochemical-degrading microorganisms. Increased levels of degradation after chronic herbicide applications is an example of rapid adaptation of soil microbes. In modern agricultural practice, it is common to apply the same pesticides to a given field year after year. There are now several well-documented cases in which this repeated use of pesticides has resulted in reduced or total loss of effectiveness of a pesticide (Alexander, 1994). In some cases this results from increased resistance of the pest to the pesticide, but in other cases, adaptation of soil microorganisms to metabolize the pesticide was the cause. One example of the latter phenomenon is the declining effectiveness of S-ethyl N,N-dipropylthiocarbamate (EPTC) as a pre-emergence herbicide (Moorman, 1988). Even a single application of EPTC to soil can result in enhanced removal of a second application compared with soil that had never been treated with EPTC (Alexander, 1994). The mechanism of enhanced biodegradation of pesticides in soil seems to be either a build up of higher populations of pesticide degrading microorganisms (e.g., Mueller et al., 1989) or selection for more efficient pesticide degrading microbes (e.g., Moorman, 1988; Scow et al., 1990b). In either case, the end result is that repeated exposure of soil to the same chemical can result in the loss of effectiveness of that chemical as a phytotoxin, whether the chemical is made by humans or plants.

20.5.3 TARGET PROXIMITY

Given the slow movement of chemicals in soil and the high probability of sorption or destruction of most organic molecules, it is likely that true allelopathic interaction can only take place when chemicals are released in close proximity to species that are susceptible to them. Thus, it is easy to envision allelochemicals being leached from litter and affecting seedling growth under and through the litter. But it is much less likely that chemicals leached from litter would affect growth of roots deeper in the soil profile. A good example of a study that considered the spatial dynamics of plant–microbe interactions was that of Romheld (1991). Romheld (1991) discusses different ways in which plants might minimize microbial breakdown of phytosiderophores by the spatial separation of the site of siderophore release and the site of maximal microbial activity. For example, siderophores are released primarily at the growing root apex (Marschner et al., 1987) before it can be heavily colonized by rhizosphere microorganisms. Thus, in effect, the plant may be able to utilize iron captured by siderophores before large populations of microorganisms can build up to destroy them (Romheld, 1991).

In the case of root-to-root interactions, proximity is probably also very important. There are several reported cases where actual root-to-root contact was necessary for allelopathic effects to become manifest. For example, Mahall and Callaway (1992) showed that when roots of different *Ambrosia dumosa* plants touched each other, growth was inhibited, whereas when roots from the same individual plant touched one another no growth inhibition was noted. This self-nonself recognition system may be important in minimizing the overlap of neighboring *Ambrosia* plants, thus decreasing intra-specific competition for nutrients (Mahall and Callaway, 1991, 1992).

The importance of root-to-root contact was expressed very early in the allelopathic literature by Massey (1925), who showed that tomatoes were only inhibited when their roots were in close proximity to the roots of Black Walnut trees (*Juglans nigra* L.). Massey (1925) wrote: "The toxic principle, it would seem, is either insoluble in the soil water or it undergoes some chemical change shortly after leaving the walnut root, thereby losing its toxicity."

The work of Massey (1925) has been largely forgotten except by several modern scientists whose results support Massey's conclusions. For example, MacDaniels and Pinnow (1976) claim that root-to-root contact is necessary for tomatoes to be inhibited by Black Walnut trees. In addition,

De Scisciolo et al. (1990) showed that juglone, the purported active agent of walnut toxicity, became rapidly unavailable when added to the soil used in their study. The rapid disappearance of juglone in soil is probably due mostly to microbial processes. Several soil bacteria have been shown to have the capacity to rapidly oxidize juglone (Rettenmaier et al., 1983; Schmidt, 1988, 1990) but De Scisciolo et al. (1990) showed that juglone disappeared even in sterile soil, although at a slower rate than in non-sterile soil.

20.6   CONCLUSION

Soil is an extremely complex system consisting of a labyrinth of mineral and organic fractions populated by countless numbers of carbon-limited microorganisms. Microbial growth and adaptation to chemical inputs has been shown to be rapid and many soil organisms are chemotactically attracted to purported allelochemicals. In addition, chemically active compounds are subject to many complexation reactions in soil that would render them ineffective as allelochemicals. Soil microbes are carbon limited and show uptake affinities for allelochemicals that are 100 to 1000 times greater than those of plants. It therefore is not surprising that organic chemicals introduced into soil move slowly and usually become unavailable to plants. Given these overwhelming constraints, future studies of allelopathy should concentrate on chemicals that are resistant to microbial breakdown and/or are released very close to their target species.

ACKNOWLEDGMENTS

We thank D. Lipson, G.M. Colores, D. Wardle, A. Dahlberg, T. Raab and W. Segal for helpful discussions and constructive comments on the manuscript.

REFERENCES

Aguilar, J. M. M., Ashby, A. M., Richards, A. J. M., Loake, G. J., Watson, M. D., and Shaw, C. H., Chemotaxis of *Rhizobium leguminosarum* biovar *phaseoli* towards flavonoid inducers of the symbiotic nodulation gene, *J. Gen. Microbiol.,* 134, 2741, 1988.

Alexander, M., *Biodegradation and Bioremediation,* Academic Press, San Diego, 1994.

Anderson, J. P. E. and Domsch, K. H., Measurement of bacterial and fungal contributions to respiration of selected agricultural and forest soils, *Can. J. Microbiol.,* 21, 314, 1975.

Armitage, J. P., Gallagher, A., and Johnston, A. W. B., Comparison of the chemotactic behavior of *Rhizobium leguminosarum* with and without the nodulation plasmid, *Mol. Microbiol.,* 2, 743, 1988.

Ashford, A. E., Allay, W. G., Peterson, C. A., and Cairney, J. W. G., Nutrient transfer and the fungus-root interface, *Aust. J. Pl. Physiol.,* 16, 85, 1989.

Azam, F., Stevenson, F. J., and Mulvaney, R. L., Chemical extractions of newly immobilized 15_N and native soil N as influenced by substrate addition rate and soil treatment, *Soil Biol. Biochem.,* 21, 715, 1989.

Batistic, L. and Mayaudon, J., Stabilization biologique dans le sol de l'acide ferulique 14_C, de l'acide vanillique 14_C et de l'acide *p*-coumarique 14_C, *Annal. l'Instit. Pasteur.* 118, 199, 1970.

Bergman K., Gulash-Hoffee, M., Hovestadt, R. E., Larosiliere, R. C., Ronco, P. G., II and Su, L., Physiology of behavioral mutants of *Rhizobium meliloti*: evidence for a dual chemotaxis pathway, *J. Bacteriol.,* 170, 3249, 1988.

Blum, U. and Shafer, S. R., Microbial populations and phenolic acids in soil, *Soil Biol. Biochem.,* 20, 793, 1988.

Bonner, J., Further investigations of toxic substances which arise from Guayule plants: relation of toxic substances to the growth of Guayule in soil, *Bot. Gaz.,* 107, 343, 1946.

Bremner, J. M. and McCarty, G. W., Effects of terpenoids on nitrification in soil, *Soil Sci. Soc. Am. J.,* 52, 1630, 1988.

Buchanan, B. A., Harper, K. T., and Fricknecht, N. C., Allelopathic effects of Bur Buttercup tissue on germination and growth of various grasses and forbs *in vitro* and in soil, *Great Basin Nat.,* 38, 90, 1978.

Cain, R. B., Aromatic metabolism by mycelial organisms: Actinomycete and fungal strategies, *in Microbial Metabolism and the Carbon Cycle,* Hagedorn, S. R., Hanson, R. S., and Kunz, D. A., Eds., Harwood Academic Publishers, Chur, Switzerland, 1988, 101.

Caetano-Anolles G., Crist-Estes, D. K., and Barner, W. D., Chemotaxis of *Rhizobium meliloti* to the plant flavone luteolin requires functional nodulation genes, *J. Bacteriol.,* 170, 3164, 1988.

Dalton, B. R., Blum, U. and Weed, S. B., Differential sorption of exogenously applied ferulic, *p*-coumaric *p*-hydroxybenzoic and vanillic acids in the soil, *Soil Sci. Soc. Amer. J.,* 53, 757, 1989.

Dao, T. H., Sorption and mineralization of plant phenolic acids in soil, *in Allelochemicals: Role in Agriculture and Forestry,* Waller, G. R., Ed., American Chemical Society Symposium Series 330, Washington, D.C., 1987, 358.

De Scisciolo, B., Leopold, D. J., and Walton, D. J., Seasonal patterns of juglone in soil beneath *Juglans nigra* (Black Walnut) and influence of *J. nigra* on understory vegetation, *J. Chem. Ecol.,* 16, 1111, 1990.

Donker, M. H., Eijsackers, H., and Heimbach, F., Eds., *Ecotoxicology of Soil Organisms,* Lewis Publishers, Boca Raton, 1994.

Donnelly, P. K., Hegde, R. S., and Fletcher, J. S., Growth of PCB-degrading bacteria on compounds from higher plants, *Chemosphere,* 28, 981, 1994.

Folsom, B. R., Chapman, P. J., and Prichard, P. H., Phenol and trichloroethylene degradation by *Pseudomonas cepacia* G4: kinetics and interactions between substrates, *Appl. Environ. Microbiol.,* 56, 1279, 1990.

Glass, A. D. M. and Bohm, B. A., The uptake of simple phenols by barley roots, *Planta,* 100, 93, 1971.

Haider, K. and Martin, J. P., Decomposition of specifically carbon-14 labeled benzoic and cinnamic acid derivatives in soil, *Soil Sci. Soc. Amer. J.,* 39, 657, 1975.

Harper, J. L., *Population Biology of Plants,* Academic Press, London, 1977.

Harwood, C. S., Rivelli, M., and Ornston, L. N., Aromatic acids are chemoattractants for *Pseudomonas putida,* *J. Bacteriol.,* 160, 622, 1984.

Heisey, R. M., Evidence for allelopathy by tree-of-heaven (*Ailanthus altissima*), *J. Chem. Ecol.,* 16, 2039, 1990.

Heisey, R. M. and Delwiche, C. C., Allelopathic effects of *Trichostema lanceolatum* (Labiatae), *J. Ecol.,* 73, 729, 1985.

Henderson, M. E. K and Farmer, V. C., Utilization by soil fungi of *p*-hydroxybenzaldehyde, ferulic acid, syringaldehyde and vanillin, *J. Gen. Microbiol.,* 12, 37, 1955.

Hess, T. F., Schmidt, S. K., and Silverstein, J., Supplemental substrate enhancement of 2,4-dinitrophenol mineralization by a bacterial consortium, *Appl. Environ. Microbiol.,* 56, 1551, 1990.

Hill, G. A. and Robinson, C. W., Substrate inhibition kinetics: phenol degradation by *Pseudomonas putida,* *Biotechnol. Bioeng.,* 17, 1211, 1975.

Hoffman, D. J., Rattner, B. A., Burton, G. A., and Cairns, J., Eds., *Handbook of Ecotoxicology,* Lewis Publishers, Boca Raton, 1995.

Höfte, M., Classes of microbial siderophores, *in Iron Chelation in Plants and Soil Microorganisms,* Barton, L. L. and Hemming, B. C., Eds., Academic Press, San Diego, 1993, 3.

Hunt, H. W., Ingham, E. R., Coleman, D. C., Elliot, E. T., and Reid, C. P. P., Nitrogen limitation of production and decomposition in prairie, mountain meadow, and pine forest, *Ecology,* 69, 1009, 1988.

Huang, P. M., Wang, M. C., and Wang, M. K., Catalytic transformation of phenolic compounds in the soil, 1999 (this volume), Chapter 18.

Inderjit and Dakshini, K. M. M., On laboratory bioassays in allelopathy, *Bot. Rev.,* 61, 28, 1995.

Jackson, L. E., Schimel, J. P., and Firestone, M. K., Short-term partitioning of ammonium and nitrate between plants and microbes in an annual grassland, *Soil Biol. Biochem.,* 21, 409, 1989.

Jones, D. L. and Darrah, P. R., Re-sorption of organic compounds by roots of *Zea mays* L. and its consequences in the rhizosphere. III. Characteristics of sugar influx and efflux, *Plant Soil* 178, 153, 1996.

Jones, K. H., Trudgill, P. W., and Hopper, D. J., Metabolism of *p*-cresol by the fungus *Aspergillus fumigatus,* *Appl. Environ. Microbiol.,* 59, 1125, 1993.

Kape, R., Parniske, M., and Werner, D., Chemotaxis and nod-gene activity of *Bradyrhizobium japonicum* in response to hydroxycinnamic acids and isoflavonoids, *Appl. Environ. Microbiol.,* 57, 316, 1991.

Kassim, G., Stott, D. E., Martin, J. P., and Haider, K., Stabilization and incorporation into biomass of phenolic and benzenoid carbons during biodegradation in soil, *Soil Sci. Soc. Am. J.,* 46, 305, 1981.

Lee, M. Y. and Kim, S. S., Purification and characterization of the far migrating anionic isoperoxidase A3 from Korean radish root, *Kor. Biochem. J.,* 23, 440, 1990.

Linton, J. D. and Stephenson, R. J., A preliminary study on growth yields in relation to the carbon and energy content of various organic growth substrates, *FEMS Microbiol. Lett.,* 3, 95, 1978.

Lockwood, J. L. and Filonow, A. B., Responses of fungi to nutrient-limiting conditions and to inhibitory substances in natural habitats, *Adv. Microbial Ecol.,* 5, 1, 1981.

Lopez de Victoria, G. and Lovell, C. R., Chemotaxis of *Azospirillum* species to aromatic compounds, *Appl. Environ. Microbiol.,* 59, 2951, 1993.

Mahall, B. E. and Callaway, R. M., Root communication among desert shrubs, *Proc. Natl. Acad. Sci. USA,* 88, 874, 1991.

Mahall, B. E. and Callaway, R. M., Root communication mechanisms and intracommunity distributions of two Mojave desert shrubs, *Ecology,* 73, 2145, 1992.

Marschner, H., Romheld, V., and Kissel, M., Localization of phytosiderophore release and iron uptake along intact barley roots, *Physiol. Plant,* 71, 157, 1987.

Massey, A. B., Antagonism of the walnuts (*Juglans nigra* L. and *J. cinerea* L.) in certain plant associations, *Phytopathology,* 15, 774, 1925.

MacDaniels, L. H. and Pinnow, D. L., Walnut toxicity, an unsolved problem, *Ann. Rep. N. Nut Growers,* 67, 114, 1976.

Michelsen, A., Schmidt, I. K., Jonasson, S., Dighton, J., Jones, H. E. and Callaghan, T. V., Inhibition of growth, and effects on nutrient uptake of arctic graminoids by leaf extracts—allelopathy or resource competition between plants and microbes?, *Oecologia,* 105, 407, 1995.

Moorman, T. B., Populations of EPTC-degrading microorganisms in soils with accelerated rates of EPTC degradation, *Weed Sci.,* 36, 96, 1988.

Mueller, J. G., Skipper, H. D., Lawerence, E. G., and Kline, E. L., Bacterial stimulation by carbamothoate herbicides, *Weed Sci.,* 37, 424, 1989.

Nair, M. G., Whitenack, C. J., and Putnam, A. R., 2,2′-oxo-1,1′-azobenzene a microbially transformed allelochemical from 2,3-benzoxazolinone, *J. Chem. Ecol.,* 16, 353, 1990.

Nor, Y. M., Phenol removal by *Eichhornia crassipes* in the presence of trace metals, *Water Res.,* 28, 1161, 1994.

Palleroni, N. J., Pseudomonaceae, *in Bergey's Manual of Determinative Bacteriology,* 9th ed., Vol. 1, Krieg, N. R. and Holt, J. G., Eds., Williams and Wilkins, Baltimore, MD, 1984, 141.

Parke, D. and Ornston, L. N., Nutritional diversity of Rhizobiaceae revealed by auxanography, *J. Gen. Microbiol.,* 130, 1743, 1984.

Parke, D., Rivelli, M., and Ornston, L. N., Chemotaxis to aromatic and hydroaromatic acids: comparison of *Bradyrhizobium japonicum* and *Rhizobium trifolii, J. Bacteriol.,* 163, 417, 1985.

Parke, D., Ornston, L. N., and Nester, E. W., Chemotaxis to plant phenolic inducers of virulence genes is constitutively expressed in the absence of the Ti plasmid in *Agrobacterium tumefaciens, J. Bacteriol.,* 169, 5336, 1987.

Peters, N. K., Frost, J. W., and Long, S. R., A plant flavone, luteolin, induces expression of *Rhizobium meliloti* nodulation genes, *Science,* 233, 977, 1986.

Rahouti, M., Seigle-Murandi, F., Steinman, R., and Eriksson, K.-E., Metabolism of ferulic acid by *Paecilomyces variottii* and *Pestalotia palmarum, Appl. Environ. Microbiol.,* 55, 2391, 1989.

Redmond, J. W., Batley, M., Djordjevic, M. A., Innes, R. W., Kuempel, P. L., and Rolfe, B. G., Flavones induce expression of nodulation genes in *Rhizobium, Nature,* 323, 632, 1986.

Reinhold, B., Hurek, T., and Frendrik, I., Strain specific chemotaxis of *Azospirillum* spp., *J. Bacteriol.,* 162, 190, 1985.

Rettenmaier, H., Kupas, U., and Lingens, F., Degradation of juglone by *Pseudomonas putida* J1, *FEMS Microbiol. Lett.,* 19, 193, 1983.

Romheld, V., The role of phytosiderophores in acquisition of iron and other micronutrients in graminaceous species: an ecological approach, *Plant Soil,* 130, 127, 1991.

Saiz-Jimenez, C., Haider, K., and Martin, J. P., Anthroquinone and phenols as intermediates in the formation of dark-colored humic acid-like pigments by *Eurotium echinulatum, Soil Sci. Soc. Am. J.,* 39, 649, 1975.

Scher, F. M., Kloepper, J. W., and Singleton, C. A., Chemotaxis of fluorescent *Pseudomonas* spp. to soybean seed exudates *in vitro* and in soil, *Can. J. Microbiol.,* 31, 570, 1985.

Schimel, J. P., Jackson, L. E., and Firestone, M. K., Spatial and temporal effects on plant-microbial competition for inorganic nitrogen in a California annual grassland, *Soil Biol. Biochem.,* 21, 1059, 1989.

Schmidt, S. K., Degradation of juglone by soil bacteria, *J. Chem. Ecol.,* 14, 1561, 1988.

Schmidt, S. K., Ecological implications of the destruction of juglone (5-hydroxy-1,4-naphthoquinone) by soil bacteria, *J. Chem. Ecol.,* 16, 3547, 1990.

Schmidt, S. K. and Gier, M. J., Dynamics of microbial populations in soil: indigenous microorganisms degrading 2,4-dinitrophenol, *Microb. Ecol.,* 18, 285, 1989.

Schmidt, S. K. and Gier, M. J., Coexisting bacterial populations responsible for multiphasic mineralization kinetics in soil, *Appl. Environ. Microbiol.,* 56, 2692, 1990.

Schmidt, S. K. and Reeves, F. B., Interference between *Salsola kali* L. seedlings: implications for plant succession, *Plant Soil,* 116, 107, 1989.

Schmidt, S. K., Scow, K. M., and Alexander, M., Kinetics of *p*-nitrophenol mineralization by a *Pseudomonas* sp.: effects of second substrates, *Appl. Environ. Microbiol.,* 53, 2617, 1987.

Scow, K. M., Simkins, S., and Alexander, M., Kinetics of mineralization of organic compounds at low concentrations in soil. *Appl. Environ. Microbiol.,* 51, 1028, 1986.

Scow, K. M., Li, D., Manilal, V. B., and Alexander, M., Mineralization of organic compounds at low concentrations by filamentous fungi, *Mycol. Res.,* 94, 793, 1990a.

Scow, K. M., Merica, R. R., and Alexander, M., Kinetics of mineralization of organic compounds at low concentrations in soil., *J. Agric. Food Chem.,* 38, 908, 1986.

Shafer, S. R. and Blum, U., Influence of phenolic acids on microbial populations in the rhizosphere of cucumber, *J. Chem. Ecol.,* 17, 369, 1991.

Shann, J. R. and Blum, U., The utilization of exogenously supplied ferulic acid in lignin biosynthesis, *Phytochemistry,* 26, 2977, 1987.

Siqueira, J. O., Nair, M. G., Hammerschmidt, R., and Safir, G. R., Significance of phenolic compounds in plant-soil-microbial systems, *Crit. Rev. Pl. Sci.,* 10, 63, 1991a.

Siqueira, J. O., Safir, G. R., and Nair, M. G., Stimulation of V.A. mycorrhiza formation and growth of white clover by flavonoid compounds, *New Phytol.,* 118, 87, 1991b.

Sparling, G. P., Ord, B. G., and Vaughan, D., Changes in microbial biomass and activity in soil amended with phenolic acids, *Soil Biol. Biochem.,* 13, 455, 1981.

Stanier, R. Y., Palleroni, N. J., and Doudoroff, M., The aerobic Pseudomonads: a taxonomic study, *J. Gen Microbiol.,* 43, 159, 1966.

Tatsumi, K., Freyer, A., Minard, R. D., and Bollag, J.-M., Enzyme-mediated coupling of 3,4-dichloroaniline and ferulic acid: a model for pollutant binding to humic materials, *Environ. Sci. Technol.,* 28, 210, 1994.

Vaughan, D., Sparling, G. P., and Ord, B. G., Amelioration of phytotoxicity of phenolic acids by some soil microbes, *Soil Biol. Biochem.,* 15, 613, 1983.

Verhagen, F. J. M., Laanbroek, H. J., and Woldendorp, J. W., Competition for ammonium between plant roots and nitrifying and heterotrophic bacteria and the effects of protozoan grazing, *Plant Soil,* 170, 241, 1995.

Wang, T. S. C., Li, S. W., and Ferng, Y. L., Catalytic polymerization of phenolic compounds by clay minerals, *Soil Sci.,* 126, 15, 1978.

Wettermark, M. H., Taylor, J. R., Rogers, M. L., and Heath, H. E., Metabolism of carbohydrate derivatives by *Pseudomonas acidovorans, J. Bacteriol.,* 138, 418, 1979.

Yang, R. D. and Humphrey, A. E., Dynamic and steady state studies of phenol biodegradation in pure and mixed cultures, *Biotechnol. Bioeng.,* 17, 1211, 1975.

Zak, D. R., Groffman, P. M., Pregitzer, K. S., Christensen, S., and Tiedje, J. M., The vernal dam: plant-microbe competition for nitrogen in northern hardwood forests, *Ecology,* 71, 651, 1990.

21 Lignin-Related Phenolic Acids in Peat Soils and Implications in Tropical Rice Sterility Problem

Takao Katase

CONTENTS

21.1   ABSTRACT

Lignin-related phenolic acids have been implicated in a rice sterility problem in Southeast Asia. Therefore, a comparison was made of phenolic constituents in peat soils from three zones: tropical Malaysia, temperate Japan, and boreal Finland. Peat samples were also collected from Sarawak, tropical Malaysia in order to make a comparison between paddy and forest peats. From peat samples were extracted the following lignin-related phenolic acids: 4-hydroxybenzoic, 4-hydroxy-3-methoxybenzoic (vanillic), 4-hydroxy-3,5-dimethoxybenzoic (syringic), *trans*-4-hydroxycinnamic (*p*-coumaric), and *trans*-4-hydroxy-3-methoxycinnamic (ferulic) acids; they were determined by gas-liquid chromatography and mass-spectroscopy. The amounts of all the phenolic acids were much greater in the tropical and temperate peat soils than in the boreal peat. Between the tropical and temperate peats, there was a marked difference in the quantitative distribution of the five phenolic acids. The ratio of the total amounts of *trans*-4-hydroxycinnamic and ferulic acids to that of 4-hydroxybenzoic, vanillic, and syringic acids was 0.12 for tropical and 2.7 for temperate peats. Between paddy fields and forest areas in Sarawak, Malaysia, the amounts of all the phenolic acids were greater in paddy peats than in the forest peats, and the ratio of hydroxycinnamic acids to hydroxybenzoic acids was larger in the former (0.10) than in the latter (0.04). After the peat soils in tropical forests were logged and followed by cultivation with rice plants, total amounts of hydroxycinnamic acids in paddy fields in Malaysia were found to be approximately twice as high as those in forest areas. On the other hand, the relationship between those of Japan was reversed; and the ratio was smaller in paddy fields than in the forest areas. It was determined that the polymerization of these hydroxycinnamic acids may cause rice sterility problem in tropical peat soils.

21.2   INTRODUCTION

Lignin-related phenolic acids: 4-hydroxybenzoic (HBA), 4-hydroxy-3-methoxybenzoic (vanillic, VNA), 4-hydroxy-3,5-dimethoxybenzoic (syringic, SRA), trans-4-hydroxycinnamic (*p*-coumaric, HCA), *trans*-4-hydroxy-3-methoxycinnamic (ferulic, FRA), and *trans*-4-hydroxy-3,5-dimethoxycinnamic acids (sinapic, SNA) occur widely in land plants (Harbone and Simmonds, 1964) and sea grasses (Zapata and MacMillion, 1979), in a free form or in combined forms. The behaviors of these phenolic acids in the natural environment seem to be closely correlated with the forms in which they exist. A free form, for example, is known to inhibit the germination of various plants or retard the growth of the plants (Börner, 1955, 1956a, b; Gortner and Kent, 1958; Henderson and Nitsche, 1962; Zenk and Müller, 1963; Guenzi and McCalla, 1966a, b; Wang et al. 1967b; Kusano and Ogata, 1974), to inhibit feeding by detritus feeders (Valiela et al., 1979), and to cause disorders in the tissues of experimental animals (Hanya et al., 1973, 1976; Takizawa, 1970). Phenolic acids have been mentioned in the allelopathic literature on land plants as major water-borne inhibitors (Müller, 1970; Rice, 1974), following the report of Börner in 1955 that a solution of 10^{-5} g/ml of HBA and HCA significantly influenced the root growth of wheat and rye (Katase, 1981a). The specificity of the *cis*-isomer of substituted cinnamic acid derivatives has been known. For example, the *cis*-forms of methyl 3,4-dimethoxycinnamate and methyl 4-hydroxy-3-methoxycinnamate were found to be the only active forms of the germination self-inhibitors from bean rust and wheat rust uredospores (Macko et al., 1972), although the active inhibitors from both spores had previously been identified as the cis- and trans-isomers of both cinnamate derivatives (Macko et al., 1970, 1971). Other investigators reported, moreover, that the *cis*- but not the *trans*-isomers of the above two cinnamates inhibit the germination of wheat-stem rust spores (Allen, 1972; Hess et al., 1975). The *cis*- forms of these compounds in soils, therefore, may become more important (Katase, 1983). These phenolic acids had already been found in water (Hanya et al., 1976) and soils (Whitehead, 1964; Morita, 1965; Guenzi and McCalla, 1966a; Wang et al., 1967a, b, Shindo and Kuwatsuka, 1978). Phenolic constituents in the temperate peat soils in Hokkaido, Japan had been investigated for a period of about

10 years (1971 to 1984) (Katase 1981a, b, 1983; Katase and Kondo, 1984a, b, 1989a). Some phenolic compounds, especially substituted hydroxycinnamic acids, are biologically active and these compounds are important in allelopathic, ecological, organic pharmacological, and medical studies.

21.3   TROPICAL RICE STERILITY PROBLEM

The lignin-related phenolic acids may have been implicated in a rice sterility problem in Southeast Asia. Total area covered by tropical peats is estimated to be approximately 32 million ha, mainly at low altitudes in the rain forest belt of Asia, Africa, and America, and accounts for approximately 15 percent of 200 million ha worldwide coverage (Driessen, 1978). Peatland in Malaysia occupies 2.4 million ha, of which approximately 1.5 million ha is in the state of Sarawak, 0.183 million ha in the Peninsula, and 0.086 million ha in Sabah. The peats in Sarawak, therefore, account for more than 60 percent of the total Malaysian peats. Polak (1975) compared Indonesian and European peats, and concluded that tropical peats are higher in lignin and lower in water-soluble compounds, cellulose, hemicelluloses, and protein, even though the content of plant nutrients in tropical peats shows much resemblance to that of European bog peats. After the death of plants, cellulose materials of the plants are utilized by brown-rot fungi, while lignin materials are decomposed by white-rot fungi. Lignin-related phenolic acids are decomposed by a laccase from *Trametes versicolor,* which belongs to white-rot fungi. The phenolic lignin degradation products may have been implicated in a rice sterility problem (Driessen, 1978). There has been little investigation of phenolic constituents in tropical peats although phenolic compounds and their condensation and polymerization products were presumed to be abundant in tropical peat (Driessen and Suhardjo, 1976). Although abiotic catalysis is a very significant reaction of the transformation of peats (Wang et al, 1986), enzymatic catalysis of the transformation of phenolic compounds was studied and discussed. The study on Malaysian peats was conducted to analyze phenolic constituents in soils taken from uncultivated areas and those from peat soils cultivated with oil palm, pineapple, and rubber tree (Katase, 1993). We describe here a dimerization product from a lignin precursor, *trans*-4-hydroxycinnamic acid by incubation with an extracellular laccase from the fungus *Trametes versicolor.* Additionally, phenolic constituents in tropical peats from Malaysia were compared with those from temperate Japan and boreal Finland (Katase, 1995), as well as between paddy and forest peats in Sarawak, Malaysia.

Most natural peats are water-saturated and need to be drained to permit the cultivation of dryland crops. Tropical peatlands are more commonly opened by individual farmers. At first, annual crops such as pineapple, cassava, gourds, corn, and beans were planted in a way that has been practiced on peat soils all over the world. Initial subsidence, however, starts directly after the lowering of the water table and can reach values of 1 m/y in deeply drained peats (Polak, 1952). This conflicts with the requirements of most crops, but makes the cultivation of wetland rice particularly attractive because it requires prolonged inundation of the land, which reduces shrinkage and disintegration of organic materials and helps to preserve the peat. Therefore, it is evident that wetland rice is best suited to the use of tropical peat areas for a long term. Unfortunately, the grain production of wetland rice on flat and stagnant deep peats is disappointingly low because of a sterility problem. Farmers in Sumatra report that more than two-thirds of the panicles of irrigated rice on deep peat remain empty (Driessen and Suhardjo, 1976). It is generally believed that paddy does not grow on peat in Malaysia (Coulter, 1950). Driessen and Suhardjo (1976) pointed out that the sterility of wetlands on deep water-saturated peat is most probably due to imperfect photosynthesis, carbohydrate translocation, or disturbance of the generative system of the rice plants. They suggested that the decisive cause is the quality of the water, based on the fact that sterility occurs predominantly in areas with water-saturated deep peats and occasionally also in areas with mineral soils that receive water from adjacent deep peat formation (Driessen, 1978). The sterility problem was presumed to be caused by noxious substances in the irrigation water (Driessen and Suhardjo, 1976). Kanapathy (1975) postulated that the cultivation of irrigated rice is unsuited to peats because of the formation of organic acids on flooding and not only because of the acidity. Rice can grow well at pH values as low as pH 3.5 (Kanapathy, 1973; Thawornwong and Diest, 1974). Kanapathy has no quantitative evidence for the postulate but his idea is supported by Kyuma's statement (personal communication) that irrigated rice is successfully grown on certain peats in Japan that are continuously leached with good quality river water, thus lowering the contents of soluble organic substances in the root zone. However, there is evidence that the organochemical composition of peats formed in the tropics differs from those under temperate and boreal climates (Polak, 1975). It is interesting to note that approximately 50,000 ha out of 2,000,000 ha of temperate peat soils in Hokkaido has been under cultivation for rice and does not have sterility problems (Kyuma, 1983).

Driessen and Suhardjo (1976) concluded that lignin-derived polyphenolic structures are highly suspect because 1) they are probably abundant in tropical peats, 2) above certain concentration, they hinder plant growth, 3) they can cause uncoupling of oxidative phosphorylation, and 4) they have a high molecular weight and are believed to inhibit enzyme-catalyzed transformations even at low concentration.

21.4 EXPERIMENTAL

21.4.1 SAMPLING SITES

21.4.1.1 Sites in Tropical, Temperate, and Boreal Areas

Tropical, temperate, and boreal peat samples were collected from Selangor in Malaysia (approximately N10° and E100°), Hokkaido in Japan (approximately N45° and E140°), and Joenseu in Finland (approximately N60° and E40°), respectively (Figure 21.1). The three sampling sites in the tropics were only approximately 2 km apart, but those of the boreal samples were much further apart. All sampling areas were uncultivated. Ferns, reeds, and *Melaleuca leucadendron* are currently present following removal of vegetated tropical forest about ten years ago. The current vegetation on temperate peats is composed of *Sphagnum* spp., *Moliniopsis japonica, Eriophorum vaginatum* L., *Carex middendorffii, Osmunda cinnamomea, Sasa amphitrica,* and *Vaccinium oxycoccus* (Katase and Kondo, 1989a); and that on boreal peat is *Sphagnum* spp., *Pinus sylvestris* L., *Paris quadrifolia*

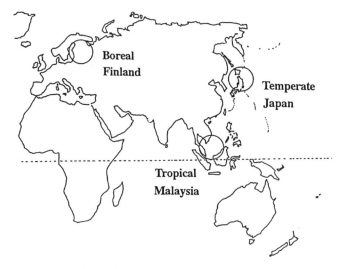

FIGURE 21.1   Map of sampling locations in tropical, temperal, and boreal zone; tropical samples were taken from Selangol, Malaysia (N10°, E100°), temprature samples from Hokkaido, Japan (N45°, E140°), and boreal samples from Joensou, Finland (N60°, E40°).

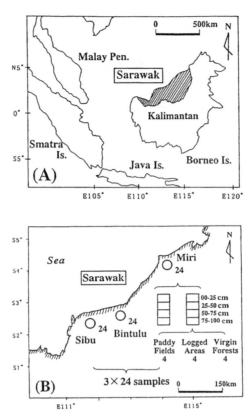

FIGURE 21.2   Location of Sarawak, Malaysia in Borneo Island, in Southern Asia (A), and sampling sites in Sarawak: Miri (Bakong and Baram), Bintulu (Tatau and Balingian) and Sibu (Naman and Rasau) (B).

L., *Rubus chamaemorus* L., *Carex* spp., *Betula nana* L., *Ledum palustre* L., *Calluna vulgaris* Hull, *Andromeda polifolia* L., *Vaccinium oxycoccus* L., *V. myrtillus*, *V. uliginosum* L., *V. vitisidaea* L., *Pleurozium schreberi*, and *Chamaedaphne calyculata* Moench (Katase, 1995).

21.4.1.2   Sites in Sarawak, Malaysia

Sarawak is situated in the northern part of Borneo Island at the north latitude between 1° and 5°, and the East longitude between 115° and 120° (Figure 21.2, A). There are four districts of peat bog areas in Sarawak. The peat samples were collected from three of the four districts: Sibu, Bintulu, and Miri; and from two sites in each district: Naman and Rasau in the Sibu district, Tatau and Balingian in the Bintulu district, and Bakong and Baram in the Miri district. At different stages of cultivation, three core samples—virgin forests, logged areas, and paddy fields—were taken from each site. Each core sample was segmented into four sections. From each district, 24 analytical samples were taken. Thus, in all, 72 samples were analyzed to determine phenolic acids of peat soils from these three districts (Figure 21.2, B).

21.4.2   Analytical Methods

21.4.2.1   Sample Preparation

Each sample was lyophilized, covered with aluminum foil, and stored at −20°C to inhibit *cis*-isomerization (Katase, 1979, 1981b). The samples were ground to fractions of less than 60 mesh with a Wiley mill for preliminary chemical analyses and then ground further to approximately 200 mesh with a microvibrational ball mill for analysis of the phenolic acids.

21.4.2.2   Total Carbon and Nitrogen Analysis

The carbon (C) and nitrogen (N) contents of the peat samples were determined with an N/C analyzer-type Sumigraph NC-80 instrument. The relative standard deviation of one determination was less than 2 percent.

21.4.2.3   Determinations of Phenolic Acids and Total Phenolic Contents

One-half gram of the pulverized peat sample was refluxed with 25 ml of ethyl acetate for 24 h, and, after removal of the ethyl acetate extract, the residual peat was refluxed with 25 ml of 2 M NaOH for 24 h. The NaOH-extracted fraction was adjusted to pH 2 by using HCl, and then extracted with ethyl acetate. This ethyl acetate extract was then made up to 5 or 50 ml volume; 0.5 ml of the ethyl acetate extract was used for determination of individual phenolic acids: HBA, VNA, SRA, HCA, and FRA by gas chromatography and mass spectroscopy (Katase, 1981a; Katase and Kondo, 1984a, b). The phenolic acids in several peat samples were determined in triplicate, and the relative standard deviation was less than 10 percent. The ethyl-acetate extract, 0.1 mL, was used for determination of the total phenolic contents by spectrophotometric method (Katase and Kondo, 1989a), and the relative standard deviation was less than 10 percent.

21.4.2.4   Identification of Reaction Products from *Trans*-4-Hydroxicinnamic Acid with Enzyme

The change in amounts of phenolic acids reacted with a laccase from *Trametes versicolor* was determined by high performance liquid chromatography (HPLC) (Katase et al., 1989b). A reaction product of HCA with the laccase was characterized by the method described previously (Katase and Bollag, 1991). One DMP unit per milliliter of the laccase was incubated with 400 mg of HCA in 2-l citrate-phosphate buffer at pH 5.5. Boiled enzymes served as controls. After 2 h incubation, the mixture was centrifuged. A 500 ml aliquot of the supernatant was extracted with 200 ml of methylene chloride. A 250 ml aliquot of the aqueous phase was then re-extracted with 200 ml of ethyl acetate. The resulting residue was separated with TLC, and the product of interest was analyzed by mass spectrometry (MS) and proton nuclear magnetic resonance spectrometry (pNMR). The molecular weight of one of the reaction products was determined by electron impact ionization using a Kratos MS 9150 double-focusing MS. Its molecular weight was confirmed by chemical ionization mass analysis on Finnigan 3200 MS. Sample introduction was by direct insertion probe. The product was confirmed with pNMR using a Bruker WM-360 instrument with acetone-D6 as solvent.

21.5   RESULTS AND DISCUSSION

21.5.1   CHARACTERISTICS OF PHENOLIC ACIDS IN PEAT SOILS

21.5.1.1   Chemical Properties of Peat Soils in Tropical Malaysia, Temperate Japan, and Boreal Finland

Ash content, total carbon, C/N ratio, and total phenolic content in tropical Malaysian, temperate Japanese, and boreal Finnish peat soils are shown in Table 21.1. The ash contents of peats in all three zones were less than 10 percent and their total carbon contents were similar. The average C/N ratios of the tropical and boreal peats were similar, but the C/N ratio of the Malaysian peats increased with soil depth, while that of the Finnish peat decreased with depth. Nitrogen in the tropical peat decreased with depth, while that of boreal peat increased. Nitrogen in the tropical peat might be more quickly utilized by microorganisms than that in the boreal peat. Nitrogen in the boreal peat accumulated in the deeper layer. The nitrogen content in the temperate peat soils was the highest of all. The difference between the temperate and boreal peats may arise from the vegetation; previous work has shown nitrogen content to be 1.5 percent for mossy peat and 2.4 percent for grassy peat (Katase and Kondo, 1984a).

TABLE 21.1
Chemical Properties of Peat Soils in Tropical Malaysia, Temperate Japan, and Boreal Finland

Depth (cm)	Ash Content (%)			Total Carbon (%)			C/N Ratio			Total Phenolics (mg/g)		
	Tropical	Temperate	Boreal	Tropical	Temperate	Boreal	Tropical	Temperate	Boreal	Tropical	Temperate	Boreal
0–9					41			17			8.8	
0–10			39 ± 1.5			45 ± 1.0			50 ± 21			13 ± 2.4
0–15	7.8 ± 2.7			46 ± 19			34 ± 4.2			13 ± 4.5		
9–22		27			40			38			8.4	
10–20			2.4 ± 0.33			43 ± 3.1			59 ± 15			15 ± 1.6
15–30	4.8 ± 1.3			50 ± 6.8			41 ± 1.6			13 ± 1.4		
20–30			2.6 ± 0.61			47 ± 4.5			41 ± 2.5			10 ± 5.5
22–34		7.7			55			26			18	
30–45	2.2 ± 0.52			58 ± 6.2			55 ± 6.2			14 ± 0.96		
34–63		3.4			57			30			19	
40–63			1.5 ± 0.76			48 ± 5.5			33 ± 7.9			15 ± 9.2
45–60	2.4 ± 0.51			58 ± 5.7			58 ± 9.1			15 ± 2.3		
60–70			4.4 ± 4.0			48 ± 5.5			37 ± 10			14 ± 2.5
63–80		4.9			58			29			23	
80–90			1.7 ± 1.5			47 ± 6.6			33 ± 4.4			16 ± 3.2
80–95		4			58			36			19	
95–110		2.7			60			32			9.9	
Mean	4.3 ± 2.7	8.3 ± 9.3	2.7 ± 1.9	52 ± 11	53 ± 8.5	46 ± 4.4	47 ± 11	27 ± 7.3	42 ± 14	14 ± 2.4	15 ± 5.8	14 ± 4.5
Samples	n = 12	n = 6	n = 18	n = 12	n = 7	n = 18	n = 12	n = 7	n = 18	n = 12	n = 7	n = 18

Tropical samples were taken from three different sites in uncultivated areas at Kelang in Selangol, Malaysia (approximately N10°, E40°). Temperate samples were taken from one site of cultivated area at Hokkaido, Japan (N45°, E140°). Boreal samples were taken from three different sites in uncultivated areas around Joensou, Finland (approximately N60°, E40°).

21.5.1.2 Relationship Between Total Phenolic Acids ($\Sigma BA + \Sigma CA$) and Their Ratio ($\Sigma CA / \Sigma BA$) in Peat Soils

The total amounts of the lignin-related phenolic acids: HBA, VNA, SRA, HCA, and FRA in tropical Malaysian, temperate Japanese, and boreal Finnish peat soils are shown in Table 21.2. The relationship between total amounts of phenolic acids [ΣBA (= HBA+VNA+SRA)+ΣCA (= HCA+FRA)] and their ratio of ΣCA to ΣBA in various peats are shown in Figure 21.3. The total amounts of phenolic acids ($\Sigma BA + \Sigma CA$) in boreal Finnish peats were much lower than in tropical and temperate peats, even though their total phenolic contents are similar (Table 21.1). This difference may be due to respective vegetation cover. Tropical and temperate peats originate mainly from woody or herbaceous plants, which are relatively rich in lignin. Boreal peats, however, come mainly from mosses. The lignin-related phenolic acids in tropical and temperate peats, therefore, appear to comprise a large portion of the total phenolic contents. The total amounts ($\Sigma BA + \Sigma CA$) in the boreal area decreased with soil depth ($r = 0.52$, $P < .01$), suggesting that phenolic acids were produced from decomposing peats. On the other hand, there was no significant relationship between $\Sigma BA + \Sigma CA$ and soil depth for tropical and temperate peats. The ratio $\Sigma CA / \Sigma BA$ in the tropical peats

TABLE 21.2
Phenolic Acids of Peat Soils in Tropical Malaysia, Temperate Japan and Boreal Finland

Depth (cm)	HBA	VNA	SRA	HCA	FRA	$\Sigma BA + \Sigma CA$	$\Sigma CA / \Sigma BA$
		Peat Soil in Selangol, Malysia ($n = 3$), mg/g soil (Ash-free base)					
0–15	3.0 ± 1.4	1.8 ± 0.5	1.1 ± 4.2	0.38 ± 0.07	0.54 ± 0.16	6.8 ± 0.16	0.16 ± 0.03
15–30	3.3 ± 2.4	1.3 ± 0.46	0.76 ± 0.3	0.22 ± 0.13	0.35 ± 0.16	5.6 ± 2.3	0.12 ± 0.04
30–45	3.8 ± 1.5	1.5 ± 1.3	1.2 ± 0.94	0.29 ± 0.34	0.55 ± 0.47	7.3 ± 3.4	0.12 ± 0.07
45–60	6.0 ± 1.2	0.95 ±0.02	0.59 ±0.18	0.16 ± 0.05	0.39 ± 0.11	8.1 ± 1.2	0.075 ± 0.02
Mean ($n = 12$)	4.0 + 1.9	1.4 + 0.72	0.89 + 0.52	0.27 + 0.18	0.46 + 0.25	6.9 + 2.3	0.12 + 0.047
		Peat Soils in Hokkaido, Japan ($n = 1$), mg/g soil (Ash-free base)					
0–9	0.24	0.98	0.49	1.6	0.89	4.2	1.47
09–22	0.26	0.76	0.29	1.7	0.85	4.1	1.60
22–34	0.41	1.10	0.96	6.3	2.10	11.0	3.30
34–63	0.26	0.95	0.81	5.8	2.0	9.9	3.90
63–80	0.21	0.73	0.61	4.1	1.20	6.8	3.50
80–95	0.16	0.37	0.25	1.6	0.51	3.0	2.74
95–110	0.22	0.33	0.24	1.2	0.77	2.7	2.50
Mean ($n = 7$)	0.25 ± 0.08	0.75 ± 0.3	0.52 ± 0.29	3.2 ± 2.2	1.2 ± 0.62	6.0 ± 3.4	2.7 ± 0.93
		Peat Soils in Joensou, Finland ($n = 3$), µg/g soil (Ash-free base)					
0–10	29 ± 6	35 ± 28	9.7 ± 6.5	11 ± 2.8	7.3 ± 5.2	94 ± 47	0.26 ± 0.07
10–20	28 ± 25	39 ± 28	40 ± 4.8	10 ± 9.2	13 ± 16	97 ± 87	0.32 ± 0.13
20–30	26 ± 14	47 ± 32	17 ± 9.5	37 ± 34	30 ± 35	160 ± 110	0.61 ± 0.45
40–50	17 ± 15	16 ± 22	4.6 ± 6.5	15 ± 2.7	9 ± 15	60 ± 79	0.15 ± 0.59
60–70	61 ± 23	58 ± 18	27 ± 18	57 ± 30	63 ± 76	270 ± 140	0.77 ± 0.59
80–90	40 ± 18	60 ± 52	31 ± 29	61 ± 40	140 ± 210	330 ± 300	0.12 ± 0.79
Mean ($n = 12$)	34 ± 21	43 ± 32	16 ± 17	32 ± 32	43 ± 91	170 ± 160	0.37 ± 0.54

Abbreviations: HBA, 4-hydroxybenzoic acid; VNA, vanillic acid; SRA, syringic acid; HCA, *trans*-4-hydroxycinnamic acid; FRA, ferulic acid; ΣBA, HBA+VNA+ΣRA; ΣCA, HCA+FRA.

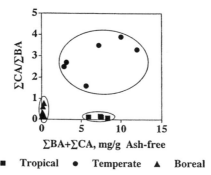

FIGURE 21.3 Relationship between phenolic acids indices, $\Sigma BA + \Sigma CA$ and $\Sigma CA/\Sigma BA$ in climatically different zones.

decreased ($r = -0.61$, $P < .01$), and increased with depth in the boreal peats with depth ($r = 0.60$, $P < .01$). This change in the ratio of $\Sigma CA/\Sigma BA$ with depth can be attributed to the microbial transformation of the hydroxycinnamic acids to hydroxybenzoic acids. *Pseudomonas fluorescens* were able to metabolize *trans*-4-hydroxycinnamic acid to 4-hydroxybenzoic acids. (Seidham et al., 1969). High microbial activity in tropical areas may quickly cause the transformation of the hydroxycinnamic acids to hydroxybenzoic acids. Katase and Bollag (1991) found that *trans*-4-hydroxycinnamic acid was transformed with oxidoreductase enzyme to measurable reaction products, one of which was identified as its dimer, *trans*-4-hydroxy-3-(*trans*-4-cinnamyloxy) cinnamicacid (see Section 21.5.2.4).

In the present study, lignin-related phenolic acids were found to differ among the tropical, temperate, and boreal peats. To understand rice-sterility problems, it would be necessary to study the phenolic constituents of paddy peat soils and irrigated water and deep water-saturated peats.

21.5.2 COMPARISON OF PHENOLIC ACIDS IN TROPICAL PEAT SOILS BETWEEN PADDY FIELDS AND FOREST AREAS

21.5.2.1 Vertical Profiles of Ash and Carbon Contents, and C/N Ratio

Vertical profiles (mean values) of ash, and organic carbon contents, and C/N ratio in peat soils, Sarawak, Malaysia are shown in Figure 21.4. The ash content of peat samples in forest areas averaged 5.0 ± 9.7 percent ($n = 24$), while that of paddy fields was 12 ± 15 percent ($n = 24$). The carbon content in forest areas averaged 49.9 ± 5.4 percent ($n = 24$), and that of paddy fields was 41.9 ± 8.5 percent ($n = 24$). The nitrogen content in forest areas averaged 2.3 ± 1.6 percent ($n = 24$), and of paddy fields 1.8 ± 1.8 percent ($n = 24$). However, the C/N ratio was larger in the paddy fields than in the forest areas. The nitrogen content in the paddy fields is relatively low.

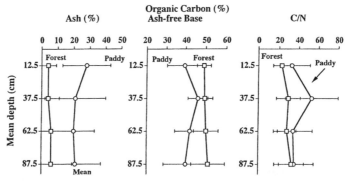

FIGURE 21.4 Vertical profile of ash, organic carbon, and C/N in peat soils, Sarawak, Malaysia ($n = 6$).

TABLE 21.3
Total Amounts of Phenolic Acids (ΣBA + ΣCA) of Peat Soils in Sarawak, Malaysia

	Depth (cm)	Mean Depth	Bintulu		Miri		Sibu		Mean ± SD	(n)
			Tatau	Balingian	Bakong	Baram	Naman	Rasau		
Virgin Forest Areas (n = 24)	0–25	12.5	1.32	0.92	2.26	5.18	2.51	3.95	2.69 ± 1.61	(n = 6)
	25–50	37.5	3.10	2.78	0.22	7.02	3.03	1.31	2.91 ± 2.31	(n = 6)
	50–75	62.5	3.42	0.37	0.21	6.65	2.95	1.47	2.51 ± 2.41	(n = 6)
	75–100	87.5	2.11	0.09	0.00	9.97	2.61	1.70	2.75 ± 3.70	(n = 6)
	Mean ± SD	—	2.49 ± 0.96	1.04 ± 1.21	0.67 ± 1.06	7.20 ± 2.01	2.78 ± 0.25	2.11 ± 1.24	2.71 ± 2.45	(n = 24)
Logged Areas (n = 3)	0–25	12.5	3.20	2.16	4.27	2.70	5.51	3.20	3.51 ± 1.20	(n = 6)
	25–50	37.5	2.97	4.33	3.04	2.50	4.19	2.18	3.41 ± 1.24	(n = 6)
	50–75	62.5	0.32	6.43	3.41	3.07	4.19	2.18	3.27 ± 2.04	(n = 6)
	75–100	87.5	—	7.60	2.80	3.36	6.84	1.55	4.43 ± 2.65	(n = 5)
	Mean ± SD	—	2.16 ± 1.60	5.13 ± 2.40	3.38 ± 0.64	2.91 ± 0.38	5.50 ± 1.08	2.27 ± 0.69	3.56 ± 1.76	(n = 23)
Paddy fields (n = 24)	0–25	12.5	2.17	7.24	6.98	4.46	3.46	4.43	4.79 ± 1.98	(n = 6)
	25–50	37.5	0	7.97	0.77	5.91	6.72	0.005	3.56 ± 3.69	(n = 6)
	50–75	62.5	0.07	5.54	7.40	8.20	3.64	0.71	4.26 ± 3.39	(n = 6)
	75–100	87.5	3.05	7.47	4.32	10.89	2.55	2.84	5.19 ± 3.33	(n = 6)
	Mean ± SD	—	1.32 ± 1.53	7.05 ± 1.05	4.87 ± 3.05	7.37 ± 2.81	4.09 ± 1.82	2.00 ± 2.02	4.45 ± 3.02	(n = 24)
Mean	00–25	12.5	2.23	3.44	4.50	4.11	3.83	3.86	3.66 ± 0.79	(n = 6)
	25–50	37.5	2.02	5.03	1.34	5.14	5.07	1.16	3.29 ± 1.98	(n = 6)
	50–75	62.5	1.27	4.11	3.67	5.97	3.59	1.45	3.35 ± 1.76	(n = 6)
	75–100	87.5	2.58	5.05	2.37	8.07	4.00	2.03	4.02 ± 2.29	(n = 6)
	Mean + SD	—	2.02 ± 0.55	4.41 ± 0.78	2.97 ± 1.40	5.83 ± 1.68	4.12 ± 0.65	2.13 ± 1.21	3.58 ± 1.70	(n = 24)

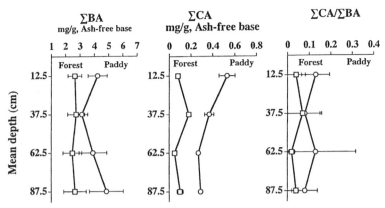

FIGURE 21.5 Vertical profiles of hydroxybenzoic acids ($\Sigma BA=HBA+VNA+SRA$), hydroxycinnamic acids ($\Sigma CA=HCA+FRA$), and their ratio, $\Sigma CA/\Sigma BA$ ($n=6$).

21.5.2.2 Vertical Profiles of Phenolic Acids in Paddy Fields and Forest Areas

Table 21.3 shows the total amounts of phenolic acids ($\Sigma BA+\Sigma CA$) of peats in Sarawak, Malaysia in different stages of cultivation: virgin forests, logged areas, and paddy fields. Figure 21.5 shows vertical profiles of hydroxybenzoic acids ($\Sigma BA=HBA+VNA+SRA$), hydroxycinnamic acids ($\Sigma CA=HCA+FRA$) and their ratio, $\Sigma CA/\Sigma BA$. All indices (ΣBA, ΣCA and $\Sigma CA/\Sigma BA$) in any layers of the profiles were larger in paddy fields than in those of forest areas. The profiles of forest areas varied little in both indices of ΣBA and ΣCA, but there was a marked variation in those of paddy fields. In contrast, their variation in ΣBA increased and ΣCA decreased beyond the 37.5 cm of depth; especially the latter decreased significantly through the layers ($r = -0.896$, $P < .05$). These profiles were presumed to indicate transformation from ΣCA to ΣBA resulting in complementary change in $\Sigma CA/\Sigma BA$ between both profiles of paddy fields and forest areas.

21.5.2.3 Comparison of Total Phenolic Acids Between Paddy Fields and Forest Areas

As shown in Table 21.3, the total amounts of phenolic acids ($\Sigma BA+\Sigma CA$) of peat soils in virgin forests and paddy fields averaged 4.5 ± 3.0 mg/g soil, ash-free base ($n = 24$), and 2.7 ± 2.5 mg/g soil, ash-free base ($n = 24$), respectively, while those of the logged areas were 3.7 ± 1.8 mg/g soil ($n = 23$). After logging, phenolic acids in peat soils would be produced by degradation of plant tissues. However, with rice cultivation, phenolic acids in fields under anaerobilc conditions might not be easily decomposed in comparison with the aerobic conditions in the logged areas.

Figure 21.6 shows relationships between the total amounts of phenolic acids ($\Sigma BA+\Sigma CA$) and their ratio ($\Sigma CA/\Sigma BA$) in soils from paddy fields and forest areas in Sarawak, Malaysia (A) as well as Japan (B). After logging, the areas were cultivated with rice and in such situations, substituted hydroxycinnamic acids (ΣCA) in peat soils in Malaysia, increased in comparison with substituted hydroxybenzoic acid (ΣBA). The total amounts of hydroxycinnamic acids (ΣCA) of paddy fields in Malaysia were found to be approximately twice as high as those of forest areas. In contrast, the relationship between these acids in Japanese fields was different, and the ratio $\Sigma CA/\Sigma BA$ was smaller in paddy fields than in forest areas. The data from Shindo et al. (1978) were recalculated for Figure 21.6. However, data from both paddy and forest areas were not from peat soils.

Based on our results, the amounts of phenolic acids and especially hydroxycinnamic acid in Japanese paddy fields were relatively smaller than those in forest areas, and those of Malaysian paddy fields were approximately twice as high as those of the virgin forest areas. However, only lignin-related phenolic acids in peat soils were not concluded to be strongly related to the rice sterility problem on wetlands in Southeast Asia. In order to clarify this problem, it will be necessary to study the phenolic constituents of paddy peats and irrigated water in much more detail.

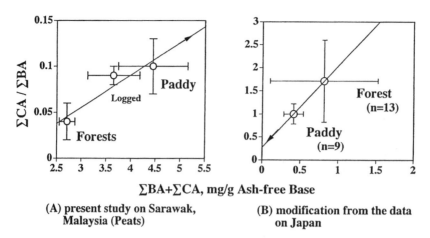

$\Sigma BA + \Sigma CA$, mg/g Ash-free Base

(A) present study on Sarawak, Malaysia (Peats)

(B) modification from the data on Japan

FIGURE 21.6 Relationships between total phenolic acids ($\Sigma BA + \Sigma CA$) and their ratio ($\Sigma CA/\Sigma BA$) of soils from paddy and forests on Sarawak, Malaysia (A), and those of Japan after Shindo et al., 1978 (B).

21.5.2.4 Identification of Reaction Products from *Trans*-4-Hydroxycinnamic Acid with Laccase

Phenolic acids are decomposed easily by enzymatic activity to various reaction products. HPLC analysis of *trans*-4-hydroxycinnamic acid, after it incubated with a laccase of *Trametes versicolor*, indicated an almost complete disappearance of the substrate and the formation of products. The results were described earlier (Katase and Bollag, 1991), and the data were represented here. Figure 21.7 shows typical HPLC chromatograms of the supernatants of a reaction mixture for a control sample (A) and a sample incubated with laccase (B). The products were named compounds, X_A, X_B, and X_C, which had retention times of 10.4, 10.9, and 13.1 minutes, respectively. The molecular weight of compound X_C determined by high-resolution mass spectrometry was 326.0728 (calculated as 326.0730), which corresponds to the molecular formula $C_{18}H_{14}O_6$. Mass fragments with m/z ratios

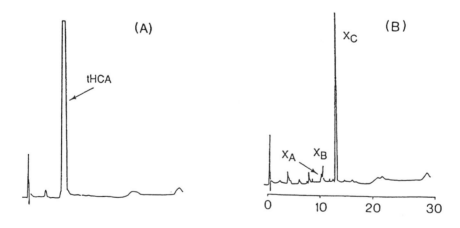

FIGURE 21.7 HPLC chromatograms of reaction mixtures of *trans*-4-hydroxycinnamic acid (tHCA) (Katase and Bollag 1991): (A) control and (B) incubated with a laccase of *Trametes versicolor*.

of 282 and 238 were assigned to the one decarboxylated (282=M−44) and the two decarboxylated (238= M−[2×44]) molecules, respectively. Decarboxylation was confirmed by the appearance of an m/z 44 fragment on the mass spectrum (Figure 21.8). Proton NMR spectroscopy supported the conclusion that compound X_C was a dimer of HCA and demonstrated that the two molecules were coupled at ortho- and O-phenolic positions. Table 21.4 shows the chemical shift and coupling constant data for product X_C. Chemical shifts of 6.88 ppm and 7.56 ppm between H_C and Hd indicated clearly their positions as illustrated in Figure 21.9. Accordingly, compound X_C was identified from the mass spectrometric data and the chemical shift and coupling constant data (Table 21.4) as *trans*-4-hydroxy-3-(*trans*-4'-cinnamyloxy)cinnamic acid. Since only small quantities of X_A and X_B were isolated, no NMR spectra could be obtained. According to our high resolution mass spectrometric results, however, X_B has a molecular weight of 282.0887 corresponding to the molecular formula $C_{17}H_{14}O_4$. Fragmentation in the mass spectra suggests that the compound has one carboxyl group and may be the one decarboxylation product. Only a low resolution mass spectrum was taken for product X_A and showed a molecular peak at m/z 242. Compound X_A may result from the removal of one carboxyl group from compound X_B. Further studies are needed to identify products X_B and X_A. The participation of laccase in natural humification processes of plant residues in soil was established by Trojanowski and Matwijow (1964). A direct interaction between the organic matter in soil

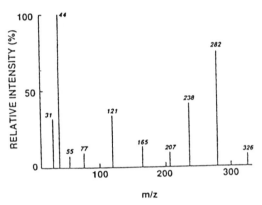

FIGURE 21.8   Mass spectrum of product Xc, resulting from incubation of *trans*-4-hydroxycinnamic acid with a laccase of *Trameters versicolor* (Katase and Bollag 1991).

TABLE 21.4
Proton NMR Chemical Shifts and Coupling Constants for Product Xc Resulting from Incubation of *trans*-4-Hydroxycinnamic Acid with a Laccase of *Trametes versicolor*

Chemical Shift	(ppm)	Type	Coupling constants (Hz)	Proton	Integral
7.75	Singlet			Hg	1H
7.65	Doublet	Jba	15.9	Hb	1H
7.56	Doublet	Jfe	8.22	Hf	1H
7.27	Doublet	Jdc	8.64	Hd	2H
6.88	Doublet	Jef	8.33	He	1H
6.85	Doublet	Jcd	8.59	Hc	2H
6.38	Doublet	Jab	15.93	Ha	1H
6.04	Doublet	Jba	7.44	Hb¢	1H
4.31	Doublet	Jab	7.44	Ha¢	1H

Signals from Hb¢ and Ha¢ of this compound were shifted to a higher magnetic field, most likely due to a ring effect of the substituted benzene ring. See Figure 21.9 for chemical structure of this compound and proton assignment (a to g).

Source: From Katase and Bollag, 1991. With permission.

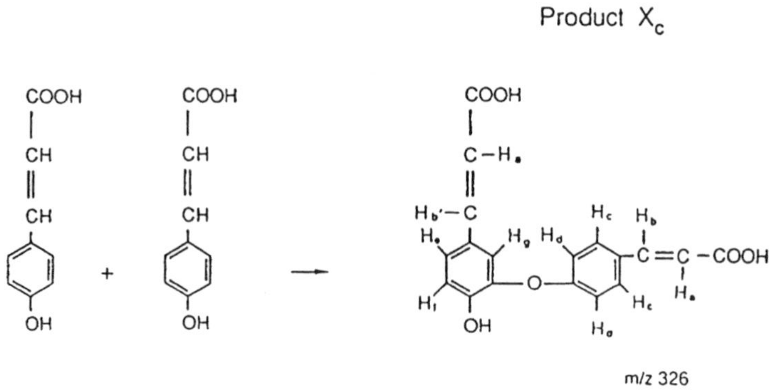

Product X_c

COOH   COOH   COOH

m/z 326

FIGURE 21.9   Scheme for the formation of a dimer of *trans*-4-hydroxycinnamic acid (product X_C) catalyzed by a laccase of *Trametes versicolor* (Katase and Bollag 1991).

and polyphenols can result in a synthetic transformation of naturally occurring phenols. Particularly, oxido-reductases (peroxidases, laccases) can catalyze coupling reactions resulting in the formation of oligomeric and polymeric products from phenols and aromatic amines (Ross and McNeilly, 1973; Mayaudon and Sarkar, 1975; Liu et al., 1981; Sjoblad and Bollag, 1981; Bollag, 1983; Leonowicz and Bollag, 1987; Simmons et al., 1987). Because intermediate and oligomerized compounds from substituted hydroxycinnamic acids are biologically and geochemically active, they should be of interest for further investigations into the presence of reaction products, such as a dimer of *trans*-4-hydroxycinnamic acid in the environment as well as in their biological activity in plants. To clarify rice-sterility problems, it will be necessary to study the phenolic constituents of paddy peat soils, as well as irrigated water and deep water-saturated peats. Only 11 percent of incubated *trans*-4-hydroxycinnamic acid was transformed to the above-mentioned three dimers (X_A, X_B, X_C); the remaining 89 percent was transformed to larger oligomers or smaller compounds. Further investigations are needed to clarify their potential fate in peat soil environment and to detect alleopathic phenolic compounds.

REFERENCES

Allen, F. J., Specificity of the *cis*-isomers of inhibitors of uredospore germination in the rust fungi, *Proc. Natl. Acad. Sci. USA,* 69, 3497, 1972.

Bollag, J-M., Synthetic reactions of aromatic compounds by fungal enzymes, *in Microbiology—1983,* Schlesinger, D., Ed., American Society for Microbiology, Washington, D.C., 1983, 203.

Börner, H., Untersuchungen öber phenolishe Verbindung aus Getreidestroh und Getreiderückständen, *Naturwiss.,* 42, 583, 1955.

Börner, H., Der papierchromatographische Nachweis von Ferulasäure in wäßrigen Extrakten von Getreidestroh und Getreideräkst, *Naturwiss.,* 43, 129, 1956.

Börner, H., Die Abgabe organishe Verbindungen aus den Karyopsen, Wurze ln und Ernteröckstönden von Roggen, Weizeln und Gerste und ihre Bedeutung bei der gegenseitigen Beeinflössung der hocheren Pflanzen, *Beitr. Biol. Pflanz.,* 32, 1, 1956b.

Coulter, J. K., Peat formation in Malaya, *Malay. Agric. J.,* 33, 63, 1950.

Driessen, P. M. and Suhardjo, H., On the defective grain formation of sawah rice on peat, *Soil Res. Inst. Bull. Bogor. Indonesia.,* 3, 20, 1976.

Driessen, P. M., Peat soils, *in* Soils and Rice, International Rice Research Institute, Ed., IRRI, Los Banos, Philippines, 1978, 763.

Gortner, W. A. and Kent, M. J., Ferulic acid and p-coumaric acid in pineapple tissue as modifiers of pineapple indoleacetic acid oxidase, *Nature,* 181, 630, 1958.

Guenzi, W. D. and McCalla, T. M., Phenolic acids in oats, wheat, sorghum and corn residues and their phytotoxycity, *Agron. J.,* 58, 303, 1966a.

Guenzi, W. D. and McCalla, T. M., Phytotoxic substances extracted from soil, *Soil Sci. Soc. Am. Proc.,* 30, 214, 1966b.

Hanya, T., Ishiwatari, R., Katase, T., Takada, T., and Nagao, K., Identification of a trace amount of organics in natural and polluted waters, *in* Trace Substances in Environmental Health-VI, Hemphill, D. D., Ed., University of Missouri, Columbia, 1973, 355.

Hanya, T., Matsumoto, G., Nagao, K., and Katase, T., The presence of *p*-coumaric and ferulic acids in natural waters and their significance in relation to environmental health, in *Trace Substances in Environmental Health-X,* Hemphill, D. D. Ed., University of Missouri, Columbia, 1976, 265.

Harborne, J. B. and Simmonds, N. W., The natural distribution of the phenolic aglycone, *in Biochemistry of Phenolic Compounds,* Harborne, J. B., Ed., Academic Press, London, 1964, 80.

Henderson, J. H. M. and Nische, J. P., Effect of certain phenolic acids on the elongation *Avena* first internodes in the presence of auxine and tryptophan, *Nature,* 195, 780, 1962.

Hess, S. L., Allen, F. J., Nelson, D. N., and Lester, H., Mode of action of methyl *cis*-ferulate, the self-inhibitor of stem rust uredospore germination, *Physiol. Plant Pathol.,* 5, 107, 1975.

Kanapathy, K., Acidity, acid sulfate soils and liming of padi fields, *Malay. Agric. J.,* 49, 154, 1973.

Kanapathy, K., Factors in the utilization of peat soils in Peninsular Malaysia, 3rd ASEAN Soil Conf., Kuala Lumpur, 1975 (cited from Kyuma, 1983).

Katase, T., Stereoisomerization of *p*-coumaric acid during analytical procedure by exposure to fluorescent light, *Bunseki Kagaku (Analyt. Chem.)* 28, 455, 1979.

Katase, T., The different forms in which *p*-coumaric acid exists in a peat soil, *Soil Sci.,* 131, 271, 1981a.

Katase, T., The different forms in which *p*-hydroxybenzoic, vanillic, and ferulic acids exist in a peat soil, *Soil Sci.,* 132, 436, 1981b.

Katase, T., The presence of *cis*-4-hydroxycinnamic acid in peat soils, *Soil Sci.,* 135, 296, 1983.

Katase, T. and Kondo, R., Distribution of some different forms of some phenolic acids in peat soils in Hokkaido, Japan. 1. *trans*-4-Hydroxycinnamic acid, *Soil Sci.,* 138, 220, 1984a.

Katase, T. and Kondo, R., Distribution of some different forms of some phenolic acids in peat soils in Hokkaido, Japan 2. 4-Hydroxybenzoic, 4-hydroxy-3-methoxybenzoic and *trans*-4-hydroxy-3-methoxycinnamic acids, *Soil Sci.,* 138, 279, 1984b.

Katase, T., Vertical profiles of *trans*- and *cis*-4-hydroxycinnamic acids and other phenolic acids in Horonobe peat soil, Japan, *Soil Sci.,* 148, 258, 1989a.

Katase,T., Hirota, S., and Bollag, J.-M., High performance liquid chromatography of dimers of *trans*-4-hydroxy-cinnamic acid and similar reaction products of some other substituted cinnamic and benzoic acids by an extracellular laccase of *Trametes versicolor, Gen. Educ. Res. Coll. Agr. Vet. Med.,* 25, 57, 1989b.

Katase, T. and Bollag, J.-M., Transformation of *trans*-4-hydroxycinnamic acids by a laccase of the fungus *Trametes versicolor*: its significance in humification, *Soil Sci.,* 151, 291, 1991.

Katase, T., Phenolic acids in tropical peats from peninsular Malaysia:occurrence and possible diagenetic behavior, *Soil Sci.,* 155, 155, 1993.

Katase, T., Phenolic acids in boreal peats from Finland and comparison with those from tropical and temperate areas, in *Plant Soil Interactions at Low pH,* Date, R. A., et al., Eds., Kluwer Academic Publishers, Netherlands, 1995, 71.

Kuprevich, V. F. and Scherbakova, T. A., Comparable enzymatic activity in diverse types of soil, in *Soil Biochemistry,* Vol. 2, Paul, E. A., and McLaren, A. D., Eds., Marcel Dekker, New York, 1971, 167.

Kusano, S. and Ogata, K., Phenolic acids in crops and their phytotoxicity, *Soil Sci. Plant Nutr.,* 45, 29, 1974.

Kyuma, K., Soils of swampy coastal areas in Southeast Asia. Part 2. Organic soils under the swamp forest, *Southeast Asia Studies,* 27, 492, 1983.

Leonowicz, A. and Bollag, J.-M., Laccases in the soil and the feasibility of their extraction, *Soil Biol. Biochem.,* 19, 237, 1987.

Liu, S-Y., Minard, R. D., and Bollag, J.-M., Oligomerization of syringic acid, a lignin-derivative, by a phenoloxidase, *Soil Sci. Soc. Am J.,* 45, 1100, 1981.

Macko, V., Staples, R. C., Gershon, H., and Renwick, J. A. A., Self-inhibitor of bean rust uredospores: methyl 3,4-dimethoxycinnamate, *Science,* 170, 539, 1970.

Macko, V., Staples, R. C., Allen, P. J., and Renwick, J. A. A., Identification of the germination self-inhibitor from wheat stem rust uredospores, *Science,* 173, 835, 1971.

Macko, V., Staples, R. C., Renwick, J. A. A., and Pirone, J., Germination self-inhibitors of rust uredospores, *Physiol. Plant Pathol.,* 2, 347, 1972.

Mayaudon, J. and Sarkar, J. M., Laccase de *Polyporus versicolor* dans le sol et la litiere, *Soil Biol. Biochem.,* 7, 31, 1975.

Morita, H., The phenolic acids in organic soil, *Can. J. Biochem.,* 43, 1277, 1965.

Müller, C. H., Phytotoxins as plant habitat variables, *Recent Adv. Phytochem.,* 3, 106, 1970.

Polak, B., Veen en veenontginning in Indonesia. MIAI. nrs 5 and 6, Vorkink, Bandung, 1952 (cited from Driessen and Suhardjo, 1976).

Polak, B., Character and occurrence of peat deposits in the Malaysian tropics, in *Modern Quaternary Research in Southeast Asia,* Bartstra, G. J. and Caspare, W. A., Eds., A. A. Belkema, Rotterdam, Netherlands, 1975, 71.

Rice, E. L., *Allelopathy,* Academic Press, New York, 1974, 353.

Ross, D. J. and McNeilly, B. A., Biochemical activities in a soil profile under hard beech forest-3. Some factors influencing the activities of polyphenol-oxidizing enzyme, *N. Z. J. Sci.,* 16, 241, 1973.

Seidham, M. M., Tome, A., and Wood, J. M., Influence of side-chain substituents on the position of cleavage of the benzene ring by *Pseudomonas fluorescences, J. Bacteriol.,* 97, 1192, 1969.

Shindo, H., and Kuwatsuka, S., Distribution of phenolic acids in soils of paddy fields and forests, *Soil Sci. Plant Nutr.,* 24, 233, 1978.

Shindo, H., Ohta, S., and Kuwatsuka, S., Distribution of phenolic acids of greenhouse and fields, *Soil Sci. Plant Nutr.,* 25, 591, 1979.

Simmons, K. E., Minard, R. D., and Bollag, J.-M., Oligomerization of an aromatic amine in the presence of oxidoreductases, *Environ. Sci. Technol.,* 21, 999, 1987.

Sjoblad, R. D. and Bollag, J.-M., Oxidative coupling of aromatic compounds by enzymes from soil microorganisms, in *Soil Biochemistry, Vol. 5,* Paul, E. A. and Ladd, J. N., Eds., Marcel Dekker, New York, 1981, 113.

Takizawa, N., *A Study on Kaschin-Beck Disease in Japan,* Ogata-shoin Book Co., Tokyo (in Japanese), 1970, 267.

Thawornwong, N. and van Diest, A., Influences of high acidity and aluminum on the growth of lowland rice, *Plant Soil,* 41, 141, 1974.

Valiera, I., Koumjian, L., Swain, T., Teal, J. M., and Hobbie, J. E., Cinnamic acid inhibition of detritus feeding, *Nature,* 280, 55, 1979.

Wang, T. S. C., Cheng, S., and Tung, H., Dynamics of soil organic acids, *Soil Sci.,* 103, 138, 1967a.

Wang, T. S. C., Yang, T., and Chung, T., Soil phenolic acids as plant growth inhibitors, *Soil Sci.,* 103, 239, 1967b.

Wang, T. S. C., Wang, M. C., and Gerng, Y. Z., Catalytic synthesis of humic substances by natural clays, silts and soils, *Soil Sci.,* 135, 350, 1983.

Whitehead, D. C., Identification of *p*-hydroxybenzoic, vanillic, *p*-coumaric and ferulic acids in soils, *Nature,* 202, 417, 1964.

Zapata, O. and McMillan, C., Phenolic acids in seagrasses, *Aquat. Bot.,* 7, 304, 1979.

Zenk, M. H. and Müller, G., *In vivo* destruction of exogenously applied indolyl-3-acetic acid as influenced by naturally occuring phenolic acid, *Nature,* 200, 761, 1963.

22 Effect of Plant Polyphenols on Nutrient Cycling and Implications for Community Structure

*Robert R. Northup, Randy A. Dahlgren, T. Mitchell Aide,
and Jess K. Zimmerman*

CONTENTS

22.1 ABSTRACT

The role of polyphenols as regulators of plant–litter–soil interactions is described in the context of two highly contrasting ecosystems. In northern California's pygmy forest, inherent conditions of the ancient soil are extremely acidic and infertile. High polyphenol concentrations in pygmy forest species regulate pathways of nitrogen cycling, create sorption capacity and complex metal cations. These attributes minimize nutrient losses from the ecosystem and ameliorate acid soil infertility factors, such as aluminum toxicity and phosphorus fixation. In this edaphic climax ecosystem, polyphenols facilitate sustained productivity on soils that otherwise are toxic, nutrient-deficient, and have a high potential for nutrient loss. This final succession sere is contrasted with *Dicranopteris pectinata* (Willd.) fern thickets on deforested sites of the wet tropics that create monospecific stands that inhibit further rain forest succession. High polyphenol concentration in the ferns may retard decomposition of leaf litter and sequester nitrogen into unavailable forms as in the pygmy forest. In these aggrading ecosystems, *Dicranopteris* remains dominant despite the rapid accumulation of organic matter and increasing levels of total soil nitrogen. Both ecosystems contradict models of plant succession based on resource availability. An important difference, however, is that *Dicranopteris* forms monospecific thickets on sites that otherwise would support a more diverse tropical rain forest community. High polyphenol production by *Dicranopteris* may allelopathically create edaphic conditions unfavorable for competing species, allowing *Dicranopteris* to remain dominant on soils that are rapidly accumulating nutrients in organic matter over time. These contrasting examples of ecosystem-scale effects of polyphenols are presented as supporting evidence for a broad hypothesis regarding the convergent evolution of tannin-rich plant communities on highly infertile soils throughout the world.

22.2 INTRODUCTION

The association between polyphenol- (i.e., tannin-) rich plant communities and nutrient-poor soils has long been recognized. On the assumption that they function as chemical defenses against other organisms (Fraenkel, 1959), production of polyphenols such as condensed tannins is believed to be selected in ecosystems where leaf longevity and low soil nutrient availability favor immobile, carbon-based defenses (Coley, 1988). Exceptionally high concentrations of these compounds in rain forest species growing on extremely infertile soils have been interpreted as reflecting natural selection for greater defense in cases where low soil nutrient availability creates high "cost" for replacing foliage lost to herbivory (Janzen, 1974). Several ecological studies, however, suggest that polyphenols are not always effective at reducing herbivory (Proctor et al., 1983; Aide and Zimmerman, 1990; Glyphis and Puttick, 1989; Balsberg Pahlson, 1989). Although differential leaf age and degree of shading result in different concentrations of condensed tannins in *Connarus turczaninowii* Tr. (Connaraceae), condensed tannin concentration showed no significant correlation with measured levels of herbivory (Aide and Zimmerman, 1990). Similarly, higher polyphenol concentrations are not correlated with reduced herbivory in *Quercus coccifera* L. and *Cistus monospeliensis* L. (Glyphis and Puttick, 1989) or *Fagus sylvatica* L. (Balsberg Pahlson, 1989). Although exceptionally high concentrations of polyphenols were observed in a rain forest growing on an acid white sand, levels of herbivory were higher than in an adjacent rain forest having lower polyphenol concentration (Proctor et al., 1983). Presumably, the large "investment" of photosynthate into polyphenol production imparts some fitness benefit to the producers. If the reduction of herbivory is not always the primary function of polyphenols, there may be some other explanation for the occurrence of polyphenol-rich plant communities on highly infertile soils throughout the world.

Soil scientists have long recognized the importance of polyphenols in pedogenic processes such as regulation of nitrogen mineralization (Handley, 1961), formation of mor-type humus (Coulson et al., 1960), and metal complexation and transport (Bloomfield, 1957). Polyphenols or their derivative, phenolic-carboxylic acids, are the dominant substrate from which stable humic substances are formed (Schnitzer et al., 1984; Stevenson, 1994). Newer analytical techniques have shown that what had been operationally defined as "lignin" is often composed mainly of polyphenols (Leary et al., 1986; Love et al., 1994; Preston, 1996). The identification of lignin as the dominant regulator of soil organic matter dynamics may be an artifact of this methodological flaw, or may result from the covariance of high lignin and high polyphenol concentrations in many species (Barry and Manley, 1986; Rittner and Reed, 1992). Although it is known that polyphenols can influence a broad range of soil properties and processes (Nicolai, 1988; Kuiters, 1990), little has been published about the potential benefits these impacts could impart to the producers (Bernays et al., 1990; Northup et al., 1995a).

An alternative explanation for the natural selection of polyphenol-rich plants in association with strongly acidic, infertile soils arises from an analysis of the role of polyphenols in plant–litter–soil interactions. Feedbacks whereby polyphenols degrade the productivity of fertile soils have been reported (Muller et al., 1987; Nicolai, 1988; Kuiters, 1990; Chapin, 1993). In contrast, relatively little has been published about feedbacks whereby polyphenols permit sustained productivity on highly acidic and infertile soils (Northup et al., 1995a). For example, by minimizing nitrogen mineralization and forming recalcitrant complexes with litter nitrogen, polyphenols shift the dominant pathway of nitrogen cycling from mineral to organic forms, minimizing N losses from the ecosystem and maximizing potential recovery of litter N by mycorrhizal symbionts (Northup et al., 1995b). The impact of polyphenols in plant–litter–soil interactions could also be beneficial to the producers if they function to create conditions that are unfavorable for competing species (Van Breemen, 1993; Chapin, 1993). This chapter presents examples from contrasting ecosystems to support a broad hypothesis regarding the occurrence of polyphenol-rich plant communities on highly infertile soils.

22.3 PYGMY FOREST ECOSYSTEMS

The "Ecological Staircase" consists of a series of coastal terraces in northern California that comprise an extreme gradient of soil acidity and fertility and support a broad range of plant communities. Continuous geologic uplift of the coast, in combination with periodic changes in sea level, created these terraces (Fox, 1976). Soils on each successive terrace initially had high fertility inherited from the geologic parent material and, through leaching and weathering processes, became progressively more infertile and acidic over time, until reaching a steady state on the oldest terraces (Jenny et al., 1969). Soils on the Ecological Staircase show a wide range of acidity and fertility: fertile and slightly acidic on the youngest terrace (T1; ~ 100,000 years old), moderately infertile and acidic on the intermediate age terrace (T2; ~ 200,000 years old), and extremely acidic and infertile on the oldest terraces (T3–T5; ~ 240,000 to > 400,000 years old) (Figure 22.1). The oldest soils

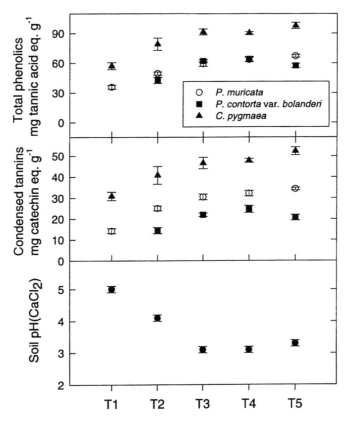

FIGURE 22.1 Concentrations of total phenolics and condensed tannins (mean $\pm$ SEM, $n = 8$–11) for *Pinus muricata* (Bishop pine), *Pinus contorta* var. *bolanderi* (Bolander pine), and *Cupressus pygmaea* (Mendocino cypress) and soil pH (CaCl$_2$) for the five sites (T1–T5) representing the soil chronosequence of the Ecological Staircase. Foliage was extracted with aqueous methanol and total phenols and condensed tannins were quantified by the Prussian Blue (Price and Butler, 1977) and vanillin methods (Broadhurst and Jones, 1978), respectively.

(T3–T5) in the sequence support a rare ecosystem of pygmy conifers ($<$ 3 m height) and Ericaceous species, dominated by endemic subspecies that are edaphic ecotypes (McMillan, 1956; Westman, 1975). Three conifer comprise much of the pygmy forest biomass; *Cupressus pygmaea* (Lemmon) Sarg. (Mendocino cypress), *Pinus contorta* var. *bolanderi* (Parl.) Vasey (Bolander pine), and *Pinus muricata* D. Don (Bishop pine)(Westman, 1975). The first two of these are endemic to the pygmy forest, and the third occurs as a distinct "blue" race of *P. muricata* on pygmy forest soils (Millar, 1989). The three conifers of the pygmy forest also grow on the younger terraces (T1 and T2) of the Ecological Staircase where they reach heights greater than 20 m.

Foliar concentrations of total phenolics and condensed tannins vary significantly ($P < .05$) in the three conifer species and are attributed to differences in soil properties (Figure 22.1). Polyphenol concentrations vary inversely with soil pH and fertility across this extreme edaphic gradient. Similarly, total phenolic and condensed tannin concentrations are very high in all Ericaceous spp. growing within the pygmy forest (Northup et al., 1998). *P. muricata* occurs on all five terraces of the Ecological Staircase and displays distinct growth habits in each soil type. In the highly infertile pygmy forest (T3–T5), approximately half of the fine roots occur in the litter layer rather than the mineral soil (Northup et al., 1995a). In contrast, few *P. muricata* roots occur in the litter layer when it grows on the most fertile (T1) soil, and approximately 25 percent of its fine roots are in the litter layer of the intermediately fertile (T2) soil. Similar variation occurs in the average age of *P. muricata* needle retention; few needles more than two years old are found on the T1 site, three- to five-year-old needles are common on the T2 site, and seven- to nine-year-old needles can be found in the pygmy forest (R. Northup, unpublished data, 1997). Differences in *P. muricata* current-year foliar polyphenol concentration (Figure 22.1) are also part of this soil-related variation in growth habit.

One consequence of this intraspecific variation in polyphenol concentrations is seen in the pattern of nitrogen release from *P. muricata* litter. With higher total phenolic and condensed tannin concentrations, a greater proportion of litter nitrogen is released as dissolved organic nitrogen (DON) rather than as mineral nitrogen (NH_4 and NO_3) (Figure 22.2). Whereas nitrate could easily be lost from the highly leached and periodically flooded soil of the pygmy forest, DON rarely leaches beyond the rooting zone (Yavitt and Fahey, 1986) and is not subject to gaseous loss through denitrification. Whereas ammonium can be consumed by a broad range of soil organisms, DON (particularly protein-tannin complexes) is difficult to utilize and cannot easily be converted to a form that might be lost from the ecosystem. This pattern of nitrogen release may constitute a nitrogen conservation mechanism in ecosystems having severe nitrogen deficiency. One potential explanation for the occurrence of polyphenol-rich plant communities on highly leached soils is that it is an adaptation to nitrogen deficiencies that mitigates further nitrogen loss.

Soil pH in the pygmy forest is extremely low (pH [$CaCl_2$] $\approx$ 3), and many soil infertility factors become more severe with decreasing pH. Aluminum toxicity, for example, can be a major limitation to the growth of many species at low pH (Roy et al., 1988). Phenolic acids leaching from the forest floor form strong 5- and 6-bond ring complexes with Al (McColl et al., 1990). The threshold concentration for toxicity of organically complexed Al is at least an order of magnitude higher than that of labile, inorganic forms of Al (Adams and Moore, 1983). Humic acids added to sand culture ameliorate Al toxicity in maize (Tan and Binger, 1986), and the greatest degree of Al detoxification was achieved with phenolic acids that form 5- and 6-bond rings with Al (Hue et al., 1986). Polyphenols are strong complexing agents for Al (Powell and Rate, 1987), particularly the ortho-phenolic group (Sikora and McBride, 1990). Thus, the enigma of exceptionally high polyphenol concentrations in tropical rain forest vegetation growing on strongly acidic soils could be a response to aluminum toxicity (Bruijnzeel and Veneklaas, 1998).

Low phosphorus availability in acidic soils can be an important limitation to ecosystem productivity. Phosphorus (P) "fixation" due to precipitation with soluble forms of aluminum, manganese, and iron, or specific sorption of phosphate to reactive surfaces are more problematic with decreasing pH. In the pygmy forest, highly weathered Al and Fe (hydr)oxides, in combination with extremely low pH, contribute to a high potential for P fixation. However, the ortho-phenolic group

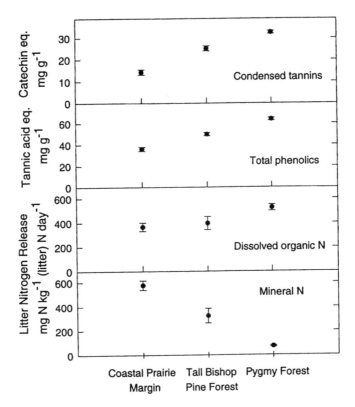

FIGURE 22.2 Foliar polyphenol concentrations (mean ± SEM) and nitrogen release (mean ± SEM) from *Pinus muricata* (Bishop pine) litter in contrasting soil environments of the Ecological Staircase. Litter was extracted with aqueous methanol and total phenols and condensed tannins were quantified by the Prussian Blue (Price and Bulter, 1977) and vanillin methods (Broadhurst and Jones, 1978), respectively. Nitrogen released was measured in the laboratory using a three-week, aerobic incubation at 25°C.

is such a strong competitor for these sites that phenolic acids can desorb "fixed" phosphate (Davis, 1982). Phenolic acids are also strong complexing agents for Al, Mn, and Fe, which lowers their solution activities and reduces the potential for precipitation with phosphate (McColl et al., 1990). Reduced reactivity toward phosphate due to complexation of Al by humic acids has been shown to increase the bioavailability of P in acid sand culture (Tan and Binger, 1986). Their capacity to complex iron and solubilize P has been suggested as a reason for natural selection of high polyphenol concentration in *Eucalyptus* growing on acidic, P-fixing soils (Hingston, 1962). Pygmy forest species are able to obtain sufficient P to sustain productivity, despite extreme conditions for potential P fixation. The high polyphenol concentrations in the pygmy forest may be part of the explanation for this ability.

Deficiency of nutrient cations (e.g, K, Ca, Mg, etc.) is another infertility factor commonly associated with acidic soils. Losses of such nutrients from highly leached soils can be minimized by adsorption to cation exchange sites. Most cation exchange capacity (CEC) in highly weathered soils arises from organic matter rather than clay minerals (Kalisz and Stone, 1980). Through retarded decomposition of organic matter, and as the dominant substrate from which humic substances are formed, polyphenols are responsible for much of the CEC in such soils (Schnitzer et al., 1984; Stevenson, 1994). Although the importance of polyphenols for creating CEC has long been recognized (e.g., Davies, 1971), overemphasis of their presumed role in antiherbivore defense has resulted in little consideration of the potential adaptive value from this impact on soil properties (Bernays et al., 1990). In the pygmy forest, CEC in the rooting zone arises exclusively from organic matter. The acidified quartz sand and highly weathered Al and Fe (hydr)oxides have very little CEC, nor can they provide any nutrient cations through additional weathering reactions. Despite these conditions, the pygmy forest soils have sufficient CEC to prevent leaching loss of the remaining nutrient cations, and sustain productivity over geologic time. Again, the high polyphenol concentration of the associated vegetation may be part of the explanation for this enigma.

22.4   *DICRANOPTERIS* FERN THICKET ECOSYSTEMS

In the Ebano Verde scientific reserve, located in a wet region of the central highlands of the Dominican Republic, thickets of *Dicranopteris pectinata* (Willd.) dominate disturbed areas where human activity (e.g., logging, slash and burn agriculture, etc.) has removed the original rain forest. These fern thickets have exceptionally low species diversity and have persisted, in some cases, for many decades. Pioneer trees usually establish on disturbed sites before the arrival of the ferns. The ferns overtopped and out-compete most of these trees for light as a result of their vine-like growth habit. Some pioneer trees are tall enough to be unaffected by light competition from the ferns; however, they appear sickly and have high rates of mortality.

Dicranopteris fern thickets at Ebano Verde typically have approximately 100-cm thick layers of litter above the mineral soil. The overwhelming majority (>85 percent) of roots are concentrated above the mineral soil in the decomposing litter (Northup, unpublished data, 1998). Laboratory analysis of *Dicranopteris* foliage reveals an exceptionally high concentration of condensed tannins (51.9 ± 2.7 mg catechin equiv/g foliage; mean ± SEM, $n = 11$). These polyphenol concentrations are equal to or greater than concentrations in conifers and Ericaceous species growing in the pygmy forest (Northup et al., 1998).

Throughout the wet tropics, this and other closely related species of fern commonly form monospecific thickets on disturbed sites, and these become a major obstacle to rain forest succession (Joachim and Kandiah, 1942; Gleissman, 1978; Walker, 1994; Cohen et al., 1995). Although the establishment of a dense tree canopy can shade out the ferns (de Ronde and Bredenkamp, 1984), it is common for these monospecific thickets to persist for decades (Guariguata, 1990; Walker, 1994; Cohen et al., 1995). *Dicranopteris* thickets can occur on strongly acidic soils with pH values as low as 3.9 (Walker, 1994); conditions conducive to aluminum toxicity, phosphorus fixation, and other infertility factors associated with low soil pH. However, where more nutrient-rich soils occur, such as in the deposition zone at the base of a landslide, faster-growing herbs and trees can form a canopy dense enough to prevent fern thicket establishment (Stewart, 1986; Guariguata, 1990).

Experiments with *Dicranopteris* litter in fern thickets show exceptionally slow rates of decomposition (Russell and Vitousek, 1997), and annual mass loss rates can be less than 16 percent (Maheswaran and Gunatilleke, 1988). Substantial accumulation of organic matter occurs despite well-drained, warm, and moist conditions that should favor rapid decomposition. Accumulation of litter is common in fern thickets and creates a physical barrier to establishment of competing seedlings (Walker, 1994). Furthermore, *Dicranopteris* litter brings about a net immobilization of N and P into the organic matter (Russell and Vitousek, 1997). This net immobilization of N and P continues even after a year of decomposition, by which time other litter types usually show net nutrient release (Maheswaran and Gunatilleke, 1988). Immobilization of N by decomposing organic matter in wet tropical forests can cause a substantial reduction in the availability of mineral N (Zimmerman et al., 1995). Retarded decomposition of *Dicranopteris* litter has been attributed to high concentrations of lignin (Russell and Vitousek, 1997). However, as previously mentioned, there are reasons to believe that decomposition is more strongly regulated by polyphenols than by lignin. The slow decomposition rates and sequestration of nutrients by *Dicranopteris* litter are consistent with the exceptionally high concentration of condensed tannin, which has been shown to hinder decomposition and result in the formation of a mor-type humus (Coulson et al., 1960).

The low vigor of taller pioneer trees in the Ebano Verde *Dicranopteris* thickets is consistent with what has been observed in comparable fern thickets elsewhere in the wet tropics (Cohen et al., 1995).

Allelopathic inhibition of other species by fern leachate has been indicated in some studies (Gleissman and Muller, 1978), but other experiments fail to confirm this finding (Walker, 1994). Potential mechanisms of allelopathy other than by direct toxicity from fern leachate have been suggested, such as induced manganese toxicity (Aragon, 1975). This possibility is intriguing given the fact that phenolic acids leaching from the litter are capable of solubilizing large amounts of manganese in the mineral soil (McColl et al., 1990) and given that the polyphenol-rich *Dicranopteris* keeps the vast majority of its own roots above the mineral soil.

Another potential mechanism of inhibition could be the immobilization of mineral N. Again there could be a role for polyphenols through retarding organic matter decomposition, prolonging N immobilization, and forming protein-tannin complexes that are very difficult to mineralize. This family of ferns is abundantly infected with vesicular arbuscular mycorrhizal (VAM) fungi (e.g., Schmid and Oberwinkler, 1995). It has been shown that ericoid and ectomycorrhizal fungi are capable of utilizing N from protein-tannin complexes, thereby "short-circuiting" the N cycle (Leake and Read, 1989; Griffiths and Caldwell, 1992). It is not yet known if VAM fungi associated with fern roots also have this capacity. However, in the absence of some capacity to utilize N from protein-tannin complexes, the fern's high polyphenol content and concentration of roots in the litter layer would appear to be maladaptive. Assuming that *Dicranopteris* does have the ability to use organic forms of N, one consequence of this would be to maintain very low levels of mineral N in the soil, thus inhibiting the performance of competing species by starving them for N.

High rates of nonsymbiotic nitrogen fixation have been measured in *Dicranopteris* fern thickets (Maheswaran and Gunatilleke, 1990; Russell and Vitousek, 1997). The energetic expense of N fixation by free-living bacteria is enormous, and the high rates measured in fern thickets are indicative of an ecosystem that has abundant organic matter but is deficient in available (mineral) N. One consequence of the N fixation occurring in *Dicranopteris* thickets is to increase total soil N over time (Walker, 1994). As aggrading ecosystems accumulating organic matter and soil N, these fern thickets would presumably be creating ideal conditions for forest succession. However, monospecific *Dicranopteris* thickets can persist nearly indefinitely (Walker, 1994). For example, *Dicranopteris* continues to dominate the understory for up to 300 years on Hawaiian lava flows (Kitayama et al., 1995). If the absolute increase in total soil N associated with these fern thickets (Walker, 1994) is due to accumulation of recalcitrant forms of N (e.g., protein-tannin complexes), available (mineral) forms of N could be too deficient for establishment of competing species.

22.5   IMPLICATIONS FOR COMMUNITY STRUCTURE

The intraspecific variation of foliar polyphenol concentration observed in conifers of the Ecological Staircase is consistent with what has been observed in other species on soil acidity gradients (e.g., Muller et al., 1987). This suggests that there has been natural selection for feedbacks between soil conditions and polyphenol production. As soils become progressively more leached and acidic, there should be greater selective pressure to produce high concentrations of polyphenols to alleviate acid soil infertility factors. The pygmy forest is an extreme example of an oligotrophic ecosystem growing on soil that is exceptionally acidic and N deficient (McMillan, 1956). The few nutrients that remain in the pygmy forest are contained entirely in the biomass or decomposing organic matter (Westman, 1975). With its high potential for N loss to leaching or denitrification, aluminum toxicity, P fixation, and nutrient cation deficiencies, there are many ways that polyphenol production in the pygmy forest can be construed as an adaptation to soil conditions (Northup et al., 1995a). In this context, intraspecific variation of polyphenol concentration would be expected along the soil gradient regardless of whether they play any role in antiherbivore defense.

The role of polyphenols in the *Dicranopteris* fern thickets is probably similar in many respects to the pygmy forest. High concentrations of condensed tannin in the vegetation facilitates the accumulation of organic matter, alters N dynamics to minimize leaching loss or denitrification, and could facilitate sustained productivity despite strongly acidic soil conditions. The most striking contrast between the two ecosystems is in regard to soil development and plant succession. The pygmy forest is an edaphic climax ecosystem on an ancient soil that represents as close to a final succession sere as can be found in nature (Westman, 1975). Despite the pygmy forest having returned to an open canopy with high light availability, and soil conditions of extreme N limitation, non-N fixing plants dominate the ecosystem. *Dicranopteris* fern thickets, on the other hand, are aggrading ecosystems that often occur on freshly exposed soil parent material. Despite rapid accumulation of organic matter and increasing levels of total soil N, succession in the fern thickets is arrested at one of its earliest seres.

Both ecosystems described in this work contradict succession models based on resource availability. According to Tilman's (1985) resource-ratio hypothesis of plant succession, conditions of high light availability and soil N limitation should favor dominance by N-fixing plants. The lack of N-fixing vascular plants in the pygmy forest may be the result of soil chemical conditions (e.g., low pH, lack of molybdenum, etc.). Free-living N-fixing bacteria are active in *Dicranopteris* thickets (Maheswaran and Gunatilleke, 1990; Russell and Vitousek, 1997) resulting in increasing quantities of total soil N (Walker, 1994). The inhibition of succession by faster-growing species following this soil enrichment also contradicts predictions of Tilman's hypothesis. However, arrested succession by monospecific thickets is consistent with the inhibition model (Connell and Slatyer, 1977), particularly given the ability of *Dicranopteris* to create a physical barrier to seedling establishment through rapid litter accumulation (Walker, 1994). Inhibition of forest succession by several polyphenol-rich Ericaceous species is also reported. Salal (*Galutheria shallon* Pursh) has been implicated as inhibiting the growth of cedar and hemlock in clearcut areas on northern Vancouver Island (Prescott and Weetman, 1994). These studies indicate that salal competes aggressively for soil nutrients because of its high below-ground biomass and the ability of its mycorrhizae to access organic forms of N that are unavailable to trees. Similarly, in clearcut areas of black spruce [*Picea mariana* (Mill.) B.S.P.] in Newfoundland, there is a permanent conversion from forest to heathland as a result of kalmia (*Kalmia angustifolia* L. var. *angustifolia*) invasion (Titus, 1995).

To account for the ability of organisms to improve their reproductive success by actively altering their environment, the concept of the "extended" phenotype has gained favor among many ecologists (Dawkins, 1982). Although the most familiar examples of this are nest construction by animals, plants can also act as ecosystem engineers (Jones et al., 1994). The adaptive value for polyphenol production can be construed in the context of the extended phenotype as arising from their many impacts on plant–litter–soil interactions (Northup et al., 1998). In the context of the two ecosystems described in this work, the extended phenotype concept is supported by the idea that high polyphenol concentrations enable plants to act as soil engineers, thereby manipulating soil conditions to their advantage. Furthermore, these contrasting ecosystems suggest two different evolutionary pathways toward the convergent evolution of polyphenol-rich plant communities, given the ability of plants to alter soil properties and nutrient cycling dynamics through polyphenol production.

Long-term stability of soil conditions on the coastal terraces of the Ecological Staircase has permitted the evolution of endemic subspecies that dominate pygmy forest biomass (Westman, 1975). Bolander pine, for example, is an edaphic ecotype of *Pinus contorta* found exclusively in the pygmy forest, and appears to have evolved there relatively recently (Aitken and Libby, 1994). One possible scenario is that a polyphenol-rich mutant, originating from the nearby tall forest, was able to colonize a pygmy forest site, taking advantage of the abundant light availability. Another possible scenario is that the mutant occurred within the tall forest as the soil became less fertile over time, and was able to remain established there when conditions became intolerable for competing individuals that required more fertile soil. In either case, accumulation of high concentrations of polyphenols would have been the only way for it to tolerate the highly acidic and infertile soil conditions.

Inhibition of forest succession by *Dicranopteris* fern thickets on soils that are accumulating organic matter and N suggests a completely different evolutionary pathway for the selection of high polyphenol concentration. Whereas the regulation of N dynamics in the pygmy forest can be construed as a mechanism to minimize losses of this extremely limited nutrient, *Dicranopteris* may produce polyphenols to exclude competing species by monopolizing the N supply. Similarly, complexation of aluminum by polyphenols can be construed as an adaptation to ameliorate toxicity in the pygmy forest, while reduction and complexation of manganese by polyphenols in the fern thickets might be a mechanism to solubilize toxic levels of manganese in the mineral soil to exclude competing species. Lastly, the retardation of organic matter decomposition by polyphenols in the pygmy forest can be construed as necessary to create sufficient CEC and sorption capacity to minimize nutrient losses in the acid quartz sand. In contrast, attenuation of organic matter decomposition may be used by *Dicranopteris* to accumulate so much litter that it creates a physical barrier to establishment of competing species. In the pygmy forest it would appear that roots are concentrated in the organic layer because the underlying mineral soil is unfavorable, since the same species will avoid placing roots in the organic layer entirely when the mineral soil is fertile. In the fern thickets, the thick accumulation of a surface litter layer and the concentration of roots in the litter layer may be yet another mechanism to exclude competition. These traits may provide a physical barrier to seedling establishment, spatially regulating nutrient cycling to keep nutrients unavailable to competing plants, and potentially allowing the ferns to allelopathically create unfavorable conditions in the underlying mineral soil without being adversely affected themselves.

This broad hypothesis is highly speculative, and much work remains to be done to test it. We do not pretend to have definitively proven anything about the evolution of polyphenol-rich plant communities, nor have we disproven the potential importance of polyphenols as an antiherbivore defense. The intention for presenting this hypothesis is to stimulate others to examine these possibilities and perhaps reconsider some of the basic assumptions that have guided chemical ecology research in recent decades. A paradigm shift toward viewing plants as soil engineers, and assessing the adaptive value of polyphenol production in this context, may help to explain patterns of phytochemistry and elucidate enigmatic aspects of plant–litter–soil interactions.

ACKNOWLEDGMENTS

We are grateful for permission to collect samples from California's Jug Handle reserve and from the Ebano Verde scientific reserve administered by the PROGRESSIO Foundation. This work was supported in part by a grants from the National Science Foundation (DEB-9527722) to the University of California-Davis and the National Aeronautic and Space Administration Institutional Research Awards for Minority Universities Program (NAGW-4059) to the University of Puerto Rico.

REFERENCES

Adams, F. and Moore, B. L., Chemical factors affecting root growth in subsoil horizons of coastal plain soils, *Soil Sci. Soc. Am. J.,* 47, 99, 1983.

Aide, T. M. and Zimmerman, J. K., Patterns of insect herbivory, growth, and survivorship in juveniles of a neotropical liana, *Ecology,* 71, 1412, 1990.

Aitken, S. N., and Libby, W. J., Evolution of the pygmy forest edaphic subspecies of *Pinus contorta* across an ecological staircase, *Evolution,* 48, 1009, 1994.

Aragon, E. L., Inhibitory effects of substances from residues and extracts of staghorn fern (*Dicranopteris linearis*), MSc thesis, University of Hawaii at Manoa, HI, 1975.

Balsberg Pahlsson, A. M., Mineral nutrients, carbohydrates and phenolic compounds in leaves of beech (*Fagus sylvatica* L.) in southern Sweden as related to environmental factors, *Tree Physiol.,* 5, 485, 1989.

Barry, T. and Manley, T., Interrelationships between the concentrations of total condensed tannin, free condensed tannin and lignin in *Lotus* sp. and their possible consequences in ruminant nutrition, *J. Sci. Food Agric.,* 37, 248, 1986.

Bernays, E. A., Cooper-Driver, G., and Bilgener, M., Herbivores and plant tannins, *Adv. Ecol. Res.,* 19, 263, 1990.

Bloomfield, C., The possible significance of polyphenols in soil formation, *J. Sci. Food Agric.,* 8, 389, 1957.

Broadhurst, R. B. and Jones, W. T., Analysis of condensed tannins using acidified vanillin, *J. Sci. Food Agric.,* 29, 788, 1979.

Bruijnzeel, L. A. and Veneklaas, E. J., Climatic conditions and tropical montane forest productivity: the fog has not lifted yet, *Ecology,* 79, 3, 1998.

Chapin, F. S., The evolutionary basis of biogeochemical soil development, *Geoderma,* 57, 223, 1993.

Cohen, A. L., Singhakumara, B. M., and Ashton, P. M., Releasing rain forest succession: a case study in the *Dicranopteris linearis* fernlands of Sri Lanka, *Restor. Ecol.,* 3, 261, 1995.

Coley, P. D., Effects of plant growth rate and leaf lifetime on the amount and type of anti-herbivore defense, *Oecologia* (Berlin), 74, 531, 1988.

Connell, J. H. and Slatyer, R. O., Mechanisms of succession in natural communities and their role in community stability and organization, *Am. Nat.,* 111, 1119, 1977.

Coulson, C. B., Davies, R. I., and Lewis, D. A., Polyphenols in plant, humus, and soil. I. Polyphenols of leaves, litter, and superficial humus from mull and mor sites, *J. Soil Sci.,* 11, 20, 1960.

Davies, R. I., Relation of polyphenols to decomposition of organic matter and to pedogenic processes, *Soil Sci.,* 111, 80, 1971.

Davis, J. A., Adsorption of natural dissolved organic matter at the oxide/water interface, *Geochimi. Cosmochimi. Acta,* 46, 2381, 1982.

Dawkins, R., *The Extended Phenotype.* Oxford University Press, Cambridge, 1982.

De Ronde, C. and Bredenkamp, B. V., The influence of *Pinus pinaster* on the spread of *Gleichenia polypodioides, South Afr. For. J.,* 131, 40, 1984.

Fox, W. W., Pygmy forest: An ecological staircase, *California Geol.,* 29, 3, 1976.

Fraenkel, G., The raison d'etre of secondary plant substances, *Science,* 129, 1466, 1959.

Gleissman, S. R., The establishment of bracken following fire in tropical habitats, *Am. Fern J.,* 68, 41, 1978.

Gliessman, S. R. and Muller, C. H., The allelopathic mechanisms of dominance in bracken (*Pteridium aquilinum*) in southern California, *J. Chem. Ecol.,* 4, 337, 1978.

Glyphis, J. P. and Puttick, G. M., Phenolics, nutrition, and insect herbivory in some Garrique and Maquis plant species, *Oecologia* (Berlin), 78, 259, 1989.

Griffiths, P. and Caldwell, B., Mycorrhizal mat communities in forest soils, *in Mycorrhizas in Ecosystems,* Read, D., Lewis, D., Fitter, A., and Alexander, I., Eds., CAB International, Wallingford, 1992, 98.

Guariguata, M. R., Landslide disturbance and forest regeneration in the upper Luquillo mountains of Puerto Rico, *J. Ecol.,* 78, 814, 1990.

Handley, W. R., Further evidence for the importance of residual leaf protein complexes in litter decomposition and the supply of nitrogen for plant growth, *Plant Soil,* 15, 37, 1961.

Hingston, F. J., Activity of polyphenolic constituents of leaves of *Eucalyptus* and other species in complexing and dissolving iron oxide, *Aust. J. Soil Res.,* 1, 63, 1962.

Hue, N. V., Craddock, G. R., and Adams, F., Effects of organic acids on aluminum toxicity in subsoils, *Soil Sci. Soc. Am. J.,* 50, 28, 1986.

Janzen, D. H., Tropical blackwater rivers, animals, and mast fruiting by the dipterocarpaceae, *Biotropica,* 6, 69, 1974.

Jenny, H., Arkley, R. J., and Schultz, A. M., The pygmy forest-podsol ecosystem and its dune associates of the Mendocino coast, *Madrono,* 20, 60, 1969.

Joachim, A. W. and Kandiah, S., Studies on Ceylon soils: The chemical and physical characteristics of the soils of adjacent contrasting vegetation formations, *Trop. Agric.,* 98, 81, 1942.

Jones, C., Lawton, J., and Shachak, M., Organisms as ecosystem engineers, *Oikos,* 69, 373, 1994.

Kalisz, P. J. and Stone, E. L., Cation exchange capacity of acid forest humus layers, *Soil Sci. Soc. Am. J.,* 44, 407, 1980.

Kitayama, K., Mueller-Dombois, D., and Vitousek, P. M., Primary succession of Hawaiian montane rain forest on a chronosequence of eight lava flows, *J. Veg. Sci.,* 6, 211, 1995.

Kuiters, A. T., Role of phenolic substances from decomposing forest litter in plant-soil interactions, *Acta Bot. Neerl.,* 39, 329, 1990.

Leake, J. R. and Read, D. J., Effects of phenolic compounds on nitrogen mobilisation by ericoid mycorrhizal systems, *Agric. Ecosyst. Environ.,* 29, 225, 1989.

Leary, G. J., Newman, R. H., and Morgan, K. R., A ^{13}C nuclear magnetic resonance study of chemical processes involved in the isolation of Klason lignin, *Holzforschung,* 40, 267, 1986.

Love, G. D., Snape, C. E., Jarvis, M. C., and Morrison, I. M., Determination of phenolic structures in flax fibre by solid-state ^{13}C-NMR, *Phytochemistry,* 35, 489, 1994.

Maheswaran, J. and Gunatilleke, I. A., Litter decomposition in a lowland rain forest and a deforested area in Sri Lanka, *Biotropica,* 20, 90, 1988.

Maheswaran, J. and Gunatilleke, I. A., Nitrogenase activity in soil and litter of a tropical lowland rain forest and an adjacent fernland in Sri Lanka, *J. Trop. Ecol.,* 6, 281, 1990.

McColl, J. G., Pohlman, A. A., Jersak, J. M., Tam, S. C., and Northup, R. R., Organics and metal solubility in California forest soils, in *Sustained Productivity of Forest Lands,* Gessel, S. P., Ed., Proceedings of the 7th North American Forest Soils Conference. Faculty of Forestry Publ., University of British Columbia, Vancouver, 1990, 178.

McMillan, C., The edaphic restriction of *Cupressus* and *Pinus* in the coast ranges of central California, *Ecol. Monogr.,* 26, 177, 1956.

Millar, C. I., Allozyme variation of Bishop pine associated with pygmy forest soils in northern California, *Can. J. For. Res.,* 19, 870, 1989.

Muller, R. N., Kalisz, P. J., and Kimmerer, T. W., Intraspecific variation in production of astringent phenolics over a vegetation-resource availability gradient, *Oecologia* (Berlin), 72, 211, 1987.

Nicolai, V., Phenolic and mineral content of leaves influences decomposition in European forest communities, *Oecologia* (Berlin), 75, 575, 1988.

Northup, R. R., Dahlgren, R. A., and Yu, Z., Intraspecific variation of conifer phenolic concentration on a marine terrace soil acidity gradient: a new interpretation, *Plant Soil,* 171, 255, 1995a.

Northup, R. R., Yu, Z., Dahlgren, R. A., and Vogt, K. A., Polyphenol control of nitrogen release from pine litter, *Nature,* 377, 227, 1995b.

Northup, R. R., Dahlgren, R. A., and McColl, J. G., Polyphenols as regulators of plant-litter-soil interactions in northern California's pygmy forest: a positive feedback?, *Biogeochemistry,* 42, 189, 1998.

Powell, H. and Rate, A. W., Aluminium-tannin equilibria: a potentiometric study, *Aust. J. Chem.,* 40, 2015, 1987.

Prescott, C. E. and Weetman, G. F., Salal cedar hemlock integrated research program: A synthesis. Faculty of Forestry, University of British Columbia, Vancouver, 1994.

Preston, C. M., Applications of NMR to soil organic matter analysis: history and prospects, *Soil Sci.,* 161, 144, 1996.

Price, M. L. and Butler, L. G., Rapid visual estimation and spectrophotometric determination of tannin content in sorghum grain, *J. Agric. Food Chem.,* 25, 1268, 1977.

Proctor, J., Anderson, J. M., Fogden, S. C., and Vallack, H. W., Ecological studies in four contrasting lowland rain forests in Gunung Mulu National Park, Sarawak. II. Litterfall, litter standing crop and preliminary observations on herbivory, *J. Ecol.,* 71, 261, 1983.

Rittner, U. and Reed, J., Phenolics and *in-vitro* degradability of protein and fiber in West African browse, *J. Sci. Food Agri.,* 58, 21, 1992.

Roy, A., Sharma, A., and Talukder, G., Some aspects of aluminum toxicity in plants, *Bot. Rev.,* 54, 145, 1988.

Russell, A. E., and Vitousek, P. M., Decomposition and potential nitrogen fixation in *Dicranopteris linearis* litter on Mauna Loa, Hawaii, *J. Trop. Ecol.,* 13, 579, 1997.

Schmid, E. and Oberwinkler, F., A light- and electron-microscopic study on a vesicular-arbuscular host-fungus interaction in gametophytes and young sporophytes of the Gleicheniaceae (Filicales), *New Phytol.,* 129, 317, 1995.

Schnitzer, M., Barr, M., and Hartenstein, R., Kinetics and characteristics of humic acids produced from simple phenols, *Soil Biol. Biochem.,* 16, 371, 1984.

Sikora, F. and McBride, M., Aluminum complexation by protocatechuic and caffeic acids as determined by ultraviolet spectrophotometry, *Soil Sci. Soc. Am. J.,* 54, 78, 1990.

Stevenson, F. J., *Humus Chemistry: Genesis, Composition, Reactions,* 2nd ed., John Wiley & Sons, New York, 1994.

Stewart, G. H., Forest dynamics and disturbance in a beech/hardwood forest, Fiordland, New Zealand, *Vegetatio,* 68, 115, 1986.

Tan, K. and Binger, A., Effect of humic acid on aluminum toxicity in corn plants, *Soil Science,* 141, 20, 1986.

Tilman, D., The resource-ratio hypothesis of plant succession, *The Am. Natur.* 125, 827, 1985.

Titus, B. B., A summary of some studies on *Kalmia angustifolia* L.: a problem species in Newfoundland forestry, Canadian Forest Service Information Report N-X-296, 1995.

Van Breemen, N., Soils as biotic constructs favouring net primary productivity, *Geoderma,* 57, 183, 1993.

Walker, L. R., Effects of fern thickets on woodland development on landslides in Puerto Rico, *J. Veg. Sci.*, 5, 525, 1994.

Westman, W. E., Edaphic climax pattern of the pygmy forest region of California, *Ecol. Monogr.*, 45, 109, 1975.

Yavitt, J. B. and Fahey, T. J., Litter decay and leaching from the forest floor in *Pinus contorta* (lodgepole pine) ecosystems, *J. Ecol.*, 74, 525, 1986.

Zimmerman, J. K., Pulliam, W. M., Lodge, D. J., Quinones-Orfila, V., Fetcher, N., Guzman-Grajales, S., Parrotta, J. A., Asbury, C. E., Walker, L. R., and Waide, R. B., Nitrogen immobilization by decomposing woody debris and the recovery of tropical wet forest from hurricane damage, *Oikos*, 72, 314, 1995.

Section IV

Biochemical, Chemical, and Physiological Aspects

23 Detoxification of Allelochemicals in Higher Plants and Enzymes Involved

Margot Schulz and A. Friebe

CONTENTS

23.1   ABSTRACT

Plants are capable to detoxify harmful molecules, such as herbicides, by absorption and conversion into biologically nonactive compounds. Molecular mechanisms involved in detoxification have been thoroughly studied with herbicides, albeit those mechanisms have certainly been developed during evolution to reduce or to compensate the reactivity of natural compounds. Detoxification or deactivation of allelochemicals, of naturally occurring growth effectors, has been only rarely investigated at the molecular level.

As a common concept, incorporated molecules are modified to increase their polarity, either by hydroxylation, dealkylation, and/or conjugation with primary metabolites, such as sugars and amino acids. The conjugation can mask the reactivity of functional groups that are responsible for biological effects. Conversion of the absorbed molecule is followed by storage within the vacuoles or by excretion into the environment. Oxidases, peroxigenases, and transferases are enzymes necessary for detoxification. These enzymes can be constitutive or they can be induced by the allelochemical, as has been shown for salicylic acid. Another strategy is to polymerize absorbed compounds, for example, ferulic acid, and to deposit them within the cell walls. In some cases, microorganisms are involved in detoxification, as it was found for hydroxamic acid degradation. The rapidity of the detoxification and the mode of enzyme induction seems to be species dependent, a fact that contributes to the difference in allelochemical tolerance among species.

23.2   INTRODUCTION

Experimental investigations of allelopathic phenomena have revealed strong species-dependent responses to allelopathically active compounds. Differences in growth stimulation or inhibition of various plant species have been found for a large variety of allelochemicals including phenolic substances (Ray and Hastings, 1992), benzoquinone (Einhellig and Souza, 1992) and anthraquinone compounds (Inoue et al., 1992), amino acids (Kato-Noguchi et al., 1994), as well as cyclic hydroxamic acids and their derivatives (Barnes and Putnam, 1987; Schulz et al., 1994). In some cases a high resistance of plants to allelochemicals was observed. For instance, the growth inhibition of several plant species by L-tryptophan, which was identified as a main compound of oat (*Avena sativa* L.) root exudates, highly differs: The allelochemical concentration for 50 percent inhibition of root growth was 0.14 mM for cockscomb (*Amaranthus caudatus* L.) and 0.15 mM for cress (*Lepidium sativum* L.) compared with 1.7 mM for the moderately resistant wheat (*Triticum aestivum* L.) (Kato-Nogchi et al., 1994).

The ability of plants to withstand increased concentrations of allelochemicals in their close surroundings is called allelopathic tolerance (Grodzinski, 1991). Allelopathic tolerance is a general feature and is a requirement for the development of characteristic plant communities. There are morphological and structural properties of plants that contribute to an increased tolerance to allelochemicals. Morphological adaptations are based on the inhibition or prevention of the uptake of biologically active compounds. Correspondingly, the permeability of the cuticles for phenolic substances varies from one to two orders of magnitude depending on their lipid composition (Shafer and Schönherr, 1985). Nevertheless, allelopathic resistance cannot be sufficiently explained by morphological properties alone, since species with similar morphological features show different allelopathic susceptibilities (Golovko and Kavelenova, 1992). Obviously, plants must have rather extensive abilities to overcome the influence of absorbed allelopathic compounds. Detoxification may represent a factor that contributes to the differential susceptibility to a particular allelochemical agent. Additional factors influencing a plant's sensitivity are, for example, the rate of uptake and translocation, susceptibility of the site of inhibition, and physical-chemical properties of the environment (e.g., soil, pH, humidity, light intensity and quality, and temperature).

General mechanisms for the detoxification of xenobiotics and herbicides and enzymes involved have long been studied and are subjects of current research (Gareis et al., 1992; Schmidt et al., 1993, 1995). The majority of enzymes involved in detoxification are constitutive. However, there are also detoxifying enzymes that are induced by absorbed compounds. We will briefly present mechanisms responsible for herbicide detoxification since similar ones have to be taken into consideration for detoxification of absorbed allelochemicals.

23.3   CONCEPTS OF HERBICIDE DETOXIFICATION

As a general concept of metabolic plant defense against xenobiotics, a three-step mechanism (Figure 23.1) has been proposed by Cole (1994) and Golovko and Kavelenova (1992). The primary metabolic reaction increases the polarity of the incorporated molecule. Most common initial biochemical reactions are hydroxylations (Thomas et al., 1964; Chkanikov, 1987; Tanaka et al., 1990a) and dealkylations (Burnet et al., 1993). These reactions are often catalyzed by cytochrome P450-dependent oxidases. The detoxification of the herbicide MCPA (4-chloro-2-methylphenoxyacetic acid), for instance, is based on methyl hydroxylation (Cole and Loughman, 1985) and the metabolization of chlorotoluron in resistant biotypes of swiss ryegrass *Lolium rigidum* Gaudin (Burnet et al., 1993) can be inhibited by application of 1-aminobenzotriazole, an inactivator of cytochrome P450 oxidases. Aside from cytochrome P450-dependent reactions, there are also alternative metabolic mechanisms, such as peroxygenase reactions. This metabolization leads to the oxidation of xenobiotics by organic peroxides without the requirement of NAD(P)H. For example, maize microsomal fractions are able to detoxify the herbicide EPTC (S-ethyl dipropyl carbamothioate) by peroxidase-catalyzed oxidation (Blee, 1991).

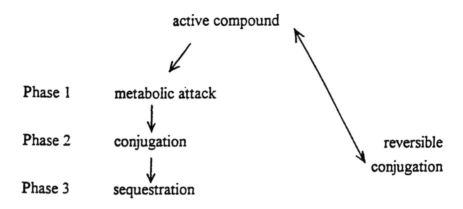

FIGURE 23.1 General detoxification mechanism for xenobiotics in plants (Cole, 1994; Grodzinski, 1991).

Following the primary metabolic reactions, the modified molecules are commonly conjugated with plant constituents such as sugars, malonic acid, or amino acids to enhance water solubility. There are numerous investigations dealing with the conjugation of exogenic phenolic compounds. The functional groups of these substances may be conjugated with monosaccharides and oligosaccharides (Balke et al., 1987) as well as amines, amino acids, and peptides (Rusness and Still, 1977). Additionally, a conjugation to membrane proteins is also possible (Ugreknelidze and Durmishidze, 1984). The products can be excreted, transported to the vacuoles, or bound to insoluble cellular constituents, for example, lignin. The conjugation with sugars is usually carried out by UDP-glycosyltransferases. Sometimes, N-glycosylation occurs as well. UDP-N-glycosyltransferases isolated from soybean are able to conjugate anilines, such as the herbicide 3,4-dichloroaniline (Sandermann et al., 1991). Glycosyl derivatives often undergo a further conjugation to malonic acid. The malonyl residue is thought to be an endogenic signal for transport processes (Mackenbrock et al., 1992).

The conjugation of xenobiotics with the tripeptide glutathione is another general defense reaction of plants. In this case, the organic compound reacts directly with endogen glutathione to the corresponding thioether. The reaction, appearing at electrophilic sites in the substrate molecule, is catalyzed by glutathione-S-transferases. Glutathione-S-transferases can be induced by exogenously applied herbicide antidotes (Mozer et al., 1983; Dean et al., 1990). This is an important mechanism for increasing herbicide resistance by herbicide safeners.

This chapter, however, deals with detoxification of allelochemicals. These natural compounds are produced by plants and can be released into the environment by passive or active processes, for example, decomposition of plant residues, leaching, evaporation, or as constituents of root exudates. Sometimes allelochemicals are converted by microbial activities, which results in more or less toxic derivatives. Other plants can absorb allelochemicals and their derivatives, which often results in significant growth reduction and reduced crop yields, unless detoxification is possible (Rice, 1984; Putnam and Tang, 1986); it seems that the capacity to detoxify harmful extrinsic molecules by masking reactive functional groups is a widespread phenomenon in the plant kingdom. Nevertheless, numerous studies substantiate inhibitory or, more rarely, stimulating effects of extrinsic natural organic molecules, often secondary compounds, on plant growth and development (Rice, 1984; Putnam and Tang, 1986; Rizvi and Rizvi, 1992). The extent of effectiveness depends on the dose and on the plant species (Weston and Putnam, 1986; Perez, 1990; Kil and Yun, 1992; Schulz et al., 1994). Thus, the latter implies differences in the ability and capacity of detoxification processes implemented on species level.

With regard to allelochemicals it is possible or has already been shown that detoxification mechanisms are similar to those described for herbicides: hydroxylations, dealkylations, or oxidations as a first step, followed by conjugation to (primary) metabolites (e.g, sugars, amino acids, peptides) as the second step (or sometimes as a first step, which depends on the chemical nature of the allelo-chemical), and, as the final step, transport to a depository, such as the vacuole or the cell wall or excretion (Table 23.1). In this chapter, we summarize by means of representative studies, mechanisms involved in allelochemical detoxification and possible reasons for species-dependent differences in allelopathic tolerance.

23.4 CONCEPTS OF ALLELOCHEMICAL DETOXIFICATION

23.4.1 OXIDATION

There are only a few studies dealing with oxidation of absorbed allelochemicals and enzymes involved in those reactions. Golovko and Kavelenova (1993) studied the effects of naturally occurring phenolic acids on peroxidase activity and isozyme patterns in allelopathically resistant and susceptible plants. Incubations with 3,5-dihydroxybenzoic acid resulted in an increased peroxidase activity in roots of rye (*Secale cereale* L.), barley (*Hordeum vulgare* L.), wheat (*Triticum aestivum* L.), and oat (*Avena sativa* L.), but resulted in an activity reduction in those of *Brassica napus* L. and *Brassica napus* var. *napus*. Changes in the enzyme activities corresponded with activity staining of the peroxidase isozymes obtained with root protein extracts from the different species after native electrophoresis. Whereas the cruciferous plant extracts prepared from incubated roots exhibited weaker staining than the untreated controls, the isozyme activities of the cereals were higher as indicated by more intensive staining, particulary in the range of low protein mobility. 3,5-dihydroxy-benzoic acid seems not to induce the synthesis of additional peroxidases, but with it, isoenzymes already present were upregulated. An increase or decrease in peroxidase activity was in agreement with allelopathical sensitivity (*Brassica napus*, *Brassica napus* var. *napus*) or tolerance (cereals, especially wheat) levels of the tested species. Unfortunately, the study does not include an approach of analyzing whether the applied phenolic acid was oxidized. Increased peroxidase level and a swelling response in cucumber (*cucomis sativus* L.) roots have been observed after application of umbelliferon (Jankay and Muller, 1976) indicating that coumarins can affect peroxidase activity in a similar way.

At present, one can only speculate about how mechanisms of phenolics trigger peroxidase activity. Phenolics are known to produce membrane depolarization and uncontrolled effluxes of inorganic ions (Glass, 1973, 1974; Harper and Balke, 1981; Balke, 1985; Macri et al., 1986; Lyn et al., 1990). Generally, stress-induced membrane depolarization is supposed to initiate membrane degradation—caused by appearing free radicals and peroxides—that can result in ion, especially K^+, effluxes leading to changes in the intracellular Ca^{2+}/K^+ ratio. According to Castillo et al. (1984), enhancement of intracellular Ca^{2+} concentration induces activated secretion of basic peroxidases and facilitates binding of the enzymes to membrane proteins. Complex interactions finally result, as well, in an activation of acidic peroxidases (Gasper et al., 1985). This process can be regarded as a putative defense mechanism that may be important for allelopathic resistance, even when oxidation of allelochemicals does not appear.

In several investigations, however, oxidation of allelochemicals has been observed. *Astilbe sinensis* Chinsis (Maxim.) Franch. (Billek and Schnook, 1967), tomato (*Lycopersicon esculentum* Mill.) (Chada and Brown, 1974), and buckwheat (*Fagopyrum esculentum* Moench.) (Schulz et al., 1993) hydroxylate salicylic acid (SA) to gentisic acid in a first detoxification step (compare: carbohydrate conjugation). Using crude extracts prepared from SA-induced buckwheat roots for hydroxylation assays, gentisic acid was only synthesized in the presence of NADH, indicating an involvement of a mono-oxygenase. The enzyme hydroxylated 22 nmol SA $\times$ min^{-1} and mg protein, 1 to 2 nmol *m*-hydroxybenzoic acid $\times$ min^{-1} and mg protein and 0.3 nmol *p*-hydroxybenzoic acid $\times$ min^{-1} and mg protein. Thus the mono-oxygenase showed a certain substrate specificity.

TABLE 23.1
Higher Plant Strategies in Detoxification and Deactivation of Allelochemicals

Allelochemical	Mechanism	Product	Species	Reference
Salicylic acid	O-glucosylation	Salicylic acid-O-β-D-glucoside	*Mallotus japonicus, Avena sativa, Vicia faba, Nicotiana tabacum, Helianthus annuus*	Tanaka et al., 1990; Enyedi et al., 1992; Klämbt, 1962; Yalpani et al., 1992 a,b; Schulz et al., 1993
Salicylic acid	Hydroxylation, glucosylation	Gentisic acid-O-β-D-glucoside	*Fagopyrum esculentum, Astilbe sinensis, Lycopersicon esculentum*	Schulz et al., 1993
Daphnetin, esculetin, m-hydroxybenzoic acid, liquiritigenin, umbelliferon	Glucosylation	Glucosides	Cultures of *Datura, Lithospermum, Perilla, Catharanthus*	Tanaka et al., 1990; Tabata et al., 1976, 1984, 1988; Suzuki et al., 1987.
p-Hydroxybenzoic acid	Glucosylation, esterfication	p-O-β-glucose-hydroxybenzoic acid, p-hydroxybenzoic acid ester	*Mallotus japonicus*	Tabata et al., 1988
Hydroquinone	Glucosylation	Arbutin	*Triticum aestivum, Datura innoxia, D. ferox, Agrostemma githago, Digitalis purpurea, Gardenia jasminoides, Antennaria micro-phylla, Euphorbia esula*	Pridham 1964, Hogan and Manners, 1990; Winter, 1961.
Resorcinol	Glucosylation	Glucosides	Numerous species of the angiosperms and gymnosperms	Pridham, 1964
Brassinolide	Glucosylation	23-β-glucoside	*Vigna radiata*	Yokota et al., 1991
BOA	Hydroxylation, glucosylation	BOA-6-O-glucoside	*Avena sativa, Triticum aestivum (Vicia faba)*	Wieland et al., 1996

TABLE 23.1 continued

Allelochemical	Mechanism	Product	Species	Reference
Castasterone	Used for brassinolide synthesis	Brassinolide	*Catharanthus roseus*	Yokota et al., 1991
Castasterone	?	Water-soluble non-glucosidic compounds	*Vigna radiata*	Yokota et al., 1991
Ferulic acid	Polymerization	Lignin	*Cucumis sativus*	Klein and Blum, 1990
Tryptophan	Involved microorganisms, conversion into IAA, in planta conjugation with amino acids	IAA-aspartat, IAA-glycin	*Triticum aestivum*	Martens and Frankenberger, 1994
BOA	Involved microorganisms, conversion into O-aminophenol and oxidative dimerization	2-amino-*3H*-phenoxazin-3-one 2-acetylamino-*3H*-phenoxazin-3-one	*Avena sativa* root colonizing microorganisms	Friebe et al., 1996

Prior to hydroxylation, decarboxylation of allelochemicals may occur. A number of benzoic acids and additional phenols have been found to be metabolized by cell suspension cultures, for example, of *Phaseolus aureus* Roxb. or *Glycine max* (L.) Merr. Demethylation, oxidative decarboxylation, and complete oxidative degradation of some applied compounds to CO_2 have been observed (Berlin et al., 1971; Harms and Haider, 1972) The purpose of these studies was, however, the elucidation of catabolic pathways and the demonstration of cell cultures as suitable systems to study catabolism of secondary plant products. A possible meaning of those reactions in allelopathic plant–plant interactions was not considered. A complete or partial oxidation of an absorbed allelochemical may present an alternative defense reaction of a plant in addition to detoxification of the compound or its derivatives by conjugation with, for example, primary metabolites, since comparable reactions have been described with intact wheat seedlings (Harms et al., 1971; Harms and Prieβ, 1973).

Oxidative dimerization was found with benzoxazolin-2-(*3H*)-one (BOA), a decomposition product of DIBOA (2,4-dihydroxy-(*2H*)-1,4-benzoxazin-3-one). Cyclic hydroxamic acids are known to be very important and highly bioactive compounds of numerous Poaceae (Niemeyer, 1988; Perez, 1990). They have been detected in several species of the genus *Agropyrum* (*A. cristatum, A. cristatum puberulum, A. desertorum, A. fragile*) (Copaja et al., 1991); and in shoot extracts and root exudates of *A. repens* seedlings: (Friebe et al., 1995). Rhizome born roots of the latter species exudes DIBOA and 2,4-dihydroxy-7-methoxy-1,4-benzoxazin-3-one (DIMBOA) in addition to a variety of further allelochemicals (Schulz et al., 1994). Moreover, cyclic hydroxamic acids are constituents of wheat, rye, and corn *Zea mays* L., but not of oat or rice (*oryza sativa* L.). Species- and dose-dependent differences in sensitivity to DIBOA have been observed by numerous authors (Barnes and Putnam, 1987; Niemeyer, 1988). A higher sensitivity of dicots to DIBOA compared with that of monocots was demonstrated by Barnes and Putnam (1987) and Schulz et al. (1994). For instance, in the presence of 500 μM DIBOA, seedling growth of *Lepidium sativum* was inhibited up to 50 percent, that of *Amaranthus retroflexus* L. and *Brassica napus* was reduced to 70 to 80 percent, whereas *Hordeum vulgare* L. showed only a 30 percent growth reduction in comparison with the controls (Figure 23.2, Table 23.2). It was suggested that some species, in particular monocots, may develop mechanisms that enable them to overcome inhibitory effects of cyclic hydroxamic acids.

When *Avena sativa* L. seedlings are incubated for 24 to 48 h with BOA, a bright orange-colored compound can be isolated from the incubation medium by ether extraction. The compound was

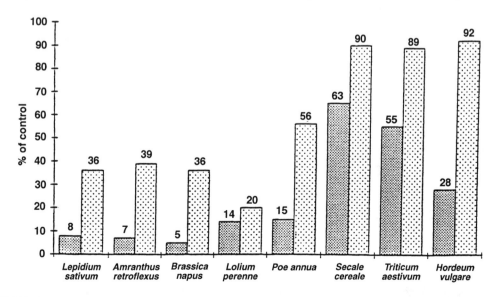

FIGURE 23.2   Effects of DIBOA (2 mM) and ferulic acid (2 mM) on seedling shoot growth of different plant species. It should be noted that shoot growth is less inhibited by DIBOA and ferulic acid than the radicle growth.

TABLE 23.2
Effects of DIBOA on Radicle Growth of Different Crops and Weeds

| | DIBOA Concentration | | | |
Species	0.05 mM	0.1 mM	0.5 mM	2 mM
Lepidium sativum	109%	101%	56%[b]	6%[b]
Amaranthus retroflexus	96%	65%[b]	20%[b]	8%[b]
Lolium perenne	95%	77%[a]	76%[a]	10%[b]
Poa annua	93%	81%[a]	51%[b]	22%[b]
Secale cereale	74%[a]	82%	50%[b]	40%[b]
Triticum aestivum	94%	72%[a]	60%[b]	40%[b]
Hordeum vulgare	95%	100%	56%[b]	27%[b]
Brassica napus	83%[a]	68%[b]	29%[b]	3%[b]

Controls = 100 percent. Significance: a, $P \leq .05$; b, $P \leq .01$ (Student's t-test).

identified by mass spectrometry, UV-spectra, melting point, and HPLC-co-chromatography as 2-amino-*3H*-phenoxazin-3-one (Figure 23.3), (Friebe et al., 1996). The substance is known to appear within 10 days in nonsterile soil after incubation with DIBOA (2,4-dihydroxy-1,4(*2H*)-benzoxazin-3-one), due to the activities of microorganisms. These organisms degrade the compound to *o*-aminophenol, an intermediate that reacts subsequently without further enzymatic catalysis, yielding 2-amino-*3H*-phenoxazin-3-one (Gagliardo and Chilton, 1992). Aminophenol appeared, as well as an intermediate, in the incubation medium with *Avena sativa*. Here 2-amino-*3H*-phenoxazin-3-one was, however, only another intermediate, since it was acetylated yielding 2-acetylamino-*3H*-phenoxazin-3-one as the final product. Both compounds are known as the natural occurring antibiotics questomycin and N-acetylquestiomycin produced by actinomycetes (Gerber and Lechevalier, 1964). When the incubation medium was supplemented with the antibiotics kanamycin, penicillin, and streptomycin, the production of the intermediates as well as that of 2-acetylamino-*3H*-phenoxazin-3-one was prevented, indicating the necessary presence of active, detoxifying microorganisms. It seems that these microorganisms colonize the roots and undertake detoxification work for the host plant. The same, putative partners-involving strategy was observed with rye and corn, but with reduced rapidity. However, in all cases studied, oxidative dimerization of *o*-aminophenol appears remarkably faster than previously found with unsterile soil.

As an additional detoxification mechanism, *Avena sativa* absorbed BOA and converted it via BOA-6-OH to BOA-6-O-glucoside, which accumulated within the roots. In addition, a third not yet identified compound was produced (Wieland et al., 1996). Moreover, a small amount of BOA-6-OH was exuded into the medium during the course of incubation. Wheat absorbed BOA quickly and synthezised the same detoxification products as *Avena sativa,* but in contrast to oat, the unidentified compound and not BOA-6-OH was exuded in low concentrations. Neither *o*-aminophenol nor 2-amino-*3H*-phenoxazin-3-one could be found in wheat incubation media. Thus, wheat seems to have a potential mechanism for the detoxification of exogenously applied BOA and is probably not

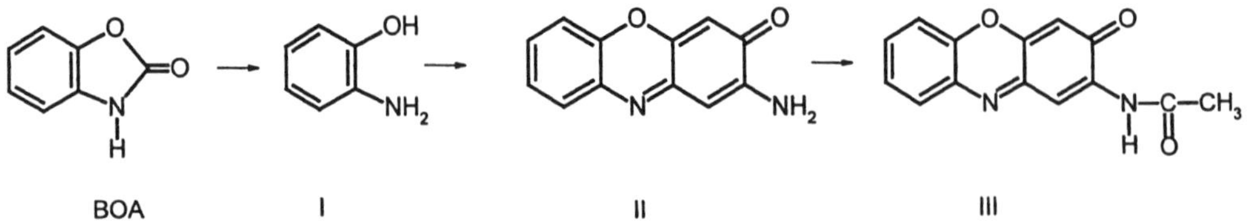

FIGURE 23.3 Microbial degradation of BOA. The intermediate *o*-aminophenol (I) is further converted to 2-acetylamino-*3H*-phenoxazin-3-one (III) via amino-*3H*-phenoxazin-3-one (II).

dependent on the additional aid of microorganisms. It has to be kept in mind that wheat, rye, and corn contain hydroxamic acids as natural constituents. The conversion of absorbed hydroxamic acids and benzoxazolinone, probably catalyzed by constitutive enzymes, may therefore be easier for plants that contain endogenous hydroxamic acids.

Vicia faba L., the only dicotyledonous species tested at present, absorbed small amounts of BOA and transformed it with low effectiveness to BOA-6-OH and its glucoside. However, detoxification appeared only with BOA concentrations no higher than 0.1 mM. o-Aminophenol and 2-amino-3H-phenoxazin-3-one could not be detected in the incubation media, even after ten days. On the contrary to the cereals and as a result of obviously underdeveloped detoxification strategies (with regard to BOA), broadbean roots blackened within 10 days when exposed to 0.5 mM BOA.

At present, bacterial or plant enzymes involved in BOA detoxification have not been investigated, but it is already thought highly likely that a species-dependent diversity in enzyme categories and enzyme properties should be expected, explaining the differences in benzoxazolinone tolerance or susceptibility. Moreover, symbioses between microorganisms and some plant species have to be taken into consideration, where the partners interact in detoxification processes. It is also possible that all organisms living in the rhizosphere, not only the symbiotic ones, are involved in detoxification. One can speculate that under certain circumstances mycorrhizas may contribute as well to allelochemical tolerance of their symbiotic partners.

23.4.2 CARBOHYDRATE CONJUGATION

A variety of phenolic compounds including simple phenolics, phenolic acids and their derivatives, flavonoids as well as, for example, anthraquinones, alkaloids, coumarins, steroids, terpenoids, and cardenolides—all described as classes of compounds with potential phytotoxic properties (Rice, 1984)—have been found to be convertable to glycosides or esters or were otherwise converted when fed to plant cell or tisssue cultures (Tabata et al., 1976; Barz et al., 1978; Barz, 1981; Berlin, 1984; Sandermann et al., 1984; Mizukami et al., 1986; Tabata et al., 1988; Koester et al., 1989; Tanaka et al., 1990; Suzuki et al., 1993; Berlin et al., 1994; Verpoorte et al., 1994). Although these investigations had with only some exceptions, biotechnological backgrounds, the results indicate that uptake and metabolization of potential and long known allelochemicals is probably a common feature not only of plant cell cultures but also in intact plants; Harms et al., (1971) described the glycosylation and esterfication of radioactively labeled phenolcarboxylic acids applied to wheat seedlings.

Tabata et al. (1988) studied the ability of ten culture strains derived from seven plant species [Perilla frutescens (L.) Britton, Catharanthus roseus G. Don, Gardenia jasminoides Ellis, Mallotus japonicus (Thumb.) Muell.-Arg., Lithospermum erythrorhizon L., Datura innoxia Mill., Bupleurum falcatum L.] to glucosylate exogenously applied phenolic compounds (coumarins: umbelliferon, esculetin, daphnetin, fraxetin; flavonoids: liquiritigenin, naringenin, baicalein; anthraquinones: alizarin, rhein, emodin; simple phenolics: salicyl alcohol, o-, m-, p-hydroxybenzoic acids). They found species- and culture strain-dependent differences in the capability to glycosylate the compounds. Only Perilla cultures accepted rhein and emodin as substrates for glycosylation. Datura, Lithospermum, Perilla, and Catharanthus cultures glycosylate daphnetin, esculetin, m-hydroxybenzoic acid, liquiritigenin, and umbelliferon. A strain of Mallotus was capable of transforming p-hydroxybenzoic acid to p-O-β-D-glucosylbenzoic acid and p-hydroxybenzoic acid glucose ester.

At present, there are only limited studies directed at molecular mechanisms responsible for effective detoxification of absorbed allelochemicals by glycosylation. Although glycosyltransferases catalyzing those reactions are certainly common enzymes (quite a number of glycosyltransferases involved in final steps of secondary product biosynthesis or biotransformation of secondary product precursors have been characterized, e.g., Koester and Barz, 1981; Schulz and Weißenböck, 1988; Kreis and May, 1990; Theurer et al., 1994), nevertheless, transferase activities that play a defined role in allelopathic tolerance of a plant are still rarely investigated.

Salicylic acid, a compound with phytotoxic properties when applied in higher concentrations (Manthe et al., 1992), functions in a low concentration as a signal molecule in plant-pathogen response and, in some species, it is known as a natural inducer of thermogenicity (Raskin, 1992; Pierpoint, 1994). The dose dependent, ambivalent properties of salicylic acid may be one reason for more intensive studies directed at mechanisms that reduce or eliminate its biological activities.

A 176-fold purified Uridine 5'-diphosphate-glucose:salicylic acid glucosyltransferase (GTase) from oat roots was found to be highly specific to the substrates UDPG (K_m = 0.28 mM) and SA (K_m = 0.16 mM). The molecular mass of the enzyme was determined to be 50 kD; its apparent iso-electric point was at pH 5.0 and UTP and, to a lower degree, the product UDP inhibited the enzyme's activity. For SA, no substrate inhibition was observed up to 1 to 6 mM. The enzyme was not a constitutive protein, but it was highly inducible by SA. Moreover, not only *de novo* synthesis of the enzyme was blocked, but existing transferase seemed to be degraded in the roots when the SA concentration of the incubation medium was lowered by successful detoxification to nontoxic levels (Yalpani et al., 1992a,b).

Vicia faba and *Fagopyrum esculentum* roots absorbed SA with triphasic kinetics. After a short first phase of passive uptake by diffusion into the apoplast, a stationary phase followed, in which duration depended on plant species (*V. faba*, 16 to 18 h; *F. esculentum*, 5 to 8 h). No obvious absorption could be observed during this phase. Finally, a third phase of active uptake occurred that correlated with the appearance of detoxification products. Like *Avena sativa*, *Vicia faba* detoxified SA by simple glycosylation and salicylic acid β-D-glucoside (GSA) accumulated in the roots (Figure 23.4). The responsible transferase was SA-inducible and highly specific for SA (K_m= ~ 0.14 M) and for the sugar donor UDPG (K_m= ~ 3.025 mM). SA concentrations higher than 600 μM resulted in substrate inhibition. The data were, however, obtained with crude protein extracts from SA-incubated roots (Schulz et al., 1993).

GSA seems to be a common detoxification product as it was also found in TMV-inoculated leaves of *Nicotiana tabacum* L. cv. Xanthi-nc, a condition, where SA acts as a signal molecule for PR-protein gene induction (Enyedi et al., 1993), in sunflower (Klämbt, 1962), and in cell suspension cultures of *Mallotus japonicus* (Tanaka et al., 1990b), when exogenously applied. For tobacco and cell cultures of *Mallotus japonicus,* inducibility of the involved glycosyl-transferases was shown.

In contrast, *Fagopyrum esculentum* oxidized SA to 2,5-dihydroxybenzoic acid (gentisic acid) and glucosylated it immediately at the 5-OH group (Figure 23.4) (Schulz et al., 1993). For this reason, free gentisic acid level remained low in the roots, whereas the glucoside accumulated. As soon as the SA concentration in the incubation medium was sufficiently lowered by active uptake and detoxification, negative effects of the phenolic acid (e.g., inhibition of the proton-efflux: K_i = 0.125 mM and K$^+$-efflux=1.4 μM/g fresh weight of the roots) disappeared. *In vitro* detoxification assays pointed to a coordinated interaction of the mono-oxygenase with the β-D-glycosyltransferase. With crude extracts prepared from SA-incubated roots, a K_m of approximately of 45 μM for gentisic acid and a K_m approximately of 2.3 mM for UDP-glucose was found; and up to 300 μM, no substrate inhibition by the phenolic compound was observed. The transferase was highly specific for gentisic acid and the sugar donor UDPG. An identical SA detoxification yielding gentisic acid

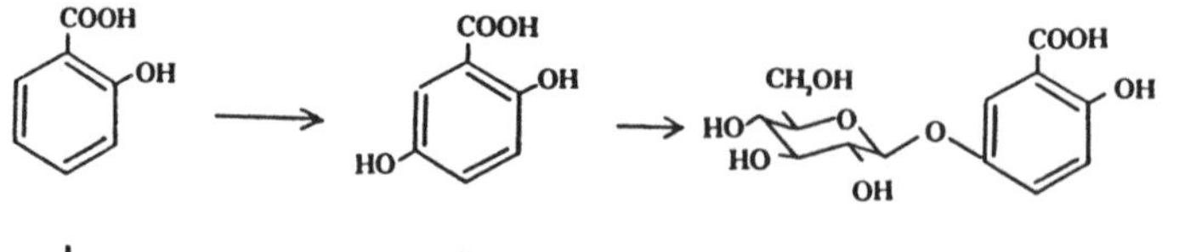

FIGURE 23.4 Detoxification of salicylic acid in buckwheat roots. The compound is first hydroxylated in the 5-position and the product gentisic acid is subsequently glucosylated. *Avena sativa* and *Vicia faba* glycosylate salicylic acid without additional modification of the molecule.

5-O-β-D-glucoside has been described for *Astilbe sinensis* (Billek and Schmook, 1967) and for tomato (Chada and Brown, 1974).

Inducibility, substrate specificity, and transient gene expression seem to be characteristic features of glycosyl transferases involved in detoxification (Balke et al., 1987), although some compounds could be immediately glycosylated after administration as was found for glycosylation of *m*- and *p*-hydroxybenzoic acid by *Mallotus japonicus* cell cultures (Tanaka et al., 1990b). In those cases, constitutive sugar transferases, probably with low specificity for phenolic substrates, may be the catalyzing enzymes. Faced with the results available in this field, one is tempted to suggest that harmful effects of SA are not of importance because plants are able to eliminate the negative biological activity of the compound by glucosylation. However, this may be a superficial consideration since first, the time period necessary for induction of detoxifying enzymes may be different for different plant species. For instance, *Avena sativa* is able to start SA detoxification quickly (within 1.5 h), buckwheat needs 5 to 8 h, and *Vicia faba* needs 16 to 18 h (Yalpani et al., 1992b; Schulz et al., 1993). Thus, broadbean is confronted with harmful SA concentrations more than ten times longer than oat.

Second, the molecular mode of enzyme induction may be different. When *Vicia faba* roots were incubated in SA-medium supplemented with 50 μg/ml cycloheximide, an inhibitor for protein systhesis, no GSA could be detected in the roots and in comparison with the controls (incubations without cycloheximide) only 20 percent SA was absorbed, indicating that the active uptake may be under control of another inducible protein. The GTase induction depends on SA concentration in the incubation medium. SA, 100 μM, resulted in a five-fold activity increase compared with the constitutive level; highest activities were achieved with 0.1 to 0.75 mM. Interestingly, concentrations above 1 mM resulted in a weaker induction, and with 4 mM SA no more inductions appeared (Schulz et al., unpublished). On the contrary, *Avena sativa* GTase was inducible to 23 times the constitute level at SA concentrations of 1 to 3.14 mM, and only 10 μM SA was necessary to elicit the induction. Obviously, *Avena sativa* metabolizes SA faster than buckwheat or broadbean, due to a faster induction of the detoxification system, and this system is less sensitive or insensitive to higher SA concentration than the one of *Vicia faba*.

There are also hints of species-dependent differences in the detoxification capacity of hydroquinone, another widespread allelochemical. The aglucon of arbutin, hydroquinone, has been known for some considerable time to be a phytotoxin to seedling growth (Schreiner and Reed, 1908; Hogan and Manners, 1990, 1991). Winter (1961) studied hydroquinone uptake by wheat and bean plants, which do not contain this compound as a normal constituent. However, both plant species absorbed hydroquinone and glycosylated the compound. Pridham (1964) showed that numerous higher plants could glycosylate hydroquinone, as it was later found in cell cultures of *Datura innoxia*, *D. ferox* L., *Agrostemma githago* L., *Digitalis purpurea*, and *Gardenia jasminoides*. Studies of Hogan and Manner (1990, 1991) indicated species-dependent differences in the effectiveness of detoxification. They co-cultured callus tissue of *Antennaria microphylla* and *Euphorbia esula* L. on the same medium and observed a significant inhibition of *E. esula* callus growth. In *A. microphylla* extracts, hydroquinone was detected, which was thought to be responsible for *E. esula* growth inhibition. Both species were certainly able to convert hydroquinone to arbutin, but *E. esula* seems to be less efficient in detoxification. Furthermore, the glycosyltransferase activity determined in cell cultures of the latter species decreased, when 1 mM hydroquinone was applied to the medium. However, the exact reasons for differences in effectiveness of detoxification are yet to be clarified at the molecular level.

Quinol and resorcinol can be glycosylated by species from almost all recent taxonomical groups of angiosperms and gymnosperms (Pridham, 1964). In all cases, UDPG was the sugar donor. Other biochemical properties of the corresponding transferases or their possible induction have not been investigated.

One of the arguments against the relevance of allelopathic effects under field or greenhouse conditions is that concentrations of phytotoxins released into the environment are too low to affect plant growth as compounds are absorbed by soil particles, if not completely degraded by microbes. Indeed, investigations on the molecular level that consider the whole complexity of manipulated or natural ecosystems are difficult to perform. In an approach to involve a substrate parameter, Politycka and Wojcik-Wojtkowiak (1991) studied sweet pepper cultivation on substrates differing in phytotoxicity level. Substrate phytotoxicity was obtained by prolonged use of the same substrate in cucumber cultivation, where phytotoxic compounds released by the plants accumulated over time or by supplementing the substrate with seven phenolic acids. When sweet pepper was used as an after-crop, phytotoxicity of the soil was eliminated, as sweet pepper absorbed the phytotoxins and detoxified the compounds by glycosylation. The glycosylated phenols accumulated first in the leaves and later on within the fruits. This result leads one to assume that under natural conditions, various plant species can probably profit by a strong detoxification capacity of other species, resulting in a diminished concentration of inhibitory allelochemicals in the soil.

There are also examples in which stimulatory compounds are glycosylated. In those cases, the term detoxification would be better replaced by the term "deactivation." Biologically highly active brassinosteroids, widely distributed in the plant kingdom, can enhance crop yields and plant growth (Rizvi and Rizvi, 1992). Evidence arose that brassinosteroids can synergistically interfere with auxins or ABA, and that they induce ethylene production (Mandava, 1988). Plants were capable of absorbing exogenously applied brassinosteroids, such as brassinolide and castasterone (Schlagnhaufer and Arteca 1991; Suzuki et al., 1993). *Vigna radiata* (L.) Wilezek hypocotyls converted 47 percent of the brassinolide fed into 23-β-D-glucopyranoside within 72 h. It was concluded that glycosylation is an important deactivation of the effector, but the catalyzing glycosyltransferase was not further investigated. Interestingly, and in contrast to *Catharanthus roseus* crown gall cells, castasterone, the biosynthetic precursor of brassinolide, was not used by *Vigna radiata* to produce brassinolide, but was mainly converted to water soluble non-glycosidic compounds (Yokota et al., 1991).

Lycopersicon esculentum metabolized an analog of brassinolide to inactive substances. After 12 h of incubation, two putative metabolites that were not further analyzed were found, the appearance of which was correlated with a reduction of ethylene over-production. Since during the first 4 h no metabolites were found, the deactivation process is probably inducible.

Glycosylation of absorbed brassinolide does not seem to be the general mechanism of deactivation since rice seedlings formed non-glycosidic polar metabolites of unknown structure. Thus, deactivation of brassinolide may point to a certain diversity of strategies realized by different species.

23.4.3 Other Conjugations

Conjugation of glutathione with herbicides catalyzed by glutathione-S-transferases, represents an irreversible mode of detoxification (Cole, 1994; Kreuz et al., 1996). Plants contain numerous glutathione-S-transferase isozymes with different functions. Some of them accept phenylpropanoids as natural substrates (Marrs, 1996). In several plants Fabaceae and in corn endogenous cinnamic acid that can be conjugated with glutathione (Edwards and Dixon, 1991). Moreover, purified corn glutathione-S-transferases with activity to *trans*-cinnamic acid were activated by supplementing the *in vitro* assays with coumaric acid or 7-hydroxycoumarin (Dean and Machota, 1993). At present, conjugation of glutathione to allelochemicals has not yet been described as a common, naturally occurring detoxification step in plants, but it is certainly worthwhile to begin a search for allelochemicals accepted as substrates for glutathione-S-transferases. Tryptophan has already been mentioned as a compound with phytotoxic properties to some species. Microorganisms in the rhizosphere of different wheat varieties can convert tryptophan into indole-3-acetic acid (IAA). The wheat roots can assimilate the phytohormone from the rooting media and most of the IAA is conjugated either with aspartic acid or glycine, although the transferases involved have not been studied (Martens and Frankenberger, 1994). This study demonstrates that molecules other than sugars can be utilized for conjugation. Amino acid transfer to allelochemicals, however, must still be demonstrated as a wide-spread detoxification step. Furthermore, in the study of Martens and Frankenberger (1994), positive interferences between higher plants and rhizosphere microorganisms have been substantiated. They are not only of importance in subsidiary detoxification or deactivation strategies, but may also play a significant role in plant development when able to change harmful molecules into useful ones (Arshad and Frankenberger, 1990).

23.4.4 INCORPORATION INTO DEPOSITORY COMPARTMENTS AND EXCRETION

Ferulic acid has often been found to inhibit plant growth (Klein and Blum, 1990). Aside from its allelopathic properties, ferulic acid is a precursor for lignin synthesis (Kindl, 1994). Shann and Blum (1987) posed the question whether plants can utilize exogenously supplied ferulic acid for lignifica-tion. As known from previous studies, cucumber absorbs ferulic acid continously over extended periods of time (Shann and Blum, 1987). Therefore, this species was chosen to elucidate the fate of absorbed ^{14}C-radiolabeled molecules. Ferulic acid treatment resulted in increased lignification and obviously, the plants used absorbed molecules for lignin synthesis. This process seems to be paral-leled by a significant increase of cell wall bound peroxidase forms (EC 1.1147) involved in the for-mation of lignin polymer. The introduction of extrinsic, absorbed allelochemicals that are phytotoxic in higher concentrations, but are also precursors or intermediates of secondary products naturally occurring in a given species or precursors of polymers like lignin into the corresponding pathways, may be a more general mechanism than previously assumed. This assumption gets some support from the work of Reinhard and co-workers (Kreis and Reinhard, 1987; Seidel and Reinhard, 1987; Seidel et al., 1990). Cell cultures of *Digitalis lanata* Ehrh., *D. purpurea* L., *D. lutea,* and *D. merto-nensis* were used for biotransformation of cardenolides, natural compounds that are charactristic sec-ondary constitutents of *Digitalis* species. When *D. lanata* cell cultures were exposed to digitoxin, this compound was transformed mainly to a series of cardiac glycosides. Although these studies were not at all associated with any aspect of allelopathy, they demonstrate the acceptance of extrin-sic molecules by cells as precursors for the biosynthesis of secondary products.

Vacuoles must be regarded as an important if not the most common depository for absorbed and subsequently conjugated allelochemicals. The transfer into the organelle seems to be trivial as vac-uoles generally store secondary products or herbicide detoxification products (Gaillard et al., 1994; Martinoia et al., 1993). Werner and Matile (1985) reported on the uptake of esculin and scopolin by barley mesophyll vacuoles. When the corresponding aglycones were applied to barley protoplasts, the compounds were first glycosylated before storing in the vaculoes. Biochemical data strongly indicate a carrier-mediated uptake into the vacuole, stimulated by ATP and inhibited by protonophores. According to Martinoia (1993), the latter suggests a proton antiport mechanism. Even though the transport of absorbed allelochemicals or their detoxification products into the vac-uole together with the molecular organization of putative carrier and translocator proteins are responsible for those events (e.g., their amino acid sequences and quaternary structures), their speci-ficity, regulation, and possible induction of encoding genes is not investigated.

Wheat and bean absorbed hydroquinone and excreted the glycosylated compound (arbutin) by an unknown mechanism into the incubation media (Winter, 1961). As already mentioned, oat roots exuded BOA-6-OH, an intermediate in BOA detoxification in low amounts, whereas wheat exuded another still unidentified metabolite. In spite of the lack of sufficient data, it cannot be denied that excretion of absorbed and then converted allelochemicals may often appear as an attempt to get rid of toxic compounds. However, this may only be transient since the metabolites can be converted again by microorganisms.

23.5   CONCLUSION

During evolution, plant species have developed strategies to detoxify allelochemicals resulting in tolerance to those compounds. Ironically, detoxification mechanisms have presently been studied extensively only with xenobiotic substances (such as herbicides) but not with natural compounds. Temporary exposure to allelochemicals may not always be a critical situation for fully developed plants, but it can be a fatal condition when occurring during germination and seedling development, the most sensitive stages in the plant's life cycle. Tolerant species in the narrow surroundings have the advantage to grow unaffected. They can acquire further capacities in competition that enables them to suppress individual plants exhibiting allelochemically induced retarded or reduced growth.

The utilization of plant detoxification strategies in agricultural management concepts is burdened with skepticism. A better understanding of the ways how plants can detoxify molecules occurring naturally in their environment will help to achieve new insights in plant–plant interactions; it is clear that microorganisms, soil conditions, and abiotic factors must be included in these studies. As a consequence, investigations in this field require interdisciplinary efforts.

ACKNOWLEDGMENTS

We thank Dr. A. Plant for critical reading and Isabel Wieland for help with the manuscript. Studies on detoxification of BOA by higher plants was performed in our laboratory, supported by the Deutsche Forschungsgemeinschaft.

REFERENCES

Arshad, M. and Frankenberger, W. T., Jr., Microbial production of plant growth regulators, in *Soil Microbial Technologies,* Metting, B., Ed., Marcel Dekker, New York, 1990, 307.

Balke, N. E., Effects of allelochemicals on mineral uptake and associated physiological processes, in *The Chemistry of Allelopathy,* Thompson, A. C., Ed., ACS Symposium Series, Washington, D.C., 1985, 161.

Balke, N. E., Davis, M., and Lee, C. C., Conjugation of allelochemicals by plants. Enzymatic glucosylation of salicylic acid by *Avena sativa,* in *Allelochemicals: Role in Agriculture and Forestry,* Waller, G. R., Ed., ACS Symposium Series 330, American Chemical Society, Washington, D.C., 1987, 214.

Barnes, P. and Putnam, A. R., Role of benzoxazinones in allelopathy of rye (*Secale cereale* L.), *J. Chem. Ecol.,* 13, 889, 1987.

Barz, W., Plant cell cultures and their biotechnological potential, *Ber. Dtsch. Bot. Ges.,* 94, 1, 1981.

Barz, W., Schlepphorst, R., Wilhelm, P., Kratzl, K., and Tengler, E., Metabolism of benzoic acids and phenols in cell suspension cultures of soybean and mung bean, *Z. Naturforsch.,* 33C, 363, 1978.

Berlin, J., Plant cell cultures—a futural source of natural products?, *Endeavour,* 8, 5, 1984.

Berlin, J., Barz, W., Harms, H., and Haider, K., Degradation of phenolic compounds in plant cell cultures, *FEBS Lett.,* 16, 141, 1971.

Berlin, J., Ruegenhagen, C., Kuzovkina, N., Fecher, L. F., and Sasse, F., Are tissue cultures of *Peganum harmala* a useful system for studying how to manipulate the formation of secondary metabolites, *Plant Tissue Organ Cult.,* 38, 289, 1994.

Billek, G. and Schmook, F. P., Zur Biosynthese der Gentisinsäure, *Monatsh. Chem.,* 98, 1651, 1967.

Blee, E., Effect of the safener dichlormid on maize peroxygenases and lipoxygenases, *Z. Naturforsch.,* 46c, 920, 1991.

Burnet, M. W. M., Loveys, B. R., Holtum, J. A. M., and Powels, S. B., A mechanism of chlorotoluron resistance in *Lolium rigidum, Planta,* 190, 182, 1993.

Castillo, F. J., Penel, C., and Greppin, H., Peroxidase release induced by ozone in *Sedum album* leaves. Involvement of Ca^{2+}, *Plant Physiol.,* 74, 846, 1984.

Chadha, K. C. and Brown, S. A., Biosynthesis of phenolic acids in tomato plants infected with *Agrobacterium tumefaciens, Can. J. Bot.,* 52, 2041, 1974.

Chkanikov, D. I., Arigidroksilirovovaniie i gluyukozilirovanie aromaticheskikh ksenobiotikov v rasteniyakh. 5 Vsesoyuznaya Simposiya po technologicheskim Soedineniyam, Tallin, Sept. 22–24, 1987, 158.

Cole, D. J. and Loughman, B. C., Factors affecting the hydroxylation and glycosylation of (4-chloro-2methyl-phenoxy) acetic acid in *Solanum tuberosum* tuber tissue, *Physiol. Veg.,* 23, 879, 1985.

Cole, J. D., Detoxification and activation of agrochemicals in plants, *Pestic. Sci.,* 42, 209, 1994.

Copaja, S. V., Barria, B. N., and Niemeyer, H. M., Hydroxamic acid content of perennial Triticeae, *Phytochemistry,* 30, 1531, 1991.

Dean, J. V., Gronwald, J. W., and Eberlein, C. V., Induction of glutathione-S-transferase isozymes in sorghum by herbicide antidotes, *Plant Physiol.,* 92, 467, 1990.

Dean, J. V. and Machota, J. H., Activation of corn glutathione-S-transferase enzymes by coumaric acid and 7-hydroxycoumarin, *Phytochemistry,* 34, 361, 1993.

Edwards, R. and Dixon, R. A., Glutathione-S-cinnamoyl transferases in plants, *Phytochemistry,* 30, 79, 1991.

Einhellig, F. A. and Souza, I. F., Phytotoxicity of sorgoleone found in grain sorghum root exudates, *J. Chem. Ecol.,* 18, 1, 1992.

Enyedi, A. J. and Raskin, I., Induction of UDP-glucose:salicylic acid glucosyltransferase activity in tobacco mosaic virus-inoculated tobacco (*Nicotiana tabacum*) leaves, *Plant Physiol.,* 101, 1375, 1993.

Friebe, A., Schulz, M., Kück, P., and Schnabl, H., Phytotoxins from shoot extracts and root exudates of *Agropyron repens* seedlings, *Phytochemistry,* 38, 1157, 1995.

Friebe, A., Wieland, I., and Schulz, M., Tolerance of *Avena sativa* to the allelochemical benzoxazolinone. Degradation of BOA by root-colonizing bacteria, *Angew. Bot.,* 70, 150, 1996.

Gagliardo, R. W. and Chilton, W. S., Soil transformation of 2(3H)-benzoxazolone of rye into phytotoxic 2-amino-3H-phenoxazin-3-one, *J. Chem. Ecol.,* 18, 1683, 1992.

Gaillard, C., Dufaud, A., Tommasini, K. K., Amrhein, N., and Martinoia, E., A herbicide antidote (safener) induces the activity of both herbicide detoxifying enzyme and of a vacuolar transporter for the detoxified herbicide, *FEBS Lett.,* 352, 219, 1994.

Gareis, C., Rivero, C., Schuphan, I., and Schmidt, B., Plant metabolism of xenobiotics. Comparison of the metabolism of 3,4-dichloroaniline in soybean excised leaves and soybean cell suspension cultures, *Z. Naturforsch.,* 47c, 823, 1992.

Gaspar, T., Penel, C., Castillo, F. J., and Greppin, H., A two-step control of basic and acidic peroxidases and its significance for growth and development, *Physiol. Plant,* 64, 418, 1985.

Gerber, N. N. and Lechevalier, M. P., Phenazines and phenoxazinones from *Waksmania aerata* sp. *nov.* and *Pseudomonas iodina, Biochemistry,* 3, 598, 1964.

Glass, A. D. M., Influence of phenolic acids on ion uptake. I. Inhibition of phosphate uptake, *Plant Physiol.,* 51, 1037, 1973.

Glass, A. D. M., Influence of phenolic acids on ion uptake. II. Inhibition of potassium absorption, *J. Exp. Bot.,* 25, 1104, 1974.

Golovko, E. A. and Kavelenova, L. M., O konzepzii allelopaticheskoi tolerantnosti rastenii, *Fiziol. Biokhim. Kul't. Rast.,* 24, 439, 1992.

Golovko, E. A. and Kavelenova, L. M., Vliyanie allelopaticheski aktivnuikh veshchestv na aktivnost i isofer-mentnuii spektr peroksidazui u allelopaticheski chustvitelnuikh i tolerantnuikh rastenii, *Fiziol. Biokhim. Kul't. Rast.,* 25, 181, 1993.

Grodzinski, A. M., Allolpatiya rasteniy i pochwoutomlenie. Izbr. trudlui, Izbrannuie trudlui, Kiew, Nauk. dumka, 1991, 431.

Harms, H. and Haider, K., Über O-Demethylierung und Decarboxylierung von Benzoesäuren in pflanzlichen Zellsuspensionskulturen, *Planta,* 105, 342, 1972.

Harms, H. and Prieß, I., Positionsspezifische O-Demethylierung von Benzoesäuren in Weizenkeimpflanzen, *Planta,* 109, 307, 1973.

Harms, H., Söchtig, and Haider, K., Aufnahme und Umwandlung von in unterschiedlichen Stellungen C^{14}-markierten Phenolcarbonsäuren in Weizenpflanzen, *Z. Pflanzenphysiologie,* 64, 437, 1971.

Harper, J. R. and Balke, N. E., Characterization of the inhibition of K^+ absorption in oat roots by salicylic acid, *Plant Physiol.,* 68, 1349, 1981.

Hogan, M. E. and Manners, G. D., Allelopathy of small everlasting (*Antennaria microphylla*) phytotoxicity to leafy spurge (*Euphorbia esula*) in tissue culture, *J. Chem. Ecol.,* 16, 913, 1990.

Hogan, M. E. and Manners, G. D., Differential allelochemical detoxification mechanism in tissue cultures of *Antennaria microphylla* and *Euphorbia esula, J. Chem. Ecol.,* 17, 167, 1991.

Inoue, M., Nishimura, H., Li, H. H., and Mizutani, J., Allelochemicals from *Polygonum sachalinense* Fr. Schm. (Polygonaceae), *J. Chem. Ecol.*, 18, 1833, 1992.

Jankay, P. and Muller, W. H., The relationships among umbelliferone, growth, and peroxidase levels in cucumber roots, *Am. J. Bot.*, 63, 126, 1976.

Kato-Noguchi, H., Mizutani, J., and Hasegawa, K., Allelopathy of oats. II. Allelochemical effect of L-tryptophan and its concentration in oat root exudates, *J. Chem. Ecol.*, 20, 315, 1994.

Kil, B. S. and Yun, K. W., Allelopathic effects of water extracts of *Artemisia princeps* var. *orientalis* on selected plant species, *J. Chem. Ecol.*, 18, 39, 1992.

Kindl, H., Biochemie der Pflanzen, 4th ed., Springer-Verlag, Berlin, 1994, 459.

Klämbt, H. D., Conversion in plants of benzoic acid to salicylic acid and its β-D-glucoside, *Nature*, 196, 491, 1962.

Klein, K. and Blum, U., Effects of soil nitrogen level on ferulic acid inhibition of cucumber leaf expansion, *J. Chem. Ecol.*, 16, 1371, 1990.

Koester, J. and Barz, W., UDP-glucose-isoflavone-7-O-glucosyl-transferase from roots of chick pea (*Cicer arietinum*), *Arch. Biochem. Biophys.*, 212, 98, 1981.

Koester, S., Upmeier, B., Komossa, D., and Barz, W., Nicotinic acid conjugation in plants and plant cell cultures of potato (*Solanum tuberosum*), *Z. Naturforsch.*, 44C, 623, 1989.

Kreis, W. and May, U., Cardenolide glucosyltransferases and glucohydrolases in leaves and cell cultures of three *Digitalis* (Scrophulariaceae) species, *J. Plant Physiol.*, 136, 247, 1990.

Kreis, W. and Reinhard, E., 12β-hydroxylation of digitoxin by suspension-cultured *Digitalis lanata* cells. Production of deacetyllanatoside C using a two stage culture method, *Planta Med.*, 54, 143, 1988.

Kreuz, K., Tommasini, R., and Martinoia, E., Old enzymes for a new job. Herbicide detoxification in plants, *Plant Physiol.*, 111, 349, 1996.

Lyn, S. W., Blum, U., Gerig, T. M., and O'Brien, T. E., Effects of mixtures of phenolic acids on phosphorous uptake by cucumber seedlings, *J. Chem. Ecol.*, 16, 2559, 1990.

Mackenbrock, U., Vogelsang, R., and Bartz, W., Isoflavone and pterocarpan malonyl glucosides and β-1,3-glucan- and chitin-hydrolases are vacular constitutents of chickpea (*Cicer arietinum*), *Z. Naturforsch.*, 47c, 815, 1992.

Macri, F., Vianello, A., and Pennazio, S., Salicylate collapsed membrane potential in pea stem mitochondria, *Physiol. Plant.*, 67, 136, 1986.

Mandava, N. B., Plant growth-promoting brassinosteroids, *Ann. Rev. Plant Physiol. Plant Mol. Biol.*, 39, 23, 1988.

Manthe, B., Schulz, M., and Schnabl, H., Effects of salicylic acid on growth and stomatal movements of *Vicia faba* L.: Evidence for salicylic acid metabolization, *J. Chem. Ecol.*, 18, 1524, 1992.

Marrs, K. A., The function and regulation of glutathione S-transferases in plants, *Annu. Rev. Physiol. Plant Mol. Biol.*, 47, 127, 1996.

Martens, D. A. and Frankenberger, W. T., Jr., Assimilation of exogenous 2-^{14}C-indole-3-acetic acid and 3-^{14}C-tryptophan exposed to the roots of three wheat varieties, *Plant Soil*, 166, 281, 1994.

Martinoia, E., Transport processes in vaculoes of higher plants, *Bot. Acta*, 105, 232, 1992.

Martinoia, E., Grill, E., Tommasini, R., Kreuz, K., and Amrhein, N., ATP-dependent glutathione S-conjugate "export" pump in the vacuolar membrane of plants, *Nature*, 346, 247, 1993.

Mizukami, H., Terao, T., Amano, A., and Ohashi, H., Glucosylation of salicylic alcohol by *Gardenia jasminoides* cell cultures, *Plant Cell Physiol.*, 27, 645, 1986.

Mozer, T. J., Tiemeier, D. C., and Jaworski, E. G., Purification and characterization of corn gluthatione-S-transferase, *Biochemistry*, 22, 1068, 1983.

Niemeyer, H. M., Hydroxamic acids (4-hydroxy-1, 4-benzoxazin-3-ones), defense chemicals in the Gramineae, *Phytochemistry*, 27, 3349, 1988.

Perez, F. J., Allelopathic effect of hydroxamic acids from cereals on *Avena sativa* and *A. fatua*, *Phytochemistry*, 29, 773, 1990.

Pierpoint, W. S., Salicylic acid and its derivatives in plants: medicines, metabolites and messenger molecules, *Adv. Bot. Res.*, 20, 163, 1994.

Politycka, B. and Wojcik-Woijtkowiak, D., Response of sweet pepper to phenols accumulated in greenhouse substrate, *Plant Soil*, 135, 275, 1991.

Pridham, J. B., The phenol glucosylation reaction in the plant kingdom, *Phytochemistry*, 3, 493, 1964.

Putnam, A. R. and Tang, C. S., *The Science of Allelopathy,* Wiley and Sons, New York, 1986, 317.

Raskin, I., Role of salicylic acid in plants, *Annu. Rev. Plant Physiol. Plant Mol. Biol.,* 43, 439, 1992.

Ray, H. and Hastings, P. J., Variation within flax (*Linum ustiatissimum*) and barley (*Hordeum vulgare*) in response to allelopathic chemicals, *Theor. Appl. Genet.,* 84, 460, 1992.

Rice, E. L., *Allelopathy,* 2nd ed., Academic Press, Orlando, FL, 1984.

Rizvi, S. J. H. and Rizvi, V., *Allelopathy. Basic and Applied Aspects,* 1st ed., Chapman & Hall, London, 1992, 480.

Rusness, D. G. and Still, G. G., Partial purification and properties of S-cysteinyl-hydroxychlorpropham transferase from *Avena sativa* L., *Pestic. Biochem. Physiol.,* 7, 220, 1977.

Sandermann, H., Use of plant cell cultures to study the metabolism of environmental chemicals, *Ecotox. Environ. Saf.,* 8, 167, 1984.

Sandermann, H., Schmitt, R., Eckey, H., and Bauknecht, T., Plant biochemistry of xenobiotics: isolation and properties of soybean O- and N-glucosyl and O- and N-malonyl transferases for chlorinated phenols and anilines, *Arch. Biochem. Biophys.,* 287, 341, 1991.

Schlagnaufer, C. D. and Arteca, R. N., The uptake and metabolism of brassinosteroid by tomato (*Lycopersicon esculentum*) plants, *J. Plant Physiol.,* 138, 191, 1991.

Schmidt, B., Rivero, C., and Thiede, B., 3,4-dichloroaniline N-glucosyl- and N-malonyltransferase activities in cell cultures and plants of soybean and wheat, *Phytochemistry,* 39, 81, 1995.

Schmidt, B., Rivero, C., Thiede, B., and Schenk, T., Metabolism of 4-nitrophenol in soybean excised leaves and suspension cultures of soybean and wheat, *J. Plant Physiol.,* 141, 641, 1993.

Schreiner, O. and Reed, H. S., The toxic action of certain organic plant constituents, *Bot. Gaz.* (Chicago), 45, 73, 1908.

Schulz, M., Friebe, A., Kück, P., Seipel, M., and Schnabl, H., Allelopathic effects of living quackgrass (*Agropyron repens* L.). Identification of inhibitory allelochemicals exuded from rhizome borne roots, *Angew. Bot.,* 68, 195, 1994.

Schulz, M., Schnabl, H., Manthe, B., Schweihofen, B., and Casser, I., Uptake and detoxification of salicylic acid by *Vicia faba* and *Fagopyrum esculentum, Phytochemistry,* 33, 291, 1993.

Schulz, M. and Weißenböck, G., Three specific UDP-glucuronate:flavone-glucuronosyl transferases from primary leaves of *Secale cereale, Phytochemistry,* 27, 1261, 1988.

Seidel, S. and Reinhard, E., Major cardenolide glycosides in embryogenic suspension cultures of *Digitalis lanata, Planta Med.,* 53, 308, 1987.

Seidel, S., Kreis, W., and Reinhard, E., Delta5-3beta-hydroxysteroid dehydrogenase/delta5-delta4-ketosteroid isomerase (3beta-HSD), a possible enzyme of cardiac gycoside biosynthesis in cell cultures of *Digitalis lanata, Plant Cell Rep.,* 10, 621, 1990.

Shafer, W. E. and Schönherr, J., Accumulation and transport of phenol, 2-nitrophenol and 4-nitrophenol in plant cuticles, *Ecoltoxicol. Environ. Safety,* 10, 239, 1985.

Shann, J. R. and Blum, U., The utilization of exogenously supplied ferulic acid in lignin biosynthesis, *Phytochemistry,* 26, 2977, 1987.

Suzuki, H., Kim, S.-K., Takahashi, N., and Yokota, T., Metabolism of castasterone and brassinolide in mung bean explant, *Phytochemistry,* 33, 1361, 1993.

Tabata, M., Umetani, Y., Ooya, M., and Tanaka, S., Glucosylation of phenolic compounds by plant cell cultures, *Phytochemistry,* 27, 809, 1988.

Tabata, M., Ikeda, F., Hiraoka, N., and Konoshima, M., Glucosylation of phenolic compounds by *Datura innoxia* cell suspension cultures, *Phytochemistry,* 15, 1225, 1976.

Tanaka, F. S., Hoffer, B. L., Shimabukuro, R. H., Wien, R. G., and Walsh, W. C., Identification of the isomeric hydroxylated metabolites of methyl 2[4-(2,4)-dichlorophenoxy]propanoate (Diclofop-Methyl) in wheat, *J. Agric. Food Chem.,* 38, 559, 1990a.

Tanaka, F. S., Hayakawa, K., Umetani, Y., and Tanaka, M., Glucosylation of isomeric hydroxybenzoic acids by cell suspension cultures of *Mallotus japonicus, Phytochemistry,* 29, 1555, 1990b.

Theurer, C., Treumann, H. J., Faust, T., May, U., and Kreis, W., Glycosylation in cardenolide biosynthesis, *Plant Cell Tissue Organ Culture,* 38, 327, 1994.

Thomas, E. W., Loughman, B. C., and Powell, R. G., Metabolic fate of 2,4-dichlorophenoxyacetic acid in the stem tissue of *Phaseolus vulgaris, Nature,* 204, 884, 1964.

Ugrekhelidze, D. and Durmishidze, S. V., Postuplenie i detoksikatsiya organicheskikh ksenobiotikov v rasteniyakh, *Tiblissi, Metsniereba,* 1984, 230.

Verpoorte, R., Plant cell biotechnology for the production of secondary metabolites, *Pure Appl. Chem.,* 66, 2307, 1994.

Werner, C. and Matile, P., Accumulation of coumarylglucosides in vacuoles of barley mesophyll protoplasts, *J. Plant Physiol.,* 118, 237, 1985.

Weston, L. A. and Putnam, A. R., Inhibition of legume seedlings growth by residues and extracts of quackgrass (*Agropyron repens*), *J. Chem. Ecol.,* 13, 403, 1986.

Wieland, I., Friebe, A., Kluge, M., Sicker, D., and Schulz, M., Detoxification of 2-(*3H*)-benzoxazolone in higher plants, First World Congress on Allelopathy, Cadiz, Spain, 1996.

Winter, A. G., New physiological and biological aspects in the interrelationships between higher plants, *Symp. Soc. Exp. Biol.,* 15, 229, 1961.

Yalpani, N., Balke, N. E., and Schulz, M., Induction of UDP-glucose:salicylic acid glucosyltransferase in oat roots, *Plant Physiol.,* 100, 1114, 1992.

Yalpani, N., Schulz, M., Davis, M. P., and Balke, N. E., Partial purification and properties of an inducible uridine 5'-diphosphate:glucose:salicylic acid glucosyltransferase from oat roots, *Plant Physiol.,* 100, 457, 1992.

Yokota, T., Ogino, Y., Suzuki, H., Takahashi, H., Saimoto, H., Fujioka, S., and Sakurai, K., Metabolism and biosynthesis of brassinosteroids, in *Brassinosteroids: Chemistry, Bioactivity, and Applications,* Cutler, H. G., Yokota, T., and Adam, G., Eds., ACS Symposium Series 474, 1991, 86.

24 Effect of Different Concentrations and Application Periods of *p*-Hydroxybenzoic Acid on Development, Yield, and Yield Components of Spring Wheat

Olaf Christen and Christiane Theuer

CONTENTS

24.1 ABSTRACT

Information on the effect of allelochemicals on grain yield of cereal crops is scarce. Therefore, the aim of our experiments was to evaluate the effect *p*-hydroxybenzoic acid had on development, yield, and yield components of spring wheat (*Triticum aestivum* L.). The development of spring wheat was considerably affected by a two- or four-week exposure to *p*-hydroxybenzoic acid immediately after germination. At growth stage 31 (tillering), the phenolic compound had significantly reduced single plant dry weight, plant length, and number of tillers per plant. The strongest inhibition was recorded after a four-week exposure to *p*-hydroxybenzoic acid. This effect diminished until the next sampling date at growth stage 76 (ripening). At maturity, the grain yield was significantly reduced by all *p*-hydroxybenzoic acid treatments. All yield components were affected. The reduced number of kernels had the largest impact on grain yield, whereas the number of tillers per plant and the thousand grain weight had smaller effects. The reduction in grain yield was more pronounced in the higher category tillers. Although the development of spring wheat was affected more by the early treatments with *p*-hydroxybenzoic acid, the larger yield reduction was recorded after an exposure at later stages during the development. We argue that this effect was caused by the great sensitivity of the cereal crop during the transition from the vegetative to the generative development. The implications of our results for the interpretation of allelopathic experiments are 1) though early stages of crop development were more sensitive to an exposure to a phytotoxic chemical, the yield reduction was more severe after the later application; and 2) more agronomic factors must be considered in allelopathic research since different tiller categories showed a distinct quantitative response and results of such experiments will be affected, for example, by the seeding rate.

24.2   INTRODUCTION

Despite substantial evidence for allelopathic effects in no-till systems, short cereal rotations or cereal monocultures, caused by phytotoxic substances released during the decomposition of plant residues, agronomy research and standard agronomy textbooks tend to ignore these results (McCalla and Norstadt, 1974; Müller-Wilmes et al., 1977; Wolf and Höflich, 1983). One of the reasons might be that the most common methods of identifying phytotoxicity have been germination or seedling growth bioassays with little or no relation to yield response under field conditions (Guenzi et al., 1967; Kimber, 1973; Leather and Einhellig, 1986; Mason-Sedun and Jessop, 1989).

Allelochemicals associated with decomposing cereal residues are mainly phenolic and short chain aliphatic acids (Chou and Patrick, 1976; Guenzi and McCalla, 1966; Tang and Waiss, 1978; Wojcik-Wojtkowiak et al., 1990). In addition, Norstadt and McCalla (1963) isolated the antibiotic patulin produced by a number of different soil fungi as another cause of toxicity problems in mulch farming.

Siqueira et al. (1991) concluded that most phenolic acids present in crop residues have also been isolated from soils; however, their concentration varies considerably depending on soil type, amount of organic matter, previous cropping, type and amount of retained residues, and extraction method. Based on the reviewed literature, Siqueira et al. (1991) suggested that the amount of p-hydroxybenzoic acid might peak at 1400 nM in the soil solution. Other phenolic acids will contribute to phytotoxic effects since in a soil environment there is normally a mixture of different phenolic acids.

Despite this evidence, the yield response of wheat to phytotoxic substances has not been adequately established. In a series of experiments, McCalla and associates (Ellis and McCalla, 1973; McCalla and Norstadt, 1974; Norstadt and McCalla, 1971) quantified the effect of a single application of the microbial product patulin on wheat grown to maturity. Christen and Lovett (1993) investigated the effect of a short-term application of p-hydroxybenzoic acid on spring barley (*Hordeum vulgare* L.) and reported yield reductions of nearly 20 percent depending on the concentration of the phenolic substance. Other experiments with short-term exposure of crops to a phytotoxic chemical were either conducted with other plant species or different chemicals, or the plants were not allowed to grow to maturity. Cochran et al. (1983) compared the influence of various concentrations of acetic acid on the average number of first stem tillers of wheat and described an almost 30 percent reduction after a four-day exposure with 10 mM acetic acid compared with the control. Lower concentrations of a short-term exposure with acetic acid did not cause long-term effects in their experiment. In contrast, Waters and Blum (1987) report that *Phaseolus* spp. bean plants recovered from two-day treatment with 1.0 or 2.0 mM ferulic acid at seedling or flowering stage. Only an exposure at pod-fill caused a significantly lower leaf area and plant dry weight at maturity due to water stress and a loss of turgidity. Using cucumber (*Cucumis sativum* L.) in a series of experiments, Blum and Rebbeck (1989) observed a rapid recovery of roots after a short-term exposure with ferulic acid, but did not relate this result to later growth, development, or yield parameters.

Agronomy research on cereal development and organogenesis with respect to reaction toward fertilizer (Langer and Liew, 1973; Thorne et al., 1988), herbicide applications (Tottman, 1977), or water stress (Christen et al., 1995) has focused on stage-specific responses and identified differences in susceptibility depending on the developmental stage of the shoot apex. In particular the transition from the vegetative to the generative stage, the so-called "double ridges stage," seems to be a period with a particularly high sensitivity toward external factors, but so far only Christen and Lovett (1993) have addressed this.

The objective of the present study was to compare the effect of an application of *p*-hydroxy-benzoic acid, as an example for a phenolic compound that has been frequently isolated from soil or soil water, on development, grain yield, and yield components of spring wheat.

24.3 METHODS

The experiment was conducted during 1995 with the spring wheat variety Devon in sand–water culture using 5-L pots filled with coarse sand. In each pot, 20 plants were transplanted after germination on tissue paper. The design was a multifactorial experiment with the following factors:

1. five concentrations of *p*-hydroxybenzoic acid (control, 0.5, 1, 2.5, and 5 mM)
2. three treatment periods (1 to 2 weeks, 1 to 4 weeks, and 3 to 4 weeks)
3. three sampling dates
4. four replicates

Therefore, this design had a total of 180 pots. The *p*-hydroxybenzoic acid (Sigma Chemical Company, St Louis, MO) was applied with a standard nutrient solution three times a week with 200 ml each time, giving a total of 600 ml per week. In order to avoid a pH effect of the acid, the pH of all solutions regardless of the *p*-hydroxybenzoic concentration was adjusted to pH 5.8 with 0.01 M Na OH. The pots were flushed after treatment periods in order to remove *p*-hydroxybenzoic acid from the solutions. The pots were kept outside and therefore the growing conditions for all treatments were similar.

The sequential harvests were conducted at GS 31 (12/6/95) and GS 76 (10/7/95), growth stages according to Zadoks et al. (1974). At these dates, single plant dry matter, plant length, and the number of tillers per plant were determined.

After the last application of *p*-hydroxybenzoic acid, the wheat was allowed to grow to maturity and was harvested on August 1 (GS 91). At maturity, all shoots having fertile ears were harvested individually and, after drying at 65°C for 48 h, the grain weight and ear structure (kernels per ear) were determined. The thousand grain weight was calculated. All data were subjected to a statistical analysis using PROC GLM (Generalized Linear Models), option LSMEANS of the SAS package (SAS 1985; Searle, 1987). The statistical procedure GLM was applied to allow for the unequal numbers of observations in the cells.

24.4 RESULTS

All three parameters, viz dry matter, plant length, and number of tillers per plant were negatively affected by the application of *p*-hydroxybenzoic acid compared with the untreated control (Table 24.1). Averaged over the different application periods, the single plant dry matter was reduced up to 43 percent following an exposure to 5 mM of the phenolic compound. The number of tillers per plant showed a smaller response because there were only 14 days between the last application and the sampling date. With respect to the different application periods, all three parameters were more affected by application during the first two or the first four weeks of the experiment. This effect occurred more pronounced at the higher concentrations, 2.5 and 5 mM.

At the sampling date GS 76, the differences in single plant dry matter (and the other parameters that are not shown) resemble the results of the first (Table 24.2). There was no difference in dry matter production with the application periods of 1 to 2 weeks and 1 to 4 weeks; however, these were significantly less than that from the 3- to 4-week period.

Averaged over the different application periods, the single plant yield of spring wheat showed a clear response to *p*-hydroxybenzoic acid (Table 24.3). Even a concentration of only 0.5 mM caused a reduction of 32 percent in single plant yield compared with the control treatment. This difference increased to 52 percent following an application of 5 mM *p*-hydroxybenzoic acid. On average, the

TABLE 24.1

Effect of Different Concentrations [mM] and Application Periods of *p*-Hydroxybenzoic Acid on Dry Matter [g] per Single Plant, Plant Length [cm], and Number of Tillers Per Plant of Spring Wheat at GS 31

	Control	0.5	1	2.5	5	∅
Dry matter per single plant						
1–2 Week	1.59[ab]	1.29[de]	1.16[ef]	1.03[fg]	0.70[h]	1.15[b]
1–4 Week	1.65[a]	1.25[de]	1.15[ef]	0.98[g]	0.68[h]	1.14[b]
3–4 Week	1.52[b]	1.72[a]	1.58[ab]	1.50[bc]	1.37[cd]	1.54[a]
∅	1.59[a]	1.42[b]	1.30[c]	1.17[d]	0.92[e]	
Single plant length						
1–2 Week	50.1[ab]	45.8[d]	43.3[e]	39.4[f]	31.9[h]	42.1[b]
1–4 Week	50.9[a]	42.3[e]	40.8[f]	36.9[g]	31.1[h]	40.4[c]
3–4 Week	48.2[c]	51.0[a]	48.9[bc]	48.1[c]	43.5[e]	47.9[a]
∅	49.7[e]	46.6[d]	44.3[c]	41.5[b]	35.5[a]	
Number of tillers per plant						
1–2 Week	6.0[a]	4.7[de]	5.0[cde]	4.5[e]	3.9[f]	4.8[b]
1–4 Week	5.7[ab]	5.1[cd]	5.1[cd]	4.6[e]	3.5[g]	4.8[b]
3–4 Week	6.0[a]	5.4[bc]	5.6[ab]	6.0[a]	5.3[bc]	5.7[a]
∅	5.9[a]	5.1[c]	5.2[b]	5.0[c]	4.2[d]	

Values followed by the same letter within the same feature are not different at $P < .05$ level.

TABLE 24.2

Effect of Different Concentrations [mM] and Application Periods of *p*-Hydroxybenzoic Acid on Dry Matter [g] per Single Plant of Spring Wheat at GS 76

	Control	0.5	1	2.5	5	∅
1–2 Week	7.97[bcd]	5.88[j]	7.20[fgh]	7.43[cfg]	7.02[gh]	7.10[b]
1–4 Week	8.01[bcd]	7.22[fgh]	6.93[h]	7.45[cfg]	6.39[i]	7.20[b]
3–4 Week	8.57[a]	8.14[abc]	7.84[cde]	8.46[ab]	7.59[dcf]	8.12[a]
∅	8.19[a]	7.08[cd]	7.32[c]	7.78[b]	7.00[d]	

Values followed by the same letter are not significantly different at $P < .05$ level.

largest yield reduction occurred after an application of p-hydroxybenzoic acid from the 1- to 4-week period. In order to explain the described differences in the single plant yield caused by the application of the phenolic compound, the different yield components have to be considered. The number of kernels per plant resembles the previously described data for the single plant yield, that is, a clear negative response to the higher concentrations of the compound as well as the strongest effect of the late application. The two other yield components—thousand grain weight and number of ears per plant—did respond less to the application. Especially the thousand grain weight showed some compensational effect after an application from the 3- to 4-week period, causing an increase in this parameter.

The previously described response of the single ear weight after an application with *p*-hydroxybenzoic acid is confirmed for all tiller categories, but the magnitude of the reaction differed between the categories (Table 24.4). On average, the highest concentration of *p*-hydroxybenzoic acid caused a yield decrease of 41 percent in the main stem tillers compared with the untreated control, whereas ears of the third category tillers suffered a 66 percent reduction in weight compared with

TABLE 24.3
Effect of Different Concentrations [mM] and Application Periods of *p*-hydroxybenzoic Acid on Single Plant Yield [g], Number of Kernels per Plant, Thousand Grain Weight [g] and Number of Ears per Plant of Spring Wheat at GS 91

	Control	0.5	1	2.5	5	Ø
Single plant yield						
1–2 Week	3.03^a	1.69^{ef}	1.75^{de}	2.31^c	2.00^d	2.16^a
1–4 Week	2.74^{ab}	2.36^c	1.51^{fg}	1.66^{ef}	1.26^g	1.91^a
3–4 Week	2.82^{ab}	1.82^{de}	1.83^{de}	1.80^{de}	0.92^h	1.84^b
Ø	2.86^a	1.96^b	1.70^c	1.92^b	1.39^d	
Number of kernels per plant						
1–2 Week	81^a	58^{de}	56^{defg}	71^b	61^{cd}	65^a
1–4 Week	76^{ab}	68^{bc}	46^{hi}	61^{cd}	45^{hi}	59^b
3–4 Week	73^{ab}	53^{efgh}	53^{efgh}	49^{ghi}	43^i	54^c
Ø	77^a	60^b	52^c	60^b	50^c	
Thousand grain weight						
1–2 Week	37^{ab}	29^e	30^d	32^d	32^d	32^b
1–4 Week	36^{ab}	35^{bc}	32^d	26^f	27^{ef}	31^b
3–4 Week	38^a	33^{cd}	35^{bc}	38^a	21^g	33^a
Ø	37^a	32^b	32^b	32^b	27^c	
Number of tillers per plant						
1–2 Week	2.3^a	1.5^f	1.7^{ed}	2.2^{ab}	2.4^a	2.0a
1–4 Week	2.1^{abc}	1.9^{cde}	1.6^{ed}	2.0^{bcd}	1.7^{ed}	1.9b
3–4 Week	2.0^{bcd}	1.8^{de}	1.9^{cde}	2.2^{ab}	1.8^{de}	1.9b
1–4 Week	76^{ab}	68^{bc}	46^{hi}	61^{cd}	45^{hi}	59^b
3–4 Week	73^{ab}	53^{efgh}	53^{efgh}	49^{ghi}	43^i	54^c
Ø	2.1^a	1.8^c	1.7^d	2.1^a	2.0^b	

Values followed by the same letter within the same feature are not significantly different at $P < .05$ level.

the untreated control. The yield of the second tillers averaged 54 percent less at the highest concentration of *p*-hydroxybenzoic acid. In general, higher tiller categories suffered more severely after an application of *p*-hydroxybenzoic acid and therefore were unable to compensate for the detrimental effects of the phenolic acid. Apart from a few exceptions, this observation can be confirmed for all application periods and different concentrations of *p*-hydroxybenzoic acid. The different components of yield showed the same behavior previously described for the single ear yield of the different tiller categories (data not shown).

24.5 DISCUSSION

The evidence presented demonstrates the potential of *p*-hydroxybenzoic acid to substantially reduce grain yields of spring wheat. We believe this to be the first report of deleterious effects of a phenolic acid application for different periods on development, grain yield, and yield components of spring wheat. These results, however, have to be interpreted very carefully in relation to yield effects observed in field experiments since the growth medium used here, sand with only negligible organic content, does not completely reflect the situation in the field. Additionally, the concentrations of *p*-hydroxybenzoic acid chosen in this experiment were generally higher than concentrations found in the soil. However, this concentration of phenolic acids in the bulk soil and the soil solution is

TABLE 24.4

Effect of Different Concentrations [mM] and Application Periods of p-Hydroxybenzoic Acid on Single Plant Yield [g] of Spring Wheat on First, Second, and Third Category Tillers at GS 91

	Control	0.5	1	2.5	5	∅
Single ear yield of first category tillers						
1–2 Week	1.69[ab]	1.42[c]	1.32[cd]	1.38[c]	1.39[c]	1.44[a]
1–4 Week	1.66[b]	1.71[ab]	1.17[e]	1.15[e]	1.01[f]	1.34[b]
3–4 Week	1.79[a]	1.19[e]	1.23[de]	1.20[de]	0.66[f]	1.21[c]
∅	1.71[a]	1.44[b]	1.24[c]	1.24[c]	1.02[d]	
Single ear yield of second category tillers						
1–2 Week	1.16[a]	0.51[de]	0.58[d]	0.86[b]	0.51[de]	0.73*
1–4 Week	1.04[a]	0.80[bc]	0.54[de]	0.60[bcd]	0.37[e]	0.67
3–4 Week	1.06[a]	0.74[bc]	0.62[bcd]	0.59[d]	0.33[f]	0.67
∅	1.09[a]	0.68[b]	0.59[c]	0.68[b]	0.40[d]	
Single ear yield of third category tillers						
1–2 Week	0.78[a]	0.28[fg]	0.49[cdef]	0.63[bc]	0.32[fg]	0.50*
1–4 Week	0.76[bc]	0.42[def]	0.46[cdef]	0.45[def]	0.24[g]	0.46
3–4 Week	0.83[a]	0.61[bcd]	0.55[cde]	0.39[efg]	0.24[g]	0.52
∅	0.79[a]	0.44[b]	0.50[b]	0.49[b]	0.27[c]	

Values followed by the same letter within the same feature are not significantly different at $P < .05$ level.
*Not significantly different at $P < .05$.

affected by various parameters, and estimates of the active part are still a subject of considerable dispute and speculation in the literature (for review see Siqueira et al., 1991). On the other hand, in the field, plants are exposed to a great number of different phenolic acids at the same developmental stage, which in combination, according to Einhellig et al. (1982), affect germination and early growth well below their separate thresholds.

An important result of this experiment is the decreasing effect of the p-hydroxybenzoic acid application on the parameters of the plant development in relation to the effect on the yield and yield components of spring wheat. This finding extends conclusions drawn by Christen and Lovett (1993) in a comparable experiment with spring barley. The reduction in dry weight, plant length, and number of tillers per plant does not completely correspond with the yield reduction, that is, in the treatment that only received p-hydroxybenzoic acid from the 3 to 4 weeks, a smaller reduction was observed during the development, but the yield was more severely affected. The finding that both yield components, that is, kernels per ear and thousand grain weight, were affected by an exposure to p-hydroxybenzoic acid in the development experiment confirms reports from Ellis and McCalla (1973), applying patulin to wheat in a similar experiment. Also, the response of the different tiller categories confirms reports by Christen and Lovett (1993) with spring barley and is in general comparable with tiller responses to stress of various kinds (Christen and Hanus, 1993; Christen et al., 1995).

Higher tiller categories proved to be more sensitive and showed a larger yield decrease. It could be argued that the development of the main stem tiller was enhanced after the application of p-hydroxybenzoic acid at double ridge and, subsequently, the higher tiller categories suffered a yield reduction due to interplant competition. Another possibility would be a larger sensitivity of the higher tiller categories at early stages of the apex development, which consequently increased the single ear weight of the main stem tiller. Both effects might also appear in combination. The so-called "sensitive stages" and, especially, the transition from the vegetative to the generative development of the apex, are solely based on experiments comparing herbicide or fertilizer treatments, and the cereal crop will not necessarily show sensitivity at similar stages toward allelochemicals.

The specific response of different tiller categories in respect to grain yield following an exposure to p-hydroxybenzoic acid clearly demonstrates the need for considering more agronomic and husbandry factors in the design of experiments investigating allelopathy. Since higher tiller categories showed a larger yield decrease after short-term stress caused by p-hydroxybenzoic acid, the yield depression in a field situation interact with seed density, thus modifying the ability of different tiller categories to compensate for the detrimental effects of allelochemicals.

The great sensitivity of cereal plants to stress at the transition of the shoot apex from the vegetative to the generative development (double ridges stage), which has been confirmed in our experiment with p-hydroxybenzoic acid, is not fully understood. The observed effects of p-hydroxybenzoic acid on wheat yield and yield components are the integrated results of specific biochemical and cytological reactions of the plant. It has been demonstrated that phenolic acids might affect processes like cell division, cell elongation, membrane permeability, and mineral uptake. Tottman (1977) investigating the effect of herbicides on wheat, argued that the yield decrease caused by a stress at double ridges is due to interference with the extreme sensitive process of primordium formation on the shoot apex. It, therefore, seems also possible that phenolics like p-hydroxybenzoic acid, which might interact with the hormone synthesis, will affect a cereal plant at this developmental stage by changing its hormonal balance.

Apart from Christen and Lovett (1993), only McCalla and Norstadt (1974) have investigated the response of a cereal crop to the application of a phytotoxic substance in the context of yield decline caused by crop residues. They compared germination bioassays with the response of wheat grown to maturity after a single application with the microbial product patulin, in order to quantify yield losses caused by phytotoxic substances. The response in tillering revealed an interaction with the medium; thus, patulin reduced the number of tillers in sand but increased tillering of wheat in soil, an effect that the authors attribute to the ability of soil microorganisms to inactivate patulin in soils. In both treatments, however, the patulin application caused a reduction in the kernel weight, due to a lower number of kernels per head. Based on further experiments, McCalla and co-workers (Ellis and McCalla, 1973; McCalla and Norstadt, 1974) concluded that spring wheat was most sensitive to an application of patulin at germination or at heading. Other application dates based on the external development of the wheat plant at the second node stage (GS 32) and the flag leaf sheath opening (GS 49) were less susceptible. Due to the different chemicals used in the bioassays, a direct comparison is restricted to the general approach; however, it is confirmed that an exposure with a phytotoxic substance during the early development of the cereal crop might cause substantial losses in grain yield.

In order to relate these findings to yield depressions described in field situations, quantitative analyses of the phenolic content in soils and/or soil water based on sequential sampling dates are required. But so far most analyses have been restricted to soil samples taken only once or twice during the growing season (Whitehead et al., 1981, 1983). Few reports have attempted to relate changes in concentrations of phytotoxic substances in the soil to the crop development in field experiments.

Müller-Wilmes et al. (1977) found the highest concentrations of phenolics in an experiment with winter barley grown in monoculture or following potatoes from early spring until ear emergence. They explain the yield depression of 10% to 23%, depending on the level of nitrogen application, to differences in the chemical composition of the phenolic compounds rather than to the actual concentration in the soil. In contrast, Wolf and Höflich (1983) compared the allelopathic potential of soil collected in a wheat monoculture with soil from wheat grown in rotation. They used a radish (*Raphanus sativus* L.) germination bioassay and reported the largest degree of toxicity in October and April. However, they did not isolate or identify the chemical nature of the effects. These results indicate a shift in phytotoxicity during the growing season, although the flow rates between the different pools of phenolics in the soil and conditions affecting these processes are little understood (Blum et al., 1991).

A quantification of the yield decrease caused by phytotoxic substances released from crop residues under field conditions is only possible if sequential soil sampling is accompanied by a detailed observation of the crop development, since differences in the apex development are most important for an understanding of yield differences in respect to environmental conditions and husbandry factors (Kirby and Appleyard, 1984; Landes and Porter, 1989). Based on the evidence presented, this is an approach that should be considered in allelopathic research.

ACKNOWLEDGMENTS

This project was financially supported by a grant of the German Research Council (Deutsche Forschungsgemeinschaft).

REFERENCES

Blum, U. and Rebbeck, J., Inhibition and recovery of cucumber roots given multiple treatments of ferulic acid in nutrient culture, *J. Chem. Ecol.,* 15, 917, 1989.

Blum, U., Wentworth, T. R., Klein, K., Worsham, A. D., King, L. D., Gerig, T. M., and Lyu, S. W. Phenolic acid content of soils from wheat-no till, wheat-conventional till, and fallow-conventional till soybean cropping systems, *J. Chem. Ecol.,* 17, 1045, 1991.

Chou, C. H. and Patrick, Z. A., Identification and phytotoxic activity of compounds produced during decomposition of corn and rye residues in soil, *J. Chem. Ecol.,* 2, 369, 1976.

Christen, O. and Hanus, H., Single ear yield from different shoot categories of winter wheat following either wheat or rapeseed, *Eur. J. Agron.,* 2, 105, 1993.

Christen, O. and Lovett, J. V., Effects of a short term *p*-hydroxybenzoic acid application on grain yield and yield components in different tiller categories of spring barley, *Plant Soil,* 151, 279, 1993.

Christen, O., Sieling, K., Richter-Harder, H., and Hanus, H., Effects of temporary water stress before anthesis on growth, development and grain yield of spring wheat, *Eur. J. Agron.,* 4, 27, 1995.

Cochran, V. L., Bikfasy, D., Elliott, L. F., and Papendick, R. I., Effect of root contact with short-chain aliphatic acids on lateral wheat growth, *Plant Soil,* 74, 369, 1983.

Einhellig, F. A., Schon, M. K., and Rasmussen, J. A., Synergistic effects of four cinnamic acid compounds on grain sorghum, *J. Plant Growth Regul.,* 1, 251, 1982.

Ellis, J. R. and McCalla, T. M., Effects of patulin and method of application on growth stages of wheat, *Appl. Microbiol.,* 25, 562, 1973.

Guenzi, W. D. and McCalla, T. M., Phytotoxic substances extracted from soil, *Soil Sci. Soc. Am. J.,* 30, 214, 1966.

Guenzi, W. D., McCalla, T. M., and Norstadt, F. A., Presence and persistence of phytotoxic substances in wheat, oat, corn and sorghum residues, *Agron. J.,* 59, 163, 1967.

Kimber, R. W. L., Phytotoxicity from plant residues. II. The effect of time of rotting of straw from grasses and legumes on the growth of wheat seedlings, *Plant Soil,* 38, 347, 1973.

Kirby, E. J. M. and Appleyard, M., *Cereal Development Guide,* 2nd ed. NAC Cereal Unit, Stoneleigh, 1984, 80.

Landes, A. and Porter, J. R., Comparison of scales used for categorising the development of wheat, barley, rye and oats, *Ann. Appl. Biol.,* 115, 343, 1989.

Langer, R. H. M. and Liew, F. K. Y., Effects of varying nitrogen supply at different stages of the reproductive phase on spikelete and grain production and on grain nitrogen in wheat, *Aust. J. Agric. Res.,* 24, 647, 1973.

Leather, G. R. and Einhellig, F. A., Bioassays in the study of allelopathy, in *The Science of Allelopathy,* Putnam, A. R. and Tang, C. S., Eds., Wiley and Sons, New York, 1986, 133.

McCalla, T. M. and Norstadt, F. A., Toxicity problems in mulch tillage, *Agric. Env.,* 1, 153, 1974.

Mason-Sedun, W. and Jessop, R. S., Differential phytotoxicity among species and cultivars of the genus *Brassica* to wheat. III. Effects of environmental factors during growth on the phytotoxicity of residue extracts, *Plant Soil,* 117, 90, 1989.

Müller-Wilmes, U., Schön, W. J., and Zoschke, M., Zur Autotoleranz der Wintergerste (*Hordeum vulgare* L.), *Z. f. Acker- und Pflanzenbau,* 145, 296, 1977.

Norstadt, F. A. and McCalla, T. M., Phytotoxic substances from a species of *Penicillium, Science,* 140, 410, 1963.

Norstadt, F. A. and McCalla, T. M., Effects of patulin on wheat grown to maturity, *Soil Sci.,* 111, 236, 1971.

SAS Institute, *SAS User's Guide, Statistics,* 5th. ed., SAS Institute, Cary, NC, 1985.

Searle, S. R., *Linear Models for Unbalanced Data,* John Wiley & Sons, New York, 1987.

Siqueira, J. O., Nair, M. G., Hammerschmidt, R., and Safir, G. R., Significance of phenolic compounds in plant-soil-microbial systems, *Crit. Rev. Plant Sci.,* 10, 63, 1991.

Tang, C. S. and Waiss, A. C., Short-chain fatty acids as growth inhibitors in decomposing wheat straw, *J. Chem. Ecol.,* 4, 225, 1978.

Thorne, G. N., Wood, D. W., and Stevenson, H. J., Effects of nitrogen supply and drought on early development of winter wheat in the field in Eastern England, *J. Agric. Sci.,* 110, 109, 1988.

Tottman, D. R., The identification of growth stages in winter wheat with reference to the application of growth-regulator herbicides, *Ann. Appl. Biol.,* 87, 213, 1977.

Waters, E. R. and Blum, U., Effects of single and multiple exposures of ferulic acid on the vegetative and reproductive growth of *Phaseolus vulgaris* BBL-290, *Am. J. Bot.,* 74, 1635, 1987.

Whitehead, D. C., Dibb, H., and Hartley, R. D., Extractant pH and the release of phenolic compounds from soils, plant roots and leaf litter, *Soil Biol. Biochem.,* 13, 343, 1981.

Whitehead, D. C., Dibb, H., and Hartley, R. D., Bound phenolic compounds in water extracts of soils, plant roots and leaf litter, *Soil Biol. Biochem.,* 15, 133, 1983.

Wolf, H. J. and Höflich, G., Phytoinhibitorische Wirkungen im Boden bei Anbau von Wintergetreide, *Zbl. Mikrobiologie, Jena,* 138, 617, 1983.

Wojcik-Wojtkowiak, D., Politycka, B., Schneider, M., and Perkowski, J., Phenolic substances as allelopathic agents arising during the degradation of rye (*Secale cereale*) tissues, *Plant Soil,* 124, 143, 1990.

Zadoks, J. C., Chang, T. T., and Konzak, T. T., A decimal code for the growth stages of cereals, *Weed Res.,* 14, 415, 1974.

25 Biochemical Effects of Allelopathic Alkaloids

M. Wink, B. Latz-Brüning, and T. Schmeller

CONTENTS

25.1 ABSTRACT

The alkaloids ajmaline, berbamine, berberine, boldine, cinchonine, cinchonidine, ergometrine, harmalin, harmin, lobeline, norharman, papaverine, quinidine, quinine, sanguinarine, and solanine affect more than one of the basic molecular targets tested (DNA intercalation, inhibition of DNA

0-8493-2116-6/99/$0.00+$.50
© 1999 by CRC Press LLC

polymerase I, protein biosynthesis, membrane stability). It is likely that these activities are responsible for the phytotoxic effects that are exhibited by these alkaloids. Since these targets are present in all cells, known allelochemical effects in microbial and animal cells can be mediated by the same compounds. In addition, the same compounds can displace specifically bound ligands from neuroreceptors, such as $alpha_1$, $alpha_2$ adrenergic receptors, serotonin receptor, and nicotinic and muscarinic acetyl choline receptors, which are only present in animals. Thus, most of the allelopathic alkaloids are compounds with a very broad activity spectrum. They obviously have evolved as multipurpose defense compounds. Since plants cannot predict or choose their competitors, infesting microorganisms, insects, and other herbivores, such preformed multipurpose compounds are certainly a means to be prepared for most situations. Considering the prominent effects of alkaloids on neurotransmission, the role of these alkaloids in plant–plant interactions might be a side-effect of their main function as defense compounds in plant–animal interactions.

25.2   INTRODUCTION

Plants compete with other plants for light, water, and nutrients. The production, accumulation and release of secondary compounds, which inhibit germination of other plants (of the same or other species) or the development of seedlings, is one of several complex strategies that have evolved to enhance the fitness of a plant producing them. A number of natural products with allelopathic properties have been reported (Rice, 1984; Waller, 1987; Rizvi and Rizvi, 1992), including many phenolics, terpenes, and alkaloids (Rice, 1984; Waller, 1987; Rizvi and Rizvi, 1992; Wink, 1993). Alkaloids with allelopathic properties include aconitine, berberine, caffeine, cinchonine, colchicine, cytisine, ergometrine, gramine, harmaline, hyoscyamine, lobeline, lupanine, narcotine, nicotine, papaverine, quinidine, quinine, salsoline, sanguinarine, sparteine, strychnine, theophylline, and yohimbine (Wink and Twardowski, 1992; Wink and Latz-Brüning, 1995). These compounds are either actively secreted into the surrounding soil via the rhizosphere or are leached out by rain from aerial parts or from detached leaves lying on the ground. Since these compounds are often rapidly degraded by soil microorganisms, allelopathic effects are not always clear-cut and appear to depend on several environmental factors (Rice, 1984; Waller, 1987; Rizvi and Rizvi, 1992).

Plants also need defenses against herbivores and microorganisms. Again, plant secondary metabolites play a very prominent role in this context (Swain, 1974; Wink, 1987, 1993; Harborne, 1993). Whereas some natural products are specifically directed toward a single group of organisms, others show a broader activity spectrum. We have found several alkaloids that inhibited feeding of insect and vertebrate herbivores or the growth of bacteria and fungi, were phytotoxic as well. Compounds that are active over a wide range of organisms should affect basic molecular targets, which are common to all of them.

In this chapter, we describe bioassays that can be used to determine interactions of alkaloids with basic molecular targets, common to all cells, such as assays for DNA intercalation, inhibition of DNA polymerase, protein biosynthesis, and membrane stability. In addition, we have developed bioassays that detect the binding of alkaloids to neuroreceptors, targets that are only present in animals and often influenced by allelochemicals, or that detect the effect of alkaloids toward reverse transcriptase, a relevant target in retroviruses. It will be shown, that there are several alkaloids that modulate both basic (present in all organisms) and advanced targets (which only exist in animals).

25.3   EXPERIMENTAL

25.3.1   ALKALOIDS

Alkaloids were either isolated and purified in our laboratory (Schmeller et al., 1994, 1995, 1997a,b) or purchased commercially (Sigma, USA; Roth, Karlsruhe). Purity of all compounds was checked by HPLC or GLC and was greater than 95 percent in all cases.

25.3.2 Assays to Determine Interactions of Alkaloids with Molecular Targets

A number of assays were established in our laboratory to determine the interaction of alkaloids with DNA and related enzymes, with biomembranes, protein biosynthesis, and neuroreceptors. These assays were all optimized and standardized in terms of linearity, reproducibility, sensitivity, and specificity. A complete documentation of these assays can be found in Latz-Brüning (1994) and Schmeller (1995).

25.3.3 Interaction of Alkaloids with DNA

25.3.3.1 Melting Point Determination

If compounds intercalate with DNA, then the melting point is shifted to higher temperatures (Maiti et al., 1982; Nandi and Maiti, 1985). *Sinapis* DNA, 70 μM, was incubated in TE-buffer (pH 7.4) with 70 μM alkaloids for 30 min at 22°C. Then the temperature was increased by 1°C/min to 90°C and the absorption was continuously determined in a spectrometer at 260 nm. Differences between two consecutive measurements were plotted to determine the kinetics of the process (Latz-Brüning, 1994).

25.3.3.2 Methylgreen Assay

Methylgreen (MG) binds to DNA and bound MG displays an absorption maximum at 642 nm, whereas free MG shows no absorption at this wavelength (Krey and Hahn, 1969; Burres et al., 1992). When alkaloids bind or intercalate with DNA, then MG is released, which can be measured as a decrease of optical density at 642 nm. DNA-methylgreen (Sigma), 70 μM, was incubated in the dark in 20 mM Tris-HCl (pH 7.4) together with up to 5 mM alkaloids. After 24 h the OD_{642} of untreated controls and treated samples was determined (Latz-Brüning, 1994).

25.3.3.3 Inhibition of DNA-Polymerase I

To determine the activity of DNA-polymerase I, we modified a nick translation assay (Sambrook et al., 1989). The assay buffer contained 50 mM Tris-HCl (pH 7.5), 10 mM $MgSO_4$, 0.1 mM DDT, 500 ng of a linearized plasmid (pUC19), 625 μM dNTPs, 0.01 μCi $\alpha^{32}P$-dCTP, 1 U DNA-polymerase I, 25 pg DNAse I, and up to 10 mM alkaloids. The reaction was started by adding DNAse I; after 15 min at 37°C, the reaction was terminated by adding 100 mM EDTA (pH 8.0). Two variations were carried out: 1) a preincubation of DNA polymerase I with alkaloids for 15 min, prior to adding plasmid DNA, and 2) a preincubation of DNA and alkaloids for 15 min before adding the enzymes in order to differentiate between alkaloidal effects on DNA polymerase I and on DNA. The incorporated radioactivity was removed from the nonincorporated $\alpha[^{32}P]$-dCTP by gelfiltration on Sephadex G 50 (Pharmacia) and measured in a liquid scintillation counter (Latz-Brüning, 1994).

25.3.3.4 Inhibition of Reverse Transcriptase (RT)

To measure the activity of reverse transcriptase, a protocol for the synthesis of cDNA was modified (Sambrook et al., 1989), and mRNA was isolated from rat liver according to standard protocol (Sambrook et al., 1989). mRNA (500 ng) and 500 ng random primer (Boehringer Mannheim) were denatured at 70°C for 5 min and immediately cooled afterward in ice-water. Then 0.3 mM dNTPs, 0.01 μCi $\alpha[^{32}P]$-dCTP, 6 U AMV reverse transcriptase (Promega) in RT buffer (50 mM Tris-HCl, pH 7.8, 10 mM $MgCl_2$, 80 mM KCl, 10 mM DTT) were added and incubated for 30 min at 42°C. The reaction was terminated by adding 100 mM EDTA; the incorporation of $\alpha[^{32}P]$-dCTP was measured as described before in the DNA-polymerase assay. Again, two preincubation strategies as previously described were used (Latz-Brüning, 1994).

25.3.3.5   Inhibition of Protein Biosynthesis

An *in vitro* reticulocyte translation assay (Boehringer Mannheim) was modified to determine an inhibition of translation by alkaloids. An assay (total volume 25 µl) contained 2 µl 12.5 × translation mix (Boehringer), 10 µl reticulocyte lysate, 200 mM K-acetate, 1.5 mM Mg-acetate, 0.25 µCi L-[4,5-^{3}H(N)]-leucine, 0.5 µg TMV-RNA (Boehringer), and up to 5 mM alkaloids (buffered to pH 7). The mixture was incubated at 30°C; reactions were terminated after 0, 10, 20, 30, and 40 min. The radiolabeled protein was precipitated by adding 200 µl ice-cold trichloroacetic acid (TCA) (50%; wt/vol) and after 30 min filtered through GF 34 filters (Schleicher-Schüll), which bind proteins. After washing the filters three times with 50 percent TCA, they were dried at 85°C. Radioactivity of the filters was determined in a liquid scintillation counter (Latz-Brüning, 1994).

25.3.3.6   Influence of Alkaloids on Membrane Permeability

Sheep erythrocytes were purified and incubated in 50 µl PBS (8 g NaCl, 0.2 g KCl, 1.44 g Na_2HPO_4, 0.24 g KH_2PO_4 in 1 l H_2O). Erythrocytes were incubated for 15 min at 10°C together with up to 5 mM alkaloids. Then erythrocytes were precipitated by centrifugation (4 min at 2000g) and the hemoglobin released from erythrocytes was determined photometrically at 543 nm (Latz-Brüning, 1994).

25.3.3.7   Membrane Preparation for Receptor-Binding Studies

Porcine brains, which were obtained within 30 min after death of the animals from a local slaughterhouse, were used to prepare receptor-rich membranes. The brains were immediately frozen in liquid N_2; 50 g brain per 200 ml ice-cold buffer (0.32 M sucrose, 10 mM potassium phosphate buffer, pH 7.0, 1 mM EDTA) were homogenized twice for 15 sec in a blender and then for 1 min with an ultraturrax. The homogenate was centrifuged three times for 15 min at 1400g and 4°C to separate cellular debris. The supernatant was spun down at 100,000g for 60 min. The resulting pellet was resuspended in buffer (as above but without sucrose). Aliquots were stored frozen at ⁻80°C. Protein content was determined by the Lowry method, using bovine serum albumin as a standard (Schmeller et al., 1994, 1995, 1997).

25.3.4   RADIO RECEPTOR BINDING ASSAYS

Binding assays (in triplicates) were performed using a rapid filtration technique, essentially as that described by Schmeller et al. (1994, 1995, 1997).

25.3.4.1   Muscarinic Receptor (mAChR)

Membrane preparations adjusted to 500 µg protein in a final volume of 500 µl buffer were incubated with [^{3}H]-quinuclidinyl benzilate (QNB) (52.3 Ci/mmol; Dupont NEN) for 1 h at 20°C in the absence and presence of alkaloids, using 20 µM atropine as a positive control substance. The incubation was stopped with 3 ml ice-cold 0.9 percent NaCl-solution and filtered (by suction) through Whatman GF/C glass fiber filters. The filters were washed three times with 3 ml 0.9 percent NaCl, placed in vials, and dried for 30 min at 60°C. Their radioactivity was measured in a liquid scintillation counter (RackBeta, Pharmacia) using Ultima-Gold (Packard) as the scintillation cocktail.

25.3.4.2   Nicotinic Receptor

[^{3}H]-nicotine (85 Ci/mmol; Amersham) was used to assay specific binding of alkaloids to the nicotinic ACh receptor (nAChR). The membrane preparation was incubated for 40 min with differing concentrations of alkaloids or 1 mM nicotine as a positive control. The GF/C filters were presoaked with polyethylene glycol 8000 (5 percent in water) for 3 h to reduce nonspecific binding of [^{3}H]-nicotine. Further procedures were the same as described above for mAChR.

25.3.4.3 Alpha$_1$ Receptor

[^{3}H]-prazosine (78 Ci/mmol; DuPont NEN) was used to assay specific binding of alkaloids to the alpha$_1$ receptor. The membrane preparation was adjusted to 400 μg in a final volume of 500 μl and incubated for 45 min at 20°C with differing concentrations of alkaloids or 400 μM phentolamine as a positive control. Further procedures were the same as described above for mAChR.

25.3.4.4 Alpha2 Receptor

[^{3}H]-yohimbine (81 Ci/mmol; DuPont NEN) was used instead of [^{3}H]-prazosine; other conditions were the same as in the alpha$_1$ receptor assay.

25.3.4.5 Serotonin2 Receptor

[^{3}H]-ketanserine (85,1 Ci/mmol; DuPont NEN) was used to assay specific binding of alkaloids to the serotonin^2 receptor (5-HT^2). The membrane preparation was adjusted to 400 μg in a final volume of 500 μl and incubated for 40 min at 20°C with differing concentrations of alkaloids or 100 μM mianserine as a positive control. Further procedures were the same as described above for mAChR.

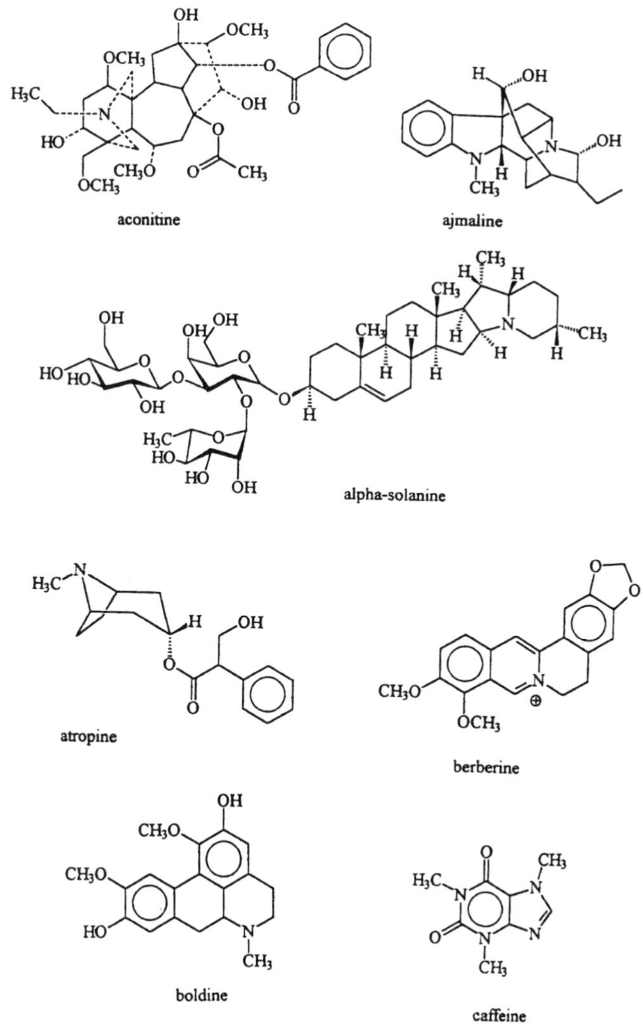

FIGURE 25.1

25.4   RESULTS AND DISCUSSION

Phytotoxic properties have been detected for the following alkaloids (see Figure 25.1): aconitine, berberine, boldine, caffeine, cinchonine, colchicine, cytisine, ergometrine, gramine, harmaline, hyoscyamine, lobeline, lupanine, narcotine, nicotine, papaverine, quinidine, quinine, salsoline, sanguinarine, sparteine, strychnine, and yohimbine (Wink and Twardowski, 1992; Wink and Latz-Brüning, 1995). The mechanisms underlying allelopathy of these compounds have hardly been studied so far (Wink and Twardowski, 1992; Wink and Latz-Brüning, 1995).

TABLE 25.1
Interaction of Alkaloids with Basic Molecular Targets. Significant Effects Are Marked in Bold (with concentrations in brackets)

	DNA Melting Temperature Increase[a] (C°)	DNA Methylgreen Release (%)	DNA DNA Pol I Inhibition (%)[b]	RNA RT Inhibition (%)[b]	Protein Biosynthesis Inhibition (%)	Membrane Stability Hemolysis (%)
aconitine	n.d.	0 (2 mM)	10 (1 mM)	3 (0.7 mM)	0 (0.2 mM)	0 (1 mM)
ajmaline	2.3	**33** (5 mM)	37 (10 mM)	**71** (10 mM)	**80** (4 mM)	0.4 (5 mM)
berbamine	13.2	**76** (1 mM)	**45** (0.6 mM)	**81** (0.5 mM)	0 (0.1 mM)	2 (1 mM)
berberine	15	**91** (0.5 mM)	**91** (1 mM)	**97** (1 mM)	**100** (1 mM)	0 (1 mM)
boldine	6	**81** (5 mM)	**69** (10 mM)	**97** (10 mM)	**30** (1 mM)	0 (5 mM)
caffeine	n.d.	0 (2 mM)	7 (10 mM)	25 (10 mM)	10 (1 mM)	0 (5 mM)
cinchonine	5	9 (5mM)	**60** (10 mM)	**72** (10 mM))	**43** (1 mM)	0 (5 mM)
cinchonidine	6	**56** (5 mM)	**55** (10 mM)	**93** (10 mM)	**90** (5 mM)	0 (5 mM)
colchicine	n.d.	0 (5 mM)	0 (5 mM)	24 (5 mM)	0 (1 mM)	0 (5 mM)
cytisine	n.d.	0 (5 mM)	14 (10 mM)	4 (10mM)	0 (1 mM)	0 (5 mM)
ergometrine	13.7	**27** (5 mM)	2 (10 mM)	**40** (10 mM)	0 (1 mM)	0 (5 mM)
gramine	n.d.	1 (5 mM)	14 (7 mM)	26 (5 mM)	0 (1 mM)	0 (5 mM)
harmalin	8.6	**100** (5 mM)	**87** (10 mM)	**90** (10 mM)	**70** (1 mM)	0 (5 mM)
harmin	16.1	**68** (1 mM)	**91** (10 mM)	**70** (0.6 mM)	**95** (1 mM)	2 (5 mM)
hyoscyamine	n.d.	5 (5 mM)	12 (10 mM)	13 (10 mM)	0 (1 mM)	0 (5 mM)
lobeline	1.5	27 (5 mM)	**46** (10 mM)	**71** (10 mM)	**50** (5 mM)	0 (5 mM)
lupanine	0	0 (5 mM)	1 (10 mM)	0 (10 mM)	10 (1 mM)	0.8 (5 mM)
narcotine	0	0 (0.5 mM)	**44** (0.7 mM)	0 (0.5 mM)	0 (0.1 mM)	5 (0.5 mM)
nicotine	n.d.	4 (5 mM)	2 (10 mM)	7 (10 mM)	20 (1 mM)	0 (5 mM)
norharman	6.2	**71** (5 mM)	**45** (7 mM)	**88** (1 mM)	**50** (0.2 mM)	3 (5 mM)
papaverine	0	1 (0.5 mM)	5 (3 mM)	**71** (2 mM)	**60** (0.8 mM)	1 (0.5 mM)
quinidine	8	**88** (5 mM)	**99** (10 mM)	**90** (10 mM)	**63** (1 mM)	0.6 (51mM)
quinine	6	**81** (5 mM)	**83** (10 mM)	**70** (5 mM)	**70** (1 mM)	0.4 (1 mM)
salsoline	n.d.	3 (5 mM)	0 (10 mM)	18 (10 mM)	**30** (1 mM)	0 (5 mM)
sanguinarine	24	**82** (0.5 mM)	**97** (0.15 mM)	**75** (0.2 mM)	10 (0.01 mM)	5 (0.1 mM)
scopolamine	0.3	0 (5 mM)	0 (10 mM)	0 (10 mM)	0 (1 mM)	0 (5 mM)
solanine	3	0 (0.5 mM)	17 (0.5 mM)	**91** (0.25 mM)	**50** (0.025 mM)	**100** (0.3 mM)
sparteine	n.d.	3 (5 mM)	24 (10 mM)	24 (10 mM)	10 (1 mM)	0.3 (5 mM)
strychnine	n.d.	3 (5 mM)	16 (10 mM)	**71** (10 mM)	20 (1 mM)	0 (5 mM)
yohimbine	n.d.	2 (1 mM)	17 (5 mM)	29 (5 mM)	**62** (1 mM)	0 (1 mM)

Abbreviation: n.d., not determined.

[a] 70 μM alkaloid solution.

[b] alkaloids were preincubated with DNA or RNA.

25.4.1 ACTIVITY OF ALKALOIDS AT BASIC MOLECULAR TARGETS

25.4.1.1 Interactions with DNA and Related Enzymes

We have analyzed whether these alkaloids affect basic molecular targets such as DNA and related processes, protein biosynthesis, or membrane stability. The interaction of alkaloids with DNA, especially intercalation, can be studied by measuring the melting temperature of DNA. Intercalating compounds increase this parameter significantly. Since the direct assay is costly and time intensive, we have used a more rapid assay in addition, which is based on the competitive release of the intercalating compound methylgreen from DNA (Burres et al., 1992). As can be seen from Table 25.1, several alkaloids increase the melting temperature of DNA; the strongest effects were detected for sanguinarine, followed by harmin, berberine, berbamine, ergometrine, harmalin, quinidine, quinine, cinchonidine, cinchonine, boldine, norharman, and possibly ajmaline. These results were corroborated by the methylgreen assay (Table 25.1). For a few alkaloids, DNA intercalation had been described before, for example for sanguinarine, berberine, or harmalin (Krey and Hahn, 1969; Maiti et al., 1982; Nandi and Maiti, 1985; Wink, 1993; Schmeller et al., 1997). DNA intercalation can affect replication and transcription, but can also lead to mutations and even cancer. It is likely that the disturbance of replication and transcription is the major mode of action of the phytotoxic compounds. Two related targets were studied, such as DNA polymerase I and reverse transcriptase (Table 25.1). Although both enzymes do not occur in plants, it is nevertheless likely that DNA and RNA polymerases of plants will be affected in a similar way. As can be seen from Table 25.1, ajmaline, berbamine, berberine, boldine, cinchonine, cinchonidine, ergometrine, harmalin, harmin, norharman, quinidine, quinine, and sanguinarine strongly affect both targets. These compounds have also shown significant effects in the DNA intercalating assays, suggesting that binding of these alkaloids to DNA or RNA is the cause of enzyme inhibition. Effects were more strongly expressed when alkaloids and nucleic acid were preincubated as compared with the situation in which alkaloids and enzymes were preincubated in the absence of nucleic acids (Latz-Brüning, 1994). Solanine, which displayed a weak increase in the melting temperature, showed a strong inhibitory effect on RT and a weaker one on DNA Pol I. Lobeline, which was only active in the methylgreen assay, showed a moderate inhibition of DNA Pol I and RT. Narcotine and papaverine, which do not intercalate DNA, inhibit DNA Pol I or RT, respectively. We assume that mechanisms other than intercalation must work in the latter instances.

25.4.1.2 Inhibition of Protein Biosynthesis

Protein biosynthesis, which is a critical and important target in all cells, was strongly affected (inhibition greater than 30 percent) by ajmaline, berberine, boldine, cinchonine, cinchonidine, harmalin, harmin, lobeline, norharman, papaverine, quinidine, quinine, salsoline, sanguinarine, solanine and yohimbine; a weaker inhibition (less than 20 percent) was observed for caffeine, lupanine, nicotine, sparteine, and strychnine (Table 25.1). Again most compounds that substantially affected DNA, DNA Pol I, and RT, were also active as inhibitors of translation. Interaction of these alkaloids with ribosomal nucleic acids, for example rRNA, tRNA, or mRNA, is likely, besides interactions with ribosomal proteins (Wink and Twardowski, 1992). Only a few intercalating compounds, such as berbamine or ergometrine were not protein biosynthesis inhibitors. On the other hand, most compounds (such as aconitine, caffeine, colchicine, cytisine, gramine, hyoscyamine, lupanine, narcotine, scopolamine, sparteine, or strychnine) that do not intercalate, do not influence translation substantially (inhibition less than 20 percent) with exceptions noted for papaverine, salsoline, and yohimbine.

25.4.1.3 Stability of Biomembranes

The integrity of biomembranes is of ultimate importance for the functioning of cells; compounds that disturb biomembranes and thus make cells leaky, are usually strong cell poisons. Natural products that exhibit these properties are either very lipophilic or amphiphilic, such as mono-, sesqui-, and diterpenes or triterpene and steroid saponins, respectively. These compounds are often phytotoxic (Wink and Twardowski, 1992; Wink and Latz-Brüning, 1995). Hemolysis in erythrocytes offers a simple *in vitro* test system to study the effect of chemicals on membrane disturbance (Latz-Brüning, 1994). Out of the series of alkaloids tested, only solanine showed a significant hemolytic activity. Since solanine has the structure of an amphiphlic saponin, this result is not surprising and is consistent with previous results (Rice, 1994; Wink 1993). Weak hemolytic properties were detected for berbamine, harmin, narcotine, norharman, and sanguinarine. Whether they contribute to allelopathy cannot be determined from these experiments.

25.4.2   OTHER TARGETS

For a few phytotoxic alkaloids, such as aconitine, caffeine, colchicine, cytisine, gramine, hyoscyamine, lupanine, nicotine, scopolamine, and sparteine, we could not detect a significant effect on the molecular targets studied (Table 25.1). We must conclude, therefore, that still other molecular targets must exist that mediate allelopathy in these and other instances.

For colchicine, it is well established that it inhibits the polymerization of tubulin to microtubules, which are important for cellular transport and cell division. Since cell division takes place during germination, the colchicine–tubulin intercations could be relevant for allelopathy.

Caffeine inhibits phosphodiesterase in animals; cAMP and cGMP and corresponding phosphodiesterase seem to exist also in plants, but their function has been discussed controversially. Whereas it is unlikely that cAMP/cGTP are part of the signal transduction (as in animals), the cyclic nucleotides might play a role in gene regulation (as in bacteria). Therefore, phosphodiesterase inhibition could be a relevant target for allelopathy; besides caffeine, theophylline (Wink and

TABLE 25.2
Interaction of Alkaloids with Neuroreceptors (Only Present in Animals).

	adrenergic receptor		serotonin receptor	acetylcholine receptor	
	alpha$_1$	alpha$_2$	5-HT$_2$	mACh	nACh
aconitine	n.a.	331	n.a.	1.3	159
berberine	3.2	0.48	1.9	1.0	35
boldine	0.5	0.09	0.67	118	11
caffeine	n.a.	n.a.	n.a.	464	n.a.
cinchonine	5.2	1.4	5.5	19.2	n.a.
cinchonidine	1.1	1.3	10.7	19.7	n.a.
colchicine	n.a.	23.8	133	347	30
cytisine	n.a.	n.a.	n.a.	398	0.14
ergometrine	4.4	0.9	1.5	2.0	178
gramine	26	8.7	8.5	677	30.7
harmalin	34	7.5	14.6	33.5	n.a.
hyoscyamine	6.1	10.1	6.0	0.005	284
lupanine	n.a.	n.a.	n.a.	118	5.3
nicotine	n.a.	n.a.	n.a.	882	0.008
quinine	5.7	2.5	6.4	4.5	n.a.
quinidine	29.7	1.3	14.4	18.4	n.a.
salsoline	115	9.8	146	n.a.	n.a.
sanguinarine	33.6	6.4	91.7	2.4	11.8
scopolamine	113	359	168	0.002	928
sparteine	n.a.	127	n.a.	21.3	330
strychnine	25.1	172	51.6	128.8	10.2

Given are concentrations (in µM) that replace 50 percent of the specifically bound ligand (IC$_{50}$); n.a., not active at 500 µM. After Schmeller et al., 1994, 1995, 1997

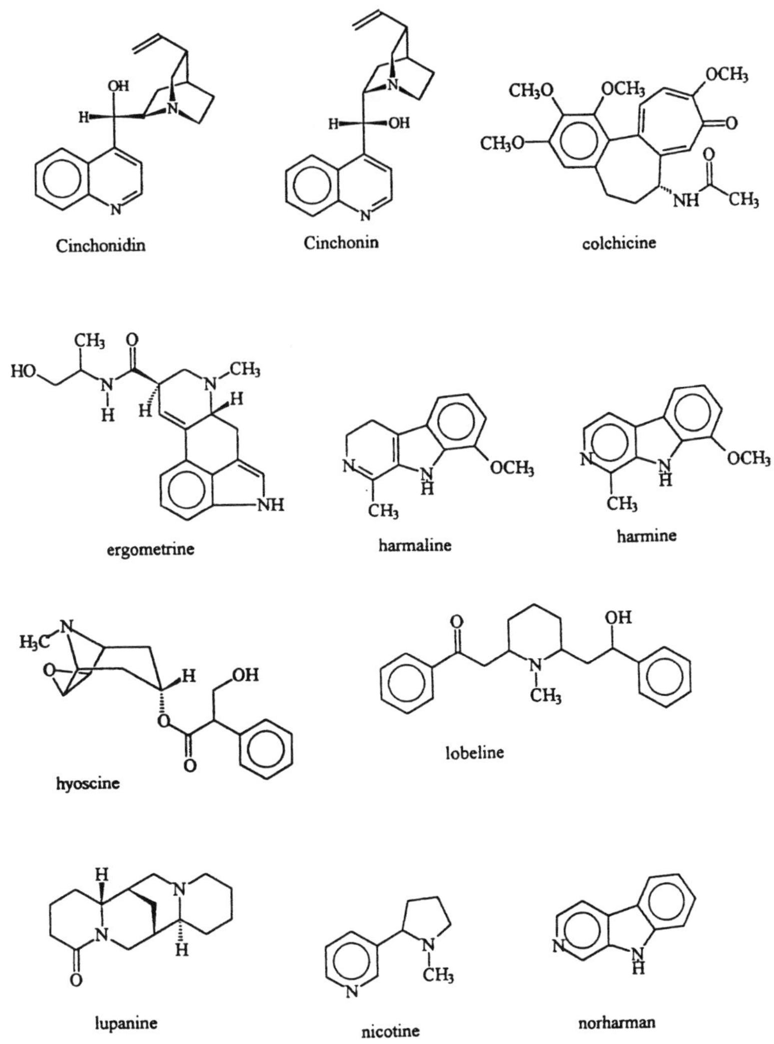

FIGURE 25.2

Twardowski, 1992) and papaverine, which are known phosphodiesterase inhibitors, exhibited allelopathic properties. Caffeine was shown to inhibit cell division in *Coffea arabica* (Friedman and Waller, 1983).

Aconitine, cytisine, lupanine, and sparteine strongly bind to acetylcholine receptors (Schmeller et al., 1994) (Table 25.2), but this target should not be relevant in plants. However, these alkaloids also affect Na^+ and K^+ channels in animals (also the allelopathic ajmaline and quinidine inhibit this target) (Wink, 1993); therefore, it needs to be established whether this target is also affected in plant cells. Gramine might interact with the metabolism of plant growth factors, since it shares structural similarities with auxins (such as IAA). Also other alkaloids might be active in this context, which needs further evaluation. See Figures 25.2 and 25.3.

25.4.3 BINDING OF ALKALOIDS TO NEURORECEPTORS

Many secondary metabolites (including alkaloids) affect neurotransmission and signal transduction (Wink, 1993), which represent extremely vulnerable targets in animals. Since any substantial interference at neuroreceptors (e.g., when natural ligands cannot bind to their receptors any longer because they are blocked by an alkaloid with structural similarity to the natural ligand) will influence neuronal signal transduction (including muscular activity) and CNS activity, the intake of at

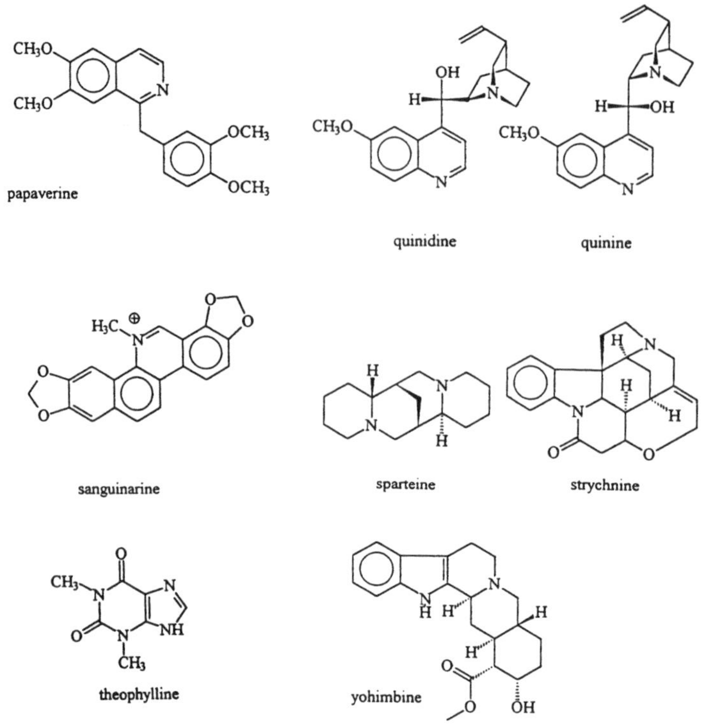

FIGURE 25.3

least a large dose of alkaloids should lead to short-term (within several hours after ingestion) physiological disturbances. These adverse effects and the bitter taste of most alkaloids should provide a clue to herbivores to avoid alkaloid-producing plants in the future (associative learning).

We have analyzed whether the alkaloids that exhibit phytotoxic properties can also affect neuroreceptors. For this purpose we have determined whether an alkaloid can displace a specifically bound ligand from a neuroreceptor, such as $alpha_1$, $alpha_2$ adrenergic receptors, serotonin receptor, and nicotinic and muscarinic acetyl choline receptors (Table 25.2). As can be seen from Table 25.2, almost every alkaloid can bind to at least one of the neuroreceptors studied, and a number of alkaloids affect several neuroreceptors.

The question is whether the inhibitor concentrations determined in *in vitro* experiments (Table 25.2) relate in any way to the *in vivo* situation. A simple calculation may help to assess this problem. Alkaloids are usually stored in high concentrations at sites that are important for growth and reproduction and can reach 1 to 2 percent of the dry weight (Wink, 1993). Assuming an alkaloid concentration of 100 mg per 100 g fresh weight and a small herbivore with a body weight of 1000 g, if this animal would ingest 100 g of an alkaloid-producing plant, it would take up 100 mg of alkaloids. Suggesting that the alkaloids are completely resorbed and equally distributed in the body, we would obtain a concentration of 100 mg alkaloids per kilogram of body weight. Taking a mean molecular weight of 400, the alkaloid concentration in our herbivore would be 250 μM (in reality non-absorption and degradation will lead to a lower value of approx. 50 to 150 μM), which would be high enough to partially or completely block the binding of acetylcholine, serotonin, or noradrenalin at their receptors (compare the IC_{50} values in Table 25.2).

25.5 CONCLUSIONS

Ajmaline, berbamine, berberine, boldine, cinchonine, cinchonidine, ergometrine, harmalin, harmin, lobeline, norharman, papaverine, quinidine, quinine, sanguinarine, and solanine affect more than one of the basic molecular targets (Table 25.1). It is likely that these interactions are responsible (at least in part) for the phytotoxic effects that are exhibited by these alkaloids (Wink and Twardowski, 1992; Wink and Latz-Brüning, 1995). Since these targets are present in all cells, known allelochemical effects in microbial and animal cells might be mediated by the same compounds.

In addition, the same compounds can bind to one or several neuroreceptors that are only present in animals. Thus most of the allelopathic alkaloids are compounds with a very broad activity spectrum. As many other secondary metabolites, they obviously have evolved as defense compounds. Because of their wide activity, we can consider them multipurpose defense compounds. Since plants cannot predict or choose their competitors, for example, infesting microorganisms, insects, and other herbivores, such preformed multipurpose compounds are certainly a means to be prepared for most situations. Considering the prominent effects of alkaloids on neurotransmission (Table 25.2), the role of these alkaloids in plant–plant interactions might be a side-effect of their main function as defense compounds in plant–animal interactions.

ACKNOWLEDGMENTS

Part of this study was supported by the Deutsche Forschungsgemeinschaft and Mittex Anlagenbau (Ravensburg).

REFERENCES

Burres, N. S., Frigo, A., Rasmussen, R. R., and McAlpine, J. B., A colorimetric microassay for the detection of agents that interact with DNA, *J. Nat. Prod.,* 55, 1582, 1992.

Friedmann, J. and Waller, G. R., Caffeine hazards and their prevention in germination of seeds of *Coffea arabica., J. Chem. Ecol.,* 9, 1099, 1983.

Harborne, J. B., *Introduction to Ecological Biochemistry,* 4th ed., Academic Press, New York, 1993.

Krey, A. K. and Hahn, F. E., Berberine: complex with DNA, *Science,* 166, 755, 1969.

Latz-Brüning, B., Molekulare Wirkmechanismen von Alkaloiden, PhD dissertation, University of Heidelberg, Germany, 1994.

Maiti, M., Nandi, R., and Chaudhuri, K., Sanguinarine: a monofunctional intercalating alkaloid, *FEBS Lett.,* 142, 280, 1982.

Nandi, R. and Maiti, M., Binding of sanguinarine to desoxyribonucleic acid of differing base composition, *Biochem. Pharmacol.,* 34, 321, 1985.

Rice, E. L., *Allelopathy,* Academic Press, Orlando, FL, 1984.

Rizvi, S. J. H. and Rizvi, V., Eds., *Allelopathy. Basic and Applied Aspects,* Chapmann & Hall, London, 1992.

Sambrook, J., Fritsch, E. F., and Maniatis, T., *Molecular Cloning: a Laboratory Manual,* Cold Spring Harbour Labs, New York, 1989.

Schmeller, T., Interaktionen von Alkaloiden mit Neurotransmitter-Rezeptoren, PhD dissertation, University of Heidelberg, Germany, 1995.

Schmeller, T., Latz-Brüning, B., and Wink, M., Biochemical activities of berberine, palmatine and sanguinarine mediating chemical defence against microorganisms and herbivores, *Phytochemistry,* 44, 257, 1997a.

Schmeller, T., El-Shazly, A., and Wink, M., Allelochemical activities of pyrrolizidine alkaloids: interactions with neuroreceptors and acetylcholine related enzymes, *J. Chem. Ecol.,* 23, 399, 1997b.

Schmeller, T., Sauerwein, M., Sporer, F., Wink, M., and Müller, W., Binding of quinolizidine alkaloids to nicotinic and muscarinic receptors, *J. Nat. Prod.,* 57, 1316, 1994.

Schmeller, T., Sporer, F., Sauerwein, M., and Wink, M., Binding of tropane alkaloids to nicotinic and muscarinic receptors, *Pharmazie,* 50, 493, 1995.

Swain, T., Secondary compounds as protective agents, *Annu. Rev. Plant Physiol.,* 28, 479, 1974.

Waller, G. R., Ed., *Allelochemicals. Role in Agriculture and Forestry,* ACS Symposium Series 330, American Chemical Society, Washington, D.C., 1987.

Wink, M., Chemical ecology of quinolizidine alkaloids, in *Allelochemicals. Role in Agriculture and Forestry,* Waller, G. R., Ed., ACS Symposium Series 330, American Chemical Society, Washington, D.C.,1987, 524.

Wink, M., Allelochemical properties and the raison d'être of alkaloids, in *The Alkaloids,* Cordell, G. A., Ed., Academic Press, Orlando, FL, 1993, 1.

Wink, M. and Twardowski, T., Allelochemical properties of alkaloids. Effects on plants, bacteria and protein biosynthesis, in *Allelopathy: Basic and Applied Aspects,* Rizvi, S. J. H. and Rizvi, V., Eds., Chapman & Hall, London, 1992, 129.

Wink, M. and Latz-Brüning, B., Allelopathic properties of alkaloids and other natural products: possible modes of action, in *Allelopathy. Organisms, Processes and Applications,* Inderjit, Dakshini, K. M. M., and Einhellig, F. A., Eds., ACS Symposium Series 582, American Chemical Society, Washington, D.C., 1995, 117.

26 Higher Plant Flavonoids: Biosynthesis and Chemical Ecology

Mark A. Berhow and Steven F. Vaughn

CONTENTS

26.1 ABSTRACT

Higher plants must protect themselves from invasion by microorganisms through physical barriers and the production of a wide range of secondary chemicals. Many of these compounds also confer selective advantages by inhibiting competing plants. The plant phenolics constitute the largest group of plant secondary compounds, primarily synthesized via the shikimic acid pathway. They constitute an extremely diverse group, many of which have been shown to be antimicrobial and/or phytotoxic. Phenolic compounds can be assigned to several chemical groups, such as simple phenolics (e.g., cinnamic and ferulic acids), lignins, coumarins, quinones, and flavonoids. Flavonoids are a biologically important and chemically diverse group that can be further divided into subgroups including anthocyanidins, flavones, flavanones, flavonols, flavanols, isoflavones, and chromones. Flavonoids may be either constitutive or induced by stimuli such as wounding or pathogenic attack. In particular, the isoflavone phytoalexins (compounds formed *de novo* after microbial invasion), produced by members of the plant family Fabaceae, have been the subject of much recent research. This article will attempt to summarize the published literature regarding the antimicrobial and phytotoxic effects of plant flavonoids and discuss their potential exploitation.

26.2 INTRODUCTION

Plants synthesize a wide variety of chemical compounds, numbering well into the hundreds of thousands, perhaps even millions. These compounds can be sorted by their chemical class and biosynthetic origin, and have been roughly separated into two main functional groups: primary or secondary metabolites. Primary metabolites consist of compounds such as carbohydrates, lipids, proteins, heme chlorophylls, and nucleic acids, which make up the physical integrity of plant cells, and are involved with the primary metabolic processes of building and maintaining living cells. Secondary metabolites have been defined historically as naturally occurring substances that do not seem to be vital to the immediate survival of the organism that produces them, and are not an essential part of the process of building and maintaining living cells. The emerging picture from recent research, however, is that these secondary metabolites play pivotal roles in the ecochemical functionality of the plant, that is, in determining the role a particular species will play in the environment. In addition, research is beginning to show a role for the these compounds at the cellular level as plant growth regulators and modulators of gene expression. Plants commit a large amount of physical resources to the synthesis and accumulation of secondary compounds.

Secondary compounds are crucial in the plant's response to stresses, such as changes in light or temperature, competition, herbivore pressure, and pathogenic attack. They also appear to be critical to a plant's ability to survive and reproduce. Plants have further adapted their use of these metabolites and incorporated them into specialized physiological functions such as reproduction and intracellular signaling. Of special interest is that these secondary metabolites are often functionally unique at the species level, and taxonomically discrete individual compounds are responsible for identical functions in separate species. Secondary metabolites have been generally grouped according to gross chemical structure and/or biosynthetic origin. These groups include terpenes, alkaloids, phenolics, and polyamines. Individual compounds from these classes have been adapted and used for many different functions by plants, as well as microbes and herbivores, which feed on them.

Research in the field of functional plant natural products has exploded in recent years (Cordell, 1995). The field of secondary metabolite interaction with other living organisms is huge and well beyond the scope of this chapter. We have chosen to provide just a few examples of the diverse functionality of just one class of plant secondary metabolites—the flavonoids. This ubiquitous class of phenolic compounds is found in nearly all vascular plants. Flavonoids are of particular interest as they appear to function in all roles in which plant secondary metabolites have been implicated (Harborne, 1994; Stafford, 1990), including that of UV light protectants, anti-oxidants, enzymatic and primary metabolite regulators, in mineral nutrition, in temperature and water stress, in pollination and seed dispersal (by their color properties), in phytohormone regulation, in pollen tube growth, and as chemoprotective agents against other plants, microbial pathogens, fungi, insects, as well as against attack by herbivores. They are, as one reviewer has pointed out, "biological molecules for useful exploitation" (Dakora, 1995). This chapter will cover all aspects of plant flavonoid interactions with other living things in the plant environment, not strictly plant–plant interactions, to cover the broadest sense of the term allelopathy. An excellent series of books has been published on plant flavonoids. These include the first comprehensive work on the subject by Geisman (1962), a series edited by Harborne (Harborne et al., 1975; Harborne and Mabry, 1982; Harborne, 1994), and work by Stafford (1990).

26.3 FLAVONOID BIOSYNTHESIS AND FUNCTIONALITY

Flavonoids have attracted the attention of humans for centuries. Certain plant flavonoids were among the first dyes used. While they have been systematically studied since around the turn of the century, only recently has their biochemical roles in plants been fully appreciated. As the activity levels of many of the enzymes in the flavonoid biosynthetic pathway are low and/or transient in nature,

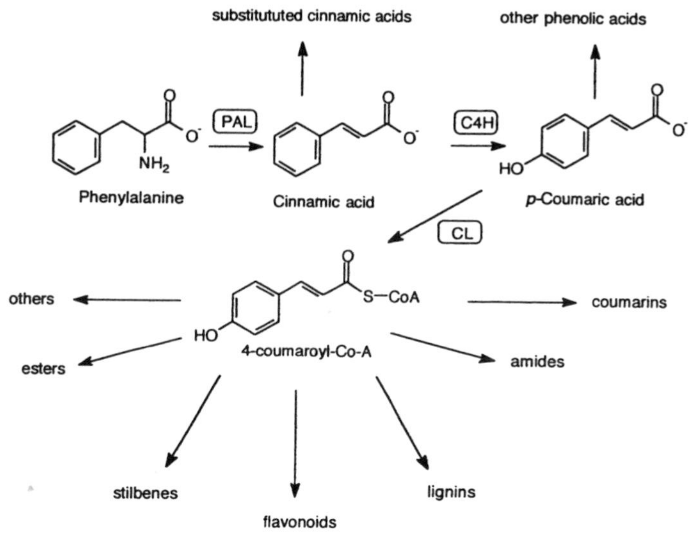

FIGURE 26.1 Biosythesis of the plant polypropanoids. Enzyme abbreviations: PAL, phenylalanine ammonia lyase; C4H, cinnamic acid 4-hydroxylase; CL, *p*-coumaroyl acid CoA ligase.

much of the early work on this pathway was conjectural in nature. Not until radiolabeled biochemicals became available in the 1960s and with the development of plant cell culture techniques in the 1970s did their physiological roles become evident.

The biosynthesis of flavonoids in plant tissues has been extensively studied in many plants and several of the biosynthetic steps have been elucidated. Several general reviews of flavonoid biosynthesis in plants have been published (Heller and Forkmann, 1993; Hahlbrock and Scheel, 1989; Stafford, 1990). Flavonoids are synthesized from phenylalanine via a biosynthetic route in plants termed the "general phenylpropanoid pathway." This important pathway generates a large number

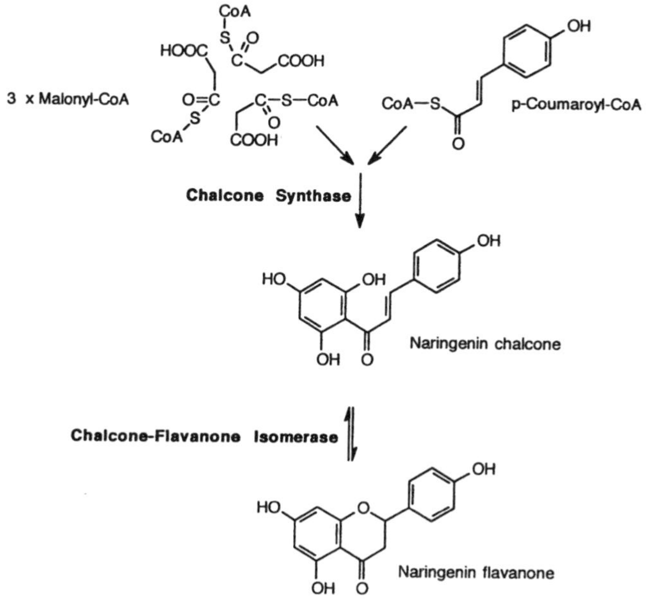

FIGURE 26.2 Biosynthesis of the flavonoid precursor naringenin.

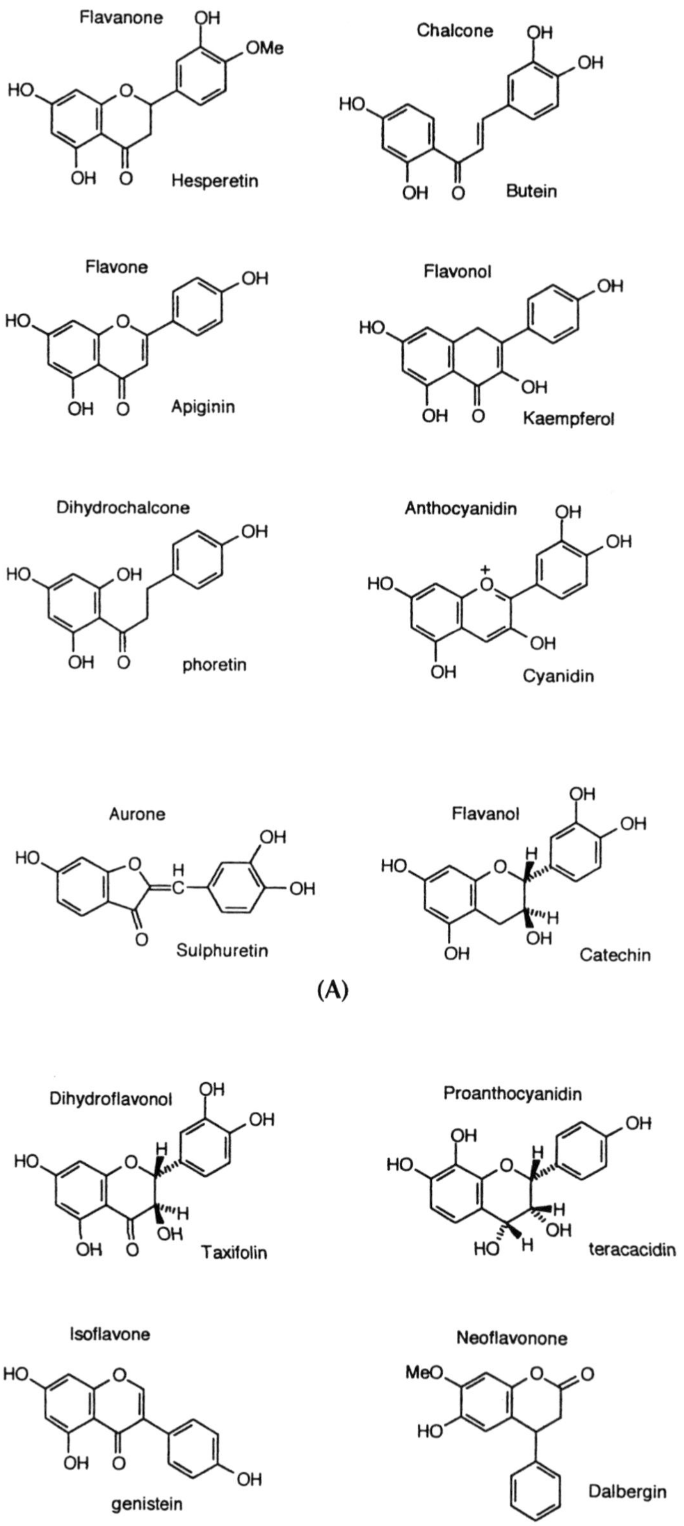

(A)

(B)

FIGURE 26.3 Chemical structures of the different classes of flavonoids.

of compounds that have in common a benzyl group structural element. Secondary metabolites derived from this pathway in addition to flavonoids include tannins, phenols, benzoic acids, stilbenes, cinnamate esters, and coumarins. A sequence of reactions converts phenylalanine into coenzyme A (CoA) derivatives of substituted cinnamic acids (Figure 26.1). These are further converted to the various metabolic compounds mentioned previously. Flavonoids are derived from the conjugation of p-coumaryl-CoA with three malonyl-CoA molecules to form naringenin chalcone, which is considered to be the precursor of all the flavonoids (Figure 26.2). Naringenin chalcone is rapidly converted to the flavanone form, naringenin. Naringenin can then be further modified enzymatically by reactions including reduction, oxidation, hydroxylation, O-methylation, O-glycosylation, C-glycosylation, acylation, sulfonation, rearrangement, and polymerization to form other flavonoids.

Flavonoids have generally been classified into 12 different subclasses by the state of oxidation and the substitution pattern at the C-2–C-3 unit (Figure 26.3). These include flavanones, flavones, flavonols, chalcones, dihydrochalcones, anthocyanidins, aurones, flavanols, dihydroflavonols, proanthocyanidins (flavan-3,4-diols), isoflavones, and neoflavones. More than 10,000 flavonoids have been identified from natural sources and others continue to be identified at a rate of more than 10 per month. Of particular interest is the use of flavonoids as taxonomic markers, because individual species within plant families often vary widely in flavonoid type and content (Seigler, 1981). In addition, individual plants within a species will produce and accumulate different flavonoids depending on several factors, such as plant growth stage, reproductive stage, the particular plant tissue involved, and the type of environmental stress or pathogenic attack involved (Dixon and Paiva, 1995).

26.3.1 CONTROL, INDUCTION, AND ACCUMULATION OF FLAVONOID

The control mechanisms for complex modifications in flavonoid biosynthesis (such as B-ring hydroxylation, methylation, or glycosylation) have been studied in a small number of cultured cell suspensions and whole plants, including parsley *(Petroselium crispum),* soybean *(Glycine max* [L.] Merr.), green bean *(Phaseolus vulgaris* L.), alfalfa *(Medicago sativa* L.), and chickpea *(Cicer arietinum* L.) systems (Hahlbrock and Scheel, 1989). Much of the recent work on the control of these pathways has been determined using new molecular biological techniques. Specfic modification of plant genomes will allow researchers to delve further into the controls operating in this complex system.

Flavonoid biosynthesis pathways are highly regulated and controlled by both normal growth and development, as well as induction by wounding or attack by pathogens. These pathways appear to be very tightly regulated and controlled by other metabolic, developmental, and stress-related factors. The enzymes in these pathways are often present as several different isozymes, indicating groups of tightly controlled metabolic chains that are driven by the formation of very specific end products (Stafford, 1990; Dixon et al., 1992; Douglas et al., 1992). Because of the wide variety of functions within the plant, it is not surprising that a large number of gene copies and isozymes for any given step in the flavonoid biosynthetic pathway have been found. Differential control may be achieved by several methods: enzyme synthesis and degradation, enzyme activation and inhibition, and enzyme compartmentalization. It has been proposed that the activities of groups of enzymes may be specially compartmentalized to yield the specific products called for by the endogenous and exogenous signals (Hrazdina and Jensen, 1992; Harrison and Dixon, 1994).

Flavonoid accumulation is integrated into a number of physiological processes in plants. The biosynthesis of new flavonoids is constitutive in plants; they are internally controlled during normal growth and development, such as new vegetative leaf growth and reproductive organ development (Heller and Forkmann, 1988). Flavonoid biosynthesis may also be induced by exogenous stimuli, such as changes in light and temperature (Hahlbrock and Scheel, 1989). It is also a part of the plant's chemical defense system. Flavonoid biosynthesis and/or modification can be triggered, or elicited, by damage to the plant caused by physical agents (wind, freezing, water stress, ozone, heavy metal ions, certain herbicides), herbivore attack (insects, grazing animals), and microbial invasion (bacteria, fungi) (Chappell and Hahlbrock, 1984; Ebel 1986; Nicholson and Hammerschmidt, 1992; Dixon et al., 1994; Bohnert et al., 1995; Dixon and Paiva, 1995). Induced compounds accumulate specifically in tissues around the wound site, often at very high concentrations. These compounds are not only toxic to the invading pests, but seem to be toxic to the plant tissues themselves; this accumulation can also be accompanied by localized areas of plant cell death, although there is some recent evidence that these two events are not necessarily tied together (Jakobeck and Lindgren, 1993). Induced accumulation of flavonoids may be in addition to constitutive levels found in unstressed plants, which also may be transformed during the induction process (hydrolysis of pre-formed glycosides, for instance) (Ebel and Grisebach, 1988; Burden and Norris, 1992). Biotic induction appears to be a complex chain of events triggered by assorted chemical breakdown products of the host-pathogen interaction—plant cell wall fragments and pathogen specific chemicals (polysaccharides such as chitosan oligomers; other bacterial and fungal cell wall fragments; specific chemicals not found in plants such as arachidonic acid and other long chain fatty acids; hydrolytic enzymes and polypeptides secreted by the pathogens; etc.) (Kobayashi et al., 1993; Darvell and Albersheim, 1984; Ebel, 1986; Lamb et al., 1989; Kuc, 1995; Negrel and Javelle, 1995). As noted previously, the physiological regulation of a plant's specific flavonoid chemistry varies depending on the specific tissue, individual plant, or species examined.

Induction has been studied extensively in cell culture systems. Phenylproponoid pathway genes are activated in cells treated with either microbes or elicitors, resulting in the formation of newly synthesized flavonoids that inhibit microbial growth. In some cases, these same biosynthetic pathways can be induced by UV light or physical damage to the cells (Liu et al., 1993a,b; Logemann et al., 1995; Negrel and Javelle, 1995; Nojiri et al., 1996). Modern genetic engineering techniques may possibly be used in plant cell cultures to produce large amounts of flavonoids either in culture or in altered plants (Zenk, 1991; Kuc, 1995). Several cell lines have already been shown to produce and accumulate flavonoids that are different than those normally found in mature plants, some of which have interesting biocidal properties (Yamamoto et al., 1993).

26.4   FLAVONOID CHEMICAL ECOLOGY

26.4.1   FLAVONOIDS IN ALLELOPATHIC INTERACTIONS

The term allelopathy was originally coined by Molisch (1937) to refer to biochemical interactions, both positive and negative, between plants and plant-associated microorganisms. Classically, this term has been used to describe the effect of extracellular compounds produced by a plant that inhibit the growth of neighboring plants. These allelochemicals can therefore be termed as a plant's chemical system for interacting with challenges from other plants in the environment (Rice, 1984). This definition has been expanded by some authors to cover compounds that are produced by plants that have inhibitory/deleterious or beneficial effects on other living things, be it other plants, insects, microorganisms, or herbivores (Waller, 1987; Waller and Chou, 1989).

Flavonoids appear to act primarily as germination and cell growth inhibitors, possibly through interference with the energy transfer system within the plant cell (Stenlid, 1970; Moreland and Novitsky, 1987, 1988). Flavones have been shown to interfere with adenoside triphosphate (ATP) formation in plant mitochondria (Stenlid, 1970, 1976). There are specific structural requirements for particular flavonoids to act as stimulators of the destruction of indoleacetic acid via IAA oxidase, which results in the inhibition of ATP formation (Stenlid, 1976). They may also act to inhibit or interfere with the mode of action of other plant hormones such as the auxins (Jacobs and Rubery, 1988; Brunn et al., 1992).

There are several recent publications concerning the allelopathic effects of flavonoids. Flavonoids are actively excreted by the roots of many plants and are released from leaf litter (Rao,

1990). Castaneda and coworkers (1992) found that flavones from *Celaenodendron mexicanum* Stand. inhibited the growth of seeds and shoots of *Amaranthus* and *Echinchloa* spp. A flavonoid from *Tithonia diversifolia* A. Gray inhibited germination of radish *(Raphinus sativa* L.*),* cucumber *(Cucumis sativus* L.*),* and onion *(Allium cepa* L.) seeds (Baruah et al., 1994). Several soybean *(Glycine max* [L.] Merr.) flavonoids were shown to have plant growth inhibitory properties (Porter et al., 1986). Flavanone, flavone, and isoflavone glycosides excreted by the weed *Pluchea lanceolata* (DC.) C. B. Clarke inhibited mustard *(Brassica spp.)* root and shoot growth (Inderjit and Dakshini, 1991, 1992), and inhibited the growth of aspargus *(Asparagus officinalis* L.) seedlings (Inderjit and Dakshini, 1995). Flavones from quackgrass *(Agropyron repens* [L.] Beauv.) inhibited radicle growth in eight crop and weed species (Weston et al., 1987). Soil accumulation of isoflavonoids released by red clover *(Trifolium pratense* L.) have also been implicated in a disease termed "red clover sickness"(Tamura et al., 1969). Flavonoids found in the seed coat of velvetleaf inhibited the germination and radicle elongation of the seeds of cress *(Lepidium sativum* L.*),* radish *(Raphinus sativa* L.*),* and soybean *(Glycine max* [L.] Merr.) (Paszkowski and Kremer, 1988).

26.4.2 PLANT–MICROBIAL INTERACTIONS

Plants produce a great number of compounds in several different chemical families that have antimicrobial properties (Grayer and Harborne, 1994). They can be either accumulated in normal, unstressed tissues or induced by the damage caused by microbial invasion of plant tissues. The term phytoalexin was first used by Müller and Börger (1941) to describe the specific chemicals that are produced by plants after they are challenged by microbial invaders. By the definition of Müller and Börger, phytoalexins are produced only in live cells after plant tissues come in contact with the microbe. Later this terminology was broadened to include not only induced chemicals, but those formed constitutively and stored in plant cells (Bailey and Mansfield, 1982; VanEtten et al., 1994). The mechanism of induction of phytoalexins has been covered in several reviews as noted above. Grayer and Harborne (1994) surveyed the published literature since 1982 concerning constitutive and induced compounds. In their two tables of over 160 examples, 17 of the active compounds were flavonoids. Where it was once thought that a certain plant family produced only a single chemical class of phytoalexins, the recent data suggest that several different classes of phytoalexins may be produced within the same family of plants (Grayer and Harborne, 1994).

26.4.3 PLANT–FUNGAL INTERACTIONS

Alfalfa *(Medicago sativa)* isoflavones were shown to inhibit the growth of *Phytophthora megasperma*, but were not as effective against four other alfalfa fungal pathogens (Blount et al., 1992). The induction of soybean *(Glycine max)* phytoalexins has been extensively studied (Ebel et al., 1984; Ebel and Grisebach, 1988; Todd and Vodkin, 1993; Uhlmann and Ebel, 1993; Miller et al., 1994). Several flavanones, flavonols, flavones, and isoflavones from soybean, including coumesterol, biochanin A, genistein, naringenin, isorhamnetin, and quercetin inhibited the growth of several *Phytophthora* species that cause root rot at concentrations of 60 to 240 micromolar (Ebel and Grisebach, 1988). Flavonoid aglycones are generally more active than glycones (Rivera et al., 1993). Flavonoid aglycones from the bark of *Prunus* species were shown to inhibit mycelial growth of the perennial canker fungus *Cytospora persoonii* (Geibel, 1995). A dimethoxyflavone from peanuts *(Arachis hypogaea* L.) inhibited the growth of *Aspergillus flavus* (Turner et al., 1975). Luteolin, eriodictyol, and 5,7-dihydroxychromone (DHC) from peanut shells lowered aflatoxin production in *Aspergillus parasiticus* cultures (DeLucca et al., 1987). DHC from peanut shells was also shown to inhibit the soil pathogens *Rhizoctonia solani* and *Sclerotium rolfsii* at biological concentrations (Vaughn, 1995). Legume seeds treated with methylisoflavans can reduce the level of infection by *Aspergillus* spp. (Weidenborner et al., 1990, 1992; Weidenborner and Jha, 1993). Flavonoids isolated from velvetleaf seed coats inhibited the growth of *Aspergillis niger, Penicillium diversum,* and *Fusarium* sp., but not *Gliocladium roseum* or *Trichoderma viride* (Paszkowski and Kremer, 1988). The inhibited fungal species are all phytopathogenic, while the latter two species are considered beneficial. The effect of the phytoalexins appears to depend on both concentration and the species of fungi involved. The accumulation of induced flavonoids, induction of suberin production, and the constitutive accumulation of gallocatechin seem to play a role in the ability of grapevine callus cultures to resist infection by downy mildew (*Plasmopara viticola*) (Dai et al., 1995a, b).

A constitutively expressed tryptophan decarboxylase gene was introduced into a potato (*Solanum tuberosum* L.) cultivar, causing an over-accumulation of tryptamine at the expense of phenylalanine and other phenolic compounds derived from phenylalanine (Yao et al., 1995). This potato cultivar had significantly lower levels of all phenolics, including lignin. The induction of phenolic phytoalexins were also drastically reduced and the transgenic cultivar tubers were much more susceptible to invasion by the late blight pathogen *Phytophthora infestans* than the untransformed cultivar.

The induction of the synthesis of the phytoalexin flavonoids kievitone, phaseollinisoflavan, and phaseollin in common bean (*Phaseolus vulgaris* L.) by root rot pathogens (*Pythium* spp.) have been studied (Liu et al., 1995). This response does not seem to be affected by the application of glyphosate (an inhibitor of an early step in aromatic amino acid synthesis), suggesting that enzymes subsequent to biosynthesis of phenylalanine are induced.

In sorghum, deoxyanthocyanidin flavonoids are specifically accumulated in special subcellular bodies in cells around wound sites on plants during invasion by the fungal pathogen *Colletotrichum graminicola*, the causal agent of sorghum anthracnose. Levels of these compounds are much higher in resistant cultivars than in the susceptible cultivars (Snyder and Nicholson, 1990). 3-Deoxyanthocyanidins from sorghum were found in higher concentrations in plant cultivars resistant to *Colletotrichum graminicola* (Tenkouano et al., 1993). It has been shown that new proteins involved in the synthesis of flavonoids in parsley cell cultures are induced by either UV-irradiation or treatment with fungal elicitors (Chappell and Hahlbrock, 1984). Other types of stress, including treatment with metals or metal salts such as cupric chloride can also induce this response (Hanawa et al., 1992). Sakuranetin, a flavanone isolated from UV-irradiated rice (*Oryza* spp.) leaves, has been shown to be an effective inhibitor of spore germination of *Pyricularia oryzae*, and was also found in higher levels in induced rice cultivars resistant to the fungi (Kodama et al., 1992). Flavan-3-ol appears to play a role in the resistance of *Malus* species to the causal agent of apple scab, *Venturia inaequalis* (Sierotzki and Gessler, 1993). Several different types of flavonoids were shown to have fungicidal activity on four fungi that infect stored grain (Weidenborner and Jha, 1993). The unsubstituted flavones were shown to have the best activity but the results were highly dependent on the fungal species tested. Some fungal pathogens also have the ability to metabolize plant phytoalexins to nontoxic forms, depending on phytoalexin concentration (Soby et al., 1996).

26.4.4 PLANT–BACTERIAL INTERACTIONS

Recent work on the mechanism of induction by bacterial attack was published by Jakobeck (1993). Flavonoids appear to play a signaling role in the induction of the accumulation of syringomycin in sweet cherry (*Prunus avium*) that is triggered by the invasion of *Pseudomonas syringae* (Mo et al., 1995). Antibacterial flavones and flavonols have been found in *Hypericum brasiliense* that were inhibitory to *Bacillus subtilis* (Rocha et al., 1995). Isorhamnetin and quercetin glucosides isolated from onion have antibacterial activity against *Pseudomonas cepacia* (Omidiji and Ehimidu, 1990). Flavonoids and conjugated polyphenols derived from flavonoids in tea inhibit the growth of food-borne bacteria including *Clostridium* spp. (Fukai et al., 1991). An acylated kempferol rhamnoside from *Pentachondra pumila* was shown to inhibit the growth of *Staphococcus aureus*. A recent paper has reported two flavones from *Artemisia giraldii* that show antibiotic activity against five bacterial and two fungal species (Zheng et al., 1996).

26.4.5 PLANT–VIRAL INTERACTIONS

There has been limited work on the effects of flavonoids on viruses. Several flavonoids that appear to be involved in resistance to viruses include flavonols in the resistance of mung beans (*Vignia radiata*) to yellow mosaic virus (Sohal and Bajaj, 1993), methoxyflavones and methoxyflavonols in the resistance of tobacco (*Nicotiana tabacum*) to the mosaic virus (French et al., 1991), and flavonols in the resistance of *Chenopodium quinoa* to potato virus X (French and Towers, 1992).

26.4.6 PLANT–INSECT INTERACTIONS

Plants produce many compounds that have been shown to deter insect feeding, acting either as metabolic poisons or feeding deterrents (Brattsten, 1986; Klocke, 1987; Nahrstedt, 1989; Jain and Tripathi, 1993). While the majority of the most active compounds are found in chemical classes other than flavonoids, several insects have been shown to be sensitive to flavonoids in feeding tests, including the fall armyworm (*Spodoptera frugiperda*) (flavones) (Gueldner et al., 1992; Wheeler et al., 1993); the spotted stalk borer (flavones) (Machocho et al., 1995); corn earworm (*Helicoverpa zea*) (flavonoid C-glycosides) (Shaver and Lukefahr, 1969; Waiss et al., 1980; Gueldner et al., 1992); Mexican bean beetle (*Epilachua varivestis*) (isoflavones from soybean) (Burden and Norris, 1992); tobacco budworm (*Heliothis virescens*) (flavonols and flavones) (Shaver and Lukefahr, 1969; Waiss et al., 1980; Hedin et al., 1988, 1992; Hedin and McCarthy, 1990); western corn rootworm (*Diabrotica virgifera*) (flavones and flavonols) (Mullin et al., 1991); pink bollworm (*Pectinophora gossypiella*) (flavonols) (Shaver and Lukefah, 1969); cabbage looper (*Trichoplusia ni*) (isoflavones) (Sharma and Norris, 1991); boll weevil (*Anthonomus grandisa*) (methoxyflavones) (Miles et al., 1993); lepidopterous larvae (*Spodoptera littoralis* and *S. exempta*) (several flavonoids and other secondary metabolites) (Simmonds et al., 1990; Pandji et al., 1993); mosquito larvae (*Streptomyces* spp.) (methoxyflavones and isoflavones) (Rao et al., 1990); *Dione juno* larvae (methoxyflavones) (Echeverri et al., 1991); spotted stalk borer (*Chilo partellus*) (three flavonoids) (Machocho et al.,1995); and pasture scarab larvae (*Costelytra zealandica* and *Heteronychus arator*) (isoflavones) (Lane et al., 1987).

As with the microbes, insects have also adapted uses for the plant flavonoids. Some insects recognize plants they feed on by the chemicals the plants produce. In the case of some butterflies, the female will determine where to lay its eggs by detecting the presence of a certain flavonoid glycosides (Brattsten, 1986; Nishida et al., 1987; Feeney et al., 1988; Nahrstedt, 1989) The flavonoids can also provide protection from fungal pathogens carried by insects (Brignolas et al., 1995).

The flavonoids may act by inhibiting important P-450–dependant enzyme systems in the insects, such as the sterol hydroxylases (Mitchell et al., 1993). Adaptation by insects include items previously mentioned and the ability to digest the offending compounds (Nahrstadt, 1989), such as transferring glutathione groups to inactivate the toxic effects of flavonoids (Yu, 1992).

26.5 OTHER FUNCTIONAL ROLES OF FLAVONOIDS

In addition to the important chemodefensive roles discussed above, the flavonoids have been implicated in a number of functional roles in both plants and animals. Flavonoids are found in nearly all plants and have probably been around nearly as long as plants themselves. The flavonoids may have developed in plants as protection from harmful UV radiation during the period when the Earth's atmosphere was developing its protective ozone layer (Kubitzki, 1987). Flavonoids have been further adapted for a number of other uses by the plants and the animals that consume them. In plants they appear to have very diverse functions, including functioning as antioxidants, superoxide radical scavengers, chelators mediating mineral uptake, enzyme inhibitors and regulators, redox cofactors, and pigmented color attractants for pollination and seed dispersal mechanisms by insects and animals (McClure, 1986).

Flavonoids have been adapted for use as host-symbiote recognition signals, such as inducing the species-specific expression of nodulation genes in free-living *Rhizobium* with legumes. This function has been extensively reviewed (Djordjevic and Weinman, 1991; Fisher and Long, 1992; Phillips, 1992; Phillips and Tsai, 1992; Gagnon et al., 1995).

Flavonoids have been shown to be utilized by plants for specific physiological functions. They have also been shown to be essential for the formation of pollen tubes in the pistil containing germinating pollen. Recent evidence suggests that phenolics may play an important role in the regulation of plant metabolism. For example, flavonoids have been shown to be naturally occurring auxin transport regulators (Jacobs and Rubery, 1988); they may be involved in the signal transduction process mediating the translation of externally perceived signals into new protein synthesis. They may also be involved either directly or indirectly in other phytohormonal systems.

26.6   SUMMARY

It is evident that flavonoids play a major role in plants. Although much basic research remains to be carried out, it is possible that many of these compounds, either as isolates or in conjunction with other compounds, may be used in either agricultural or pharmaceutical roles. The use of flavonoids and other plant phenolics as biopesticides, either as an applied pesticide or through genetically engineered plants, has been widely discussed (Swain, 1977; Bell, 1981; Friend and Rathmell, 1984; Clark et al., 1985; Cutler et al., 1986; McClure, 1986; Harborne, 1986, 1988; Laks and Pruner, 1988; Waller, 1989; Dixon and Lamb, 1990; Waterman, 1990; Anderson, 1991; Nicholson and Hammerschmidt, 1992; Ryals et al., 1994; Dakora, 1995). Developing an understanding of the distribution of flavonoids in plants will provide an assessment of the diversity of these compounds in the plant kingdom; developing a better understanding of their functional roles in plants may lead to new biologically compatible pesticides that could be controlled or triggered by manipulation of the flavonoid pathway in the plant.

Researchers doing basic work on the mechanism of plant defense and secondary metabolite accumulation are optimistic about the practical applications resulting from this line of research (Chasen, 1994). However, the use of naturally occurring secondary metabolites as alternatives for pest control is not without pitfalls. A number of reviewers have pointed out problems that could arise from the increased use of natural products either directly applied in sufficiently effective levels or from genetically altered plants producing higher levels of these natural pesticides (Kuc, 1995). Although there is significant emerging evidence that the natural levels of flavonoids have human health benefits, relatively little is known about the chronic effects of these compounds in higher doses (Middleton, 1988; Fenwick et al., 1990; Pathak et al., 1991). Conventional breeding programs have produced potato, celery, and melon varieties that were more resistant to storage pathogens and insects, but could not be commercially released because of higher levels of alkaloids, psoralens, or cucurbitacins, which were either toxic to humans, produced allergenic responses, or were inedible due to increased bitterness (Fenwick et al., 1990). While it may be possible to offset some of these effects, potential underlying biochemical problems must be considered when embarking on a plan to implement natural biopesticides commercially. Above all, a full understanding of the underlying mechanisms involved in the biochemistry both in the plants and in the human diet must be considered when carrying out this type of research. More effort is required in very basic research in this field, thus integrating the information to gain a full understanding of the complex interactions between plants, pests, and human physiology in order to yield usable biopesticide strategies.

REFERENCES

Anderson, A. J., Phytoalexins and plant resistance, in *Mycotoxins and Phytoalexins,* Sharma, R. P. and Salunkhe, D. K., Eds., CRC Press, Boca Raton, LA, 1991, 569.

Bailey, J. A. and Mansfield, J. W., *Phytoalexins,* John Wiley & Sons, New York, 1982.

Baruah, N. C., Sarma, J. C., Barua, N. C., Sarma, S. and Sharma, R. P., Germination and growth inhibitory sesquiterpene lactones and a flavone from *Tithonia diversifolia, Phytochemistry,* 36, 29, 1994.

Bell, E. A., The physiological role(s) of secondary (natural) products, in *Secondary Plant Products,* Stumpf, P. K. and Conn, E. E., Eds., Academic Press, New York, 1981, 1.

Blount, J. W., Dixon, R. A., and Paiva, N. L., Stress responses in alfalfa (*Medicago sativa* L.) XVI. Antifungal activity of medicarpin and its biosynthetic precursors; implications for the genetic manipulation of stress metabolites, *Physiol. Mol. Plant Pathol.,* 41, 333, 1992.

Bohnert, H. J., Nelson, D. E., and Jensen, R. G., Adaptations to environmental stresses, *Plant Cell,* 7, 1089, 1995.

Brattsten, L. B., Fate of ingested plant allelochemicals in herbivorous insects, in *Molecular Aspects of Insect–Plant Interactions,* Brattsten, L. B. and Ahmand, S., Eds., John Wiley & Sons, New York, 1986, 211.

Brignolas, F., Lacroix, B., Lieutier, F., Sauvard, D., Drouet, A., Claudot, A.-C., Yart, A., Berryman, A. A., and Christiansen, E., Induced responses in phenolic metabolism in two Norway spruce clones after wounding and inoculations with *Ophiostoma polonicium,* a bark beetle-associated fungus, *Plant Physiol.,* 109, 821, 1995.

Brunn, S. A., Muday, G. K., and Haworth, P., Auxin transport and the interactions of phytotropins, *Plant Physiol.,* 98, 101, 1992.

Burden, B. J. and Norris, D. M., Role of the isoflavonoid coumestrol in the constitutive antixenosic properties of "Davis" soybeans against an oligophagous insect, the Mexico bean beetle, *J. Chem. Ecol.,* 18, 1069, 1992.

Castaneda, P., Garcia, M. R., Hernandez, B. E., Torres, B. A., Anaya, A. L., and Mata, R., Effects of some compounds isolated from *Celaenodendron mexicanum* Standl (Euphorbiaceae) on seeds and phytopathogenic fungi, *J Chem. Ecol.,* 18, 1025, 1992.

Chappell, J. and Hahlbrock, K., Transcription of plant defense genes in response to UV light or fungal elicitor, *Nature,* 311, 76, 1984.

Chasen, R., Phytochemical forecasting, *Plant Cell,* 6, 3, 1994.

Chou, C. and Waller, G. R., *Phytochemical Ecology: Alleochemicals, Mycotoxins, and Insect Phermones and Allomones, Proceedings of Symposium held in 1988 at Taipei, Taiwan.* Institute of Botany, Academia Sinica, Taipei, Taiwan, 1989.

Clark, A. M., Hufford, C. D., El-Feraly, F. S., and McChesney, J. D., Antimicrobial agents from plants: a model for studies of allelopathic agents?, in *The Chemistry of Allelopathy,* Thompson, A. C., Ed., American Chemical Society, Washington, D.C., 1985, 327.

Cordell, G. A., Changing strategies in natural products chemistry, *Phytochemistry,* 40, 1585, 1995.

Cutler, H. G., Severson, R. F., Cole, P. D., Jackson, D. M., and Johnson, A. W., Secondary metabolites from higher plants: their possible role as biololgical control agents, in *Natural Resistance of Plants to Pests,* Green, M. B., and Hedin, P. A., Eds., American Chemical Society, Washington, D.C., 1986, 178.

Dai, G. H., Andary, C., Mondolot-Cosson, L., and Boubals, D., Histochemical studies on the interaction between three species of grapevine, *Vitis vinifera, V. rupestris* and *V. rotundifolia* and the downy mildew fungus, *Plasmopara viticola, Physiol. Mol. Plant Pathol.,* 46, 177, 1995a.

Dai, G. H., Andary, C., Mondolot-Cosson, L., and Boubals, D., Involvement of phenolic compounds in the resistance of grapevine callus to downy mildew *(Plasmopara viticola), Eur. J. Plant Pathol.,* 101, 541, 1995b.

Dakora, F. D., Plant flavonoids: biological molecules for useful exploitation, *Aust. J. Plant Physiol.,* 22, 87, 1995.

Darvell, A. G. and Albersheim, P., Phytoalexins and their elicitors—a defenses against microbial action in plants, *Ann. Rev. Plant Physiol.,* 35, 243, 1984.

DeLucca, A. J. I., Palmgren, M. S., and Daigle, D. J., Depression of aflotoxin production by flavonoid-type compounds from peanut shells, *Phytopathology,* 77, 1560, 1987.

Dixon, R. A. and Lamb, C. J., Molecular communication in interactions between plants and microbial pathogens, *Ann. Rev. Plant Physiol. Plant Mol. Biol.,* 41, 339, 1990.

Dixon, R. A., Choudhary, A. D., Dalkin, K., Edwards, R., Fahrendorf, T., Gowri, G., Harrison, M. J., Lamb, C. J., Loake, G. J., Maxwell, C. A., Orr, J., and Paiva, N. L., Molecular biology of stress-induced phenylpropanoid and isoflavonoid biosynthesis in alfalfa, in *Phenolic Metabolism in Plants,* Stafford, H. A., and Ibrahim, R. K., Eds., Plenum Press, New York, 1992, 91.

Dixon, R. A., Harrison, M. J., and Lamb, C. J., Early events in the activation of plant defense responses, *Ann. Rev. Phytopath.,* 32, 479, 1994.

Dixon, R. A. and Paiva, N. L., Stress-induced phenylpropanoid metabolism, *Plant Cell,* 7, 1085, 1995.

Djordjevic, M. A. and Weinman, J. J., Factors determining host recognition in the clover-*Rhizobium* symbiosis, *Aust. J. Plant Physiol.,* 18, 543, 1991.

Douglas, C. J., Ellard, M., Hauffe, K. D., Molitor, E., d. Sá, M. M., Reinold, S., Subramaniam, R., and Williams, F., General phenylpropanoid metabolism: regulation by environmental and developmental signals, in *Phenolic Metabolism in Plants,* Stafford, H. A., and Ibrahim, R. K., Eds., Plenum Press, New York, 1992, 63.

Ebel, J., Phytoalexin synthesis: the biochemical analysis of the induction process, *Ann. Rev. Phytopathol.,* 24, 235, 1986.

Ebel, J. and Grisebach, H., Defense strategies of soybean against the fungus *Phytophthora megasperma* f. sp. *glycinea*: a molecular analysis, *Trends Biochem. Sci.,* 13, 1988.

Ebel, J., Schmidt, W. E., and Loyal, R., Phytoalexin synthesis in soybean cells: elicitor induction of phenylalanine ammonia-lyase and chalcone synthase mRNAs and correlation with phytoalexin accumulation, *Arch Biochem Biophys.,* 232, 240, 1984.

Echeverri, F., Cardona, G., Torres, F., Pelaez, C., Quinones, W., and Renteria, E., Ermanin: an insect deterrent flavonoid from *Passiflora foetida* resin, *Phytochemistry,* 30, 153, 1991.

Feeney, P., Sachdev, K., Rosenberry, L., and C. M., Luteolin 7-O-(6″-O-malonyl)-β-D-glucose and trans-chlorogenic acid: oviposition stimulants for the Black Swallowtail butterfly, *Phytochemistry* 27, 3439, 1988.

Fenwick, G. R., Johnson, I. T., and Hedley, C. L., Toxicity of disease-resistant plant strains, *Trends Food Sci. Tech.,* 1, 23, 1990.

Fisher, R. F. and Long, S. R., *Rhizobium*–plant signal exchange, *Nature,* 357, 655, 1992.

French, C. J., Elder, M., Leggett, F., Ibrahim, R. K., and Towers, G. H. N., Flavonoids inhibit infectivity of tobacco mosaic virus, *Can. J. Plant Pathol.,* 13, 1, 1991.

French, C. J. and Towers, G. H. N., Inhibition of infectivity of potato virus X by flavonoids, *Phytochemistry,* 31, 3017, 1992.

Friend, J. and Rathmell, W. G., *Phytoalexins, Fourth International Symposium on Antibiotics and Agriculture,* University of Hull, Butterworth, London, England, 1984.

Fukai, K., Ishigami, T., and Hara, Y., Antibacterial activity of tea polyphenols against pytopathogenic bacteria, *Agric. Biol. Chem.,* 55, 1895, 1991.

Gagnon, H., Grandmaison, J., and Ibrahim, R. K., Phytochemical and immunochemical evidence for the accumulation of 2′-hydroxylupalbigenin in lupin nodules and bacteriods, *Mol. Plant-Microbe Interactions,* 8, 131, 1995.

Geibel, M., Sensitivity of the fungus *Cytospora persoonii* to the flavonoids of *Prunus cerasus, Phytochemistry,* 38, 599, 1995.

Geissman, T. A., *The Chemistry of Flavonoid Compounds,* Pergamon Press, New York, 1962.

Grayer, R. J. and Harborne, J. B., A survey of antifungal compounds from higher plants, 1982–1993, *Phytochemistry,* 37, 19, 1994.

Gueldner, R. C., Snook, M. E., Widstrom, N. W., and Wiseman, B. R., TLC screen for maysin, chlorogenic acid, and other possible resistance factors to the fall armyworm and the corn earworm in *Zea mays, J. Agric. Food Chem.,* 40, 1211, 1992.

Hahlbrock, K. and Scheel, D., Physiology and biochemistry of phenylpropanoid metabolism, *Ann. Rev. Plant Physiol. Plant Mol. Biol.,* 40, 347, 1989.

Hanawa, F., Tahara, S., and Mizutani, J., Antifungal stress compounds from *Veratrum grandiflorum* leaves treated with cupric chloride, *Phytochemistry,* 31, 3005, 1992.

Harborne, J. B., The role of phytoalexins in natural plant resistance, in *Natural Resistance of Plants to Pests,* Green, M. B., and Hedin, P. A., Eds., American Chemical Society, Washington, D.C., 1986, 22.

Harborne, J. B., Flavonoids in the environment: structure-activity relationships, in *Plant Flavonoids in Biology and Medicine II,* Cody, V., Middleton, E., Harborne, J. B., and Beretz, A., Eds., Alan R. Liss, New York, 1988, 17.

Harborne, J. B., *The Flavonoids: Advances in Research Since 1986*, Chapman and Hall, London, 1994.

Harborne, J. B. and Mabry, T. J., *The Flavonoids: Advances in Research,* Chapman and Hall, London, 1982.

Harborne, J. B., Mabry, T. J., and Mabry, H., *The Flavonoids*, Academic Press, New York, 1975.

Harrison, M. J. and Dixon, R. A., Spatial patterns of expression of flavonoid/isoflavonoid pathway genes during interactions between roots of *Medicago truncatula* and the mycorrhizal fungus *Glomus versiforme, Plant J.,* 6, 9, 1994.

Hedin, P. A., Jenkins, J. N., Thompson, A. C., McCarty, J. C., Jr., Smith, D. H., Parrott, W. L., and Shepherd, R. L., Effects of bioregulators on flavonoids, insect resistance, and yield of seed cotton, *J. Agric. Food Chem.*, 36, 1055, 1988.

Hedin, P. A., Jenkins, J. N., and Parrott, W. L., Evaluation of flavonoids in *Gossypium arboreum* (L.) cottons as potential source of resistance to tobacco budworm, *J. Chem. Ecol.*, 18, 105, 1992.

Hedin, P. A. and McCarthy, J. C., Possible roles of cotton bud sugars and terpenoids in oviposition by the boll weevil, *J. Chem. Ecol.*, 16, 757, 1990.

Heller, W. and Forkmann, G., Biosynthesis of flavonoids, in *The Flavonoids, Advances in Research Since 1986*, Harborne, J. B., Ed., Chapman and Hall, London, 1993, 499.

Hrazdina, G. and Jensen, R. A., Spatial organization of enzymes in plant metabolic pathways, *Annu. Rev. of Plant Physiol. and Plant Mol. Biol.*, 43, 241, 1992.

Inderjit and Dakshini, K. M. M., Hesperetin 7-rutinoside (hesperidin) and taxifolin 3-arabinoside as germination and growth inhibitors in soils associated with the weed, *Pluchea lanceolata* (DC) C.B. Clarke (Asteraceae), *J Chem. Ecol.*, 17, 1585, 1991.

Inderjit and Dakshini, K. M. M., Formononetin 7-O-glucoside (ononin), an additional growth inhibitor in soils associated with the weed, *Pluchea lanceolata* (DC) C.B. Clarke (Asteraceae), *J. Chem. Ecol.*, 18, 713, 1992.

Inderjit and Dakshini, K. M. M., Alleopathic effect of *Pluchea lanceolata* (Asteraceae) on characteristics of four soils and tomato and mustard growth, *Am. J. Bot.*, 81, 799, 1994.

Inderjit and Dakshini, K. M. M., Quercetin and quercitrin from *Pluchea lanceolata* and their effect on growth of asparagus bean, in *Alleopathy: Organisms, Processes, and Applications*, Inderjit, Dakshini, K. M. M., and Einhelling, F. A., Eds., ACS Symposium Series 582, American Chemical Society, Washington, D.C., 1995.

Jacobs, M. and Rubery, P. H., Naturally occuring auxin transport regulators, *Science*, 241, 346, 1988.

Jain, D. C. and Tripathi, A. K., Potential of natural products as insect antifeedants, *Phytother. Res.*, 7, 327, 1993.

Jakobeck, J. L. and Lindgren, P. B., Generalized induction of defense responses in bean is not correlated with the induction of the hypersensitive response, *Plant Cell*, 5, 49, 1993.

Klocke, J. A., Natural compounds useful in insect control, in *Alleochemicals: Role in Agriculture and Forestry*, Waller, G. R., Eds., American Chemical Society, Washington, D.C., 1987, 397.

Kobayashi, A., Tai, A., Kanzaki, H., and Kawazu, K., Elicitor-active oligosaccharides from algal laminaran stimulate the production of antifungal compounds in alfalfa, *Z. Naturfors. C.*, 48, 575, 1993.

Kodama, O., Miyakawa, J., Akatsuka, T., and Kiyosawa, S., Sakuranetin, a flavanone phytoalexin from ultraviolet-irradiated rice leaves, *Phytochemistry*, 31, 3807, 1992.

Kubitzki, K., Phenylpropanoid metabloism in relation to land plant origion and diversification, *J. Plant Physiol.*, 131, 17, 1987.

Kuc, J., Phytoalexins, stress metabolism, and disease resistance in plants, *Ann. Rev. Phytopathol.* 33, 275, 1995.

Laks, P. E. and Pruner, M. S., Flavonoid biocides: structure/activity relations of flavonoid phytoalexin analogues, *Phytochemistry*, 28, 87, 1988.

Lamb, C. J., Lawton, M. A., Dron, M., and Dixon, R. A., Signals and transduction mechanisims for activation of plant defenses against microbial attack, *Cell*, 56, 215, 1989.

Lane, G. A., Sutherland, O. R. W., and Skipp, R. A., Isoflavonoids as insect feeding deterrents and antifungal components from root of *Lupinus angustifolius*, *J. Chem. Ecol.*, 13, 771, 1987.

Liu, L., Punja, Z. K., and Rahe, J. E., Effect of *Pythium* spp. and glyphosate on phytoalexin production and exudation by bean (*Phaseolus vulgaris* L.) roots grown in different media, *Physiol. Mol. Plant Pathol.*, 47, 1995.

Liu, Q., Dixon, R. A., and Mabry, T. J., Additional flavonoids from elecitor-treated cell cultures of *Cephalocereus senilis*, *Phytochemistry*, 34, 167, 1993a.

Liu, Q., Markham, K. R., Paré, P. W., Dixon, R. A., and Mabry, T. J., Flavonoids from elicitor-treated cell suspension cultures of *Cephalocereus senilis*, *Phytochemistry*, 32, 925, 1993b.

Logemann, E., Wu, S.-C., Schroder, J., Schmetzer, E., Somssich, I. E., and Hahlbrock, K., Gene activation by UV light, fungal elicitor or fungal infection in *Petroselinum crispum* is correlated with repression of cell cycle-related genes, *Plant J.*, 8, 865, 1995.

Machocho, A. K., Lwande, W., Jondiko, J. I., Moreka, L. V. C., and Hassanali, A., Three new flavonoids from the root of *Tephrosia emoroides* and their antifeedant activity against the larvae of the spotted stalk borer *Chilo partellus* swinhoe, *Int. J. Pharmacogn.*, 33, 222, 1995.

McClure, J. W., Physiology of flavonoids in plants, in *Plant Flavonoids in Biology and Medicine*, Cody, V., Middleton, E., and Harborne, J. B., Eds., Alan R. Liss, New York, 1986, 77.

Middleton, E. J., Plant flavonoid effects on mammalian cell systems, in *Herbs, Spices and Medicinal Plants,* Craker, L. E., and Simon, J. E., Eds., Oryx Press, Phoenix, AZ, 1988, 103.

Miles, D. H., Tunsuwan, K., Chittawong, V., Kokpol, U., Choudhary, M. I., and Clardy, J., Boll weevil antifeedants from *Arundo donax, Phytochemistry,* 34, 1277, 1993.

Miller, K. J., Hadley, J. A., and Gustine, D. L., Cyclic β-1,6-1,3-glucans of *Bradyrhizobium japonicum* USDA 110 elicit isoflavonoid production in the soybean *(Glycine max)* host, *Plant Physiol.,* 104, 917, 1994.

Mitchell, M. J., Keogh, D. P., Crooks, J. R., and Smith, S. L., Effects of plant flavonoids and other allelochemicals on insect cytochrome P-450 dependent steroid hydroxylase activity, *Insect Biochem. Mol. Biol.,* 23, 65, 1993.

Mo, Y.-Y., Geibel, M., Bonsall, R. F., and Gross, D. C., Analysis of sweet cherry (*Prunus avium* L.) leaves for plant signal molecules that activate the syrB gene required for the synthesis of the phytotoxin, syringomycin, by *Pseudomonas syringae* pv. *syringe, Plant Physiol.,* 107, 603, 1995.

Molisch, J., *Der Einfluss einer Pflanze auf die andere-Allelopathie,* Jena, Germany, 1937.

Moreland, D. E. and Novitsky, W. P., Effects of phenolic acids, coumarins, and flavonoids onisolated choroplasts and mitochondria, in *Allelochemicals: Role in Agriculture and Forestry,* Waller, G. R., Ed., ACS Symposium Series 330, American Chemical Society, Washington, D.C., 1987.

Moreland, D. E. and Novitsky, W. P., Interference by flavone and flavonols with chloroplast mediated electron transport and phosphorylation, *Phytochemistry,* 27, 3359, 1988.

Müller, K. O. and Börger, H., *Arb. Biol. Abt. (Ansl.-Reichstanst.), Berlin,* 23, 189, 1941.

Mullin, C. A., Alfatafa, A. A., Harman, J. L., Everett, S. L., and Serino, A. A., Feeding and toxic effects of floral sesquiterpene lactones, diterpenes, and phenolics from sunflower (*Helianthus annuus* L.) on western corn rootworm, *J. Agric. Food Chem.,* 39, 2293, 1991.

Nahrstedt, A., The significance of secondary metabolites for the interactions between plants and insects, *Planta Medica,* 55, 333, 1989.

Negrel, J. and Javelle, F., Induction of phenylpropanoid and tyramine metabolism in pectinase- or pronaseelicited cell suspension cultures of tobacco *(Nicotiana tabacum), Physiol. Plant.,* 95, 569, 1995.

Nicholson, R. L. and Hammerschmidt, R., Phenolic compounds and their role in disease resistance, *Ann. Rev. Phytopathol.,* 30, 369, 1992.

Nishida, R., Obsungi, T., Kokubo, S., and Fukami, H., Oviposition stimulents of a citrus-feeding swallowtail butterfly, *Papilio xuthus* L., *Experientia,* 43, 342, 1987.

Nojiri, H., Sugimori, M., Yamane, H., Nishimura, Y., Yamada, A., Shibuya, N., Kodama, O., Murofushi, N., and Omori, T., Involvement of jasmonic acid in elicitor-induced phyoalexin production in suspension-cultured rice cells, *Plant Physiol.,* 110, 387, 1996.

Omidiji, O. and Ehimidu, J., Changes in the content of antibacterial isorhamnetin 3-glucoside and quercetin 3-glucoside following inoculation of onoin (*Allium cepa* L. cv. Red Creole) with *Pseudomonas cepacia, Physiol. Mol. Plant Pathol.,* 37, 281, 1990.

Pandji, C., Grimm, C., Wray, V., Witte, L., and Proksch, P., Insecticidal constituents from four species of the Zingiberaceae, *Phytochemistry,* 43, 415, 1993.

Paszkowski, W. L. and Kremer, R. J., Biological activity and tentative idenfitication of flavonoid components in velvetleaf (*Abutilon theophrasti* Medik.) seed coats, *J. Chem. Ecol.,* 14, 1573, 1988.

Pathak, D., Pathak, K., and Singla, A. K., Flavonoids as medicinal agents: recent advances, *Fitoterapia,* 62, 371, 1991.

Phillips, D. A., Flavonoids: plant signals to soil microbes, in *Phenolic Metabolism in Plants,* Stafford, H. A., and Ibrahim, R. K., Eds., Plenum Press, New York, 1992, 201.

Phillips, D. A. and Tsai, S. M., Flavonoids as plant signals to rhizosphere microbes, *Mycorrhiza,* 1, 55, 1992.

Porter, P. M., Banwart, W. L., and Hassett, J. J., Phenolic acids and flavonoids in soybean root and leaf extracts, *Environ. Exp. Bot.,* 26, 65, 1986.

Rao, A. S., Root flavonoids, *Bot. Rev.,* 56, 90, 1990.

Rao, K. V., Chattopadhyay, S. K., and Reddy, G. C., Flavonoids with mosquito larval toxicity, *J. Agric. Food Chem.,* 38, 1427, 1990.

Rice, E. L., *Allelopathy,* Academic Press, Orlando, FL, 1984.

Rivera Vargas, L. I., Schmitthenner, A. F., and Graham, T. L., Soybean flavonoid effects on and metabolism by *Phytophthora sojae, Phytochemistry,* 32, 851, 1993.

Rocha, L., Marston, A., Potterat, O., Kaplan, M. A. C., Stoeckli-Evans, H., and Hostettmann, K., Antibacterial phloroglucinols and flavonoids from *Hypericum brasiliense, Phytochemistry,* 40, 1447, 1995.

Ryals, J., Uknes, S., and Ward, E., Systematic acquired resistance, *Plant Physiol.,* 104, 1109, 1994.

Seigler, D. S., Secondary metabolites and plant systematics, *in Secondary Plant Products,* Stumpf, P. K. and Conn, E. E., Eds., Academic Press, New York, 1981, 139.

Sharma, H. C. and Norris, D. M., Chemical basis of resistance in soyabean to cabbage looper, *Trichoplusia ni, J. Sci. Food Agric.,* 55, 353, 1991.

Shaver, T. N. and Lukefahr, M. J., Effect of flavonoid pigments and gossypol on growth and development of the bollworm, tobacco budworm, and pink bollworm, *J. Econ. Entom.,* 62, 643, 1969.

Sierotzki, H. and Gessler, C., Flavan-3-ol content and the resistance of *Malus x domestica* to *Venturia inaequalis* (Cke.) Wint., *Physiol. Mol. Plant Pathol.,* 42, 291, 1993.

Simmonds, M. S. J., Blaney, W. M., Monache, F. D., and Bettolo, G. B. M., Insect antifeedant activity associated with compounds isolated from species of *Lonchocarpus* and *Tephrosia, J Chem. Ecol.,* 16, 365, 1990.

Snyder, B. A. and Nicholson, R. L., Synthesis of phytoalexins in sorghum as a site-specific response to fungal ingress, *Science,* 248, 1637, 1990.

Soby, S., Caldera, S., Bates, R., and VanEtten, H., Detoxification of the phytoalexins maackiain and medicarpin by fungal pathogens of alfalfa, *Phytochemistry,* 41, 759, 1996.

Sohal, B. S. and Bajaj, K. L., Effects of yellow mosaic virus on polyphenol metabolism in resistant and susceptible mungbean (*Vigna radiata* L. Wilczek) leaves, *Biochem. Physiol. Pflanzen,* 188, 419, 1993.

Stafford, H. A., *Flavonoid Metabolism,* CRC Press, Boca Raton, FL, 1990.

Stenlid, G., Flavonoids as inhibitors of the formation of adenosine triphosphate in plant mitocondria, *Phytochemistry,* 9, 2251, 1970.

Stenlid, G., Effects of substituents in the A-ring on the physiological activity of flavones, *Phytochemistry,* 15, 911, 1976.

Swain, T., Secondary compounds as protective agents, *Ann. Rev. Plant Physiol.,* 28, 479, 1977.

Tamura, S., Chang, C.-F., Suzuki, A., and Kumai, S., Chemical studies on "clover sickness." I. Isolation and structural elucidation of two new isoflavonoids in red clover. *Agric. Biol. Chem.,* 33, 391, 1969.

Tenkouano, A., Miller, F. R., Hart, G. A., Frederiksen, R. A., and Nicholson, R. L., Phytoalexin assay in juvenile sorghum: an aid to breeding for anthracnose resistance, *Crop Sci.,* 33, 243, 1993.

Todd, J. J. and Vodkin, L. O., Pigmented soybean *(Glycine max)* seed coats accumulate proanthocyanidins during development, *Plant Physiol.,* 102, 663, 1993.

Turner, R. B., Lindsey, D. L., Davis, D. D., and Bishop, R. D., Isolation and identification of 5,7 dimethoxyisoflavone, an inhibitor of *Aspergillus flavus* from peanuts, *Mycopathologia,* 57, 39, 1975.

Uhlmann, A. and Ebel, J., Molecular cloning and expression of 4-coumarate:coenzyme A ligase, an enzyme involved in the resistance response of soybean *(Glycine max* L.) against pathogen attack, *Plant Physiol.,* 102, 1147, 1993.

VanEtten, H. D., Mansfield, J. W., Bailey, J. A., and Farmer, E. E., Two classes of plant antibiotics: phytoalexins versus "phytoanticipins," *Plant Cell,* 6, 1191, 1994.

Vaughn, S. F., Phytotoxic and antimicrobial activity of 5,7-dihyrochromone from peanut shells, *J. Chem. Ecol.,* 21, 107, 1995.

Waiss, A. C. J., Chan, B. G., Ellinger, C. A., Dreyer, D. L., Binder, R. G., and Gueldner, R. C., Insect growth inhibitors in crop plants: control of lepidopterous larval pests, *Bull. Entom. Soc. of Am.,* 27, 217, 1980.

Waller, G. R., *Allelochemicals: Role in Agriculture and Forestry,* ACS Symposium Series 330, American Chemical Society, Washington, D.C., 1987.

Waller, G. R., Biochemical frontiers of allelopathy, *Biologia Plantarum,* 31, 418, 1989.

Waterman, P. G., Searching for bioactive compounds: various strategies, *J. Natur. Prod.,* 53, 13, 1990.

Weidenborner, M., Hindorf, H., Jha, H. C., and Tsotsonos, P., Antifungal activity of flavonoids against storage fungi of the genus *Aspergillus, Phytochemistry,* 29, 1103, 1990.

Weidenborner, M., Hindorf, H., Weltzien, H. C., and Jha, H. C., An effective treatment of legume seeds wioth flavonoids and isoflavonoids against storage fungi of the genus *Aspergillus, Seed Sci. Technol.,* 20, 447, 1992.

Weidenborner, M. and Jha, H. C., Antifungal activity of flavonoids and their mixtures against different fungi occurring on grain, *Pestic. Sci.,* 38, 347, 1993.

Weston, L. A., Burke, B. A., and Putnam, A. R., Isolation, characterization and activity of phytotoxic compounds from quackgrass (*Agropyron repens* (L.) Beauv.), *J. Chem. Ecol.,* 13, 403, 1987.

Wheeler, G. S., Slansky, F., Jr., and Yu, S. J., Fall armyworm sensitivity to flavone: limited role of constitutive and induced detoxifying enzyme activity, *J. Chem. Ecol.,* 19, 645, 1993.

Yamamoto, H., Yan, K., Ieda, K., Tanaka, T., Iinuma, M., and Mizuno, M., Flavonol glycosides production in cell suspension cultures of *Vancouveria hexandra, Phytochemistry,* 33, 841, 1993.

Yao, K., De Luca, V., and Brisson, N., Creation of a metabolic sink for tryptophan alters the phenylpropanoid pathway and the susceptibility of potato to *Phytophthora infestans, Plant Cell,* 7, 1787, 1995.

Yu, S. J., Plant-allelochemical-adapted glutathione transferases in Lepidoptera, *in Molecular Mechanisms of Insecticide Resistance,* Mullin, C.A. and Scott, J. G., Eds., ACS Symposium Series No. 505, American Chemical Society, Washington, D.C., 1992, 174.

Zenk, M. H., Chasing the enzymes of secondary metabolism: plant cell cultures as a pot of gold, *Phytochemistry,* 30, 3861, 1991.

Zheng, W. F., Tan, R. X., Yang, L., and Liu, Z. L., Two flavones from *Artemisia giraldii* and their antimicrobial activity, *Planta Medica,* 62, 160, 1996.

27 Allelochemical Function of Coumarins on the Plant Surface

Alicja M. Zobel

CONTENTS

27.1 ABSTRACT

Plant coumarins play an important role in allelochemical interactions. The objective of this review is to discuss the allelochemical interactions of coumarins on the plant surfaces. The localization of coumarin on roots, shoots, seeds, and fruits has been discussed. Mechanisms of allelopathic interactions of coumarins have been outlined.

27.2 INTRODUCTION

Allelochemicals are compounds that, when synthesized by a plant, can influence the physiology of other species both by inhibiting and stimulating biochemical processes (Rice, 1984; Narwal, 1995). Such compounds can also influence the plant producing them, and if the influence is negative, the effect is known as autotoxity (Kil and Lee, 1987; Hegazy et al., 1990). Some such chemicals of plant origin can stimulate growth of the plant producing them (Schmidt and Reeves, 1989). The direct influence of one organism on another requires physical transference of allelochemicals (Read and Jensen, 1989), which would normally be synthesized at some defined stage of phenotypical development. The influence can also be indirect if the original allelochemical is metabolized by soil microbes to some other physiologically active product (DeScisciolo et al., 1990; Levitt and Lovett, 1984). Plants may have a mechanism to detoxify allelochemicals (Hogan and Manners, 1991).

Phytoalexins, on the other hand, are compounds that may exist in healthy plants in small quantities, but increase dramatically as a result of microbial infection (Beier and Oertli, 1983; Surico et al., 1987). Phytoalexins and allelochemicals are located in distinct compartments of a plant's tissue (Zobel, 1993, 1996)—one filled with an aqueous solution (vacuole) and the second with air (intercellular spaces and plant surface). The first compartment, inside the cell, contains predominantly water-soluble compounds and is surrounded by the tonoplast, a membrane that may be destroyed by precipitation with phenolic compounds if they are continuously synthesized under stress conditions

0-8493-2116-6/99/$0.00+$.50

(Zobel and Nighswander, 1991) on the endoplasmic reticulum (Hrazdina et al., 1987). We have found an example of this in the case of simulated acid rain and salt spray on pine needles (Zobel and Nighswander, 1990, 1991; Zobel, 1996).

The second compartment, filled with air, consists of both intracellular spaces and the surface of the plant, each rather neglected in biochemistry. We must realize that biochemical processes take place in the cell cytoplasm and to some extent in the cell wall, but on the surface of cells (adjacent to the intercellular spaces and covered by epicuticular waxes) there exist only uncatalyzed chemical and physical processes (Zobel, 1993, 1996). Every plant deposits in its vacuoles a large number of secondary metabolites from among the hundreds it may elaborate (Harborne, 1985), some of which can be relatively inactive in that compartment (e.g., in bound form as a glucoside), but are activated when released from the vacuoles (Murray et al., 1982). This could explain why some allelochemicals become toxic only after the death of a plant or of a particular tissue (White et al., 1989).

On the surface of parenchyma cells phenolic compounds, covering or impregnating the walls of cells surrounding intracellular spaces, will form a defense barrier against microbial attack (Zobel and Brown, 1989; Zobel and March, 1993; Zobel et al., 1994b). This answers the intriguing question of why plants are not generally vulnerable to rotting, even though they possess numerous stoma in the epidermis that allow bacteria and fungi to easily penetrate the plant body with air exchange, not to mention continual mechanical damage with risk of infection, as, for example, during the mowing of lawns.

On the surface of a plant, Baker (1982) has reported that the aerial parts are covered by the cuticle and epicuticular waxes, which have phenolic compounds that he did not identify. Waxes could form a water-resistant barrier (Esau, 1977) and be responsible for some dispersion of radiation falling on the plant surface (Martin et al., 1991; Cen and Bornman, 1993; Waterman and Mole, 1994). Phenolic compounds embedded in waxes, if present in sufficient concentration to inhibit microorganisms, could form a shield against bacteria and germinating fungal spores. Because they absorb ultraviolet (UV) rays, they could likewise be effective against such radiation. Some, at least, can convert this absorbed energy to lower-energy, longer wavelength radiation in the visible range (Murray et al., 1982; Zobel et al., 1993).

Up to now, two large groups of secondary metabolites have been found to be represented in epicuticular waxes on the surface of aerial parts of plants: flavonoids (Wollenweber and Dietz, 1981) and coumarins (Zobel and Brown, 1989). Over 400 species known to contain flavonoids had at least some of these compounds on their surfaces (Wollenweber, 1986). Extruded flavonoids sometimes amount to 3 percent of the total flavonoid concentration, as in the case of *Primula*. Furanocoumarins were first found to exist on *Daucus carota* L. leaves, in very small concentrations, when the relationship between plants and insects was investigated (Ashwood-Smith et al., 1983; Städler and Buser, 1984). These compounds were removed by washing, for up to 60 seconds, with hexane, alcohols, or other organic solvents, based on techniques developed by Tulloch (1987) to remove waxes. We used an improved method of brief dipping in almost boiling water (Zobel and Brown, 1988); because the resulting high temperature melts the surface waxes, we were able to remove up to 100 times more furanocoumarins from the *Ruta graveolens* surface than by the use of organic solvents (Zobel and Brown, 1988, 1989, 1990a).

Many species of herbs and medicinal plants contain furanocoumarins on their surfaces. Several other coumarins were also identified on the surface along with furanocoumarins (Zobel et al., 1991c), but to date many of the peaks detected on chromatograms have not been identified. Because of a possible synergistic or additive effect between different compounds, further investigations of this phenomenon are needed.

Most recently we have isolated and identified furanocoumarins located on the root surfaces of three plant species (unpublished data). Protection against bacteria and UV could be one of their roles on the aerial plant parts, but the question of their roles on underground organs has not yet been addressed. Most likely it would be an antimicrobial role, because coumarins are known to be antifungal (Chakraborty et al., 1957; Towers, 1987b), antibacterial (Towers, 1987a), and even antiviral (Towers, 1986).

Allelopathic potentials would be enhanced by greater extrusion of compounds to the root surface, along with qualitative differences, if these chemicals were toxic to other plants, as they would inevitably be washed off the surface into the soil. It is important to distinguish between allelopathy and competition of plants for resources (Weidenhamer et al., 1989), as the former must involve compounds extruded into soil. We have now proved that a protective shield of phenolic compounds is not confined to shoots. Roots of *Ruta*, *Pastinaca*, and *Angelica* contain furanocoumarins identified as xanthotoxin, bergapten, and imperatorin, in different proportions depending on the species (unpublished data). Autotoxicity is most likely nonexistent in the case of shoots because the cuticle and epicuticular wax prevent water, and possibly dissolved substances in it, from penetrating back into the protoplast of the epidermal cells. With roots, which contain very little if any waxes and absorb water readily, autotoxity can be a problem if surface deposits are in high concentrations. In the case of *Ruta graveolens* L., we found that concentrations of compounds deposited on the root surface were lower than on leaves of the same seedling, although the percentage of extrusion was high (*ca.*, 80 percent). Such compounds, after extrusion to the plant surface, can have several fates: degradation with inactivation; modification to a more active form; or, if they are volatile, disappearance.

In the course of evolution, when the first organisms appeared on Earth's surface they faced several problems after they had managed to prevent desiccation. Conditions on land were harsher than those in water because of direct UV irradiation and high oxygen concentration. Desiccation could be avoided by covering the parts exposed to wind with waxes, and only opening holes at some convenient times for gas exchange. Oxygen can be used by plant cells during the night and in the photorespiration process during the day; but in light, the additional oxygen from photosynthesis leads to dramatically increased oxygen concentration in the ground cytoplasm, which could be dangerous to cells if no peroxisomes existed. Thus, antioxidants have been necessary for plants from the very beginning of land colonization. Coumarins can be antioxidants and scavengers of free radicals *in vitro* (Paya et al., 1993). Only recently have we realized how important a role antioxidants play in our diet (Stavric et al., 1992); a similar role may exist in plants.

The next dangerous environmental factor was UV radiation, which had to be prevented from penetrating into the cells or, if that was found to be impossible, any damage, direct and indirect, had to be repaired. Recently we have found that UV can actually be utilized by autofluorescing furanocoumarins (Zobel and Brown, 1993; Zobel et al., 1994a); substitution of hydroxyl groups in the benzene ring rendered such aromatic products able to absorb UV and emit longer wavelength radiation (Zobel and March, 1993; Zobel, 1997). A greater variety of phenolic compounds in any plant would clearly increase the range of its UV absorption capacities. Over 100,000 secondary metabolites have now been identified (Harborne, 1985, 1988, 1991), with new ones discovered at the rate of two to three a day, and the vast majority of these absorb UV, each having its own characteristic wavelength of maximum absorption. But after a compound has absorbed UV, which is high-energy radiation, it acquires excitation energy, and to revert to the ground state this energy must be discharged. This can be accomplished by emission of heat or longer wavelength radiation, but if this is not done there is an unavoidable increase in reactivity, and free radicals may be formed. One way to overcome the high reactivity of such radicals would be for the plant to produce compounds exhibiting autofluorescence, such as, for instance, furanocoumarins (Zobel and March, 1993; Zobel et al., 1993).

But evolving plants went further and began to utilize UV, possibly as early as they started to use visible light in assimilation by cryptochromes (Manunelli, 1989), about which little is still known (Senger and Schmidt, 1986). We have postulated (Zobel and Brown, 1993; Zobel, 1996) that autofluorescing furanocoumarins may be transducers of UV into lower-wavelength radiation—that is, usable light, allowing *Ruta graveolens* plants not only to survive under 366 nm radiation for as long as four weeks, but even to synthesize more furanocoumarins than control plants in light.

Another function of UV is in photomorphogenesis (Salisbury and Ross, 1992), where it participates in the action of cryptochromes. These pigments have not yet been fully characterized but are known to be responsible for the blue light syndrome (Senger, 1980; Senger and Lipson, 1987) and UV radiation effects. They respond to differences in wavelength of as little as 10 nm, causing, for instance, a rosette structure in alpine plants by shortening internodes.

Only recently has more attention been paid to characterization of different flavonoid and coumarin aglycones (Zobel, 1986; Zobel and Lynch, 1995) as special filters of short wave radiation, although screening of UV through the epidermis has been investigated (Gislefoss et al., 1992). As we might expect, plants have evolved a mechanism for protection against UV radiation above 280 nm, which penetrates the Earth's atmosphere, but what about shorter radiation? Is it more damaging only because of its higher energy, or as well by reason of the fact that plants have had no experience with UV-C and UV-B? Detached leaves of *Ruta* plants exposed to continuous 254 nm radiation showed a dramatic decline in their content of xanthotoxin, bergapten, and psoralen during the first one to two days, but after four days the concentration of these compounds returned to control levels, and after seven days it even exceeded by several times the control concentration (Zobel et al., 1994a). Thus, we have observed a process of bouncing back or recovery that has been observed in other plants as well.

Much more attention should be paid to this recovery process, which has been only recently investigated, and which, in the author's opinion, is responsible for the stable morphology of plants in spite of all mutagens affecting them. Major genetic information has remained intact and is still expressed in phenotypic reactions during plant ontogenesis (Zobel, 1989a,b). A recovery mechanism would be very beneficial to any plant; otherwise continuous production of phenolic compounds after biosynthesis had been switched on, as an immediate, direct response to stress, could lead to uncontrolled deposition of phenolic compounds in a vacuole, and subsequent death of a cell. We might suggest that such a response is the basis of the reaction of pine needles to salt and simulated acid spray in which necrosis, not visible macroscopically, appeared within parenchyma cells beneath the epidermis (Zobel and Nighswander, 1990). After the first four to seven days, cells under the stoma contained more phenolic compounds than those located farther from possible pollutants; after 28 days these closer cells, which then contained degraded cytoplasm, had died. Plants have not produced a mechanism of switching off production of defense compounds because these were relatively new (in evolution) stresses.

We now postulate that a relatively new stress, pollution by humans (Zobel and Nighswander, 1991; Zobel, 1997), is leading only to switching on biosynthesis of phenolic compounds without a later switching off. Plants at the current stage of evolution have not yet developed genes for switching off production of phenolic compounds. They may be able to respond to so many different stresses because of the existence of several isoenzymes of each enzyme involved in biosynthesis of phenolic compounds, along with several genes for their production. Such an enzyme, for instance, is chalcone synthase, which has several such isoenzymes and genes (An et al., 1993). In the course of evolution, chalcone synthase would be responsible for switching from biosynthesis of aromatic amino acids, which are primary plant products, to production of aromatic compounds, which are secondary products.

Nowadays, both marine and terrestrial plants contain hundreds of different natural products (Porter and Targett, 1988; Harborne, 1991) that could account for possible rapid adaptation of marine plants to land conditions by enhancing biosynthesis of natural metabolites. Thus, over the course of evolution, any nonspecific response to stress would be to initiate biosynthesis of some phenolic compounds. Then a second set of genes, or other mechanism of control, for example, feedback control by phenolic compounds, would come into operation. Addition of feedback inhibition would be needed for switching this response off and letting the plant recover before it could over-produce phenolic compounds and increase concentrations to the point of killing itself. This suggestion of self control is probably valid, in our view (Wronka et al., 1995), because we found phenolic compounds, precipitated by caffeine during fixation for the electron microscope, in the nucleus associated with chromatin, but not in RNA. That, together with the fact that genes of *Rhizobium* and other symbiotic bacteria are regulated by phenolic compounds of the host plant (Peters and Verma, 1990), raises the question, which still needs to be addressed, of whether the cell's own genes can be regulated by phenolic compounds in the same cell.

Plants that lack a mechanism for switching off the genes of phenolic compound production, or lack cellular control of such production, must continuously produce these natural products and eventually commit suicide. If so, the extreme manifestation of allelopathy, autotoxity, would exist. Thus, total lack or even low efficiency of a control mechanism may be responsible for autotoxity of compounds extruded by any part of one plant and affecting growth after having entered a root of that plant. That process would depend on the concentration of compounds, and we should look more closely at the phenomenon of stimulation and inhibition by the same molecule (Einhellig and Leather, 1988) depending, for instance, on concentration. Very little has been investigated on this subject.

27.3   LOCALIZATION OF COUMARINS ON SHOOTS

We found several such events in plants containing coumarins while investigating the localization of furanocoumarins and evaluating their role in particular species, organs, or even tissues. All plant species investigated, approximately 20, from families of Rutaceae, Umbelliferae, and Leguminosae, had coumarins on their surfaces. On the surface of *Ruta graveolens* leaves there were changes in concentration of psoralen, xanthotoxin, and bergapten during the vegetative period, depending on environmental conditions (Zobel, 1991; Zobel and Brown, 1995) and the position of the leaf on the plant shoot. Younger leaves, third from the shoot apex, had higher concentrations than the third from the bottom of the branch (Zobel and Brown, 1991a). During the senescing process in *R. graveolens*, yellow leaves contained less furanocoumarin than green and dry leaves—only about 10 percent of the concentration in young leaves on the same branch. Do the younger leaves need more protection and thus a higher concentration?

During the vegetative period leaves of *Angelica archangelica* L. and *Heracleum mantegazianum* Somm. & Lev. (Zobel and Brown, 1991b) showed a similar trend of changes with maturation, and younger leaves had more furanocoumarins than older leaves. *Helacleum lanatum* Michx. (Zobel and Brown, 1990b) concentrations changed with temperature, and after five days of temperatures reaching 36°C, the concentrations were manifold higher than that at the beginning, with exceptionally high extrusion. After two nights of freezing temperatures, concentrations of psoralen, xanthotoxin, and bergapten were increased up to 100 times. Such plants would thus be more toxic at these times, indicating that there are differences in the potential for allelopathy depending on environmental conditions.

UV radiation enhanced not only production of allelochemicals, but even more the extrusion process, which seems to be the first switched reaction in response to this stress (Zobel et al., 1994a, b). *Ruta graveolens* plants exposed to 366 and 254 nm had furanocoumarin concentrations several times as high as did plants growing in normal light, and after four weeks under 366 nm their leaves contained cells that, under examination in an electron microscope, did not show morphological changes. The chloroplasts possessed starch grains, strongly suggestive of continuing assimilation. Our conclusion would be that transduction of UV into visible light, as noted previously, produces autofluorescence (Zobel et al., 1993) that might be strong enough to sustain sugar production, and could be responsible for the well-being of this particular species. We saw only sporadically damaged epidermal cells surrounded by undamaged ones in germinating roots of *Brassica napus* (unpublished data) in UV radiation.

Extrusion of furanocoumarins to the surface may have another purpose—avoidance of autotoxity. As they are so dangerous that they cannot be kept in vacuoles, aglycones have to be located outside the plasmalemma. Such an explanation of avoidance of autotoxity by suspension culture cells was first given by Matern et al. (1988). We have supported this view since our finding that callus cells have furanocoumarins on their surface, where they can be both protective compounds and waste products due to overproduction. Furanocoumarin concentration decreases during the senescence process, suggesting that these compounds are not waste products, but could be re-utilized (Zobel and Brown, 1991a). When extruded to the leaf surface they can become even more toxic, as was found for allelochemicals changed by biotic and abiotic factors (Pal, 1994; Weidenhamer et al., 1989).

27.4 LOCALIZATION ON SEEDS AND FRUITS

Seeds and fruits of many plant species contain coumarins (Murray et al., 1982), which have been found in substantial amounts on the surface (Zobel and Brown, 1991c). In the case of *Angelica archangelica* Linn. (Zobel and Brown, 1991b), furanocoumarins were situated on the surface of fruits, within dead tissue of fruit, and on the surface of seed coats and within them, as well as on the surface and in the interior of the embryo (Zobel and Brown, 1991b; Zobel et al., 1991a).

Different proportions of furanocoumarins but similar trends of localization occurred in case of *Psoralea*, which contained furanocoumarin crystals on the embryo surface itself (Zobel et al., 1991a). We have suggested that these furanocoumarins can act as autoinhibitors of germination, thus accounting for difficulties often met in attempts to germinate embryos of some *Umbelliferae* (Zobel and Brown, 1991c). Each dry fruit has on its surface different concentrations of these very potent germination inhibitors, which varied tenfold among different individual fruits and which could keep embryos dormant, preventing germination. Winter thaws and spring rains cause dissolution and removal of these compounds, allowing the embryo to germinate. We found that xanthotoxin at 100 ppm concentration inhibits mitoses (Podbielkowska et al., 1994) and oxygen uptake by isolated mitochondria but stimulates oxygen uptake at 10 ppm. Thus, as the xanthotoxin concentration in seeds gradually diminishes, it could cross the threshold of stimulation. An additional benefit would be to decrease palatability to herbivores, as in case of other allelochemicals (Dhaliwal et al., 1990). There are very few animals, apart from a few specialist insects, that consume seeds of *Rutaceae* and fruits of *Umbelliferae* (Nitao, 1989, 1990; Berenbaum, 1995).

But after seeds and fruits have been deposited in the soil, leakage from the surface and from dead tissues may remove coumarins, which would then begin to act as allelochemicals.

27.5 LOCALIZATION OF COUMARINS ON ROOTS

Xanthotoxin has been found as crystals on stored roots of parsnips (Ceska et al., 1986). We found that roots of three-month-old seedlings of *Ruta graveolens* L., *Pastinaca sativa* L., and *Angelica archangelica* had xanthotoxin and bergapten on their surface (Table 27.1). When seedlings of *R. graveolens* (Table 27.2) were irradiated by 366 nm UV, the roots reacted by lowering the total concentration of bergapten, xanthotoxin, and imperatorin by over 50 percent. Interior concentrations did not change, but surface concentrations of xanthotoxin and bergapten declined severalfold. When *Angelica archangelica* seedlings (unpublished data) were irradiated by 366 nm UV the opposite reaction was observed: total concentration doubled, with drastically increased extrusion and a parallel decline of interior concentrations. These two species reacted differently: thus, *Ruta* decreased extrusion and maintained interior concentration, but *Angelica* increased extrusion and decreased interior concentration. In *Ruta* something as yet unexplained happened to these compounds in the soil, where they were released immediately from the root surface. In *Angelica* higher deposits observed at first on the root surface remained there, but ultimately, of course, they would have been released into the soil. Roots of other plants were found to extrude scopoletin under stress conditions (unpublished data). When such compounds are extruded to the surface of roots they have the potential to become involved in communication between plants, and in the relationship among plants, soil, insects, and microbes.

TABLE 27.1

Xanthotoxin (X) and Bergapten (B) on the Surface of Roots of Seedlings of *Ruta graveolens*, *Angelica archangelica*, and *Pastinaca sativa*

Species	Surface Deposits[a]		
	X	B	X+B
Ruta graveolens	8.55 ± 0.95	13.6 ± 1.6	22.15 ± 2.15
Pastinaca sativa	2.30 ± 0.24	1.80 ± 0.2	4.10
Angelica archangelica	0.16 ± 0.01	0.10 ± 0.01	0.26

[a] Fresh weight, μg/g.

TABLE 27.2

Concentrations of Xanthotoxin (X), Bergapten (B), and Imperatorin (I) on and Inside Roots of *Ruta graveolens* Seedlings due to UV-A Radiation

	Roots[a]				% on Surface			
	X	B	I	X+B+I	X	B	I	X+B+I
Control surface	8.55 ± 0.95	13.6 ± 1.6	traces	22.15	89	81	0	62
UV surface	traces	2.37 ± 0.5	traces	2.37	0	40	0	15

[a] Fresh weight, μ/g fresh weight ± standard deviation.

27.6 MECHANISM OF REACTION OF COUMARINS IN ALLELOCHEMISTRY

High concentrations of coumarins, reaching milligrams per gram fresh weight, are found on plant leaves, roots, seeds, and fruits, and could be washed into the soil by rain. After tissue decay, they could also be transferred there from the cell interior. Even though they are present in dried leaves (Zobel and Brown, 1991a) in relatively lower concentrations than in living tissues (10 percent of the concentration in young leaves), they still account for large amounts (700 μg/g). Additionally, when they are together in the soil, and cell compartmentation is no longer a factor, some synergistic reactions may occur; Berenbaum and Neal (1985) found that xanthotoxin was synergistically activated by myristicin. Once in the soil, allelochemicals undergo abiotic and biotic changes (Pal, 1994), in the latter case because microbes can actively transform them to either more or less toxic compounds. There is also a reciprocal action: these compounds, both before and after such transformations, can influence microbes, because they may have antibacterial, antifungal, and even antibacteriophagic properties. Xanthotoxin concentrations of 75 to 100 ppm are enough to kill several bacteria, but on root surfaces of three species investigated (Zobel et al., 1991b), concentrations reached only 22 ppm, suggesting that no individual compound was bactericidal. Thus, on roots the protective chemical mechanism must involve at least several coumarins and possibly other phenolics that could interact in their effects. Investigating reactions of coumarin and several of its methyl, methoxyl, hydroxyl, and furano derivatives, we found that these compounds are antimitotic, reacting individually at concentrations of 50 to 200 ppm (Podbielkowska et al., 1994; Keightly et al., 1996). At the same time, except for coumarin itself and 4-hydroxycoumarin, the investigated coumarins caused chromosomal aberrations.

Coumarins caused changes in cell membrane permeability, decreased oxygen uptake, and degradation of oxidative phosphorylation, and altered the activity of enzymes and the cell structure (Podbielkowska et al., 1995). Mitochondria acquired a structure similar to that observed in cells deprived of oxygen or due to anticancer agents (Podbielkowska et al., 1981). Substantial areas of cytoplasm were separated by endomembranes and either contained a high density of ribosomes or became degraded and formed autophagic vacuoles (Podbielkowska et al., 1995).

Such strongly reactive compounds must be segregated from the cell cytoplasm where they are synthesized. Glycosides, which are relatively nontoxic, can be tolerated in vacuoles, but the more toxic aglycones are extruded outside the plasma membrane. Such extruded compounds can act immediately once in the soil, but any present in the vacuoles will have input into allelochemistry of the soil only after decay of the plant. More investigations are needed on reactivity of allelochemicals in soil and the contribution of these compounds to coexistence of plants, microbes, and animals during their coevolution.

REFERENCES

An, C., Ichinose, Y., Yamada, T., Tanaka, Y., Shiraiski, T., and Oku, H., Organization of the genes encoding chalcone synthase in *Pisum sativum, Plant Mol. Biol.,* 21, 789, 1993.

Ashwood-Smith, M. J., Poulton, G. A., Ceska, O., Lin, M., and Furniss, E., An ultrasensitive bioassay for the detection of furanocoumarins and other photosensitizing molecules, *Photochem. Photobiol.,* 38, 113, 1983.

Baker, E. A., Chemistry and morphology of plant epicuticular waxes, *in The Plant Cuticle,* Cultler, D. F., Alvin, K. L. and Price, C. E., Eds., Academic Press, London, 1982, 139.

Beier, R. C. and Oertli, E. H., Psoralen and other linear furanocoumarins as phytoalexins in celery (*Apium graveolens*), *Phytochemistry,* 22, 2595, 1983.

Berenbaum, M. and Neal, J.J., Synergism between myristicin and xanthotoxin, a naturally occurring plant toxicant, *J. Chem. Ecol.,* 11, 1349, 1985.

Cen, Y. P. and Bornman, J. F., The effect of exposure of enhanced UV-B radiation on the penetration of monochromatic polychromatic UV-B radiation in leaves of *Brassica napus, Physiol. Plant,* 87, 249, 1993.

Ceska, O., Chaudhary, S., Warrington, P., Poulton, G., and Ashwood-Smith, M. J., Naturally occurring crystals of photocarcinogenic furanocoumarins on the surface of parsnip roots sold as food, *Experientia,* 42, 1302, 1986.

Chakraborty, D. P., Das Gupta, A., and Bose, P. H., On the antifungal action of some natural occurring coumarins, *Ann. Biochem. Exp. Med.,* 17, 59, 1957.

DeSciscicolo, B., Leopold, D. J., and Walton, D. C., Seasonal patterns of juglone in soil beneath *Juglans nigra* (black walnut) and influence of *J. nigra,* on understory vegetation, *J. Chem. Ecol.,* 16, 1111, 1990.

Dhaliwal, G. S., Pathak, M. D., and Vega, C. R., Effect of a rice allelochemical on insect pests, predators and plant pathogens, *J. Insect. Sci.,* 3, 136, 1990.

Einhellig, F. A. and Leather, G. R., Potentials for exploiting allelopathy to enhance crop production, *J. Chem. Ecol.,* 14, 1829, 1988.

Esau, K., *Anatomy of Seed Plants,* 2nd ed., John Wiley & Sons, London, 1977.

Gislefoss, J. S., Kieldtradt, B., and Bakken, A. K., Optical properties of the epidermis of leek epidermis (*Allium ampelosparsum* L.) and cabbage (*Brassica oleracea* L.) after enhanced ultraviolet radiation, *Acta Agric. Scand.,* 42, 173, 1992.

Harborne, J. B., Phenolics and plant defense, *Ann. Proc. Phytochem. Soc. Europe,* 25, 393, 1985.

Harborne, J. B., *The Flavonoids: Advances in Research,* Vol. 2, Chapman and Hall, London, 1988.

Harborne, J. B., Flavonoids as pigments, in *Herbivores: Their Interaction with Secondary Plant Metabolities,* 2nd ed., Rosential, G. A. and Berenbaum, M. R., Eds., Academic Press, New York, 1991, 389.

Hegazy, A. K., Mansour, K. S., and Abdel-Hady, N. F., Allelopathic and autotoxic effects of *Anastatica hierochuntica* L., *J. Chem. Ecol.,* 16, 2183, 1990.

Hogan, M. E. and Manners, G. D., Differential allelochemical detoxification mechanism in tissue cultures of *Antennaria microphylla* and *Euphorbia esula, J. Chem. Ecol.,* 17, 167, 1991.

Hrazdina, G., Zobel, A. M., and Hoch, H. C., Biochemical, immunological, and immunochemical evidence for the association of chalcone synthase with endoplasmic reticulum membranes, *Proc. Natl. Acad. Sci. USA,* 84, 8966, 1987.

Keightly, A. M., Dobrzynska, M., Podbielkowska, M., Renke, K., and Zobel, A. M., Coumarin and its 4- and 7-substituted derivatives as retardants of mitoses in *Allium* root promeristem, *Int. J. Pharm.,* 34, 105, 1996.

Kil, B. S. and Lee, S. Y., Allelopathic effects of *Chrysanthemum morifolium* on germination and growth of several herbaceous plants, *J. Chem. Ecol.,* 13, 299, 1987.

Levitt, J. and Lovett, J. V., Activity of allelochemicals of *Datura stramonium* L. (Thornapple) in contrasting soil types, *Plant Soil,* 79, 181, 1984.

Manunelli, A. Z., Interaction between cryptochrome and phytochrome in higher plant photomorgenesis, *Am. J. Bot.,* 76, 143, 1989.

Martin, G., Myers, D. A., and Vogelmann, T. C., Characterization of plant epidermal lens effects by a surface replication technique, *J. Exp. Bot.,* 42, 581, 1991.

Matern, U., Strasser, H., Wendorff, H., and Hameski, D., Coumarins and furanocoumarins, in *Cell Culture and Somatic Cell Genetics in Plants,* Vol. 5, Vasil, I. and Constabel, F., Eds., Academic Press, New York, 1988, 2.

Murray, R. D. H., Méndez, J. and Brown, S. A., *The Natural Coumarins: Occurrence, Chemistry and Biochemistry,* John Wiley & Sons, Chichester, England, 1982.

Narwal, S. S., *Allelopathy in Crop Production,* Scientific Publishers, Jodhpur, 1995.

Nitao, J. K., Enzymatic adaptation in a specialist herbivore for feeding on furanocoumarin-containing plants, *Ecology,* 70, 629, 1989.

Nitao, J. K., Metabolism and excretion of the furanocoumarin xanthotoxin by parsnip webworm, *Depressaria pastinacella, J. Chem. Ecol.,* 16, 417, 1990.

Pal, S., Role of abiotic and biotic catalysts in the transformation of phenolic compounds through oxidative coupling reaction, *Soil Biol. Biochem.,* 26, 813, 1994.

Paya, M., Halliwell, B., and Hoult, J., Peroxyl radical scavenging by a series of coumarins, *Free Rad. Res. Comms.,* 17, 293, 1993.

Peters, N. K. and Verma, D., Phenolic compounds as regulators of gene expression in plant–microbe interactions, *Mol. Plant. Microbe Interact.,* 5, 33, 1990.

Podbielkowski, M., Waleza, M., and Zobel, A. M., Influence of methotrexate on respiration and ultrastructure of meristematic cells from *Allium cepa* roots, *Acta Soc. Bot. Pol.,* 50, 563, 1981.

Podbielkowski, M., Kupidlowska, E., Waleza, M., Dobrzynska, K., Louis, S. A., Keightly, A., and Zobel, A. M., Coumarins as antimitotics, *Int. J. Pharm.,* 32, 262, 1994.

Podbielkowski, M., Waleza, M., Dobrzynska, K., and Zobel, A. M., Reaction of coumarin and its derivatives on ultrastructure ATP-ases and acid phosphatases in meristematic cells of *Allium cepa* roots, *Int. J. Pharm.,* 34, 105, 1996.

Porter, J. W. and Targett, N. M., Allelochemical interactions between sponges and corals, *Biol. Bull. Mar. Biol. Lab. Woods-Hole,* 175, 230, 1988.

Read, J. J. and Jensen, E. H., Phytotoxicity of water-soluble substances from alfalfa and barley soil extracts on four crop species, *J. Chem. Ecol.,* 15, 619, 1989.

Rice, E. L., *Allelopathy,* Academic Press, Orlando, FL, 1984.

Salisbury, F. B. and Ross, C. W., Photomorphogenesis, in *Plant Physiology,* 4th ed., Wadsworth Publishing, 1992, 438.

Schmidt, S. K. and Reeves, F. B., Interference between *Salsola kali* L. seedlings: implications for plant succession, *Plant Soil,* 116, 107, 1989.

Senger, H., *The Blue Light Syndrome,* Springer-Verlag, Berlin, 1980.

Senger, H. and Lipson, E. D., *Problems and Prospects of Blue and Ultraviolet Light Effects in Phytochrome and Photoregulation in Plants,* Academic Press, New York, 1987.

Senger, H. and Schmidt, W., Cryptochrome and UV receptors, in *Photomorphogenesis in Plants,* Kendrich, R. E. and Kronenberg, G. H., Eds., Martinus Nishoff, Boston, 1986, 137.

Städler, E. and Buser, H. R., Defense chemicals in the leaf surface wax synergistically stimulate oviposition by a phytophagous insect, *Experimentia,* 40, 1157, 1984.

Starvic, B., Matula, T. I., Klassen, R., Downie, R. H., and Wood, R. J., Effect of flavonoids on mutagenicity and bioavailability of xenobiotics in foods, in *Phenolic Compounds in Food and Their Effects on Health. II. Antioxidants and Cancer Prevention,* Huang, M., Ho, C., and Lee, C., Eds., ACS Symposium Series 507, Washington, D.C., 1992, 239.

Surico, G., Varvano, L., and Solfrizzo, M., Linear furanocoumarins accumulation in celery plants infected with *Erwinia carotovera* pathovar *carotovera, J. Agric. Food Chem.,* 35, 406, 1987.

Towers, G. H. N., Induction of cross-links in viral DNA by naturally occurring photosensitizers, *Photochem. Photobiol.,* 44, 187, 1986.

Towers, G. H. N., Fungicidal activity of naturally occurring photosensitizers, *Am. Chem. Soc. Symp. Ser.,* 339, 231, 1987a.

Towers, G. H. N., Comparative anti-bacteriophage activity of naturally occurring photosentizers, *Plant Med.,* 53, 536, 1987b.

Tulloch, A. P., Epicuticular waves of *Abies balsamea* and *Picea glauca:* occurrence of long-chain methyl esters, *Phytochemistry,* 26, 1041, 1987.

Waterman, P. G. and Mole, S., *Analysis of Phenolic Plant Metabolites,* Blackwell Scientific, London, 1994.

Weidenhamer, J. D., Hartnett, D. C., and Romeo, J. T., Density-dependent phytotoxicity: distinguishing resource competition and allelopathic interference in plants, *J. Appl. Ecol.,* 26, 613, 1989.

White, R. H., Worsham, A. D., and Blum, U., Allelopathic potential of legume debris and aqueous extracts, *Weed Sci.,* 37, 674, 1989.

Wollenweber, E., Flavonoid aglycones in leaf exudate constituents in higher plants, in *Studies in Organic Chemistry, Flavonoids, Bioflavonoids,* Farkas, L., Gabor, M., and Kallay, F., Eds., Elsevier, Amsterdam, 1986, 155.

Wollenweber, E. and Dietz, U. H., Occurrence and distribution of free flavonoid aglycones in plants, *Phytochemistry,* 20, 869, 1981.

Wronka, M., Kuras, M., Tykarska, T., Podstolski, A., and Zobel, A. M., Inhibition of the production of phenolic compounds in *Brassica napus* by 2-amino-oxyacetic acid, *Ann. Bot.,* 75, 319, 1995.

Zobel, A. M., Sites of localization of phenolics in tannin coenocytes in *Sambucus racemosa* L., *Ann. Bot.,* 57, 801, 1986.

Zobel, A. M., Origin of nodes and internodes in plant shoots. I. Transverse zonation of apical parts of the shoot, *Ann. Bot.,* 63, 201, 1989a.

Zobel, A. M., Origin of nodes and internodes in plant shoots. II. Models of origin of nodes and internodes, *Ann. Bot.,* 63, 209, 1989b.

Zobel, A. M., Effect of the change from field to greenhouse environment on the linear furanocoumarin levels of *Ruta chalepensis, J. Chem. Ecol.,* 17, 21, 1991.

Zobel, A. M., Phenolic compounds: an answer by the plant to air pollution, Proceedings of Second Princess Chulabhorn Meeting, Thailand, 1993.

Zobel, A. M., Phenolic compounds as bioindicators of air pollution, in Yunas and Iqbal, Eds., John Wiley & Sons, Chichester, England, 1996, 100.

Zobel, A. M., Coumarins in fruit and vegetables, in *Phytochemistry of Fruit and Vegetables,* Thomas-Barberan, F. A., Wiley and Sons, New York, 1997, 173.

Zobel, A. M. and Brown, S. A., Determination of furanocoumarins on the leaf surface of *Ruta graveolens* with an improved extraction technique, *J. Nat. Prod.,* 51, 941, 1988.

Zobel, A. M. and Brown, S. A., Histological localization of furanocoumarins in *Ruta graveolens, Can. J. Bot.,* 67, 915, 1989.

Zobel, A. M. and Brown, S. A., Dermatitis-inducing furanocoumarins on the leaf surfaces of eight species of rutaceous and umbelliferous plants, *J. Chem. Ecol.,* 16, 693, 1990.

Zobel, A. M. and Brown, S. A., Seasonal changes of furanocoumarins in leaves of *Heracleum lanatum, J. Chem. Ecol.,* 16, 1623, 1990b.

Zobel, A. M. and Brown, S. A., Psoralens in senescing leaves of *Ruta graveolens, J. Chem. Ecol.,* 17, 1801, 1991a.

Zobel, A. M. and Brown, S. A., Furanocoumarin concentrations in fruits and seeds of *Angelica archangelica, Enviro. Exp. Bot.,* 31, 447, 1991b.

Zobel, A. M. and Brown, S. A., Psoralens on the surface of seeds of *Rutaceae* and fruits of *Umbelliferae* and *Leguminosae, Can. J. Bot.,* 69, 485, 1991c.

Zobel, A. M. and Brown, S. A., Influence of low-intensity ultraviolet radiation on extrusion of furanocoumarins to the leaf surface, *J. Chem. Ecol.,* 19, 939, 1993.

Zobel, A. M. and Brown, S. A., Coumarins in the interactions between the plant and its environment, *Allelopathy J.,* 2, 9, 1995.

Zobel, A. M. and Lynch, J. M., Production of anthocyanins and other flavonoids in *Acer saccharum* and *Acer platanoides* in response to stress, Proceedings of Meeting on Air Pollution, September 1994, Athens, Greece, 1995, 210.

Zobel, A. M. and March, R. E., Autofluorescence reveals different histological localizations of furanocoumarins in fruits of some *Umbelliferae* and *Leguminosae, Ann. Bot.,* 71, 251, 1993.

Zobel, A. M. and Nighswander, J. E., Accumulation of phenolic compounds in the necrotic areas of Austrian and red pine needles due to salt spray, *Ann. Bot.,* 66, 629, 1990.

Zobel, A. M. and Nighswander, J. E., Accumulation of phenolic compounds in the necrotic areas of Austrian and red pine needles due to sulphuric acid spray as a bioindicator of air pollution, *New Phytologist,* 117, 565, 1991.

Zobel, A. M., Brown, S. A., and March, R. E., Histological localization of psoralens in fruits of *Psoralea bituminosa, Can. J. Bot.,* 69, 1673, 1991a.

Zobel, A. M., Mwiraria, K. and Louis, S. A., Two mechanisms of action of psoralen and xanthotoxin on mitotically active cells, Joint Meeting of I.S.C.E. and PSNA, August 11–15, 1990, Quebec, 1991b.

Zobel, A. M., Wang, J., March, R. E., and Brown, S. A., Occurrence of other coumarins with psoralen, xanthotoxin, and bergapten on leaf surfaces of seven plant species, *J. Chem. Ecol.,* 17, 1859, 1991c.

Zobel, A. M., Sandstrom, T., Nighswander, J. E., and Dudka, S., Uptake of metals by aquatic plants and changes in phenolic compounds, *Heavy Metals Environ.,* 1, 210, 1993.

Zobel, A. M., Chen, Y., and Brown, S. A., Influence of UV on furanocoumarins in *Ruta graveolens* leaves, *Acta Hort.,* 381, 355, 1994a.

Zobel, A. M., Crellin, J., Brown, S. A., and Glowniak, K., Concentrations of furanocoumarins under stress conditions and their histological localization, *Acta Hort.,* 381, 510, 1994b.

28 Ecological Significance of Plant Saponins

Wieslaw A. Oleszek, Robert E. Hoagland, and Robert M. Zablotowicz

CONTENTS

28.1 ABSTRACT

Saponins are a group of steroid or triterpene natural products occurring in many plant families. Their structural diversity is very broad due to the large number of substituent functional groups and sugar moieties. This divergence of chemistry is responsible for the wide array of specific biological, chemical, and physical properties displayed by saponins. Differential synthesis and/or accumulation of saponins and their aglycones is observed in different plant tissues and organs depending upon species or genotype, age, and environmental conditions. Some of these compounds may be exuded from plant tissues or be released during mechanical injury or decay of vegetative material, thereby posing a potential for allelopathic interactions. The ecological/physiological roles of this group of compounds has been defined with regard to certain plant–microbe, plant–plant, or plant–insect interactions. However, with most of these compounds, their role is hypothetical and their modes of biological action are not totally understood. The purified compounds can elicit changes in phenomena such as membrane permeability, leakage, and hemolysis at the cellular level in plants, animals, and microorganisms. Some saponins can affect (inhibit or promote) whole plant parameters such as germination and the development of roots and shoots. A function of certain saponins as protectants from phytopathogen infection and insect predation suggests that they may have fungicidal and insecticidal roles in nature. Fungi have historically been used as indicator organisms to test saponin toxicity and recent studies have demonstrated that certain genera of soil and rhizosphere bacteria are also sensitive to these compounds. The most widely studied class of saponins are those isolated from alfalfa (*Medicago sativa* L.) and these compounds are the major focus of this review. More specifically, this review discusses the ecological significance of saponins, especially their allelopathic potential and action on some plants, animals, microbes, pathogens, and insects relevant to agricultural production.

0-8493-2116-6/99/$0.00+$.50
© 1999 by CRC Press LLC

28.2 INTRODUCTION

Saponins are a group of compounds with a steroidal or triterpene skeleton (aglycone) substituted with carboxyl, carbonyl, or hydroxyl functional groups. These functional groups create significant diversity in the structure of saponins even within different organs of the same species. Additionally, this structural diversity is greatly increased by the substitution of some functional groups possessing sugar chains. Sugar chains are composed of a combination of glucose, galactose, arabinose, rhamnose, xylose, uronic acids, and in some rare cases, apiose. Sugars may be attached to the aglycone either as one-, two-, or three-sided chains. The terms monodesmoside, bidesmoside, or tridesmoside, respectively, have been given to these saponins (Greek desmos, chain) (Wulff, 1968). As a consequence of different aglycone structure and sugar substitution, structurally divergent saponins may elicit different reactions in living organisms such as microbes, plants, and animals. Compared with other groups of secondary compounds (e.g., phenolics, terpenoids, or alkaloids), the ecological function of saponins is much less well-known. This chapter discusses some possible, ecological roles of saponins, and, in particular, will focus on saponin glycosides of alfalfa (*Medicago sativa* L.). At present, evidence for these roles is predominantly circumstantial.

28.3 EXUDATION OF SAPONINS FROM PLANT ROOTS

28.3.1 SOYASAPOGENOLS

There is limited information on the ability of saponins to pass from the roots of plants into the environment. Most saponins are rather large molecules with membrane affinity, and thus it seems quite doubtful that they can easily pass the membrane barrier. Although some reports suggest this possibility, the mechanism of exudation has not been fully explained. The work of Lynn (1985) documented that root exudates from sericea lespedeza [*Lespedeza cuneata* (Dumont) G. Don] contain two triterpene compounds. These were identified as soyasapogenols B and E. It was suggested that these two compounds were capable of inducing differentiation of the haustorium in *Agalinis purpurea*. The idea of a synergistic effect with some unspecified compounds was also introduced. Moreover, it was shown that soyasapogenol A isolated from the roots of *Lactuca cuneata* stimulated haustorial formation (Chang and Lynn, 1986). In this respect, soyasapogenols were thought to act in the same way as xenognisine A and its related isoflavone xenognisine B. The exudation rate was established at 42.8 and 56.7 pmol per root per day for soyasapogenol B and E, respectively, while the ratio of soyasapogenol B/E in the roots was 6:1. It was concluded that this selective exudation may either represent an active aspect of root metabolism, or reflect cellular compartmentalization of secondary metabolites and their passive leakage from the root. It is noteworthy to emphasize that genistein, a chemical signal that activates the nodulating genes of the symbiotic nitrogen-fixing microorganism (*Rhizobium leguminosarum*) (Zaat et al., 1987), was found in roots in amounts comparable with soyasapogenol B, but was exuded at a rate two orders of magnitude lower than the soyasapogenols.

The natural distribution of soyasapogenol B and E still has not been completely clarified. Earlier work on soybean [*Glycine max* (L.) Merr.] and alfalfa (Kitagawa et al., 1976) has shown that soyasapogenol B is a natural aglycone, and that a series of glycosidic forms of this sapogenin can be found in plants. Soyasapogenol E was found together with soyasapogenols C, D, and F in hydrolysates of soyasapogenol B glycosides when hydrolysis was performed in the presence of water. It is unclear if soyasapogenol E is a natural aglycone or simply an artifact. There has been one glycoside of soyasapogenol E identified in alfalfa (Kitagawa et al., 1988) from a total of about 30 triterpene glycosides reported. Recent work by Massiot et al. (1992), Kudou et al. (1993), and Tsurumi et al. (1992), indicates that series B saponins, particularly soyasaponin I (a major saponin of soyasapogenol B in alfalfa and soybean seeds) are not the genuine compounds. Their DDMP (2,3-dihydro-2,5-dihydroxy-6-methyl-4H-pyrane-4-one) conjugates, soyasaponin VI (chromosaponin), seem to be naturally occurring forms, with soyasaponin I appearing as an artifact. These conjugates can generate soyasapogenols B or E, depending on enzymatic hydrolysis or temperature treatment (Figure 28.1). The extraction or purification procedures used strongly influence the structure of the resulting compound, and the isolated product may be an artifact. It is impossible to say whether soyasapogenol B or E are the genuine *in vivo* products. DDMP conjugated saponins are readily soluble in water, while free soyasaponin I precipitates from alcohol-water solutions. Therefore, DDMP conjugates of saponins may be of ecological significance by allowing transport of saponins across epidermal membranes due to greater solubility. The lack of activity of soyasaponin I in routine biological tests (hemolysis, *Trichoderma viride*, and rat intestine integrity), (Oleszek, 1996) demonstrates its low sterol/membrane affinity. In such a case, solubility may be the most crucial parameter in the ability of the saponin to cross the membrane. DDMP conjugates have been reported in many kinds of legumes such as Indian potato (*Apios tuberosa* Moench.), chickpea (*Cicer arietinum* L.), scarlet runner bean (*Phaseolus coccineus* L.), kidney bean (*Phaseolus vulgaris* L.), pea (*Pisum sativum* L.), mung bean [*Vigna mungo* (L.) Hepper], and cowpea [*Vigna sinensis* (L.) Hassk] (Kudou et al., 1994). Soyasapogenol B has been identified in hydrolysates of saponins from seeds of 32 species of the genus *Medicago* (Jurzysta et al., 1992).

Soyasapogenol saponins may be widely distributed in the legumes as DDMP conjugated forms. Their physiological/allelopathic function needs to be examined more thoroughly. However, in most of the traditional tests, soyasapogenol-based saponins show marginal activity. Their high concentrations in the physiologically active tissues of soybean seedlings (Shimoyamada et al., 1990) and pea (Tsurumi et al., 1992), may indicate their importance to the plant. High concentration in the root tips and hairs may be of allelopathic significance since root hairs have a short life span and may release saponins to the environment when they decay. Thus, release of saponins into the environment may not be a mechanism of passive leakage or active exudation, but simply of root hair death and decay.

Some released saponins can act as plant growth promoters. In *in vitro* germination tests performed with wheat (*Triticum aestivum* L.), soyasaponin I (100 to 500 ppm) was the only one of several triterpene glycosides that stimulated seedling root growth (Oleszek, 1993). Medicagenic acid (Ma), hederagenin, and zanhic acid glycosides (Figure 28.2) were inhibitory at the same concentrations (Table 28.1). Similarly, Tsurumi and Tsujino (1995) reported that chromosaponin I stimulated lettuce root growth. These observations were supported by experiments performed by Waller et al. (1995b), which showed that mung bean saponins (predominantly soyasaponin I) enhanced the growth of new mung bean plants when added to the soil.

The above considerations need further experimental support to be fully documented. However, since soyasapogenol-based saponins occur widely in Papilionaceae, show haustorium-inducing activity, and promote the growth of plants, there may be an important allelopathic significance to their presence in root tissue. Evaluation of the role of these compounds in plant/rhizobium interactions would be a promising area of research.

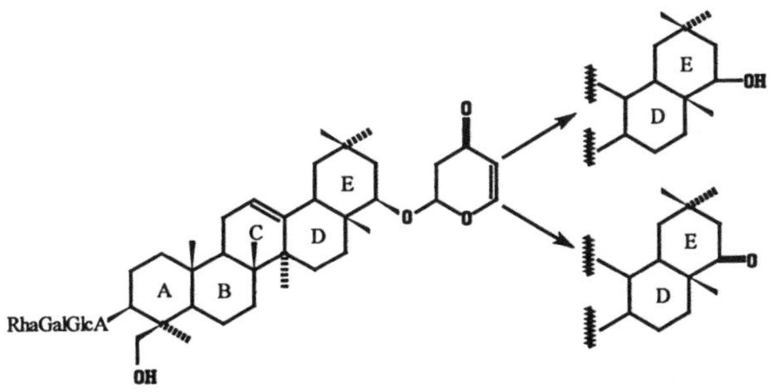

FIGURE 28.1 Soyasapogenol B and E formation during hydrolysis of chromosaponin.

TABLE 28.1
The Growth of Roots of Wheat Seedlings (% Control) Treated with Individual Alfalfa Saponins

Concentration	Wheat Root Length (% Control)										
Saponins (ppm)	MaNa	1	2	3	4	5	6	7	8	9	LSD
100	29!	108	97	55	76	37!	97	34!	82	43	7.7
200	22!	111	69	35!	54	27!	114	28!	63	28!	8.7
300	22!	111	58	26!	57	22!	113	21!	44	27!	7.3
400	12!	98	44!	21!	49	16!	131	20!	34!	21!	6.2
500	16!	43!	36!	13!	46	14!	124	15!	30!	17!	7.2
LSD	6.2	5.6	5.5	9.2	14.4	4.3	16.2	5.7	8.3	2.9	

Linear regression equation coefficients

x coeff. $\times 10^{-3}$	−33	−144	−147	−85	−62	−57	+73	−45	−134	−60
Constant	29	137	105	56	74	40	94	37	91	45
(95% CI)	(6)	(31)	(10)	(11)	(16)	(5)	(19)	(5)	(11)	(6)
R	−0.80	−0.78	−0.97	−0.96	−0.72	−0.95	+0.87	−0.91	−0.96	−0.93

R, correlation coefficient (significant at $P < .05$) for linear regression where x corresponds to saponin concentration and y to seedling root length; !, treatment where the browning of root tips was observed; LSD, least significant difference at $P < .005$. MaNa, medicagenic acid sodium salt; 1 - 3GlcA Ma; 2 - 3GlcA,28AraRhaXyl Ma; 3 - 3Glc Ma; 4 - 3Glc,28Glc Ma ; 5 - 3Glc,28AraRhaXyl Ma; 6 - 3GlcAGalRha Soyasap. B; 7 - AraGlcAra Hederagenin; 8 - 3 - GlcGlcGlc,23Ara,28AraRha Xyl Api zanhic acid; 9 - 3GlcGlcGlc zanhic acid.

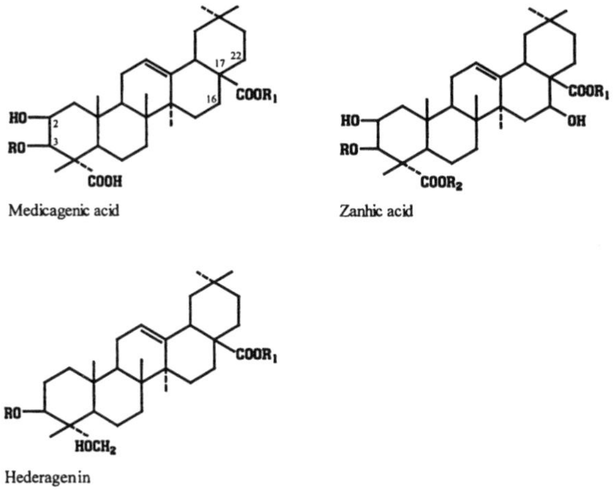

Medicagenic acid

Zanhic acid

Hederagenin

FIGURE 28.2 Chemical formulas of aglycones of alfalfa saponins tested (see also Table 28.1).

28.3.2 ALFALFA SAPONINS

Of all the saponins studied as allelopathic agents, root saponins from alfalfa have received the most attention. As early as 1954, Mamedov observed yield reductions when cotton (*Gossypium hirsutum* L.) was grown in the field after an alfalfa crop. He also showed that the fungus *Rhizoctonia ader-chaldii*, which is very harmful to cotton seedlings, grew vigorously on alfalfa residues. He concluded that it was not allelopathy, but rather a soil-borne disease responsible for the cotton yield reduction. One year later, Mishustin and Naumova (1955) reported that alfalfa root saponins were toxic to cotton and reduced cotton seed germination. They further suggested that saponins were released from alfalfa roots during the vegetative period, and accumulated in the soil in quantities high enough to harm succeeding cotton plants. To test this hypothesis, these authors compared the hemolytic indices of soil extracts and alfalfa root extracts. Interestingly, the hemolytic index of the soil extract taken from a two-year-old alfalfa stand was only two times lower than for the root extract. The average saponin concentration in alfalfa roots is about 4 to 5% plant dry weight. Theoretically, if the hemolytic activity of soil extracts was totally due to alfalfa saponins, the concentration of saponins in soils should be 1 to 2% of the soil dry weight, which is rather unlikely. Thus, the proof for the existence of pure allelopathy in alfalfa stands was quite weak. Many authors have followed this idea, quoting Mishustin and Naumova as a classical paper. The evidence to prove that saponins are really exuded into the environment from the alfalfa root system, has yet to appear. Birk (1969) reported that alfalfa saponins could be isolated from the dry and sieved soil where alfalfa had been grown, but the origin of these saponins (exudates, small lateral roots, etc.) was not explained.

Despite divergent opinions on the exudation of saponins from plant parts, there is evidence that saponins may harm plants when introduced in pure forms or as dried plant material to the growing medium or into the soil. Experiments supporting this observation are presented in Table 28.2. The symptoms of the influence of saponins on plants were similar in most experiments. Germination of seeds was least affected, but seedling growth was retarded depending on the concentration of the preparation. In general, roots showed higher sensitivity than coleoptiles. The tested plant species showed dose-dependent activities; that is, for most species there was growth stimulation at saponin concentrations ranging from 10 to 100 ppm. Exceptions were barnyardgrass [*Echinochloa crus-galli* (L.) Beauv.] and cheat (*Bromus secalinus* L.), the growth of which was inhibited even at 10 ppm (Waller et al., 1993, 1995a). Soil environment strongly modified the activity of saponins, and much higher concentrations were needed to inhibit plant growth in clay loam than in sand (Oleszek and Jurzysta, 1987). Soil microflora were able to decompose alfalfa saponins quickly. Some of the compounds were bound or sorbed to the organic soil complex, and in this state they could not affect plant growth due to reduced bioavailability. Many saponins and sapogenins were shown to have little

TABLE 28.2
Inhibition of Seed Germination and Seedling Growth by Alfalfa Preparations

Alfalfa Sample	Species Tested	Parameter Measured	References
root saponin	cotton	germination	Mishustin and Naumova, 1955
seed extract	mustard	germination	Bastek et al., 1962
root extract	grasses	seedling growth	Lawrance and Kichler, 1962
top extract	corn, wheat	germination, seedling growth	Guenzi et al., 1964; Kimber, 1973
top & root extract	tomato, wheat, barley	seedling growth	Stachurska-Bac, 1965
alfalfa meal, saponin mixture	cotton	germination, radicle growth	Pedersen, 1965
alfalfa soil	cotton	germination	Shany et al., 1970
seed saponin mix	barley, oat, wheat, rye	germination, seedling growth	Jurzysta, 1970
powdered roots	wheat	germination, seedling growth	Oleszek et al., 1987
root saponin mix	wheat	germination	Oleszek and Jurzysta, 1987
powdered roots		seedling growth	Oleszek and Jurzysta, 1987
soil extracts	barley, wheat, radish, alfalfa, clover	seedling growth	Read & Jansen, 1989
single root saponin	wheat, cotton	germination, seedling growth	Tarikov et al., 1988
root saponin mix	wheat, cheat	seedling growth	Wyman-Simpson et al., 1991
root saponin mix	dandelion, coffeeweed, pigweed, wheat, barnyardgrass	germination, seedling growth	Waller et al., 1993, 1995a
single alfalfa saponins	wheat	germination, seedling growth	Oleszek, 1993

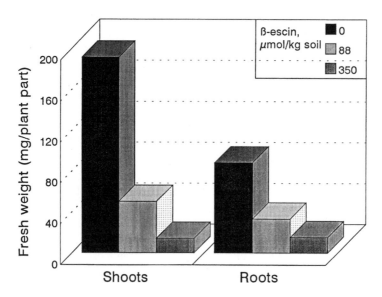

FIGURE 28.3   Effects of two concentrations of β-escin on barnyardgrass shoot and root fresh weight accumulation 11 days after treatment.

effect on seed germination of several weed and crop species (Hoagland et al., 1996a). However, seedling growth of some species was dramatically inhibited. For example, barnyardgrass and wheat were sensitive to β-escin [from horse chestnut (*Aesculus hippocastanum* L.)]. β-Escin strongly inhibited the emergence of barnyardgrass from a Bosket sandy loam soil; and after emergence, the growth of roots and shoots was severely stunted (Figure 28.3) (Hoagland et al., 1996a). Growth parameters were reduced by certain saponins (betulin, escin, glycyrrhetinic acid, and oleandrin) in different crop and weed species using several different assays (Hoagland et al., 1996a). However, biological activity varied depending on species and compound. Other related compounds, amasterol (Roy et al., 1982), chondrillasterol (Fischer, 1991; Bradow, 1985), and various terpenes and sterols (Nord and Van Atta, 1960; Duke, 1991; Duke and Kenyon, 1993) also have phytotoxic activity on a variety of plant species.

Studies of the activity of single saponins showed some correlation between compound structure and phytotoxic activity (Oleszek, 1993) (Table 28.1). All the glycosides of medicagenic acid, hederagenin, and zanhic acid were inhibitory. Soyasaponin I, in contrast, was slightly stimulatory to wheat seedling growth. Among the group of medicagenic acid glycosides, those having glucose in the C-3 position (**3, 4, 5**) showed higher activity than similar compounds substituted with glucuronic acid (**1, 2**). Zanhic acid glucosides (**8, 9**) were more active than medicagenic acid glucuronides, and monodesmosides were more active than bi- or tridesmosides. Browning symptoms were observed in meristematic areas of seedling roots. Similar browning of roots and increased anthocyanin accumulation in the stems of barnyardgrass seedlings treated with β-escin have been reported (Hoagland et al., 1996a).

28.4   INSECTICIDAL ACTIVITY

Various plant metabolites are known to be produced by plants for defense against herbivores, but saponins are not well recognized with respect to their insecticidal activity (Hostettmann and Marston, 1995). The most spectacular examples of the research performed on saponins in relation to insect feeding and development are presented in Table 28.3. Most studies evaluated alfalfa saponins and their activity against alfalfa pests, but some *in vitro* and field experiments were performed on other insects. These studies were aimed at possible practical application of saponins with agronomic

TABLE 28.3
Insecticidal Activity of Saponins of Different Origin

Saponin	Insect	Effective Concentration	Reference
alfalfa root	*Tribolium castaneum*	0.1%	Birk, 1969
alfalfa	*Empoasca fabae*	0.01-0.1%	Roof et al., 1972
Yucca saponin	*Empoasca fabae*	0.01-0.1%	Roof et al. 1972
Baker's saponin	*Acyrthosiphon pisum*	0.1%	Horber, 1972
alfalfa saponin	*Melolontha vulgaris*	residues	Horber, 1972
alfalfa extracts	*Costelytra zealandica*	1.5 mg	Sutherland et al.,1982
camellidin II	*Eurema hecabe mandarina*	—	Numata et al., 1987
alfalfa saponin	*Tenebrio molitor*	0.1%	Pracros, 1988
peduncloside	*Spodoptera litura*	0.2%	Nakatani et al., 1989
rotungenoside	*Spodoptera litura*	0.2%	Nakatani et al., 1989
alfalfa saponins	*Labesia botrana*	0.01%	Tava et al., 1992
alfalfa saponins	*Adoxophyes orana*	0.01%	Tava et al., 1992
alfalfa saponins	*Tetranychus urticae*	0.1%	Puszkar et al., 1994
alfalfa saponins	*Phorodon humuli*	0.1%	Puszkar et al., 1994
alfalfa saponins	*Leptinotarsa decemlineata*	0.5%	Waligura and Krzymañska, 1994
alfalfa saponins	*Ostrinia nubilalis*	10 mg/g	Nozzolillo et al., 1997

potential as insecticides. Thus, the activities against the Colorado potato beetle (*Leptinotarsa decemlineata*) (Waligura and Krzymanska, 1994) and hop (*Humulus* spp.) pests, including the spider mite (*Tetrarychus urticae*) and hop aphid (*Phorodon humuli* Schrank)(Puszkar et al., 1994) were tested. Spraying the leaves of plants with 0.1 to 0.2% of saponin solution reduced the number of spider mites approximately 85 to 90% and this activity persisted seven days after treatment. A structure-dependent effect on insects was shown; that is, prosapogenins obtained by alkaline hydrolysis of the purified mixture of alfalfa root saponins were much more active than genuine saponins and the medicagenic acid sodium salt itself. Prosapogenin, and to a lesser degree medicagenic acid, also killed a number of spider mite eggs. This clearly shows that monodesmosidic saponins obtained during alkaline hydrolysis are much more potent against insects than bidesmosidies that occur in genuine saponin mixtures. With respect to hop aphids, crude saponins were the most inhibitory. They reduced aphid number by 50 percent two weeks after treatment at a concentration of 0.2%.

Generally, effective concentrations of saponins are not much different among insects tested. It must be pointed out that in a number of examples, the saponin tested was very poorly characterized, thus comparison of the data is very difficult. Many of the insects were inhibited at saponin concentrations ranging from 0.01 to 0.1 percent, whether or not they were pests of alfalfa, indicating that saponins are not very species-specific with respect to insecticidal activity.

28.5 ANTIMICROBIAL ACTIVITY

The antifungal activity of saponins has been the subject of much research, primarily for pharmaceutical and agricultural purposes (Hostettmann and Marston, 1995; Grayer and Harborne, 1994; Gruiz, 1996). Agricultural interest focuses on two functions of saponins: their importance as plant protectants, that is, as possible fungicides in environmentally friendly agriculture, and their ecological or allelopathic interactions upon beneficial soil microbes and crop plant roots.

The function of saponins as protectants is based on the speculation that saponins may act as constitutive antifungal compounds (prohibitins). The best recognized cases are the oat (*Avena sativa* L.) root triterpene saponins (avenacins A-1, A-2, B-1 and B-2) that provide protection against

Gaeunamomyces graminis var *tritici* (*Ggt*), and the glycoalkaloid α-tomatine that protects tomato from fungal pathogens (Osbourn et al., 1996). Based on these two examples and some other extensive studies, several fundamental principles can be noted:

1. Toxic effects of saponins may be ascribed to their ability to complex with fungal pathogen membrane sterols; this effect is relatively nonspecific.
2. Antifungal activity is strongly attributed to the structure of the aglycone, but sugar substitution can modify this activity.
3. Saponins may be toxic in either the aglycone or glycosidic form.
4. Bidesmosidic and tridesmosidic forms are less active than corresponding monodesmosides; many exceptions can be found.
5. Fungi can detoxify saponins by the production of saponin-detoxifying enzymes (usually hydrolases); the ability of a pathogenic fungus to detoxify plant saponin may determine its host range (Bowyer et al., 1995).

Some saponins exhibit fungicidal activity at very low concentrations. Medicagenic acid 3-O-glucoside inhibited *T. viride* growth 100 percent at a concentration as low as 1.6 ppm (Oleszek et al., 1990). This compound showed high inhibitory activities against medically important yeasts at a concentration range of 3 to 15 ppm (Polacheck et al., 1986). The same range (5 to 10 ppm) was reported by Osbourn et al. (1996) for the activity of avenacin A-1 against *Ggt*, but a different isolate of the same fungus pathogenic to oats (*Gga*) required concentrations greater than 50 ppm for 100 percent inhibition. In recent research, activity of alfalfa root saponins against soil-borne fungal pathogens of cereals, *Ggt* and *Cephalosporium gramineum*, was determined (Martyniuk et al.,1995). Preparations such as the total root saponin mixture, the prosapogenin mixture obtained by alkaline hydrolysis of the root saponin mixture, medicagenic acid (Ma), MaNa, 3Glc Ma, and 3Glc,28Glc Ma were tested *in vitro* (Figures 28.4 and 28.5). The data show that the mixture of root saponins was completely inactive against *Ggt*, but was inhibitory to *C. gramineum,* providing 80 percent inhibition at 100 ppm. The most potent inhibitors of *Ggt* radial growth on a culture plate were MaNa (ID$_{50}$=3.5 ppm), 3Glc Ma (ID$_{50}$=7 ppm), and prosapogenins (ID$_{50}$=8 ppm). These substances completely inhibited

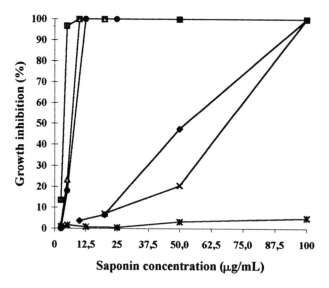

FIGURE 28.4   Inhibition by alfalfa root saponins and their derivatives of *Gaeumannomyces graminis* var *tritici* growth on corn meal agra (data adapted from Martyniuk et al., 1995).
(—◆— Ma, —⊟— Ma Na, —△— 3Glc Ma, —✕— 3,28Glc Ma, —✳— crude saponins, —●— prosapogenins

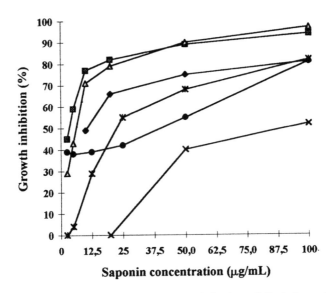

FIGURE 28.5 Inhibition by alfalfa root saponins and their derivatives of *Cephalosporium gramineum* growth on corn meal agra (data adapted from Martyniuk et al., 1995).
(──◆── Ma, ──☐── Ma Na, ──△── 3Glc Ma, ──✕── 3,28Glc Ma, ──✳── crude saponins, ──●── prosapogenins

linear growth of these pathogens at about 10 ppm. The first two compounds were also the most inhibitory to *C. gramineum*. These results indicate that some alfalfa saponins have the potential to inhibit mycelial growth and the reproduction of fungi pathogenic in cereals, but more research is needed prior to their field application. Zentmyer and Thompson (1967) demonstrated that efficacy of alfalfa saponins against *Phytophthora cinnamomi in vitro* differed significantly from that determined in soil.

Relatively little is known about the effects of saponins on bacteria, especially terrestrial and plant-associated species. The general consensus has been that the major effect of saponins on microorganisms was interaction with membrane sterols. Thus, bacteria (and also Phycomycetes, e.g., *Phytophthora*) should not be sensitive to saponins since their membranes are low in sterols. Recent studies comparing medicagenic acid-sensitive and -resistant *Trichoderma viride* strains (Gruiz, 1996) indicated that fatty acid composition is as important as sterol content with regard to saponin sensitivity. We found that bacterial growth was inhibited by the saponin β-escin, and that the effect was strain- as well as species-specific (Zablotowicz et al., 1996). Certain members of the Rhizobiaceae family (*Agrobacterium tumefaciens*, *Rhizobium meliloti*, and *Bradyrhizobium japonicum*) were the most sensitive to β-escin (Figure 28.6) (Zablotowicz et al., 1996). These effects were not as great as growth inhibition caused by β-escin in fungi such as *Trichoderma viride* (Gruiz, 1996). Structurally divergent saponins and aglycones can have differential effects on different genera of bacteria. One of the major physiological effects of saponins (and certain aglycones) on bacteria is the leakage of protein and certain enzymes from their cells (Zablotowicz et al., 1996; Hoagland et al., 1996b). Extracellular protein leakage elicited by saponins and certain aglycones from alfalfa and other plants on three species of rhizobacteria are summarized in Table 28.4 (data adapted from Zablotowicz et al., 1996; Hoagland et al., 1996b). All saponins with a glucose substituent at the C3 carbon caused leakage in all strains tested. Soyasaponin I, with a three sugar moiety (glucuronic acid–galactose–rhamnose) at the C3 carbon, elicited leakage only in the *Curtobacterium* strain. Of the aglycones, medicagenic acid and glycyrrhetic acid caused leakage, while betulin and hecogenin did not. It is of interest that medicagenic acid with glucose at both the C3 and C28 carbon was not as active as glucose substituted only at the C3 carbon, while the aglycone was more active than the 3Glc,28Glc Ma. Differential effects of various alfalfa saponins were found on enzymatic activities

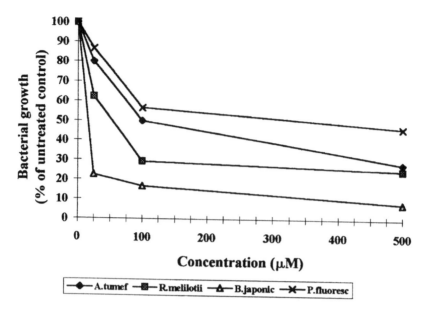

FIGURE 28.6   Effects of various concentrations of β-escin on the growth of pure cultures of strains of *Agrobacterium tumefaciens* (♦), *Bradyrhizobium japonicum* (Δ), *Pseudomonas fluorescens* (X), *Rhizobium melilotii* (■).

(i.e., FDA-esterase and TTC-dehydrogenase activities) in the *Bacillus thuringiensis* strain, but not in the *Pseudomonas fluorescens* strain (Table 28.5). *P. fluorescens* strain RA2 was uneffected; but in cells of *B. thuringiensis* UA404, 3,28NaMa and 3Glc,28NaMa reduced both TTC-dehydrogenase and FDA hydrolysis more than 95%. The 3Glc,28Glc Ma had no effect on these parameters. These data indicate that bacterial membranes can be affected by certain saponins and their aglycones, resulting in a significant loss of vital activity, especially in certain Gram-positive genera such as *Bacillus*. The ecological impact of saponins produced by crop plants and weeds on microbial populations and on certain microbial processes in soil warrants further study.

TABLE 28.4
Effect of Saponins and Aglycones from Alfalfa and Other Plants on Extracellular Protein Leakage of Three Bacterial Strains, 24 h after Treatment (500 μm)[a]

Treatment	*Pseudomonas fluorescens* RA-2	*Curtobacterium flaccumfaciens* JM-1011	*Bacillus thuringiensis* UZ404
	Protein, % of untreated control		
β-Escin	146[b]	141	168[b]
Betulin	64[b]	100	100
Hecogenin	100	106	84
Glycyrrhetic acid	185[b]	177[b]	1180[b]
3Glc,28Glc Ma	134[b]	225[b]	167[b]
3, 28 Na Ma	146[b]	492[b]	756[b]
3Glc,28 Na Ma	186[b]	766[b]	684[b]
Soyasaponin I	82[b]	200[b]	42[b]

[a]Protein was calculated on a mg/ml basis. [b]Indicates value is significantly different from untreated control value for that bacterial strain at the 95 percent confidence level.

TABLE 28.5
Effect of Alfalfa Saponins on Dehydrogenase (TTC) and Esterase (FDA) Activities of *P. fluorescens* (RA-2) and *B. thuringiensis* (UZ404)

| | nmol/ml culture/h | | | |
| | RA-2 | | UZ404 | |
Treatment	TTC	FDA	TTC	FDA
Control	56[a]	552[a]	465[a]	4536[a]
3Glc,28Glc Ma	58[a]	560[a]	557[a]	4070[a]
3, 28 Na Ma	60[a]	586[a]	9[b]	187[b]
3Glc,28Na Ma	60[a]	581[a]	13[b]	212[b]

Abbreviations: TTC, triphenyltetrazolium chloride; FDA, fluorescein diacetate.

[a,b] Values within a column followed by the same letter do not differ significantly at the 95 percent confidence level.

28.6 ECOLOGICAL FUNCTION OF ALFALFA SAPONINS

According to some authorities, allelopathic substances may only be secondarily functional in plants, having arisen initially as plant responses to herbivore pressure (Whittaker and Feeny, 1971). Distribution of particular saponins in alfalfa plants may suggest some ecological functions for these compounds. There are 12 recognized taxonomic sections within the genus *Medicago* (Small and Jomphe, 1989). Comprehensive analyses of the aglycone composition in seeds of different species of *Medicago* showed considerable difference among species (Jurzysta et al., 1992). A high frequency of hemolytic seeds are found in the sections *Spirocarpos*, *Lupularia*, and *Geocarpa* (Jurzysta et al., 1988). These elevated frequencies correlate with differences in seed biology of most species of these sections. In *Spirocarpos*, the effective propagule is the fruit itself, which is indehicent, retaining all of its seeds together for one year or more, until weathering allows moisture to penetrate and germination to occur. Seeds never escape from the fruit, and while more than one seed may germinate within the fruit, normally only one survives to produce a mature plant. Hemolytic saponins may protect seeds from insect and microbe attacks and also may increase competitiveness for survival among the seedlings within a fruit and with surrounding competitive seedlings. In the sections, *Lupularia* and *Geocarpa*, fruits are also the effective propagules. Since these fruits contain only one seed, protection from microbes and insects is of vital significance.

The *Medicago* species that tend to scatter their seeds (e.g., *M. sativa*) do not possess hemolytic seeds, but biologically active saponins are rapidly synthesized soon after the germination process begins. According to some reports, the concentration of saponins in alfalfa sprouts (measured with biological tests) may reach levels of 8.7 to 9.5 percent dry weight (Oakenfull and Sidhu, 1989; Górski et al., 1991). These concentrations are overestimates since biological tests are very sensitive to some single, highly active saponins like 3Glc Ma. Small variations in the concentration of this compound may simulate fundamental changes in total saponin concentration. In the juvenile seedling stage, this predominant saponin is 3Glc Ma, which is highly active in the *T. viride* bioassay. In fact, up to the third day of germination, seedlings contain soyasaponin I at 2 mg/g dry matter. Synthesis of medicagenic acid glycosides at levels detectable by HPLC techniques begins three days after germination (Oleszek, 1998). Zanhic acid tridesmoside was detected 12 days after germination began. In general, there is an increase in saponins from 2 mg/g during early germination, to 6 to 8 mg/g in 16-day-old seedlings, which is ten-fold less than previously reported. The high content of biologically active saponins at the beginning of seedling development might be advantageous for plant establishment by increasing seedling competitiveness and disease resistance.

TABLE 28.6
Distribution of Saponins in Mature Alfalfa Plant

Compound	Saponin Content (% in dry matter)		Activity		
	Root	Shoot	Antifungal	Allelopathic	Other
3Glc Ma	0.25	—	very high	very high	PD moderate
3Glc,28Glc Ma	0.39	—	high	high	PD inactive
3Glc,28AraRhaXyl Ma	0.40	—	high	very high	PD moderate
3GlcA,28AraRhaXyl Ma	0.70	0.37	moderate	moderate	PD moderate
3GlcA,28AraRha Ma	—	0.11	—	—	—
3AraGlcAra Hederagenin	0.11	—	very high	very high	—
3GlcAGalRha Soayapogenol B	0.05	0.26	inactive	stimulation	PD inactive
3GlcGlcGlc,23Ara 28Ara RhaXylApi Zannic acid	—	0.30	very low	low	PD very high very bitter astringent

Abbreviation: PD, potential difference across the rat intestine.

The distribution of saponins in mature alfalfa plants suggests the adaptive importance of these compounds. The concentration and composition of saponins in roots and aerial parts of alfalfa are presented in Table 28.6. These data clearly show that alfalfa roots contain glycosides with antifungal and allelopathic activities (Nowacka and Oleszek, 1994). The distribution of these compounds inside the root strongly suggests their potential for protection. Most of the saponins can be found in root epidermal tissues, while tissues toward the root axis were void of saponins (Pedersen, 1975; Quazi, 1976). Aerial parts of alfalfa do not contain saponins with antifungal or allelopathic activities, but rather house the saponins that play important roles against herbivores. Thus, aerial plant parts contain predominantly 3GlcA,28AraRhaXyl-medicagenic acid and zanhic acid tridesmoside, which has a strong bitter taste, astringent activity, and reacts with intestinal membranes, altering their permeability (Oleszek et al., 1992, 1994). These characteristics may evoke repellent properties by these compounds, providing protection against herbivores. It is unclear whether saponin distribution in roots and foliage is the result of localization of saponin synthesis and/or of restricted transport in the plant, or whether the protective function of saponins is a secondary phenomenon.

REFERENCES

Bastek, A., Kosik, J., Rut, O. and Slebodzinska, A., Studies on the allelopathy in crops and the meaning of the phenomenon in agriculture, *Zesz. Nauk WSR Wroclaw,* (in Polish) 46, 109, 1962.

Birk, Y., Saponins, *in Toxic Constituents of Foodstuffs,* Liener, I. E., Ed., Academic Press, New York, 1969, 169.

Bowyer, P., Clark, B. R., Lunness, P., Daniels, M. J. and Osbourn, A. E., Host range of a plant pathogenic fungus determined by a saponin detoxifying enzyme, *Science,* 267, 371, 1995.

Bradow, J. M., Germination regulation by *Amaranthus palmeri* and *Ambrosia artemissifolia, in The Chemistry of Allelopathy, Biochemical Interactions Among Plants,* Thompson, A. C., Ed., ACS Symposium Series No. 268, American Chemical Society, Washington, D.C., 1985, 285.

Chang, M. and Lynn, D. G., The haustorium and the chemistry of host recognition in parasitic angiosperms, *J. Chem. Ecol.,* 12, 561, 1986.

Duke, S. O., Plant terpenoids as pesticides, *in Handbook of Natural Toxins, Vol. 6, Toxicology of Plant and Fungal Compounds,* Keeler, R. F. and Ti, A. T., Eds., Marcel Dekker, New York, 1991, 269.

Duke, S. O. and Kenyon, W. H., Peroxidizing activity determined by cellular leakage, *in Target Assays for Modern Herbicides and Related Phytotoxic Compounds,* Böger, P. and Sandmann, G., Eds., Lewis Publishers, Boca Raton, FL, 1993, 61.

Fischer, N. H., Plant terpenoids as allelopathic agents, *in Ecological Chemistry and Biochemistry of Plant Terpenoids,* Harborne, J. B. and Tomas-Barberan, F. A., Eds., Claredon Press, New York, 1991, 377.

Górski, P. M., Miersch, J. and Ploszynski, M., Producion and biological activity of saponins and canavanine in alfalfa seedlings, *J. Chem. Ecol.,* 17, 1135, 1991.

Grayer, R. J. and Harborne, J. B., A survey of antifungal compounds from higher plants, 1982–1993, *Phytochemistry,* 37, 19, 1994.

Gruiz, K., Fungitoxic activity of saponins: practical use and fundamental principles, *in Saponins Used in Food and Agriculture,* Waller, G. R. and Yamasaki, K., Eds., Plenum Publishing, New York, 1996, 527.

Guenzi, W. D., Kehr, W. R. and McCalla, T. M., Water-soluble phytotoxic substances in alfalfa forage: variation with variety, cutting, year and stage of growth, *Agron. J.,* 56, 499, 1964.

Hoagland, R. E., Zablotowicz, R. M. and Reddy, K. N., Studies of the phytotoxicity of saponins on weed and crop plants, *in Saponins Used in Food and Agriculture,* Waller, G. R. and Yamasaki, K., Eds., Plenum Publishing, New York, 1996a, 57.

Hoagland, R. E., Zablotowicz, R. M., Oleszek, W. and Jurzysta, M., Effect of alfalfa saponins on rhizosphere bacteria, *Phytopathology,* 86, S97, 1996b.

Horber, E., Alfalfa saponins significant in resistance to some insects, *in Insect and Mite Nutrition,* North Holland, Amsterdam, 1972, 611.

Hostettmann, K. and Marston, A., *Saponins,* Cambridge University Press, Cambridge, England, 1995.

Jurzysta, M., Effect of saponins from seeds of lucerne on germination and growth of cereal seedlings, *Zesz. Nauk. UMK Torun,* 13, 253, 1970.

Jurzysta, M., Small, E. and Nozzolillo, C., Hemolysis, a synapomorphic discriminator of an expanded genus *Medicago* (Leguminosae), *Taxon,* 37, 354, 1988.

Jurzysta, M., Burda, S., Oleszek, W., Ploszynski, M., Small, E. and Nozzolillo, C., Chemical composition of seed saponins as a guide to the classification of *Medicago* species, *Can. J. Bot.,* 70, 1384, 1992.

Kimber, R. W. L., Phytotoxicity from plant residues. 2. The effect of time of rotting of straw from some grasses and legumes on the growth of wheat seedlings, *Plant Soil,* 38, 347, 1973.

Kitagawa, I., Yoshikawa, M. and Yosioka, I., Saponin and sapogenol XIII. Structures of three soybean saponins: soyasaponin I, soyasaponin II and soyasaponin III, *Chem. Pharm. Bull.,* 24, 131, 1976.

Kitagawa, I., Taniyama, T., Murakami, T. and Yoshikawa, M., Saponin and sapogenol. XLVI. On the constituents in aerial parts of American alfalfa, *Medicago sativa* L. The structure of dehydrosoyasaponin I, *Yakugaku Zassi,* 108, 547, 1988.

Kudou, S., Tonomura, M., Tsukamoto, C., Uchida, T., Sakabe, T., Tamura, N. and Okubo, K., Isolation and structural elucidation of DDMP-conjugated soyasaponins as genuine saponins from soybean seeds, *Biosci. Biotech. Biochem.,* 57, 546, 1993.

Kudou, S., Tonomura, M., Sukamoto, C., Uchida, T., Yoshikoshi, M. and Okubo, K., Structural eludication and physiological properties of genuine soybean saponins, *in Food Phytochemicals for Cancer Prevention. I. Fruits and Vegetables,* Huang, M. T., Toshihiko, O., Ho, C. T. and Rosen, R. T., Eds., ACS Symposium Series No. 546, American Chemical Society, Washington, D.C., 1994, 340.

Lawrance, T. and Kichler, M. R., The effect of fourteen root extracts upon germination and seedling length of fifteen plant species, *Can. J. Plant Sci.,* 42, 308, 1962

Lynn, D. G., The involvement of allelochemicals in the host selecion of parasitic angiosperms, *in The Chemistry of Allelopathy; Biochemical Interactions Among Plants,* Thompson, A. C., Ed., ACS Symposium Series No 268, American Chemical Society, Washington, D.C., 1985, 55.

Mamedov, U., Poor germination of cotton seeds in the soils taken from under the perennial grasses, *Tr. Azerbaj. C-H Inst.,* 1, 37, 1954.

Martyniuk, S., Jurzysta, M., Bialy, Z. and Wróblewska, B., Saponins as inhibitors of cereal pathogens: *Gaeumannomyces graminis* var *tritici* and *Cephalosporium gramineum,* Proceedings of the 11th International Symposium, *Modern fungicides and antifungal compounds,* Reinhardsbrunn/Friedrichroda, 1995, 193.

Massiot, G., Lavaud, C., Benkhaled, M. and Le Men-Olivier, L., Soyasaponin VI, a new maltol conjugate from alfalfa and soybean, *J. Nat. Prod.,* 55, 1339, 1992.

Mishustin, E. H. and Naumova, A. N., Secretion of toxic substances by alfalfa and their effect on cotton and soil microflora, *Izvestia Ak. Nauk CCCP,* 6, 3, 1955.

Nakatani, M., Hatanaka, S., Komura, H., Kubota, T. and Hase, T., The structure of rotungenoside, a new bitter triterpene glucoside from *Illex rotunda, Bull. Chem. Soc. Jpn.,* 62, 469, 1989.

Nord, E. C. and Van Atta, G. R., Saponin—a seed germination inhibitor, *Forest Sci.,* 6, 53, 1960.

Nowacka, J. and Oleszek, W., Determination of alfalfa (*Medicago saiva*) saponins by high-performance liquid chromatography, *J. Agric. Food Chem.,* 42, 727, 1994.

Nozzolillo, C., Arnason J. T., Campos, F., Donskov, N. and Jurzysta, M., Alfalfa leaf saponins and insect resistance, *J. Chem. Ecol.,* 23, 995, 1997.

Numata, A., Kitajima, A. and Katsuno, T., An antifeedant for the yellow butterfly larvae in *Camellia japonica*: a revised structure of camellidin II, *Chem. Pharm. Bull.,* 35, 3948, 1987.

Oakenfull, D. and Sidhu, S., Saponins, *in Toxicants of Plant Origin. Vol. II. Glycosides,* Cheeke, P. R., Ed., CRC Press, Boca Raton, FL, 1989, 97.

Oleszek, W., Allelopathic potentials of alfalfa (*Medicago sativa*) saponins: their relation to the antifungal and haemolytic activities, *J. Chem. Ecol.,* 19, 1063, 1993.

Oleszek, W., Alfalfa saponins: structure, biological activity and chemotaxonomy, *in Saponins Used in Food and Agriculture,* Waller, G. R. and Yamasaki, K., Plenum Publishing, New York, 1996, 155.

Oleszek, W., The composition and concentration of saponins in alfalfa (*Medicago sativa* L.) seedlings, *J. Agric. Food Chem.,* 1998, 46,960.

Oleszek, W. and Jurzysta, M., The allelopathic potential of alfalfa root medicagenic acid glycosides and their fate in soil environments, *Plant Soil,* 98, 67, 1987.

Oleszek, W., Jurzysta, M. and Górski, P., Alfalfa saponins—the allelopathic agent, *in Allelopathy: Basic and Applied Aspects,* Rizvi, S. J. and Rizvi, V., Eds., Chapmann and Hall, London, 1992, 151.

Oleszek, W., Jurzysta, M., Górski, P., Burda, S. and Ploszynski, M., Studies on *Medicago lupulina* saponins. 6. Some chemical characteristics and biological activity of root saponins, *Acta. Soc. Bot. Pol.,* 56, 119, 1987.

Oleszek, W., Price, K. R., Colquhoun, I. J., Jurzysta, M., Polszynski, M. and Fenwick, G. R., Isolation and identification of alfalfa root saponins: their activity in relation to a fungal bioassay, *J. Agric. Food Chem.,* 38, 1810, 1990.

Oleszek, W., Jurzysta, M., Ploszynski, M., Colquhoun, I. J., Price, K. R. and Fenwick, G. R., Zanhic acid tridesmoside and other dominant saponins from alfalfa (*Medicago sativa*) aerial parts, *J. Agric. Food Chem.,* 40, 191, 1992.

Oleszek, W., Nowacka, J., Gee, J. M., Wortley, G. M. and Johnson, I. T., Effects of some purified alfalfa (*Medicago sativa*) saponins on transmural potential difference in mammalian small intestine, *J. Sci. Food Agric.,* 65, 35, 1994.

Osbourn, A. E., Bowyer, P. and Daniels, M. J., Saponin detoxification by plant pathogenic fungi, *in Saponins Used in Traditional and Modern Medicine,* Waller, G. R. and Yamasaki, K., Eds., Plenum Publishing, New York, 1996, 547.

Pedersen, M. W., Effect of alfalfa saponin on cotton seed germination. *Agron. J.,* 57, 516, 1965.

Pedersen, M. W., Relative quantity and biological activity of saponins in germinated seeds, roots and foliage of alfalfa, *Crop Sci.,* 15, 541, 1975.

Polacheck, U., Zehavi, M., Naim, M., Levy, M. and Evron, R., Activity of compound G2 isolated from alfalfa roots against medically important yeasts, *Antimicrob. Agents Chemother.,* 30, 290, 1986.

Pracros, P., Mesure de l'activité des saponines de la luzerne par le larvesdu ver de farine: *Tenebrio molitor* L. (Coléoptére, Tenebrionidae). Recherche des fractions de saponines responsables des effects antinutritionels observés, *Agronomie,* 8, 793, 1988.

Puszkar, L., Jastrzebski, A., Jurzysta, M. and Bialy, Z., Alfalfa saponins—a chance in integrated hop protection. *Proceedings of the XXXIV Meeting IOR,* Poznan, Poland, 1994, 255.

Quazi, H. M., Distributions of saponins in roots of lucerne (*Medicago sativa* L.), *NZ J. Agric. Res.,* 19, 347, 1976.

Read, J. J. and Jensen, E. H., Phytotoxicity of water soluble substances from alfalfa and barley soil extracts of four crop species, *J. Chem Ecol.,* 15, 619, 1989.

Roof, M., Horber, E. and Sorensen, E. L., Bioassay technique for the potato leafhopper. *Empoasca fabae* (Harris), *Proceedings of the North Central Branch Entomology Society American Meeting,* Kansas City, KA, 1972, 140.

Roy, S., Dutta, A. K. and Chakraborty, D. P., Amasterol, an ecogsone precurssor and a growth inhibitor from *Amaranthus viridis, Phytochemistry,* 21, 2417, 1982.

Shany, S., Birk, Y., Gestetner, B. and Bondi, A., Lucerne saponins. III. Effect of lucerne saponins on larval growth and their detoxification by various sterols, *J. Sci. Food Agric.,* 21, 508, 1970.

Shimoyamada, M., Kudo, S., Okubo, K., Yamauchi, F. and Harada, K., Distributions of saponin constituents in some varieties of soybean plants, *Agric. Biol. Chem.,* 54, 77, 1990.

Small, E. and Jamphe, M., A synopsis of the genus *Medicago* (Leguminosae), *Can. J. Bot.,* 67, 3260, 1989.

Stachurska-Bac, A. and Szuwalska, Z., Influence of some plant extracts on the growth of tomato, wheat and barley plants, *Zesz. Nauk WSR Wroclaw, Roln. XIX,* 60, 164, 1965.

Sutherland, O. R. W., Hutchins, R. F. N. and Greenfield, W. J., Effect of lucerne saponins and *Lotus* condensed tannins on survival of grass grub, *Costelytra zealandica, NZ J. Zoology,* 9, 511, 1982.

Tarikov, C., Timbekova, A. E., Abubakirov, M. K. and Koblov, R. K., Growth regulating activity of triterpene glycosides from alfalfa (*Medicago sativa*) roots, *Uzbek. Biol. J.,* 6, 24, 1988.

Tava, A., Forti, D. and Odoardi, M., Alfalfa saponins: isolation, chemical characterization and biological activity against insects, *Proceedings of the X International Conference EUCARPIA*, Lodi, Italy, 1992, 283.

Tsurumi, S. and Tsujino, Y., Chromosaponin I stimulates the growth of lettuce roots, *Physiol. Plant.,* 93, 785, 1995.

Tsurumi, S., Takagi, T. and Hashimoto, T., A γ-pyronyl-triterpenoid saponin from *Pisum sativum, Phytochemistry,* 31, 2435, 1992.

Waligura, D. and Krzymanska, J., The influence of secondary plant substances: glucosinolaes, alkaloids and saponins on the feeding of Colorado potato beetle (*Leptinotarsa decemlineata* Say), *Proceedings of the XXXIV Session IOR*, Poznan, Poland, 1994, 9.

Waller, G. R., Jurzysta, M. and Thorne, R. L. Z., Allelopathic activity of root saponins from alfalfa (*Medicago sativa* L.) on weeds and wheat, *Bot. Bull. Acad. Sin.,* 34, 1, 1993.

Waller, G. R., Jurzysta, M. and Thorne, R. L. Z., Root saponins from alfalfa (*Medicago sativa* L.) and their allelopathic activity on weeds and wheat, *Allelopathy J.,* 2, 21, 1995a.

Waller, G. R., Cheng, C. S., Chang-Hung C., Kim, D., Yang, C. F., Huang, S. C. and Lin, Y. F., Allelopathic activity of naturally occurring compounds from mung bean (*Vigna radiata*) and their surrounding soil, in *Allelopathy: Organisms, Processes and Applications,* Inderjit, Dakshini, K. M. M. and Einhellig, F. A., Eds., ACS Symposium Series No 582, American Chemical Society, Washington, D.C., 1995b, 242.

Whittaker, R. H. and Feeny, P. P., Allelochemicals: chemical interactions between species, *Science,* 171, 757, 1971.

Wyman-Simpson, C. L., Waller, G. R., Jurzysta, M., McPherson, J. K. and Young, C. C., Biological activity and chemical isolation of root saponins of six cultivars of alfalfa (*Medicago sativa* L.), *Plant Soil,* 135, 83, 1991.

Wulff, G., Neuere Enwicklungen auf dem Saponingebiet, *Dtsch. Apoth. Ztg.,* 108, 797, 1968.

Zablotowicz, R. M., Hoagland, R. E. and Wagner, S. C., Effects of saponins on the growth and activity of rhizosphere bacteria, in *Saponins Used in Food and Agriculture,* Waller, G. R. and Yamasaki, K., Eds., Plenum Publishing, New York, 1996, 83.

Zaat, S. A. J., Wijffelman, C. A., Spaink, H. P., Brussel, A. A. N., Okker, R. J. H. and Lugtenberg, B. J. J., Induction of the *nogA* promoter of the *Rhizobium leguminosarum* Sym plasmid pRL1JI by plant flavanones and flavones, *J. Bacteriol.,* 169, 198, 1987.

Zentmyer, G. A. and Thompson, C. R., The effect of saponins from alfalfa on *Phytophthora cinnamomi* in relation to control of root rot of avocado, *Phytopathology,* 57, 1278, 1967.

29 Allelopathic Potential of Grain Sorghum (*Sorghum bicolor* [L.] Moench) and Related Species

Leslie A. Weston, Chandrashekhar I. Nimbal, and Philippe Jeandet

CONTENTS

29.1 ABSTRACT

Sorghum is an economically important forage, grain, and cover crop that has shown strong allelopathic activity in many field, greenhouse, and laboratory studies. The related hybrid cover crop, sorghum-sudangrass hybrid (SSH), and johnsongrass, a noxious weed, have also exhibited similar interference with crop and weed growth. All of these species contain a number of interesting and well-characterized phytochemicals including dhurrin and its decomposition products, phenolic acids, as well as the exceptionally active long chain hydroquinone, sorgoleone. Studies concerning the mode and site of action of these chemicals on higher plant growth have shown that these products contribute to decreased plant growth and inhibition of respiration and photosynthesis in higher plant systems.

29.2 INTRODUCTION

The change in composition of plant communities as a result of allelochemical production has been well documented in past literature (Rice, 1984). However, many of the observations made with regard to allelopathic interactions have concerned the harmful effects of weeds or perennial species on other crops or weeds. Although allelopathy has often been associated with adverse impacts of higher plants in agricultural systems, (Putnam and Weston, 1986), recent investigators also suggest that allelopathic traits of crop species be further utilized in agroecosystems to aid in weed suppression (Einhellig and Leather, 1988; Putnam and DeFrank, 1983; Putnam et al., 1983; Weston, 1990; Worsham, 1989).

Worsham and Lewis (1985) suggested that reduced weed control is currently the most important factor limiting the acceptance of no-tillage production methods. No-tillage production can be defined as production utilizing soil conserving tillage techniques that result in a crop residue

0-8493-2116-6/99/$0.00+$.50

remaining on the soil surface, or a soil surface that is left largely undisturbed during the growing season of the subsequent crop. Therefore, identification of cover crops that provide weed suppressive activity may increase acceptance and utilization of these soil-conserving tillage techniques. Cover crops are typically utilized in no-tillage production systems because they provide surface residue that may decrease soil erosion and increase water retention (Ebelhar et al., 1984). In many cases, increases in yield and labor efficiency have also resulted when minimum tillage techniques were adopted in agronomic cropping systems (Hayes, 1982; Young, 1982). Significant increases in available nitrogen and soil organic matter have also been reported over time in no-tillage systems (Ebelhar et al., 1984; Utomo et al., 1990). In states such as Kentucky, it is estimated that no-tillage production is utilized on nearly 50% of the acreage of corn (*Zea mays* L.) and soybean (*Glycine max* L. Merr.), and the trend toward no-tillage production is increasing.

In the United States, sorghum is cultivated on more than 600,000 ha for grain, with greater than 350,000 ha used for forage, green manures, and in the production of sorghum syrup (Heath et al., 1985). Sorghum is often chosen as a summer annual cover crop because of its rapid growth and ability to suppress weeds (Forney et al., 1985). Sorghum sudangrass hybrids, (SSH), (*Sorghum bicolor* (L.) Moench × *Sorghum sudanese* (Piper) Stapf) are also produced as forages and cover crops in tropical regions around the world. Past studies with SSH have shown that it is quite weed suppressive under field situations (Forney et al., 1985), and has demonstrated allelopathic activity in container media (Geneve and Weston, 1988; Iyer, 1980) and in root leachates (Forney and Foy, 1985).

Johnsongrass, *Sorghum halepense* (L.) Pers., is a related species and occurs as a noxious weed in many areas of the world. Originally native to Southern Asia and India, it is now one of the world's ten worst weed species. It is particularly difficult to control with chemicals or cultural practices since it is a perennial with a rhizomatous growth habit (Warwick et al., 1984). Nicollier et al., (1983) observed strong inhibitory activity of selected compounds extracted from johnsongrass. A significant body of literature currently documents the weed suppressive and allelopathic potential of related sorghum species. This chapter will address the studies describing this interference, the secondary products involved in these interactions, and their role in allelopathic interference mechanisms.

29.3   ALLELOPATHIC INTERACTIONS OF *SORGHUM BICOLOR*

In the field, sorghum can be produced as a summer annual cover crop, or utilized as a forage or grain crop. When sorghum residues are allowed to remain on the soil surface in no-tillage systems, or when they are tilled into the soil as green manures, they have shown strong weed suppressive potential. Overland (1966) described sorghum as a smother crop used to suppress weed populations over time. Guenzi and McCalla (1966) estimated the levels of five phenolic acids that are produced by decomposing sorghum residues and indicated that sorghum produced a high level of *p*-coumaric acid, in comparison with corn, oat (*Avena sativa* L.), and wheat (*Triticum aestivum* L.). These authors showed in later studies that toxicity from high concentrations of phenolics could persist for over half a year in field conditions (Guenzi et al., 1967), but it is difficult to estimate what levels would typically be produced by a sorghum cover crop. In orchards, Putnam (1986) showed that grain sorghum was useful in controlling weeds over time. Residues from dead sorghum covers inhibited weed growth by more than 60% in orchard cropping systems for several weeks after cover crop kill. In annual cropping systems, planting into a dead cover crop of sorghum resulted in almost complete control of certain weeds (Putnam and DeFrank, 1983). These authors attributed part of the weed suppression observed by these residues to allelopathy. Einhellig and Rasmussen (1989) also noted the strong weed suppressive potential of grain sorghum residues during subsequent establishment of row crops in Nebraska. Density of annual weeds, particularly the broadleaf weeds, was reduced by up to two to four times that found in the stubble of other crops one year following establishment of sorghum as a summer crop. Weed populations following sorghum residues were compared with those occurring in plots following corn or soybean and were two to four times less.

When sorghum residues or living plants are extracted or assayed directly, they contain a variety of water-soluble substances that can inhibit germination or seedling growth. Allelopathic effects of germinating sorghum seeds have been examined in some detail. Panasiuk et al. (1986) showed that germinating seeds of sorghum were phytotoxic to ten types of germinating weed seeds in petri dish assays. In recent studies, we observed the reduction in radicle length of several weed species including velvetleaf (*Abutilon theophrasti* Medikus), green foxtail [*Setaria viridis* (L.) Beauv.], and smooth pigweed (*Amaranthus hybridus* L.), in the presence of germinating sorghum seedlings (Hoffman et al., 1996). Optimal levels or close to optimal levels of edaphic factors such as light, temperature, nutrients, and water were provided and closely controlled to minimize any competitive effects, even in the presence of germinating sorghum (Hoffman et al., 1996). Allelopathic activity is thus observed with crop residues in the field and germinating sorghum seedlings.

In the greenhouse, residues of sorghum as well as living sorghum plants have been evaluated for allelopathic or weed suppressive activity. When weeds were interplanted with sorghum and grown under greenhouse conditions, the inhibitory effect of sorghum on some weeds was evident even after two months of growth (Panasuik et al., 1986). When sorghum was grown in proximity to weeds, significant reductions in dry matter of weeds were observed, in comparison with those produced in monocultures. In other greenhouse experiments, Putnam and DeFrank (1983) observed that the growth of purslane (*Portulaca oleracea* L.), redroot pigweed (*Amaranthus retroflexus* L.), and other small seeded species was greatly reduced (by three to five times that of the no-tillage control) when sorghum residues were present on the surface of muck or loamy sand soils (Table 29.1). However, when residues were incorporated into soil, no apparent inhibition of weed growth was observed.

Sorghum shoots produce and release a variety of chemicals, some in high quantities, which are phytotoxic to crops and weeds. Sorghums produce and release cyanogenic glycosides, and a number of phenolic breakdown products of these glycosides also contribute to short-term plant growth suppression (Einhellig and Rasmussen, 1989; Guenzi and McCalla, 1966) (Figure 29.1). The toxicity of sorghum to livestock has long been associated with the release of HCN from dhurrin (*p*-hydroxy-(S)-mandelonitrile B-D-glucopyranoside), which is present in sorghum herbage. Selection for sorghum lines containing low levels of HCN has been successful (Lamb et al., 1987). The presence of HCN has long been associated with reduced respiration and electron transport in mammalian and higher plant systems. Lehle and Putnam (1983) also found inhibitory activity in extracts of sorghum herbage and suggested that phytotoxicity may result from numerous compounds with diverse chemical composition, which remain unidentified.

Although sorghum herbage or shoot biomass can be produced in remarkably large quantities on a per plant or per hectare basis, sorghum shoot residues generally contain high levels of water and carbohydrates, leading to rapid decompositon under typical cropping situations. After cover crop kill, or tillage of shoot residues into the soil, decompostion of the shoot residue proceeds, and stubble is often negligible after 4 to 8 weeks. It is difficult to estimate how long phenolics produced by shoots may persist, but most estimates have shown that soil activity of these released phenolics tends to be limited unless large quantities are produced and released over time (Guenzi and McCalla, 1966).

Sorghum roots and the components they produce have been closely studied recently, due to the high level of activity associated with these allelochemicals. Forney and Foy (1985) found that rhizosphere products of hydroponically grown Sudex were more phytotoxic than compounds from other plant parts and toxicity increased with increasing plant age up to six weeks. They also noted a yellow pigmentation associated with the rhizosphere product, suggestive of anthocyanins. Netzley and Butler (1986) isolated sorgoleone, the oxidized quinone form of a hydrophobic *p*-benzoquinone derivative, from sorghum root exudates (Figure 29.1). Sorgoleone is the major constituent of sorghum root exudates and is easily extracted from roots in methylene chloride or ethanol. Both quinone and hydroquinone forms of the molecule are exuded by living root systems (Fate et al., 1990). The hydroquinone form is rapidly converted to the quinone form because it is unstable in the

TABLE 29.1

Total Weed Biomass as Influenced by Sorghum Residues and Simulated Tillage on a Muck and Loamy Sand Soil

Tillage	Sorghum Residue	Total Weed Biomass (g/flat)[a]	
		Houghton Muck	Spink's Loamy Sand
−	+	3.3	1.0
−	−	10.8	5.7
+	+	18.4	11.3
+	−	14.7	10.2

[a] F value for the interaction of tillage vs. none $\times$ sorghum vs. none is significant at $P < .05$

Source: From Putnam and DeFrank, 1983. With permission.

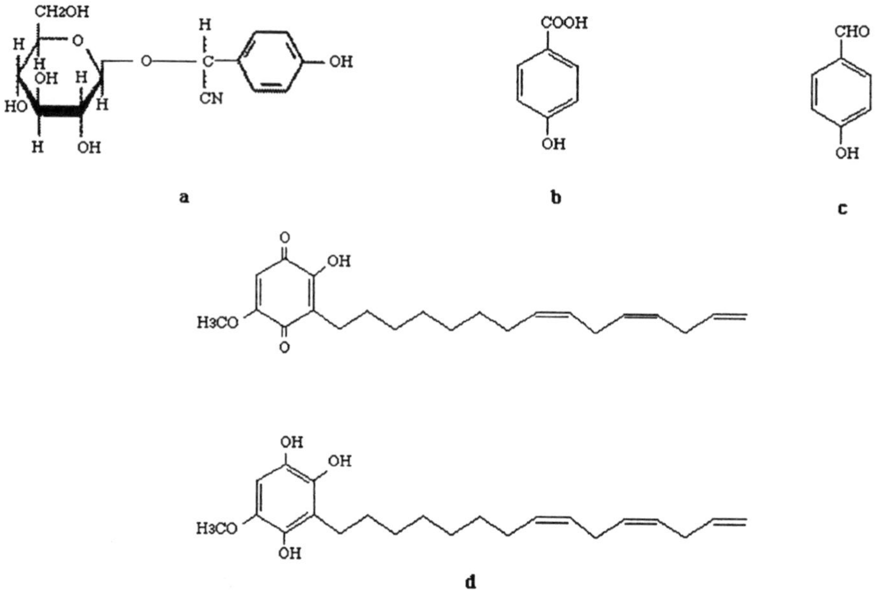

FIGURE 29.1   Structures of important phytotoxic compounds identified in shoot or root extracts of *Sorghum* spp. a) dhurrin, b) *p*-hydroxybenzoic acid, c) *p*-hydroxybenzaldehyde, d) sorgoleone as the quinone form and reduced form.

soil environment. In recent studies, Nimbal et al. (1996a) showed that sorgoleone was produced in large quantities by a variety of sorghum accessions, including *S. bicolor* and *S. sudanense*. The production of sorgoleone by five-day-old sorghum seedlings was evaluated by HPLC analysis (Table 29.2). These experiments showed that accessions produced variable quantities of sorgoleone, with certain accessions producing greater than 1% sorgoleone, or 10 mg per g fresh weight root tissue. Interestingly, the exudate of sorghum seedling roots often contains greater than 85% pure sorgoleone. This was first shown by Netzley and Butler (1986), who also indicated that sorgoleone showed some phytotoxic activity to crop plants. They were primarily interested in sorgoleone and, in particular, its hydroquinone form because of their potent activity as germination stimulants for witchweed (*Striga asiatica* L. Kuntz) (Netzley et al., 1988).

Sorgoleone likely accounts for a significant portion of the allelopathic effects of *S. bicolor* (Einhellig and Souza, 1992; Nimbal et al., 1996a). Einhellig and coworkers (1993) first showed that sorgoleone possesses remarkable phytotoxic activity in different growth assays. Our recent studies have confirmed this in several assay systems. At concentrations of less than 250 mM, sorgoleone

TABLE 29.2
Sorgoleone Production in Various Sorghum Genotypes[a]

Sorghum Genotype	Percent Germination	RFW (g)	Amount of Sorgoleone (mg)	Sorgoleone per Unit RFW (mg/g)	Percent Purity
RTX 433	65.3 ± 4.62	0.15 ± 0.03	0.10 ± 0.00	0.67	91.8
RBN 9040	42.6 ± 1.33	0.13 ± 0.01	1.00 ± 0.25	7.70	89.5
IS 3723C	82.6 ± 1.33	0.29 ± 0.01	1.23 ± 0.18	4.24	85.7
IS 8266C	88.0 ± 0.00	0.22 ± 0.01	1.10 ± 0.10	5.00	87.2
BN 122	94.7 ± 1.33	0.35 ± 0.01	2.00 ± 0.10	5.71	75.9
RTX 7078	56.0 ± 4.00	0.15 ± 0.01	1.80 ± 0.21	12.0	98.7
B. Martin	97.3 ± 2.67	0.32 ± 0.03	1.73 ± 0.13	5.40	78.2
IS 8160C	92.0 ±4.62	0.22 ± 0.03	2.00 ± 0.15	9.10	83.8
RTX 415	66.7 ± 6.67	0.20 ± 0.02	1.87 ± 0.03	9.35	92.4
RTX 430	24.0 ± 4.62	0.08 ± 0.04	0.30 ± 0.10	3.75	94.9
IS1 318C	94.7 ± 5.33	0.17 ± 0.03	2.43 ± 0.47	14.20	87.0
RTX 700	85.3 ± 5.81	0.25 ± 0.02	2.30 ± 0.32	9.20	78.9
Greenleaf	57.3 ± 0.57	0.06 ± 0.01	0.73 ± 0.47	11.4	99.1
IS 7333C	81.3 ± 1.33	0.27 ± 0.01	1.63 ± 0.27	6.00	83.4
EH - Sart	81.3 ± 4.80	0.30 ± 0.04	1.60 ± 0.21	5.33	87.1
IS 5893C	88.0 ± 4.62	0.19 ± 0.02	0.33 ± 0.07	1.74	82.4
BTX 3042	94.7 ± 1.33	0.19 ± 0.02	0.30 ± 0.06	1.58	77.7
B. Redlan	81.3 ± 1.33	0.15 ± 0.03	2.67 ± 1.31	17.80	80.2
RN 97	98.7 ± 1.33	0.25 ± 0.00	1.70 ± 1.15	6.80	81.6
Piper	76.0 ± 3.60	0.17 ± 0.01	0.17 ± 0.06	1.00	98.6
IS 1269C	60.0 ± 2.00	0.20 ± 0.03	0.77 ± 0.15	1.10	83.2
IS 7041C	96.0 ± 2.31	0.37 ± 0.02	0.60 ± 0.31	1.62	87.9
IS 1098C	53.3 ± 8.74	0.08 ± 0.01	0.20 ± 0.10	2.50	88.2
IS 12611C	84.0 ± 0.00	0.27 ± 0.02	0.67 ± 0.03	2.48	84.1
B. Wheatland	89.3 ± 1.33	0.33 ± 0.06	0.83 ± 0.23	2.50	78.3

Abbreviation: RFW, root fresh weight.

[a]The data are means (and SD) of three replicates of 25 seedlings each.

appears to have limited effects on seed germination and seedling growth. Einhellig and Souza (1992) showed some inhibition of seedling tef [*Eragrostis tef* (Zncc.) Trotter], whereas Nimbal et al. (1996a) found that aqueous solutions of sorgoleone had limited effects on germination or radicle elongation of a variety of crop and weed species even at concentrations as high as 2 mM. Sorgoleone is hydrophobic in nature and can be difficult to maintain in aqueous solutions, thus contributing to the difficulty of conducting these whole plant assays. After examining several systems, sorgoleone was successfully dissolved in minute quantities of acetone and then diluted with Hoagland's solution, followed by sonication. In this manner, it was apparent that sorgoleone was quite phytotoxic to velvetleaf, barnyardgrass [*Echinochloa crus-galli* (L.) Beauv.], and large crabgrass [*Digitaria sanguinalis* (L.) Scop.] seedlings (Einhellig and Souza, 1992; Nimbal et al., 1996a), with tolerance exhibited by the large-seeded species, ivyleaf morning glory [*Ipomoea hederacea* (L.) Jacq.] (Nimbal et al., 1996a) (Table 29.3). These experiments provide strong evidence that sorgoleone is phytotoxic at micromolar concentrations and is selective among plant species.

Sorgoleone causes tissue bleaching in sensitive photosynthetic tissues as well as interference with respiratory metabolism, and its mode of action in higher plants is thought to be related to inhibition of electron transport. Low concentrations of sorgoleone interfered with photosynthesis and reduced O_2 evolution by photosynthesizing leaf discs (Einhellig et al., 1993). Rasmussen et al. (1992) found that sorgoleone inhibited O_2 uptake (state III and state IV respiration) in isolated

TABLE 29.3
Effect of Sorgoleone on the Growth of 15-day-old Weed Seedlings Under Hydroponic Culture

Sorgoleone Concentration (μM)	Shoot Dry Weight (g)				Root Dry Weight (g)			
	Large Crabgrass	Barnyardgrass	Velvetleaf	Ivyleaf Morning-Glory	Large Crabgrass	Barnyardgrass	Velvetleaf	Ivyleaf Morning-Glory
0	0.11	0.33	0.13	0.27	0.20	0.17	0.06	0.07
10	0.06	0.31	0.15	0.25	0.20	0.13	0.05	0.08
50	0.04	0.27	0.09	0.28	0.20	0.09	0.04	0.09
100	0.04	0.27	0.08	0.25	0.20	0.08	0.04	0.09
200	0.02	0.18	0.06	0.26	0.20	0.08	0.03	0.10
GR_{50} [a]	~10	~200	~200	>200	>200	~100	~200	>200

[a] GR_{50} is the concentration of sorgoleone required for 50 percent growth inhibition. GR_{50} values were estimated by linear regression analysis.

Source: From Nimbal et al., 1996a. With permission.

soybean and corn mitochondria, perhaps by blocking electron flow in the mitochondrial electron transport system at concentrations of less than 0.4 mM. The hydrophobic nature of sorgoleone and its action on mitochondria also suggested it might strongly interfere with photosynthetic electron transport. In studies recently conducted by Nimbal et al. (1996b), the inhibition of photosynthetic electron transport in isolated thylakoids of spinach (*Spinacea oleracea* L.) and cell cultures off common groundsel (*Senecio vulgaris* L.), redroot pigweed and potato (*Solanum tuberosum* L.) was investigated. Sorgoleone was equally or more active than diuron (DCMU), with an I_{50} at 0.09 µM. Sorgoleone competed effectively with atrazine, in a manner similar to that of diuron (Figure 29.2) and metribuzin (data not shown), two potent photosynthetic inhibitor herbicides. Competitive binding of sorgoleone was not observed in atrazine-resistant pigweed or potato thylakoids. We conclude from these studies evaluating herbicide binding that sorgoleone is a remarkably potent inhibitor of photosynthetic electron transport and binds to a Q_B niche of the D1 protein in a manner similar to other PSII inhibitors, such as diuron.

In further studies, the effects of sorgoleone on photosynthetic electron transport in O_2-evolving chloroplast thylakoids as well as Triton-X-100 prepared photosystem II (PS II) membranes were evaluated (Gonzalez et al., 1998). The Hill activity of the thylakoids proved to be at least as sensitive to sorgoleone as it was to diuron (DCMU), a potent herbicidal inhibitor of PSII. However, a photosystem I (PSI) partial reaction was not affected by a tenfold greater concentration of sorgoleone than is required for complete inhibition of Hill activity. Measurements of flash-induced Chl *a* variable fluorescence showed that sorgoleone neither dissipated excitation energy nor diminished the amplitude of Chl *a* variable fluorescence. However, sorgoleone inhibited the decay of variable fluorescence as effectively as diuron, which blocks the oxidation of the PSII secondary electron acceptor, Q_B, by displacing Q_B from the D1 protein. Additionally, sorgoleone competitively inhibited the binding of ^{14}C-atrazine to the Q_B locus. Increasing durations of trypsin proteolysis of the Q_B binding niche caused parallel losses of inhibition of O_2 evolution from sorgoleone and diuron, as well as from bromoxynil, a benzonitrile type herbicide also binding to the Q_B locus. Sorgoleone is therefore a potent inhibitor of electron transfers between Q_A and Q_B at the reducing side of PSII, and thus is potentially a very effective herbicide.

Recent studies evaluating the interaction of sorgoleone within the Qb binding pocket using 3D computer imaging analysis were conducted. The electrostatic charge distribution of the sorgoleone molecule is markedly similar to that of other PSII inhibitors, thereby allowing for similar tight binding within the Qb binding site and strong competitive inhibition with other moleculess binding at this site. Remaining studies must now determine the actual significance of sorgoleone to sorghum allelopathy, especially in a natural soil setting.

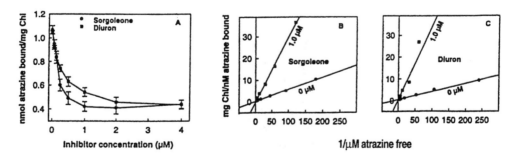

FIGURE 29.2   Competitive binding of labeled atrazine vs. sorgoleone or diuron to spinach thylakoids. Inhibitor concentration was varied while keeping atrazine concentration at 0.4 µM (A). Double reciprocal plots for competitive binding in the presence or absence of 1.0 µM sorgoleone (B) and diuron (C). Data adapted with permission from Nimbal et al. 1996b.

Recently, we closely evaluated the chemical constituents of root exudates of *Sorghum bicolor* cv. Pioneer 8333. Root exudates of five-day-old sorghum seedlings were evaluated using mass spectrometry (CI, EI, FAB, GC-MS with TMS derivatization, and electrospray MS) and ^{13}C and ^{1}H NMR spectroscopy techniques. Consistent with the report of Fate et al. (1990), we discovered that, in general, greater than 80% of newly extracted exudates consisted of the quinone form of the molecule, with a smaller percentage in the reduced hydroquinone form. In addition, we have characterized five or more constituents of the root exudate that are present to a lesser extent. One structure appears to be novel because of different substituents on the main ring. The others are also closely related with slight modifications in the bis-allylic methylene bonding pattern of the side chain. Structurally, these compounds are likely the result of sorghum biosynthesis, rather than artifacts of the extraction process (Jan St. Pyrek, personal communication; Ogawa and Natori, 1968; Thomson, 1971). Fate et al. (1990) also reported the presence of several closely related hydroquinones in freshly collected root exudates.

Recently, we began close examination of the stability of sorgoleone in soil. We discovered that sorgoleone has stability, not only in the soil but also in chemical degradation experiments. In the soil, we developed methods for extraction and quantification over time, using methanolic extraction, filtration, and HPLC analysis, similar to those used by Kelley et al., 1994. Sorgoleone concentration decreases over time in sterile and non-sterile systems (unpublished data), but the parent compound can be recovered and quantified from the soil medium. Currently, we are attempting isolation and evaluation of sorgoleone and its major metabolites in soil. At least one to two polar metabolites appear prevalent in all systems. In whole plant assays, when sorgoleone was added directly to a sand/Maury silt loam soil (50/50 vol/vol) system, lettuce (*Lactuca sativa* L.) growth reduction was positively correlated with increasing sorgoleone concentration. Growth reduction and chlorosis was noted even at the lowest concentration incorporated into soil (10 ppm), after 14 d growth following planting of pregerminated seeds into a treated sand/Maury silt loam medium (Nimbal and Weston, 1996). In this way, bioassays for sorgoleone concentration can be developed as well as studies evaluating species tolerance to sorgoleone over time, in conditions simulating field settings.

In chemical degradation experiments, sorgoleone exhibited considerable stability compared with other natural products of phenolic structure, when exposed to degradative reagents. It may be that the bis-allylic methylene bonding pattern (with three double bonds in the side chain) or antioxidant activity associated with the presence of quinone functions are responsible for the stability and persistence of this compound over time in the laboratory or in the field. Very little is known about the specific functional roles of root exudates such as sorgoleone within the soil rhizosphere. Related quinones occurring in plant tissues, such as irisoquin, have also been shown to be potent metabolic toxins (Ogawa and Natori, 1968). It is certainly possible that such a metabolically expensive process as the extensive hydroquinone exudation by *Sorghum* spp. is an ecologically active defense mechanism.

29.4 ALLELOPATHIC INTERACTIONS OF *SORGHUM BICOLOR* × *SORGHUM SUDANENSE* (SSH) AND JOHNSONGRASS (*SORGHUM HALEPENSE*)

SSH exhibits great potential to produce large quantities of biomass rapidly when selected for use as a summer annual cover crop. Typically, SSH achieves greater height and biomass than the related *Sorghum bicolor* in a shorter period of time. SSH has been evaluated in the field and greenhouse for weed suppressive potential. In the field, Forney et al. (1985) obtained excellent suppression of broadleaf species when Sudex was used as a green manure crop preceding alfalfa (*Medicago sativa* L.) establishment. Iyer et al. (1980) demonstrated that Sudex incorporated into a container soil medium inhibited the growth of *Pinus* seedlings. In studies conducted with Eastern redbud (*Cercis canadensis* L.), seedling growth was significantly reduced when seedlings were co-cultivated with living SSH and when SSH leaf tissue was incorporated into the growth medium (Geneve and Weston, 1988). Growth reductions could not be reversed with increased fertilizer rates. Redbud growth decreased linearly with increasing amounts of fresh or dried SSH incorporated into the growth medium. Interestingly, when redbud seedlings were planted into living SSH in which shoots were repeatedly trimmed to short heights to minimize shoot competition effects, growth of redbud seedlings over a seven-week period was completely inhibited. In comparison, when shoot residues were incorporated, up to 25 percent decrease in growth was observed. This indicates that living root systems of SSH may be particularly important in the allelopathic interference mechanism.

In past experiments, we evaluated the phytotoxicity and chemical constituents produced by SSH shoots cv. FFR 201 (Weston et al., 1989). Allelopathic potential, as measured by radicle elongation of herbaceous indicator species, decreased with increasing Sudex age. Greatest potential allelopathic activity of sudex shoot tissue was observed when Sudex was collected at seven days of age. Small-seeded broadleaf species were more inhibited than were grass species (Table 29.4 and Weston et. al, 1989). Two major phytoinhibitors were isolated from aqueous extracts of Sudex shoot material by partitioning with diethyl ether, followed by thin layer and liquid column chromatography. Phytoinhibitors were identified as the enzymatic breakdown products of the cyanogenic glycoside dhurrin and included *p*-hydroxybenzoic acid and *p*-hydroxybenzaldehyde (Figure 29.1). The specific activity of these compounds in seed bioassays ranged from 100 to 140 ppm, typical for many related phenolics. SSH tissue seven days old possessed a greater percentage of these inhibitors on a per gram basis than did older Sudex tissue.

Allelopathy in johnsongrass shoot residues has been attributed in the past to phenolic acids, primarily *p*-hydroxybenzoic acid, released upon degradation of dhurrin and also to the HCN produced (Abdul Wahab and Rice, 1967) (Figure 29.1). Nicollier et al. (1983) assessed the activity of dhurrin and other compounds and concluded that significant inhibition occurred as a result of other phenolic compounds in johnsongrass extracts.

Forney and Foy (1985) also discovered that root leachates from hydroponically grown Sudex inhibited the growth of both monocot and dicot seedlings. Purification of collected Sudex phytotoxins was not attempted. Specifically, when chloroform soluble products were collected from the leachates of SSH, johnsongrass, rye (*Secale cereale* L.), and cucumber (*Cucumis sativus* L.), the exudates from johnsongrass were most toxic, followed by SSH and rye as well as cucumber. The acidic fraction of the chloroform extracts of johnsongrass was the most toxic overall. The products collected from the johnsongrass rhizospheres were similar in color to those collected from SSH rhizospheres.

TABLE 29.4

Influences of Age of SSH Shoot Residue upon Radicle Elongation after 96 h and the Quantities of *p*-Hydroxybenzoic Acid and Aldehyde Obtained after Partitioning and HPLC Quantification of 100 g of SSH Shoot Residues of Various Ages

	Radicle Length (mm)		Ether Extract	*p*-Hydroxy-benzoic Acid	*p*-Hydroxy-benzaldehyde
Residue Age (Days)	Tomato	Foxtail Millet	mg/100 g	mg/100 g	mg/100 g
Control	9.1	35.8	—	—	—
7	0.7	19.5	230	136.6	2.6
14	2.4	21.2	113	37.1	2.1
21	7.7	26.4	75	13.7	1.7
28	7.8	36.4	55	8.1	1.6
LSD (0.05)	4.3	7.3			

Data adapted from Weston et al., 1989.

Our work with various sorghum accessions has shown that SSH accessions produce sorgoleone, and often in high quantities (Nimbal et al., 1996a). Due to similarities in toxicity and appearance of johnsongrass and SSH root exudates, this exceptionally toxic allelochemical is likely also produced by johnsongrass rhizomes or roots. Using HPLC and TLC, we have recently determined that johnsongrass produces considerable quantities of sorgoleone and related components.

In conclusion, sorghum is an economically important forage, grain, and cover crop that has shown strong allelopathic activity and potential for activity in many field, greenhouse, and laboratory studies. The related hybrid cover crop, SSH, and johnsongrass, a noxious weed, have also exhibited similar interference with crop and weed growth. These species all contain a number of interesting and well-characterized phytochemicals including dhurrin and its breakdown products, HCN, and phenolic acids, and the exceptionally active long chain hydroquinone, sorgoleone. Studies concerning the mode and site of action of these chemicals upon higher plant growth have shown that these products contribute to decreased overall plant growth and reductions in respiration and photosynthetic efficiency in higher plant systems. It is possible that in the future, the development of sorghum cover crops with enhanced weed suppressive potential could allow for long-term suppression of weeds in agroecosystems. A knowledge of how these products are produced in related plant species, metabolized, and finally degraded in soil may enable us to pursue these goals.

REFERENCES

Abdul-Wahab, A. S. and Rice, E. L., Plant inhibition by johnsongrass and its possible significance in old-field successicon, *Bull. Torrey Bot. Club*, 94, 486, 1967.

Ebelhar, S. A., Frye, W. W., and Blevins, R. L., Nitrogen from legume cover crops for no-tillage corn, *Agron. J.*, 76, 51, 1984.

Einhellig, F. A. and Leather, G. R., Potentials for exploiting allelopathy to enhance crop production, *J. Chem. Ecol.*, 14, 1829, 1988.

Einhellig, F. A. and Rasmussen, J. A., Prior cropping with grain sorghum inhibits weeds, *J. Chem. Ecol.*, 15, 951, 1989.

Einhellig, F. A., Rasmussen, J. A., Hejl, A., and Souza, I. F., Effects of root exudate sorgoleone on photosynthetics, *J. Chem. Ecol.*, 19, 369, 1993.

Einhellig, F. A. and Souza, J. F., Allelopathic activity of sorgoleone, *J. Chem. Ecol.*, 18, 1, 1992.

Fate, G., Chang, M., and Lynn, D. G., Control of germination in *Striga asiatica:* chemistry of spatial definition, *Plant Physiol.*, 93, 201, 1990.

Forney, D. R. and Foy, C. L., Phytotoxicity of products from rhizospheres of a sorghum-sudangrass hybrid (*S. bicolor* × *S. sudanenase*), *Weed Sci.*, 33, 597, 1985.

Forney, D. R., Foy, C. L., and Wolf, D. D., Weed suppression in no-till alfalfa (*Medicago sativa*) by prior cropping of summer annual forage grasses, *Weed Sci.*, 33, 490, 1985.

Geneve, R. L. and Weston, L. A., Growth reduction of eastern redbud (*Cercis canadensis* L.) seedlings caused by interaction with a sorghum-sudangrass hybrid (sudex), *J. Environ. Hortic.*, 6, 24, 1988.

Gonzalez, V., Nimbal, C. I., Weston, L. A., and Cheniae, G. M., Inhibition of a photosystem II electron transfer reaction by sorgoleone, a natural product, *J. Agric. Food Chem.*, 1998, 45, 1415–1421.

Guenzi, W. D. and McCalla, T. M., Phenolic acids in rats, wheat, sorghum and corn residues and their phytotoxicity, *Agron. J.*, 58, 303, 1966.

Guenzi, W. D., McCalla, T. M., and Nordstadt, F. A., Presence and persistence of phytotoxic substances in wheat, oat, corn and sorghum residues, *Agron. J.*, 59, 163, 1967.

Hayes, W. A., *Minimum Tillage Farming*, No-Till Farmer, Brookfield, WI, 1982.

Heath, M. F., Barnes, R. F., and Metcalfe, D. S., *Summer Annual Grasses in Forages: The Science of Grassland Agriculture*, Iowa State University Press, 1985, 278.

Hoffman, M. L., Weston, L. A., Snyder, J. C., and Regnier, E. E., Interference mechanisms between geminating seeds and between seedlings: bioassays using cover crop and weed species, *Weed Sci.*, 44, 579, 1996.

Iyer, J. G., Wilde, S. A., and Corey, R. B., Green manure of sorghum-sudan: it's toxicity to pine seedlings, *Tree Planter's Notes*, 31, 11, 1980.

Kelley, W. T., Coffey, D., and Mueller, T. C., Liquid chromatographic determination of phenolic acids in soil, *J. Assoc. Anal. Chem. Int.*, 77, 805, 1994.

Lamb, J. F. S., Haskins, F. A., Gorz, H. J., and Vogel, K. P., Inheritance of seedling hydrocyanic acid potential and seed weight in sorghum-sudangrass crosses, *Crop Sci.*, 27, 522, 1987.

Lehle, F. R. and Putnam, A. R., Allelopathic potential of sorghum (*Sorghum bicolor*): isolation of seed germination inhibitors, *J. Chem. Ecol.*, 9, 1223, 1983.

Netzley, D. H. and Butler, L. G., Roots of sorghum exude hydrophobic droplets containing biologically active components, *Crop Sci.*, 26, 776, 1986.

Netzley, D. H., Reopel, J. L., Ejeta, G., and Butter, L., Germination stimulants of witchweed (*Striga asiatica*) from hydrophobic root exudate of Sorghum (*Sorghum bicolor*), *Weed Sci.*, 36, 441, 1988.

Nicollier, J. F., Pope, D. F., and Thompson, A. C., Biological activity of dhurrin and other compounds from Johnsongrass (*Sorghum halepense*), *J. Agric. Food Chem.*, 31, 744, 1983.

Nimbal, C. I., Pedersen, J., Yerkes, C. N., Weston, L. A., and Weller, S. C., Phytotoxicity and distribution of sorgoleone in grain sorghum germplasm, *J. Agric. Food Chem.*, 44, 1343, 1996a.

Nimbal, C. I., Yerkes, C. N., Weston, L. A., and Weller, S. C., Herbicidal activity and site of action of the natural product sorgoleone, *Pestic. Biochem. Physiol.*, 54, 73, 1996b.

Nimbal, C. I. and Weston, L. A., Mode of action of sorgoleone, a natural product isolated from *S. bicolor*, First World Congress on Allelopathy, Abstracts, Cadiz, Spain, 1996.

Ogawa, H. and Natori, S., Hydroxybenzoquinones from myrsinaceae plants. II. Distribution among myrsinaceae plants in Japan, *Phytochemistry*, 7, 773, 1968.

Overland, L., The role of allelopathic substances in "smother crop" barley, *Am. J. Bot.*, 53, 423, 1966.

Panasuik, O., Bills, D. D., and Leather, G. R., Allelopathic influence of *S. bicolor* on weeds during germination and early development of seedlings, *J. Chem. Ecol.*, 12, 1533, 1986.

Putnam, A. R., Allelopathy: can it be managed to benefit horticulture?, *Hort. Sci.*, 21, 411, 1986.

Putnam, A. R. and DeFrank, J., Use of phytotoxic plant residues for selective weed control, *Crop Protect.*, 2, 173, 1983.

Putnam, A. R. and Weston, L. A., Adverse impacts of allelopathy in agricultural systems, in *The Science of Allelopathy*, Putnam, A. R. and Tang, L. S., Eds., John Wiley & Sons, New York, 1986, 43.

Putnam, A. R., DeFrank, J., and Barnes, J. P., Exploration of allelopathy for weed control in annual and perennial cropping systems, *J. Chem. Ecol.*, 9, 1001, 1983.

Rasmussen, J. A., Heil, A. M., Einhellig, F. A., and Thomas, T. A., Sorgoleone from root exudate inhibits mitochondrial functions, *J. Chem. Ecol.*, 18, 197, 1992.

Rice, E. L., *Allelopathy*, 2nd ed., Academic Press, Orlando, FL, 1984.

Thomson, R. H., *Naturally Occurring Quinones*, Academic Press, New York, 1971, 9.

Utomo, M., Frye, W. W., and Blevins, R. L., Sustaining soil nitrogen for corn using hairy vetch cover crops, *Agron. J.*, 82, 979, 1990.

Warwick, S. I., Thompson, B. K., and Black, L. D., Population variation in *Sorghum halpense*, Johnsongrass, at the northern limits of its range, *Can. J. Bot.*, 62, 1781, 1984.

Weston, L. A., Cover crop and herbicide influence on row crop seedling establishment in no-tillage culture, *Weed Sci.*, 38, 166, 1990.

Weston, L. A., Harmon, R., and Mueller, S., Allelopathic potential of sorghum-sudangrass hybrid (sudex), *J. Chem. Ecol.*, 15, 1855, 1989.

Worsham, A. D., Use of allelopathic cover crops for weed suppression, *Alternatives in Pest Management Workshop*, University of Illinois, IL, 1989.

Worsham, A. D. and Lewis, W. M., Weed management: key to no-tillage crop production, *in Proceedings of the 1985 S. Region No-Till Conference*, Hargrove, W. L. and Buswell, F. C., Eds., Griffin, GA, 1985, 177.

Young, H. M., Jr., *No-Tillage Farming*, No-Till Farmer, Brookfield, WI, 1982.

30 An Integrated View of Allelochemicals Amid Multiple Stresses

F. A. Einhellig

CONTENTS

30.1 ABSTRACT

The focus of this chapter is on building the conceptual framework for allelopathy as an integrated component among many interactive stresses. Case studies of allelochemical inhibition identify multiple compounds that contribute to the process. A noninhibitory concentration of a specific compound inhibits growth when this compound acts additively or synergistically with other allelochemicals that are present, and this joint action is the normal situation. Abiotic and biotic stresses in the plant environment will enhance the physiological effects of allelochemicals. Data show that the concentration of ferulic acid required to inhibit seedling growth is lower when there is an associated heat stress, mild moisture stress, or herbicide involvement. Sensitivity to allelopathy is also greater when there is infection by pathogens, insect herbivory, nutrient deficiency, certain soil organics, and a variety of other factors such as plant–plant competition. For the donor plant, many of these same stresses cause an increase in tissue levels of allelochemicals and, subsequently, the toxicity of the leachates and residues from these plants. Thus, in a holistic perspective, the relative presence and activity of allelochemicals depend on multiple stresses in a natural community or agroecosystem. Recognizing stress interactions involving allelochemicals is essential for explanations of allelopathy, and this information has implications for capitalizing on allelopathy in crop production systems.

30.2   INTRODUCTION

By definition and investigative understanding, allelopathy is a process involving chemical interactions among plants in certain natural communities and agroecosystems. As with other facets of ecology, allelopathy is not an isolated event, but in the broad scope it involves a complex set of interactions. Hence, it is not surprising that the more we dissect the process of allelopathy to focus on and study specific events, the less these components resemble reality. This should not imply that it is unnecessary to simplify the system and limit the variables in investigations that seek to provide insights on allelopathy. However, in the interpretation of such data, one must recognize that the whole complex of interactions may not have been considered. The goal of this chapter is to show that multiple stress factors in the plant environment, both abiotic conditions and stresses of biotic origin, alter the quantity of allelochemicals produced and the impact allelochemicals have on other plants.

It has long been recognized that the plant environment is a dynamic complex of interrelationships that include both interactions between components of the physical environment and those that directly involve plants (Billings, 1952). The plant environment is almost always less than ideal, typically having components that diverge far enough from the optimum to be a measurable stress that may reduce productivity. While moisture and temperature conditions are among the more common stresses, a variety of interactions with associated plant species and other organisms add to the stress dynamics. Plants are rarely exposed to a single stress; adjustments to several concurrent stresses at one time are required. Also, the duration and intensity of stress and its associated growth detriment or damage vary markedly within short time-frames as well as during the life cycle of a plant. Discounting the dimension of time, a holistic model can be used to illustrate the variety of plant-stress interactions that include allelopathy (Figure 30.1). These dynamics include stress enhancement of allelochemical production and the effects associated stresses have on the action of allelochemicals. The model moves toward a more integrated perspective on plant stress than earlier descriptions as a stress triangle (Einhellig, 1989, 1996).

30.3   ALLELOCHEMICAL COMPLEXES

30.3.1   THE DIVERSITY OF COMPOUNDS

Allelopathy rarely, if ever, results from the action of a single compound. Every in-depth analysis of chemicals arising from an allelopathic plant has shown more than one compound with the capability for affecting growth. Often, studies have focused only on allelochemicals of one chemical class, yet regularly these investigations show the presence of several compounds. Phenolic acids have been the most frequently reported chemicals implicated in allelopathy, and multiple phenolic acids are

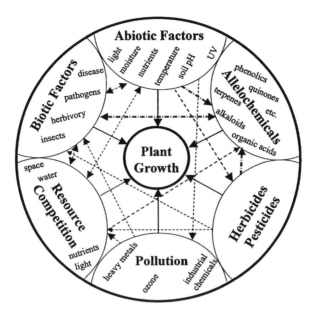

FIGURE 30.1 Paradigm of the complex interactions between environmental factors and plant growth. Solid lines show direct plant effects; dash lines show factor interactions; double-dash lines show known interactions with allelochemicals. Interactive dynamics also occur among factors within an arc.

consistently reported (Inderjit, 1996). A single example will suffice: Chou and Lin (1976) connected autotoxicity of rice (*Oryza sativa* L.) to *p*-hydroxybenzoic, *p*-coumaric, vanillic, ferulic, and *o*-hydroxyphenylacetic acids, and additional investigations have supported the role of phenolic acids in allelopathic effects of rice (Chou et al., 1981; Rice et al., 1981).

Allelopathic case studies applying analytical techniques to isolate and identify more than one class of allelochemicals typically report contributions from several different classes of compounds. Several examples chosen for compound diversity illustrate this reality. Chou and Waller (1980) noted that coffee (*Coffea arabica* L.) allelopathy involved both phenolic acids and phytotoxic purine alkaloids. Vegetational patterns arising from allelopathic effects of prostrate knotweed (*Polygonum aviculare* L.) were first associated with several phenolics, but additional studies showed 14-22 carbon, long-chain fatty acids were part of the toxicity (Alsaadawi et al., 1983). Interference by *Parthenium hysterophorus* L., a well-studied aggressive weed, is due to the combination of sesquiterpene lactones and phenolics (Kanchan and Jayachandra, 1980). Parthenin and coronopilin are prominent among the sesquiterpene lactones; phenolic acids include caffeic, vanillic, *p*-coumaric, anisic, *p*-hydroxybenzoic, chlorogenic, and ferulic. Bradow and Connick (1988a,b) isolated and tested an array of methyl ketone and aliphatic alcohol and aldehyde inhibitors arising from Palmer amaranth (*Amaranthus palmeri* S. Wats.). Autotoxicity of alfalfa (*Medicago sativa* L.) has been attributed to triterpenoid glycosides known as saponins (Oleszek, 1993), with Wyman-Simpson et al. (1991) reporting an average of 14 different saponins per cultivar. However, alfalfa seedlings also contain biologically active canavanine and non-protein amino acids (Gorski et al., 1991).

It is not unusual to have multiple compounds representing more than two chemical classes implicated in allelopathy. A variety of cinnamic and benzoic acids, tannins, flavonoids, several cyanogenic glycosides, and *p*-benzoquinones known as sorgoleone cause *Sorghum* allelopathy (Einhellig, 1995b). To date, more than 30 compounds have been identified with allelopathy in the Florida scrub community, including a large number of mono-, sesqui-, di-, and triterpenoids, several types of flavonoids, acetophenone, and hydrocinnamic acid (Fisher et al., 1988, 1994). Some of the allelochemicals in this scrub community are generated by photochemical activation and degradation of the original plant-produced compounds after their release into the environment.

30.3.2 ADDITIVE OR SYNERGISTIC EFFECTS

The axiom that the whole is greater than the sum of its parts also holds true for allelopathy since almost all allelochemical chemicals are found in the environment in very low concentrations. Tests conducted with single allelochemicals at their estimated concentration in the environment often show either no effects on growth or a very minor impact. However, inhibition by the total allelochemical complex may be significant. A recent illustration is the study by Lydon et al. (1997) on wormwood (*Artemisia annua* L.) inhibition. Artemesinin is recognized as the major allelochemical in wormwood. However, the Lydon et al. (1997) study of seedling emergence, growth, and survival in soil pots that had incorporation of either pure artemisinin or leaf-tissue extracts showed that the effects of treatment with extracted levels of pure artemisinin alone were much less than those caused by the total leaf extract. This example is one of many illustrating that the total allelopathic activity from a plant is not explained by a single compound even when other contributors remain unidentified.

It is now common place for investigators to run bioassays with a mixture of allelochemicals (Bradow and Connick, 1988a,b; Blum et al., 1992; Boufalis and Pellissier, 1994; Fisher et al., 1994; Weidenhamer et al., 1994; Veronneau et al., 1997), thereby highlighting the joint action of compounds. However, only a few studies have incorporated the necessary controls to establish the relative contribution of each compound. Analyses to determine additive or synergistic effects in allelochemical mixtures are complicated by the fact that any one allelochemical in the environment is almost always below the inhibition-threshold concentration. We know little about effects of subthreshold allelochemical exposures on plant physiology (Figure 30.2). Relatively low concentrations are often slightly stimulatory, in contrast to inhibitory effects by the same compound at higher levels. Combination experiments are further complicated in that the inhibition thresholds vary among compounds, types of bioassays, selectivity of the bioassay plants, and environmental conditions. The seedling growth-inhibition threshold for many coumarins, cinnamic and benzoic acids, flavonoids, monoterpenes, and sesquiterpene lactones is in the 100 to 1000 μM range, but some allelochemicals have thresholds below this range (Einhellig, 1995b). Inhibition thresholds in seed germination bioassays are often higher than for seedling growth.

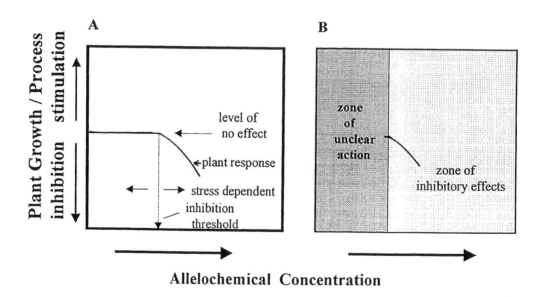

Allelochemical Concentration

FIGURE 30.2 Diagrammatic representation showing (A) an inhibition threshold that varies according to the compound, the species or process, and other environmental stress; and (B) subthreshold, shaded zone of unknown physiological action.

30.3.2.1   Similar Compounds

The cinnamic and benzoic acids appear to have a similar mode of action, altering membrane functions, which results in an array of changes in plant metabolism (Einhellig, 1995b). Their effects include changes in nutrient uptake and retention and overall plant–water relationships. Hence, it is not surprising that combinations of these compounds have at least an additive effect.

I first designed experiments to evaluate the effects of combinations of phenolic compounds by starting with concentrations of the singly applied allelochemical at a concentration at or below the inhibition threshold (i.e., no effect). Treatments with mixtures of two, three, or four compounds at these concentrations resulted in significant inhibition of seed germination and seedling growth (Einhellig and Rasmussen, 1978; Rasmussen and Einhellig, 1977, 1979; Einhellig et al., 1982). Subsequent work showed an equimolar combination containing 50 µM each of ten phenolic acids approximated the inhibition on velvetleaf (*Abutilon theophrasti* Medic.) seedling growth of 500 µM of any one compound (Einhellig, 1989). Under the controlled environmental conditions of these experiments, 50 µM of each phenolic acid alone had no effect and the growth-inhibition threshold for each compound was approximately 250 µM. The data clearly show that an allelochemical concentration that appears to be inconsequential by itself can have a significant effect in concert action with other compounds.

Primarily through the use of cucumber (*Cucumis sativus* L. cv. Early Green Cluster) growth bioassays, Blum and other scientists in his laboratory have shown that the components in a mixture of phenolics typically have an additive effect, but as their individual concentrations get higher, some antagonistic interactions can occur (Blum et al., 1984, 1985, 1989; Lyu et al., 1990; Gerig and Blum, 1991, 1993). Some useful models, such as those by Colby (1967) and most recently Gerig et al. (1989), provide a perspective to evaluate the joint action of compounds. However, the predictive use of these models is uncertain when extrapolating to interactions of concentrations of potential phytotoxins that are well below the inhibition threshold. What we do know is that the joint action of low levels of allelochemicals in the field can result in growth inhibitory effects (Einhellig, 1989, 1996; Blum, 1996).

Investigating interactions among volatile monoterpenes, Asplund (1969) showed dramatic synergism inhibition from a mixture of (+)pulegone and (−)-camphor at one-hundredth that of concentrations required for phytotoxicity of the single compounds. This unique finding suggests the need for evaluating interactions among compounds other than phenolic acids. Wallace and Whitehead (1980) found interactive effects among short-chain organic acids, but not the same synergistic as that found by Asplund.

30.3.2.2   Different Chemical Classes

I used grain sorghum (*Sorghum bicolor* [L.] Moench) seedling growth as a bioassay to analyze effects of a three-component mixture containing a phenolic acid (salicylic acid), a coumarin (umbelliferone), and a flavonoid (rutin) (Einhellig, 1996). Seedling growth was slightly inhibited by 150 µM of each compound alone, whereas the mixture of the three reduced plant growth to 34 percent of control plants. This combination effect was statistically the same as 450 µM of either salicylic acid or umbelliferone, yet 450 µM rutin was much less inhibitory than the mixture. Hence, these data show slightly more than an additive effect for salicylic acid, umbelliferone, and rutin. They also exemplify some of the complications in interaction studies, as the toxicity of all three compounds was the same at 150 µM but it was quite different at 450 µM.

The joint inhibitory action of salicylic acid, umbelliferone, and rutin does not mean that the three allelochemicals have the same mechanism of action. Current information suggests their mechanisms of interference with physiological processes are probably different (Einhellig, 1995b). However, all three chemicals are a stress on the receiving plant, and clearly, their joint action illustrates the predominate pattern of allelopathic activity.

30.4   STRESS PARTNERS IN ALLELOPATHY

It is imperative that the conceptual framework for allelopathy includes the realization that almost every other environmental stress interacts with the activity of allelochemicals, and the evidence is that any simultaneous stress enhances allelopathic effects. Moisture and nutrient deficits, extremes in temperature and irradiance, and disease and predation damage are common natural stresses. The presence of herbicides and other industrial chemicals are additional stresses in the environment (Rama Devi et al., 1997).

30.4.1   NATURAL ABIOTIC FACTORS

30.4.1.1   Moisture Stress

Allelopathy often has been documented in moisture-stressed environments such as the arid old-fields of Oklahoma, saline agricultural land in Pakistan, and the California chaparral with its long, dry summer and fall (Rice, 1984; Mahmood et al., 1989). Allelopathy may play a more important role in arid environments because the chemicals persist in the environment for longer periods than in humid environments. In addition, the effects of allelopathy may be magnified in arid environments by the combined impact of moisture and allelochemical stress. In a three-year field study in northeastern Nebraska, we found grain sorghum significantly suppressed weeds, and this was primarily due to allelopathy (Einhellig and Rasmussen, 1989). Low rainfall is a major crop limitation in this region of Nebraska, and I believe that moisture stress enhanced the allelopathic effects of grain sorghum. Weidenhamer et al. (1989) suggested that low moisture and extremely high temperatures contributed to allelopathy of the Florida scrub. To the extent that some of the flavonoids and polyphenols may be active in allelopathy, a similar inference is evident from data showing that a number of climatic stresses induce flavonoid synthesis and accumulation in plants (Chaves et al., 1997; Chaves and Escudaro, this volume).

Using osmotic stress to simulate moisture stress, I designed experiments to quantitatively assess the impact of stress combinations between ferulic acid (FA) and reductions in water potential (ψ) of the plant environment. The bioassay system was grain sorghum seedlings grown in nutrient culture manipulated for ψ and allelochemical stress. Even a minor osmotic stress enhanced inhibition of FA in a 12-block matrix including all combinations of two osmotic conditions (nutrient media and -0.2 MPa ψ) with three levels of ferulic acid (0, 100, and 250 μM) (Einhellig, 1987). The data showed moisture-stressed plants were more sensitive to FA, as this allelochemical at 100 μM had no effect on growth, but in the -0.2 MPa environment 100 μM FA significantly suppressed seedling growth below that of plants grown in -0.2 MPa without FA. Experiments were conducted using several osmotic agents, and the same trends were recorded. In a 16-block matrix using polyethylene glycol, 150 μM FA and ψ at 0.15 MPa resulted in 96 and 95 percent of control seedling growth, respectively. However, grain sorghum seedlings in the combination of these two stresses had 59 percent of the dry weight of untreated plants after 11 days (Einhellig, 1989).

Additional experiments investigated grain sorghum seed germination under combinations of osmotic and FA stress. The threshold concentration of FA required to inhibit germination of grain sorghum (2500 μM) is more than tenfold above the level for inhibition of seedling growth; similarly, germination is less affected by minor reductions in ψ. However, I found grain sorghum seed germination was delayed (evaluated by a germination index) more by a combination of 0.3 MPa ψ and 2500 μM FA than either treatment alone (Einhellig, 1989).

Duke et al. (1983) concluded that the mechanism of action of water stress was similar to that in phenolic acids, and there is strong evidence that phenolic acids alter plant water balance (Einhellig, 1995b). Hence, it is not surprising that plants grown in environments with regular moisture stress will be more susceptible to allelopathy than those in more optimum moisture conditions. This assumes that there is sufficient moisture to leach allelochemicals from the producing plant or plant residue. In contrast, there are situations where allelopathic cover crops may be less effective in suppressing weeds because there is not enough rainfall during a critical period to leach allelochemicals from the mulch into the soil.

30.4.1.2   Heat

High temperature and moisture are often, but not always, companion conditions. Effects from allelopathic residues and aqueous extracts appear to be temperature sensitive (Bhowmik and Doll, 1983; Steinsiek et al., 1982). Several investigations showed higher temperature environments resulted in greater inhibition from specific allelochemicals, such as gramine and some phenolic acids (Glass, 1976; Hanson et al., 1983).

Einhellig and Echrich (1984) reported the quantitative assessment of temperature–allelochemical interactions on grain sorghum and soybean (*Glycine max* [L.] Merr.). Temperatures in a range normal to the field environment were used, and plants were grown in a glass house under full sun. The data showed temperatures at the high end of the optimum range for these species reduced the concentration of ferulic acid required to inhibit seedling growth. The growth-inhibition threshold for grain sorghum was 200 μM FA at 37°C, compared with 400 μM at 29°C. The higher of these two temperatures was above the optimum for the C-3 soybean species. However, the same pattern held for soybean. In a relatively hot environment (34°C), 100 μM FA inhibited soybean growth, but at 23°C, 250 μM FA was required to inhibit growth over the 10-d treatment period.

30.4.1.3   Light Environment

Variations in irradiance received by plants has typically, and rightly, been considered as part of competitive interference. However, Bhowmik and Doll (1983) reported that photosynthetic photon flux densities (PPFD) altered the allelopathic effects of weed residues on crops. At 30/20°C day/night temperatures, the inhibitory effects of redroot pigweed (*Amaranthus retroflexus* L.) and yellow foxtail (*Setaria glauca* [L.] Beauv.) residues on corn (*Zea mays* L.) were less when plants were grown with moderate PPFD (380 to 570 μmol photons m^{-2} sec^{-1}) as compared with 760 μmol photon m^{-2} sec^{-1}. The data showed a number of interactive effects involving temperature, irradiance, and phytotoxicity of the weed residues.

30.4.1.4   Nutrient Deficiencies

Some of the strongest evidence for allelopathy comes from nutrient-deficient environments, such as Oklahoma old-field succession and the Florida scrub community (Rice, 1984; Weidenhamer et al., 1989). Yet the case for nutrient stress interacting with allelochemical stress needs further clarification. Williamson et al. (1992) reported that hydrocinnamic acid was more toxic to little bluestem (*Schizachyrium scoparium* [Michx.] Nash) under low-N and low-K conditions. Similarly, *p*-coumaric and vanillic acids had their greatest effects on barley (*Hordeum vulgare* L.) growth under nitrogen and phosphorus deficient conditions (Stowe and Osborn, 1980). Hence, a number of researchers have suggested that increasing fertilization will overcome allelopathy.

To test the interaction of phosphorus level with allelopathy, I grew grain sorghum seedlings in a 20-block matrix of four levels of ferulic acid across five levels of phosphorus. The data showed low phosphorus did not enhance the action of ferulic acid, nor did high phosphorus overcome ferulic acid inhibition (Einhellig, 1989). This outcome is consistent with findings of Bhowmik and Doll (1984) showing that adding nitrogen and phosphorus supplements to various weed residues did not overcome inhibition from the weed residue. In contrast, Hall et al. (1983) reported nutrient enrichment reduced phytotoxicity of sunflower (*Helianthus annuus* L.). As will be discussed later, perhaps the larger role of nutrient stress on allelopathy is that of increasing the quantity of allelochemicals produced by plants (Armstrong et al., 1970; Lehman and Rice, 1972; Koeppe et al., 1976). Also, growth under high-nutrient environments may increase the toxicity of the subsequent residue (Luu et al., 1982; Mwaja et al., 1995).

30.4.1.5   Soil Organics

Blum has done an excellent job of developing the perspective that nonallelochemical organic compounds in the soil will influence the action of phenolic acids. The presence of abundant soil methionine lowered the inhibition threshold for p-coumaric acid on morning-glory (*Ipomoea hederacea* [L.] Jacquin) (Blum et al., 1993). Increases of glucose in the soil and noninhibitory levels of phenylalanine lowered the amount of p-coumaric acid required to inhibit shoot and seedling biomass (Pue et al., 1995). The data indicate glucose and phenylalanine slowed the rate of degradation of p-coumaric metabolism by microbes, thus increasing the effectiveness of a given concentration of inhibitor. Soils contain a variety of other organic molecules that may modify the effectiveness of allelochemicals, and certainly interactions involving pH are mitigating factors (Lehman and Blum, 1997).

30.4.2   DIRECT BIOLOGICAL STRESS

30.4.2.1   Disease Infections

Interactions among plants and their pathogens involve allelochemicals in at least three ways. First, there is substantial evidence for microbial toxins being the mode of damage to plants by many disease organisms (Ayers, 1991; Rice, 1995). Second, a rapid build up in secondary plant compounds is a long-recognized protective and resistance response to infections (Swain, 1977). Third, interactive effects of allelopathy and disease can result in an enhancement of both the pathogenesis and injury from the allelochemicals. Overlying all three of these situations is the additive deleterious effect that occurs from the combined action of pathogens and abiotic stress, since each reduces a plant's capacity for response.

An early evidence of interactions between allelopathy and disease was the recognition that pathogenesis of root rot fungi was greater in fields having decomposing, allelopathic residues (Patrick et al., 1964). In studies in the United Kingdom on problems associated with decomposing straw and weed residues, Lynch (1987) concluded that residues decomposing under relatively dry conditions promote *Fusarium* populations, which act in concert with deleterious action of acetic and other short-chain organic acids. Asparagus (*Asparagus officinalis* L.) regeneration problems illustrate the clearest evidence that disease can synergize effects of allelopathy (Hartung and Stephens, 1983; Peirce and Colby, 1987). Allelopathic effects of asparagus residue and leachate on a new crop of asparagus were positively correlated with the severity of infection by *Fusarium* spp. Both studies concluded that it was the combined action of allelochemicals and infection that contributed to slow growth of the new crop of asparagus.

30.4.2.2   Herbivores

There is abundant evidence that insect feeding and other herbivore damage elicit a redirection of plant metabolism to defense compounds. Since some of these compounds—certain phenolics, flavonoids, alkaloids—are the same compounds active in allelopathy, herbivory has an indirect effect on the relative amount of allelochemicals that may be present in the environment. At the same time, abiotic stress can either positively or negatively alter the suitability of the plant to insect feeding (Jones and Coleman, 1991).

I know of no definitive studies to quantitatively evaluate the simultaneous action of allelopathy and insect damage on plants. It is logical that the two sources of stress will have an additive detriment to plant growth, but this suggestion needs further investigation.

30.4.3   COMPETITIVE INTERFERENCE

Although we have little concrete information on how interplant competition affects allelopathy, the relevance of this stress seems predictable. Competition is responsible for increasing moisture stress, nutrient deficiency, or an unfavorable irradiance environment. As noted, all of these stresses amplify production of certain allelochemicals, and they enhance the deleterious action of compounds. However, the interactive role of competitive interference with allelopathy is complicated by how competition alters the relative availability of allelochemicals to an individual plant. Data from Weidenhamer et al. (1989) and Weidenhamer (1996) support the hypothesis that allelopathic effects are density dependent. The potential for one plant or species to be negatively affected by allelochemicals is likely to be changed by the extent to which competitive species remove those same compounds from the immediate environment, and the time frame in which this occurs.

30.4.4   ANTHROPOGENIC SOURCES

30.4.4.1   Herbicides

The potential joint action of herbicides and allelochemicals should be of major interest. As a mainstay of modern agriculture, herbicides are a common factor in agroecosystems. It is valuable to understand the potential for herbicides to act in concert with allelopathic residues and natural action of allelopathic crops on weeds. Likewise, while many herbicides are considered selective in their action, they still impart some crop stress. Herbicide carryover and unwanted transport of herbicides also result in crop damage, and this may be occurring in the same time frame as allelopathic interference in the crop environment.

I have reported laboratory studies showing interactive inhibition from several herbicides and allelochemicals (Einhellig, 1987, 1989, 1996; Einhellig and Leather, 1988). Each experiment was a matrix of treatments with seedlings grown in nutrient culture amended with either an herbicide, allelochemical, or various combinations of the two. The threshold toxicity for herbicides is typically one or more orders of magnitude below that for allelochemicals, and problems arise because of the killing effect of herbicides at most points above their inhibition threshold. The difficulties of these experiments are compounded by abiotic conditions affecting herbicide and allelochemical activity. Nevertheless, in separate studies with trifluralin [2,6-dinitro-N,N-dipropyl-4-(trifluoromethyl)benzenamine], alachlor [2-chloro-N-(2,6-diethylphenyl)-N-(methoxymethyl) acetamide], and atrazine [6-chloro-N-ethyl-N'-(1-methylethyl)-1,3,5-triazine-2,4-diamine], the data showed at least an additive effect from near-threshold concentrations of the herbicide in combination with a phenolic acid.

A combination of 10 µg L^{-1} (10 ppb) atrazine and 250 µM ferulic acid reduced oat *(Avena sativa* L.) seedling growth 47 percent over 10-d treatments whereas separately their effects were 9 and 31 percent inhibition, respectively (Einhellig, 1987). Using grain sorghum as the bioassay seedlings, plants in a combination of 0.1 mg L^{-1} trifluralin combined with 40 mg L^{-1} ferulic acid had only 47 percent of control dry weight, whereas their separate treatments resulted in 65 and 70 percent of controls (Einhellig, 1989). Alachlor, which is commonly used to control many annual grasses, was tested in a 16-block matrix of four levels of herbicide and ferulic acid (Einhellig, 1996). Seedlings in either 50 mg L^{-1} alachlor or 100 µM ferulic acid achieved 85 to 92 percent of control plant growth by the end of the experiment, whereas those in the combination of these two stresses were at 63 percent.

The herbicide–phenolic acid interaction data has practical implications for agricultural management practices. In the future, allelopathic crops and residues may be used in combination with herbicides as planned weed control (Einhellig and Leather, 1988; Einhellig, 1995a).

30.4.4.2   Other Chemicals

Human activities have resulted in many chemical stresses on plants in addition to herbicides, some anticipated and others not. Use of pesticides may directly impact plant metabolism. Ozone and other atmospheric pollutants, concentration of heavy metals, and the wide distribution of many industrial molecules challenge plant functions. How these conditions interact with allelochemical stress remains an open question.

30.5   ENVIRONMENTAL STIMULATION OF ALLELOCHEMICAL PRODUCTION

30.5.1   ALLELOCHEMICALS IN PLANT TISSUE

The literature on both plant defense and abiotic stress adjustments gives abundant evidence of the quantitative plasticity of secondary compounds in plants (Timmermann et al., 1984; Gershenzon, 1984). A number of commercial bioregulators, such as mepiquat chloride and chlorequat chloride, also induce changes in allelochemical production (Hedin, 1990). The most frequent stress-induced response is an overall increase in secondary compounds, albeit it is not predictable which compounds will be elevated—and therein lies the challenge. Several chemicals implicated in allelopathy have been documented to follow the pattern of increasing under stress, and this increase in tissue content must ultimately result in a greater chance for chemical interference to occur in a field situation. Space allows only a few selected examples leading to the generalization postulated in Figure 30.3, and I want to clearly state that every allelopathic chemical does not respond in the same way.

Moisture deficits have been most often studied with investigators typically only evaluating effects on one compound or class of allelochemicals. Reported increases in allelochemicals in moisture-stressed plants include chlorogenic acid (del Moral, 1972), total phenolics (Balakumar et al., 1993; Kumar et al., 1991), flavonoids (Chaves and Escudero, this volume), hydroxamic acids (Richardson and Bacon, 1993), and monoterpenes (Gilmore, 1977; Kainulainen et al., 1992). Depending on the level of stress, tissue, and compounds investigated, stressed plants may have several times the tissue concentration of one of these allelochemicals when compared with unstressed plants.

Work at the University of Oklahoma in the 1960s and early 1970s showed that plant content of the coumarins and the chlorogenic acids were subject to environmental conditions (Wender, 1970; Armstrong et al., 1970; Lehman and Rice, 1972; Koeppe et al., 1969, 1976). The focus was on tobacco (Nicotiana tabacum L.) and sunflower. Tissue levels of these phenolics were modified by certain herbicides, mineral deficiency (low N, K, Mg, S), low temperature, and UV radiation. Results varied according to the stress, plant age, and tissue type, but an increased allelochemical content was the common denominator. At one extreme, Dieterman et al. (1964) found that sunflowers sprayed with 2,4-D (2,4-dichlorophenphenoxyacetic acid) had significant increases in scopoletin, ayapin, and scopolin, with the latter being 30 times the level in untreated plants.

Tissue levels of allelopathic compounds and phytotoxicity are subject to seasonal changes and geographic location of the plants according to the associated stress. Examples include the phytotoxicity of bracken (Pteridium aquilinum [L.] Kuhn), quinolizidine alkaloid content of lupine (Lupinus argenteus L. argenteus), and polyphenols of certain species of the Mediterranean (Dolling et al., 1994; Carey and Wink, 1994; Chaves and Escudero, 1998). High temperature can be an elevating factor, as shown by temperature-induced increases of hydroxamic acids (Gianoli and Niemeyer,

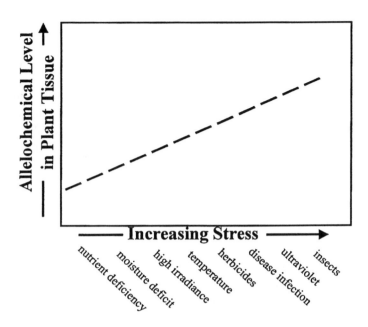

FIGURE 30.3 Generalized interrelationship between stress and overall allelochemical concentration in plants. Stress effects on concentration of specific compounds will vary. Some of the more common stress are shown.

1997). Hence, more stressful environments are likely to have situations of allelopathic interference. Although it is not a plant stress, some instances of CO_2 enrichment result in an increase in plant concentrations of phenols, terpenes, and other secondary metabolites (Penuelas et al., 1996; Penuelas and Llusia, 1997). The data in these reports are variable, but even when terpene and phenolic increases are relatively small, the quantity of these allelochemicals per plant increased significantly under elevated CO_2 treatments mainly through increased plant biomass.

30.5.2 LEACHATE AND EXUDATE PHYTOTOXICITY

Ultimately, higher plant concentration of allelochemicals must result in larger releases into the environment, and a few studies have made a direct connection of these situations to allelopathic inhibition. Sunflowers grown with limited phosphorus leached more phenolics (Koeppe et al., 1976), and residues from plants that were nutrient-limited were more inhibitory to redroot pigweed germination in the soil (Hall et al., 1983). Sterling et al. (1987) found that velvetleaf grown under high-temperature stress produced exudates from the glandular trichomes that were twice as toxic as exudates during moderate temperature. Water stress decreased the volume of trichome exudate and increased its phytotoxicity. Drought stress of purple nutsedge (*Cyperus rotundus* L.) resulted in an increased release of toxic secondary metabolites to the rhizosphere (Tang et al., 1995). This was postulated as the probable reason for greater interference of purple nutsedge on sweet corn yield under limited soil-moisture conditions.

Sorgoleone, a *p*-benzoquinone exuded by the roots of grain sorghum, is a potent inhibitor of photosynthesis (Einhellig et al., 1993; Nimbal et al., 1996b). Nimbal et al. (1996a) reported that sorgoleone is often produced at levels greater than 1 percent of the seedling dry weight, but various grain sorghum genotypes differed more than 25-fold in the quantity of sorgoleone produced. In my laboratory, we found some variation in the amount of sorgoleone exudate collected from the same seed source (Einhellig and Souza, 1992). Although no quantification was carried out, it appeared that roots under dry conditions produced a higher quantity of sorgoleone.

30.6   A UNIFYING PERSPECTIVE

Current evidence strongly supports the position that multiple stress conditions interact with allelopathy, and to disregard the impact of these biotic and abiotic stresses leaves a gap in our understanding of chemical interactions among plants. The diagrammatic representation in Figure 30.1 is presented as a unifying perspective on interactive factors. The data show that allelopathy results from the joint action of multiple compounds in which individual compounds are often present in concentrations below their inhibition threshold. The complex of inhibitors arising from an allelopathic plant may include compounds from different chemical classes and biosynthethic pathways. Almost every form of environmental stress that plants may encounter results in elevated production and release of one or more classes of these allelochemicals.

There also is growing evidence supporting the conjecture that inhibitory action of allelochemicals is enhanced by the concert action of other stress conditions. However, the number of investigations evaluating this hypothesis is relatively small, and only a few of the array of known allelochemicals have been included in these studies. Certainly more work is needed that analyzes interactive effects between allelochemicals and specific additional stresses in the plant environment. There is an absence of data at the next level of complexity, which is the study of the action of allelochemicals in concert with multiple other stresses. We also have a decided lack of information on how the degree of various stress combinations influences the extent of damage to plant growth or productivity. It is possible that at very low stress levels there will be compensatory metabolic adjustments that avoid plant damage. However, as noted previously, allelopathy is often found in plant communities where moisture, temperature, and other environmental conditions are very unfavorable. Since multiple stress associations are commonplace, recognition of the probability for deleterious additive and sometimes synergistic action of stresses helps clarify the processes causing allelopathy in field situations.

REFERENCES

Alsaadawi, I. S., Rice, E. L., and Karns, T. K. B., Allelopathic effects of *Polygonum aviculare* L. III. Isolation, characterization, and biological activities of phytotoxins other than phenols, *J. Chem. Ecol.*, 9, 761, 1983.

Armstrong, G. M., Rohrbaugh, L. M., Rice, E. L., and Wender, S. H., The effects of nitrogen deficiency on the concentration of caffeoylquinic acids and scopolin in tobacco, *Phytochemistry*, 9, 945, 1970.

Ayres, P. G., Growth responses induced by pathogens and other stresses, in *Response of Plants to Multiple Stresses*, Mooney, H. A., Winner, W. E., and Pell, E. J., Eds., Academic Press, San Diego, CA, 1991, 227.

Asplund, R. O., Some quantitative aspects of phytotoxicity of monoterpenes, *Weed Sci.*, 17, 454, 1969.

Balakumar, T., Vincent, V. H. B., and Paliwal, K., On the interaction of UV-B radiation (280–315 nm) with water stress in crop plants, *Physiol. Plant.*, 87, 217, 1993.

Bhowmik, P. C., and Doll, J. D., Growth analysis of corn and soybean response to allelopathic effects of weed residues at various temperatures and photosynthetic photon flux densities, *J. Chem. Ecol.*, 9, 1263, 1983.

Bhowmik, P. C., and Doll, J. D., Allelopathic effects of annual weed residues on growth and nutrient uptake of corn and soybean, *Agron. J.*, 76, 383, 1984.

Billings, W. D., The environmental complex in relation to plant growth and distribution, *Quart. Rev. Biol.*, 27, 251, 1952.

Blum, U., Allelopathic interactions involving phenolic acids, *J. Nematol.*, 28, 259, 1996.

Blum, U., Dalton, B. R., and Rawlings, J. O., Effects of phenolic acid and some of its microbial metabolic products on radicle growth of cucumber, *J. Chem. Ecol.*, 10, 1169, 1984.

Blum, U., Dalton, B. R., and Shann, J. R., Effects of various mixtures of ferulic acid and some of its microbial metabolic products on cucumber leaf expansion and dry matter in nutrient culture, *J. Chem. Ecol.*, 11, 619, 1985.

Blum, U., Gerig, T. M., and Weed, S. B., Effects of mixtures of phenolic acids on leaf area expansion of cucumber seedlings grown in different pH Portsmouth A_1 soil materials, *J. Chem. Ecol.*, 15, 2413, 1989.

Blum, U., Gerig, T. M., Worsham, A. D., Holappa, L. D., and King, L. D., Allelopathic activity of wheat-conventional and wheat-no-till soils: Development of soil extract bioassays, *J. Chem. Ecol.*, 18, 2191, 1992.

Blum, U., Gerig, T. M., Worsham, A. D., and King, L. D., Modification of allelopathic effects of *p*-coumaric acid on morningglory seedling biomass by glucose, methionine, and nitrate, *J. Chem. Ecol.*, 19, 2791, 1993.

Boufalis, A., and Pellissier, F., Allelopathic effects of phenolic mixtures on respiration of two spruce mycorrhizal fungi, *J. Chem. Ecol.*, 20, 2283, 1994.

Bradow, J. M., and Connick, W. J., Jr., Volatile methyl ketone seed-germination inhibitors from *Amaranthus palmeri* S. Wats. residues, *J. Chem. Ecol.*, 14, 1617, 1988a.

Bradow, J. M., and Connick, W. J., Jr., Seed germination inhibition by volatile alcohols and other compounds associated with *Amaranthus palmeri* residues, *J. Chem. Ecol.*, 14, 1633, 1988b.

Carey, D. B., and Wink, M., Elevation variation of quinolizidine alkaloid contents of lupine (*Lupinus argenteus*) of the Rocky Mountains, *J. Chem. Ecol.*, 20, 849, 1994.

Chaves, N., and Escudero, J. C., Seasonal variation of flavonoid synthesis induced by ecological factors, 1999 (this volume), Chapter 17.

Chaves, N., Escudero, J. C., and Guterrez-Merino, C., Role of ecological variables in the seasonal variation of flavonoid content of *Cistus ladanifer* exudate, *J. Chem. Ecol.*, 23, 2577, 1997.

Chou, C. H. and Lin, H. J., Autointoxication of mechanism of *Oryza sativa*. I. Phytotoxic effects of decomposing rice residues in soil, *J. Chem. Ecol.*, 2, 353, 1976.

Chou, C. H. and Waller, G. R., Isolation and identification by mass spectrometry of phytotoxins in *Coffea arabica*, *Bot. Bull. Academia Sinica*, 21, 25, 1980.

Chou, C. H., Chiang, Y. C., and Cheng, H. H., Autointoxication mechanism of *Oryza sativa*. III. Effect of temperature on phytotoxin production during rice straw decomposition in soil, *J. Chem. Ecol.*, 7, 741, 1981.

Colby, S. R., Calculating synergistic and antagonistic response of herbicide combinations, *Weeds*, 14, 20, 1967.

del Moral, R., On the variability of chlorogenic acid concentration, *Oecologia*, 9, 289, 1972.

Dieterman, L. J., Lin, C. Y., Rohrbaugh, L. M., and Wender, S. H., Accumulation of ayapin and scopolin in sunflower plants treated with 2,4-dichlorophenoxyacetic acid, *Arch. Biochem. Biophys.*, 106, 274, 1964.

Dolling, A., Zackrisson, O., and Nilsson, M. C., Seasonal variation in phytotoxicity of bracken (*Pteridium aquilinum* (L.) Kuhn), *J. Chem. Ecol.*, 20, 3163, 1994.

Duke, S. O., Williams, R. D., and Markhart, A. H., III, Interactions of moisture stress and three phenolic compounds on lettuce seed germination, *Ann. Bot.* (London), 52, 923, 1983.

Einhellig, F. A., Interactions among allelochemicals and other stress factors of the plant environment, in *Allelochemicals: Role in Agriculture Land Forestry*, Waller, G. R., Ed., ACS Symposium Series 330, American Chemical Society, Washington, D.C., 1987, 343.

Einhellig, F. A., Interactive effects of allelochemicals and environmental stress, in *Phytochemical Ecology: Allelochemicals, Mycotoxins and Insect Pheromones and Allomones*, Chou, C. H. and Waller, G. R., Eds., Institute of Botany, Academia Sinica Monograph Series No. 9, Taipei, ROC, 1989, 101.

Einhellig, F. A., Allelopathy: current status and future goals, in *Allelopathy: Organisms, Processes, and Applications*, Inderjit, Dakshini, K. M. M., and Einhellig, F. A., Eds., ACS Symposium Series 583, American Chemical Society, Washington, D.C., 1995a, 1.

Einhellig, F. A., Mechanism of action of allelochemicals in allelopathy, in *Allelopathy: Organisms, Processes, and Applications*, Inderjit, Dakshini, K. M. M., and Einhellig, F. A., Eds., ACS Symposium Series 582, American Chemical Society, Washington, D.C., 1995b, 96.

Einhellig, F. A., Interactions involving allelopathy in cropping systems, *Agron. J.*, 88, 886, 1996.

Einhellig, F. A., and Eckrich, P. C., Interactions of temperature and ferulic acid stress on grain sorghum and soybeans, *J. Chem. Ecol.*, 10, 161, 1984.

Einhellig, F. A., and Leather, G. R., Potentials for exploiting allelopathy to enhance crop production, *J. Chem. Ecol.*, 14, 1829, 1988.

Einhellig, F. A., and Rasmussen, J. A., Synergistic inhibitory effects of vanillic and *p*-hydroxybenzoic acids on radish and grain sorghum, *J. Chem. Ecol.*, 4, 425, 1978.

Einhellig, F. A., and Rasmussen, J. A., Prior cropping with grain sorghum inhibits weeds, *J. Chem. Ecol.*, 15, 951, 1989.

Einhellig, F. A., and Souza, I. F., Phytotoxicity of sorgoleone found in grain sorghum root exudates, *J. Chem. Ecol.*, 18, 1, 1992.

Einhellig, F. A., Schon, M. K., and Rasmussen, J. A., Synergistic inhibitory effects of four cinnamic acid compounds on grain sorghum, *J. Plant Growth Regul.*, 1, 251, 1982.

Einhellig, F. A., Rasmussen, J. A., Hejl, A. H., and Souza, I. F., Effects of root exudate sorgoleone on photosynthesis, *J. Chem. Ecol.*, 19, 369, 1993.

Fisher, N. H., Tanrisever, N., and Williamson, G. B., Allelopathy in the Florida scrub community as a model for natural herbicide actions, in *Biologically Active Natural Products: Potential Use in Agriculture,* Cutler, H. G., Ed., ACS Symposium 380, American Chemical Society, Washington, D.C., 1988, 233.

Fisher, N. H., Williamson, G. B., Weidenhamer, J. D., and Richardson, D. R., In search of allelopathy in the Florida scrub: the role of terpenoids, *J. Chem. Ecol.,* 20, 1355, 1994.

Gerig, T. M., and Blum, U., Effects of mixtures of four phenolic acids on leaf area expansion of cucumber seedlings grown in portsmouth B_1 soil materials, *J. Chem. Ecol.,* 17, 29, 1991.

Gerig, T. M., and Blum, U., Modification of an inhibitory curve to account for effects of a second compound, *J. Chem. Ecol.,* 19, 2783, 1993.

Gerig, T. M., Blum, U., and Meier, K., Statistical analysis of the joint inhibitory action of similar compounds, *J. Chem. Ecol.,* 15, 2403, 1989.

Gianoli, E., and Niemeyer, H. M., Environmental effects on the accumulation of hydroxamic acids in wheat seedlings: the importance of plant growth rate, *J. Chem. Ecol.,* 23, 543, 1997.

Gilmore, A. R., Effects of soil moisture stress on monoterpenes in loblolly pine, *J. Chem. Ecol.,* 3, 667, 1977.

Glass, A. D. M., The allelopathic potential of phenolic acids associated with the rhizosphere of *Pteridium aquilinum, Can. J. Bot.,* 54, 2440, 1976.

Gorski, P. M., Miersch, J., and Ploszynski, M., Production and biological activity of saponins and canavanine in alfalfa seedlings, *J. Chem. Ecol.,* 17, 1135, 1991.

Hall, A. B., Blum, U., and Fites, R. C., Stress modification of allelopathy of *Helianthus annus* L. debris on seed germination, *Am. J. Bot.,* 69, 776, 1983.

Hanson, A. D., Ditz, K. M., Singletary, G. W., and Leland, T. J., Gramine accumulation in leaves of barley grown under high temperature stress, *Plant Physiol.,* 71, 896, 1983.

Hartung, A. C., and Stephens, C. T., Effects of allelopathic substances produced by asparagus on incidence and severity of asparagus decline due to *Fusarium* crown rot, *J. Chem. Ecol.,* 8, 1163, 1983.

Hedin, P. A., Bioregulator-induced changes in allelochemicals and their effects on plant resistance to pests. *Crit. Rev. Plant Sci.,* 9, 371, 1990.

Inderjit, Plant phenolics in allelopathy, *Bot. Rev.,* 62, 186, 1996.

Jones, C. G., and Coleman, J. S., Plant stress and insect herbivory: toward an integrated perspective, in *Response of Plants to Multiple Stresses,* Mooney, H. A., Winner, W. E. and Pell, E. J., Eds., Academic Press, San Diego, CA, 1991, 249.

Kanchan, S. D. and Jayachandra, Allelopathic effects of *Parthenium hysterophorus* L. IV. Identification of inhibitors, *Plant Soil,* 55, 67, 1980.

Kainulainen, P., Oksanen, J., Palomaki, V., Hokopainen, J. K., and Hokopainen, T., Effect of drought and waterlogging stress on needle monoterpenes of *Picea abies, Can. J. Bot.,* 70, 1613, 1992.

Koeppe, D. E., Rohrbaugh, L. M., and Wender, S. H., The effect of varying U.V. intensities on the concentration of scopolin and caffeoylquinic acids on tobacco and sunflower, *Phytochemistry,* 8, 889, 1969.

Koeppe, D. E., Southwick, L. M., and Bittell, J. E., The relationships of tissue chlorogenic acid concentrations and leaching of phenolics from sunflowers grown under varying phosphate nutrient conditions, *Can. J. Bot.,* 54, 593, 1976.

Kumar, S. S., Nalwad, U. G., and Barsarkar, P. W., Influence of moisture stress on the accumulation of phenols in marigold (*Tagetes erecta* L.), *Geobios,* 18, 165, 1991.

Lehman, M. E., and Blum, U., Cover crop debris effects on weed emergence as modified by environmental factors, *Allelopathy J.,* 4, 69, 1997.

Lehman, R. H., and Rice, E. L., Effect of deficiencies of nitrogen, potassium and sulfur on chlorogenic acids and scopolin in sunflower, *Am. Midl. Natur.,* 87, 71, 1972.

Luu, K. T., Matches, A. G., and Peters, E. J., Allelopathic effects of tall fescue on birdsfoot trefoil as influenced by N fertilization and seasonal change, *Agron. J.,* 74, 805, 1982.

Lydon, J., Teasdale, J. R., and Chen, P. K., Allelopathic activity of annual wormwood (*Artemisia annua*) and the role of artemisinin, *Weed Sci.,* 45, 807, 1997.

Lynch, J. M., Allelopathy involving microorganisms: case histories from the United Kingdom, in *Allelochemicals: Role in Agriculture and Forestry,* Waller, G. R., Ed., ACS Symposium Series 330, American Chemical Society, Washington, D.C., 1987, 44.

Lyu, S. W., Blum, U., Gerig, T. M., and Brien, T. E., Effects of mixtures of phenolic acids on phosphorous uptake by cucumber seedlings, *J. Chem. Ecol.,* 16, 2559, 1990.

Mahmood, K., Malik, K. A., Sheikh, K. H., and Lodhi, M. A. K., Allelopathy in saline agricultural land: vegetation successional changes and patch dynamics, *J. Chem. Ecol.,* 15, 565, 1989.

Mwaja, V. N., Masiunas, J. B., and Weston, L. A., Effects of fertility on biomass, phytotoxicity, and allelochemical content of cereal rye, *J. Chem. Ecol.*, 21, 81, 1995.

Nimbal, C. I., Pedersen, J. F., Yerkes, C. N., Weston, L. A., and Weller, S. C., Phytotoxicity and distribution of sorgoleone in grain sorghum germplasm, *J. Agric. Food Chem.*, 44, 1343, 1996a.

Nimbal, C. I., Yerkes, C. N., Weston, L. A., and Weller, S. C., Herbicidal activity and site of action of the natural product sorgoleone, *Pestic. Biochem. Physiol.*, 54, 73, 1996b.

Oleszek, W., Allelopathic potentials of alfalfa (*Medicago sativa*) saponins: their relation to antifungal and hemolytic activities, *J. Chem. Ecol.*, 19, 1063, 1993.

Patrick, Z. A., Toussoun, T. A., and Koch, L. W., Effects of crop-residue decomposition products on plant roots, *Annu. Rev. Phytopathol.*, 2, 267, 1964.

Peirce, L. C., and Colby, L. W., Interactions of asparagus root filtrate with *Fusarium oxysporum* f. sp. *asparagi*, *J. Am. Soc. Hortic. Sci.*, 112, 35, 1987.

Penuelas, J., Estiarte, M., Kimball, B. A., Idso, S. B., Pinter, P. J., Jr., Wall, G. W., Garcia, R. L., Hansaker, D. J., Lamorte, R. L., and Hendrix, D. L., Variety of responses of plant phenolic concentrations to CO_2 enrichment, *J. Exp. Bot.*, 47, 1463, 1996.

Penuelas, J., and Llusia, J., Effects of carbon dioxide, water supply, and seasonality on terpene content and emission by *Rosmarinus officinalis*, *J. Chem. Ecol.*, 23, 979, 1997.

Pue, K. J., Blum, U., Gerig, T. M., and Shafer, S. R., Mechanism by which noninhibitory concentrations of glucose increase inhibitory activity of *p*-coumaric acid on morning-glory seedling biomass accumulation, *J. Chem. Ecol.*, 21, 833, 1995.

Rama Devi, S., Pellissier, F., and Prasad, M. N. V., Allelochemicals, in *Plant Ecophysiology,* Prasad, M. N. V., Ed., John Wiley & Sons, New York, 1997, 253.

Rasmussen, J. A. and Einhellig, F. A., Synergistic inhibitory effects of *p*-coumaric and ferulic acids on germination and growth of grain sorghum, *J. Chem. Ecol.*, 3, 197, 1977.

Rasmussen, J. A. and Einhellig, F. A., Inhibitory effects of combinations of three phenolic acids on grain sorghum germination, *Plant Sci. Lett.*, 14, 69, 1979.

Rice, E. L., Allelopathy, 2nd ed., Academic Press, Orlando, FL, 1984, 422.

Rice, E. L., Biological control of weeds and plant diseases: advances in applied allelopathy, University of Oklahoma Press, Norman, OK, 1995, 439.

Rice, E. L., Lin, C. V., and Huang, C. Y., Effects of decomposing rice straw on growth of and nitrogen fixation by *Rhizobium*, *J. Chem. Ecol.*, 7, 333, 1981.

Richardson, M. D., and Bacon, C. W., Cyclic hydroxamic acid accumulation in corn seedlings exposed to reduced water potentials before, during, and after germination, *J. Chem. Ecol.*, 19, 1613, 1993.

Steinsiek, J. W., Oliver, L. R., and Collins, F. C., Allelopathic potential of wheat (*Triticum aestivum*) straw on selected weed species, *Weed Sci.*, 30, 495, 1982.

Sterling, T. M., Houtz, R. L., and Putnam, A. R., Phytotoxic exudates from velvetleaf (*Abutilon theophrasti*) glandular trichomes, *Am. J. Bot.*, 74, 453, 1987.

Stowe, L. G., and Osborn, A., The influence of nitrogen and phosphorus levels on the phytotoxicity of phenolic compounds, *Can. J. Bot.*, 58, 1149, 1980.

Swain, T., Secondary compounds as protective agents, *Ann. Rev. Plant Physiol.*, 28, 479, 1977.

Tang, C. S., Cai, W. F., Kohl, K., and Nishimoto, R. K., Plant stress and allelopathy, in *Allelopathy: Organisms, Processes, and Applications,* Inderjit, Dakshini, K. M. M., and Einhellig, F. A., Eds., ACS Symposium Series 582, American Chemical Society, Washington, D.C., 1995, 142.

Timmermann, B. N., Steelink, C., and Loewus, F. A., Eds., Phytochemical adaptations to stress, *Rec. Adv. Phytochem.*, 18, 1984.

Veronneau, H., Greer, A. F., Daigle, S., and Vincent, G., Use of mixtures of allelochemicals to compare bioassays using red maple, pin cherry, and American elm, *J. Chem. Ecol.*, 23, 1101, 1997.

Wallace, J. M., and Whitehead, L. C., Adverse synergistic effects between acetic, propionic, butyric and valeric acids on the growth of wheat seedlings roots, *Soil Biol. Biochem.*, 12, 445, 1980.

Weidenhamer, J. D., Distinguishing resource competition and chemical interference: overcoming the methodological impasse, *Agron. J.*, 88, 866, 1996.

Weidenhamer, J. D., Harrtnett, D. C., and Romeo, J. T., Density-dependent phytotoxicity: distinguishing resource competition and allelopathic interference in plants, *J. Appl. Ecol.*, 26, 613, 1989.

Weidenhamer, J. D., Menelaou, M., Macias, F. A., Fischer, N. H., Richardson, D. R., and Williamson, G. B., Allelopathic potential of menthofuran monoterpenes from *Calamintha ashei*, *J. Chem. Ecol.*, 20, 3345, 1994.

Wender, S. H., Effects of some environmental stress factors on certain phenolic compounds in tobacco, *Rec. Adv. Phytochem.*, 3, 1, 1970.

Williamson, G. B., Obee, E. M., and Weidenhamer, J. D., Inhibition of *Schizachyrium scoparium* (Poaceae) by the allelochemical hydrocinnamic acid, *J. Chem. Ecol.*, 18, 2095, 1992.

Wyman-Simpson, C. L., Waller, G. R., Jurzysta, M., McPherson, J. K., and Young, C. C., Biological activity and chemical isolation of root saponins of six cultivars of alfalfa (*Medicago sativa* L.), *Plant Soil*, 135, 83, 1991.

Section V

Biological Control of Plant Disease
and Weeds: Applied Aspects

31 Potentially Useful Natural Product Herbicides from Microorganisms

Horace G. Cutler

CONTENTS

31.1 ABSTRACT

A number of microbial metabolites, and their templates, have potential as natural product herbicides. These include simple structures such as α-methylene-β-alanine and *cis*-2-amino-1-hydroxycyclobutane-1-acetic acid. The medium-sized structures include the nucleoside formycin, herbicidins A, B, E, F, and G; gougerotin; deoxyguanosine; coaristeromycin; 5′-deoxytoyocamycin; coformycin; and adenine 9-β-arabinofuranoside. Other nitrogen-containing metabolites are hydantocidin; 5-hydroxy-9-methylstreptimidone; 9-methylstreptimidone; phthoxazolin; and ustiloxins A, B, C, D, and E. Polyoxygenated and ring- oxygenated metabolite examples include arabenoic acid; herboxidiene; 2-butyryl-3,5-dihydroxy-cyclohex-2-ene-1-one; 2-acetyl-3,5-dihydroxy-cyclohex-2-ene-1-one; cladospirone bisepoxide; dipoxin α, η, ξ, and δ; (−)-canadensolide; sporothoriolide; 4-*epi*-ethiosolide; and pironetin. Though rare, chlorinated metabolites are also of interest and 4-chlorothreonine is examined with its biological activity.

31.2 INTRODUCTION

In previous reviews (Cutler, 1987, 1988a,b,c, 1991, 1992a,b, 1995; Edwards et al., 1988), we have shown that biologically active secondary metabolites of microbial origin offer a rich, diverse source of novel chemical structures and templates, which may be potentially useful as agrochemicals. As the search for utilitarian, readily biodegradable compounds continues, the literature continues to yield a kaleidoscope of structurally different and significant compounds. However, the primary discovery of a new metabolite is often slanted toward one discipline, for example, pharmaceuticals, because that is where the most profit can be made. But during the exploration process, microbial zoopathogens, phytopathogens, and plants may be included in bioassay screens on a gratuitous basis so that the initial intention may undergo some metamorphosis and instead of a medicinal, something agrochemical may be the apparent end product. In any event, a pharmaceutical may have use as a medicinal, herbicide, or agricultural fungicide. The purpose of this review is to highlight certain natural product structures from microorganisms that have herbicide potential.

31.3   SIMPLE STRUCTURES

Relatively simple natural product structures are nearly always of interest to both the biochemist and synthetic chemist. To the biochemist, these structures are often similar to compounds commonly found in the various metabolic pathways and they are, in a sense, mimics that can be manipulated to disrupt biochemical sequences, as with the amino acids (*vide infra* 4-chlorothreonine). Likewise, the synthetic chemist especially understands that these sorts of compounds can be synthesized in sufficient quantity for testing in a multitude of bioassays, whereas natural products from the source may only be available in very limited quantities.

Streptomyces sp. A12701 has yielded α-methylene-β-alanine (Figure 31.1). Initial fermentation was carried out in liquid culture consisting of tryptone-yeast extract, and this was used to seed a secondary shake liquid medium consisting of dextrin, glucose, soybean flour, yeast extract, and $CaCO_3$. Clean up, on reversed phase C_{18} and Sephadex LH-20, eventually yielded 15 mg of pure metabolite, which was bioassayed against *Arabidopsis thaliana* L. (Heynh.) in an agar-based titration study. At 50 μg/ml there was growth retardation, while higher concentrations fully inhibited seed germination. Small scale post-emergence screens indicated that there was slight activity against velvetleaf (*Abutilon theophrasti* Medik.) at 5.6 kg/ha (Isaac et al., 1991a). The discovery of this compound is of interest from at least two aspects. First, the simplicity of the structure and its relationship to the herbicide glycine phosphate indicates that there may be other congeners that possess herbicidal activity. Second, this is the first discovery of this particular secondary metabolite in a microorganism. It had earlier been isolated from the sponge *Fasciospongia cavernosa* (Kashman et al., 1973), collected in the Red Sea, and from a different sponge, *Spongia* cf. *zimocca* that originated in Hawaii (Yunker and Schener, 1978). In spite of the very obvious relationship to glycine phosphate, the substance was not examined for herbicidal activity until almost 20 years after its discovery (Isaac et al., 1991a), leaving one to wonder just how many other natural products remain in a chemical limbo. It is also of interest to note that the identical compound occurs both in a microorganism and two sponges, suggesting that during the evolutionary process the mechanism for biosynthesizing the metabolite has not been deleted. The question then remains as to the possibility of finding α-methylene-β-alanine in higher plants.

Another relatively simple structure is *cis*-2-amino-1-hydroxycyclobutane-1-acetic acid (Figure 31.2), which has been isolated from shake liquid fermentation. The constituents of the medium were identical to those used for the production of α-methylene-β-alanine. Following a rigorous fraction-

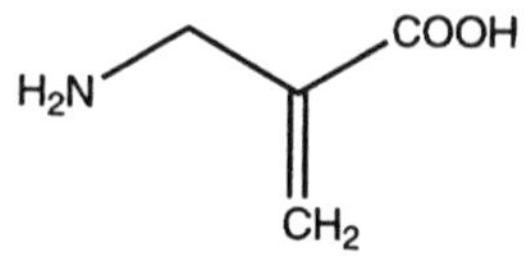

FIGURE 31.1   α-Methylene-β-alanine (*Streptomyces* sp. A12701).

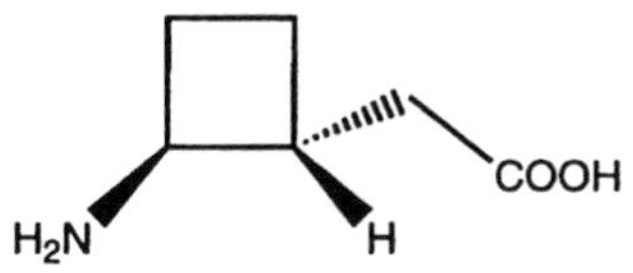

FIGURE 31.2   *cis*-2-Amino-1-hydroxycyclobutane-1-acetic acid (*Streptomyces rochei* A13018).

ation by column chromatography, a final yield of 12 mg of pure metabolite was obtained. The optically active form of the compound was assayed against *Arabidopsis thaliana* in an agar-based titration bioassay and 10 μg/ml induced some chlorosis, although 50 μg/ml resulted in bleaching and limited growth retardation. The racemic mixture, when applied to morningglory (*Ipomoea* sp.) and Indian mustard (*Brassica juncea* [L.] Czern. & Coss.) at rates of 11.2 kg/ha in post emergence assays gave severe chlorosis. Further experiments indicated that the metabolite inhibited the biosynthesis or incorporation of sulfur-containing amino acids by plants and that the chlorotic effect could be reversed by adding L-cysteine or L-methionine (Ayer et al., 1991).

It should be emphasized that *cis*-2-amino-1-hydroxycyclobutane-1-acetic acid is a component of the dipeptide 1*S*,2*S*-1-hydroxy-2-[(*S*)-valylamino] cyclobutane-1-acetic acid (Figure 31.3) (Pruess et al., 1974), which, like the former compound, exhibits antibacterial activity. The close structural relationship between the two compounds, and the biological activity exhibited by both, led to the synthesis of 25 analogs of *cis*-2-amino-1-hydroxycyclobutane-1-acetic acid, wherein there were substitutions on the carboxylic acid function {*O-tert*-butyl, OCH_3, $O(CH_2)CH_3$, and $NHCH(CH_3)_2$, NEt2, NBn2 amides, exocyclic methylene ($C(CH_3)_2$, hydroxyl (OCH_3, $OCH_2CH=CH_2$), and amino (NBn2) substitutions} (Ayer et al., 1991). Of these, a singular compound possessing both the $NBn_2$and $CONBn_2$ gave better activity that resulted in growth retardation and necrosis (Ayer et al., 1991). Examination of the original parent structure jogs the memory concerning a previously reported metabolite, moniliformin, 3-hydroxycyclobut-3-ene-1,2-diol, isolated from solid-state fermentation of *Fusarium moniliforme*, which possesses the cyclobutene structure. It had potent herbicidal activity (Cole et al., 1973) and locked mitosis, in corn (*Zea mays* L.) roots at c-metaphase, similar to colchicine (Steyer and Cutler, 1984). It would appear that the cyclobutane or cyclobutene structure has the potential to yield some bioactive derivatives.

31.4   MEDIUM-SIZED STRUCTURES

31.4.1   NUCLEOSIDES

A set of compounds occurring naturally that control growth and development in organisms are the nucleosides; one may speculate as to the number of possible permutations that nature may make using these templates. In addition, the question may be posed as to how many of these derivatives

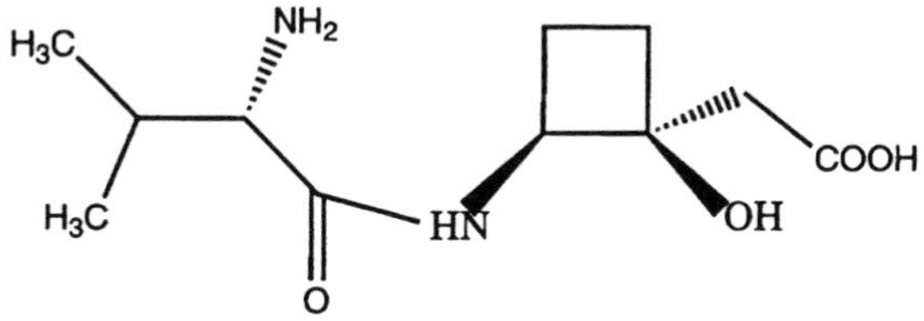

FIGURE 31.3   (1*S*,2*S*)-1-Hydroxy-2-[(*S*)-valylamino]cyclobutane-1-acetic acid.

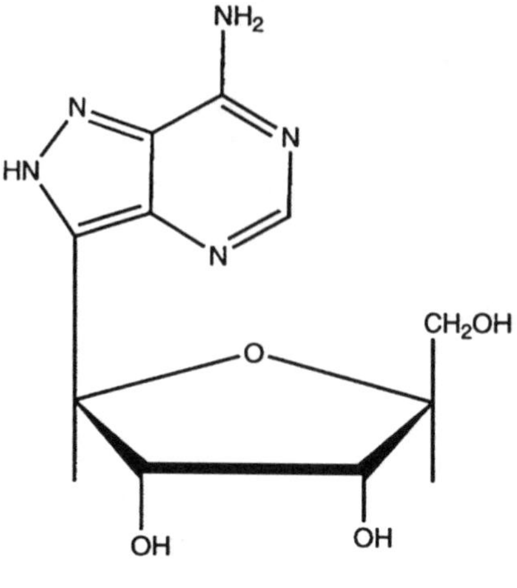

FIGURE 31.4 Formycin (*Nocardia interforma*).

have any utilitarian biological activity. Among those reported are formycin (Figure 31.4), isolated from *Nocardia interforma*, which was shown to have herbicidal activity (Hori et al., 1964). Later, a series named herbicidins A, B, E, F, and G (Figure 31.5), unique purine analogs, were found in *Streptomyces saganonensis* No. 4075 and also exhibited specific herbicidal activity (Arai et al., 1976; Haneishi et al., 1976; Takiguchi et al., 1979a,b). While there was some initial confusion relative to the absolute structures, which is not uncommon when very small amounts of a natural product are available for analysis, the problem has been corrected (Terahara et al., 1982). But of this family, only herbicidins A and B have been intensively examined for herbicidal activity on both monocotyledonous and dicotyledonous plants. Assays at concentrations that ranged from 37.5 to 300 mg/l showed that barnyardgrass (*Echinochloa crus-galli* [L.] Beauv.), goosegrass (*Eleusine indica* [L.] Gaertn.), mannagrass (*Glyceria septentrionalis* A.S.Hitchc.), and green panicum (*Panicum* sp.), were relatively susceptible, while rice *(Oryza sativa* L.) was not affected, showing a certain herbicidal selectivity for these compounds. On the other hand, herbicidin A has been experimentally used to control the phytopathogen *Xanthomonas oryzae* in rice with good success. In addition, both her-

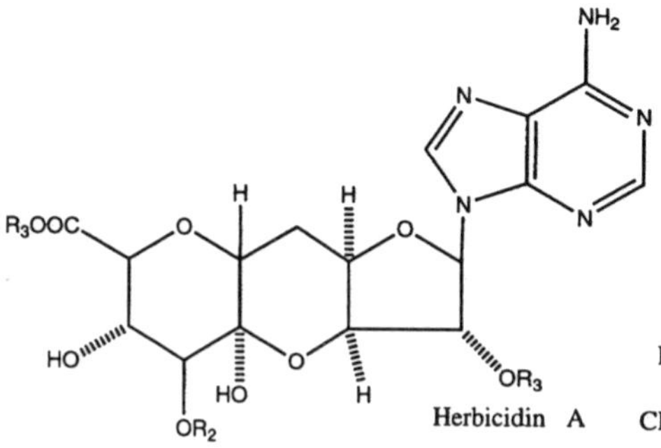

	R_1	R_2	R_3
Herbicidin A	CH_3	$CO(CH_2OH)C=CHCH_3$	CH_3
B	CH_3	H	CH_3
C	CH_3	$COCH(CH_3)_2$	CH_3
F	CH_3	$CO(CH_2)C=CHCH_3$	CH_3
G	H	$CO(CH_3)C=CHCH-3$	H

FIGURE 31.5 Herbicidins A, B, E, F, and G (*Streptomyces saganonensis*).

bicidins A and B were toxic to the dicotyledonous plants common purslane (*Portulaca oleracea* L.), *Achyranthes* sp., white goosefoot (*Chenopodium* sp.), smartweed (*Polygonum* sp.), wild amaranth (*Amaranthus* sp.), Asiatic dayflower (*Commelina communis* L.), tomato (*Lycopersicon esculentum* Mill.), and radish (*Raphanus sativus* L.), at rates of 30 to 300 mg/l. Herbicidins C and E inhibited germination of Chinese cabbage (*Brassica pekinensis* L.) and the MICs were 12.5 and 25 μg/ml, respectively (Takiguchi et al., 1979a); but against the Cyanobacterium *Anacystis nidulans* M-6, the MICs were 100 μg/ml and 100 to 200 μg/ml, respectively. They were not active against an array of 12 microorganisms, 11 of which were of medical interest and the twelfth, *Pyricularia oryzae,* a pathogen of rice. In contrast, both herbicidins F and G are active against *Trichophyton rubrum* (MIC 6.25 μg/ml), *T. asteroides* (MIC 6.25 μg/ml), *T. mentagrophytes* (6.25 to 12.5 μg/ml), *Botrytis cinerea* (MIC 12.5 μg/ml), and *Blastomyces brasiliensis* (MIC 12.5 to 25 μg/ml) (Takiguchi et al., 1979b).

Gougerotin (Figure 31.6) has been produced in liquid shake culture of *Streptomyces* sp. No. 179 and was isolated in a relatively large quantity (400 mg) after a two-step chromatography procedure (Kanzaki et al., 1962). The original isolation was reported in 1962 from *Streptomyces gougerotii* (Murao and Hayashi, 1983) and, additionally, it was shown to be a strong protein synthesis inhibitor in both prokaryotes and eukaryotes (Yukioka, 1975). However, 20 years elapsed before the report of its herbicidal activity was disclosed (Kanzaki et al., 1962). The measure of this activity was obtained with rice (cv. Tan-ginbozu) measuring second leaf sheath length. Concentrations of gougerotin from 0.03 to 0.3 mM were inhibitory and specifically, the metabolite reduced growth by 10 and 60 percent at 0.03 and 0.1 mM while 100% inhibition was induced with 0.3 mM (Kanzaki et al., 1962). At levels between 0.03 and 0.1 mM, the effects of gougerotin could be reversed by GA_3, but this was not the case with the 0.3 mM concentration of the metabolite (Kanzaki et al., 1962).

Recently, five natural product nucleosides were isolated from microbial sources and demonstrated herbicidal activity. The first, deoxyguanosine (Figure 31.7) was isolated from cell-free filtrates of *Thermoactinomycete* sp. A6019. Approximately 1 mg was isolated, but it was sufficient for testing against *Lemna minor* L. where it was highly phytotoxic at 100 μg/ml. Coaristeromycin (Figure 31.8) has been isolated from the fermentation of *Streptomyces* sp. A6308, in addition to aristeromycin (Figure 31.9). Of these, coaristeromycin exhibited herbicidal activity against yellow nutsedge (*Cyperus esculentus* L.), johnsongrass (*Sorghum halepense* [L.] Pers.), barnyardgrass and Indian mustard at 6 kg/ha (Isaac et al., 1991b). No data were supplied for aristeromycin, implying, but not proving, that the carbonyl at C6 is essential for biological activity in this particular series.

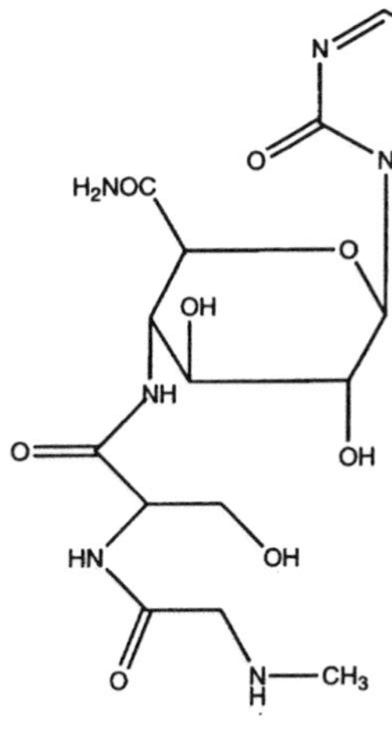

FIGURE 31.6 Gougerotin (*Streptomyces* sp.).

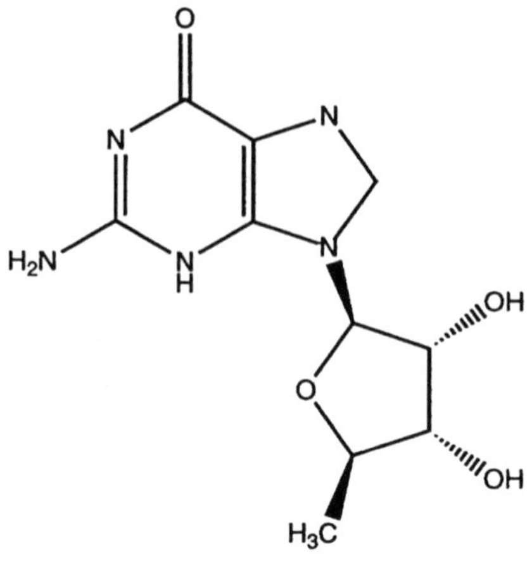

FIGURE 31.7   5-Deoxyguanosine (*Thermoactinomycete* sp. A6109).

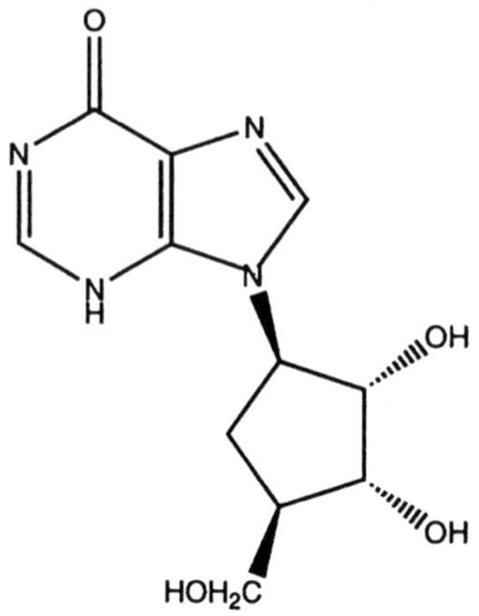

FIGURE 31.8   Coaristeromycin (*Streptomyces* sp. A6308).

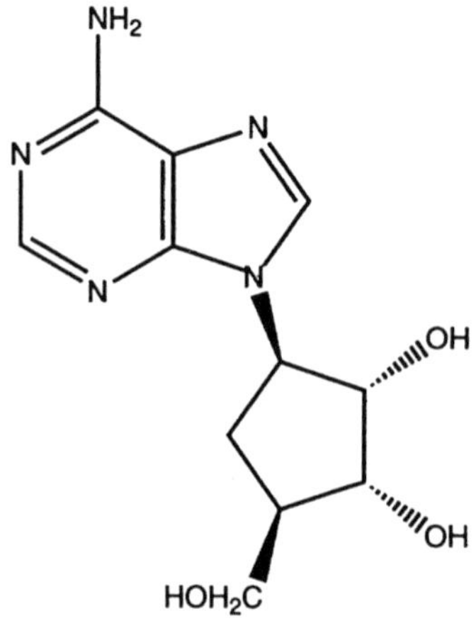

FIGURE 31.9   Aristeromycin (*Streptomyces* sp. A6308).

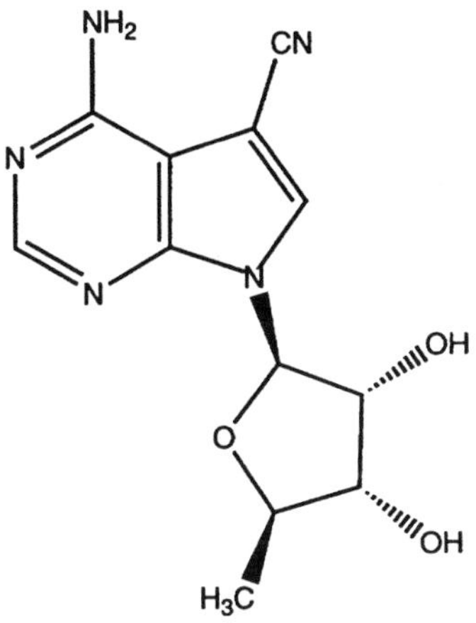

FIGURE 31.10 5′-Deoxytoyocamycin (*Streptomyces* sp. A14345).

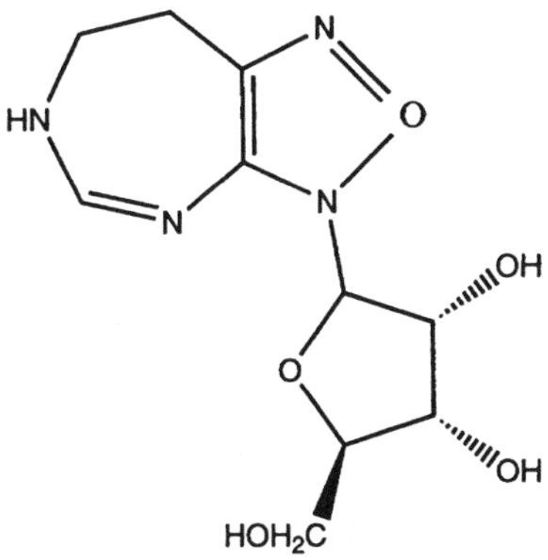

FIGURE 31.11 Coformycin (*Actinomycete* A990).

5′-Deoxytoyocamycin (Figure 31.10) was isolated from *Streptomyces* sp. A14345 in a large amount (5.2 mg) from 20 ml of culture filtrate and it was observed to cause bleaching in *L. minor* at rates of 10 μg/ml. Another close relative, coformycin (Figure 31.11), that possesses an additional ring carbon, exhibited broader herbicidal activity than the previously mentioned structures. While only 2.7 mg of the metabolite was recovered from 1 l of culture filtrate, it was sufficient to demonstrate herbicidal activity against seedlings of johnsongrass, barnyardgrass, morning-glory (*Ipomoea* sp.), and crabgrass (*Digitaria* sp.) at 6 kg/ha (Isaac et al., 1991a). Adenine 9-β-D-arabinofuranoside (Figure 31.12), an isomer of aristeromycin, was found in the fermentation broth of *Actinoplane* sp. A9222 and it inhibited germination of *A. thaliana* at rates of 25 μg/ml (Isaac et al., 1991b).

Of the six nucleosides just described, 5′-deoxyguanosine, coaristeromycin, and 5′-deoxytoyocamycin were novel natural products. They had, however, been previously synthesized but had not been evaluated for phytotoxic properties (Marumoto et al., 1976; Wang et al., 1977; McGee and Martin, 1986). On the other hand, coformycin and adenine 9-β-D-arabinofuranoside had been isolated as natural products but had not been tested on plants. This, of course, leads to the observation that there are numerous natural products that have been isolated, but their biological activities have not been examined in diverse and important bioassays. Furthermore, there appear to be synthetic structures, based on natural product templates, that have suffered the same fate. A reasonably priced index of secondary metabolite structures and their synthetic mimics outlining their biological properties, or lack thereof, would be an invaluable asset to secondary metabolite researchers.

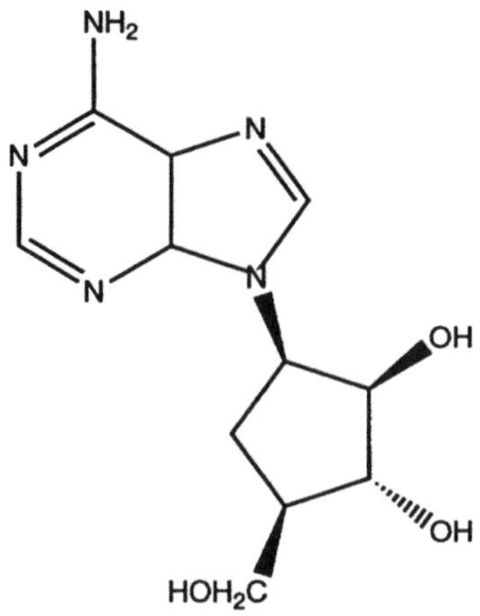

FIGURE 31.12   Adenine 9-β-D-arabinofuranoside (*Actinoplane* sp. A9222).

31.4.2   NITROGEN-CONTAINING METABOLITES

A very detailed examination has been made of hydantocidin (Figure 31.13), a metabolite obtained
from the fermentation of *Streptomyces hygroscopicus* SANK 63584, wherein 300 l of culture broth
finally yielded 37 mg of highly purified material. While the compound did not show any antifungal
or antibacterial activity, it did have potent herbicidal effects on both monocotyledonous and
dicotyledonous plants. In greenhouse screens at application rates of 500 mg/l on two-week-old
seedlings, hydantocidin was more active than bialaphos and approximately the same as glycine
phosphate. Those monocotyledonous plants that were greatly affected by the metabolite were, barnyardgrass, blackgrass (*Alopecurus myosyroides* Hyds.), large crabgrass (*Digitaria sanguinalis* [L.]
Scop.), giant foxtail (*Setaria faberi* Herrm.), green foxtail (*Setaria viridis* [L.] Beauv.), johnsongrass, and wild oat (*Avena fatua* L.). Dicotylendonous plants that were either killed, or highly
necrotized were common cocklebur (*Xanthium strumarium* L.), jimsonweed (*Haplopappus pluri-
florus* [Gray] Hall), common lambsquarters (*Chenopodium album* L.), tall morning-glory (*Ipomoea
purpurea* [L.] Roth), black nightshade (*Solanum nigrum* L.), redroot pigweed (*Amaranthus
retroflexus* L.), prickly sida (*Sida spinosa* L.), common ragweed (*Ambrosia artemisiifolia* L.), velvetleaf (*Abutilon theophrasti* Medik.), and wild mustard (*Sinapis arvensis* L.). However, bermudagrass (*Cynodon dactylon* [L.] Pers.) was hardly affected by hydantoin, while glycine phosphate and
bialaphos were moderately active. Quackgrass (*Elytrigia repens* (L.) Nevski, formerly *Agropyron
repens* [L.] Beauv.) was marginally affected by the metabolite, which proved to be equal to glycine
phosphate and more active than bialaphos. Horsenettle (*Solanum carolinense* L.), purple nutsedge
(*Cyperus rotundus* L.), and yellow nutsedge (*C. esculentus* L.) were completely controlled by hydantocidin (Nakajima et al., 1991). While many natural products have a tendency to be target specific,
hydantocidin has broad, potent herbicidal activity. The question remains unanswered about the specific activity of the various isomers, but certainly, the structure-activity examination would yield a
wealth of information.

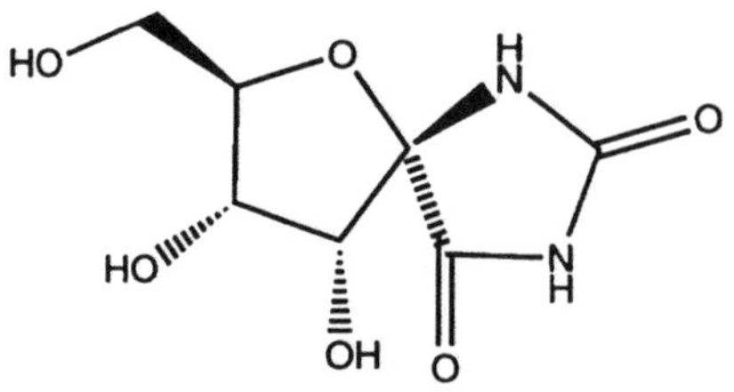

FIGURE 31.13   Hydantocidin (*Streptomyces hygroscopicus* SANK 63584).

Liquid-shake fermentation of *Streptomyces* sp. HIL Y-9065403 has yielded two glutarimide metabolites. The first, 5-hydroxy-9-methylstreptimidone (Figure 31.14), which exhibited herbicidal activity, and the second, 9-methylstreptimidone (Figure 31.14), a known antibiotic (Saito et al., 1974; Chatterjee et al., 1995). Only 27 mg of the 5-hydroxy species was isolated from 17 l of culture filtrate, but this sufficed to obtain both the physical data and some preliminary herbicide screening. Initial assays, in model systems, indicated that 100 percent growth inhibition was obtained at rates of 2 mg/l in *Lemna gibba* and *Avena sativa* with 9-methylstreptimidone, but it gave very limited control of phytopathogens *in vitro* (Chatterjee et al., 1995). The 5-hydroxy-9-methylstreptimidone congener possessed weak herbicidal activity. Oddly, the culture filtrate exhibited herbicidal activity against *L. gibba* and *A. sativa,* but the final product was only weakly active. Three questions arise as to this apparent paradox. First, is it possible that the two compounds synergise to elicit the herbicidal response? Second, is there another metabolite that is more potent but was not recovered? Third, is some co-factor necessary to enhance the activity of the molecule? The 5-hydroxy offers tantalizing possibilities with respect to synthetic modification because there are several readily available functional groups for derivatization.

A secondary metabolite that has undergone several synthetic manipulations is the herbicide CL 22T, phthoxazolin (Figure 31.15), originally isolated from *Streptomyces* sp. OM-57114 by Omura et al. (1990), in Japan, and also by French researchers from *Streptomyces griseoauranticus* (Fabre et al., 1988; Legendre and Arman, 1989). Initially, the Japanese investigation was prompted by the search for specific inhibitors of cellulose biosynthesis, with special emphasis on the phytopathogen *Phytophthora parasitica*, which was used as a bioassay system because it was known to possess cellulose in the cell wall. Since neither *Candida albicans* nor *Pyricularia oryzae* contain cellulose, these two species were also used as a bioassay system for cell free cultures. Thus, by a process of

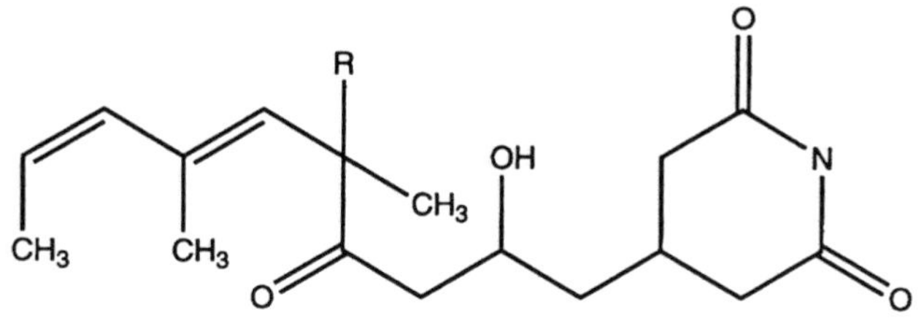

	R
5-Hydroxy-9-methylstreptimidone	OH
9-Methylstreptimidone	H

FIGURE 31.14 Streptimidones (*Streptomyces* sp. HIL Y-9065403).

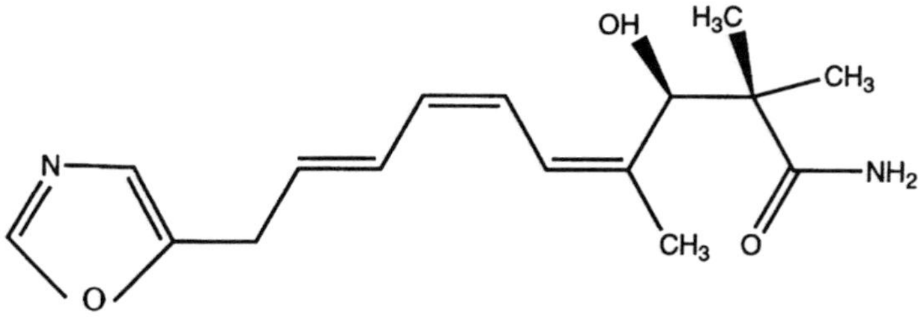

FIGURE 31.15 CL 22T, Phthoxazolin (*Streptomyces griseoauranticus*).

elimination, only cellulose biosynthesis inhibitors were discovered. Some 20,000 soil organisms were examined for this singular property until one was found that was highly active against *P. parasitica* and, in addition, gave strong herbicidal activity against radish seedlings (Omura et al., 1990). Furthermore, the culture filtrate inhibited the incorporation of ^{14}C glucose into the alkali-insoluble fraction of *Acetobacter xylinum*, in resting cells. The organism normally produces extracellular cellulose from glucose (Omura et al., 1990).

The production of the metabolite was worked from 30 l of liquid culture and approximately half this amount gave 17 mg of pure phthoxazolin, the trivial name being a combination of the genus *Phytophthora* and the oxazole moiety. Rates of 100 and 10 μg/ml inhibited radish seed 100 and 30 percent, respectively, and the MIC was established as 25 μg/test tube [5 radish seeds were incubated at 27°C for 3 days under light on metabolite-supplemented wet cotton in test tubes] (Omura et al., 1990). The claim was made that phthoxazolin has potential as a herbicide, plant growth regulator, and as a tool for studying cellulose biosynthesis (Omura et al., 1990).

Extensive derivitazation of phthoxazolin was carried out by French workers seven years after its original disclosure (Legendre et al., 1995), and the biological activity of 34 products were compared with the parent material in the radish seedling bioassay. Most of the products were derivatives of the OH function at the C3 position, though certain singular derivatives included the exocyclic NH_2 and N in the oxazole ring. The IC_{20}, the minimal concentration needed to inhibit 20 percent of radish seedling growth, was 0.5 μg/ml for phthoxazolin; 2 μg/ml for the $Si(CH_2)_2$ derivative of the OH function; 5 μg/ml for the closed ring structure $-O-C=O-NH-C(CH_3)_2-C(R)H-$; 20 μg/ml for the OCH_3 derivative; and 50 μg/ml for the decarboxylated species. All the other derivatives had IC_{20} values in excess of 100 μg/ml, or 125 μg/ml. As is so often the case, nature seems to evolve the most biologically potent structure and the synthetic derivatives are disappointing. However, this is not always the case and the statement should only be accepted as a general rule of thumb.

Another series of fairly complex small nitrogen-containing molecules are ustiloxins A, B, C, D, and E (Figure 31.16) that have been extracted from false smut balls produced by the phytopathogen *Ustilaginoidea virens* on rice panicles. While it was initially demonstrated that water extracts of smut balls were toxic to rabbits (Suwa, 1915), later work showed that aqueous extracts induced abnormal swelling in rice seedlings, very similar to the response obtained by the rice pathogen *Rhizopus chinensis* (Iwasaki et al., 1984). In fact, the latter produces rhizoxin, which inhibits microtubule formation, and it was anticipated that smut balls would produce the same toxin. However, the cyclic peptide ustiloxin A was initially isolated from water extracts of *Ustilaginoidea virens* (Koiso et al., 1992) and, later, ustiloxins B, C, and D were also discovered (Koiso et al., 1994). Both ustiloxin A and B induced abnormal swelling of rice seedling roots at 100 and 10 μg/ml, but unfortunately there was not enough of either ustiloxin C or D to complete experiments in plants (Koiso et al., 1994). Other data strongly suggest that these compounds interfere with tubulin formation. Of interest is the point that these metabolites are closely related, in part of their structure, to the hexapeptide myco-

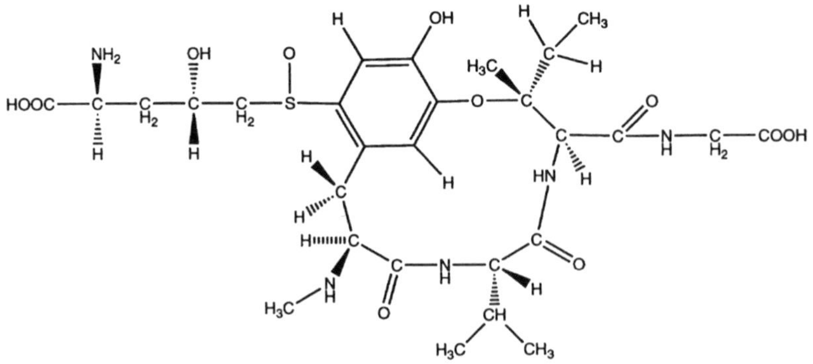

FIGURE 31.16 (A) Ustiloxin A; (B) Ustiloxin B; (C) Ustiloxin C; (D) Ustiloxin D *(Ustilaginoidea virens)*.

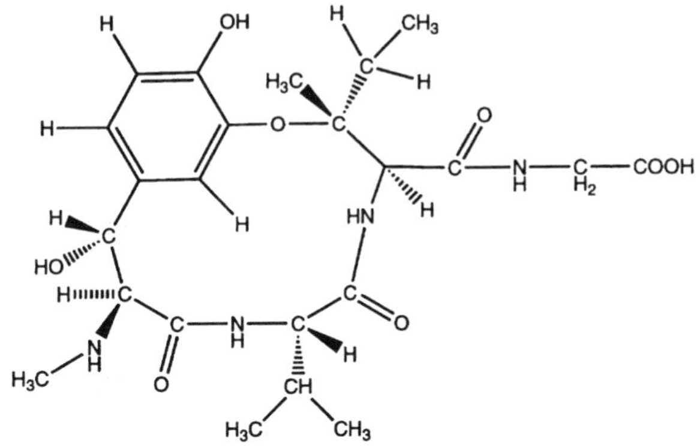

FIGURE 31.16 Continued

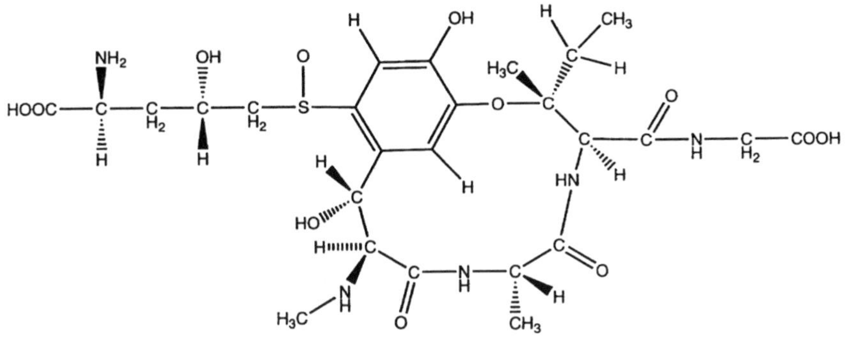

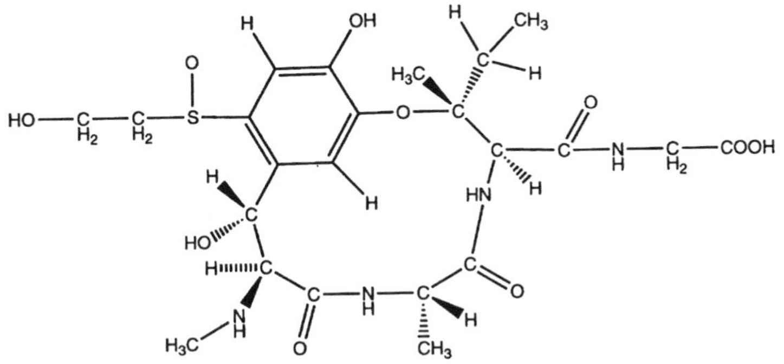

toxin phomopsin A, isolated from *Phomopsis leptostromiformis* (Culvenor et al., 1989), which has been shown to inhibit brain tubulin isomerization.

31.4.3 POLYOXYGENATED AND RING-OXYGENATED METABOLITES

An undisclosed fungus, allotted the code F6286, has produced an α, β-unsaturated pentenoic acid, trivially named arabenoic acid (Figure 31.17), in the amount of 64 mg of pure material from 700 ml of culture filtrate. The absolute structure is (*E*)-5-hydroxy-3-methoxy-2-pentenoic acid. Using an agar-based titration bioassay, with *Arabidopsis thaliana*, the compound inhibited growth at rates of 50 μg/ml and there was complete inhibition at higher rates. In post-emergence application, the metabolite was effective against velvetleaf, common cocklebur, and Indian mustard at rates of

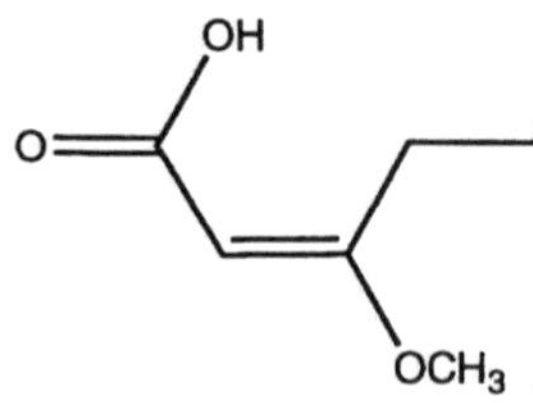

FIGURE 31.17   Arabenoic acid (F-6286, unidentified).

11.2 kg/ha. In the presence of diazomethane and acid, the compound formed the cyclic structure 5,6-dihydro-4-methoxy-2*H*-pyran-2-one, a toxin produced by *Penicillium italicum* (Isaac et al., 1991c). But the parent compound was not shown to be an artifact of the latter and, in addition, the toxin was not active against *A. thaliana* at rates of 50 µg/ml.

Streptomyces chromofuscus A7847 produces the novel structure herboxidiene (Figure 31.18), which proved to be a selective and potent herbicide against a number of weeds, but did not affect wheat (*Triticum aestivum* L.). The culture medium was manipulated to produce 155 mg/l of the metabolite, ensuring that a number of bioassays could be carried out on greenhouse-grown corn, rice (*Orzya sativa* L.), soybean (*Glycine max* L.), annual morning-glory (*Ipomoea* sp.), wheat (*Triticum aestivum* L.), oilseed rape (*Brassica napus* L.), wild buckwheat (*Polygonum convolvulus* L.), and hemp sesbania (*Sesbania exaltata* [Raf.] ex A. W. Hill). With rates of 5.592 kg/ha, herboxidiene was 100 percent inhibitory to rice. At 1.118 kg/ha, these figures remained the same except for soybean (75 percent inhibition) and rice (also 75 percent inhibition). At concentrations down to 0.017 kg/ha, these figures dropped, as might be expected, and apart from rice, soybean, and wheat, rates of 0.069 kg/ha effectively controlled the other plant species tested (Miller-Wideman et al., 1992).

Sometimes trivial names for biologically active natural products become reduced to code numbers. Such was the case with the novel chlorosis-inducing compounds AB5046A and AB5046B, which were chemically identified as 2-butyryl-3,5-dihydroxy-cyclohex-2-ene-1-one and 2-acetyl-

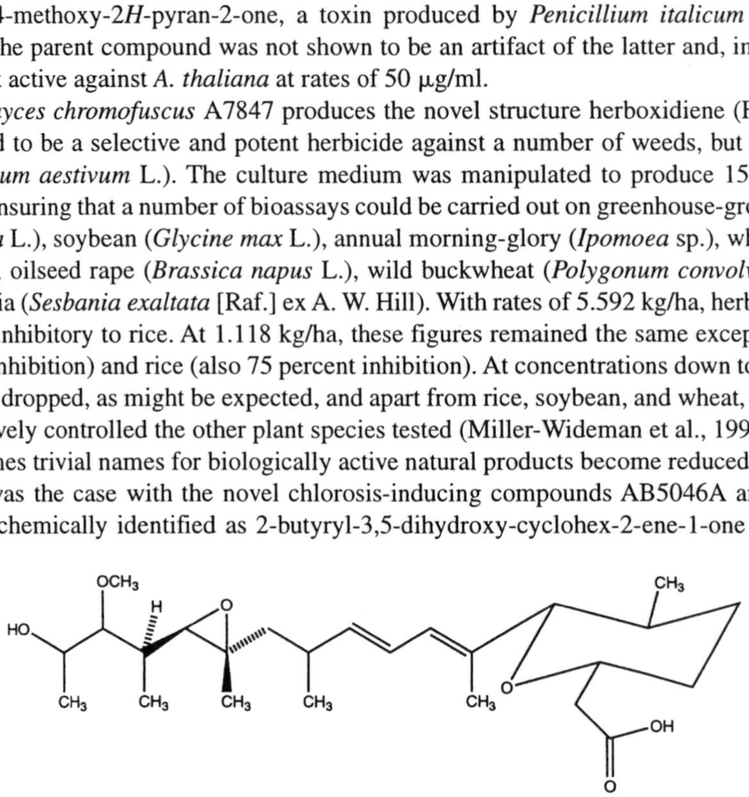

FIGURE 31.18   Herboxidiene (*Streptomyces chromofuscus* A7847).

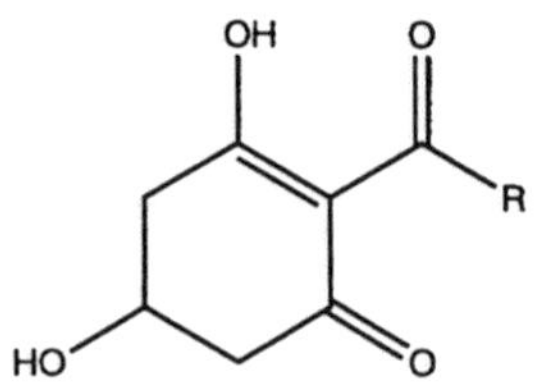

	R
AB 5046 A :	$CH_2CH_2CH_3$
AB 5046 B :	CH_3

FIGURE 31.19   AB 5046 A and B (*Nodulisporium* sp.).

3,5-dihydroxy-cyclohex-2-ene-1-one, respectively (Figure 31.19). These were metabolites of a *Nodulisporium* sp., found in soil samples, that were grown on liquid shake culture. In this particular case, the yields were relatively high so that 10 l of medium gave rise to 1.45 g of AB5046A and 200 mg of AB5046B. The effects of the metabolites used as sodium salts against Japanese barnyard millet (*Echinochloa utilis*) were dramatic, and concentrations of both chemical species at 50 and 100 mg/l caused complete chlorosis, manifested by totally white leaves, without inhibition of seed germination. But at concentrations of 12.5 to 25 mg/l, the activity of AB5046B was only half that of its congener (Igarashi et al., 1993). Other monocotyledonous plants bioassayed included large crabgrass (*Digitaria sanguinalis* [L.] Scop.), green foxtail (*Setaria viridis* [L.] Beauv.), rice, flatsedge (*Cyperus iria* [L.]), and the dicotyledonous plants green gram (*Phaseolus aureus* L.), Chinese radish (*Raphanus sativus* L.), hairy beggarticks (*Bidens pilosa* L.), and livid amaranth (*Amaranthus lividus* L.). Of these, all the monocotyledonous species were susceptible to AB5046A from 6.25 to 12.5 mg/l, while the dicotyledonous plants, with the exception of livid amaranth, which behaved like the monocotyledons, were resistant (Igarashi et al., 1993). The IC_{50}, the amount necessary to induce a 50 percent chlorosis with AB5046A, was calculated to be approximately 0.094 mM. Neither compound induced toxicity in mice when administered orally at 300 mg/kg (Igarashi et al., 1993). Of parallel interest is the information that compounds with this skeleton, such as 3,5-dihydroxy-2-dodecanoyl-cyclohex-2-ene-1-one (Oliver et al., 1990) and 3,6-dihydroxy-2-[1-oxo-10 (*E*)-tetradecenyl] cyclohex-2-ene-1-one (Lusby et al., 1987) have been discovered in insects, but their herbicidal activity has not been reported. The discovery of AB5046A and B in *Nodulisporium* forges an interesting link, in terms of biosynthetic pathways, between microorganisms and insects. Furthermore, the functional groups on both AB5046A and B that are available for derivatization suggest that other utilitarian compounds may be synthesized.

Cladospirone bisepoxide (Figure 31.20) (Petersen et al., 1994) has been so named because its genesis was thought to be a *Cladosporium* sp. Accessed as F24'707, the organism was later identified, by personal communication, as belonging to the Coelomycetes because of the formation of pycnidia. This compound, too, has been isolated in a relatively large quantity with amounts up to 1.5 g/l being obtained from shake cultures; yields of 1.16 g/l have been obtained at the bioreactor level. The metabolite induced both herbicidal and antimicrobial activity (Petersen et al., 1994). With respect to the former, garden cress (*Lepidium sativum* L.) seeds were sown on filter papers and treated with concentrations of 5, 25, 50, and 100 μg of metabolite per 20 mm disk. At 24 and 48 hours following treatment, inhibition, shoot growth, and chlorophyll production were noted, depending on the concentration used. Root tip growth was greatly reduced by rates of 25 and 50 μg and, furthermore,

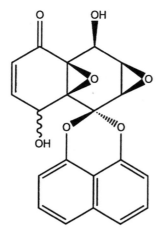

FIGURE 31.20 Cladospirone bisepoxide (unidentified Coelomycete).

roots appeared to be very intertwined. Higher concentrations completely inhibited growth for 48 hours, but at lower concentrations seeds behaved exactly like the controls (Petersen et al., 1994).

Those microorganisms that were examined were *Bacillus subtilis* ETH 2016, *Staphylococcus aureus* ETH 2070, *Sarcina lutea* ATCC9341, *Escherichia coli* ETH 2018, *Pseudomonas aeruginosa* ATCC 10145, *Candida albicans* ETH 6370, *Sacchromyces cerevisiae* ETH 108, *Botrytis cinerea* Tü 157, *Pyricularia oryzae* AC 164, *Mucor miehei* Tü 284, and *Paecilomyces varioti* Tü 137: 50 μg of the metabolite was added to each paper disk. Of agronomic importance was the finding that both *Botrytis cinerea* and *Pyricularia oryzae* were controlled by cladospirone bisepoxide. Those medically significant organisms that were inhibited were *Bacillus subtilis, Straphylococcus aureus, Escherichia coli,* and *Sacchromyces cerevisiae.*

The spironaphthodioxines with their substituted decaline groups are a series of novel structures that include diepoxin, and α, η, ξ, and σ (Schlingman et al., 1993). A fungus found growing on a tree trunk in Panama yielded the antibiotic MK 3018 (Anonymous, 1988) and the closely related preussomerin A, isolated from the fungus *Preussia isomera* (Weber et al., 1990).

Occasionally an old structure reappears in the literature along with new relatives. Such an example is (−)-canadensolide (Figure 31.21), which, of itself, has a fascinating history. In 1978, McCorkindale and co-workers (1968) disclosed the biosynthetic pathway for this metabolite, a dilactone isolated from *Penicillium canadense,* which had been shown to have antifungal properties. What makes the story intriguing is that McCorkindale astutely surmised that hexylitaconic acid might be a precursor and, subsequently, synthesized the ^{14}C isotopic species. This, when introduced into fermentation cultures of *P. canadense,* became incorporated into ^{14}C canadensolide and proved, unequivocally, that hexylitaconic acid was the necessary precursor (McCorkindale et al., 1978). The complete conformation of (−)-canadensolide (McCorkindale et al., 1978) was rectified (Kato et al., 1971, 1975) and confirmed by synthesis (Anderson and Fraser-Reid, 1978, 1985). Oddly, hexylitaconic acid was discovered, six years after the (−)-canadensolide saga, in *Aspergillus niger* (Isogai et al., 1984) as a biologically active natural product and it proved to be a potent growth promoter in rice at 20 mg/l, but it inhibited growth at 100, 200, and 500 mg/l. In turn, hexylitaconic acid proved

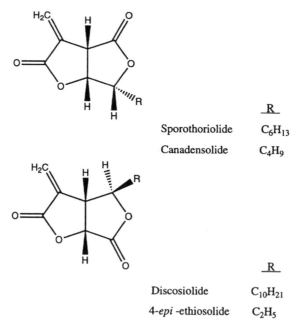

	R
Sporothoriolide	C_6H_{13}
Canadensolide	C_4H_9

	R
Discosiolide	$C_{10}H_{21}$
4-*epi*-ethiosolide	C_2H_5

FIGURE 31.21 Furofurandiones (*Sporothrix* sp., *Discosia* sp., *Pezicula livida*).

to be similar in structure to another root growth stimulator, radiclonic acid, from an unidentified fungus (Sassa et al., 1973a,b).

Compounds related to ($-$)-canadensolide are sporothriolide, from *Sporothrix* sp. Strain 700, discosiolide, from *Discosia* sp. Strain 1290, and 4-*epi*-ethiosolide (Figure 31.21) from *Pezicula livida*. These are all furofuranosides. By way of colorful history, and to show that there are some esoteric collection sites, *Sporothrix* sp. had been isolated from the slag heap of a goldmine in West Borneo. The *Discosia* sp. originated from a Teneriffe soil sample collected at 1100 m in a foggy wooded location; *Pezicula livida* was an endophyte on *Betula lenta*. At concentrations of 4 mg/ml for sporothorolide, germination of *Lepidium sativum* L. and *Medicago sativa* L. were both inhibited 95 percent relative to controls. Discosiolide was not inhibitory to either species at 5.1 mg/ml, and 4-*epi*-ethiosolide completely inhibited both at 20 mg/ml, relative to controls. From the synthetic perspective, dihydrosporothriolide in which the exocyclic double bond (CH_2) is saturated, completely inhibited *M. sativa*, but it had no effect on *L. sativum* (Krohn et al., 1994). This is yet another example of a case in which a less than 1 percent addition to the molecular weight completely alters both the specific activity and target specifity. And in terms of Avagadro's number, relative to specific activity, the mathematics become intriguing.

In addition to the herbicidal activity, sporothriolide showed promising fungicidal control of *Botrytis cinerea* that infected pepper seedlings when treated with solutions containing 500 mg/l. Again, all three metabolites gave some measure of control against *Bacillus megatarium, E. coli, Ustilago violacea, Mycotypha microspora,* and *Eurotium repens*. But, dihydrosporothriolide was not active against either the bacteria or *E. repens*. Related compounds include ($-$)-avenaciolide (Hughes, 1978), in which the side chain is C_8H_{17}. It has shown activity against *Botrytis allii,* and the synthetic racemic mixture of sporothriolide, for which a patent has been issued, states that the molecule cures abscesses. This set of compounds has enough functional groups to make the candidates very suitable for derivitization and, possibly, the production of biologically active homologs and analogs. It must be stated that discosiolide, in which the side chain is $C_{10}H_{21}$, as opposed to 4-*epi*-ethiosolide, with a C_2H_5 side chain, does not possess herbicidal activity, but the 4-*epi*-compound does. The question then arises as to what the necessary optimum chain length is to elicit biological activity with this molecule.

One of the major problems that occurs in cereal crops is lodging, which results in harvest reductions. A few synthetic plant growth regulators have been developed to inhibit stem growth in these crops so that the resulting dwarf plants are not prone to wind damage while, at the same time, seed head formation and yield are not compromised. One of the more interesting natural products, isolated from the culture broth of *Streptomyces* sp. NK10958, that has been used experimentally to control lodging is pironetin (Figure 31.22), which was shown to have the structure (5*R*, 6*R*)-5-ethyl-5,6-dihydro-6-[(*E*)-(2*R*,3*S*,4*R*,5*S*)-2-hydroxy-4-methoxy-3,5-dimethyl-7-nonenyl]-2*H*-pyran-2-one. The fermentation, carried out in aerated, agitated liquid gave a final yield, following purification, of 8 g from 200 l. This included not only separation from the liquid, but also extraction of the mycelial cake with methanol. Pironetin was formulated into a 10 percent emulsifiable concentrate, consisting of 39% xylene, 39 percent isopropanol, and 12 percent surfactant, and evaluated on rice seedlings in greenhouse assays that incorporated the known antilodging agents paclobutrazol and inabenfide. Rice seedlings that were in the 2.5-leaf development stage were transplanted into pots filled with paddy diluvium soil, fertilized, and puddled with water; a bioassay known as the Wagner pot test. These were flooded to a depth of 3.5 cm and placed in a greenhouse; they were then treated, using a submerged treatment, at 21, 15, 9, or 5 days prior to heading with rates of 1.8 g/d for paclobutrazol, 15 g/d for inabenfide, and at 10, 5, 2.5, 0.63, and 0.16 g/d for pironetin. Pironetin induced a 14 to 23% decrease in stem length with rates of 10 g/d, and there were no deleterious effects on rice yields when the metabolite was applied at 9 or 5 days prior to heading. There were slight reductions in yield when pironetin was applied 21 or 12 days before heading, but they were not statistically significant. It was concluded that the microbial metabolite was superior to the two commercial plant growth regulators tested (Kobayashi et al., 1994a).

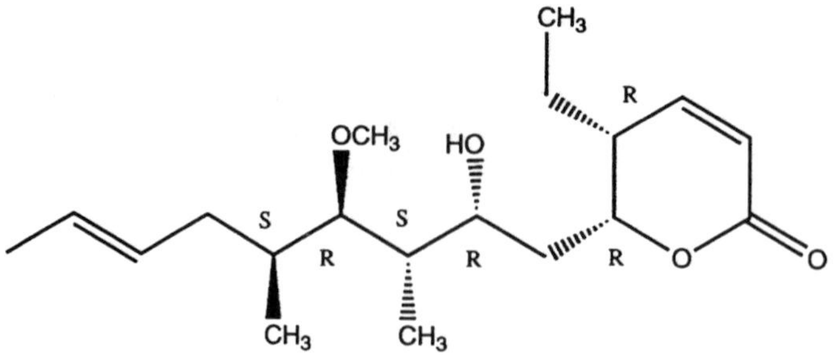

FIGURE 31.22 Pironetin (*Streptomyces* sp.).

When the absolute structure was reported (Kobayashi et al., 1994b), reference was made to the observation that part of the structure of pironetin, the 4-ethyl-5-substituted-α,β-unsaturated-δ-lactone, forms an integral portion of phosphazomycin C, which has been reported to have both anti-fungal and antitumor properties (Tomiya et al., 1990), but which has not been tested for plant growth- regulating properties. Likewise, kazusamycin is a 4-methyl-5-substituted-α,β-unsaturated-δ-lactone, which exhibits antitumor properties (Komiyama et al., 1995), but has not been examined as a plant growth regulator. As a terminal statement, Kobayashi et al., (1994b) state in their report concerning the structure of pironetin that the cytotoxicity of the metabolite to tumor cell lines will be revealed. It is significant that other pyrones have shown marked biological activity, for example, 6-pentyl-α-pyrone, which has potent fungicidal properties (Cutler et al., 1986). It seems highly probable that derivatives of pironetin will have some specific biological activities and that the template itself will be an important source for the genesis of biorational compounds of commercial value.

31.4.4 Chlorinated Metabolites

This brief overview started with a simple molecule, α-methylene-β-alanine, and it seems only fitting that it should end with a relatively uncomplicated structure, albeit a chlorinated compound, 4-chlorothreonine (Figure 31.23). Production of the metabolite was by *Streptomyces* sp. OH-5092 in aerated, agitated liquid and 40 l finally yielded 140 mg of pure substance after considerable work-up. Halogenated microbial secondary metabolites are relatively rare, the notable exceptions being the ochratoxins, the syringostatins (Isogai et al., 1990; Fukuchi et al., 1992a), syringomycin (Fukuchi et al., 1990, 1992b), and griseofulvin. Presumably, as marine microorganisms are increasingly examined, it may be that the numbers of these types of compounds will increase and, perhaps, there may be iodated and brominated congeners of considerable interest.

4-Chlorothreonine was tested against radish and sorghum (*Sorghum vulgare* [L.] Moench) and seeds were treated at rates of 30 and 120 μg/tube; bialaphos was included as an internal standard in the bioassay. Growth inhibition, relative to controls, was measured after four days and the metabolite inhibited radish seedling growth 30 and 70 percent at 30 and 120 μg/tube, respectively, but bialaphos inhibited 40 and 90% at each of the concentrations. Sorghum was inhibited 30 and 80% with rates of 30 and 120 μg/tube, respectively, while bialophos inhibited 50 and 80 percent at these

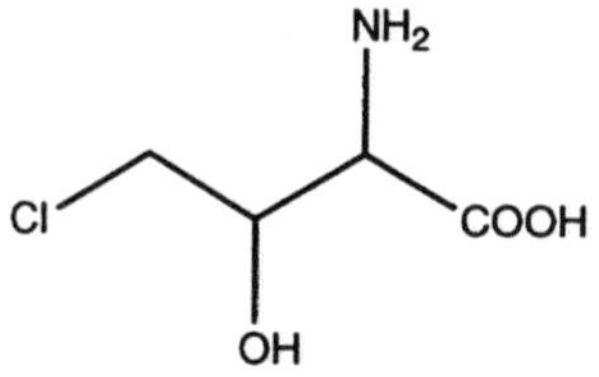

FIGURE 31.23 4-Chlorothreonine (*Streptomyces* sp.).

concentrations. Furthermore, 4-chlorothreonine inhibited *Candida albicans* in *in vitro* assays, and 50 µg/paper disk induced an inhibition zone of 21 mm. This inhibition could be reversed by the addition of 250 µg/disk of either L-alanine, proline, threonine, or DL-γ-aminobutyric acid. Consequently, it has been proposed, though not yet proved, that the herbicidal mode of action may be due to the inhibition of amino acid metabolism. Of paramount importance is the role of 4-chlorothreonine as a key intermediate in the β-lactam biosynthetic pathway and, therefore, the compound is important to the antibiotic and pharmaceutical industry (Yoshida et al., 1994).

31.5 CONCLUSION

The example of 4-chlorothreonine is important because, in a sense, it offers a philosophical paradigm by bringing a number of practical questions into focus. First, was the research conducted with the goal of finding a new agrochemical? Second, was the intention to discover a new antibiotic? Third, was the intention the elucidation of a key intermediate in β-lactam antibiotic production? If the answer is yes to all three questions, then there are excellent prospects for the production of biodegradable natural product herbicides that already occur in the ecosystem. Any other answer, with exception to the first question, indicates that some serious re-evaluations need to be considered so that natural products, some of which may have already been discovered but which lay dormant, can be commercially developed to benefit agriculture without compromising the environment.

REFERENCES

Anderson, R. C. and Fraser-Reid, B., An asymmetric synthesis of naturally occurring canadensolide, *Tetrahedron Lett.,* 1978, 3233, 1978.

Anderson, R. C. and Fraser-Reid, B., Synthesis of bis-γ-lactones from "diacetone glucose". V. Optically active canadensolide, *J. Org. Chem.,* 50, 4786, 1985.

Anonymous, Japanese Patent No. 1294-686A (Mitsubishi Kasei Corp.) priority appl. No. JP 120717 of 19. 05, Derwent patent Abstr. 90-012996/02, 1988.

Arai, M., Haneishi, T., Kitahara, N., Enokita, R., Kawakubo, K., and Kondo, Y., Herbicidins A and B, two new antibiotics with herbicidal activity. I. Producing organism and biological activities, *J. Antibiot.,* 29, 863, 1976.

Ayer, S. W., Isaac, B. G., Luchsinger, K., Makkar, N., Tran, M., and Stonard, R., *cis*-2-amino-1-hydroxy-cyclobutane-1-acetic acid, a herbicidal antimetabolite produced by *Streptomyces rochei* A13018, *J. Antibiot.,* 44, 1460, 1991.

Chatterjee, S., Vijayakumar, E. K. S., Chatterjee, S., Blumbach, J., and Ganguli, B. N., 5-Hydroxy-9-methyl-streptimidone, a new glutarimide from a *Streptomyces* sp. HIL Y-9065403, *J. Antibiot.,* 48, 271, 1995.

Cole, R. J., Kirksey, J. W., Cutler, H. G., Doupnik, B. L., and Peckham, J. C., from *Fusarium moniliforme.* Effects on plants and animals, *Science,* 179, 1324, 1973.

Culvenor, C. C. J., Edgar, J. A., MacKey, M. F., Gorst-Allman, C. G., Marasas, W. F. O., Steyn, P. S., Vleggaar, R., and Wessels, P. L., Structure elucidation and absolute configuration of phomopsin A, a hexapeptide mycotoxin produced by *Phomopsis leptostromiformis, Tetrahedron,* 45, 2351, 1989.

Cutler, H. G., Japanese contributions to development of allelochemicals, in *Allelochemicals: Role in Agriculture and Forestry,* Waller, G. R., Ed., ACS Symposium Series No. 330, American Chemical Society, Washington, D.C., 1987, 23.

Cutler, H. G., Perspectives on discovery of microbial phytotoxins with herbicidal activity, *Weed Technol.,* 2, 525, 1988a.

Cutler, H. G., Unusual plant growth regulators from microorganisms, *Crit. Rev. Plant Sci.,* 6, 323, 1988b.

Cutler, H. G., Natural products and their potential in agriculture; a personal overview, in *Biologically Active Natural Products,* Cutler, H. G., Ed., ACS Symposium Series No. 380, American Chemical Society, Washington, D.C., 1988c, 1.

Cutler, H. G., Phytotoxins of microbial origin, in *Handbook of Natural Toxins,* Keeler, R. F. and Tu, A. T., Eds., Marcel Dekker, New York, 1991, 411.

Cutler, H. G., Effects of natural products from microorganisms on higher plants, in *Plant Biochemical Regulators,* Gausman, H. W., Ed., Marcel Dekker, New York, 1992a, 113.

Cutler, H. G., Herbicidal compounds from higher plants, in *Phytochemical Resources for Medicine and Agriculture,* Nigg, H. N. and Seigler, D., Eds., Plenum Press, New York, 1992b, 205.

Cutler, H. G., Microbial natural products that affect plants, phytopathogens, and certain other microorganisms, *Crit. Rev. Plant Sci.,* 14, 413, 1995.

Cutler, H. G., Cox, R. H., Crumley, F. G., and Cole, P. D., 6-Pentyl-α-pyrone from *Trichoderma harzianum:* plant growth inhibitory, and antimicrobial properties, *Agric. Biol. Chem.,* 50, 2943, 1986.

Edwards, J. V., Dailey, O. D., Bland, J. M., and Cutler, H. G., Approaches to structure/function relationship, in *Biologically Active Natural Products,* Cutler, H. G., Ed., ACS Symposium Series No. 380, American Chemical Society, Washington, D.C., 1988, 35.

Fabre, B. E., Arman, E., Etienne, G., Legendre, F., and Tiraby, G., A simple screening method for insecticidal substances from Actinomycetes, *J. Antibiot.,* 41, 212, 1988.

Fukuchi, N., Isogai, A., Yamashita, S., Suyama, K., Takemoto, J. Y., and Suzuki, A., Structure of phytotoxin syringomycin produced by a sugarcane isolate of *Pseudomonas syringae* pv. *syringae, Tetrahedron Lett.,* 31, 1589-1, 1990.

Fukuchi, N., Isogai, A., Nakayama, J., Yamashita, S., Suyama, K., and Suzuki, A., Isolation and structural elucidation of syringostatins, phytotoxins produced by *Pseudomonas syringae* pv. *syringae* lilac isolate, *J. Chem. Soc. Perkin Trans.,* 11, 875, 1992a.

Fukuchi, N., Isogai, A., Nakayama, J., Takayama, S., Yamashita, S., Suyama, K., Takemoto, J. Y., and Suzuki, A., Structure and stereochemistry of three phytotoxins, syringomycin, syringotoxin and syringostatin, produced by *Pseudomonas syringae* pv. *Syringae, J. Chem. Soc. Perkin Trans.,* 1, 1149, 1992b.

Haneishi, T., Terahara, A., Kayamori, H., Yabe, J., and Arai, M., Herbicidins A and B, two new antibiotics with herbicidal activity. II. Fermentation, isolation and physico-chemical characterization, *J. Antibiot.,* 29, 870, 1976.

Hori, M., Ito, E., Takita, T., Koyama, G., Takeuchi, T., and Umezawa, H., A new antibiotic, formycin, *J. Antibiot.,* (Ser. A.) 17, 96, 1964.

Hughes, D. L., The crystal structure of (−)-avenaciolide, *Acta Crystallogr.,* 34, 3674, 1978.

Igarashi, M., Tetsuka, Y., Mimura, A., Takahashi, A., Tamamura, T., Sato, K., Naganawa, H., and Takeuchi, T., AB5046A and B, novel chlorosis-inducing substances from *Nodulisporium* sp., *J. Antibiot.,* 46, 1843, 1993.

Isogai, A., Washizu, M., Kondo, K., Murakoshi, S., and Suzuki, A., Isolation and identification of (+)-hexylitaconic acid as a plant growth regulator, *Agric. Biol. Chem.,* 48, 2607, 1984.

Isogai, A., Fukuchi, N., Yamashita, S., Suyama, K., and Suzuki, A., Structure of syringostatins A and B, novel phytotoxins produced by *Pseudomonas syringae* pv. *syringae* isolated from lilac blight, *Tetrahedron Lett.,* 31, 695, 1990.

Isaac, B. G., Ayer, S. W., and Stonard, R. J., The isolation of α-methylene-β-alanine, a herbicidal microbial metabolite, *J. Antibiot.,* 44, 795, 1991a.

Isaac, B., Ayer, S. W., Letendre, L. J., and Stonard, R. J., Herbicidal nucleosides from microbial sources, *J. Antibiot.,* 44, 729, 1991b.

Isaac, B., Ayer, S. W., and Stonard, R. J., Arabenoic acid, a natural herbicide of fungal origin, *J. Antibiot.,* 44, 793, 1991c.

Iwasaki, S., Kobayashi, H., Furukawa, J., Namikoshi, M., Okuda, S., Sato, Z., Matsuda, I., and Noda, T., Studies on macrocyclic lactone antibiotics. VII. Structure of a phytotoxin "Rhizoxin" produced by *Rhizopus chinensis, J. Antibiot.,* 37, 354, 1984.

Kanzaki, T., Higashide, E., Yamamoto, H., Shibata, M., Nakazawa, K., Iwasaki, H., Takewaka, T., and Miyake, A., Gougerotin, a new antibacterial antibiotic, *J. Antibiot.* (Ser. A), 15, 93, 1962.

Kashman, Y., Fishelson, L., and Ne'eman, I., *N*-acyl-2-methylene-β-alanine methyl esters from the sponge *Fasciospongia cavernosa, Tetrahedron,* 29, 3655, 1973.

Kato, M., Tanaka, R., and Yoshikoshi, A., Synthesis of (±)-canadensolide and its C-5 epimer. Revision of the stereochemistry, *J. Chem. Soc. Chem. Comm.,* 1561, 1971.

Kato, M., Kageyama, M., and Yoshikoshi, A., Synthetic study of (±)-canadensolide and related dilactones, *J. Org. Chem.,* 40, 1932, 1975.

Kobayashi, S., Tsuchiya, K., Harada, T., Nishide, M., Kurokawa, T., Nakagawa, T., Shimada, N., and Kobayashi, K., Pironetin, a novel plant growth regulator produced by *Streptomyces* sp. NK10958. I. Taxonomy, production, isolation and preliminary characterization, *J. Antibiot.,* 47, 697, 1994a.

Kobayashi, S., Tsuchiya, K., Kurokawa, T., Nakagawa, T., Shimada, N., and Iitaka, Y., Pironetin, a novel plant growth regulator produced by *Streptomyces* sp. NK10958. II. Structural elucidation, *J. Antibiot.*, 47, 703, 1994b.

Koiso, Y., Natori, M., Iwasaki, S., Sato, S., Sonoda, R., Fujita, Y., Yaegashi, H., and Sato, Z., Ustiloxin: a phytotoxin and mycotoxin from false smut balls on rice panicles, *Tetrahedron Lett.*, 33, 4157, 1992.

Koiso, Y., Li, Y., Iwasaki, S., Hanaoka, K., Kobayashi, T., Sonoda, R., Fujita, Y., Yaegashi, H., and Sato, Z., Ustiloxins, antimitotic cyclic peptides from false smut balls on rice panicles caused by *Ustilaginoidea virens*, 47, 765, 1994.

Komiyama, K., Okada, K., Oka, H., Tomisaka, S., Miyano, T., Funayama, S., and Umezawa, I., Structural study of a new antitumor antibiotic, kazusamycin, *J. Antibiot.*, 38, 220, 1985.

Krohn, K., Ludewig, K., Aust, H. J., Draeger, S., and Schulz, B., Biologically active metabolites from fungi. III. Sporothriolide, discosiolide, and 4-*epi*-ethisolide-new furofurandiones from *Sporothrix* sp., *Discosia* sp., and *Pezicula livida*, *J. Antibiot.*, 47, 113, 1994.

Legendre, F. and Arman, E., Nouveaux composés herbicides, le CLT22T et ses dérivés, procédés de préparation de ces composés, composition les contenant et procédés de traitment des desherbage les utilisant, French Patent 89,08615, June 28, 1989.

Legendre, F., Maturano, M. D., Etienne, G., Klaebe, A., and Tiraby, G., Synthesis and biological activity of derivatives of the herbicidal metabolite CL22T (Phthoxazolin), *J. Antibiot.*, 48, 341, 1995.

Lusby, W. R., Oliver, J. E., Neal, J. W., Jr., and Heath, R. R., Isolation and identification of the major component of setal exudate from *Corythucha ciliata*, *J. Nat. Prod.*, 50, 1126, 1987.

Marumoto, R., Yoshioka, Y., Furukawa, Y., and Honjo, M., Synthesis of aristeromycin analogs, *Chem. Pharm. Bull.*, 24, 2624, 1976.

McCorkindale, N. J., Wright, J. L. C., Brian, P. W., Clarke, S. M., and Hutchinson, S. A., Canadensolide-an antifungal metabolite of *Penicillium canadense*, *Tetrahedron Lett.*, 727, 1968.

McCorkindale, N. J., Blackstock, W. P., Johnston, G. A., Roy, T. P., and Troke, J. A., *11th. Int. Sym. Chem. Nat. Prod.* (IUPAC), Vol. 1, 1978, 151.

McGee, D. P. and Martin, J. C., Acyclic nucleoside analogs: methods for the preparation of 2',3'-secoguanosine, 5'-deoxy-2',3'-secoguanosine and *(R,S)*-9-[1-(2-hydroxyethoxy)-2-hydroxyethyl] guanine, *Can. J. Chem.*, 64, 1885, 1986.

Miller-Wideman, M., Makkar, N., Tran, M., Isaac, B., Biest, N., and Stonard, R., Herboxidiene, a new herbicidal substance from *Streptomyces chromofuscus* A7847, *J. Antibiot.*, 45, 914, 1992.

Murao, S. and Hayashi, H., Gougerotin, as a plant growth inhibitor, from *Streptomyces* sp. No. 179, *Agric. Biol. Chem.*, 47, 1135, 1983.

Nakajima, M., Itoi, K., Takamatsu, Y., Kinoshita, T., Okazaki, T., Kawakubo, K., Shindo, M., Honma, T., Tohjigamori, M., and Haneishi, T., Hydantocidin: a new compound with herbicidal activity from *Streptomyces hygroscopicus*, *J. Antibiot.*, 44, 293, 1991.

Oliver, J. E., Lusby, W. R., and Neal, J. W., Jr., Exocrine secretions of the andromeda lace bug *Stephanitis takeyai* (Hemiptera: Tingidae), *J. Chem. Ecol.*, 16, 2243, 1990.

Omura, S., Tanaka, Y., Kanaya, I., Shinose, M., and Takahashi, Y., Phthoxazolin, a specific inhibitor of cellulose biosynthesis, produced by a strain of *Streptomyces* sp., *J. Antibiot.*, 43, 1034, 1990.

Petersen, F., Moerker, T., Vanzanella, F., and Peter, H. H., Production of cladospirone bisepoxide, a new fungal metabolite, *J. Antibiot.*, 47, 1098, 1994.

Pruess, D. L., Scannel, J. P., Blount, J. F., Ax, H. A., Kellett, M., Williams, T. H., and Stempel, A., Antimetabolites produced by microorganisms. XI. 1-*(S)*-hydroxy-2-*(S,S)*-valylamino-cyclobutane-1-acetic acid, *J. Antibiot.*, 27, 754, 1974.

Saito, N., Kitame, F., Kikuchi, M., and Ishida, N., Studies on a new antiviral antibiotic, 9-methylstreptimidone. I. Physiochemical and biological properties, *J. Antibiot.*, 27, 206, 1974.

Sassa, T., Tomizuka, K., Ikeda, M., and Miura, Y., Isolation of a new root growth-stimulating substance from a fungus, *Agric. Biol. Chem.*, 37, 1221, 1973a.

Sassa, T., Takemura, T., Ikeda, M., and Miura, Y., Structure of radiclonic acid, a new plant growth-regulator produced by a fungus, *Tetrahedron Lett.*, 2333, 1973b.

Schlingman, G., West, R. R., Milne, L., Pearce, C. J., and Carter, G. T., Diepoxins, novel fungal metabolites with antibiotic activity, *Tetrahedron Lett.*, 34, 7225, 1993.

Steyer, C. H. and Cutler, H. G., Effects of moniliformin on mitosis in maize (*Zea mays* L.), *Plant Cell Physiol.*, 25, 1077, 1984.

Suwa, M., Study on ina-kouji. I, *JAMAS (Igaku Chuo Zasshi, Japan)*, 13, 661, 1915.

Takiguchi, Y., Yoshikawa, H., Terahara, A., Torikata, A., and Terao, M., Herbicidins C and E, two new nucleoside antibiotics, *J. Antibiot.*, 32, 857, 1979a.

Takiguchi, Y., Yoshikawa, H., Terahara, A., Torikata, A., and Terao, M., Herbicidins F and G, two new nucleoside antibiotics, *J. Antibiot.*, 32, 862, 1979b.

Terahara, A., Haneishi, T., Arai, M., Hata, T., Kuwano, H., and Tamura, C., The revised structure of herbicidins, *J. Antibiot.*, 35, 1711, 1982.

Tomiya, T., Uramoto, M., and Isono, K., Isolation and structure of phosphazomycin C, *J. Antibiot.*, 43, 118, 1990.

Wang, Y., Hogencamp, H. P. C., Long, R. A., Revankar, G. R., and Robins, R. K., A convenient synthesis of 5'-deoxynuleosides, *Carbohyd. Res.*, 59, 449, 1977.

Weber, H. A., Baenziger, N. C., and Gloer, J. B., Structure of preussomerin A: an unusual new antifungal metabolite from the coprophilous fungus *Preussia isomera*, *J. Am. Chem. Soc.*, 112, 6718, 1990.

Yoshida, H., Arai, N., Sugoh, M., Iwabuchi, J., Shiomi, K., Shinose, M., Tanaka, Y., and Omura, S., 4-Chlorothreonine, a herbicidal antimetabolite produced by *Streptomyces* sp. OH-5093, *J. Antibiot.*, 47, 1165, 1994.

Yukioka, M., *Antibiotics*, Vol. 3, Corcoran, J. W. and Hahn, F. G., Eds., Springer Verlag, Berlin, 1975.

Yunker, M. B. and Schener, P. J., Alpha-oxygenated fatty acids occurring as amides of 2-methylene-β-alanine in a marine sponge, *Tetrahedron Lett.*, 4651, 1978.

32 Allelopathic Interactions in the Biological Control of Postharvest Diseases of Fruit

Carolee T. Bull

CONTENTS

32.1 ABSTRACT

Allelopathic interactions occur among microorganisms when biochemicals produced by one organism influence another. Microbial antagonists used to control diseases of plants, produce secondary metabolites that influence plant pathogens. These secondary metabolites influence plant disease by antibiosis, nutrient competition, parasitism, or inducing resistance in the plant. Research has been conducted on the mechanisms by which antagonists of postharvest pathogens control diseases. Although there is evidence that antibiotics, degradative enzymes, and nutrient sequestering molecules may be involved in biological control of postharvest diseases, conclusive demonstrations of their roles are lacking. Molecular genetic tools have been effectively used in research on biological control of soilborne plant pathogens. Use of these tools will greatly advance our understanding of how microbial inoculants control postharvest diseases of plants.

32.2   INTRODUCTION

Primary metabolism is the interrelated series of enzymatic reactions providing living organisms with energy, synthetic intermediates, and key macromolecules such as DNA and protein. These reactions unify all forms of life in that they all use the same pathways of primary metabolism. In contrast, secondary metabolism is the part of metabolism that is not essential for life and plays no part in the growth of the organism. Products of secondary metabolism have no explicit role in the maintenance or reproduction of the producing organism (Singleton and Sainsbury, 1987). Unlike the products of primary metabolism, the enzymatic reactions and the products of secondary metabolism are heterogeneous. Secondary metabolism results in the production of a variety of biological compounds produced by relatively few organisms.

Microorganisms produce a plethora of secondary metabolites, many of which are released into the environment where they influence other organisms. Allelopathy is the historical term used to describe biochemical interactions, both positive and negative, between all organisms, including microorganisms (Rice, 1995). In general, the term allelopathy has only been applied to negative interactions between plants but as stated, allelopathic interactions also occur among microorganisms. Plant pathologists have exploited allelopathic interactions among microorganisms to favor healthy plants and inhibit organisms causing disease. Secondary metabolites produced by antagonistic microorganisms influence soilborne plant pathogens and mediate disease control (Handelsman and Stabb, 1996). Biological control is the use of one organism to control a disease caused by another. The term allelopathy is rarely applied to the biological control of plant pathogens, although secondary metabolites produced by antagonists have negative effects on pathogens. Rice (1995) states, "One of the most exciting areas [of allelopathy research] is in phytopathology, where much progress has been made concerning . . . the use of microorganisms in biological disease control . . . ." Instead of discussing allelopathy, plant pathologists have chosen to describe the interactions more specifically and do not discuss the general concept of allelopathy in relation to plant disease.

In his treatment of the topic of allelopathic interactions among plant-associated microorganisms, Rice (1995) describes many biological control systems. He recognized that the mechanism by which disease control was mediated was not demonstrated for many of the systems. He predicted that more information would soon be available. His prediction was true for all fields of biological control. In particular, many advances have recently been made in understanding how microorganisms mediate the control of postharvest diseases of fruit. In this chapter, I do not attempt to review all of the literature on the biological control of postharvest diseases, but describe some examples that may not be considered classical allelopathy but illustrate the effects of chemicals produced by biological control agents on plant disease. In addition, I suggest approaches to further our understanding of mechanisms involved in biological control of postharvest diseases. Other reviews will be helpful for those interested in the general advances in the field of biological control of postharvest diseases (Janisiewicz, 1988, 1998; Pusey, 1989; Viñas, 1995; Wilson et al., 1991; Wisniewski and Wilson, 1992).

32.3   BIOLOGICAL CONTROL OF POSTHARVEST DISEASES

Postharvest diseases of plants are economically very important and can result in losses of up to 50 percent of fruits and vegetables in some regions (Ceponis et al., 1986; Jeffries and Jeger, 1990). Postharvest decays result in millions of dollars in losses in the citrus industry alone (Eckert and Eaks, 1989). Postharvest diseases develop during harvesting, storage, grading, packing, transport, and until the product is consumed. In addition to the cost of growing and producing a crop, there are postharvest handling, storage, and transport costs associated with harvested products that are not associated with losses due to field diseases. Numerous synthetic chemicals have been used to combat postharvest losses. Additional measures and alternatives to these chemicals are currently being developed because of the high level of resistant isolates present in some fungal populations (Brown, 1977; Delp, 1980; Eckert and Ogawa, 1985), the very high cost of developing and registering new chemicals, stricter regulations on use of chemicals, and the public's increasing awareness and perception of dietary risks and environmental impacts. Viable alternatives to fungicides, including physical treatments, careful handling techniques, packing-house sanitation, treatments to increase fruit resistance, and biological control have been suggested (Smilanick and Denis-Arrue, 1992).

Biological control of plant pathogens is an emerging alternative to synthetic chemicals in several crop-disease systems. Control of soilborne plant pathogens is in the forefront of biological control research due to a long history of research and development beginning more than half a century ago. Currently, there are 22 products registered with the U.S. Environmental Protection Agency (US EPA) as biopesticides against soilborne plant pathogens (Cook et al., 1996). Although biological control of postharvest diseases has only been studied for the last two decades, there are three commercially available biopesticides registered with the US EPA (Cook et al., 1996).

Postharvest diseases represent an excellent opportunity for the success of biological control. Application of biological control products can easily be integrated into existing packing-house operations. Additionally, environmental conditions can be altered during postharvest handling and storage to increase the efficacy of the biological control products.

Although environmental conditions can rarely be manipulated to increase biological control of soilborne and foliar plant pathogens, tools and information can be gleaned from research on these systems to advance biological control of postharvest diseases. In particular, molecular genetic methods have the potential to expand our understanding of interactions between microorganisms and the mechanisms by which postharvest biological control operates. The putative mechanisms by which postharvest diseases are controlled have been reviewed (Droby and Chalutz, 1994; Janisiewicz, 1998; Wilson et al., 1991, 1994). To date, conclusive evidence implicating particular mechanisms has not been reported.

32.4   MECHANISMS BY WHICH MICROORGANISMS CONTROL POSTHARVEST DISEASES

Disease control using biological agents has not been as consistent as control with synthetic chemicals. Improvements in biological control will rely on: 1) selecting better organisms, 2) improving existing organisms, 3) applying the organisms under optimal conditions for their success, 4) preconditioning the pathogen population to be more susceptible to antagonists, 5) optimizing formulations of biological control products to enhance disease control characteristics, and 6) better integrating the use of biological control agents biorational disease control measures. More information about how biological control of postharvest pathogens is mediated is needed in order to plan strategies for improved control (Bull et al., 1997; Droby and Chalutz, 1994; Wisniewski and Wilson, 1992). Antibiosis, nutrient competition, mycoparistism, induced resistance, and pre-emptive exclusion have been reported as possible mechanisms by which antagonistic microorganisms control plant diseases (Droby and Chalutz, 1994; Janisiewicz and Roitman, 1988; Smilanick and Denis-Arrue, 1992; Wilson et al., 1991, 1994; Wisniewski et al., 1991). Here I discuss the involvement of chemicals, produced by biological control agents, in disease control.

32.5   ANTIBIOSIS

In their natural environment, microorganisms multiply when nutrients are available and environmental conditions are favorable. During an exponential growth phase, they use available nutrients to produce the essential metabolites needed for growth and multiplication. When nutrients become limited, growth slows and they enter a stationary phase. Microorganisms produce a variety of secondary metabolites at the end of exponential growth and during this stationary phase (Vining, 1990).

Production of secondary metabolites by antagonists of soilborne pathogens is regulated by environmental signals and genetic regulators (Corbell and Loper, 1995; Laville et al., 1992; Sarniguet et al., 1995). Nutrient requirements for the production of secondary metabolites are specific. Minor changes in laboratory media often result in changes in the quality and quantity of secondary metabolites produced (Gutterson, 1990; James and Gutterson, 1986; Kraus and Loper, 1995; Nowak-Thompson et al., 1994; Roitman et al., 1990; Shanahan et al., 1992; Walker et al., 1996). For example, the quantity of the antibiotics 2,4-diacetylphloroglucinol, pyoluteorin, and pyrrolnitrin produced by the antagonist *Pseudomonas fluorescens* Pf-5 were affected by the carbon source used in culture media (Nowak-Thompson et al., 1994). Environmental factors, including the presence and type of carbon, probably influence production of secondary metabolites by microbes on plant surfaces. Some secondary metabolites have no apparent effect on the producing organisms but are deleterious to other microorganisms.

Plant pathologists use the term antibiosis to describe the production of compounds by a biological control agent that directly inhibits the growth or normal activity of a pathogen (Fravel, 1988; Handelsman and Stabb, 1996; Handelsmen and Parke, 1989; Gutterson 1990). Antibiosis is therefore a clear example of allelopathy. Antibiotics and bacteriocins are the most well-studied metabolic products involved in antibiosis by antagonists used to control plant pathogens. Antibiotics are low molecular weight compounds that are produced by microbes and have deleterious effects on growth or metabolic activities of other microorganisms (Fravel, 1988). The involvement of antibiotics in disease control of soilborne pathogens has been demonstrated repeatedly (Fravel, 1988; Gutterson, 1990; Handelsman and Stabb, 1996). In contrast, few antagonists of postharvest diseases have been shown to produce antibiotics *in vitro*. The role of antibiotics in biological control of postharvest diseases has not been investigated to the same extent as their role in controlling soilborne plant pathogens.

32.5.1 ANTIBIOTIC PRODUCTION BY BACTERIAL ANTAGONISTS

Strains of *Bacillus subtilis* (Cohn) Prazmowski have been shown to control a number of postharvest diseases of apple, peach, nectarine, plum, cherry, apricot, and grape (Pusey, 1989; Pusey and Wilson, 1984). This bacterium produces compounds *in vitro* that are inhibitory to plant pathogens (Gutter and Littauer, 1953; Pusey and Wilson, 1984; Singh and Deverall, 1984; Utkhede and Sholberg, 1986). Scientists have investigated the roles these inhibitory compounds play in disease control. *Bacillus subtilis* strain B-3 controls brown rot on stone fruit caused by *Monilinia fructicola* (Wint.) Honey (Pusey and Wilson, 1984). In addition to inhibiting the pathogen *in vitro*, autoclaved spent-cell filtrate protected fruit (Pusey and Wilson, 1984). Other known antagonists from various collections did not show *in vitro* inhibition of *M. fructicola* and did not control the disease on peach. Iturin peptides with antifungal activities were isolated from strain B-3 (Gueldner et al., 1988; McKeen et al., 1986). These compounds have a wide spectrum of antifungal activity against plant pathogenic fungi *in vitro* but inhibit few bacteria (Pusey, 1989). These data suggested that antibiosis may be the major mechanism by which this antagonist operates.

In addition to strains of *B. subtilis*, fluorescent pseudomonads have been used to control postharvest diseases. Strains of *Pseudomonas cepacia* (Ex Burkholder) Palleroni & Holmes produce antifungal compounds that are implicated in control of postharvest diseases (Huang et al., 1993; Janisiewicz and Roitman, 1988; Roitman et al., 1990; Smilanick and Denis-Arrue, 1992; Smilanick et al., 1993). *Pseudomonas cepacia* strain LT412 reduces disease incidence of gray mold (*Botrytis cinerea* Pers.:Fr.), blue mold (*Penicillium expansum* Lk. & Thom.) on apple (Janisiewicz and Roitman, 1988), green mold (*P. digitatum* [Pers.:Fr.] Sacc.) on citrus (Huang et al., 1993; Smilanick and Denis-Arrue, 1992), and brown rot (*M. fructicola*) of peach and nectarine (Smilanick et al., 1993), and inhibits these fungi *in vitro*. Pyrrolnitrin produced by strain LT412 also inhibits these pathogens *in vitro* (Janisiewicz and Roitman, 1988; Roitman et al., 1990; Smilanick and Denis-Arrue, 1992).

Botrytis cinerea can cause disease on foliar and fruit tissues of strawberry. *Pseudomonas antimicrobica* inhibits *B. cinerea in vitro* and inhibited conidial germination on strawberry leaves (Walker et al., 1996). Additionally, cell-free extracts from this organism inhibited the pathogen. The nature of the antibiotic(s) has not been described. However, inhibition of the pathogen was media dependent, indicating that the quality and quantity of nutrients available to this organism influences antibiotic production.

Pseudomonas syringae van Hall strains ESC-10 and ESC-11 are the active ingredients in two of the commercially available products for biological control of postharvest diseases. These organisms are related to the pathogenic bacterium *P. syringae* pv. *syringae* van Hall; however, for strains ESC-10 and ESC-11 no host has been found (Smilanick et al., 1996). A common attribute of *P. syringae* pv. *syringae* strains is the production of cycliclipodepsipeptides, which are virulence determinants and inhibit fungi *in vitro*. Strains ESC-10 and ESC-11 produce the cycliclipodepsipeptide, syringomycin E. Both strains and purified syrinogmycin E inhibit *P. digitatum in vitro* (Bull et al., 1998; Wadsworth et al., 1996).

32.5.2 PURIFIED ANTIBIOTICS IN DISEASE CONTROL

Although antibiotic production *in vitro* has been demonstrated for a number of antagonists the role of antibiotic production in disease control has not been exhaustively investigated. Demonstrating antibiotic production and antibiosis *in vitro* is the initial step in determining if antibiosis is involved in biological control. However, *in vitro* inhibition is not always directly correlated to disease control (Fravel, 1988). A second step in demonstrating antibiosis is to provide evidence that purified antibiotics have the same antifungal spectrum as the antagonist and control disease when applied to plant surfaces. This was an important approach, which led to useful information for scientists studying biocontrol of soilborne diseases during the 1980s (Howell and Stipanovick, 1979, 1980). Scientists studying postharvest biocontrol have also used this approach. As stated, *P. cepacia* produces pyrrolnitrin *in vitro* and controls postharvest diseases caused by *B. cinerea* and *P. expansum*. Pyrrolnitrin isolated from *P. cepacia* controlled decay caused by these pathogens on apple when applied at a rate of 0.01 mg/ml. Also, purified syringomycin E from *P. syringae* strains ESC-10 and ESC-11, which inhibited *P. digitatum in vitro,* controlled green mold on lemons caused by this pathogen (Bull et al., 1998; Wadsworth et al., 1996).

Control of disease following application of purified antibiotic to plant surfaces indicates that antibiosis may be involved in disease control, but this is not conclusive. Molecular tools developed for the study of other plant–microbe interactions can be used to define the role of antibiosis in biocontrol of postharvest diseases.

32.5.3 USE OF MUTANTS TO STUDY ANTIBIOSIS

Handelsman and Parke (1989) insightfully discuss what criteria need to be met in order to demonstrate antibiosis by biological control agents. They define what are called by some scientists, "the Koch's postulates for demonstration of antibiosis." In short, antibiotic strains are mutagenized to produce variants that no longer produce antibiotics. The mutants are then complemented for antibiotic production by cloned antibiotic biosynthesis genes on recombinant plasmids. Demonstration of antibiosis is complete when the wild type and constructed strains are tested for biological control and there is a constant correlation between disease control and antibiotic production. This is the most rigorous demonstration of antibiosis currently available.

Our own work has demonstrated the need for such studies to understand the role of antibiosis in biological control. Syringomycin E produced by *P. syringae* strains ESC-10 and ESC-11 controls *P. digitatum* when applied directly into wounds after the wounds are inoculated with pathogen (Bull et al., 1998). The level of control achieved by applying syringomycin E to lemon [*citrus limon* (L.) Burm. f.)] is equivalent to the level achieved by applying the biological control agents. However, mutants that no longer produce syringomycin E controlled green mold on lemon to the same extent as the wild-type syringomycin E producers. These studies demonstrated that although purified syringomycin E can control green mold, production of this compound is not needed for disease control by the antagonists. A complicating factor is that in the syringomycin mutants, a second antifungal compound, syringopeptin, is made in increased amounts and may control the disease in the absence of syringomycin E (Bull and Smilanick, unpublished observations).

A second powerful approach using mutational analysis involves generating antibiotic resistant mutants of the pathogen. The pathogen and isogenic mutants resistant to the antibiotic in question are tested for amenability to biological control. There is compelling evidence that the antibiotic plays an important role in disease control if the resistant mutant is no longer controlled by the antagonist producing the antibiotic. Smilanick and Denis-Arrue (1992) used this approach to demonstrate that although *P. cepacia* produces pyrrolnitrin, which inhibits the pathogen *P. digitatum in vitro,* biological control is mediated by some other mechanism. Pyrrolnitrin resistant mutants of *P. digitatum* were still controlled by *P. cepacia.*

To date, only a few correlations between antibiotic production *in vitro* and biological control have been made. No definitive evidence is available demonstrating the role of antibiosis in biological control of postharvest diseases. In two cases, molecular genetic techniques have been used to define the role of antibiosis in the control of postharvest diseases (Smilanick and Denis-Arrue, 1992; Bull and Smilanick, unpublished observations). In both examples, production of antibiotics was not needed for the biological control of the pathogens being studied. In addition to the lack of genetic evidence for the role of antibiosis in the biological control of postharvest pathogens, antibiotic production *in situ* by these strains has not been demonstrated. Several antibiotics that are important in biological control of soilborne diseases have been isolated from the rhizosphere after application of antibiotic producing strains (Keel et al., 1992; Thomashow and Weller, 1988). It is not known if iturins, pyrrolnitrin, and syringomycin E are produced *in situ* in wounds or on the surface of fruit.

32.5.4   ANTIBIOTIC PRODUCTION BY YEAST ANTAGONISTS

In addition to bacterial biological control agents, yeasts also control postharvest diseases (Arras, 1996; Chalutz and Wilson, 1990; Chand-Goyal and Spotts, 1996; Filonow et al., 1996; Lurie et al., 1995; Roberts, 1990; Suzzi et al., 1995; Wilson and Chalutz, 1989; Wisniewski and Wilson, 1992). Demonstration of antibiotic production by bacterial biological control agents has historical precedence. In contrast, antibiotics have rarely been isolated from yeasts. Yeasts are the organisms of choice for some scientists because they propose that yeasts are less likely to produce antibiotics. There are, however, examples of antibiotic production by yeasts (Takesako et al., 1991; Da Silva and Pascholati, 1992; MacWilliam, 1959; McCormack et al., 1994).

The common consumable yeast *Saccharomyces cerevisiae* Meyen has been used to protect corn plants in greenhouse experiments from *Colletotrichum graminicola* (Ces.) Wils. (Da Silva and Pascholati, 1992). In addition, cell-free filtrates reduced the development of this pathogen on corn leaves while autoclaved cell-free filtrates did not. This indicated that heat-liable compounds may be involved in the antagonistic activity of *S. cerevisiae* against *C. graminicola* on corn.

In many reports, other authors have indicated that yeasts do not produce antibiotics after co-inoculating a yeast and one indicator strain on one medium. As stated before, secondary metabolite production by microorganisms is media dependent *in vitro* and presumably dependent on the nutritional environment *in situ*. In many cases, bacteria must first be incubated for several days before challenging them with the indicator organism in order to demonstrate antibiosis. For example, antibiotic production by *P. syringae* strain ESC-10 was not detected until the antagonist was incubated for five days or more on a special medium containing plant signal molecules (Bull et al., 1998; Janisiewicz and Marchi, 1992). Synthetic media cannot be presumed to mimic the environment on plants in which biological antagonists are required to operate.

A good example of the need to explore more than one medium and to incubate producing strains for some period of time prior to introduction of the pathogen comes from work with an avirulent strain of the yeast-like fungus, *Geotrichum candidum* Link. Initially, production of antifungal compounds by *G. candidum* strain avir was evaluated by coinoculating the organism with *Penicillium digitatum* on potato dextrose agar (PDA). No zone of inhibition was observed on PDA and it was assumed that *G. candidum* avir produced no secondary metabolites inhibitory to *P. digitatum* (C. Eayre, personal communication). Subsequently, nine media were tested and *G. candidum* strain avir was grown for five days before being challenged with the pathogen. The avir strain produced a zone of inhibition on one of the nine media tested (Eayre, Skaria, and Bull, unpublished results). It can therefore be concluded that *G. candidum* strain avir produces an antifungal compound(s) on at least one medium. The nature of the antifungal compound(s) has not been determined.

While in most studies authors have not fully evaluated the production of antifungal compounds by yeasts, Filonow and coworkers (1996) conducted an extensive evaluation of antibiotic production by yeasts. Several different media were used. No inhibition was observed *in vitro* even when concentrated broth from cultures was used.

It is never completely clear that the lack of demonstration of antibiotic production in culture is evidence that antibiotics are not produced *in situ*. The lack of production of an antibiotic on a single medium against a single indicator strain does not indicate that antibiotics are not produced by the antagonists. The physiology of microorganisms under laboratory conditions may be very different from those on plant and fruit surfaces. Antibiotic production by *P. syringae* strain ESC-10 was not detected unless plant signal molecules were incorporated into the medium (Bull et al., 1998). A negative result in agar plate bioassays only indicates that no compounds effective against a given target organism are produced by the antagonist on the given media tested.

32.5.5 BACTERIOCIN PRODUCTION BY BACTERIAL ANTAGONISTS

Bacteriocins are antibiotics with bactericidal effects only on strains closely related to the producing bacterium (Vidaver, 1976, 1983). Strains of antagonistic bacteria that produce bacteriocins may be useful as biocontrol agents for suppression of closely related phytopathogens because related bacteria may occupy similar niches on plant surfaces. An antagonist that inhabits the infection court of a target pathogen may be an optimal biological control agent (Cook and Baker, 1983). Although biological control of bacterial pathogens by closely related antagonists that produce bacteriocins seems promising, examples of effective biological control operating by this mechanisms are few. Nevertheless, *Agrobacterium radiobacter* (Beijerink & Van Delden) Conn strain K-84, one of the most successful biological control agents, elicits biological control in part through the production of a bacteriocin. Production of the bacteriocin agrocin 84 by *A. radiobacter* is responsible largely for the biological control of *A. tumefaciens* (Smith & Thownsend) Conn (Cooksey and Moore, 1982a,b). It is unlikely that bacteriocins will be exploited for the control of postharvest diseases of fruit because the majority of diseases are caused by fungal pathogens; however, this may be an appropriate approach to control postharvest diseases of vegetables that are often caused by bacteria. In addition, it is useful to understand that non-pathogenic organisms that occupy a similar niche as the target pathogen may be exploited for control of postharvest diseases.

32.6 PARASITISM

Parasitism is an interaction involving physical contact between the organisms and is beneficial to one organism at the expense of the other. Biological control of plant pathogens is thought to be mediated, in some cases, by antagonistic microbes that parasitize plant pathogens and directly reduce the pathogens inoculum density (Droby et al., 1993; Elad, 1996; Wisniewski et al., 1991). Mycoparasitism is implicated in biological control of fungal diseases by the antagonist *Trichoderma harzianum* Rifai. Accumulative evidence from several lines of research suggest that biological control by *T. harzianum* may be mediated by parasitism (Bélanger et al., 1995; Elad, 1996; Handelsman and Parke, 1989). Parasitism is not in itself an example of allelopathy; however, in some cases the antagonists produce enzymes that degrade the structural integrity of the plant pathogens and is therefore included in this chapter.

Several lines of evidence indicate that parasitism may be involved in biological control: 1) reduction in inoculum levels, 2) direct contact between the antagonist and the pathogen, 3) damage to the pathogen at points of direct contact, 4) isolation of cell wall degrading enzymes from the antagonist, and 5) demonstration that enzymes purified from the biological control agents damage the pathogen or reduce disease. Although parasitism has been discussed widely as a mechanism by which antagonists control soilborne fungal diseases, parasitism has only been implicated in a few instances in the biological control of postharvest diseases. In 1991, Wisniewski and coworkers suggested that parasitism was involved in the control of postharvest diseases. *Pichia guilliemondii* Wickerham isolate 87 controls diseases caused by *B. cinerea* and *P. expansum* on apple. Using electron microscopy, strong attachment of *Pichia guilliemondii* to *B. cinerea* hyphae followed by pitting and collapse of the pathogens hyphae indicated that mycoparasitism might be involved. In addition, the enzyme β-(1-3) glucanase was isolated from the antagonist grown *in vitro* and on cell walls from the pathogen.

Others are using advanced molecular genetic techniques to investigate the role of cell wall degrading enzymes in the biological control of plant pathogens. *Pichia anomola* (Hansen) Kurtzman strain K also controls gray mold on apple caused by *B. cinerea*. As with *P. guilliemondii*, *P. anomola* produces degredative enzymes. The enzyme exo-β-1,3-glucanase is produced in culture and on *B. cinerea* cell wall preparations. The purified enzyme inhibits conidia germination and germ tube growth of the target pathogen. In addition, exo-β-1,3-glucanase activity was detected on apple treated with the antagonist (Jijakli and Lepoivre, 1998). DNA sequences suspected to be involved in production of this enzyme have been identified. Their role in enzyme production and disease control is being studied. A consistent relationship between the presence of the genes, enzyme production, and disease control would indicate that this enzyme plays an important role in disease control. No correlation was found between segregation of *PAEXG*1a and *PAEXG*2a, two DNA sequences believed to be involved in synthesis of the enzyme, and enzyme or biological control activity (Grevesse et al., 1998). To date, no segregants devoid of *in vitro* enzyme production have been found. Additional detailed genetic analysis will provide conclusive evidence needed to determine if exo-β-1,3-glucanase is involved in disease control.

32.7 INDUCED RESISTANCE

Plant resistance to virulent pathogens can be increased by "immunizing" the plant through inoculation with less virulent or non-pathogenic fungi, bacteria, or viruses. Through chemical interactions the plant's own defenses are stimulated to prevent disease (Kuc and Strobel, 1992; Sequiera, 1983). Phytoalexin accumulation and development of localized or systemic acquired resistance can result from inoculation with biological control agents. Because chemicals produced by the biological control agents may be involved in induction of these defense responses, they are discussed here. Reviews discussing the possible implementation of induced resistance as a measure to reduce postharvest disease are available (Droby and Chalutz, 1994; Elad, 1996; Wilson et al., 1994).

Several authors have demonstrated that antagonists increase accumulation of antimicrobial phytoalexins in harvested fruit. Phytoalexin accumulation is stimulated by microbial invasion and chemical or mechanical damage. The antagonistic yeast, *Candida famata* (Harrison) Meyer and Yarrow strain F35 controlled green mold of citrus caused by *Penicillium digitatum* and induced accumulation of two phytoalexins, scopoletin and scoparone (Arras, 1996). Fruit inoculated with the antagonist accumulated up to 12 times more of the antimicrobial compounds at the inoculation site than did untreated fruit (Arras, 1996). Other biological antagonists also induced accumulation of these compounds (Rodov et al., 1994). Like other phytoalexins, scoparone and scopoletin inhibit fungal growth and development (Rodov et al., 1994; Kim et al., 1991) indicating that they may be involved in controlling the pathogen when accumulated in adequate concentrations. The effects of phytoalexin accumulation in control of postharvest diseases may be very localized as biological control agents were not able to induce resistance in fruit when they were applied a short distance from the pathogen (Elad, 1996).

Ethylene production is also induced by biological control agents. On citrus, inoculation of wounds with yeast results in ethylene production but ethylene is not produced by the yeast *in vitro* (Droby and Chalutz, 1994). Ethylene is also implicated in resistance mechanisms in citrus. Production of ethylene by fruit treated with antagonists indicates that induction of defense responses may be involved in disease suppression.

The best evidence to date of the involvement of induced resistance in postharvest biological control of disease on fruit, involves the induction of systemic acquired resistance in melons. The bacterial antagonist *Serratia marescens* was used to treat seeds and roots of muskmelon. After the fruit developed, they were placed on trays of inoculum to ensure that all fruit had similar inoculation pressure. Fruit harvested from plants with treated seeds and roots had significantly lower levels of Fusarium fruit rot than did fruit from non-treated plants (Eayre et al., 1997). A systemic response to the biological control agents was responsible for the reduced incidence of disease after harvest; the fruit never came in contact with the biological control agents. These data are the first indications that antagonists induce the plant's defense system to control postharvest diseases. Further work in this area will define and exploit induced resistance for biological control of postharvest diseases.

32.8   NUTRIENT COMPETITION

As with parasitism, nutrient competition is not intuitively a good example of allelopathic interactions among microorganisms. In many cases, no clear compound is being produced, which affects the growth and well-being of a second organism. There are some examples, however, of compounds produced by the antagonist that mediate nutrient competition and are involved in biological control. Siderophores are high affinity iron chelators that are produced by microbes under iron-limiting conditions (Neilands, 1981, 1982). When iron is limiting, siderophores excreted into the environment bind ferric ions and shuttle iron to the producing strain. Specific outer membrane proteins are required for the uptake of the ferric-siderophore complex (Neilands, 1982). Siderophores sequester available iron from the environment, thus making iron limiting to microbes that cannot utilize a prevalent ferric-siderophore (Neilands, 1982). Biological control of soilborne diseases is believed to involve microbial siderophores that mediate competition for iron (Loper and Buyer, 1991; Loper and Ishimaru, 1991). Siderophore production by antagonists can therefore be considered to play a role in an allelopathic relationship between microorganisms involved in biological control. Although nutrient competition has been implicated as a mechanism by which antagonists control postharvest diseases, competition for iron has not been studied in these systems. The commercially available antagonists, *Pseudomonas syringae* strains ESC-10 and ESC-11, produce siderophores, as do all fluorescent pseudomonads. In soil systems, biologically available iron is limiting and competition may therefore affect the ability of a pathogen to cause disease. The level of biologically available iron in harvested fruit is unknown and may not be limiting. No research on the role of siderophore production in the control of postharvest diseases of fruit has been reported.

Competition for nutrients, other than those for which specific shuttle vectors are produced by microbes, is very difficult to study. In many cases, it is the mechanism that is suggested by default due to the lack of evidence for other mechanisms (Handelsman and Parke, 1989; Paulitz, 1990).

Competition for nutrients has been suggested to function in the biological control of postharvest diseases (Filonow et al., 1996; Janisiewicz, 1996; Mercier and Wilson, 1994; Roberts, 1990; Droby et al., 1989).

32.9 CONCLUSION

As Rice (1995) suggested, many interactions between plant pathogens and biological control agents are allelopathic. In his treatment of the subject, he described all systems that had been studied at the time. He correctly assumed that many of these interactions were not allelopathic. While other authors may not agree that parasitism, induced resistance, and competition for nutrients meet the classical definition of allelopathy, I have chosen to include them here. The broad discussion that Rice (1995) made has set the stage for including many different chemical interactions between microorganisms within the topic of allelopathy. This may be justifiable in that the term allelopathy may have slightly different nuances in plant–plant verses microbe–microbe interactions. For example, competition between plants would not be considered allelopathy, but competition between microbes for iron through the production of siderophores might be.

Antibiosis is clearly one allelopathic phenomenon involved in biological control of plant pathogens. In most cases, there is only correlative evidence that production of secondary metabolites mediates biological control of postharvest diseases. Conclusive evidence for the involvement of antibiosis is lacking for the majority of systems. The mechanisms by which antagonists control soilborne and foliar diseases have been studied for much longer than mechanisms involved in biological control of postharvest diseases. Molecular genetic methods have been used to demonstrate the roles antibiosis, parasitism, induced resistance, and nutrient competition play in controlling soilborne and foliar diseases. These tools have helped to demonstrate that in general one mechanism does not operate at the exclusion of the others. Increased application of molecular tools to understand biological control of postharvest diseases is needed to elucidate the mechanisms involved in disease control. Understanding these mechanisms should help us to select and deploy biological antagonists for optimal control of postharvest diseases.

ACKNOWLEDGMENTS

Special thanks to Drs. L. Houck and W. J. Janisiewicz for critical review of this manuscript and to Stephanie Gularte for help in manuscript preparation.

REFERENCES

Arras, G., Mode of action of an isolate of *Candida famata* in biological control of *Penicillium digitatum* in orange fruits, *Postharvest Biol. Technol.*, 8, 191, 1996.

Bélanger, R. R., Dufour, N., Caron, J., and Benhamou, N., Chronological events associated with the antagonistic properties of *Trichoderma harzianum* against *Botrytis cinerea*: indirect evidence for sequential role of antibiosis and parasitism, *Biocontrol Sci. Technol.*, 5, 41, 1995.

Brown, G., Application of benzimidazole fungicides for citrus decay control, *Proc. Int. Soc. Citric,* 1, 273, 1977.

Bull, C. T., Wadsworth, M. L., Sorensen, K. N., Takemoto, J. Y., Austin, R. K., and Smilanick, J. L., Syringomycin E produced by biological control agents controls green mold on lemons, *Biol. Control,* 12, 89, 1998.

Bull, C. T., Stack, J. P., and Smilanick, J. L., *Pseudomonas syringae* strains ESC-10 and ESC-11 survive in wounds on citrus and control green and blue mold of citrus, *Biol. Control,* 8, 81, 1997.

Ceponis, M. J., Cappellini, R. A., and Lightner, G. W., Disorders in citrus shipments to the New York market, 1972–1984, *Plant Dis.,* 70, 1162, 1986.

Chalutz, E. and Wilson, C. L., Postharvest biocontrol of green and blue mold and sour rot of citrus fruit by *Debaryomyces hansenii, Plant Dis.,* 74, 134, 1990.

Chand-Goyal, T. and Spotts, R. A., Control of postharvest pear diseases using natural saprophytic yeast colonists and their combination with a low dosage of thiabendazole, *Postharvest Biol. Technol.,* 7, 51, 1996.

Cook, R. J. and Baker, K. F., *The Nature and Practice of Biological Control of Plant Pathogens,* American Phytopathological Society, St. Paul, MN, 1983.

Cook, R. J., Bruckart, W. L., Coulson, J. R., Goettel, M. S., Humber, R. A., Lumsden, R. D., Maddox, J. V., McManus, M. L., Moore, L., Meyer, S. F., Quimby, P. C., Jr., Stack, J. P., and Vaughn, J. L., Safety of microorganisms intended for pest and plant disease control: a framework for scientific evaluation, *Biol. Control,* 7, 333, 1996.

Cooksey, D. A. and Moore, L. W., Biological control of crown gall with an agrocin mutant of *Agrobacterium radiobacter, Phytopathology,* 72, 919, 1982a.

Cooksey, D. A. and Moore, L. W., High frequency spontaneous mutation to agrocin 84 resistance in *Agrobacterium tumefaciens* and *A. rhizogenes, Physiol. Plant Pathol.,* 20, 129, 1982b.

Corbell, N. and Loper, J. E., A global regulator of secondary metabolite production in *Pseudomonas fluorescens* Pf-5, *J. Bacteriol.,* 177, 6230, 1995.

Da Silva, S. R. and Pascholati, S. F., *Saccharomyces cerevisiae* protects maize plants, under greenhouse conditions, against *Colletotrichum graminicola, Zeitschrift Pflanzenkrankheiten Pflanzenschutz,* 99, 159, 1992.

Delp, C. J., Coping with resistance to plant disease, *Plant Dis.,* 64, 652, 1980.

Droby, S. and Chalutz, E., Mode of action of biocontrol agents of postharvest diseases, in *Biological Control of Postharvest Diseases,* Wilson, C. L., and Wisniewski, M. E., Eds., CRC Press, Boca Raton, FL, 1994, 63.

Droby, S., Hofstein, R., Wilson, C. L., Wisniewski, M., Fridlender, B., Cohen, L., Weiss, B., Daus, A., Timar, D., and Chalutz, E., Pilot testing of *Pichia guilliermondii:* a biocontrol agent of postharvest diseases of citrus fruit, *Biol. Control,* 3, 47, 1993.

Droby, S., Chalutz, E., Wilson, C. L., and Wisniewski, M., Characterization of the biocontrol activity of *Debaryomyces hansenii* in the control of *Penicillium digitatum* on grapefruit, *Can. J. Microbiol.,* 35, 794, 1989.

Eayre, C. G., Kloepper, J. W., and Tuzun, S., Plant growth promoting rhizobacteria induce systemic resistance to Fusarium fruit rot in muskmelons, *Phytopathology,* 87, S27, 1997.

Eckert, J. W. and Ogawa, J. M., The chemical control of postharvest diseases: subtropical and tropical fruits, *Ann. Rev. Phytopathol.,* 23, 421, 1985.

Eckert, J. W. and Eaks, I. L., Postharvest disorders and diseases of citrus fruits, in *The Citrus Industry,* Vol. 4, Reuther, W., Calavan, E. C., and Carman, G. E., Eds., University of California Press, Berkeley, CA, 1989, 179.

Elad, Y., Mechanisms involved in the biological control of *Botrytis cinerea* incited diseases, *Eur. J. Plant Pathol.,* 102, 719, 1996.

Filonow, A. B., Vishniac, H. S., Anderson, J. A., and Janisiewicz, W. J., Biological control of *Botrytis cinerea* in apple by yeasts from various habitats and their putative mechanisms of antagonism, *Biol. Control,* 7, 212, 1996.

Fravel, D. R., Role of antibiosis in the biocontrol of plant diseases, *Ann. Rev. Phytopathol.,* 26, 75, 1988.

Grevesse, C., Jijakli, M. H., Duterme, O., Colinet, D., and Lepoivre, P., Preliminary study of exo-β-1,3-glucanase encoding genes in relation to the protective activity of *Pichia anomala* strain K against *Botrytis cinerea* on postharvest apples, Molecular Approaches in Biological Control, September 15–18, Delémont, Switzerland, 1998.

Gueldner, R. C., Reilly, C. C., Pusey, P. L., Costello, C. E., Arrendale, R. F., Cox, R. H., Himmelsbach, D. S., Crumley, F. G., and Cutler, H. G., Isolation and identification of iturins as antifungal peptides in biological control of peach brown rot with *Bacillus subtilis, J. Agric. Food Chem.,* 36, 366, 1988.

Gutter, Y. and Littauer, F., Antagonistic action of *Bacillus subtilis* against citrus fruit pathogens, *Bull. Res. Council Israel,* 3, 192, 1953.

Gutterson, N., Microbial fungicides: recent approaches to elucidating mechanisms, *Crit. Rev. Biotechnol.,* 10, 69, 1990.

Handelsman, J. and Parke, J. L., Mechanisms in biocontrol of soilborne plant pathogens, in *Plant-Microbe Interactions, Vol. 3, Molecular Genetic Perspectives,* Kosuge, T. and Nester, E. W., Eds., McGraw-Hill, New York, 1989, 27.

Handelsman, J. and Stabb, E. V., Biocontrol of soilborne plant pathogens, *Plant Cell,* 8, 1855, 1996.

Howell, C. R. and Stipanovic, R. D., Control of *Rhizoctonia solani* on cotton seedlings with *Pseudomonas fluorescens* and with an antibiotic produced by the bacterium, *Phytopathology,* 69, 480, 1979.

Howell, C. R. and Stipanovic, R. D., Suppression of *Pythium ultimum*-induced damping-off of cotton seedlings by *Pseudomonas fluorescens* and its antibiotic, pyoluteorin, *Phytopathology,* 70, 712, 1980.

Huang, Y. B., Deverall, J., and Morris, S. C., Effect of *Pseudomonas cepacia* on postharvest biocontrol of infection by *Penicillium digitatum* and on wound responses of citrus fruit, *Australasian Plant Pathol.,* 22, 84, 1993.

James, D. W., Jr. and Gutterson, N. I., Multiple antibiotics produced by *Pseudomonas fluorescens* HV37a and their differential regulation by glucose, *Appl. Environ. Microbiol.,* 52, 1183, 1986.

Janisiewicz, W., Biological control of diseases of fruits, in *Biocontrol of Plant Diseases,* Mukerji, K. G. and Garg, K. L., Eds., CRC Press, Boca Raton, FL, 1988, 153.

Janisiewicz, W., Biocontrol of postharvest diseases of temperate fruits, in *Plant-Microbe Interactions and Biological Control,* Boland, G. J., and Kuykendall, L. D., Eds., Marcel Dekker, New York, 1998, 171.

Janisiewicz, W., Ecological diversity, niche overlap, and coexistence of antagonists used in developing mixtures for biocontrol of postharvest diseases of apples, *Phytopathology,* 86, 437, 1996.

Janisiewicz, W. J. and Marchi, A., Control of storage rots on various pear cultivars with a saprophytic strain of *Pseudomonas syringae, Plant Dis.,* 76, 555, 1992.

Janisiewicz, W. J. and Bors, B., Development of a microbial community of bacterial and yeast antagonists to control wound-invading postharvest pathogens of fruits, *Appl. Environ. Microbiol.,* 61, 3261, 1995.

Janisiewicz, W. J. and Roitman, J., Biological control of blue mold and gray mold on apple and pear with *Pseudomonas cepacia, Phytopathology,* 78, 1697, 1988.

Jeffries, P. and Jeger, M. J., The biological control of postharvest diseases of fruit, *Postharvest News Info.,* 1, 365, 1990.

Jijakli, M. H. and Lepoivre, P., Characterization of an Exo-β-1,3-glucanase produced by *Pichia anomala* strain K, antagonist of *Botrytis cinerea* on apples, *Phytopathology,* 88, 335, 1998.

Keel, C., Schnider, U., Maurhofer, M., Voisard, C., Laville, J., Burger, P., Wirthner, P., Haas, D., and Défago, G., Suppression of root diseases by *Pseudomonas fluorescens* CHA0: importance of the bacterial secondary metabolite 2,4-diacetylphloroglucinol, *Mol. Plant-Microbe Interact.,* 5, 4, 1992.

Kim, J. J., Ben-Yehoshua, S., Shapiro, B., Henis, Y., and Carmeli, S., Accumulation of scoparone in heat-treated lemon fruit inoculated with *Penicillium digitatum* Sacc, *Plant Physiol.,* 97, 880, 1991.

Kraus, J. and Loper, J. E., Characterization of a genomic region required for production of the antibiotic pyoluteorin by the biological control agent *Pseudomonas fluorescens* Pf-5, *Appl. Environ. Microbiol.,* 61, 849, 1995.

Kuc, J. and Strobel, N. E., Induced resistance using pathogens and nonpathogens, in *Biological Control of Plant Diseases—Progress and Challenges for the Future,* Tjamos, E. C., Papavizas, G. C., and Cook, R. J., Eds., Plenum Press, New York, 1992, 295.

Laville, J., Voisard, C., Keel, C., Maurhofer, M., Défago, G., and Haas, D., Global control in *Pseudomonas fluorescens* mediating antibiotic synthesis and suppression of black root rot of tobacco, *Proc. Natl. Acad. Sci. USA,* 89, 1562, 1992.

Loper, J. E. and Buyer, J. S., Siderophores in microbial interactions on plant surfaces, *Mol. Plant-Microbe Interact.,* 4, 5, 1991.

Loper, J. E., and Ishimaru, C. A., Factors influencing siderophore-mediated biocontrol activity of rhizosphere *Pseudomonas* spp, in *The Rhizosphere and Plant Growth,* Keister, D. L., and Cregan, P. B., Eds., Kluwer Academic, Netherlands, 1991, 253.

Lurie, S., Droby, S., Chalupowicz, L., and Chalutz, E., Efficacy of *Candida oleophila* strain 182 in preventing *Penicillium expansum* infection of nectarine fruits, *Phytoparasitica,* 23, 231, 1995.

MacWilliam, I. C., A survey of the antibiotic powers of yeasts, *J. Gen. Microbiol.,* 21, 410, 1959.

McCormack, P. J., Wildman, H. G., and Jeffries, P., Production of antibacterial compounds by phylloplane-inhabiting yeasts and yeastlike fungi, *Appl. Environ. Microbiol.,* 60, 927, 1994.

McKeen, C. D., Reilly, C. C., and Pusey, P. L., Production and partial characterization of antifungal substances antagonistic to *Monilinia fructicola* from *Bacillus subtilis, Phytopathology,* 76, 136, 1986.

Mercier, J. and Wilson, C. L., Colonization of apple wounds by naturally occurring microflora and introduced *Candida oleophila* and their effect on infection by *Botrytis cinerea* during storage, *Biol. Control,* 4, 138, 1994.

Nielands, J. B., Microbial iron compounds, *Annu. Rev. Biochem.,* 50, 715, 1981.

Nielands, J. B., Microbial envelope proteins related to iron, *Annu. Rev. Microbiol.,* 36, 285, 1982.

Nowak-Thompson, B., Gould, S. J., Kraus, J., and Loper, J. E., Production of 2,4-diacetylphloroglucinol by the biocontrol agent *Pseudomonas fluorescens* Pf-5, *Can. J. Microbiol.*, 40, 1064, 1994.

Paulitz, T. C., Biochemical and ecological aspects of competition in biological control, in *New Directions in Biological Control: Alternatives for Suppressing Agricultural Pests and Diseases,* Baker, R. R. and Dunn, P. E., Eds., Alan R. Liss, New York, 1990, 713.

Pusey, P. L. and Wilson, C. L., Postharvest biological control of stone fruit brown rot by *Bacillus subtilis, Plant Dis.*, 68, 753, 1984.

Pusey, P. L., Use of *Bacillus subtilis* and related organisms as biofungicides, *Pestic. Sci.*, 27, 133, 1989.

Rice, E. L., *Biological Control of Weeds and Plant Diseases: Advances in Applied Allelopathy,* University of Oklahoma Press, Norman, OK, 1995.

Roberts, R. G., Postharvest biological control of gray mold of apple by *Cryptococcus laurentii, Phytopathology,* 80, 526, 1990.

Rodov, V., Ben-Yehoshua, S., Fang, D., D'hallewin, G., and Castia, T., Accumulation of phytoalexins scoparone and scopoletin in citrus fruits subjected to various postharvest treatments, *ACTA Hortic.*, 381, 517, 1994.

Roitman, J. N., Mahoney, N. E., and Janisiewicz, W. J., Production and composition of phenylpyrrole metabolites produced by *Pseudomonas cepacia, Appl. Microbial. Biotechnol.*, 34, 381, 1990.

Sarniguet, A., Kraus, J., Henkels, M. D., Muehlchen, A. M., and Loper, J. E., The sigma factor σ^s affects antibiotic production and biological control activity of *Pseudomonas fluorescens* Pf-5, *Proc. Natl. Acad. Sci. USA*, 92, 12255, 1995.

Sequeira, L., Mechanisms of induced resistance in plants, *Annu. Rev. Microbiol.*, 37, 51, 1983.

Shanahan, P., O'Sullivan, D. J., Simpson, P., Glennon, J. D., and O'Gara, F., Isolation of 2, 4-diacetylphloroglucinol from a fluorescent pseudomonad and investigation of physiological parameters influencing its production, *Appl. Environ. Microbiol.*, 58, 353, 1992.

Singh, V. and Deverall, B. J., *Bacillus subtilis* as a control agent against fungal pathogens of citrus fruit, *Trans. Br. Mycol. Soc.*, 83, 487, 1984.

Singleton, P. and Sainsbury, D., *Dictionary of Microbiology and Molecular Biology,* 2nd ed., John Wiley & Sons, New York, 1987, 797.

Smilanick, J. L. and Denis-Arrue, R., Control of green mold of lemons with *Pseudomonas* species, *Plant Dis.*, 76, 481, 1992.

Smilanick, J. L., Denis-Arrue, R., Bosch, J. R., Gonzalez, A. R., Henson, D., and Janisiewicz, W. J., Control of postharvest brown rot of nectarines and peaches by *Pseudomonas* species, *Crop Protection,* 12, 513, 1993.

Smilanick, J. L., Gouin-Behe, C. C., Margosan, D. A., Bull, C. T., and Mackey, B. E., Virulence on citrus of *Pseudomonas syringae* strains that control postharvest green mold of citrus fruit, *Plant Dis.*, 80, 1123, 1996.

Suzzi, G., Romano, P., Ponti, I., and Montuschi, C., Natural wine yeasts as biocontrol agents, *J. Appl. Bacteriol.*, 78, 304, 1995.

Takesako, K., Ikai, K., Haruna, F., Endo, M., Shimanaka, K., Sono, E., Nakamura, T., and Kato, I., Aureobasidins, new antifunal antibiotics taxonomy, fermentation, isolation, and properties, *J. Antibiot.*, 44, 919, 1991.

Thomashow, L. S. and Weller, D. M., Role of a phenazine antibiotic from *Pseudomonas fluorescens* in biological control of *Gaeumannomyces graminis* var. *tritici, J. Bacteriol.*, 170, 3499, 1988.

Utkhede, R. S. and Sholberg, P. L., *In vitro* inhibition of plant pathogens by *Bacillus subtilis* and *Enterobacter aerogenes* and *in vivo* control of two postharvest cherry diseases, *Can. J. Microbiol.*, 325, 963, 1986.

Vidaver, A. K., Prospects for control of phytopathogenic bacteria by bacteriophages and bacteriocins, *Annu. Rev. Phytopathol.*, 14, 451, 1976.

Vidaver, A. K., Bacteriocins: the lure and the reality, *Plant Dis.*, 67, 471, 1983.

Viñas, I., Contol biológico en la conservación de frutas de pepita, *Microbiol. Sem.*, 11, 115, 1995.

Vining, L. C., Functions of secondary metabolites, *Annu. Rev. Microbiol.*, 44, 395, 1990.

Wadsworth, M. L., Bull, C. T., and Smilanick, J. L., Preliminary characterization of antifungal compounds produced by biological control agents *Pseudomonas syringae* strains ESC-10 and ESC-11, *Phytopathology,* 86, S112, 1996.

Walker, R. K., Emslie, K. A., and Allan, E. J., Bioassay methods for the detection of antifungal activity by *Pseudomonas antimicrobica* against the grey mould pathogen *Botrytis cinerea, J. Appl. Bacteriol.*, 81, 531, 1996.

Wilson, C. L. and Chalutz, E., Postharvest biological control of *Penicillium* rots of citrus with antagonistic yeasts and bacteria, *Scientia Hort.,* 40, 105, 1989.

Wilson, C. L., Wisniewski, M. E., Biles, C. L., McLaughlin, R., Chalutz, E., and Droby, S., Biological control of post-harvest diseases of fruits and vegetables: alternatives to synthetic fungicides, *Crop Protect.,* 10, 172, 1991.

Wilson, C. L., El Ghaouth, A., Chalutz, E., Droby, S., Stevens, C., Lu, J. Y., Khan, V., and Arul, J., Potential of induced resistance to control postharvest diseases of fruits and vegetables, *Plant Dis.,* 78, 837, 1994.

Wisniewski, M. E. and Wilson, C. L., Biological control of postharvest diseases of fruits and vegetables: recent advances, *Hort. Science,* 27, 94, 1992.

Wisniewski, M., Biles, C., Droby, S., McLaughlin, R., Wilson, C., and Chalutz, E., Mode of action of postharvest biocontrol yeast, *Pichia guilliermondii.* I. Characterization of attachment to *Botrytis cinerea, Physiol. Mol. Plant Pathol.,* 39, 245, 1991.

33 Potential of Cultivar Sunflowers (*Helianthus annuus* L.) as a Source of Natural Herbicide Templates

Francisco A. Macías, Rosa M. Varela, Ascensión Torres, and José M. G. Molinillo

CONTENTS

33.1 ABSTRACT

A study was conducted to exploit the potential of cultivar sunflowers (*Helianthus annuus* L.) as sources of natural herbicide templates. Germination and growth bioassay of 26 cultivar sunflower varieties using two dicotyledonous species (*Lactuca sativa* L. and *Lepidium sativum* L.) and two monocotyledonous species (*Hordeum vulgare* L. and *Triticum aestivum* L.) were carried out. To select appropriate plant developmental stages with significant profiles of biological activity, four different stages were selected. In addition to triterpenes, steroids, diterpenes, chalcones, lignans, and sesquiterpene lactones with germacranolide and guaianolide skeletons, four new families of sesquiterpene (heliannuol), were isolated, identified, and demonstrated to have potential use as natural herbicide templates.

33.2   INTRODUCTION

The concept that some crop plants may be allelopathic to common crop land weeds is receiving greater attention as an alternative weed control strategy (Worsham, 1989). Rice (1984) defined allelopathy as "the direct or indirect harmful or beneficial effect by one plant on another through chemical compounds that escape into the environment." A large number of studies have considered the role of allelopathy in agriculture (Worsham, 1989; Altieri, 1988; Einhellig and Leather, 1988; Putnam, 1988a; Tauscher, 1988). Most of this research has concentrated on determining the effect of decomposing crop residues on succeeding crops (Cochran et al., 1977), on inhibition of crop production by weeds (Einhellig and Rasmussen, 1973; Rasmussen and Einhellig, 1975), and on crop to crop interactions (Menges and Tamez, 1981; Schon and Einhellig, 1982). Research on the possible allelopathic effects of crops on weeds is relatively recent (Putnam et al., 1983; Einhellig and Rasmussen, 1989; González et al., 1992) and sparse.

Cultivation of sunflower is predominantly to produce oil and plays an important role in southern parts of Europe. Biochemical investigations on sunflower (*Helianthus annuus* L.) have revealed that this species is a rich source of sesquiterpenoids (Gershenzon et al., 1982; Spring et al., 1981; Macías et al., 1993b, 1996a) and other plant secondary metabolites with a wide spectrum of biological activities (Wilson and Rice, 1968); nevertheless, little is known about the function of these compounds. Sunflowers are known to inhibit weeds (Irons and Burnside, 1982; Leather, 1982, 1983, 1987; Varela, 1992; Macías et al., 1993a, 1997), but the chemistry involved in this process is still unresolved.

The objectives of the present study were to study the potential that sunflower cultivars have as sources of allelopathic compounds and to establish a correlation between structure of allelopathic compounds and their potential as natural herbicides and herbicide templates.

The bioassay results of 26 sunflower cultivars used for crop production in the Andalusian region (Spain) against two dicotyledonous species (lettuce and cress) and two monocotyledonous species (barley and wheat) are presented in Figures 33.1–33.8 and discussed. These bioassays were performed at four different developmental stages in order to establish which variety showed a more significant profile of activity and to determine the best stage for using the plant material (fresh leaves) without injuring the plant used for crop production. As a result of these bioassays, *Helianthus annuus* L. var SH-222 and VYP were the first two selected, and they were selected during the third plant development stage (plant 1.2 m tall with flowers, one month before harvest).

The subsequent bioassays, with fractions obtained from the first chromatographic separation, were conducted following the isolation of the active principles (Figures 33.9–33.11). From the moderately polar bioactive fractions, in addition to triterpenes, steroids, diterpenes, chalcones, and lignans, 12 sesquiterpene lactones with germacranolide and guaianolide skeletons, four members of a new family of sesquiterpenes (heliannuols), were isolated, identified, and biologically tested as allelopathic agents with potential use as natural herbicide templates (Figures 33.12–33.15).

33.3   MATERIAL AND METHODS

33.3.1   PLANT MATERIAL

Fresh leaves of 26 cultivars *H. annuus* were collected in August 1991 from four different developmental stages: 1) first stage, 15-20 cm tall plants; 2) second stage, 50 cm tall plants; 3) third stage, 1.0 m tall plants, with flowers (1 month before harvest); and 4) fourth stage, 1.0-1.5 m tall plants, with flowers, close to the harvest. The cultivars were numbered as follows: **1** (SH-25), **2** (Heliandalus), **3** (SH-26), **4** (Sirio-G-100), **5** (AB-E-353), 6 (Tornasol), **7** (Alhama), **8** (Hysun-33), **9** (Florida-2000), **10** (Rustiflor), **11** (SH-222), **12** (Tesoro-92), **13** (Viki), **14** (Maribel), **15** (Feria), **16** (AS-545), **17** (Ariflor), **18** (AB-P-113), **19** (VYP), **20** (Sungro-380), **21** (Lotus-985),

22 (Peredovick), **23** (Florasol), **24** (Solre-2), **25** (Toledo), and **26** (Monro-45). They were provided by Rancho de la Merced, Agricultural Research Station, Junta de Andalucía, Jerez, Spain, and were used in bioassays to test for effects on the germination and growth response of two dicotyledonous (lettuce and cress) and two monocotyledonous species (barley and wheat).

33.33.2 SEED GERMINATION BIOASSAYS

Seeds of lettuce, barley, and wheat were obtained from Rancho La Merced, Junta de Andalucía, Jerez, Cádiz, Spain. All undersized and damaged seeds were discarded and the assay seeds were selected for uniformity.

Bioassays have been optimized as follows: bioassays consisted of germinating 25 lettuce seeds and cress for 5 days (three for germination and two for root and shoot growth) and 2.5 days, respectively, and 5 wheat and barley seeds for 3 days, in the dark, at 25°C, in 9-cm plastic Petri dishes containing a 9-cm sheet of Whatman no. 1 filter paper, with 10 ml of a test or control solution for lettuce and cress and 5 ml for barley and wheat. There were 3 replicates for lettuce and cress and 19 for barley and wheat of each treatment and of concomitant controls. The number of seeds per replicate, time, pH, and temperature of germination were chosen in agreement with a number of preliminary experiments, varying the number of seeds, volume of test solution per dish, and the incubation period. To evaluate stimulatory effects of lettuce germination, we optimized the bioassay to obtain an average of 55 to 60 percent of germination of the control, using variety *nigra*. We use variety *romana* for the rest of the evaluated parameters.

Test solutions of water extracts were prepared by diluting the aqueous leaf extract (3 L H_2O:1 kg fresh plant) to 1:10, 1:20, and 1:40 (vol/vol; extract/H_2O) using deionized H_2O and for the fractions by diluting the appropriate amount of each to obtain a similar concentration of 1:10, 1:20, and 1:40 relative to the original aqueous extract. Test solutions of pure products (10^{-4} M) were prepared using deionized water and test solutions 10^{-5} to 10^{-9} M were obtained making a dilution series. Solutions of water insoluble products (10^{-4} M) were prepared using DMSO (0.1 percent vol/vol) as a carrier. Parallel controls consisted of deionized H_2O with DMSO.

The pH was adjusted to 6.0 before the bioassay using MES (2-[N-morpholino]ethanesulfonic acid, 10 mM). The osmotic pressure values were measured on a Vapor Pressure Osmometer WESCOR 5500 and ranged between 30 and 38 m osmolar. These values are low enough to consider their influence as negligible.

The germination, root, and shoot length values were tested by the Mann-Whitney's test, and the 0.01 level of confidence was accepted.

33.3.3 EXTRACTION AND ISOLATION OF ACTIVE PRINCIPLES FROM *HELIANTHUUS ANNUUS* VAR. SH-222 AND VYP

Fresh leaves of *H. annuus* var SH-222 (Semillas Pacífico, Spain) and VYP (KOIPESOL, Spain) were collected during the third plant development stage. This collection period was selected on the basis of bioactivity exhibited by the aqueous leaf extracts of four plant development stages (Macías et al., 1993). Six kilograms of fresh leaves was soaked in H_2O (wt/vol, 1:3) for 24 h at 25°C, in the dark. The H_2O extracts were re-extracted ($\times 8$) with 0.5 L of CH_2Cl_2 for each 1.0 L of H_2O, and the combined extracts were dried over Na_2SO_4 and evaporated *in vacuo* to yield 16 g of crude extract, termed H_2O-CH_2Cl_2 extract. The H_2O-CH_2Cl_2 extract was separated by column chromatography on silica gel using hexane-EtOAc mixtures of increasing polarity yielding 170 fractions of 50 mL each, which were combined to give 28 (SH-222) and 30 (VYP) fractions after comparison by TLC.

Following bioactive levels, fractions **A, C, D, E, F, I, P,** and **R** from SH-222 (Macías et al., 1993a, b, 1994) were selected and chromatographed using MERCK-HITACHI L-6200A HPLC with Hibar Si60 (Merck) column, RI and/or UV detector at 254 nm, variable hexane:EtOAc mixtures as eluent, and 4 mL min^{-1} flow rate yielding: heliannuol B (**2**), kaur-16-en-19-oic ac. (**21**), β-sitosterol (**19**), and stigmaste-7,24(28)-dien-3-ol (**20**) from fraction **A**; heliannuol B (**2**) from fraction **C**, heliannuols B (**2**) and D (**4**) from fraction **D**, **E**, and **F**; heliannuol A (**1**) and 1β,6α-dihydroxy-eudesm-4(15)-ene (**5**) from fraction **I** and **J**; heliannuols A (**1**) and C (**3**), *E*-isoferulaldehyde (**6**), and 1β,6α-dihydroxy-eudesm-4(15)-ene (**5**) from fraction **P**; and annuolides A-E (**7-11**), heliannoic ac. (**24**), *E*-isoferulaldehyde (**6**) and five bisnorsesquiterpenes from fraction **R**.

Using the same criteria as above, fractions **B, D, F, K, L, N, O, Q**, and **S** from VYP (Macías et al., 1996, 1997, 1994) were selected and chromatographed using the indicated system, yielding: kaur-16-en-19-oic ac. (**21**) from fraction **B**; compound 21 and α-amyrin (**26**) from fraction **D**; β-sitosterol (**19**), stigmasta-7,24(28)-dien-3-ol (**20**), 21, angeloylgrandifloric ac. (**22**), and grandifloric ac. (**23**) from fraction **F**; heliannuols B (**2**) and D (**4**) from fraction **K**; heliannuols A (**1**) and C (**3**) from fraction **L**; 1β,6α-dihydroxy-eudesm-4(15)-ene (**5**) and diterpene 24 from fraction **N**; annuolides A (**7**) and C (**9**), pinoresinol (**27**), kukulkanin B (**28**), and tambulin (**29**) from fraction **O**; sesquiterpene lactone **12**, annuolide F (**13**), helivypolides A-B (**14-15**), leptocarpine (**16**), heliangine (**17**), and 2α-hydroxy-8β-angeloyloxy-costunolide (**18**) from fraction **Q**; and a bisnorsesquiterpene with skeleton **25** from fraction **S**.

33.4   RESULTS AND DISCUSSION

The experimental results of 26 cultivar sunflower varieties used for crop production in the Andalusian region (Spain) on the germination, radical length, and shoot length of lettuce during four different plant development stages are summarized in Figures 33.1–33.4. The numbers are expressed as percents of the control, hence 0 is identical to the control, a positive value represents stimulation, and a negative value represents inhibition.

In general, a sudden change in the profile of activity shown by aqueous extracts of fresh leaves on germination of lettuce is observed between the first plant developmental stage and subsequent stages. Whereas, for this first stage stimulation of germination (average of 50 percent) is typical of the profile of activity, inhibition is typical for the other three. There is no clear correlation between concentration and the level of activity observed, whereas the maximum activity is a 1:10 dilution for varieties **1, 8, 12, 18, 19**, and **26**, and dilution 1:20 for varieties **7, 11, 13, 15, 16**, and **21**. From this first stage, varieties **7** (1:20), **8** (1:10), **12** (1:10), **16** (1:20), **18** (1:10, 1:40), **21** (1:20), and **22** (1:40) exhibit a stimulatory effect on germination of 70 percent as well as **13** (1:20) with an 80 percent stimulation (Figure 33.1).

This observed stimulatory effect on germination is exploited by farmers who grow sunflower crops. At the initiation of cultivation, a few sunflower seeds are grown together until they are 20 to 30 cm tall plants then, in a process called "sunflower castration," only one plant is left to develop. This process agrees with our bioassay results, because from the second plant development stage until harvest, there is a strong inhibitory effect on germination and root length. Significant root and shoot inhibition was observed in varieties **10** (1:20, 1:40), **23** (1:40), **25** (1:40), and **26** (1:10, 1:40).

As mentioned above, during the second plant development stage, signs of activity on germination became negative. In general, a strong inhibitory effect on germination and root length, and a stimulatory effect on shoot length was found.

In the second stage (Figure 33.2) the inhibitory effects on seed germination were observed for varieties **1, 6, 10, 11, 15, 19, 21, 25**, and **26**, while a significant reduction in root length was recorded for varieties **4, 5, 8, 15, 19, 20**, and **22**. In general, stimulatory effects on shoot length were observed, except for varieties **4** (1:40) and **15** (1:10, 1:20).

The third stage was characterized by the strongest inhibitory effect on germination (Figure 33.3), particularly for 1:10 dilutions, where an average of −60 percent (1:10) for germination and root length was found. Varieties **6, 9, 11, 12, 16, 17, 19, 23**, and **25** should be emphasized. The effect on the shoot length was, however, stimulatory.

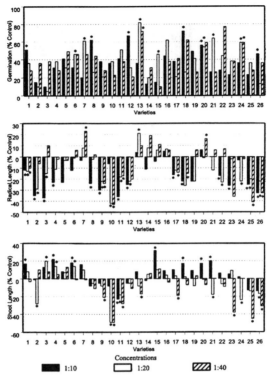

FIGURE 33.1   Effect on the germination and growth of *Lactuca sativa* L. of 26 cultivar sunflower varieties in the first plant development stage.

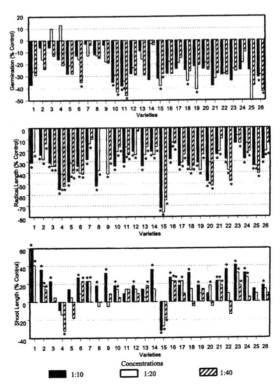

FIGURE 33.2   Effect on the germination and growth of *Lactuca sativa* L. of 26 cultivar sunflower varieties in the second plant development stage.

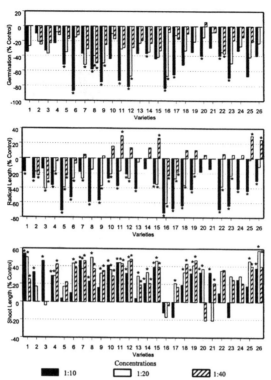

FIGURE 33.3   Effect on the germination and growth of *Lactuca sativa* L. of 26 cultivar sunflower varieties in the third plant development stage.

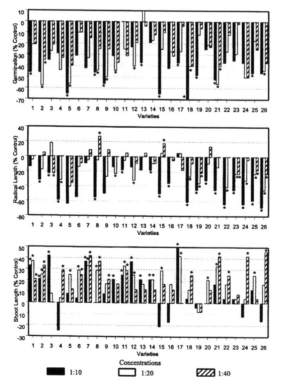

FIGURE 33.4   Effect on the germination and growth of *Lactuca sativa* L. of 26 cultivar sunflower varieties in the fourth plant development stage.

The fourth stage exhibited a profile similar in activity as that observed in the third stage, but of a lesser degree (Figure 33.4). There was a concentration-dependent inhibition in seed germination and root length. However, stimulatory effects on the shoot length were enhanced with dilution.

As result of this previous bioassay conducted during four different plant development stages, we can conclude that the most interesting stage, from the allelopathic activity point of view, is the third stage. To better evaluate this particular allelopathic activity, we carried out another bioassay in which we evaluated the bioactivity against a dicotyledonous species cress, (very sensitive for root length evaluation) and two monocotyledonous species (barley and wheat) using 13 varieties selected from the third plant development stage on the bases of their activity profile shown in the previous bioassay (Figures 33.5–33.7).

No significant allelopathic effects were observed on seed germination of cress (Figure 33.5), except for varieties **5** (1:40) and **6** (1:10), where inhibitory effects of −31 and −39 percent, respectively, were found. Nevertheless, strong inhibitory effects on the growth is observed with an average of −60 percent on root length, and is particularly high for varieties **6, 11, 25,** and **26** at 1:10 dilution, with an average of −80 percent on shoot length, which was especially intense for varieties **6, 9, 11, 19, 22,** and **25** at 1:10 dilution. The observed effects were concentration dependent but were still high at a 1:40 dilution, with average values of −30 percent on root length and −50 percent on shoot length.

In general, no significant inhibition or promotion in seed germination, root length, and shoot length was recorded in either of the monocotiledonous species (Figures 33.6, 33.7). Only germination of barley was affected by varieties **5, 6, 11, 22,** and **26** with an average of −40 percent at higher concentrations (1:10). The effect on growth was, in general, of little stimulation and significance except for varieties **10, 11, 12,** and **26** in which inhibitions of an average of −30 percent on shoot length was observed for barley.

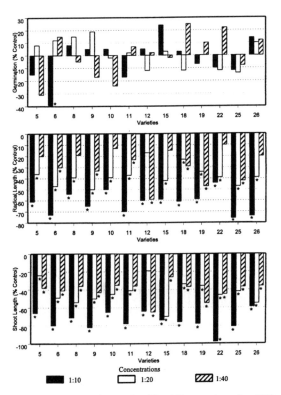

FIGURE 33.5 Effect on the germination and growth of *Lepidium sativum* L. of 13 selected cultivar sunflower varieties in the third plant development stage.

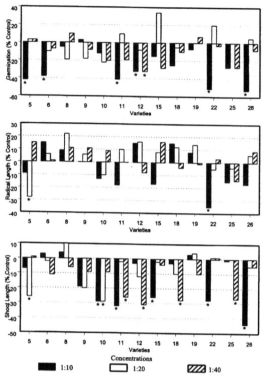

FIGURE 33.6 Effect on the germination and growth of *Hordeum vulgare* L. of 13 selected cultivar sunflower varieties in the third plant development stage.

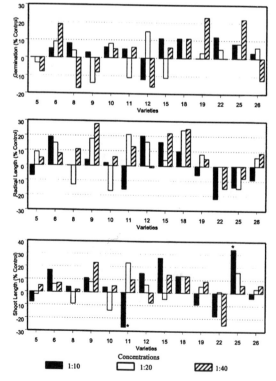

FIGURE 33.7 Effect on the germination and growth of *Triticum aestivum* L. of 13 selected cultivar sunflower varieties in the third plant development stage.

Based on the above bioassays, two sunflower varieties, SH-222 (**11**) and VYP (**19**), were the two varieties first selected; the third plant development stage was chosen as the time to carry out the chemical analysis to evaluate their allelopathic potential.

The subsequent bioassays against lettuce, with fractions obtained from the first chromatographic separation of the H_2O-CH_2Cl_2 crude extract from the fresh leaves of varieties SH-222 and VYP (Figure 33.8) during the third stage, monitored the isolation of active principles.

Following bioassay directed fractionation, fractions **A, C, D, E, F, I, P,** and **R** from SH-222 (**11**) were the first to be chemically analyzed. These yielded the four first members of the novel family of sesquiterpenes, heliannuols, and heliannuol A (**1**) (Macías et al., 1996) from fractions **I, J,** and **O**; heliannuol B (**2**) from fractions **A, C, D, E** and **F**; heliannuol C (**3**) from fraction **P**; and heliannuol D (**4**) (Macías et al., 1994) from fractions **D, E,** and **F**; as well as the diterpene **21** and the sterols **19** and **20** from fraction **A**; 1β,6α-dihydroxy-eudesm-4(15)-ene (**5**) from fraction **I, J,** and **P**; *E*-isoferulaldehyde (**6**) from fraction **P** and **R**; five new guaianolides named annuolides A-E (**7-11**) (Macías et al., 1993a), the diterpene 17-hydroxy-16-*ent*-kauran-19-oic acid (**24**), and five compounds with the bisnorsesquiterpene skeleton **25** from fraction **R** (Figures 33.9, 33.10). Similar bisnorsesquiterpenes have been found from *Helianthus heterophyllus*, and their allelopathic activities have been demostrated (Herz and Bruno, 1986; Kato et al., 1977).

The second selected variety was VYP (**19**) from which, following bioassay directed fractionation, fractions **B, D, F, K, L, N, O, Q,** and **S** were selected and chemically analyzed to yield: heliannuols A-D (**1–4**) (Macías et al., 1994) from fractions **K** and **L**; 1β,6α-dihydroxy-eudesm-4(15)-ene (**5**) from fraction **N**; α-amyrin (**26**) from fraction **D**; β-sitosterol (**19**) and stigmasta-7,24(28)-dien-3-ol (**20**) from fraction **F**; four kaurenic diterpenes, kaur-16-en-19-oic ac. (**21**) from fractions **B** and **D**, angeloylgrandifloric ac. (**22**) and grandifloric ac. (**23**) from fraction **F**, and diterpene **24** from fraction **N** (Macías et al., 1997); nine sesquiterpene lactones, annuolides A (**7**) and C (**9**), guaianolide **12**, annuolide F (**13**), helivypolides A and B (**14–15**), leptocarpine (**16**), heliangine (**17**), and

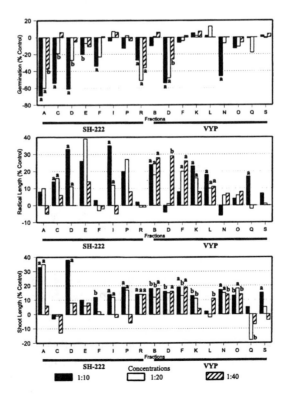

FIGURE 33.8 Effect of fractions **A-N** from SH-222 leaves aqueous extract on the germination, root and shoot length of *Lactuca sativa* L.

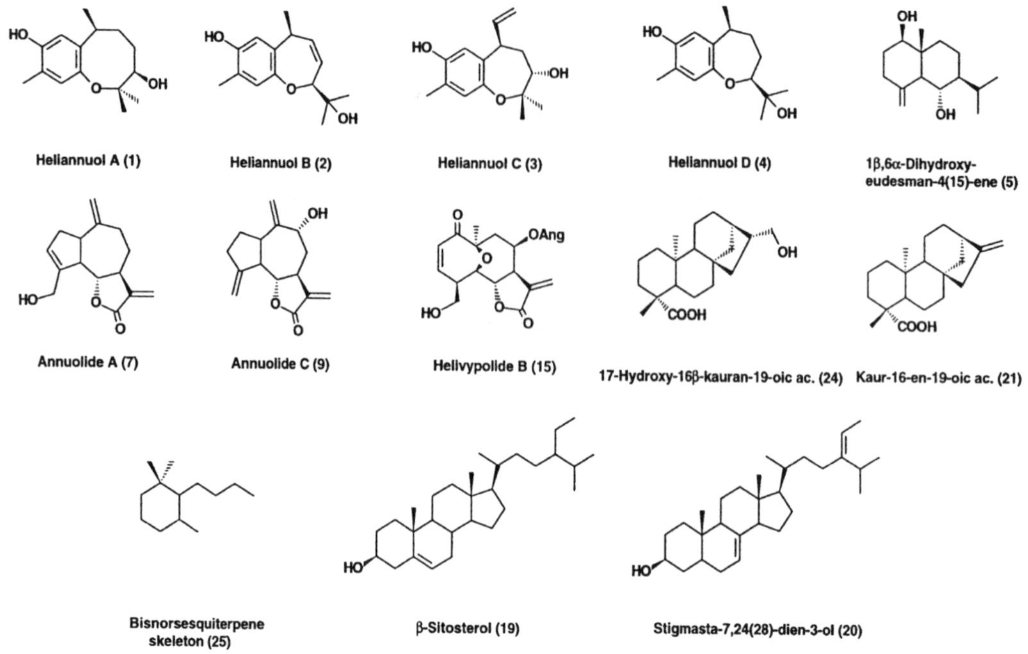

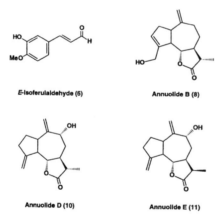

Heliannuol A (1)   Heliannuol B (2)   Heliannuol C (3)   Heliannuol D (4)   1β,6α-Dihydroxy-
                                                                                                eudesman-4(15)-ene (5)

Annuolide A (7)   Annuolide C (9)   Helivypolide B (15)   17-Hydroxy-16β-kauran-19-oic ac. (24)   Kaur-16-en-19-oic ac. (21)

Bisnorsesquiterpene   β-Sitosterol (19)   Stigmasta-7,24(28)-dien-3-ol (20)
skeleton (25)

FIGURE 33.9   Allelopathic compounds from cultivar sunflower varieties SH-222 and VYP.

E-Isoferulaldehyde (6)   Annuolide B (8)

Annuolide D (10)   Annuolide E (11)

FIGURE 33.10   Allelopathic compounds isolated only from cultivar sunflower varieties SH-222.

2α-hydroxy-8β-angeloyloxy-costunolide (**18**) from fraction **Q** (Macías et al., 1996); pinoresinol (**27**), kukulkanin B (**28**), and tambulin (**29**) from fraction **O**, and the monoterpene **25** from fraction **S** (Macías et al., 1997) (Figures 33.9–33.11).

To evaluate their potential allelopathic activity and to obtain information about the specific requirements needed for their bioactivity, the effect of aqueous solutions from 10^{-4} to 10^{-9} M of compounds **1** through **29,** were evaluated on root and shoot lengths of lettuce, barley, wheat, cress, tomato (*Lycopersicon esculentum* Mill), and onion (*Allium cepa* L.) seedlings. Most relevant results are presented in Figures 33.12–33.15 and Tables 33.1–33.4.

As observed with fractions **A, D,** and **F** from variety SH-222 (**11**) and **K, L,** and **N** from variety VYP (**19,** Figure 33.8) from which **1** through **6** were isolated, the novel family of sesquiterpenes: heliannuols A-D (**1-4**) showed (Table 33.1, Figures 33.12, 33.13) a homogeneous inhibitory profile of activity for heliannuol **A** (**1**) with an average of -40 percent inhibition on the germination of lettuce from $10^{-4}-10^{-9}$ M and a small effect on the germination and growth of barley seeds, whereas

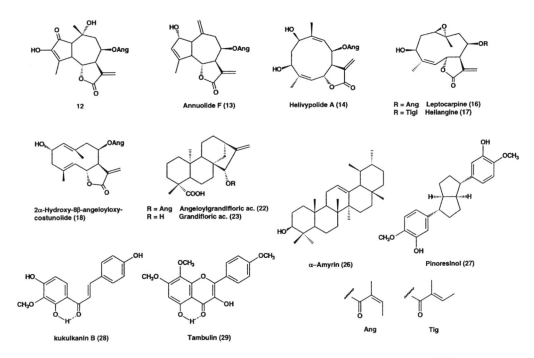

FIGURE 33.11 Allelopathic compounds isolated only from cultivar sunflower varieties VYP.

heliannuol D (**4**) showed a strong stimulation on the germination of lettuce (average 50 percent) as well as inhibition on root and shoot length with averages of −22 percent and −30 percent, respectively. The effects on germination and growth of heliannuols B (**2**) and C (**3**) as well as compound **6** are, in general, of low or no significance, except in monocotyledons in which the same root and shoot stimulation has been found particularly for **2** and **6**. The sesquiterpene **5** shows a profile of activity on the growth of lettuce very similar to heliannuol D (**4**), with special emphasis on root length with an inhibitory effect of −39 percent at 10^{-4} M. Further research has shown a differential activity with compounds **2** and **4** against growth of two additional seeds. Heliannuol B (**2**) has a powerful inhibitory effect on shoot length of cress (−60 percent, 10^{-4} M; −40 percent, 10^{-5} M; −30 percent, 10^{-6} M; −40 percent, 10^{-7} M; −38 percent, 10^{-8} M), that was not observed over root growth. Heliannuol D (**4**) showed a similar inhibitory activity on root length (−40 percent, 10^{-3} M; −50 percent, 10^{-4} M; −40 percent, 10^{-5} M; −50 percent, 10^{-6} M) and shoot length (−45 percent, 10^{-3} M; −40 percent, 10^{-4} M; −35 percent, 10^{-5} M) of onion seeds.

Several investigators have studied the regulatory activities of sesquiterpene lactones on seed germination and plant growth, and reported that the activity is affected by the conformation of the molecules, and the accessibility of groups that can be alkylated as a methylene lactone moiety (Fischer, 1986; Fischer et al., 1989; Macías et al., 1992).

As observed with the fraction from which guaianolides were isolated, **8, 9,** and **11** showed (Figures 33.10, 33.14) a high inhibitory activity on the germination of lettuce seeds in high and low concentrations (10^{-5} M **8:** −71 percent; 10^{-6} M **9:** −62 percent; 10^{-7} M **11:** −43 percent) except for compounds **12** and **13** in which stimulatory effects were found (average of 40 percent). The effects on root and shoot length are, in general, of no significance except in compounds **12** and **13** in which inhibitory effects on shoot length are an average of −35 percent and −25 percent, respectively, −30 percent on root length for **12** has been found. Seed germination and growth of barley seeds (Figure 33.15) are, in general, of no significance except in **7** and **10**. There were stimulatory effects on germination by **7** (10^{-5} M, 27 percent) and **10** (10^{-5} M, 17 percent; 10^{-6} M, 23 percent) (Macías et al., 1993a).

TABLE 33.1

Statistical Results of Allelopathic Bioassays (using *Lactuca sativa* L. and *Hordeum vulgare* L.) of Compounds 1–6*

Compound	Germination (%)						Radical length (%)						Shoot length (%)					
	10^{-4}M	10^{-5}M	10^{-6}M	10^{-7}M	10^{-8}M	10^{-9}M	10^{-4}M	10^{-5}M	10^{-6}M	10^{-7}M	10^{-8}M	10^{-9}M	10^{-4}M	10^{-5}M	10^{-6}M	10^{-7}M	10^{-8}M	10^{-9}M
L. sativa																		
1	−36[a]	−5	−36[a]	−32[a]	−29[b]	−62[a]	−2	−6[b]	−8[b]	−3	−1	−11	4	−5	13[a]	1	8[b]	0
2	−5	−10	−25[b]	−22[b]	−5	1	3	−1	−2	−7	1	−4	−1	−1	−2	0	−1	2
3	−53[a]	6	−18	2	24[b]	16	0	9[a]	−7[b]	−4	0	1	−11[b]	−6	−8[b]	−3	−7[b]	−9[b]
4	30[b]	80[a]	68[a]	62[a]	57[a]	35[b]	−23[a]	−17[b]	−6	−21[b]	−11	−12	−20[b]	−23[a]	−25[a]	−38[a]	−31[a]	−27[a]
5	−60[a]	−20	−8	−10	11	18	−38[a]	−10[a]	−18[a]	−16[a]	−23[a]	−14[a]	−2	−9[b]	−12[a]	−12[a]	−18[a]	−10[a]
6	−43[a]	−12	−3	−41	2	4	−14[a]	−2	−4	−4	7[b]	2	−10[b]	−1	1	−1	3	−3
H. vulgare																		
1	−16	−3	6	−9	3	−10	10[b]	11[a]	19	7	13[a]	5	−2	0	3	−5	3	1
2	—	4	−12	−27[b]	−21[b]	−19[b]	—	16[a]	13	6	5	14[a]	—	30[a]	25[a]	2	23[a]	1
3	—	−3	−3	−22	−6	−1	—	12[a]	−4	−2	3	−5	—	−13[a]	−1	16[a]	15[a]	4
4	−5	−14	9	7	−34[b]	−9	−20[a]	−14[b]	−8	−8	−17[a]	−12	−4	−10	−7	−4	−21[a]	−28[a]
5	—	−6	−21[b]	3	0	−18	—	3	0	−10[b]	6	0	—	8	−10	−2	4	−5
6	−12	0	−3	−22	−5	−18[a]	21[a]	11[a]	5	8[b]	9[b]	8[b]	17[a]	12[b]	9	8	−1	3

* Values are expressed as a percentage of the control; if it is not indicated, $P > .05$ for Mann-Whitney test.

[a] Values are significantly different, $P < .01$

[b] Values are significantly different, $.01 < P < .05$.

TABLE 33.2
Statistical Results of Allelopathic Bioassays (using *Lactuca sativa* L.) of Compounds 7–13, 16, and 17*

Compound	Germination (%)						Radical length (%)						Shoot length (%)					
	10^{-4}M	10^{-5}M	10^{-6}M	10^{-7}M	10^{-8}M	10^{-9}M	10^{-4}M	10^{-5}M	10^{-6}M	10^{-7}M	10^{-8}M	10^{-9}M	10^{-4}M	10^{-5}M	10^{-6}M	10^{-7}M	10^{-8}M	10^{-9}M
7	−20	−59[a]	−31[a]	−29	−7	−41[a]	−8	−14[b]	−13[a]	−3	−7[b]	−2	−10[b]	−11	−8	−6	−4	−5
8	−64[a]	−71[a]	−53[a]	−38[a]	−34	−12	−23[a]	−18[a]	−10[b]	−1	−12[a]	0	1	−6	6	2	2	2
9	−59[a]	−43[a]	−62[a]	−22	−47[a]	−55[a]	−10[b]	−6	−19[a]	−22[a]	−6	−16[a]	−4	−3	−5	−10	−8	−2
10	−57[b]	−7	−51[a]	—	−22	−17	−28[a]	−2	−12[a]	—	0	−8	−11[b]	−6	−2	—	−3	−1
11	−38[a]	−43[a]	−25	−41[a]	−22	−22	−10[b]	−9[b]	−5	−5	−6	2	7[b]	8	4	4	3	1
12	—	−42[b]	25	33	13	46[a]	—	−33[a]	−14[b]	−24[b]	−25[b]	−29[a]	—	−34[a]	−28[a]	−34[a]	−32[a]	−34[a]
13	—	38[b]	25	17	21	63[a]	—	−18	−11	−25[b]	−20[b]	−15[b]	—	−24[b]	−15[b]	−30[a]	−19[a]	−19[a]
16	−50	63	50	42[b]	0	0	3	−11	−17[b]	−27[a]	−24[b]	−37[a]	−24[b]	−20[a]	−24[a]	−27[a]	−21[b]	−36[a]
17	—	−78[a]	13	21	0	−4	—	−47[b]	−60[a]	−45[a]	−13	−28	—	−27	−35[a]	−18	−7	−21

* Values are expressed as a percentage of the control; if it is not indicated, $P < .05$ for Mann-Whitney test.

[a] Values are significantly different, $P < .01$.

[b] Values are significantly different, $.01 < P < .05$.

TABLE 33.3
Statistical Results of Allelopathic Bioassays (using *Lepidium sativum* L. and *Lycopersicon esculentum* L.) of Compounds 13–18*

Compound	Germination (%)						Radical length (%)						Shoot length (%)					
	10^{-4}M	10^{-5}M	10^{-6}M	10^{-7}M	10^{-8}M	10^{-9}M	10^{-4}M	10^{-5}M	10^{-6}M	10^{-7}M	10^{-8}M	10^{-9}M	10^{-4}M	10^{-5}M	10^{-6}M	10^{-7}M	10^{-8}M	10^{-9}M
Lepidium sativum																		
13	—	2	19	11	2	14	—	8	0	9	-17[b]	13	—	2	4	-10	-10	12
14	—	0	27	-13	-2	23	—	-2	-13	5	-5	11	—	18	5	-1	8	28[b]
15	—	-4	51[b]	22	-4	-9	—	-4	-6	11	11	0	—	-10	-8	0	8	10
16	-14	8	0	-25	19	-35[b]	30[b]	-6	10	-6	23[b]	-2	12	-6	2	-6	27[b]	-9
17	2	43	11	-20	14	-18	19	0	11	-3	8	-8	18	-10	-5	-7	7	-10
18	—	6	46[b]	4	-10	28[b]	—	-2	-13	7	-1	11	—	-3	-14	-9	-2	2
L. esculentum																		
13	—	-3	4	-8	-3	-3	—	-7	-4	-4	-4	-6	—	-17[b]	-16[b]	-18[b]	-10	-7
14	—	-17	-3	6	0	-13	—	-4	-3	-7	-3	-7[b]	—	-8	-15[b]	-10	-2	-12
15	—	1	8	-6	-18	-11	—	10[b]	3	-4	3	-20[a]	—	-10	-16[b]	-7	-7	-16[b]
16	0	11[b]	-3	-4	-14	-1	-22[a]	12[b]	-17[a]	-1	-7	-10	-22[a]	-8	-13	6	2	-8
17	—	0	-4	0	3	-1	—	-12	-7	-6	4	-8	—	-21[b]	-5	-11	5	-8
18	—	-1	15[b]	10	-1	-9	—	5	5	-8	-3	-7	—	3	-13[b]	-4	-4	-3

* Values are expressed as a percentage of the control; if it is not indicated, $P > .05$ for Mann-Whitney test.

[a] Values are significantly different, $P < .01$.

[b] Values are significantly different, $.01 < P < .05$.

TABLE 33.4
Statistical Results of Allelopathic Bioassays (using *Hordeum vulgare* L. and *Triticum aestivum* L.) of Compounds 7–10, 13–16, and 18[*]

Compound	Germination (%)						Radical length (%)						Shoot length (%)					
	10^{-4}M	10^{-5}M	10^{-6}M	10^{-7}M	10^{-8}M	10^{-9}M	10^{-4}M	10^{-5}M	10^{-6}M	10^{-7}M	10^{-8}M	10^{-9}M	10^{-4}M	10^{-5}M	10^{-6}M	10^{-7}M	10^{-8}M	10^{-9}M
H. vulgare																		
7	7	27[b]	8	3	3	−17	−19[a]	−7	−11[b]	−4	−10	−4	−10	−19[a]	−17[a]	−6	−4	7
8	—	−21[b]	−10	−6	—	−15	—	19[a]	9[b]	8	—	−2	—	18[a]	16[a]	5	—	14[b]
9	—	−10	11	−5	7	−14	—	−2	−3	−2	−4	−7	—	−12[b]	−2	10	−2	11
10	—	17	23[b]	−3	2	2	—	9[b]	−5	0	5	−1	—	−4	−12[b]	5	−1	−1
13	—	—	−37	−11	37	4	—	—	4	−3	−14	13	—	—	8	−1	−13	−2
14	—	24	19	22	21	−9	—	−6	−10	14	−16	2	—	−8	−18[b]	2	−31[b]	−14
15	—	—	22	11	42[a]	−9	—	—	1	10	−2	16	—	—	−2	−3	−16	4
16	—	−37	−11	16	13	−4	—	−8	−1	−3	15	14	—	−8	−6	−15	−4	−5
18	—	—	7	−30	42	30	—	—	5	9	−5	10	—	—	−2	17	−7	−3
T. aestivum																		
13	—	—	−2	−15	−19	−23[b]	—	—	−4	25[b]	−11	12	—	—	−10	17	−12	14
14	—	—	2	−43	−51[b]	−53[a]	—	—	−10	−9	−27[b]	−9	—	—	−9	−15	−19	−5
15	—	—	−35	−30	−48[b]	−47[b]	—	—	−22[b]	−1	−22[b]	−9	—	—	−24[b]	−13	−13	−7
16	—	4	−6	−2	−11	−9	—	−9	2	−11	17	−1	—	−4	2	−4	26[b]	10
18	—	—	−2	−9	−29	−14	—	—	−19[b]	10	−7	4	—	—	−19	8	−10	12

[*]Values are expressed as a percentage of the control; if it is not indicated, $P > .05$ for Mann-Whitney test.

[a]Values are significantly different, $P < .01$.

[b]Values are significantly different, $.01 < P < .05$.

The germacranolides **14** through **18** provoked stimulatory effects on the germination specifically by **12** (average 40 percent), **13** (average 50 percent), and **16** (average 55 percent) except at high concentrations in which an inhibitory effect was observed (average −40 percent) (Figure 33.14). The interesting effects were those on the root and shoot length with averages of −30 (**16**) and −50 percent (**17**) for root length and −30 (**16**) and −35 (**17**) for shoot length. The effects on the germination and growth of barley are again, in general, of no significance except for in germination, where stimulatory effects of averages 23 (**14**) and 35 percent (**15**) have been found.

The above findings suggest that the heliannuols **1** through **4**, the guaianolides **7** through **13**, and the germacranolides **14** through **18** are likely to be significantly involved in the allelopathic action of cultivar sunflowers, with certain specificity on some dicotyledonous species.

The inhibitory effects on seed germination of some dicotyledonous species and the stimulatory effects on seed germination of some monocotyledonous species suggest that these heliannuols, guaianolides, and germacranolides are excellent candidates to be used as pre-emergent herbicides on monocotyledonous crops at very low doses (10^{-4} to to 10^{-9} M), where even some stimulation on the germination of the monocotyledons can be expected. These suggestions are in agreement with field observations in southern parts of Mexico where corn (*Zea mays* L.) fields support an important number of *Helianthus maximiliani* (Gershenzon and Mabry, 1984) and *H. microcephalus* (Gao et al., 1987) plants with positive effects on the crop, from which guaianolides similar to annuolides **A** (**7**) and **D** (**10**) have been isolated.

There are excellent reviews about the potential use of allelochemicals as herbicides, from higher plants and microbes (Einhellig and Leather, 1988; Duke, 1986a; Duke and Lydon, 1987; Duke, 1986b; Cutler, 1988), or from plants (Putnam, 1988b; Vaughn and Vaughn, 1988). Recently, the potential of allelopathy in the search for natural herbicide models has been reviewed (Macías, 1995).

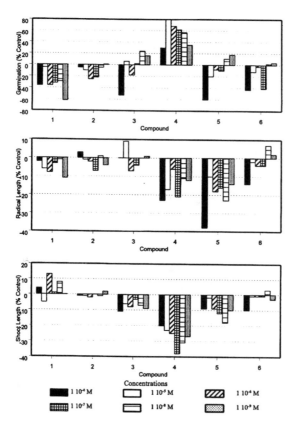

FIGURE 33.12   Effect of compounds **1-6** on the germination, root and shoot length of *Lactuca sativa* L.

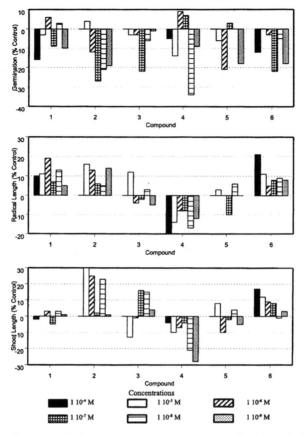

FIGURE 33.13   Effect of compounds **1-6** on the germination, root and shoot length of *Hordeum vulgare* L.

Most potential natural allelochemicals are terpenoids: monoterpenes, sesquiterpenes, sesquiterpene lactones, and triterpenes, and fatty acids with an activity range of 0.25 to 10^5 ppb. Their potential uses are discussed on the basis of their stability in the soil (some of the degradation products are more active than their precursors, particularly phototoxins), the environmental safety (biodegradability), the site of action, the accessibility (knowledge of the location of active principles within the plant might be crucial to determine the real possibilities of practical applications in agriculture), and the level of activity (to be successful, natural phytotoxins might be active at lower concentrations in comparison with synthetic herbicides).

The normal tested range of activity for allelochemical is 10^{-4} to 10^{-9} M with approximately 0.1 to 10^4 ppb; while we can consider compounds to be good candidates for natural herbicide templates when the range is between 10^{-5} and 10^{-7} M with approximately 10 to 10^3 ppb. If we compare this with the recent low-dose generation of synthetic herbicides, characterized by selective weed control at rates of < 100 g/ha at approximately < 100 ppb (soil weight basis, distributed 10 cm deep), use of these types of natural products become attractive. Therefore, the heliannuols, guaianolides, and germacranolides isolated from cultivar sunflowers with activity levels on dicotyledons at ranges between 10^{-5} and 10^{-9} M with approximately 0.1 to 103 ppb are excellent candidates as natural herbicides. However, more chemical analyses need to be done and it is necessary to conduct a systematic study of the mechanism involved in the allelopathic activity of these terpenoids. However, we can conclude that, based on their bioactivity, it is possible to use fresh leaves of cultivar sunflowers as sources of allelopathic agents with potential use as natural herbicide templates.

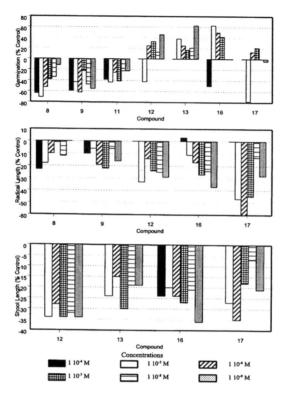

FIGURE 33.14 Effect of selected sesquiterpene lactones on the germination, root and shoot length of *Lactuca sativa* L.

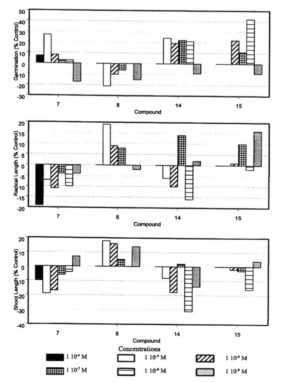

FIGURE 33.15 Effect of selected sesquiterpene lactones on the germination, root and shoot length of *Hordeum vulgare* L.

ACKNOWLEDGMENTS

This research was supported by the Dirección General de Investigación Científica y Técnica, Spain (DGICYT; Project No. PB95-1231). We thank Dr. Alberto García de Luján y Gil de Bernabé, and Miguel Lara Benítez "Rancho de la Merced" Junta de Andalucía, Jerez (Spain) for providing plant material. R.M.V. acknowledges a fellowship from Junta de Andalucía.

REFERENCES

Altieri, M. A., The impact, uses, and ecological role of weeds in agroecosystems, in *Weed Management in Agroecosystems: Ecological Approaches,* Altieri, M. A. and Liebman, M., Eds., CRC Press, Boca Raton, FL, 1988, 1.

Cochran, V. L., Elliot, L. F., and Papendick, R. I., The production of phytotoxins from surface crop residues, *Soil Sci. Soc. Am. J.,* 41, 903, 1977.

Cutler, H. G., Perspectives on discovery of microbial phytotoxins with herbicidal activity, *Weed Technol.,* 2, 525, 1988.

Duke, S. O., Naturally occurring chemical compounds as herbicides, *Rev. Weed Sci.,* 2, 15, 1986a.

Duke, S. O., Microbially produced phytotoxins as herbicides. A perspective, in *The Science of Allelopathy,* Putnam, A. R. and Tang, C. S., Eds., John Wiley & Sons, New York, 1986b, 287.

Duke, S. O. and Lydon, J., Herbicides from natural compounds, *Weed Technol.,* 1, 122, 1987.

Einhellig, F. A. and Rasmussen, J. A., Allelopathic effects of *Rumex crispus* on *Amaranthus retroflexus,* grain sorghum and field corn, *Am. Midl. Nat.,* 90, 79, 1973.

Einhellig, F. A. and Leather, G. R., Potentials for exploiting allelopathy to enhance crop production, *J. Chem. Ecol.,* 14, 1829, 1988.

Einhellig, F. A. and Rasmussen, J. A., Prior cropping with grain sorghum inhibits weeds, *J. Chem. Ecol.,* 4, 425, 1989.

Fischer, N. H., The function of mono and sesquiterpenes as plant germination and growth regulators, in *The Science of Allelopathy,* Putnam, A. R. and Tang, C. S., Eds., John Wiley & Sons, New York, 1986, 203.

Fischer, N. H., Weidenhamer, J. D., and Bradow, J. M., Inhibition and promotion of germination by several sesquiterpenes, *J. Chem. Ecol.,* 15, 1785, 1989.

Gao, F., Wang, H. and Mabry, T. J., Sesquiterpene lactones and flavonoids from *Helianthus* species, *J. Nat. Prod.,* 50, 23, 1987.

Gershenzon, J., Mabry, T. J., Sesquiterpene lactones from a Texas populations of *Helianthus maximilliani,* *Phytochemistry,* 23, 1959, 1984.

Gershenzon J., Ohno N., and Mabry, T. J., The terpenoid chemistry of sunflowers (*Helianthus*), *Rev. Latinoamer. Quím.,* 12, 53, 1982.

González, L., Souto, X. C., Bolaño, J. C., and Reigosa, M. J., Allelopathic potential of different accesions of *Caspium annuum* L. Application to the management of weeds. Proceeding of SEMh (Sociedad Española de Malherbología) Congress, 1992, 367.

Herz, W. and Bruno, M., Heliangolides, Kauranes and other constituents of *Helianthus heterophyllus,* *Phytochemystry,* 25, 1913, 1986.

Irons, S. M. and Burnside, O. C., Competitive and allelopathic effects of sunflower (*Helianthus annuus*), *Weed Sci.,* 30, 372, 1982.

Kato, T., Tsunakawa, M., Sasaki, N., Aizawa, H., Fujita, K., Kitahara, Y., and Takahashi, N., Growth and germination inhibitors in rice husks, *Phytochemistry,* 16, 45, 1977.

Leather, G. R., Weed control using allelopathic crop plants, *J. Chem. Ecol.,* 9, 983, 1982.

Leather, G. R., Sunflowers (*Helianthus annuus*) are allelopathic to weeds, *Weed Sci.,* 31, 37, 1983.

Leather, G. R., Weed control using allelopathic sunflowers and herbicide, *Plant Soil,* 98, 17, 1987.

Macías, F. A., Allelopathy in the search of natural herbicide models, in *Allelopathy. Organisms, Processes and Applications,* Inderjit, Dakshini, K. M. M., and Einhellig, F. A., Eds., ACS Symposium Series, American Chemical Society, Washington, D.C., 1994, 310.

Macías, F. A., Galindo, J. C. G., and Massanet, G. M., Potential allelopathic activity of several sesquiterpene lactone models, *Phytochemistry,* 31, 1969, 1992.

Macías, F. A., Molinillo, J. M. G., Varela, R. M., and Torres, A., Structural elucidation and chemistry of a novel family of bioactive sesquiterpenes: heliannuols, *J. Org. Chem.,* 59, 8261, 1994.

Macías, F. A., Molinillo, J. M. G., Torres, A., and Varela, R. M., Potential allelopathic sesquiterpene lactones from sunflower leaves, *Phytochemistry,* 43, 1205, 1996.

Macías, F. A., Molinillo, J. M. G., Torres, A. and Varela, R. M. Bioactive norsesquiterpene from *Helianthus annuus* with potential allelopathic activity, *Phytochemistry,* 44, 1997.

Macías, F. A., Varela, R. M., Torres A., and Molinillo, J. M. G., Potential allelopathic guaianolides from cultivar sunflower leaves, var. SH-222, *Phytochemistry,* 34, 669, 1993a.

Macías, F. A., Varela, R. M., Torres, A., and Molinillo, J. M. G., Novel sesquiterpene from bioactive fractions of cultivar sunflowers, *Tetrahedron Lett.,* 34, 1999, 1993b.

Menges, R. M. and Tamez, S., Common sunflower (*Helianthus annuus*) interference in onions (*Allium cepa*), *Weed Sci.,* 29, 641, 1981.

Putnam, A. R., Allelopathy: problems and opportunities in weed management, in *Weed Management in Agroecosystems: Ecological Approaches,* Altieri, M. A. and Liebman, M., Eds., CRC Press, Boca Raton, FL, 77, 1988a.

Putnam, A. R., Allelochemicals from plants as herbicides, *Weed Technol.,* 2, 510, 1988b.

Putnam, A. R. and Duke, W. B., Allelopathy in agroecosystems, *Annu. Rev. Phytopathol.,* 16, 431, 1978.

Putnam, A. R., Defrank, J., and Barnes, J. P., Exploitation of allelopathy for weed control in annual and perennial cropping systems, *J. Chem. Ecol.,* 9, 1001, 1983.

Rasmussen, J. A. and Einhellig, F. A., Noncompetitive effects of common milkweed, *Asclepias syriaca* L., on germination and growth of grain sorghum, *Am. Midl. Nat.,* 94, 478, 1975.

Rice, E. L., *Allelopathy,* 2nd ed. Academic Press, New York, 1984.

Schon, M. K. and Einhellig, F. A., Allelopathic effects of cultivated sunflower on grain sorghum, *Bot. Gaz.,* 143, 505, 1982.

Spring, O., Albert, K., and Gradmann, W., Annuitrin, a new biologically active germacranolide from *Helianthus annuus, Phytochemistry,* 20, 1883, 1981.

Tauscher, B., Allelochemicals-eine interdisziplinäre Herausforderung, *Z. Pflkrankh. Pflschutz, Sonderh.,* 11, 15, 1988.

Varela, R. M., Allelopathic studies on cultivar sunflowers, Master dissertation, University of Cádiz, Spain, 1992.

Vaughn, M. A., and Vaughn, K. C., Mitotic disrupters from higher plants and their potential uses as herbicides, *Weed Technol.,* 2, 533, 1988.

Wilson, R. E and Rice, E. L., Allelopathy as expressed by *Helianthus annuus* and its role in old-field succession, *Bull. Torrey Bot. Club,* 95, 432, 1968.

Worsham, A. D., Current and potential techniques using allelopathy as an aid in weed management, in *Phytochemical Ecology: Allelochemicals, Mycotoxins and Insect Pheromones and Allomones,* Chou, C. H. and Waller, G. R., Eds., Institute of Botany, Academia Sinica Monograph Series No. 9, Taipei, Taiwan, 275, 1989.

34 Biological Properties of Rue (*Ruta graveolens* L.): Potential Use in Sustainable Agricultural Systems

Giovanni Aliotta and Gennaro Cafiero

CONTENTS

34.1   ABSTRACT

Rue (*Ruta graveolens* L.) is an ancient medicinal plant with various uses including insect control and food preservation. Several secondary metabolites such as alkaloids, coumarins, flavonoids, ketones, terpenoids, and organic acids have been identified from different parts of rue plant. The upper green leaves, however, are particularly rich in these compounds. Furanocoumarins, from rue leaves, play a defensive role in protecting the plant against insects, pathogens, and herbivores, and they affect allelopathically the germination of crop and weed seeds. The rue allelochemicals, 5-methoxypsoralen, 8-methoxypsoralen, and 4-hydroxycoumarin, delayed and decreased the germination of radish (*Raphanus sativus* L.) seeds more in the light than in the darkness.

These coumarins in the presence of light inhibit water uptake, swelling of seed coat, and endosperm and cell elongation of radicle in radish seeds. The potential use of rue infusion as a possible bioherbicide against germination of weed seeds was explored. In pot culture studies, the effect of 5 and 10 percent leaf infusion of rue on seed germination of a major weed purslane (*Portulaca oleracea* L.) was investigated. The application of 5 percent infusion three times at five-day intervals, significantly checked the growth of this weed. This is considered as a major breakthrough because herbicide control of purslane in irrigated summer crops is very difficult.

34.2 INTRODUCTION

Crops free from weeds and pests have been the dream of farmers since the origin, of agriculture. In general, weeds are unwanted companions in crop production because of their negative interference. Knowledge of weed and crop biology, their origin, and their interactions may help in developing an ecological approach for the control of weeds in sustainable agricultural systems (Rice, 1984; Zimdahl, 1993). We learn a lot by studying natural defenses against diseases and pests, as well as from the traditional farmers who have developed the arts of survival (Harlan, 1992).

A weed can be defined as a generally unwanted organism that thrives in habitats disturbed by humans (Harlan and de Wet, 1965; Zimdahl, 1993). The word "crop" covers all that is harvested. Aside from crops, many weeds are adapted to people-managed habitats, that is, croplands. Moreover, crops and weeds are derived from a common progenitor; for example, there are weeds (wild races) and cultivated races of many plants such as barley (*Hordeum vulgare* L.), oat (*Avena sativa* L.), carrot (*Daucus carota* L.), lettuce (*Lactuca sativa* L.), tomato (*Lycopersicon esculentum* Mill.), and rice (*Oryza sativa* L.). That is why agricultural practices favorable to crops are also favorable to weeds (Harlan, 1992; Inderjit and Dakshini, 1996).

The use of natural plant products in comparison with synthetic chemicals is considered safer for the environment because of their biodegradability. Compared with long persistence, non-target toxicity, pollutive, carcinogenic, and mutagenic activity of synthetic agrochemicals, natural plant products are biodegradable and likely to be recycled through nature. The increased need for the most cost-effective, efficacious, selective, and environmentally safer agrochemicals (herbicides) is being suggested by several trends in agriculture and agrochemistry. Further, the increased incidence of weed resistance to important herbicides classes such as s-triazines (Gressel, 1985) and dinitroanilines (Mudge et al., 1984) enhances the need for the discovery of new herbicides. It is therefore the most current need of the industry to isolate, identify, and characterize natural plant products and explore the possibility of their potential use as herbicides and in the biocontrol of weeds and plant diseases. Discovery of natural plant products with potential use as herbicides is complicated by three factors: complicated chemical structures, extremely small amounts of compound available for bioassays, and rediscovery of known compounds. Moreover, some of the most important natural phytotoxins such as artemisinin, hypericin, and benzoic acid do not exhibit selective phytoxicity. Not all allelopathic compounds are good candidates for biocontrol. For example, certain phenolic compound such as benzoic acid derivatives, coumarins, flavonoids, and quinones are biologically active at ranges between 10^{-2} and 10^{-5} M. However, good candidates for natural herbicides should have activity ranges between 10^{-5} and 10^{-7} M. Thus, many terpenoids, sesquiterpenes, and triterpenoids are better candidates for natural herbicides (Macias, 1995).

In our allelopathic studies to isolate potential bioherbicides and algicides from plants, we have worked on poisonous plants rich in active principles because of their potential role in the interactions between plants and other organisms (Aiello et al., 1992; Aliotta et al., 1989, 1994, 1995).

In this chapter, we present an overview of research on rue (*Ruta graveolens* L. Rutaceae, Figure 34.1), an ancient medicinal plant, an evergreen shrub native of Southern Europe. Its leaves are bitter in taste and emit a powerful odor. It is cited in the ancient herbals, folklore, alchemy, and even in demonology. Furthermore, we shall also discuss the potential of rue infusion and its allelochemicals in weed management.

34.3 RUE ALLELOPATHIC COMPOUNDS

Several secondary metabolites from rue are listed in Table 34.1. Allelopathic compounds isolated from different parts of the rue plant belong to coumarins (coumarin, bergapten, psoralen, umbelliferone, xanthotoxin), flavonoids (quercetin, rutin), organic acids (salicylic acid), and terpenoids (cineole, limonene, pinene) (Aliotta et al., 1994; Duke, 1986; Harborne, 1967; Murray et al., 1982).

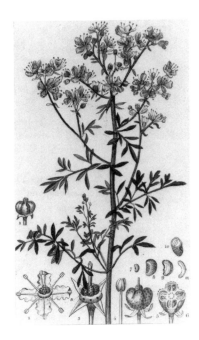

FIGURE 34.1   *Ruta graveolens* L.

TABLE 34.1
Secondary Chemical Constituents of *Ruta graveolens* L. Adapted from Murray et al., 1982

Compound	Plant Part Examined
Alkaloids	
Arborinine	
g-Fagarine	
Graveoline	
Graveolinine	
Kokusaginine	
6-Methoxy-dictamine	
Rutacridone	
Skimmianine	
Coumarins	
Bergapten*[a]	Stems, leaves, cell cultures
(-)-Byakangelicin	Roots
Chalepensin	Roots
Coumarin*[a]	Leaves
Daphnoretin	Aerial parts
Daphnoretin methyl ether	
Daphnorin	Roots
Gravelliferone	Roots
Gravelliferone methyl ether	Roots
Herniarin	Cell cultures
Isoimperatorin	Roots and stems
Isopimpinellin	Cell cultures
Isorutarin	Cell cultures, roots
Marmesin	Cell cultures, roots
Marmesinin	Roots

8-Methoxy-gravelliferone
Pangeline
Psoralen[a]   Cell cultures, roots,
                                                          stems, leaves

Rutacultin   Roots
Rutamarin   Aerial parts, roots, stems
Rutamarin alcohol   Roots
Rutaretin   Leaves (?)
Rutarin   Aerial parts, roots
Scopoletin   Cell cultures
Suberenon   Roots
Umbelliferone[a]   Cell cultures, roots
Xanthotoxin[a]   Celle cultures, stems, leaves
Xanthyletin   Roots

Flavonoids

Quercetin[a]   Leaves
Rutin[a]   Leaves

Ketones

Methyl-nonyl-ketone   Aerial parts
Methyl-heptyl-ketone   Aerial parts

Organic acids

Anisic acid
Caprinic acid
Caprylic acid
Oenanthylic acid
Plagonic acid
Salicylic acid[a]

Terpenoids

Cineole[a]   Aerial parts
Limonene[a]   Aerial parts
Pinene[a]   Aerial parts
Guaiacol   Aerial parts

[a]Allelochemicals

Zobel and Brown (1988) studied the localization of the three dominant furanocoumarins of rue: psoralen, xanthotoxin, and bergapten, with an improved extraction technique. This technique showed
that furanocoumarins are deposited mainly on the leaf surface in a thick epicuticular layer of wax in
amounts of approximately ca. 1 μg/g fresh weight. Concentrations of the three furanocoumarins
were also measured on the surface and within mature whole leaves of rue that contained green, yellow, and dry yellow leaves. Upper green leaves contained higher concentrations of these coumarins
than lower green leaves, green leaves contained several times as much as yellow leaves, and dry
leaves contained even smaller amounts than yellow ones. Xanthotoxin was always the predominant
furanocoumarin (Zobel and Brown, 1991). Zobel and Brown (1995) have argued that the presence
of coumarins on the surface of plant cells indicates their important ecological role both in plant protection and in communication with its environment. Their location on the leaves may render the plant
unpalatable to herbivores, provide defense against insect attack, control oviposition, and protect
from bacteria and fungal spores. Removal from the surface may inhibit the germination of other
species. When present on the embryo surface, coumarins may cause autoinhibition and their leaching in water during the spring permits the seeds to germinate. Environmental stresses such as high
altitude, extreme temperature, change in the UV level, and atmospheric pollution influence the production of coumarins and their extrusion to the surface.

Allelopathic interference of the flavonoid, quercetin, have also been demonstrated by Harborne et al. (1964) and Inderjit and Dakshini (1995). The aqueous extract of the weed *Pluchea lanceolata* (DC.) C. B. Clarke (Asteraceae) containing quercetin affected the seedling growth of the legume, asparagus bean (*Vigna unguiculata* (L.) Walp. var. *sesquipedalis*). It was not detected from the weed-associated soil probably because of microbial decomposition by fungi such as the species of *Aspergillus*. The quercetin isolated from velvetleaf (*Abutilon theophrasti* Medik.) seed coats significantly inhibited the germination and radicle growth of cress, radish, and soybean (*Glycine max* L.) (Paszkowsky and Kemer, 1988). Quercetin and other flavonoids influence the light-induced electron transport and phosphorylation of isolated spinach (*Spinacea oleracea* L.) chloroplast, affecting both ATP generation and electron transport pathways (Moreland and Novitsky, 1988). Finally, quercetin significantly inhibits soluble enzymes such as glutathione-S-transferases involved in the metabolism of plant allelochemicals (Berenbaum and Neal, 1987).

Einhellig (1995a) has suggested that salicylic acid is the most growth inhibitory and it appears to be an endogenous signal in systemic activation of plant defenses after a localized exposure to certain viral and fungal pathogens.

Various monoterpenes such as 1,8-cineole, limonene, and pinene, have been suggested for their allelopathic potential (Fischer, 1986; Muller, 1965; Muller and Muller, 1964; Muller et al., 1964; Weidenhamer et al., 1993)

The additive affects of rue allelopathic compounds has been demonstrated in its allelopathic interference. Under field conditions, allelopathic interference is more likely due to additive activities of allelopathic compounds than that of a single compound (Einhellig, 1995a)

34.4   RUE AND SEED GERMINATION

A common test of the allelopathic potential of a plant *in vitro* is to determine the effects of its aqueous leachate on the seed germination and subsequent radicle growth. Seed is an ideal tool because it is a dispersal unit of life formed by three parts that are genetically different: seed coat, endosperm, and an embryo able to sprout out in a new plant if exposed to the right conditions (Evenari, 1980). It is known that allelochemicals detected in rue are either seed germination regulators or inhibitors (Murray et al., 1982). However, clear insight into the precise morphological and physiological perturbations caused by these substances on test seeds are not known owing to the difficulty in determining the specific site of their action. Besides, many allelopathic studies have been carried out *in vitro,* and this is why our knowledge is limited about the fate of the allelochemicals in the soil (Inderjit and Dakshini, 1994).

We have selected rue for allelopathic studies because 1) our previous studies established coumarins as potent allelochemicals (Aliotta et al., 1992), 2) the presence of large amounts of coumarins on the leaf surface of rue and their easy extraction through leaching make them ideal (Zobel and Brown, 1988), and 3) rue and basil never grow together nor near one another (Grieve, 1967).

Coumarin and phenylpropanoids were tested for their biological activity on seed germination and subsequent root growth in light and darkness (Aliotta et al., 1993). The coumarin was found to be the most potent inhibitor, with some exceptions—phenylpropanoids with a carboxylic group in the side chain inhibited root growth. Microscopic observations of root treated with coumarin suggested that this substance inhibits the elongation of cells of the differentiating zone of the root. Besides, it is known that the coumarins and phenolic compounds interfere with many vital plant processes, including cell division, mineral uptake, stomatal function, water balance, respiration, photosynthesis, protein and chlorophyll synthesis, and phytochrome activity (Einhellig, 1995b).

We therefore, investigated the allelopathic properties of rue on crop and weed seeds focusing the attention to the hilum-micropylar region, where normally a radicle protrudes.

34.5 RUE *VERSUS* RADISH GERMINATION

We tested rue leachate for its allelopathic activity *in vitro* on radish germination in light and darkness (Aliotta et al., 1994). This delayed the onset and decreased the rate of germination in the light rather than in darkness. Three active pure compounds were isolated and identified as coumarins: 5-methoxypsoralen (5-MOP), 8-methoxypsoralen (8-MOP), and 4-hydroxycoumarin. Their concentrations in the infusion were 10^{-4} M, 2×10^{-4} M, and 0.4×10^{-5} M, respectively.

To ascertain whether the effect of coumarins on radish germination was at the level of the embryo or was mediated by the seed coat and endosperm, these latter were removed and embryos were tested for their germination in light in the presence of each coumarin. It appeared that each coumarin inhibited the seeds with their coat, but did not significantly inhibit seeds without a coat and endosperm. When the most of the seeds soaked in water were germinating, treated seeds were dormant and different uptake of water into these seeds was evident (Figure 34.2).

In radish seed, three layers of cells may be recognized in the seed coat of the mature seed: the epidermis (formed by compressed cells), the palisade layer (which represents cells more or less isodiametrical or radially elongated), and the inner parenchyma (pigmented layer one cell thick) (Vaughan and Whitehouse, 1971). The endosperm persists as a well-formed aleurone layer intimately associated with the seed coat. The hyalin layer covers the embryo (Figure 34.3A), which is folded with cotyledons against the radicle.

The hilum-micropylar region of the seed was highly specialized and characterized by a flaking epidermis, thickened aleurone cells, and a hyalin layer. Moreover, this area represents a marker for comparison between seeds (Figure 34.3, B). Stereo-microscopic observations of moistened excised seeds, cut in half along the two orthogonal planes of their major axis (Figure 34.4), showed that the section of hilum-micropylar region of control seed presents two bands of different colors; the external band is black and the inner is grey. This latter does not appear in the treated seed. SEM and light microscope observations of the hilum-micropylar region showed that the hilum was more evident in the control than in the treated seed (Figure 34.5, A and C).

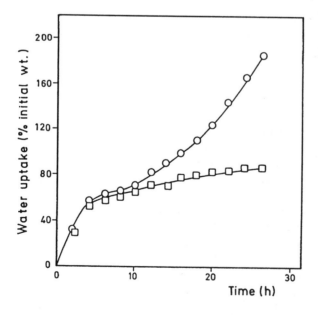

FIGURE 34.2 Water uptake by radish seeds in presence (□) and absence (○) of 5-MOP 2×10^{-4} M (on a percent of initial weight of seeds). After Aliotta et al., 1994

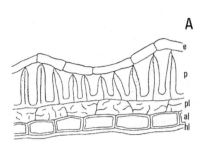
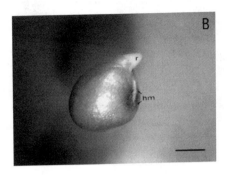

FIGURE 34.3 Graphic showing the structural features of a seed coat and endosperm of radish seed (A), according to Vaughan and Whitehouse, 1971 (A). Stereomicrograph of a germinating seed of radish showing the hilum micropylar region and the radicle (B). Abbreviations: e, epidermis; p, palisade; pl, pigment layer; al, aleurone layer; hl, hyalin layer; r, radicle; hm, hilum-micropylar region. Bar = 1 mm. After Aliotta et al., 1994

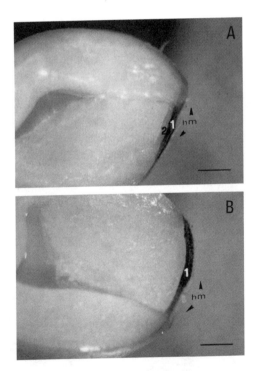

FIGURE 34.4 Stereomicrographs of radish seeds, moistened (A) and 5-MOP–treated (B), 18 h after sowing. Cut seeds excised along the two orthogonal planes of their major axis. The hilum-micropylar region shows two bands in (A) and one in (B). Bar = 0.5 mm. After Aliotta et al., 1994

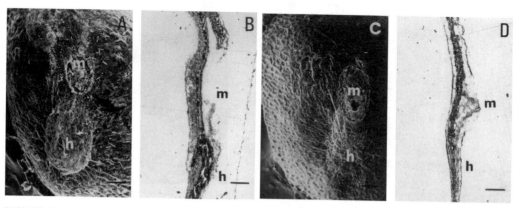

FIGURE 34.5 SEM (A, C) and light microscope (B, D) micrographs of the hilum-micropylar region of radish; 18 h after sowing, the hilum is more evident in water-moistened seed (A, B), than in 5-MOP–treated seed (C, D). Abbreviations: m, micropyle; h, hilum. Bar = 150 μm. After Aliotta et al., 1994

This aspect was confirmed by the comparison of the hilum-micropylar semi-thin section of control and treated seeds under the light microscope, where the embryos were removed and the seed coats and endosperms were prepared for transmission electron microscopic (TEM) observations (Figure 34.5, B and D). The hilum of the control seed is thicker and pigmented; the tested pigment layer is more evident, and there are more layers of aleurone cells that are filled by light-dense bodies (Figure 34.6, A and B). In this respect it was interesting to compare the TEM ultrastructure of the seed coat and endosperm of control and 5-MOP–treated seeds (Figure 34.7, A and B), as shown under the light microscope. Moreover, the palisade layer of treated seed appears thicker than in the control. TEM comparison between aleuronic cells of the control (Figure 34.8, A) and treated cells (Figure 34.8, B), reveals that the cells of the control are healthy with some evident organelles—nucleous, rough endoplasmic reticulum, plastid, plasmodesmata, conspicous constrictions, protein bodies, and lipid droplets. By contrast, cells of the treated seed resemble those in the dried quiescent seed (Figure 34.8, C). The differences observed between moistened and 5-MOP–treated radish seeds represent useful signals to establish whether the water uptake of the seed will culminate in radicle emergence.

These findings suggest that in radish seeds, 5-MOP in the presence of light provides, either directly or indirectly, a critical signal that inhibits water uptake into the seed, switches off the swelling of the seed coat epidermis as well as of the cell of the endosperm aleurone layer and the related process of cell elongation in the embryo. Thus, it is possible that germination in the radish

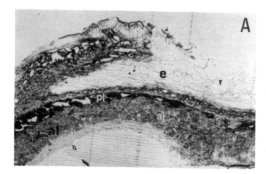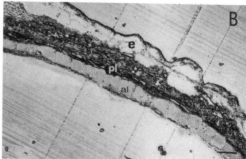

FIGURE 34.6 Semi-thin light microscope section of hilum-micropylar region of water-moistened seed, 18 h after sowing (A) and 5-MOP–treated radish seed (B). Bar = 150 μm. Abbreviations: e, epidermis; pl, pigment layer; al, aleurone layer. After Aliotta et al., 1994

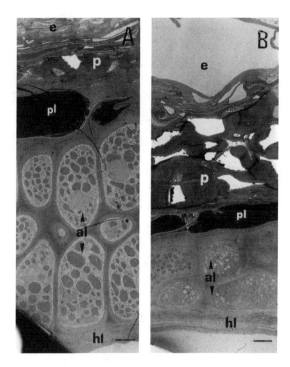

FIGURE 34.7   TEM comparison between cells of seed coats of seed and endosperm of radish 18 h after sowing in water (A) and in presence of the 5-MOP (B). Control shows a thin layer of palisade cells, and more developed pigmented and aleuronic layers. Bar = 30 μm. After Aliotta et al., 1994

results from a combination of lowered resistance of the swollen seed coat, endosperm, and of the related elongation on embryo cells. These effects due to the active coumarins are prevented by a preliminary removal of seed coats or by darkness. A similar pattern has been observed with coumarin by Aliotta et al. (1992, 1993).

34.6   RUE *VERSUS* WEEDS GERMINATION

On the basis of results obtained in radish seeds, we began to use the rue leachates as a possible bioherbicide against germination of weeds in an agricultural soil (Aliotta et al., 1995).

Greenhouse experiments in pots containing a soil differently perfused with aqueous extracts of rue leaves were carried out in the summer of 1994. In treated pots, weeds emerged later than in the control. In all observations, the presence of purslane was always recorded for about 95 percent of total weeds. Purple nutsedge (*Cyperus rotundus* L.) was noticed for less than 5 percent, while a very small number of redroot pigweed (*Amaranthus retroflexus* L.), common lambsquarters (*Chenopodium album* L.), bermudagrass (*Cynodon dactylon* L.), and dwarf spurge (*Euphorbia exigua* L.) were recorded.

Therefore, the following considerations were referred to purslane only, a serious weed of 45 crops in 81 countries, considered the ninth most common weed on Earth (Holm et al., 1977). It has been estimated that a plowed layer of the soil cropped with maize (*Zea mays* L.) contains about 220,000 purslane seeds per m^2 (Leguizamon and Cruz, 1981). The treatments consisted of three concentrations of rue leachates (1, 5, and 10 percent) and three applications, repeated every five days, were compared. After two weeks, weeds were weekly classified and counted (for a month), without moving them from the soil. Treatments of 1 percent infusion (one and two applications), 5 percent infusion (one application), and 10 percent infusion (one application) were not significantly different

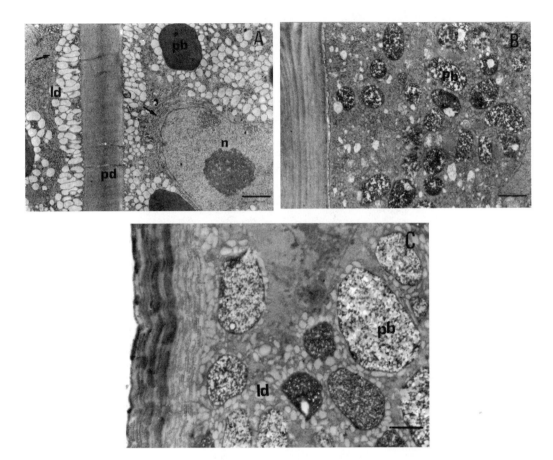

FIGURE 34.8 TEM comparison of aleuronic cells of dried radish seed (C), water-moistened (A) and 5-MOP–treated seeds (B), 18 h after sowing. Aleuronic cells of the control shows well-developed organelles: nucleus (n), rough endoplasmic reticulum (→), plasmodesmata (pd), protein bodies (pb) and lipid droplets (ld). Different profiles appear in dried (C) and 5-MOP-treated seeds (B). Bar = 5 μm. After Aliotta et al., 1994

from the control (Table 34.2). However, the other five treatments differed from the control or from the above treatments. The most effective weed control was performed by the rue extracts at the concentration of 5 and 10 percent with three applications.

Preliminary light and electronic microscopic observations were carried out to ascertain the effect of rue infusion on purslane germination (Figure 34.9). This investigation revealed that the purslane seed is lenticular, the seed coat is bitegmic, and the embryo is curved around the starchy hard perisperm that represents the seed storage reserves. Furthermore, the inner tegument (tegmen) of the control seed appears to be pierced and the radicle protruded (Figure 34.9, A). By contrast, in treated dormant seed, the tegmen appears intact and not expanded (Figure 34.9, B). Comparison between a tegmen cell of the control and that of the treated seed (Figure 34.9, C and D), shows that the cell of the control is healthy, and cytoplasmic structures are well differentiated. Tegmen cell of the treated seed also appears less differentiated. These observations suggest that tegmen, the outermost living structure of the purslane seed, is the primary target of rue leachates. This is in accord with the similar role as that of the aleuronic layer in a radish seed, where the seed coat is dead.

34.7   EXPERIMENTAL CONSIDERATIONS

To exploit the possibility of rue infusion in biocontrol of purslane, the most appropriate treatment has to synthesize the lowest number of applications as well as the lowest concentration of the extract.

TABLE 34.2
Effect of Different Treatments with Rue Infusion on Purslane Germination

Treatments (%Rue Infusion)	No. Of Times	Weeks			
		1	2	3	4
1	1	68a	68a	64a	61a
1	2	43a	43a	43a	41a
1	3	19b	11b	10b	11b
5	1	57a	55a	50a	46a
5	2	19b	16b	13b	10b
5	3	3d	1d	1d	1c
10	1	47a	52a	51a	45a
10	2	9c	4c	3c	4b
10	3	1d	1d	2cd	1c
Control		61a	61a	57a	57a

Different letters correspond to different statistical values; Student-Newman-Keuls test, P = .05 value refers to the average of purslane plants/pot.

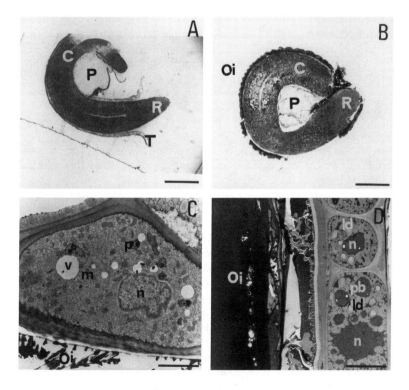

FIGURE 34.9   Light microscope micrographs of the control and treated purslane seeds after 30 h of sowing, respectively (A, B), and transmission electron micrographs (C, D) showing the tegmen cells of control and treated seeds at the same time. (A, B) C, cotyledones; Oi, outer integument; P, perisperm; R, radicle; T, tegmen (inner integument. Bar = 0.1 mm. (C, D) m, mitochondria; n, nucleus; Oi, outer integument; p, plastids; v, vacuoles; Bar = 5 mm. After Aliotta et al., 1995

The concentration of 5 percent, when applied two times, might be the most suitable. It would be of interest to investigate the reason for which, concentration being equal, allelopathic activity is less effective with one application such as for the treatment of 10 percent compared with 5 percent with two applications; higher concentrations of rue infusion could be investigated in order to find the concentration allowing the most efficacy when applied only once.

These results are important considering that purslane is difficult to control by chemical herbicides in irrigated summer crops. In conclusion, we emphasize the important properties of rue allelopathic compounds that are responsible for a broad spectrum of biological activities, often mediated by light, on different organisms. Generally, they induce an inhibitory growth delaying, modifying, and locking cellular activity of endosperm, embryo, and seed germination. Such effects have been extensively verified more with animals than in plant cells. Our studies suggest that the outermost living structures of the seed (i.e., seed coat or endosperm) are the primary target of allelochemicals.

Considerable research remains to be done to ascertain the potential of the simple rue infusion, as a better cost-effective tool for weed managements in crops (e.g., its biodegradability and selectivity). Generally, only about one out of every 10,000 chemicals screened enters the second phase of testing, which means the stage of extensive field evaluation (Han, 1987). Furthermore, it is important to study the effects of rue infusions, if any, on chemical, physical and biological soil properties. Nevertheless, the properties of rue open up a promising avenue of research.

REFERENCES

Aiello, R., Aliotta, G., Molinaro, A., Pinto, G., and Pollio, A., Anti-algal agents isolated from *Arum italicum, Planta Med.,* 58, 652, 1992.

Aliotta, G., De Napoli, L., and Piccialli, G., Inhibition of radish germination by *Anagallis arvensis* L. extract, *Giorn. Bot. Ital.,* 123, 291, 1989.

Aliotta, G., Fuggi, A., and Strumia, S., Coat-imposed dormancy by coumarin in radish seeds: the influence of light, *Giorn. Bot. Ital.,* 126, 631, 1992.

Aliotta, G., Cafiero, G., Fiorentino, A., and Strumia, S., Inhibition of radish germination and root growth by coumarin and phenylpropanoids, *J. Chem. Ecol.,* 19, 175, 1993.

Aliotta, G., Cafiero, G., De Feo, V., and Sacchi, R., Potential allelochemicals from *Ruta graveolens* L. and their action on radish seeds, *J. Chem. Ecol.,* 20, 2761, 1994.

Aliotta, G., Cafiero, G., De Feo, V., Palumbo, A. D., and Strumia, S., Inhibition of weeds germination by a simple infusion of rue, Proceedings of XXII Annual Meeting of the Plant Growth Regulator Society of America 93–97, Minneapolis, MN, 1995.

Berenbaum, C. H., and Neal, J. J., Interactions among allelochemicals and insect resistance in crop plants, *in Allelochemicals: Role in Agriculture and Forestry,* Waller, G. R., Ed., ACS Symposium Series 330, American Chemical Society, Washington, D.C., 1987, 416.

Duke, J. A., *Handbook of Medicinal Herbs,* CRC Press, Boca Raton, FL, 1986, 677.

Einhellig, F. A., Allelopathy: current status and future goals, *in Allelopathy: Organisms, Processes and Applications,* Inderjit, Dakshini, K. M. M., and Einhellig, F. A., Eds., ACS Symposium Series 582, American Chemical Society, Washington, D.C., 1995a, 1.

Einhellig, F. A., Mechanism of action of allelochemicals in allelopathy, *in Allelopathy: Organisms, Processes and Applications,* Inderjit, Dakshini, K. M. M., and Einhellig, F. A., Eds., ACS Symposium Series 582, American Chemical Society, Washington, D.C., 1995b, 96.

Evenari, M., The history of germination research and the lesson it contains for today, *Israel Bot.,* 29, 4, 1980.

Fischer, N. H., The function of mono and sesquiterpenes as plant germination and growth regulators, *in The Science of Allelopathy,* Putnam, A. R., and Tang, C. S., Eds., John Wiley & Sons, New York, 203, 1986.

Gressel, J., Herbicide tolerance and resistance: Alteration of site of activity, *in Weed Physiology: Vol. II. Herbicide Physiology,* Duke, S. O., Ed., CRC Press, Boca Baton, FL, 1985, 159.

Grieve, M., *A Modern Herbal,* Hafner Publishing, New York, 1967, 888.

Han, S. K., Potential industrial application of allelochemicals and their problems, *in Allelochemicals: Role in Agriculture and Forestry,* Waller, G. R., Ed., ACS Symposium Series 330, American Chemical Society, Washington, D. C., 1987, 449.

Harborne, J. B. and Simmonds, N. W., Phenolic glycosides and their natural distribution, *in Biochemistry of Phenolic Compounds,* Harbone, J. B., Ed., Academic Press, New York, 1964, 77.

Harborne, J. B., *Comparative Biochemistry of the Flavonoids,* Academic Press, London, 1967.

Harlan, J. R. and deWet, J. M. J. Some thoughts about weeds, *Econ. Bot.,* 19, 16, 1965.

Harlan, J. R., *Crops & Man.* American Societies of Agronomy and Crop Sciences, Madison, WI, 1992, 269.

Holm, L. G., Plucknett, D. L., Pancho, J. V., and Herberger, J. P., *The World's Worst Weeds Distribution and Biology,* University Press of Hawaii, Honolulu, 1977, 609.

Inderjit and Dakshini, K. M. M., Allelopathic effect of *Pluchea lanceolata* (Asteraceae) on characteristics of four soils and tomato and mustard growth, *Am. J. Bot.,* 81, 799, 1994.

Inderjit and Dakshini, K. M. M., Quercitin and quercitrin from *Pluchea lanceolata* and their effect on growth of asparagus bean, *in Allelopathy: Organisms, Processes and Applications,* Inderjit, Dakshini, K. M. M., and Einhellig, F. A., Eds., ACS Symposium Series 582, American Chemical Society, Washington, D.C., 1995, 86.

Inderjit and Dakshini, K. M. M., The allelopathic potential of *Pluchea lanceolata*: effect of cultivation, *Weed Sci.,* 44, 393, 1996.

Inderjit, Dakshini, K. M. M., and Einhellig, F. A., *Allelopathy: Organisms, Processes and Applications,* ACS Symposium Series 582, American Chemical Society, Washington, D.C., 1995, 381.

Leguizamon, E. S. and Cruz, P. A., Poblacion de semillas en perfil arable de suelos sometidos a distinto manejo, *R. Cien. Agrop.,* 2, 83, 1981.

Macías, F. A., Allelopathy in the search for natural herbicide models, *in Allelopathy: Organisms, Processes, and Applications,* Inderjit, Dakshini, K. M. M., and Einhellig, F. A., Eds., ACS Symposium Series 582, American Chemical Society, Washington, D.C., 1995, 310.

Moreland, D. E. and Novitsky, W. P., Interference by flavone and flavonols with chloroplast-mediated electron transport and phosphorylation, *Phytochemistry,* 27, 3359, 1988.

Mudge, L. L., Gossett, B. J., and Murphy, T. R., Resistence of goosegrass (*Eleusine indica*) to dinitrocniline herbicides, *Weed Sci.,* 32, 591, 1984.

Muller, C. H., Inhibitory terpenes volatized from *Salvia* shrubs, *Bull. Torrey Bot. Club,* 92, 38, 1965.

Muller, W. H. and Muller, C. H., Volatile growth inhibitors produced by *Salvia leucophylla, Bull. Torrey Bot. Club,* 91, 327, 1964.

Muller, C. H., Muller, W. H., and Haines, B. L., Volatile growth inhibitors produced by aromatic shrubs, *Science,* 143, 471, 1964.

Murray, R. D. M., Mendez, J., and Brown, S. A., *The Natural Coumarins: Occurrence, Chemistry and Biochemistry,* Wiley, Chichester, England, 1982, 701.

Rice, E. L., *Allelopathy,* 2nd ed., Academic Press, Orlando, FL, 1984, 422.

Paszkowsky, W. L. and Kemer, R. J., Biological activity and tentative identification of flavonoid components in velvetleaf (*Abutilon theophrasti* Medicus) seed coats, *J. Chem. Ecol.,* 14, 1573, 1988.

Vaughan, J. G., and Whitehouse, J. M., Seed structure and taxonomy of the Cruciferae, *Bot. F. Linn. Soc.,* 64, 383, 1971.

Weidenhamer, J. D., Macías, F. A., Fisher, N. H., Williamson, G. B., Just how insoluble are monoterpenes, *J. Chem. Ecol.,* 19, 1799, 1993.

Zimdahl, R. L., *The Fundamentals of Weed Science,* Academic Press, New York, 1993.

Zobel, A. M. and Brown, S. A., Determination of furanocoumarins on the leaf surface of *Ruta graveolens* with an improved extraction technique, *J. Nat. Prod.,* 51, 941, 1988.

Zobel, A. M. and Brown, S. A., Psoralens in senescing leaves of *Ruta graveolens, J. Chem. Ecol.,* 17, 1801, 1991.

Zobel, A. M. and Brown, S. A., Coumarins in the interactions between the plant and its environment, *Allel. J.,* 2, 9, 1995.

Index

A

AB5046A and AB5046B, 508–509
Abscisic acid, 326
Abutilon theophrasti (velvetleaf), 159, 469, 498, 504, 507, 555
ACC, 159
Acer campestre, 104
Acer negundo, 104
Acetic acid compounds, 258, 402, 481, 498–499
Acetobacter diazotrophicus, 158
Acetobacter xylinum, 506
Acetophenone, 256
Acetylcholine, pollen cholinesterase assay, 122
N-Acetylquestiomycin, 390
Acetyltropine, 229
Achillea millefolium, 102, 103
Achillea spp., 107–108, 113
Achyranthes spp., 501
Acidic fog, 328
Acidic rain, 440
Acidic soils, 29, 369–377, See also Soil pH
Acinetobacter calcoaceticus, 260
Aconitine, 412, 416, 418, 419
Acorus gramineus, 51
Acremonium strictum, 229
Acridone alkaloids, 113
Actinidia arguata, 317
Actinolite, 289
Actinoplane spp., 503
Acyrthosiphon pisum, 457
Adaptive response
    insects and flavonoids, 431
    soil microbes vs. allelochemicals, 328, 347
Additive effects, 41, 482–483, See also
            Synergistic effects
    environmental effects on flavonoid synthesis, 268, 275
    herbicides and allelochemicals, 6
Adenine-ß-methylstreptimidone, 497
Adenine 9-ß-D-arabinofuranoside, 503
Adenostoma fasciculatum, 149, 150
Adoxophyes orana, 457
Adrenaline, 122
Adsorption-desorption processes, 61, 341
Aesculus, 314
Aesculus hippocastanum, 104, 105, 112, 119, 269, 316, 326, 456
Agalinis purpurea, 452
Agamosperm pollen allelopathy, 132
Agar media, 135, See Media
Aglycones
    compartmentalization, 446
    soyaspogenols, 452–453, See also Saponins

Agrobacterium radiobacter, 523
Agrobacterium tumefaciens, 523
Agrobacterium spp., 344
Agrocin, 523
Agropyron repens, 36, 429
Agropyrum spp., 389
Agrostemma githago, 393
Agrostis stolonifera, 136, 141
Air pollution, 328–329
Ajmaline, 411, 416, 417, 419, 420
Alachlor, 40, 487
Albite, 289
Alcaligenes faecalis, 158
Alfalfa, See *Medicago sativa*
Algae, See also Plankton
    allelochemicals from benthic cyanobacteria, 182–186
    allelochemical transfer, 184
    allelopathy from submersed macrophytes, 186–194
    antibiotics, 173–174
    Australian studies, 229–230
    bacterial allelopathic effects, 159
    benthic and littoral habitats, 180, 182
    competition from benthic cyanobacteria, 179–180
    herbivore deterrents, 230
    hydrophyte interactions, 46, 51
    UV light effects, 273
Algal blooms, See under Plankton
Algicidal hydrolyzable polyphenols, 189–194
    enzyme inhibition activity, 191–192
    excretion, 190–191
    nutrient effects, 192–193
Alizarin, 391
Alkaline phosphatase inhibition, 191–192
Alkaloids, 411–421
    Australian plants, 227–229
    biosynthetic pathways, 79–80
    concentration, 420
    diurnal variations, 79–80
    DNA interactions, 413–414, 417
    enzyme inhibition, 413, 417, 418
    intact cell microspectrofluorimetry, 100, 103, 104, 108, 113
    membrane permeability and, 414, 417
    multi-functionality, 412
    neuroreceptor binding, 414–417, 419–420
    plant growth factors and, 419
    protein biosynthesis inhibition, 414, 417
    relevance of in *vitro* experiments, 420
    slimes, 116
Allelochemical interactions, 4, See Allelopathy
Allelochemicals, 256, See also Alkaloids; Phenolic
            acids and compounds; Polyphenols; Tannins;
            specific compounds